# Scientific Computation

C. Canuto M. Y. Hussaini
A. Quarteroni T. A. Zang

# Spectral Methods

## Fundamentals in Single Domains

With 106 Figures and 19 Tables

Claudio Canuto
Dipartimento di Matematica
Politecnico di Torino
Corso Duca degli Abruzzi, 24
10129 Torino, Italy
e-mail: claudio.canuto@polito.it

M. Yousuff Hussaini
School of Computational Science
Florida State University
Tallahassee, FL 32306-4120, USA
e-mail: myh@csit.fsu.edu

Alfio Quarteroni
SB-IACS-CMCS, EPFL
1015 Lausanne, Switzerland
and
MOX, Politecnico di Milano
Piazza Leonardo da Vinci, 32
20133 Milano, Italy
e-mail: alfio.quarteroni@epfl.ch

Thomas A. Zang, Jr.
NASA Langley Research Center*
Mail Stop 449
Hampton, VA 23681-2199, USA
e-mail: Thomas.A.Zang@nasa.gov

* *This does not constitute an endorsement of this work by either the U.S. Government or the NASA Langley Research Center.*

*Cover picture*: See Fig. 4.4 (*left*)

ISBN 978-3-642-06800-3 e-ISBN 978-3-540-30726-6

ISSN 1434-8322



Springer is a part of Springer Science+Business Media

springer.com


Printed in Germany



Cover design: *Erich Kirchner* Heidelberg

# Preface

As a tool for large-scale computations in fluid dynamics, spectral methods were originally proposed in 1944 by Blinova, first implemented in 1954 by Silberman, virtually abandoned in the mid-1960s, resurrected in 1969–70 by Orszag and by Eliason, Machenhauer and Rasmussen, developed for specialized applications in the 1970s, endowed with the first mathematical foundations by the seminal work of Gottlieb and Orszag in 1977, extended to a broader class of problems and thoroughly analyzed in the 1980s, and entered the mainstream of scientific computation in the 1990s. Two decades ago when we wrote *Spectral Methods in Fluid Dynamics* (1988) both the subject and the authors were barely past their adolescence. As the field and the authors are now in their middle age, the time seems ripe for a more mature discussion of the field, accounting for the main contributions of the intervening years. Motivated by the many favorable comments we have received and the continuing interest in the first book (which will be referred to as CHQZ1), yet desiring to present a more modern perspective, we embarked on a project which has resulted in this book (referred to as CHQZ2) and its companion book (Canuto, Hussaini, Quarteroni and Zang (2007), referred to as CHQZ3). These, like our first text on this subject, are books about spectral methods for partial differential equations – when to use them, how to implement them, and what can be learned from their rigorous theory.

The original promoters of spectral methods were meteorologists studying global weather modeling and fluid dynamicists investigating isotropic turbulence. The converts who were inspired by the successes of these pioneers remained, for the most part, confined to these and closely related fields throughout the 1970s. During that decade spectral methods appeared to be well-suited only for problems governed by ordinary differential equations or by partial differential equations with (mostly) periodic boundary conditions. And, of course, the solution itself needed to be smooth.

Both the theory and the algorithms of classical (single-domain) spectral methods for smooth problems were already reasonably mature in the mid-1980s. On the theoretical side, approximation theory results were available for periodic and nonperiodic problems, stability and convergence analyses were in-hand for steady and unsteady linear problems, and detailed numerical analyses had been produced for a variety of methods for fluid dynamics applications, and particularly for the incompressible Navier-Stokes equations. Open issues included discontinuous problems (with compressible flows of particular interest), convergence analysis of iterative methods, artificial outflow boundary conditions, and rigorous analysis of time discretizations. On the algorithms front, explicit methods for fully periodic problems were routine, efficient direct solution methods were available for several important constant-coefficient implicit equations, numerous efficient algorithms were available for incompressible flows with at most one nonperiodic direction, and shock-fitting methods had been developed for compressible flows. Numerous approaches were being tried for discontinuous problems, especially for shock capturing in compressible flows. Rapid developments were taking place in iterative methods for implicit equations. The extension of spectral methods to problems in complex geometries through multidomain spectral approaches was proceeding explosively.

Singular progress has indeed been made over the past two decades in extending spectral methods to arbitrary geometries, enabling what some would consider the mathematical nirvana of a method of arbitrarily high order capable of application to problems on an arbitrary geometry. In this respect, the trajectory of spectral methods over the past 20 years has been converging towards that of *hp* finite-element methods.

This process of migration from single-domain to multidomain spectral methods has required the injection of novel mathematical tools, and stimulated original investigation directions. Mathematics has had a profound impact on the correct design and interpretation of the methods, and in some cases it has inspired the development of discontinuous spectral methods (such as the mortar method and the discontinuous Galerkin method) even for problems with continuous solutions. On the other hand, since in general a geometrically complex computational domain is split into polygonal or polyhedral subdomains (or elements), tensor-product domains are no longer a prerequisite for spectral methods, with the development of spectral bases on triangles and tetrahedra.

One of the most pronounced changes is that the strong form of differential equations has lost its primacy as the anchor for the discretization of the problem. Multidomain spectral methods are more easily and reliably approached, both algorithmically and theoretically, from weak formulations of the differential equations. Moreover, the use of many subdomains has motivated the use of moderate polynomial degrees in every subdomain – small from the perspective of classical spectral methods, but large from the perspectives of finite-difference and finite-element methods. From a theoretical viewpoint, new error estimates have been established for which the roles of the local polynomial degree and the geometrical size of the local elements are both captured. From an algorithmic point of view, the role of matrices has been addressed in great detail, corresponding to the increased interest in small and moderate values of $N$ and on techniques of matrix assembly. Exploitation of advanced linear algebra tools for sparse, ill-conditioned systems has become of paramount importance.

In spite of this major change of perspective, the new multidomain spectral methods still enjoy some of the most distinguishing (and desirable) features of "classical" spectral methods – Gaussian integration formulas, low dispersion, and ease of preconditioning by low-order discretization matrices.

Over the past twenty years the appeal of spectral methods for applications such as computational fluid dynamics has expanded, commensurate with the erosion of most of the obstacles to their wider application. Beyond the specific techniques, the culture of high-order methods has entered the background knowledge of numerical analysts. Spectral methods have been traditional in academic instruction since the 1990s and began to penetrate industrial applications this decade. In fact, spectral methods are successfully used nowadays for widely diverse applications, such as wave propagation (for acoustic, elastic, seismic and electromagnetic waves), solid and structural analysis, marine engineering, biomechanics, astrophysics, and even financial engineering. Their principal appeal in the academic research environment still relies on their superior rate of convergence, which makes them an ideal virtual lab. In the industrial (extra-academic) environment, spectral-based codes are appreciated, and often preferred, owing to the low dissipation and dispersion errors, the neat way to treat boundary conditions, and, today, the availability of efficient algebraic solvers that allow a favorable trade-off between accuracy and computational cost.

The basics of classical spectral methods remain essential for current research on the frontiers of both the algorithms and the theory of spectral methods. At the same time, multidomain spectral methods have already warranted books in their own right. Our objectives with the current two books are to modernize our thorough discussion of classical spectral methods, accounting for advances in the theory and more extensive application experience in the fluid dynamics arena, while summarizing the current state of multidomain spectral methods from the perspective of classical spectral methods. The major methodological developments in classical spectral methods during the past two decades have been the emergence of the Galerkin with numerical integration (G-NI) approach, the decline of the tau method to a niche role, improved treatment of boundary conditions, the adaptation of advanced direct and iterative methods to spectral discretizations also thanks to a better insight into the mathematical basis of preconditioning, the development of more sophisticated tools to control spurious high-frequency oscillations without losing the formal accuracy of the method, and the formulation of spectral discretizations on triangles (in two dimensions) and tetrahedra (in three dimensions). From the applications perspective in fluid dynamics, new algorithms have been produced for compressible linear and secondary stability, for parabolized stability equations, for velocity-vorticity formulations of incompressible flow, and for large-eddy simulations, along with refinement of spectral shock-fitting methods. Moreover, the once intense debate over the impact of aliasing errors has settled down to polite differences of opinion.

While a significant amount of material in the two new books has been retained from portions of our earlier text, CHQZ1, the majority of the material is new. The most consistent augmentation is that all chapters are enhanced by the addition of material for the G-NI method. The added material has necessitated publishing this new work as two separate books. The rationale for the division of the material between the two books is that we furnish in this first book, CHQZ2, a comprehensive discussion of the generic aspects of classical spectral methods, while the second book, CHQZ3, focuses on applications to fluid dynamics and on multidomain spectral methods.

Chapters 1–4 of the present book are of general interest. Chapter 1 provides a motivational introduction to spectral methods, as well as a preview of the more sophisticated single-domain applications in fluid dynamics presented in the second book. Chapter 2 contains a thorough discussion of classical orthogonal expansions, supplemented with a basic description of spectral approximations on triangles and tetrahedra. Chapter 3 provides a comprehensive guide to spectral discretizations in space for partial differential equations in one space dimension, using the Burgers equation model problem for illustrative purposes. A discussion of boundary conditions for hyperbolic equations, and detailed prescriptions for the construction of mass and stiffness matrices for elliptic problems are also given. Chapter 4 focuses on solution techniques for the algebraic systems generated by spectral methods. In addition to a number of now classical results, the chapter offers a thorough investigation of modern direct and iterative methods, as befits the extensive developments that have transpired in the past two decades. A large number of original numerical examples are presented in these two chapters. Chapters 5–7 focus on the mathematical theory of classical spectral methods. Chapter 5 consists of a review of those results from approximation theory which are pertinent to the theoretical analysis of spectral methods. Most of them are classical; however a few of them are newer, as they highlight the dependence on both polynomial degree and geometrical parameters for both tensor-product domains and simplicial domains (triangles and tetrahedra). Chapter 6 is the focal point of this book regarding the theory of spectral methods. The fundamental stability and convergence results are established for all kinds of numerical spectral approximations (Galerkin, tau, collocation, and G-NI) to linear partial differential equations, both steady and unsteady. Finally, Chap. 7 addresses the theoretical analysis of spectral approximations to a family of partial differential equations that can be regarded as the building blocks of mathematical modelling in continuum mechanics in general, and in fluid dynamics in particular. It places particular emphasis on the Poisson equation, singularly perturbed elliptic equations that govern advection-diffusion and reaction-diffusion processes featuring sharp boundary layers, the heat equation, hyperbolic equations and systems, and the steady Burgers equation. Moreover, it addresses the eigenvalue analysis of matrices produced by spectral approximations, and illustrates recent techniques to resolve the Gibbs phenomenon for discontinuous solutions through filtering, singularity detection and spectral reconstruction techniques. The first book ends with four Appendices surveying several algorithmic and theoretical numerical analysis topics that are not specific to spectral methods, but of sufficient utility to some readers to warrant inclusion. In Appendix A we review some basic notations and theorems from functional analysis. Appendix B reviews the fast Fourier transform and some adaptations that are particularly useful to Fourier and Chebyshev methods. Appendix C is a gentle introduction to iterative methods and lists several specific iterative algorithms that have been exploited in spectral methods, while Appendix D describes some basic concepts, specific numerical schemes, and stability regions for those temporal discretizations that have been favored by the spectral methods community.

In our second book (Canuto, Hussaini, Quarteroni and Zang (2007)), Chap. 1 covers the basic equations of fluid mechanics. Chapter 2 is solely devoted to spectral algorithms for analyses of linear and nonlinear stability of fluid flows. Applications to compressible flows and to parabolized stability equations post-date our earlier book. Chapter 3, on algorithms for incompressible flows, has a sharp emphasis on those algorithms that remained in reasonably extensive use post-1990 and provides a modern discussion of solution techniques for problems with two or more nonperiodic directions. Chapter 4, on algorithms for hyperbolic systems and compressible flows, emphasizes algorithms for enforcing boundary conditions, methods for computing homogeneous, compressible flows, and an improved approach to shock fitting. Chapter 5 introduces the main strategies to construct spectral approximations in complex domains, and in particular the spectral-element method, the mortar-element method, the spectral discontinuous Galerkin method, as well as the more traditional patching collocation method. Their theoretical properties are analyzed, and their algebraic aspects are investigated. Chapter 6 illustrates solution strategies based on domain decomposition techniques for the spectral discretizations investigated in Chap. 5. Both Schur-based and Schwarz-based iterative solvers and preconditioners are considered, and their computational advantages (in particular, their property of scalability with respect to the number of subdomains) are illustrated. Our project closes in the same manner in which it began, with a survey of representative large-scale applications of (this time multidomain) spectral methods.

Whereas with our first text we made a valiant effort to provide comprehensive coverage of all available spectral methods (at least for fluid dynamics applications) and to provide a bibliography that encompassed all extant references to spectral methods, here we acknowledge the practical impossibility of such an ambition in the face of all the work that has since transpired in the field. We still aim to provide comprehensive coverage of general methodology.

However, our coverage of particular algorithms is necessarily representative rather than complete. Our aim is to focus on those algorithms that have stood the test of time in fluid dynamical applications, as assessed by how widely they have been used in the past two decades. But our knowledge in this area is certainly not exhaustive, and others would no doubt have made somewhat different choices. In our citations we enforce a strong preference for archival publications. We recognize that many developments appeared earlier (in some cases many years earlier) in pre-prints or conference publications. But we only cite non-archival sources when no archival reference is available.

The many numerical examples produced expressly for these books have all been run on desktop computers (under both Linux and Macintosh operating systems), usually in 64-bit arithmetic with the standard IEEE precision of $2^{-52} \approx 2 \times 10^{-16}$. A half-dozen or so different computers were employed, with clock speeds on the order of 1–3 GHz; some of these computers had two CPUs. The workhorse languages were Matlab and Fortran, with no special effort devoted to fine-tuning the performance of the codes. The reader will certainly appreciate that the occasional timings presented here are meant solely to provide a rough comparison between the costs of alternative algorithms and should not be construed as representing a definitive verdict on the efficiency of the methods.

Nowadays, considerable software for spectral methods is freely available on the web, ranging from libraries of basic spectral operations all the way to complete spectral codes for Navier-Stokes (and other complex) applications. Due to the highly dynamic nature of these postings, we have chosen not to list them in the text (except to acknowledge codes that we have used here for numerical examples), but to maintain a reasonably current list of such sources on the Web site (http://www.dimat.polito.it/chqz/) for this and the companion text. There is always the possibility that this site itself may need to be moved due to unforeseen circumstances; in that event one should check the Springer site for the link to the detailed book Web site.

The authors are grateful to Dr. Wolf Beiglböck, Dr. Ramon Khanna and the Springer staff for their patience while waiting for our long overdue manuscript. The authors are pleased to acknowledge the many discussions and helpful comments on the manuscript that have been provided by colleagues such as Paola Gervasio, David Kopriva, Giovanni Monegato, Luca Pavarino and Andrea Toselli. The technical support of Paola Gervasio and Marco Discacciati in running numerical tests, preparing figures and tables, typing and editing a significant part of the whole manuscript is gratefully acknowledged. Thanks are also due to Stefano Berrone and Sophie Fosson for providing further technical support, and to Susan Greenwalt for her administrative support of this project. We appreciate the generosity of those individuals who have given us permission to reprint figures from their work in these texts. The authors are grateful to the Politecnico di Torino, the Florida State University, the Ecole Polytechnique Fédérale de Lausanne and the Politecnico di Milano for their facilitation of this endeavor. One of us (MYH) is particularly grateful to Provost Lawrence Abele of Florida State University for his encouragement and support for this project. Finally, we are most appreciative of the support and understanding we have received from our wives (Manuelita, Khamar, Fulvia and Ann) and children (Arianna, Susanna, Moin, Nadia, Marzia and Silvia) during this project.

| | |
|---|---|
| Torino, Italy | *Claudio Canuto* |
| Tallahassee, Florida | *M. Yousuff Hussaini* |
| Lausanne, Switzerland and Milano, Italy | *Alfio Quarteroni* |
| Carrollton, Virginia | *Thomas A. Zang* |

February, 2006

# Contents

# List of Figures

# List of Tables

# 1. Introduction

## 1.1 Historical Background

Spectral methods are a class of spatial discretizations for differential equations. The key components for their formulation are the trial functions (also called the expansion or approximating functions) and the test functions (also known as weight functions). The trial functions, which are linear combinations of suitable trial basis functions, are used to provide the approximate representation of the solution. The test functions are used to ensure that the differential equation and perhaps some boundary conditions are satisfied as closely as possible by the truncated series expansion. This is achieved by minimizing, with respect to a suitable norm, the residual produced by using the truncated expansion instead of the exact solution. The residual accounts for the differential equation and sometimes the boundary conditions, either explicitly or implicitly. For this reason they may be viewed as a special case of the method of weighted residuals (Finlayson and Scriven (1966)). An equivalent requirement is that the residual satisfy a suitable orthogonality condition with respect to each of the test functions. From this perspective, spectral methods may be viewed as a special case of Petrov-Galerkin methods (Zienkiewicz and Cheung (1967), Babuška and Aziz (1972)).

The choice of the trial functions is one of the features that distinguishes the early versions of spectral methods from finite-element and finite-difference methods. The trial basis functions for what can now be called *classical spectral methods* – spectral methods on a single tensor-product domain – are global, infinitely differentiable and nearly orthogonal, i.e. the matrix consisting of their inner products has very small bandwidth; in many cases this matrix is diagonal. (Typically the trial basis functions for classical spectral methods are tensor products of the eigenfunctions of singular Sturm-Liouville problems). In contrast, for the $h$ version of finite-element methods, the domain is divided into small elements, and low-order trial functions are specified in each element. The trial basis functions for finite-element methods are thus local in character and still nearly orthogonal, but not infinitely differentiable. They are thus well suited for handling complex geometries. Finite-difference methods are typically viewed from a pointwise approximation perspective rather than from a trial function/test function perspective. However, when appropriately translated into a trial function/test function formulation, the finite-difference trial basis functions are likewise local.

The choice of test functions distinguishes between the three earliest types of spectral schemes, namely, the Galerkin, collocation, and tau versions. In the Galerkin (1915) approach, the test functions are the same as the trial functions. They are, therefore, infinitely smooth functions that individually satisfy some or all of the boundary conditions. The differential equation is enforced by requiring that the integral of the residual times each test function be zero, after some integration-by-parts, accounting in the process for any remaining boundary conditions. In the collocation approach the test functions are translated Dirac delta-functions centered at special, so-called collocation points. This approach requires the differential equation to be satisfied exactly at the collocation points. Spectral tau methods are similar to Galerkin methods in the way the differential equation is enforced. However, none of the test functions need satisfy the boundary conditions. Hence, a supplementary set of equations is used to apply the boundary conditions.

The collocation approach appears to have been first used by Slater (1934) and by Kantorovic (1934) in specific applications. Frazer, Jones and Skan (1937) developed it as a general method for solving ordinary differential equations. They used a variety of trial functions and an arbitrary distribution of collocation points. The work of Lanczos (1938) established for the first time that a proper choice of trial functions and distribution of collocation points is crucial to the accuracy of the solution. Perhaps he should be credited with laying down the foundation of the orthogonal collocation method. This method was revived by Clenshaw (1957), Clenshaw and Norton (1963) and Wright (1964). These studies involved the application of Chebyshev polynomial expansions to initial-value problems. Villadsen and Stewart (1967) developed this method for boundary-value problems.

The earliest applications of the spectral collocation method to partial differential equations were made for spatially periodic problems by Kreiss and Oliger (1972) (who called it the Fourier method) and Orszag (1972) (who termed it pseudospectral). This approach is especially attractive because of the ease with which it can be applied to variable-coefficient and even nonlinear problems. The essential details will be furnished below.

The Galerkin approach enjoys the esthetically pleasing feature that the trial functions and the test functions are the same, and the discretization is derived from a weak form of the mathematical problem. Finite-element methods customarily use this approach. Moreover, the first serious application of spectral methods to PDE's – that of Silberman (1954) for meteorological modeling – was a Galerkin method. However, spectral Galerkin methods only became practical for high resolution calculations of such nonlinear problems after Orszag (1969, 1970) and Eliasen, Machenhauer and Rasmussen (1970) developed transform methods for evaluating the convolution sums arising from quadratic nonlinearities. (Nonlinear terms also increase the cost of finite-element methods, but not nearly as much as they do for spectral Galerkin methods.) For problems containing more complicated nonlinear terms, high-resolution spectral Galerkin methods remain impractical.

The tau approach is a modification of the Galerkin method that is applicable to problems with nonperiodic boundary conditions. It may be viewed as a special case of the so-called Petrov-Galerkin method. Lanczos (1938) developed the spectral tau method, and Orszag's (1971b) application of the Chebyshev tau method to produce highly accurate solutions to fluid dynamics linear stability problems inspired considerable use of this technique, not just for computing eigenvalues but also for solving constant-coefficient problems or subproblems, e.g., for semi-implicit time-stepping algorithms.

In the middle 1980's newer spectral methods, which combined the Galerkin approach with Gaussian quadrature formulas, came into common use. These methods share with the Galerkin approach the weak enforcement of the differential equation and of certain boundary conditions. In their original version the unknowns are the values of the solution at the quadrature points, as in a collocation method. We shall refer to such approaches as Galerkin with numerical integration, or G-NI, methods.

The first unifying mathematical assessment of the theory of spectral methods was provided in the monograph by Gottlieb and Orszag (1977). The theory was extended to cover a large variety of problems, such as variable-coefficient and nonlinear equations. A sound approximation theory for the polynomial families used in spectral methods was developed. In his monograph Mercier (1981) advanced the understanding of the role of Gaussian quadrature points for orthogonal polynomials as collocation points for spectral methods, as had originally been observed in 1979 by Gottlieb. Stability and convergence analyses for spectral methods were produced for a variety of approaches. The theoretical analysis of spectral methods in terms of weak formulations proved very successful. As a matter of fact, this opened the door to the use of functional analysis techniques to handle complex problems and to obtain the sharpest results. Application developments were equally extensive, and by the late 1980's spectral methods had become the predominant numerical tool for basic flow physics investigations of transition and turbulence. All in all, the 10 years that followed were extremely fruitful for the theoretical development and the application deployment of spectral methods.

Developments of the first five years that followed Gottlieb and Orszag (1977) were reviewed in the symposium proceedings edited by Voigt, Gottlieb and Hussaini (1984). Indeed, that very symposium in 1982 inspired the youthful incarnations of the present authors to produce their first text on this subject (Canuto, Hussaini, Quarteroni and Zang (1988)). Subsequently, numerous other texts and review articles on various aspects of spectral methods appeared. Boyd (1989, and especially the 2001 second edition) contains a wealth of detail and advice on spectral algorithms and is an especially good reference for problems on unbounded domains and in cylindrical and spherical coordinate systems. A sound reference for the theoretical aspects of spectral methods for elliptic equations was provided by Bernardi and Maday (1992b, 1997). Funaro (1992) and Guo (1998) discussed the approximation of differential equations by polynomial expansions. Fornberg (1996) is a guide for the practical application of spectral collocation methods, and it contains illustrative examples, heuristic explanations, basic Fortran code segments, and a succinct chapter on applications to turbulent flows and weather prediction. Trefethen (2000) is a lively introduction to spectral collocation methods and includes copious examples in Matlab. Focused applications of spectral methods on particular classes of problems were provided by Tadmor (1998) and Gottlieb and Hesthaven (2001) for first-order hyperbolic problems, by Cohen (2002) for wave equations, and by Bernardi, Dauge and Maday (1999) for problems in axisymmetric domains. Peyret (2002) provided a rather comprehensive discussion of Fourier and Chebyshev spectral methods for the solution of the incompressible Navier-Stokes equations, specifically in the primitive equations and vorticity-streamfunction formulations.

By the late 1980's classical spectral methods were reasonably mature, and the research focus had clearly shifted to the use of high-order methods for problems on complex domains. We shall refer to this class of spectral methods generically as *multidomain spectral methods* or as *spectral methods in arbitrary geometries*. The 1988 book by the present authors closed with an overview of this then nascent subject. Funaro (1997) treats spectral-element methods in the context of elliptic boundary-value problems, especially convection-dominated flows, and includes a multidomain treatment for complex geometry. The first comprehensive texts on spectral methods in complex domains appeared around the year 2000. Karniadakis and Sherwin (1999) provides a unified framework for spectral-element methods (as introduced by Patera (1984)) and *hp* finite-element methods (see, for example, Babuška, Szabó and Katz (1981)). It includes structured and unstructured domains, and applications to both incompressible and compressible flows. The Deville, Fischer and Mund (2002) text focuses on high-order methods in physical space (collocation and spectral-element methods) with applications to incompressible flows. Its coverage of the implementation details of such methods on vector and parallel computers distinguishes it from other books on the subject. Although specifically devoted to the *hp*-version of finite-element methods, the book by Schwab (1998) provides many useful theoretical results about the approximation properties of high-order polynomials in complex domains.

The present book is focused on the fundamentals of spectral methods on simple domains. A companion book (Canuto, Hussaini, Quarteroni and Zang (2007)) discusses specific spectral algorithms for fluid dynamics applications and describes the evolution of spectral methods to complex domains. We shall refer to the companion book as CHQZ3. Citations in the present text that refer to specific material in the companion book will have the format

CHQZ3, Chap. x or CHQZ3, Sect. x.y. For example, a reference such as CHQZ3, Chap. 1 refers to Chapter 1 of Canuto, Hussaini, Quarteroni and Zang (2007).

## 1.2 Some Examples of Spectral Methods

Spectral methods are distinguished not only by the fundamental type of the method (Galerkin, collocation, Galerkin with numerical integration, or tau), but also by the particular choice of the trial functions. The most frequently used trial functions are trigonometric polynomials, Chebyshev polynomials, and Legendre polynomials. In this section we shall illustrate the basic principles of each method and the basic properties of each set of polynomials by examining in detail one particular spectral method on each of several different types of differential equations. Each of these examples will be reconsidered in Chap. 6 from a rigorous theoretical point of view.

### 1.2.1 A Fourier Galerkin Method for the Wave Equation

Many evolution equations can be written as

$$\frac{\partial u}{\partial t} = \mathcal{M}(u) , \tag{1.2.1}$$

where $u(\mathbf{x}, t)$ is the solution, and $\mathcal{M}(u)$ is an operator (linear or nonlinear) that contains all the spatial derivatives of $u$. Equation (1.2.1) must be coupled with an initial condition $u(\mathbf{x}, 0)$ and suitable boundary conditions.

For simplicity suppose that there is only one spatial dimension, that the spatial domain is $(0, 2\pi)$, and that the boundary conditions are periodic. Most often spectral methods are used only for the spatial discretization. The approximate solution is represented as

$$u^N(x,t) = \sum_{k=-N/2}^{N/2} a_k(t)\phi_k(x) . \tag{1.2.2}$$

The $\phi_k$ are the trial functions, whereas the $a_k$ are the expansion coefficients. In general, $u^N$ will not satisfy (1.2.1), i.e., the residual

$$\frac{\partial u^N}{\partial t} - \mathcal{M}(u^N)$$

will not vanish everywhere. The approximation is obtained by selecting a set of test functions $\psi_k$ and by requiring that

$$\int_0^{2\pi} \left[ \frac{\partial u^N}{\partial t} - \mathcal{M}(u^N) \right] \psi_k(x)\, \mathrm{d}x = 0 , \tag{1.2.3}$$

for $k = -N/2, \ldots, N/2$, where the test functions determine the weights of the residual. In this sense the approximation is obtained by a method of weighted residuals. Most often the numerical analysis community describes discretizations of differential equations formulated by integral expressions such as (1.2.3) (possibly after applying integration-by-parts) as discrete *weak formulations*. This more common terminology is the one that we follow in this text. The alternative, discrete *strong formulation* is characterized by enforcing that the approximate representation of the solution, e.g., (1.2.2), satisfy the differential equation exactly at a discrete set of points. Finite-difference methods use a strong formulation, as do spectral collocation methods – see the example in Sect. 1.2.2. A more comprehensive discussion of alternative formulations of differential problems is provided in Sect. 3.2.

The most straightforward spectral method for a problem with periodic boundary conditions is based on trigonometric polynomials:

$$\phi_k(x) = e^{ikx} , \tag{1.2.4}$$

$$\psi_k(x) = \frac{1}{2\pi} e^{-ikx} . \tag{1.2.5}$$

Note that the trial functions and the test functions are essentially the same, and that they satisfy the (bi-)orthonormality condition

$$\int_0^{2\pi} \phi_k(x)\psi_l(x)\, \mathrm{d}x = \delta_{kl} . \tag{1.2.6}$$

If this were merely an approximation problem, then (1.2.2) would be the truncated Fourier series of the known function $u(x,t)$ with

$$a_k(t) = \int_0^{2\pi} u(x,t)\psi_k(x)\, \mathrm{d}x \tag{1.2.7}$$

being simply the familiar Fourier coefficients. For the partial differential equation (PDE), however, $u(x,t)$ is not known; the approximation (1.2.2) is determined by (1.2.3).

For the linear hyperbolic problem

$$\frac{\partial u}{\partial t} - \frac{\partial u}{\partial x} = 0 , \tag{1.2.8}$$

i.e., for

$$\mathcal{M}(u) = \frac{\partial u}{\partial x} , \tag{1.2.9}$$

condition (1.2.3) becomes

$$\frac{1}{2\pi}\int_0^{2\pi} \left[ \left( \frac{\partial}{\partial t} - \frac{\partial}{\partial x} \right) \sum_{l=-N/2}^{N/2} a_l(t) e^{ilx} \right] e^{-ikx}\, \mathrm{d}x = 0 .$$

The next two steps are the analytical (spatial) differentiation of the trial functions:

$$\frac{1}{2\pi}\int_0^{2\pi}\left[\sum_{l=-N/2}^{N/2}\left(\frac{\mathrm{d}a_l}{\mathrm{d}t}-ila_l\right)e^{ilx}\right]e^{-ikx}\,\mathrm{d}x=0\;,$$

and the analytical integration of this expression, which produces the dynamical equations

$$\frac{\mathrm{d}a_k}{\mathrm{d}t}-ika_k=0\;,\qquad k=-N/2,\ldots,N/2\;. \tag{1.2.10}$$

The initial conditions for this system of ordinary differential equations (ODEs) are the coefficients for the expansion of the initial condition. For this Galerkin approximation,

$$a_k(0)=\int_0^{2\pi}u(x,0)\psi_k(x)\,\mathrm{d}x\;. \tag{1.2.11}$$

For the strict Galerkin method, integrals such as those that appear in (1.2.11) should be computed analytically. For the simple example problem of this subsection this integration can indeed be performed analytically. For more complicated problems, however, numerical quadratures are performed. This is discussed further in Sect. 1.2.3.

We shall use the initial condition

$$u(x,0)=\sin(\pi\cos x) \tag{1.2.12}$$

to illustrate the accuracy of the Fourier Galerkin method for (1.2.8). The exact solution,

$$u(x,t)=\sin[\pi\cos(x+t)]\;, \tag{1.2.13}$$

has the Fourier expansion

$$u(x,t)=\sum_{k=-\infty}^{\infty}a_k(t)e^{ikx}\;, \tag{1.2.14}$$

where the Fourier coefficients are

$$a_k(t)=\sin\left(\frac{k\pi}{2}\right)J_k(\pi)e^{ikt}\;, \tag{1.2.15}$$

and $J_k(t)$ is the Bessel function of order $k$.

The asymptotic properties of the Bessel functions imply that

$$k^p a_k(t)\to 0\quad\text{as}\quad k\to\infty \tag{1.2.16}$$

for all positive integers $p$. As a result, the truncated Fourier series,

$$u^N(x,t) = \sum_{k=-N/2}^{N/2} a_k(t) e^{ikx} , \tag{1.2.17}$$

converges faster than any finite power of $1/N$. This property is often referred to as spectral convergence.

An illustration of the superior accuracy available from the spectral method for this problem is provided in Fig. 1.1. Shown in the figure are the maximum errors after one period at $t = 2\pi$ for the spectral Galerkin method, a second-order finite-difference method, an (explicit) fourth-order finite-difference method, a fourth-order compact method, and a sixth-order compact method. The integer $N$ denotes the degree of the expansion (1.2.17) for the Fourier Galerkin method and the number of grid points for the finite-difference and compact methods. The time discretization was the classical fourth-order Runge-Kutta method and the exact initial Fourier coefficients were used for the spectral method. In all cases the time-step was chosen so small that the temporal discretization error was negligible. (Appendix D furnishes the formulas (and stability regions) for commonly used time discretizations. The familiar formula for the classical fourth-order Runge-Kutta methods is given in (D.2.17).)

The second-order and fourth-order finite-difference methods used here and elsewhere in this book for examples are the standard central-difference methods with 3-point and 5-point explicit stencils, respectively. The fourth-order and sixth-order compact methods used in our examples are the classical 3-point Padé approximations (see, for example, Collatz (1966) and Lele (1992))

$$u'_{j-1} + 4u'_j + u'_{j+1} = \frac{3}{\Delta x}(u_{j+1} - u_{j-1}) \tag{1.2.18}$$

and

$$u'_{j-1} + 3u'_j + u'_{j+1} = \frac{7}{3\Delta x}(u_{j+1} - u_{j-1}) + \frac{1}{12\Delta x}(u_{j+2} - u_{j-2}) , \tag{1.2.19}$$

respectively, where $\Delta x$ is the grid spacing and $u'_j$ denotes the approximation to the first derivative at $x_j = j\Delta x$. Of course, when nonperiodic boundary conditions are present, special stencils are needed for points at, and sometimes also adjacent to, the boundary.

Figure 1.2 compares these various numerical solutions for $N = 16$ with the exact answer. Note that the major errors in the finite-difference solutions are ones of *phase* rather than *amplitude*. In many problems the very low phase error of spectral methods is a significant advantage.

Because the solution is infinitely smooth, the convergence of the spectral method on this problem is more rapid than any finite power of $1/N$. Actually, since the solution is analytic, convergence is exponentially fast. (The errors for the $N \geq 64$ spectral results are so small that they are swamped by the round-off error of these calculations. Unless otherwise noted, all numerical examples presented in this book were performed in 64-bit arithmetic.)

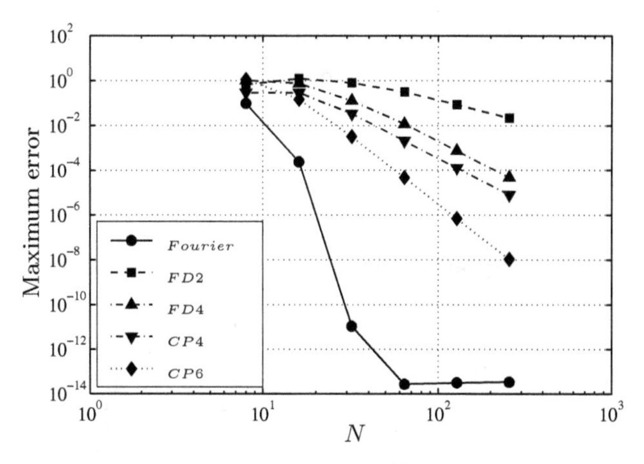


**Fig. 1.1.** Maximum errors for the linear hyperbolic problem at $t = 2\pi$ for Fourier Galerkin and several finite-difference schemes

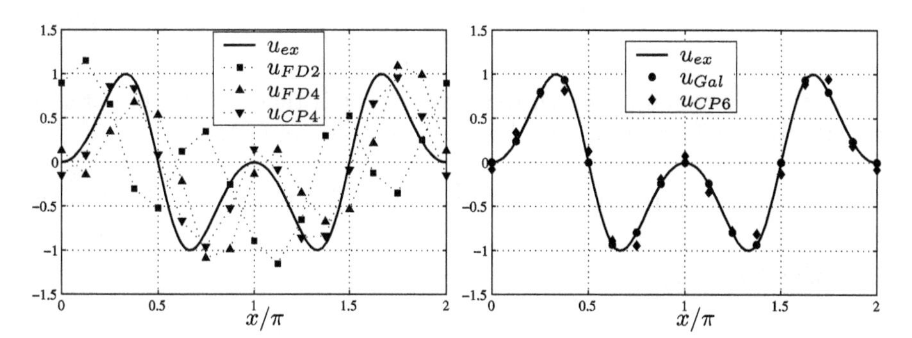


**Fig. 1.2.** Numerical solutions for the linear hyperbolic problem at $t = 2\pi$ for $N = 16$ for Fourier Galerkin and several finite-difference schemes

In most practical applications the benefit of the spectral method is not the extraordinary accuracy available for large $N$ but rather the small size of $N$ necessary for a moderately accurate solution.

### 1.2.2 A Chebyshev Collocation Method for the Heat Equation

Fourier series, despite their simplicity and familiarity, are not always a good choice for the trial functions. In fact, for reasons that will be explored in the next chapter, Fourier series are only advisable for problems with periodic boundary conditions. A more versatile set of trial functions is composed of the Chebyshev polynomials. These are defined on $[-1, 1]$ by

$$T_k(x) = \cos(k \cos^{-1} x) , \tag{1.2.20}$$

for $k = 0, 1, \ldots$.

Let us focus on the linear heat equation

$$\frac{\partial u}{\partial t} - \frac{\partial^2 u}{\partial x^2} = 0 , \tag{1.2.21}$$

i.e.,

$$\mathcal{M}(u) = \frac{\partial^2 u}{\partial x^2} , \tag{1.2.22}$$

on $(-1, 1)$ with homogeneous Dirichlet boundary conditions,

$$u(-1, t) = 0, \qquad u(1, t) = 0 . \tag{1.2.23}$$

Choosing the trial functions

$$\phi_k(x) = T_k(x) , \qquad k = 0, 1, \ldots, N , \tag{1.2.24}$$

the approximate solution has the representation

$$u^N(x, t) = \sum_{k=0}^{N} a_k(t)\phi_k(x) . \tag{1.2.25}$$

In the collocation approach the requirement is that (1.2.21) be satisfied exactly by (1.2.25) at a set of collocation points $x_j$ in $(-1, 1)$:

$$\left.\frac{\partial u^N}{\partial t} - \mathcal{M}(u^N)\right|_{x=x_j} = 0 , \qquad j = 1, \ldots, N-1 . \tag{1.2.26}$$

The boundary conditions

$$u^N(-1, t) = 0, \qquad u^N(1, t) = 0 \tag{1.2.27}$$

and the initial condition

$$u^N(x_k, 0) = u(x_k, 0) , \qquad k = 0, \ldots, N , \tag{1.2.28}$$

accompany (1.2.26).

Equations (1.2.26) are based on the strong formulation of the differential equation, since the approximate solution is required to satisfy the differential equation exactly at a set of discrete points, in this case called the collocation points. One can formally obtain the same equations starting from a weak formulation of the problem by taking as test functions the (shifted) Dirac delta-functions (distributions)

$$\psi_j(x) = \delta(x - x_j) , \qquad j = 1, \ldots, N-1 , \tag{1.2.29}$$

and enforcing the conditions

$$\int_{-1}^{1} \left[\frac{\partial u^N}{\partial t} - \mathcal{M}(u^N)\right] \psi_j(x) \,\mathrm{d}x = 0 , \qquad j = 1, \ldots, N-1 \tag{1.2.30}$$

(where the integral should really be interpreted as a duality; see (A.10)).

A particularly convenient choice for the collocation points $x_j$ is

$$x_j = \cos \frac{\pi j}{N} \, . \tag{1.2.31}$$

Not only does this choice produce highly accurate approximations, but it also is economical. Note that

$$\phi_k(x_j) = \cos \frac{\pi j k}{N} \, . \tag{1.2.32}$$

This enables the Fast Fourier Transform (FFT) to be employed in the evaluation of $\mathcal{M}(u^N)|_{x=x_j}$, as is discussed in Sect. 2.4.

For the particular initial condition

$$u(x,0) = \sin \pi x \, , \tag{1.2.33}$$

the exact solution is

$$u(x,t) = e^{-\pi^2 t} \sin \pi x \, . \tag{1.2.34}$$

It has the infinite Chebyshev expansion

$$u(x,t) = \sum_{k=0}^{\infty} b_k(t) T_k(x) \, , \tag{1.2.35}$$

where

$$b_k(t) = \frac{2}{c_k} \sin\left(\frac{k\pi}{2}\right) J_k(\pi) e^{-\pi^2 t} \, , \tag{1.2.36}$$

with

$$c_k = \begin{cases} 2 \, , & k = 0 \, , \\ 1 \, , & k \geq 1 \, . \end{cases} \tag{1.2.37}$$

Because of the rapidly decaying $J_k(\pi)$ factor, the truncated series converges at an exponential rate. A well-designed collocation method will do the same. (Since the finite series (1.2.25) is not simply the truncation of the infinite series (1.2.35) at order $N$, the expansion coefficients $a_k(t)$ and $b_k(t)$ are not identical.)

Unlike a Galerkin method, which in its conventional version is usually implemented in terms of the expansion coefficients $a_k(t)$, a collocation method is implemented in terms of the nodal values $u_j(t) = u^N(x_j, t)$. Indeed, in addition to (1.2.25), we have the expansion

$$u^N(x,t) = \sum_{j=0}^{N} u_j(t) \phi_j(x),$$

where now $\phi_j$ denote the discrete (shifted) delta-functions, i.e., the unique $N$-th degree polynomials satisfying $\phi_j(x_i) = \delta_{ij}$ for $0 \leq i, j \leq N$.

(These particular functions will be more commonly denoted by the symbol $\psi_j$ in the sequel and referred to as characteristic Lagrange polynomials; see, e.g., (1.2.55)). The expansion coefficients are used only in an intermediate step, namely, in the analytic differentiation (with respect to $x$) of (1.2.25). The details of this step, which will be derived in Sect. 2.4, follow.

The expansion coefficients are given by

$$a_k(t) = \frac{2}{N\bar{c}_k} \sum_{l=0}^{N} \bar{c}_l^{\,-1} u_l(t) \cos \frac{\pi l k}{N}\,, \qquad k = 0, 1, \ldots, N\,, \tag{1.2.38}$$

where

$$\bar{c}_k = \begin{cases} 2\,, & k = 0 \text{ or } N\,, \\ 1\,, & 1 \le k \le N-1 \end{cases}. \tag{1.2.39}$$

The exact derivative of (1.2.25) is

$$\frac{\partial^2 u^N}{\partial x^2}(t) = \sum_{k=0}^{N} a_k^{(2)}(t) T_k(x)\,, \tag{1.2.40}$$

where

$$\begin{aligned} & a_{N+1}^{(1)}(t) = 0, \qquad a_N^{(1)}(t) = 0\,, \\ & \bar{c}_k a_k^{(1)}(t) = a_{k+2}^{(1)}(t) + 2(k+1) a_{k+1}(t)\,, \qquad k = N-1, N-2, \ldots, 0\,, \end{aligned} \tag{1.2.41}$$

and

$$\begin{aligned} & a_{N+1}^{(2)}(t) = 0, \qquad a_N^{(2)}(t) = 0\,, \\ & \bar{c}_k a_k^{(2)}(t) = a_{k+2}^{(2)}(t) + 2(k+1) a_{k+1}^{(1)}(t)\,, \qquad k = N-1, N-2, \ldots, 0\,. \end{aligned} \tag{1.2.42}$$

The coefficients $a_k^{(2)}$ obviously depend linearly on the nodal values $u_l$; hence, there exists a matrix $D_N^2$ such that

$$\left.\frac{\partial^2 u^N}{\partial x^2}(t)\right|_{x=x_j} = \sum_{k=0}^{N} a_k^{(2)}(t) \cos \frac{\pi j k}{N} = \sum_{l=0}^{N} (D_N^2)_{jl} u_l(t) \tag{1.2.43}$$

(see Sect. 2.4.2 for more details). By (1.2.27), we actually have $u_0(t) = u_N(t) = 0$. Substituting the above expression into (1.2.26), we end up with a system of ordinary differential equations for the nodal unknowns:

$$\frac{\mathrm{d}u_j}{\mathrm{d}t}(t) = \sum_{l=0}^{N} (D_N^2)_{jl} u_l(t), \qquad j = 1, \ldots, N-1. \tag{1.2.44}$$

Supplemented by the initial conditions (1.2.28), the preceding system of ordinary differential equations for the nodal values of the solution is readily integrated in time.

The maximum errors at $t = 1$ in the numerical solutions for a Chebyshev collocation method, a second-order finite-difference method and a fourth-order compact method are given in Fig. 1.3, along with the maximum errors for the truncated Chebyshev series of the exact solution at $t = 1$. The Chebyshev method used the $N + 1$ non-uniformly distributed collocation points (1.2.31), whereas the finite-difference methods used $N + 1$ uniformly distributed points. The maximum errors have been normalized with respect to the maximum value of the exact solution at $t = 1$. The fourth-order scheme is the classical 3-point Padé approximation,

$$u''_{i-1} + 10u''_i + u''_{i+1} = \frac{12}{(\Delta x)^2}(u_{i-1} - 2u_i + u_{i+1}) \,, \qquad i = 1, \ldots, N-1 \,, \tag{1.2.45}$$

supplemented with a compact, third-order approximation at the boundary points (see Lele (1992)), e.g.,

$$u''_0 + 11u''_1 = \frac{1}{(\Delta x)^2}(13u_0 - 27u_1 + 15u_2 - u_3) \,, \qquad i = 0 \,. \tag{1.2.46}$$

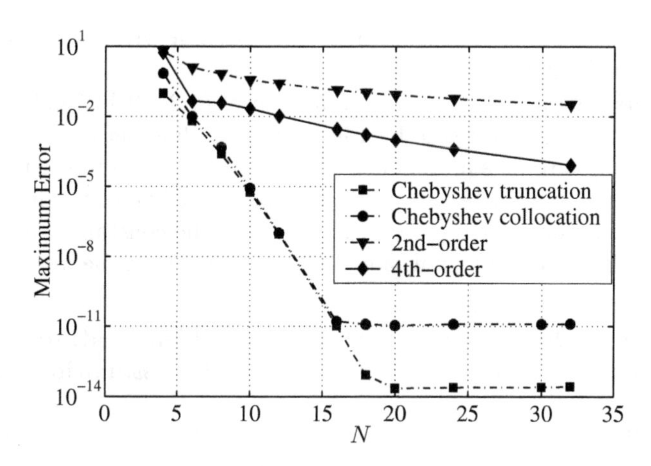


**Fig. 1.3.** Maximum errors for the heat equation problem at $t = 1$ for Chebyshev collocation and several finite-difference schemes. The Chebyshev truncation result is shown for comparison

Before leaving this example, we consider a more general equation than (1.2.21), namely,

$$\frac{\partial u}{\partial t} - \frac{\partial}{\partial x}\left(\kappa \frac{\partial u}{\partial x}\right) = 0, \tag{1.2.47}$$

where the conductivity coefficient $\kappa$ varies in $(-1,1)$ and may even depend on the solution $u$. In this case, it is not convenient to apply the collocation scheme (1.2.26) to equation (1.2.47) directly, as this would require the exact differentiation of the heat flux $\mathcal{F}(u^N) = \kappa \frac{\partial u^N}{\partial x}$. Instead, one first computes the nodal values $F_l(t) = \mathcal{F}(u^N)(x_l)$, $l = 0, \dots, N$, of this flux, then applies a transformation similar to (1.2.38), and follows that with a differentiation of the flux as in (1.2.41); the resulting expansion of the derivative is then evaluated at the collocation points. This process amounts to differentiating exactly the numerical flux $\mathcal{F}^N(u^N) = I_N(\mathcal{F}(u^N))$, which is obtained by interpolating the flux $\mathcal{F}(u^N)$ at the collocation points by a global $N$-degree algebraic polynomial. (Here and in the rest of the book, $I_N$ is a general symbol that denotes an interpolation operator.) The resulting collocation scheme reads as follows:

$$\frac{\partial u^N}{\partial t} - \frac{\partial}{\partial x} I_N\left(\kappa \frac{\partial u^N}{\partial x}\right)\bigg|_{x=x_j} = 0\,, \qquad j = 1, \dots, N-1\,. \tag{1.2.48}$$

Equivalently, we have

$$\frac{\mathrm{d}u_j}{\mathrm{d}t}(t) = \sum_{l=0}^{N} (D_N)_{jl} F_l(t), \qquad j = 1, \dots, N-1, \tag{1.2.49}$$

where $D_N$ is the Chebyshev collocation derivative matrix, which is discussed in detail in Sect. 2.4.2.

The approach used for the discretization of (1.2.47) highlights a general strategy that is adopted for collocation methods: differentiation is applied to a function only after the argument of the function is interpolated by a global polynomial at a suitable set of collocation points. Obviously, when the argument is itself a polynomial of degree $\leq N$, as in the constant-coefficient heat equation (1.2.21), the interpolation returns the value of the argument.

### 1.2.3 A Legendre Galerkin with Numerical Integration (G-NI) Method for the Advection-Diffusion-Reaction Equation

Spectral methods are also applicable to time-independent equations. The general boundary-value problem is given by the equation

$$\mathcal{M}(u) = f \tag{1.2.50}$$

to be solved in a specified domain, along with the boundary conditions

$$\mathcal{B}(u) = 0\,. \tag{1.2.51}$$

As a first example, we consider the one-dimensional advection-diffusion-reaction equation

$$\mathcal{M}(u) = \frac{\mathrm{d}\mathcal{F}(u)}{\mathrm{d}x} + \gamma u = f, \tag{1.2.52}$$

where the advection-diffusion flux is defined as

$$\mathcal{F}(u) = -\nu \frac{\mathrm{d}u}{\mathrm{d}x} + \beta u.$$

The domain for the equation is $(-1, 1)$, and the boundary conditions are

$$\mathcal{B}_1(u) = u(-1) = 0, \tag{1.2.53a}$$

$$\mathcal{B}_2(u) = \mathcal{F}(u)(1) + g = 0. \tag{1.2.53b}$$

We assume that the coefficients $\nu$, $\beta$ and $\gamma$ as well as the data $f$ may vary in the domain, and that the diffusion coefficient satisfies $\nu \geq \bar{\nu}$ for some constant $\bar{\nu} > 0$.

Trial and test functions are defined as follows. Consider the $N$-th degree Legendre orthogonal polynomial $L_N(x)$. (A detailed discussion of the properties of Legendre polynomials is furnished in Sect. 2.3.) The polynomial $L_N$ has $N-1$ extrema $x_j$, i.e., $L'_N(x_j) = 0$, for $j = 1, \ldots, N-1$; they belong to the interval $(-1, 1)$. Adding the boundary points $x_0 = -1$ and $x_N = 1$, we obtain $N+1$ points, which are high-precision quadrature nodes (they are termed the *Legendre Gauss-Lobatto nodes*); indeed, there exist weights $w_j$ such that the quadrature formula

$$\int_{-1}^{1} p(x)\,\mathrm{d}x \sim \sum_{j=0}^{N} p(x_j) w_j \tag{1.2.54}$$

is exact for all polynomials $p$ of degree $\leq 2N-1$. Based on these nodes, we now introduce the *characteristic Lagrange polynomials*

$$\psi_j(x) = \frac{1}{N(N+1)} \frac{(1-x^2)}{(x_j - x)} \frac{L'_N(x)}{L_N(x_j)}, \qquad j = 0, \ldots, N, \tag{1.2.55}$$

which are discrete (shifted) delta-functions, i.e., they are $N$-th degree polynomials which approximate the (shifted) Dirac delta-functions $\delta(x - x_j)$, as they satisfy

$$\psi_j(x_k) = \delta_{jk}, \qquad j, k = 0, \ldots, N. \tag{1.2.56}$$

In view of the boundary condition (1.2.53a), we drop $\psi_0$. The remaining functions $\psi_j$, $j = 1, \ldots, N$, will be our trial and test functions. The approximate solution is sought in the form

$$u^N(x) = \sum_{l=1}^{N} u_l \psi_l(x). \tag{1.2.57}$$

Note that the coefficients in the expansion are precisely the values of $u^N$ at the nodes $u_l = u^N(x_l)$, $l = 1, \ldots, N$.

In order to arrive at the equations which uniquely define $u^N$, we have to go back to the exact solution $u$ of our boundary-value problem. We shall derive a set of integral conditions satisfied by the exact solution (which constitute the weak formulation of the problem). The same integral conditions are enforced on the discrete solution. To this end, consider (1.2.52), multiply both sides by any test function $\psi_j$ and integrate over the interval $(-1, 1)$; we obtain the equations

$$\int_{-1}^{1} \frac{\mathrm{d}\mathcal{F}(u)}{\mathrm{d}x} \psi_j \,\mathrm{d}x + \int_{-1}^{1} \gamma u \, \psi_j \,\mathrm{d}x = \int_{-1}^{1} f \, \psi_j \,\mathrm{d}x, \quad j = 1, \ldots, N. \tag{1.2.58}$$

Integrating the first term by parts, we get

$$\begin{aligned} \int_{-1}^{1} \frac{\mathrm{d}\mathcal{F}(u)}{\mathrm{d}x} \psi_j \,\mathrm{d}x &= -\int_{-1}^{1} \mathcal{F}(u) \frac{\mathrm{d}\psi_j}{\mathrm{d}x} \,\mathrm{d}x + [\mathcal{F}(u) \, \psi_j]_{-1}^{1} \\ &= -\int_{-1}^{1} \mathcal{F}(u) \frac{\mathrm{d}\psi_j}{\mathrm{d}x} \,\mathrm{d}x - g \, \delta_{jN}, \end{aligned}$$

where we have used the boundary condition (1.2.53b), as well as the relations (1.2.56). Thus, recalling the definition of the flux $\mathcal{F}(u)$, we see that $u$ satisfies

$$\begin{aligned} \int_{-1}^{1} \nu \frac{\mathrm{d}u}{\mathrm{d}x} \frac{\mathrm{d}\psi_j}{\mathrm{d}x} \,\mathrm{d}x - \int_{-1}^{1} \beta u \frac{\mathrm{d}\psi_j}{\mathrm{d}x} \,\mathrm{d}x + \int_{-1}^{1} \gamma u \, \psi_j \,\mathrm{d}x \\ = \int_{-1}^{1} f \, \psi_j \,\mathrm{d}x + g \, \delta_{jN}, \qquad j = 1, \ldots, N. \end{aligned} \tag{1.2.59}$$

This is precisely the set of equations which we ask to be satisfied by $u^N$ as well. If we replace $u$ by $u^N$ in (1.2.59), we obtain the numerical scheme

$$\begin{aligned} \int_{-1}^{1} \nu \frac{\mathrm{d}u^N}{\mathrm{d}x} \frac{\mathrm{d}\psi_j}{\mathrm{d}x} \,\mathrm{d}x - \int_{-1}^{1} \beta u^N \frac{\mathrm{d}\psi_j}{\mathrm{d}x} \,\mathrm{d}x + \int_{-1}^{1} \gamma u^N \, \psi_j \,\mathrm{d}x \\ = \int_{-1}^{1} f \, \psi_j \,\mathrm{d}x + g \, \delta_{jN}, \qquad j = 1, \ldots, N. \end{aligned} \tag{1.2.60}$$

Note that $u^N$ satisfies (1.2.53a) exactly; conversely, (1.2.53b) is not enforced directly on $u^N$, yet it has been incorporated into (1.2.59). We say that we enforce this boundary condition in a *weak*, or *natural*, manner.

Since the integrals in (1.2.59) are evaluated exactly, we have obtained a *pure Galerkin* scheme. However, only in special situations (e.g., constant coefficients and data) can the integrals above be computed analytically. Otherwise, we have to resort to numerical integration, in which case the natural choice is the quadrature formula (1.2.54). In this way, we obtain the following modified scheme, which we term the *Galerkin with numerical integration scheme*, or in short, the G-NI scheme:

$$\sum_{k=0}^{N}\left(\nu\frac{\mathrm{d}u^N}{\mathrm{d}x}\frac{\mathrm{d}\psi_j}{\mathrm{d}x}\right)(x_k)\,w_k-\sum_{k=0}^{N}\left(\beta u^N\frac{\mathrm{d}\psi_j}{\mathrm{d}x}\right)(x_k)\,w_k+\sum_{k=0}^{N}(\gamma u^N\psi_j)(x_k)\,w_k$$
$$=\sum_{k=0}^{N}(f\,\psi_j)(x_k)\,w_k+g\,\delta_{jN},\quad j=1,\dots,N. \tag{1.2.61}$$

Inserting the expansion (1.2.57) for $u^N$, we can rephrase this scheme as a system $K\mathbf{u}=\mathbf{b}$ of $N$ algebraic equations in the unknowns $u_l$; in particular, they are

$$\sum_{l=1}^{N}K_{jl}u_l=b_j,\qquad j=1,\dots,N, \tag{1.2.62}$$

where the matrix entries are

$$K_{jl}=\sum_{k=0}^{N}\left(\nu\frac{\mathrm{d}\psi_l}{\mathrm{d}x}\frac{\mathrm{d}\psi_j}{\mathrm{d}x}\right)(x_k)\,w_k-\left(\beta\frac{\mathrm{d}\psi_j}{\mathrm{d}x}\right)(x_l)\,w_l+\gamma(x_j)w_j\delta_{lj}\,,$$

and the right-hand side components are

$$b_j=f(x_j)w_j+g\,\delta_{jN}.$$

Efficient solution techniques for such a system are described in Sect. 4.2.

The G-NI scheme can be given a pointwise, or collocation-like, interpretation, which serves to highlight the effect of the weak enforcement of the boundary condition (1.2.53b). To this end, we denote by $I_N\varphi$ the $N$-th degree algebraic polynomial that interpolates a function $\varphi$ at the Gauss-Lobatto nodes $x_j,\ j=0,\dots,N$; this allows us to introduce the numerical flux

$$\mathcal{F}^N(u^N)=I_N(\mathcal{F}(u^N)).$$

The two first sums in (1.2.61) can be written as

$$\sum_{k=0}^{N}\left(\nu\frac{\mathrm{d}u^N}{\mathrm{d}x}\frac{\mathrm{d}\psi_j}{\mathrm{d}x}\right)(x_k)\,w_k-\sum_{k=0}^{N}\left(\beta u^N\frac{\mathrm{d}\psi_j}{\mathrm{d}x}\right)(x_k)\,w_k=$$
$$=-\sum_{k=0}^{N}\left(\mathcal{F}(u^N)\frac{\mathrm{d}\psi_j}{\mathrm{d}x}\right)(x_k)\,w_k=-\sum_{k=0}^{N}\left(\mathcal{F}^N(u^N)\frac{\mathrm{d}\psi_j}{\mathrm{d}x}\right)(x_k)\,w_k.$$

Now it is crucial to observe that both the terms $\mathcal{F}^N(u^N)\dfrac{\mathrm{d}\psi_j}{\mathrm{d}x}$ and $\dfrac{\mathrm{d}\mathcal{F}^N(u^N)}{\mathrm{d}x}\psi_j$ are polynomials of degree $\le 2N-1$; hence, they can be integrated exactly by the quadrature formula (1.2.54). Thus, we are allowed to counter-integrate by parts in the last sum appearing above, obtaining

$$
\begin{aligned}
-\sum_{k=0}^{N}\Big(\mathcal{F}^N(u^N)\frac{\mathrm{d}\psi_j}{\mathrm{d}x}\Big)(x_k)\,w_k &= -\int_{-1}^{1}\mathcal{F}^N(u^N)\,\frac{\mathrm{d}\psi_j}{\mathrm{d}x}\,\mathrm{d}x \\
&= \int_{-1}^{1}\frac{\mathrm{d}\mathcal{F}^N(u^N)}{\mathrm{d}x}\,\psi_j\,\mathrm{d}x - [\mathcal{F}^N(u^N)\,\psi_j]_{-1}^{1} \\
&= \sum_{k=0}^{N}\Big(\frac{\mathrm{d}\mathcal{F}^N(u^N)}{\mathrm{d}x}\,\psi_j\Big)(x_k)\,w_k - \mathcal{F}(u^N)(1)\,\psi_j(1)\,.
\end{aligned}
$$

If we insert this expression into (1.2.61) and use the relations (1.2.56), we obtain the following equivalent formulation of the G-NI scheme:

$$
\Big(\frac{\mathrm{d}\mathcal{F}^N(u^N)}{\mathrm{d}x}+\gamma u^N\Big)(x_j)w_j - \mathcal{F}(u^N)(1)\delta_{jN} = f(x_j)w_j + g\delta_{jN}, \quad j=1,\dots,N. \tag{1.2.63}
$$

For $j=1,\dots,N-1$, this is simply

$$
\frac{\mathrm{d}\mathcal{F}^N(u^N)}{\mathrm{d}x}+\gamma u^N - f\Big|_{x=x_j} = 0, \tag{1.2.64}
$$

i.e., at the internal quadrature points we are collocating the differential equation after replacing the exact flux $\mathcal{F}(u^N)$ by the numerical one $\mathcal{F}^N(u^N)$. For $j=N$ we get

$$
\frac{\mathrm{d}\mathcal{F}^N(u^N)}{\mathrm{d}x}+\gamma u^N - f\Big|_{x=1} - \frac{1}{w_N}\,(\mathcal{F}(u^N)+g)\big|_{x=1} = 0, \tag{1.2.65}
$$

i.e., at $x=1$ we are collocating a particular linear combination of the discrete form of the differential equation and the boundary condition. Since $1/w_N$ grows like $N^2$ as $N\to\infty$ (see Sect. 2.3.1), (1.2.65) shows that the boundary condition is approximately fulfilled in a more and more accurate way as the equation residual $\mathcal{M}^N(u^N)-f\big|_{x=1}$ gets smaller and smaller for $N\to\infty$ (recall that the residual vanishes at all internal nodes, see (1.2.64)).

The example addressed above is indeed a paradigm for a general class of second-order steady problems. The G-NI discretization consists of collocating the differential equation (with numerical flux) at the internal Gauss Lobatto nodes; Dirichlet boundary conditions (i.e., conditions involving only pointwise values of the unknown function) are fulfilled exactly at the boundary points, whereas Neumann or Neumann-like boundary conditions (i.e., conditions involving also the first derivative(s) of the unknown function) are enforced via an intrinsically (and unambiguously) defined penalty method.

The accuracy of the G-NI method is illustrated by the following example. We consider the problem (1.2.50)–(1.2.53) in the interval $(-1,1)$ with $\nu=1$, $\beta(x)=\cos(\pi/4\cdot(1+x))$ and $\gamma=1$. The right-hand side $f(x)$ and the datum $g$ are computed so that the exact solution is

$$
u(x)=\cos(3\pi(1+x))\,\sin(\pi/5\cdot(x+0.5))+\sin(\pi/10)\,. \tag{1.2.66}
$$

For several values of $N$, we denote by $u^N$ the G-NI solution ($N$ is the polynomial degree) and by $u^p$ ($p = 1,2,3$) the (piecewise-polynomial) finite-element solution corresponding to a subdivision in subintervals of equal size. In all cases, $N+1$ denotes the total number of nodal values. In Fig. 1.4 (left) we plot the maximum error of the solution, while on the right we plot the absolute error of the boundary flux $|(\nu \frac{du^p}{dx}(1) + \beta u^p(1)) - g|$ for $p = 1,2,3,N$. The two errors exhibit a similar decay with respect to $N$. In particular, the boundary condition at $x = 1$ is fulfilled with spectral accuracy.

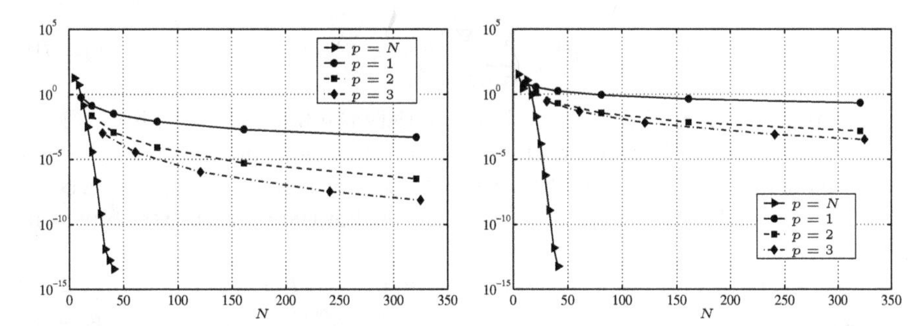


**Fig. 1.4.** Comparison between the accuracy of the G-NI solution (corresponding to the curve $p = N$) and the finite-element solutions of order $p = 1,2$ and 3 versus $N$ which represents the total number of nodal values. The maximum error between the numerical solution and the exact one $u(x) = \cos(3\pi(1+x)) \cdot \sin(\pi/5 \cdot (x+0.5)) + \sin(\pi/10)$ (*left*) and the absolute value of the error on the flux at $x = 1$ (*right*)

### 1.2.4 A Legendre Tau Method for the Poisson Equation

Our second example of a steady boundary-value problem is the Poisson equation on $(-1,1) \times (-1,1)$, with homogeneous Dirichlet boundary conditions. The choice of $\mathcal{M}$ and $\mathcal{B}$ in (1.2.50) and (1.2.51) is as follows:

$$\mathcal{M}(u) = -\left(\frac{\partial^2 u}{\partial x^2} + \frac{\partial^2 u}{\partial y^2}\right), \tag{1.2.67}$$

$$\mathcal{B}_1(u) = u(x,-1), \tag{1.2.68a}$$

$$\mathcal{B}_2(u) = u(x,+1), \tag{1.2.68b}$$

$$\mathcal{B}_3(u) = u(-1,y), \tag{1.2.68c}$$

$$\mathcal{B}_4(u) = u(+1,y). \tag{1.2.68d}$$

(We prefer to use the negative sign in second-derivative operators such as (1.2.67) so that $\mathcal{M}(u)$ is a positive, rather than a negative, operator. Although this might be disconcerting to some, it does simplify the discussion of the mathematical properties of the operator and its numerical approximations. For example, some spectral approximations to (1.2.67)–(1.2.68) yield symmetric and positive-definite matrices, albeit not the particular approximation discussed in the present subsection. This will become clearer in due course, particularly in Chaps. 4, 6 and 7.)

Both Legendre and Chebyshev polynomials are suitable trial functions. A two-dimensional Legendre expansion is produced by the tensor-product choice

$$\phi_{kl}(x,y) = L_k(x)L_l(y)\,, \qquad k,l = 0,1,\dots,N\,, \tag{1.2.69}$$

where $L_k$ is the Legendre polynomial of degree $k$. The approximate solution is

$$u^N(x,y) = \sum_{k=0}^{N}\sum_{l=0}^{N} a_{kl}L_k(x)L_l(y)\,. \tag{1.2.70}$$

Note that the trial functions do not satisfy the boundary conditions individually. (In most Galerkin methods the trial functions do satisfy the boundary conditions.) In this case two separate sets of test functions are used to enforce the PDE and the boundary conditions. For the PDE the test functions are

$$\psi_{kl}(x,y) = Q_k(x)Q_l(y)\,, \qquad k = 0,1,\dots,N-2\,, \tag{1.2.71}$$

where

$$Q_k(x) = \frac{2k+1}{2}L_k(x)\,; \tag{1.2.72}$$

for the boundary conditions they are

$$\chi_k^i(x) = Q_k(x)\,, \qquad \begin{array}{l} i = 1,2\,, \\ k = 0,1,\dots,N\,, \end{array} \tag{1.2.73a}$$

(1.2.73b)

$$\chi_l^i(y) = Q_l(y)\,, \qquad \begin{array}{l} i = 3,4\,, \\ l = 0,1,\dots,N\,. \end{array} \tag{1.2.73c}$$

The integral conditions for the differential equations are

$$\int_{-1}^{1} \mathrm{d}y \int_{-1}^{1} \mathcal{M}(u^N)\psi_{kl}(x,y)\,\mathrm{d}x = 0\,, \qquad k,\ l = 0,1,\dots,N-2\,, \tag{1.2.74}$$

while the equations for the boundary conditions are

$$\int_{-1}^{1} \mathcal{B}_i(u^N)\chi_k^i(x)\,\mathrm{d}x = 0\,, \qquad \begin{array}{l} i = 1,2\,, \\ k = 0,1,\dots,N\,, \end{array} \tag{1.2.75a}$$

$$\int_{-1}^{1} \mathcal{B}_i(u^N)\chi_l^i(y)\,\mathrm{d}y = 0\,, \qquad \begin{array}{l} i = 3,4\,, \\ l = 0,1,\dots,N\,. \end{array} \tag{1.2.75b}$$

Four of the conditions in (1.2.75) are linearly dependent upon the others; in effect the boundary conditions at each of the four corner points have been applied twice. For the Poisson equation the above integrals may be performed analytically. The result is

$$-(a_{kl}^{(2,0)} + a_{kl}^{(0,2)}) = f_{kl} \,, \qquad k,\ l = 0, 1, \ldots, N-2 \,, \tag{1.2.76}$$

$$\sum_{k=0}^{N} a_{kl} = 0 \,, \quad \sum_{k=0}^{N} (-1)^k a_{kl} = 0 \,, \qquad l = 0, 1, \ldots, N \,, \tag{1.2.77a}$$

$$\sum_{l=0}^{N} a_{kl} = 0 \,, \quad \sum_{l=0}^{N} (-1)^l a_{kl} = 0 \,, \qquad k = 0, 1, \ldots, N \,, \tag{1.2.77b}$$

where

$$f_{kl} = \int_{-1}^{1} \mathrm{d}y \int_{-1}^{1} f(x,y) \psi_{kl}(x,y) \, \mathrm{d}x \,, \tag{1.2.78}$$

$$a_{kl}^{(2,0)} = \left(k + \tfrac{1}{2}\right) \sum_{\substack{p=k+2 \\ p+k \text{ even}}}^{N} [p(p+1) - k(k+1)] a_{pl} \,, \tag{1.2.79a}$$

$$a_{kl}^{(0,2)} = \left(l + \tfrac{1}{2}\right) \sum_{\substack{q=l+2 \\ q+l \text{ even}}}^{N} [q(q+1) - l(l+1) a_{kq}] \,. \tag{1.2.79b}$$

These last two expressions represent the expansions of $\partial^2 u^N / \partial x^2$ and $\partial^2 u^N / \partial y^2$, respectively, in terms of the trial functions.

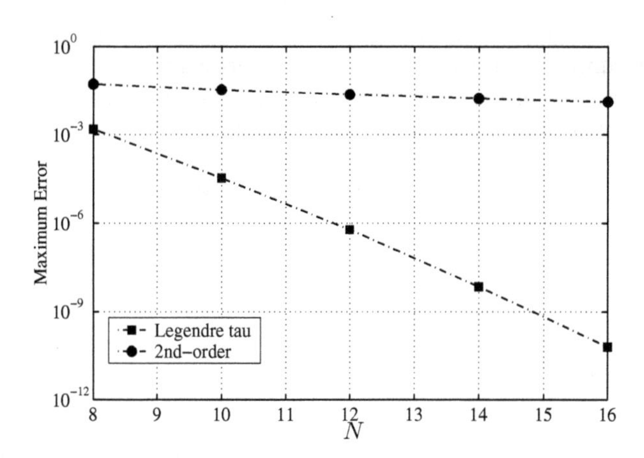


**Fig. 1.5.** Maximum errors for the Poisson problem for Legendre tau and second-order finite-difference schemes

The Legendre tau approximation to the Poisson equation consists of (1.2.76) and (1.2.77). An efficient scheme for the solution of these equations is provided in Sect. 4.1.

The specific example that will be used to illustrate the accuracy of this method is

$$f(x,y) = 2\pi^2 \sin \pi x \, \sin \pi y \,, \tag{1.2.80}$$

which corresponds to the analytic solution

$$u(x,y) = \sin \pi x \, \sin \pi y \,. \tag{1.2.81}$$

The results are given in Fig. 1.5 along with results for a second-order finite-difference scheme. The integer $N$ denotes the degree of the expansion (1.2.70) in each dimension for the Legendre tau method and the number of uniform intervals in each dimension for the finite-difference method.

### 1.2.5 Basic Aspects of Galerkin, Collocation, G-NI and Tau Methods

The Galerkin, collocation, G-NI and tau methods are more general than suggested by any of the above examples. In a broad sense, pure Galerkin and tau methods are implemented in terms of the expansion coefficients, whereas collocation methods and G-NI (Galerkin with numerical integration) methods are implemented in terms of the physical space values of the unknown function. The first example illustrated only one of the key aspects of Galerkin methods – the test functions are the same as the trial functions. The other important aspect is that the trial functions must individually satisfy all or part of the boundary conditions (the remaining ones are enforced weakly within the integral conditions). In the case of periodic boundary conditions the trigonometric polynomials automatically satisfy these requirements. Otherwise, simple linear combinations of the orthogonal polynomials will usually suffice. For example, an obvious choice of trial functions for a Chebyshev Galerkin approximation to the fourth example is

$$\phi_k(x) = \begin{cases} T_0(x) - T_k(x) \,, & k \text{ even} \geq 2 \,, \\ T_1(x) - T_k(x) \,, & k \text{ odd} \geq 3 \,; \end{cases}$$

a computationally more efficient choice (see Sect. 2.3.3) is provided by

$$\phi_k(x) = T_{k-2}(x) - T_k(x) \,, \qquad k \geq 2 \,.$$

On the other hand, for the tau method the trial functions do not individually satisfy the boundary conditions. Thus, some equations are needed to ensure that the global expansion satisfies the boundary conditions. Some of the integral equations corresponding to the highest order test functions are dropped in favor of these boundary condition equations.

The collocation method uses the values of the function at certain physical points as the fundamental representation; the expansion functions are employed solely for evaluating derivatives (and only when a fast transform is available and convenient). The collocation points for both the differential equations and the boundary conditions are usually the same as the physical grid points. The most effective choice for the grid points are those that correspond to quadrature formulas of maximum precision.

The Galerkin with numerical integration (G-NI) method aims at preserving the advantages of both Galerkin and collocation methods. Integrals appearing in the weak formulation of the problem are efficiently approximated by the quadrature formulas mentioned above. Usually, the solution is again represented in physical space through its values at a selected set of nodes. In most cases, as in the example in Sect. 1.2.3, the nodes that serve to represent the solution coincide with the nodes that are used for quadrature. Some exceptions are discussed in later chapters. Certain boundary conditions (for instance, those involving derivatives for second-order operators) are imposed weakly, through a penalty approach that naturally stems from the weak formulation of the problem.

## 1.3 Three-Dimensional Applications in Fluids: A Look Ahead

Chapters 2–4 of CHQZ3 are devoted to the details of spectral algorithms for investigations of instability, transition and turbulence in fluid flows. The simplest class of flows, termed *laminar flow*, comprises those flows in which the motion is quite regular and predictable, even though possibly unsteady. (Plane Poiseuille flow, discussed in CHQZ3, Sects. 1.3, 2.3 and 3.4, is one example of a laminar flow.) Laminar flows are either stable or unstable. In somewhat oversimplified terms, linearly stable flows are those in which all sufficiently small perturbations to the mean flow decay, whereas unstable flows are those in which some small perturbations grow. Many flows start out as laminar, become unstable (in space or time), and eventually undergo a transition to turbulent flow. The complex category of *turbulent flow* is described by Hinze (1975) as

> *"Turbulent fluid motion is an irregular condition of flow in which the various quantities show a random variation with time and space coordinates, so that statistically distinct average values can be discerned."*

In this section we illustrate some representative flow physics results from many of the principal fully spectral algorithms that we discuss in Chaps. 2–4.

Turbulent flows contain a wide range of length scales, bounded above by the geometric dimension of the flow field and bounded below by the dissipative action of the molecular viscosity (see, for instance, Tennekes and Lumley (1972, Chap. 3)). The ratio of the macroscopic (largest) *integral length scale* $L$ to the microscopic (smallest) length $\eta$ (usually known as the *Kolmogorov length scale*) is

$$\frac{L}{\eta} = \mathrm{Re}^{3/4} ,$$

where the Reynolds number Re is

$$\mathrm{Re} = \frac{uL}{\nu} , \tag{1.3.1}$$

with $\nu$ denoting the kinematic viscosity and $u = \left(\overline{u'^2}/3\right)^{1/2}$, where $u'$ is the fluctuating velocity, and the bar denotes time averaging. To resolve these scales, $N$ mesh points would be needed in each direction, where

$$N = c_1 \frac{L}{\eta} .$$

(A summary of nondimensionalization in general and Reynolds numbers in particular is provided in CHQZ3, Sect. 1.1.4.)

Two simple classes of turbulent flows are homogeneous turbulence, for which the flow properties are invariant with respect to translations, and isotropic turbulence, for which the flow properties everywhere are invariant with respect to rotations. (Isotropic turbulence is necessarily homogeneous.) For the simulation of homogeneous turbulence with a spectral method, it is appropriate to take $c_1 = 2$; for a fourth-order scheme $c_1$ would be about 6 and for a second-order scheme about 24. (These estimates are based on the typical requirement of 0.1% or better accuracy per period, using estimates such as those by Kreiss and Oliger (1972), and conclusions from the channel flow computations presented in CHQZ3, Sect. 1.3.) The ratio of the time scales of the macroscopic and microscopic motions is $T/t = \sqrt{\mathrm{Re}}$. Consequently, the number of time-steps required to describe the flow during the characteristic period (or *temporal scale*) of the physically significant events is

$$N_{Ts} = c_0 \sqrt{\mathrm{Re}} , \tag{1.3.2}$$

where the multiplicative factor, $c_0$, is between 100 and 1000 depending on the time-stepping algorithm and the time interval needed to obtain reasonable statistics for the flow. Now, the number of operations required to update the solution per time-step of a multistep scheme such as Adams-Bashforth or per stage of a multistage scheme such as Runge-Kutta is

$$c_2 N^3 \log_2 N + c_3 N^3 ,$$

where, for the spectral method, $c_2 = 45$, $c_3 = 35$, for the fourth-order spatial method, $c_2 = 17$, $c_3 = 120$, and for the second-order spatial method, $c_2 = 17$, $c_3 = 60$. (For the finite-difference methods, this assumes that the convection term is treated explicitly, the diffusion term is treated implicitly, a Poisson equation is solved for the pressure, and that the implicit equations for the finite-difference method are solved exactly using FFTs. See CHQZ3, Sect. 3.3 for the details of the spectral algorithm.) Thus, for homogeneous turbulence simulations, the storage requirement is roughly proportional to

$$4c_1^3 \mathrm{Re}^{9/4} , \tag{1.3.3}$$

and the total number of operations is approximately

$$c_0\, c_1^3 \mathrm{Re}^{11/4} \left[ c_2 \log_2 (c_1 \mathrm{Re}^{3/4}) + c_3 \right] . \tag{1.3.4}$$

The estimates above provide the resolution requirements for computations in which all the scales of the flow are resolved numerically. Such a computation is known as a *direct numerical simulation* (DNS). Many of the examples that follow are from DNS computations.

The original Orszag and Patterson (1972) computations were performed in an era in which the fastest supercomputer had a speed of roughly 1 MFlop ($10^6$ floating point operations per second). Using a typical value of $c_0 = 500$, the computer time required then for one realization of homogeneous turbulence by a spectral method was, according to (1.3.4), about 10 hours for their $\mathrm{Re} = 45$ cases. (Their computations used $N = 32$ modes in each direction.) For sustained performances typical of the fastest supercomputers circa 1980 (100 MFlop), the computer time required for one realization of homogeneous turbulence by a spectral method is 6 minutes for $\mathrm{Re} = 45$ and 2 years for $\mathrm{Re} = 3000$ (for the Brachet et al. (1983) case mentioned below, although they were able to save a factor of 64 by exploiting symmetries). Assuming a sustained performance of 1 TFlop ($10^{12}$ floating point operations per second, typical of the very fastest supercomputers circa 2000), the computer time required for one realization of homogeneous turbulence by a spectral method is about 10 hours for $\mathrm{Re} = 3000$, and about 4 months for $\mathrm{Re} = 40,000$ (for the Kaneda and Ishihara (2006) results mentioned below).

Spectral methods have been singularly successful for this problem since the corresponding requirements for a fourth-order finite-difference method are typically a factor of 10 longer in time and a factor of 20 larger in storage. Second-order finite-difference methods require more than 3 orders of magnitude more resources than spectral methods on this problem. Moreover, Fourier functions arise naturally in the theoretical analysis of homogeneous turbulence, and they are the natural choice of trial functions for spectral methods. Thus, the spectral methods, apart from their computational efficiency, have the added advantage of readily permitting one to monitor and diagnose nonlinear interactions which contribute to resonance effects, energy transfer, dissipation and other dynamic features. Furthermore, if there are any symmetries underlying a problem, and symmetry-breaking phenomena are precluded, spectral methods permit unique exploitation of these symmetries. (Since the finite-difference methods cannot benefit from the symmetries exploited by Brachet et al. (1983), even the fourth-order method is nearly a thousand times less efficient than the spectral method in this case.) These advantages in computational efficiency are so compelling that they have motivated many flow physics research groups to adopt spectral methods despite their additional complexity rather than simply waiting for increased computational power to make their desired computations feasible. These advantages have also inspired many numerical analysts to develop more efficient spectral methods and to provide their firm theoretical foundation.

Much theoretical work on homogeneous turbulence has focused on the details of the inertial range, which is the range of scales of motion (well observed experimentally) that are not directly affected by the energy maintenance and dissipation mechanisms (Mestayer et al. (1970)) and that possess an energy spectrum exhibiting a scaling behavior (Grant, Stewart, and Moilliet (1962)):

$$E(k,t) = k^{-m}$$

where $k$ is the magnitude of the wavenumber vector and $m$ is close to $5/3$. The spectrum with $m = 5/3$ is the famous Kolmogorov spectrum. The huge Reynolds numbers required to produce an extended inertial range are experimentally accessible only in geophysical flows such as planetary boundary layers and tidal channels.

The pioneering simulations of isotropic turbulence by Orszag and Patterson (1972) evolved over the subsequent decade-and-a-half to the first numerically computed three-dimensional inertial range by Brachet et al. (1983). (See CHQZ3, Sects. 3.3.1 and 3.3.2 for details on this Fourier Galerkin algorithm.) The Reynolds number was 3000 and, of course, crude by experimental standards. This calculation of the Taylor-Green vortex was feasible only because the symmetries of the problem were fully exploitable with the spectral method to obtain an effective resolution of $256^3$, i.e., the equivalent of $N = 256$ modes in each spatial direction. Among the salient results of this study is the physical insight gained into the behavior of turbulence at high Reynolds number, including the formation of an inertial range and the geometry of the regions of high vorticity.

Two decades later Kaneda and Ishihara (2006) (see also Yokokawa et al. (2002)) exploited 512 nodes of the Earth Simulator (then the world's fastest computer) to perform isotropic turbulence simulations using a very similar, Fourier spectral algorithm on grids as large as $4096^3$. (The sustained speed was as fast as 16 TFlop.) Figure 1.6 illustrates the regions of intense vorticity in 1/64 of the volume of their $2048^3$ simulation for Re $= 16,135$. The macroscopic scale $L$ is approximately 80% the size of one edge of the figure, and the microscopic scale $\eta$ is 0.06% of the edge length. Among the many results obtained from their high-resolution simulations was convincing evidence that the scaled energy spectrum (where the wavenumber is scaled by the inverse of the Kolmogorov length scale $\eta = (\nu^3/\bar{\epsilon})^{1/4}$, with $\nu$ the viscosity and $\bar{\epsilon}$ the average dissipation rate) is not the classical Kolmogorov result of $k^{-5/3}$, but rather $k^{-m}$ with $m \simeq 5/3 - 0.10$.

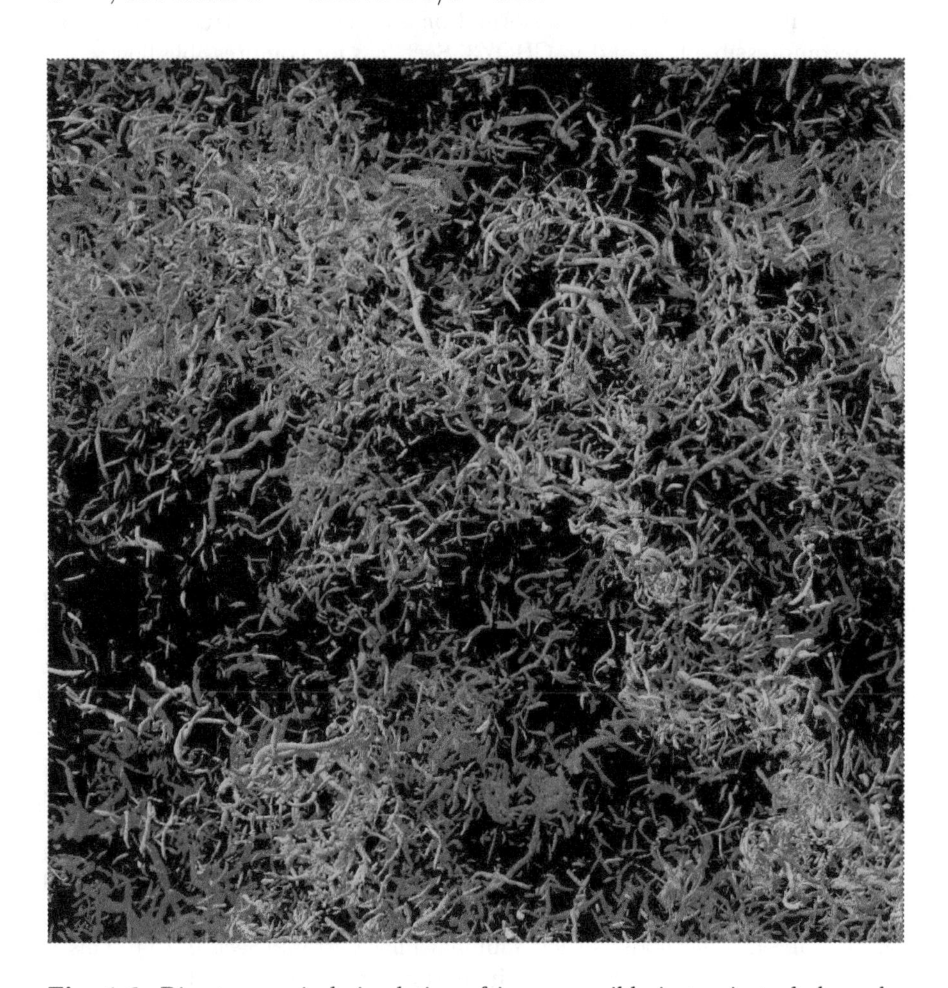

**Fig. 1.6.** Direct numerical simulation of incompressible isotropic turbulence by Kaneda and Ishihara (2006) on a $2048^3$ grid. The figure shows the regions of intense vorticity in a subdomain with 1/4 the length in each coordinate direction of the full domain [Reprinted with kind permission by the authors]

Rogallo (1977) developed a transformation that permits Fourier spectral methods to be used for homogeneous turbulence flows, such as flows with uniform shear. Blaisdell, Mansour and Reynolds (1993) used the extension of this transformation to the compressible case to simulate compressible, homogeneous turbulence in uniform shear on $192^3$ grids ($N = 192$ grid points in each spatial direction) using a Fourier collocation method. (In this example, as in all the examples cited in this section for inhomogeneous flows, the $y$ direction is the direction of inhomogeneity.) Figure 1.7 illustrates the coalescence of sound waves that is responsible for enhanced turbulence production in compressible flows. The Rogallo transformation is described in CHQZ3, Sect. 3.3.3 for incompressible flow and in CHQZ3, Sect. 4.3 for compressible flow.

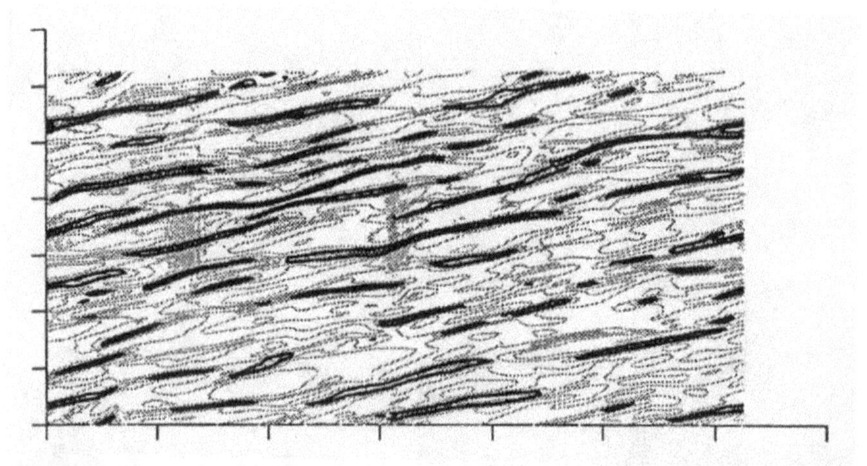

**Fig. 1.7.** Two-dimensional slice illustrating contours of the pressure field from a compressible homogeneous turbulence DNS by Blaisdell and Zeman (1992) [Reprinted with permission from G.A. Blaisdell, O. Zeman (1992); Center for Turbulence Research, Stanford University/NASA Ames Research Center]

The applications cited above were all for problems with no physical boundaries. Spectral algorithms for problems with solid boundaries are more subtle, largely because a pure Fourier method is no longer appropriate. It was not until the late 1970's that reliable Fourier-Chebyshev algorithms were applied to the simplest wall-bounded flows (Orszag and Kells (1980), Kleiser and Schumann (1980)). The principal advantage of such spectral methods over finite-difference methods is their minimal phase errors (Sect. 1.2.1). This is especially important in numerical simulations of instability and transition to turbulence, because such simulations must follow the evolution and nonlinear interaction of waves through several characteristic periods. Since phase errors are cumulative, a method that admits phase errors of even a few percent per period is unacceptable.

Kleiser and Schumann (1984) devised an influential algorithm for plane channel flow using two Fourier directions and one Chebyshev direction. This algorithm was later used by Gilbert and Kleiser (1990) for the first simulation of the complete transition to turbulence process in a wall-bounded flow using a $128^3$ grid. Figure 1.8 illustrates the evolution of one of the principal diagnostics of a transitional flow – the wall-normal shear of the streamwise velocity $\partial u/\partial y$. The ordinate in the top part of the figure is the Reynolds number based on the wall shear velocity; it is given by $\mathrm{Re}_\tau = \sqrt{\frac{1}{\nu}\frac{\partial \bar{u}}{\partial y}}\ h$, where $h$ is the channel half-width and $\bar{u}(y,t)$ is the average over $x$ and $z$ of the streamwise velocity. The bottom part of the figure illustrates the evolution of the vertical shear at the spanwise station containing the peak shear. These detailed results compared very favorably with the vibrating ribbon experiments of Nishioka, Asai and Iida (1980). The $t = 136$ frame was already computed by Kleiser and Schumann (1984) at lower resolution. (The Kleiser-Schumann algorithm is given in detail in CHQZ3, Sect. 3.4.1.)

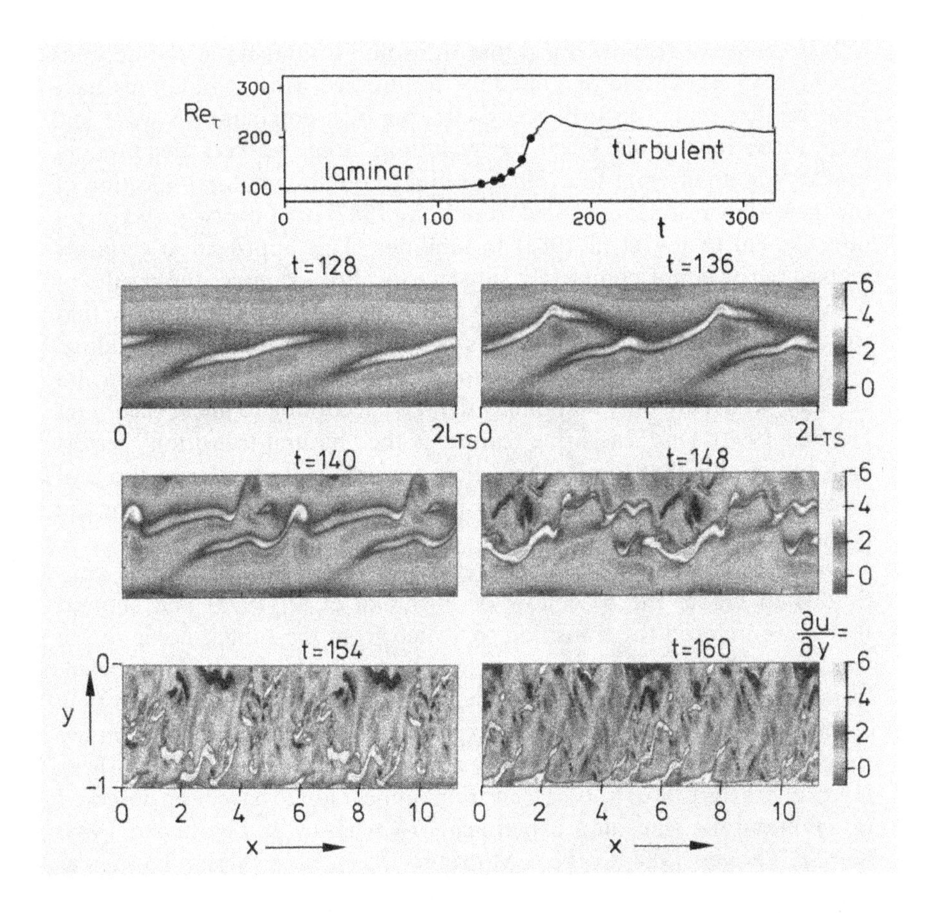


**Fig. 1.8.** DNS of transition to turbulence in plane channel flow by Gilbert and Kleiser (1990). The top figure illustrates the evolution in time of the Reynolds number based on wall friction velocity. The remaining frames illustrate the shear, $\partial u/\partial y$, in the bottom half of the channel in a two-dimensional slice at the spanwise ($z$) location containing the maximum shear [Reprinted with permission from N. Gilbert, L. Kleiser (1990); © 1990, Taylor and Francis Group]

Another widely-used algorithm, this one based on the vorticity-velocity equations, was originally developed by Kim, Moin and Moser (1987) for plane channel flow (see CHQZ3, Sect. 3.4.1). Figure 1.9 shows results from Rogers and Moser (1992) using the adaptation of this algorithm to incompressible, free shear layers; Fourier series are employed in the two homogeneous directions ($x$ and $z$) and Jacobi polynomials (see Sect. 2.5) in the $y$ direction. This figure, based on computations on a $64 \times 128 \times 64$ grid, illustrates several aspects of the vorticity from a simulation that is most representative of experiments on vortex roll-up in mixing layers. The thin, shaded surfaces correspond to the rib vortices (large component of vorticity normal to the spanwise direction), the cross-hatched surfaces denote the "cups" (regions of strong spanwise vorticity) that are critical to free shear layer transition, and the lines are vortex lines that comprise the rib vortices.

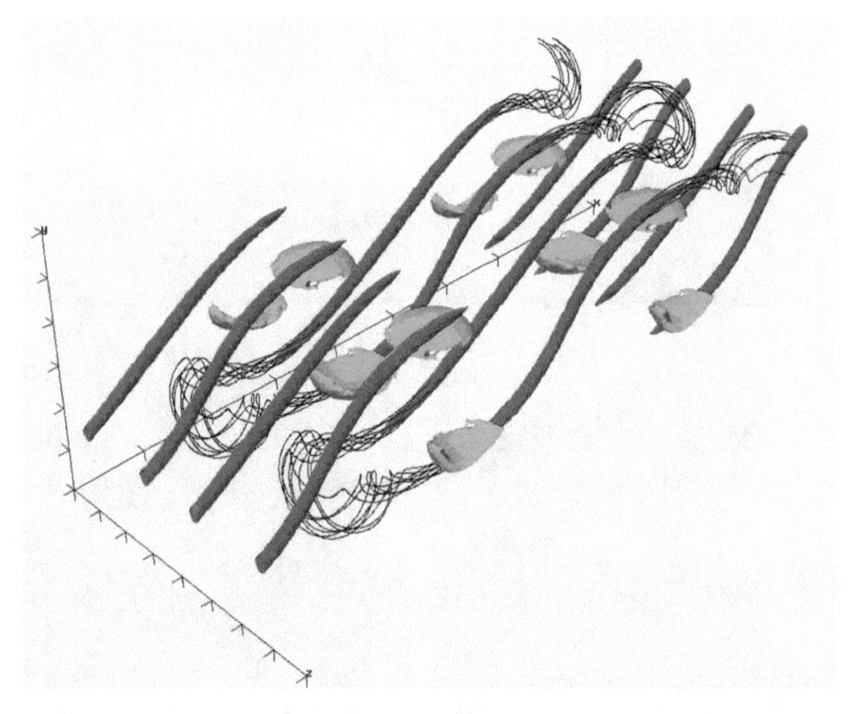

**Fig. 1.9.** DNS of vortex rollup in an incompressible free shear layer by Rogers and Moser (1992). The surfaces denote two types of regions of strong vorticity and the lines are vortex lines [Reprinted with permission from M.M. Rogers, R.D. Moser (1992); © 1992, Cambridge University Press]

Orszag and Kells (1980) and Orszag and Patera (1983) pioneered the use of splitting methods for wall-bounded flows. Figure 1.10 illustrates results from a later version of a splitting method, due to Zang and Hussaini (1986),

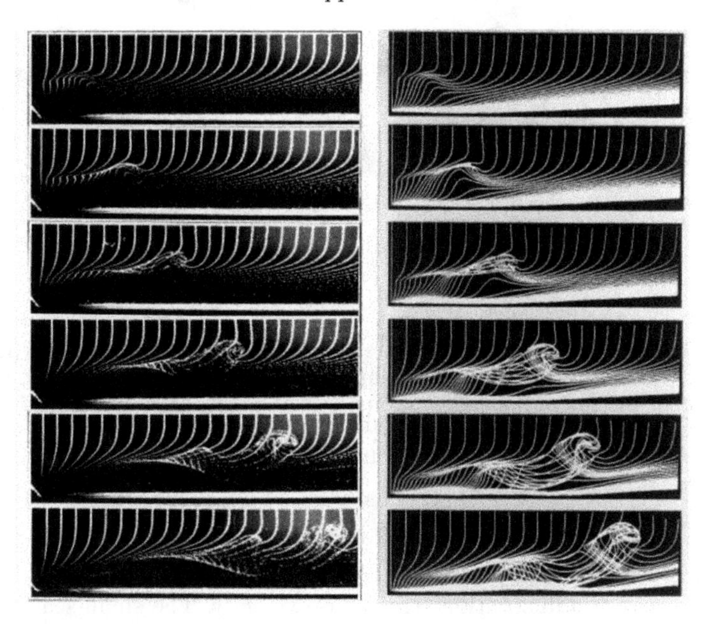

**Fig. 1.10.** Comparison of hydrogen bubble flow visualizations (*left*) of incompressible flat plate boundary-layer transition with DNS results of Zang, Hussaini and Erlebacher (*right*) [Reprinted with permission from T.A. Zang, M.Y. Hussaini (1987); © 1987 ASME]

applied to transition in a simplified version of flow past a flat plate. (The simplification invokes the parallel flow approximation that is discussed in CHQZ3, Sects. 2.3.2 and 3.4.5.) The left half of the figure is taken from the experiments of Hama and Nutant (1963) who used a hydrogen bubble flow visualization technique to illustrate the strongly nonlinear stage of transition. The right half of the figure, from Zang, Hussaini and Erlebacher (see Zang, Krist, Erlebacher and Hussaini (1987) and Zang and Hussaini (1987)), shows how well this phenomena was reproduced in the numerical computations using a $128 \times 144 \times 288$ grid. These authors demonstrated that the fine details of the vortex roll-ups were not present in the streamwise symmetry plane but only appeared in a streamwise plane displaced by a small fraction of the spanwise wavelength from the symmetric plane. (Details of the splitting algorithms are provided in CHQZ3, Sect. 3.4.2.)

This same splitting algorithm – the Zang-Hussaini version – was used by Scotti and Piomelli (2001) in their $64^3$ large-eddy simulations of pulsating channel flow. *Large-eddy simulation* (LES) is one method of accounting for the effects of turbulence by solving an augmented set of equations on a grid much coarser than for a DNS. (See CHQZ3, Sect. 1.1.3 for a summary of LES and Sagaut (2005) for a thorough discussion of the subject.) Figure 1.11 illustrates

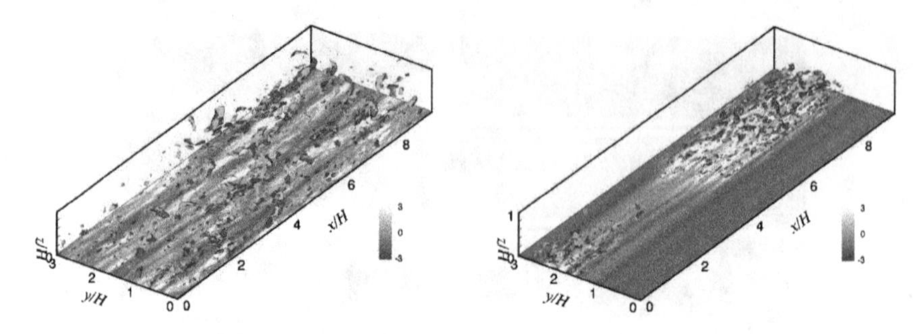

**Fig. 1.11.** Turbulent fluctuations near the bottom wall in incompressible pulsating channel flow from the LES computations of Scotti and Piomelli (2001). The left frame is near the end of the acceleration phase and the right frame is at the middle of the deceleration phase of the cycle [Reprinted with permission from A. Scotti, U. Piomelli (2001); © 2001, American Institute of Physics]

the flow structures at a fully turbulent phase of the oscillation (left half of the figure) and at a relaminarization phase (right half). The solid surface is a contour of the fluctuating streamwise velocity. The small-scale surfaces are contours of a measure of the coherent vorticity due to rotational motions. Note that the grid used for this large-eddy simulation was significantly coarser than that used in many of the examples above for transitional and turbulent flows. This illustrates a major attraction of the LES approach. The smaller grid permits wide parameter studies to be performed as opposed to the one-of-a-kind simulations typical of direct numerical simulations for such flows. Scotti and Piomelli did parametric studies using LES to characterize the detailed physics of such pulsating flows.

Figure 1.12 illustrates results from three additional classes of spectral algorithms. The physical problem is the study of the instability of flow past a flat plate. Unlike the computation of Zang, Hussaini and Erlebacher, shown above in Fig. 1.10, where the parallel flow approximation was used to study the temporal instability of this important physical problem, the results in Fig. 1.12 were for the unadulterated, spatial instability of the nonparallel flow past a flat plate. This problem requires the resolution of 10's or 100's of wavelengths in the streamwise direction (and has challenging outflow boundary conditions) rather than the mere 1 or 2 wavelengths in $x$ that are needed in the parallel flow approximation. The direct numerical simulation results used Spalart's (1988) ingenious *fringe method*, which permits a highly accurate approximation to be obtained with a Fourier approximation in $x$. (See CHQZ3, Sect. 3.6.1 for the details.) These two-dimensional DNS computations required approximately 4 points per wavelength in $x$ and no more than 40 Jacobi polynomials in $y$. The *parabolized stability equations* (PSE) *method* solve a much more economical set of equations using a marching method in $x$, a low-order Fourier expansion in $z$ and a Chebyshev collocation method in $y$ with $N \leq 40$. (See CHQZ3, Sects. 2.4.1 and 2.5.2 for PSE algorithms.)

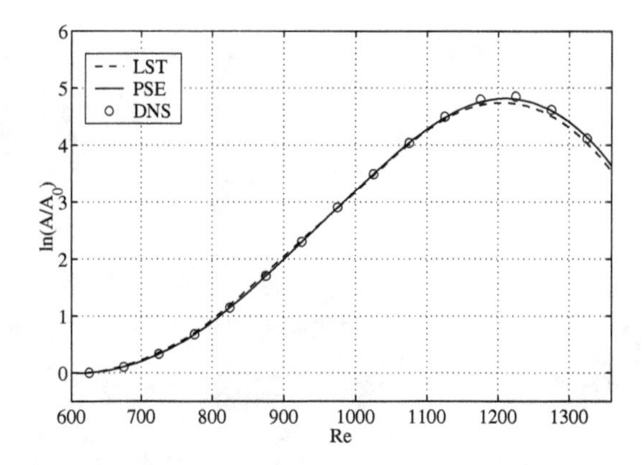


**Fig. 1.12.** Evolution of the spatial instability of an incompressible flat-plate boundary layer by Bertolotti, Herbert and Spalart (1992). Results are shown for direct numerical simulation (DNS), parabolized stability equations (PSE) and linear stability theory (LST) using the parallel flow approximation [Adapted with permission from F.P. Bertolotti, Th. Herbert, P.R. Spalart (1992); © 1992, Cambridge University Press]

The figure compares the spatial development of the maximum streamwise velocity perturbation as computed by the DNS and by the PSE; also shown for comparison are results of linear stability theory (LST) using the parallel flow approximation. (Spectral algorithms for linear stability are discussed in CHQZ3, Sect. 2.3.) The results of the PSE method agree well with the DNS results and are far cheaper. Hence, the PSE is far better suited to parametric studies.

Simulations of much later stages of transition in spatially developing flows have also been performed with both PSE and DNS techniques utilizing spectral methods. The spatial simulation of oblique transition in a boundary layer on a $1200 \times 64 \times 96$ grid by Berlin, Wiegel and Henningson (1999) is a prime example of a high-resolution DNS using the fringe method with a Fourier-Chebyshev algorithm. Figure 1.13 illustrates a comparison of their numerical results with flow visualizations of their experiment on transition in a boundary layer. (The algorithm uses components discussed in CHQZ3, Sects. 3.4.1, 3.4.4 and 3.6.1.)

In addition to the DNS, LES and PSE computations emphasized in the examples so far, spectral methods have also excelled in computations of eigenvalue problems. Indeed, Orszag's (1971b) demonstration of the power of Chebyshev spectral methods for discretizing the eigenvalue problems arising in linear stability analyses inspired many subsequent workers to adopt spectral methods for such problems in both incompressible and compressible flows. Eventually, in the 1990's computer resources were adequate for solving such problems with two or even three directions treated as inhomogeneous. An example of a large-scale eigenvalue problem solved by Theofilis (2000),

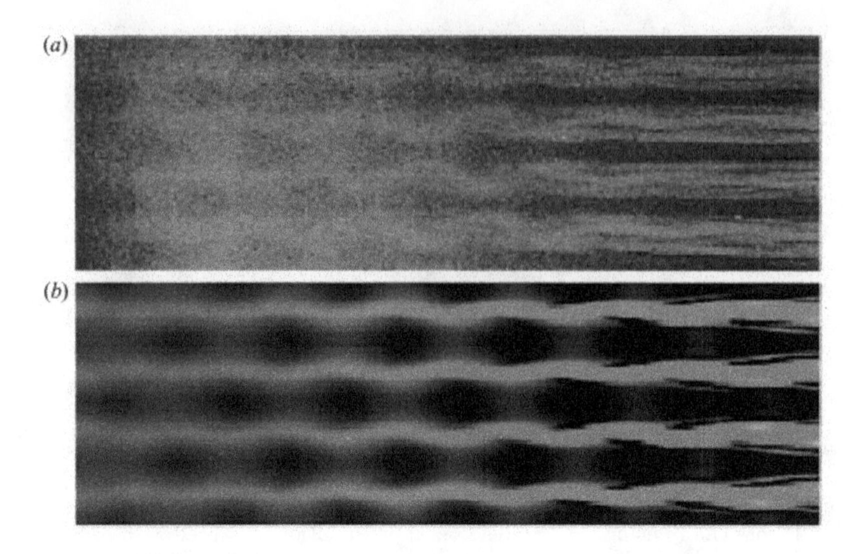


**Fig. 1.13.** Streamwise velocity flow visualizations of incompressible boundary-layer transition by Berlin, Wiegel and Henningson (1999): experiment (**a**) and spatial computation (**b**) [Reprinted with permission from S. Berlin, M. Wiegel, D.S. Henningson (1999); © 1999, Cambridge University Press]

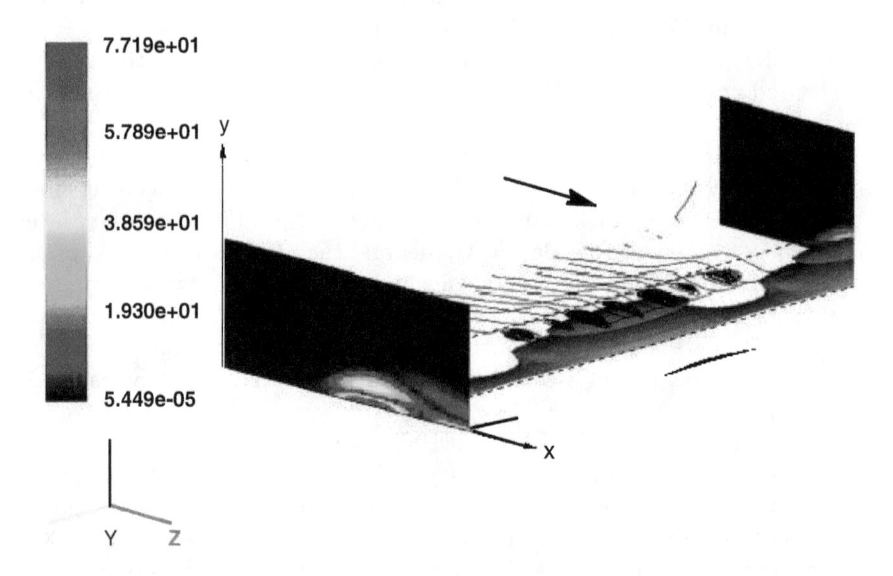


**Fig. 1.14.** Isosurface of disturbance vorticity of the primary instability of an incompressible separation bubble by Theofilis (2000) [Reprinted with permission from Springer-Verlag Berlin, Heidelberg 2006]

who used two Chebyshev directions and one Fourier direction, is given in Fig. 1.14. Spectral algorithms for discretizing the eigenvalue problems of fluid dynamical linear stability are described in much of CHQZ3, Chap. 2.

This list is by no means exhaustive and certainly neglects applications in related disciplines such as meteorology, oceanography, plasma physics and general relativity. Many of the components of algorithms mentioned above have been analyzed theoretically. The essential elements of the numerical analysis are provided in Chap. 7. Rigorous error estimates for some incompressible Navier-Stokes algorithms are reviewed in CHQZ3, Chap. 3.

The examples in this section have been confined to those using classical spectral methods. We noted earlier in this section that fourth-order methods require a factor of 10 more computational resources than spectral methods. The desire to handle problems in complex domains with greater than fourth-order accuracy has motivated the development of higher order methods using domain decomposition. Chapters 5 and 6 of the companion book (CHQZ3) survey spectral methods in complex domains. Chapters 2–7 of this book and Chaps. 1–4 of CHQZ3 are devoted to classical spectral methods.

# 2. Polynomial Approximation

The expansion of a function $u$ in terms of an infinite sequence of orthogonal functions $\{\phi_k\}$, e.g., $u = \sum_{k=-\infty}^{\infty} \hat{u}_k \phi_k$ or $u = \sum_{k=0}^{\infty} \hat{u}_k \phi_k$, underlies many numerical methods of approximation. The accuracy of the approximations and the efficiency of their implementation influence decisively the domain of applicability of these methods in scientific computations.

The most familiar approximation results are those for periodic functions expanded in Fourier series. The $k$-th coefficient of the expansion decays faster than any inverse power of $k$ when the function is infinitely smooth and all its derivatives are periodic as well. In practice this decay is not exhibited until there are enough coefficients to represent all the essential structures of the function. The subsequent rapid decay of the coefficients implies that the Fourier series truncated after just a few more terms represents an exceedingly good approximation of the function. This characteristic is usually referred to as *spectral accuracy* of the Fourier method.

The property of spectral accuracy is also attainable for smooth but nonperiodic functions provided that the expansion functions are chosen properly. It is not necessarily true that the coefficients of the expansion of a smooth function in terms of any orthogonal smooth basis decay faster than algebraically – usually spectral accuracy is attained only when the function exhibits very special boundary behavior. However, the eigenfunctions of a singular Sturm-Liouville operator allow spectral accuracy in the expansion of any smooth function. No *a priori* restriction on the boundary behavior is required. Moreover, since the eigenfunctions of the most common singular Sturm-Liouville problems are polynomials, such systems are a natural extension of the Fourier system for the approximation of nonperiodic functions.

The expansion in terms of an orthogonal system introduces a linear transformation between $u$ and the sequence of its expansion coefficients $\{\hat{u}_k\}$. This is usually called the *transform* of $u$ between physical space and transform (or wavenumber) space. If the system is complete in a suitable Hilbert space, this transform can be inverted. Hence, functions can be described both through their values in physical space and through their coefficients in transform space.

The expansion coefficients depend on (almost) all the values of $u$ in physical space, and they can rarely be computed exactly. A finite number of approximate expansion coefficients can be easily computed using the values of $u$ at a finite number of selected points, usually the nodes of high-precision quadrature formulas. This procedure defines a *discrete transform* between the set of values of $u$ at the quadrature points and the set of approximate, or *discrete*, coefficients. With a proper choice of the quadrature formulas, the finite series defined by the discrete transform is actually the interpolant of $u$ at the quadrature nodes. If the properties of accuracy (in particular the spectral accuracy) are retained by replacing the finite transform with the discrete transform, then the interpolant series can be used instead of the truncated series to approximate functions.

For some of the most common orthogonal systems (Fourier and Chebyshev polynomials) the discrete transform can be computed in a "fast" way, i.e., with an operation count with leading term $(5/2)N \log_2 N$, where $N$ is the number of polynomials, rather than with the $2N^2$ operations required by a matrix-vector multiplication.

In this chapter we shall describe in detail those orthogonal systems for which spectral accuracy is guaranteed. Some of their approximation properties will be surveyed, and practical indications on how to use the approximating functions will be given. A rigorous description of the approximation properties is postponed to Chapter 5.

The first five sections in this chapter are devoted to one-dimensional approximation. Multidimensional approximations on a Cartesian domain (i.e., a Cartesian product of intervals) are constructed by the familiar tensor-product approach. They are considered in Sect. 2.8; some specific formulas are also given in Sect. 5.8. Finally, several approximations in non-Cartesian domains such as triangles and hexahedra are briefly surveyed in Sect. 2.9.

The technical definitions of the integrals, Hilbert spaces, and norms used in the analysis of spectral methods are provided in Appendix A. They are referenced within the text by the label of that section in the appendix in which they are discussed.

We concentrate on discussing the details of spectral approximations in Cartesian coordinates on bounded domains. The texts by Fornberg (1996) and Boyd (2001) each furnish an entire chapter on spectral methods in spherical and polar coordinates. Boyd (2001) also devotes an entire chapter to spectral methods on unbounded domains. We do, however, make some brief comments on polar coordinate systems in Sect. 3.9. We also cover the fundamentals of expansions in Laguerre polynomials (for semi-infinite intervals) and in Hermite polynomials (on the infinite interval), as well as the basics of spectral approximations on triangles (in two dimensions) and tetrahedra, prisms and pyramids (in three dimensions). Mappings for both bounded and unbounded domains are also outlined.

Our discussion of spectral approximations is confined to applications to deterministic problems. Many years ago Wiener (1930) proposed the use of expansions in multidimensional Hermite polynomials for approximating

Gaussian stochastic processes. In the late 1980's such expansions began to be used for large-scale computations of stochastic structural dynamics problems; see Ghanem and Spanos (1991). Xiu and Karniadakis (2002) extended this approach to a wide set of orthogonal polynomial expansions and have focused their subsequent work on fluid dynamics applications. A key aspect of this approach is that truncation for multidimensional expansions is not accomplished in the manner that is customary for the tensor-product expansions used in deterministic spectral methods. The interested reader should consult these basic references and keep abreast of this rapidly developing class of numerical methods, which is commonly referred to as *polynomial chaos*.

## 2.1 The Fourier System

### 2.1.1 The Continuous Fourier Expansion

The set of functions

$$\phi_k(x) = e^{ikx} \tag{2.1.1}$$

is an orthogonal system over the interval $(0, 2\pi)$:

$$\int_0^{2\pi} \phi_k(x)\overline{\phi_l(x)}\,\mathrm{d}x = 2\pi\delta_{kl} = \begin{cases} 0 & \text{if } k \neq l\,, \\ 2\pi & \text{if } k = l\,. \end{cases} \tag{2.1.2}$$

(The overline on $\phi_l(x)$ denotes its complex conjugate.) For a complex-valued function $u$ defined on $(0, 2\pi)$, we introduce the *Fourier coefficients* of $u$:

$$\hat{u}_k = \frac{1}{2\pi}\int_0^{2\pi} u(x)e^{-ikx}\,\mathrm{d}x\,, \qquad k = 0, \pm1, \pm2, \ldots\,. \tag{2.1.3}$$

The integrals in (2.1.3) exist if $u$ is Riemann integrable (see (A.8), i.e., Sect. 8 of Appendix A), which is ensured, for instance, if $u$ is bounded and piecewise continuous in $(0, 2\pi)$. More generally, the Fourier coefficients are defined for any function that is integrable in the sense of Lebesgue (see (A.9)).

The relation (2.1.3) associates with $u$ a sequence of complex numbers called the *Fourier transform* of $u$. It is possible as well to introduce a *Fourier cosine transform* and a *Fourier sine transform* of $u$, respectively, through the formulas

$$a_k = \frac{1}{2\pi}\int_0^{2\pi} u(x)\cos kx\,\mathrm{d}x\,, \qquad k = 0, \pm1, \pm2, \ldots\,, \tag{2.1.4}$$

and

$$b_k = \frac{1}{2\pi}\int_0^{2\pi} u(x)\sin kx\,\mathrm{d}x\,, \qquad k = 0, \pm1, \pm2, \ldots\,. \tag{2.1.5}$$

The three Fourier transforms of $u$ are related by the formula $\hat{u}_k = a_k - ib_k$ for $k = 0, \pm 1, \pm 2, \ldots$. Moreover, if $u$ is a real valued function, $a_k$ and $b_k$ are real numbers, and $\hat{u}_{-k} = \overline{\hat{u}_k}$.

The Fourier series of the function $u$ is defined as

$$Su = \sum_{k=-\infty}^{\infty} \hat{u}_k \phi_k \,. \tag{2.1.6}$$

It represents the formal expansion of $u$ in terms of the Fourier orthogonal system. In order to make this expansion rigorous, one has to cope with three problems:

(i) When and in what sense is the series convergent?
(ii) What is the relation between the series and the function, $u$?
(iii) How rapidly does the series converge?

The basic issue is how $u$ is approximated by the sequence of trigonometric polynomials

$$P_N u(x) = \sum_{k=-N/2}^{N/2-1} \hat{u}_k e^{ikx} \,, \tag{2.1.7}$$

as $N$ tends to $\infty$. Theoretical discussions of truncated (or finite) Fourier series are customarily given for

$$P_N u(x) = \sum_{k=-N}^{N} \hat{u}_k e^{ikx} \tag{2.1.8}$$

rather than for (2.1.7). We have chosen to use the (mathematically unconventional) form (2.1.7) because it corresponds directly to the way spectral methods are actually programmed. In most cases, the most important characterization of the approximation is the number of degrees of freedom. Equation (2.1.7) corresponds to $N$ degrees of freedom and is preferred by us for this reason. We shall refer to $P_N u$ as the $N$-th order truncated Fourier series of $u$.

Points (i), (ii) and (iii) have been subjected to a thorough mathematical investigation. See, for example, Zygmund (1959). We review here only those basic results relevant to the application of spectral methods to partial differential equations.

We recall the following results about the convergence of the Fourier series. Hereafter, a function $u$ defined in $(0, 2\pi)$ will be called periodic if $u(0^+)$ and $u(2\pi^-)$ exist and are equal.

(a) If $u$ is continuous, periodic, and of bounded variation on $[0, 2\pi]$ (see (A.8)), then $Su$ is uniformly convergent to $u$, i.e.,

$$\max_{x \in [0,2\pi]} |u(x) - P_N u(x)| \to 0 \quad \text{as } N \to \infty \,.$$

(b) If $u$ is of bounded variation on $[0, 2\pi]$, then $P_N u(x)$ converges pointwise to $(u(x^+) + u(x^-))/2$ for any $x \in [0, 2\pi]$ (here $u(0^-) = u(2\pi^-)$).
(c) If $u$ is continuous and periodic, then its Fourier series does not necessarily converge at every point $x \in [0, 2\pi]$.

A full characterization of the functions for which the Fourier series is everywhere pointwise convergent is not known. However, a full characterization is available within the framework of Lebesgue integration for convergence in the mean. The series $Su$ is said to be convergent in the mean (or $L^2$-convergent) to $u$ if

$$\int_0^{2\pi} |u(x) - P_N u(x)|^2 \, dx \longrightarrow 0 \quad \text{as } N \longrightarrow \infty . \tag{2.1.9}$$

Clearly, the convergence in the mean can be defined for square-integrable functions. Integrability can be intended in the Riemann sense, but the most general results require that the integral in (2.1.9) be defined according to Lebesgue. Henceforth, we assume that $u \in L^2(0, 2\pi)$, where $L^2(0, 2\pi)$ is the space of (classes) of the Lebesgue-measurable functions $u : (0, 2\pi) \to \mathbb{C}$ such that $|u|^2$ is Lebesgue-integrable over $(0, 2\pi)$ (see (A.9)). $L^2(0, 2\pi)$ is a complex Hilbert space (see (A.1)) with inner product

$$(u, v) = \int_0^{2\pi} u(x)\overline{v(x)} \, dx \tag{2.1.10}$$

and norm

$$\|u\| = \left( \int_0^{2\pi} |u(x)|^2 \, dx \right)^{1/2} . \tag{2.1.11}$$

Let $S_N$ be the space of the trigonometric polynomials of degree $N/2$, defined as

$$S_N = \text{span}\{e^{ikx} \mid -N/2 \leq k \leq N/2 - 1\} . \tag{2.1.12}$$

Then by the orthogonality relation (2.1.2) one has

$$(P_N u, v) = (u, v) \quad \text{for all } v \in S_N . \tag{2.1.13}$$

This shows that $P_N u$ is the orthogonal projection of $u$ upon the space of the trigonometric polynomials of degree $N/2$. Equivalently, $P_N u$ is the closest element to $u$ in $S_N$ with respect to the norm (2.1.11).

Functions in $L^2(0, 2\pi)$ can be characterized in terms of their Fourier coefficients, according to the Riesz theorem, in the following sense. If $u \in L^2(0, 2\pi)$, then its Fourier series converges to $u$ in the sense of (2.1.9), and

$$\|u\|^2 = 2\pi \sum_{k=-\infty}^{\infty} |\hat{u}_k|^2 \quad \text{(Parseval identity)} . \tag{2.1.14}$$

(In particular, the numerical series on the right-hand side is convergent.) Conversely, for any complex sequence $\{c_k\}$, $k = 0, \pm 1, \ldots$, such that $\sum_{k=-\infty}^{\infty} |c_k|^2 < \infty$, there exists a unique function $u \in L^2(0, 2\pi)$ such that its Fourier coefficients are precisely the $c_k$'s for any $k$. Thus, for any function $u \in L^2(0, 2\pi)$ we can write

$$u = \sum_{k=-\infty}^{\infty} \hat{u}_k \phi_k \,, \tag{2.1.15}$$

where the equality has to be intended between two functions in $L^2(0, 2\pi)$. The Riesz theorem states that the finite Fourier transform is an isomorphism between $L^2(0, 2\pi)$ and the space $l^2$ of complex sequences $\{c_k\}$, $k = 0, \pm 1, \pm 2, \ldots$ , such that $\sum_{k=-\infty}^{\infty} |c_k|^2 < \infty$.

The $L^2$-convergence does not imply the pointwise convergence of $P_N u$ to $u$ at all points of $[0, 2\pi]$. However, a nontrivial result by Carleson (1966) asserts that $P_N u(x)$ converges to $u(x)$ as $N \to \infty$ for any $x$ outside a set of zero measure in $[0, 2\pi]$.

We deal now with the problem of the rate of convergence of the Fourier series. Hereafter, we set

$$\sum_{|k| \gtrsim N/2} \equiv \sum_{\substack{k<-N/2 \\ k \geq N/2}} \,.$$

First of all, note that by the Parseval identity one has

$$\|u - P_N u\| = \left( 2\pi \sum_{|k| \gtrsim N/2} |\hat{u}_k|^2 \right)^{1/2} \,. \tag{2.1.16}$$

On the other hand, if $u$ is sufficiently smooth, then

$$\max_{0 \leq x \leq 2\pi} |u(x) - P_N u(x)| \leq \sum_{|k| \gtrsim N/2} |\hat{u}_k| \,. \tag{2.1.17}$$

This shows that the size of the error created by replacing $u$ with its $N$-th order truncated Fourier series depends upon how fast the Fourier coefficients of $u$ decay to zero. This in turn depends on the regularity of $u$ in the domain $(0, 2\pi)$ and on the periodicity properties of $u$. Indeed, if $u$ is continuously differentiable in $[0, 2\pi]$, then, for $k \neq 0$,

$$\begin{aligned} 2\pi \hat{u}_k &= \int_0^{2\pi} u(x) e^{-ikx} \, dx \\ &= \frac{-1}{ik}(u(2\pi^-) - u(0^+)) + \frac{1}{ik} \int_0^{2\pi} u'(x) e^{-ikx} \, dx \,. \end{aligned} \tag{2.1.18}$$

Hence,

$$\hat{u}_k = O(k^{-1}) \,. \tag{2.1.19}$$

If now $u'$ is itself continuously differentiable in $[0, 2\pi]$, the last integral in (2.1.18) is $2\pi$ times the $k$-th Fourier coefficient of $u'$; hence, it decays like $k^{-1}$. It follows that $\hat{u}_k = O(k^{-2})$ if and only if $u(2\pi^-) = u(0^+)$. Iterating this argument, one proves that *if $u$ is $m$-times continuously differentiable in $[0, 2\pi]$ $(m \geq 1)$, and if $u^{(j)}$ is periodic for all $j \leq m-2$, then*

$$\hat{u}_k = O(k^{-m}) \,, \qquad k = \pm 1, \pm 2, \ldots \,. \tag{2.1.20}$$

(The symbol $u^{(j)}$ denotes the $j$-th derivative of $u$.) The same result holds if $u$ is $(m-1)$-times differentiable almost everywhere in $(0, 2\pi)$, with its $(m-1)$-th derivative of bounded variation in $[0, 2\pi]$, and if $u^{(j)}$ is periodic for all $j \leq m-2$. In this case the integral on the right-hand side of (2.1.18) has to be replaced by the Riemann-Stieltjes integral $\int_0^{2\pi} e^{-ikx}\, du(x)$ (see (A.8)).

As a corollary of (2.1.20), we conclude that *the $k$-th Fourier coefficient of a function which is infinitely differentiable and periodic with all its derivatives on $[0, 2\pi]$ decays faster than any negative power of $k$.*

***Examples***

(1) The function

$$u(x) = \begin{cases} 1\,, & \dfrac{\pi}{2} < x \leq \dfrac{3\pi}{2}\,, \\[2ex] 0\,, & 0 < x \leq \dfrac{\pi}{2},\ \dfrac{3\pi}{2} < x \leq 2\pi \end{cases} \tag{2.1.21}$$

is of bounded variation in $[0, 2\pi]$. Its Fourier coefficients are

$$\hat{u}_k = \begin{cases} \pi & \text{if } k = 0\,, \\ 0 & \text{if } k \neq 0,\ \text{even}\,, \\ \dfrac{(-1)^{(k-1)/2}}{k} & \text{if } k \neq 0,\ \text{odd}\,. \end{cases}$$

Several truncated Fourier series for this function are illustrated in Fig. 2.1(a). The pointwise convergence is linear and the series is not uniformly convergent. A more detailed discussion of the convergence is given in Sect. 2.1.4.

(2) The function

$$u(x) = \sin(x/2) \tag{2.1.22}$$

is infinitely differentiable in $[0, 2\pi]$, but $u'(0^+) \neq u'(2\pi^-)$. Its Fourier coefficients are

$$\hat{u}_k = \frac{2}{\pi} \frac{1}{1 - 4k^2} \,.$$

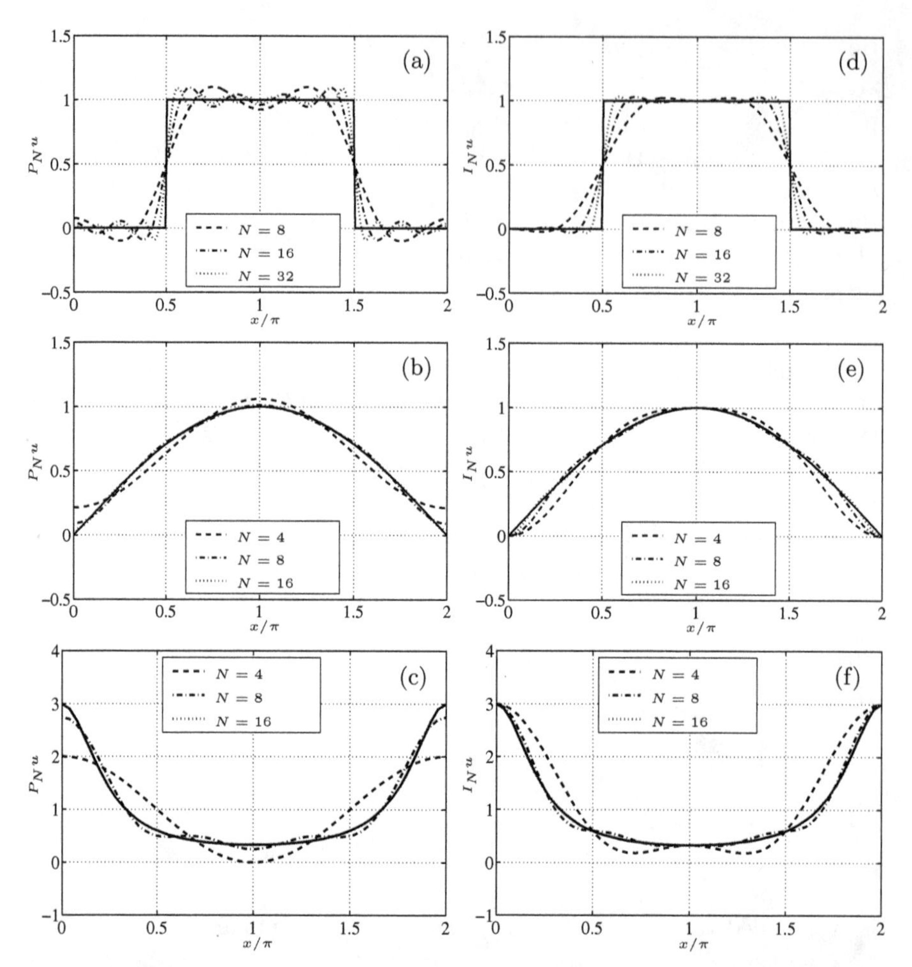


**Fig. 2.1.** Trigonometric approximations to the square wave ((a) and (d)), to $u(x) = \sin(x/2)$ ((b) and (e)) and to $u(x) = 3/(5-4\cos x)$ ((c) and (f)). Parts (**a**), (**b**), and (**c**) display truncated Fourier series. Parts (**d**), (**e**), and (**f**) display Fourier interpolating polynomials. The exact function is denoted by the solid curve

The truncated series for this function are shown in Fig. 2.1(b). The convergence is quadratic except at the endpoints. Here it is linear and monotonic, which is an obvious consequence of the coefficients decaying quadratically with the same sign.

(3) The function

$$u(x) = \frac{3}{5-4\cos x} \tag{2.1.23}$$

is infinitely differentiable and periodic with all its derivatives in $[0, 2\pi]$. Its Fourier coefficients are

$$\hat{u}_k = 2^{-|k|} , \qquad k = 0, \pm 1, \dots .$$

Note that $u$ is actually real analytic on the real axis. This results in the exponential decay of its Fourier coefficients. The rapid convergence is evident in Fig. 2.1(c). Note that the truncated series for $N = 16$ is virtually indistinguishable from the function itself.

We should stress that the asymptotic rate of decay of the Fourier coefficients does not convey the whole story of the error made in a given approximation, If a series has a finite rate of decay, $\hat{u}_k = O(k^{-m})$, then this decay is observed only for $k >$ some $k_0$. Should the series be truncated below $k_0$, then the approximation will be quite poor indeed. Even for an infinitely differentiable function there is some minimum acceptable $k_0$, and truncations below this level yield thoroughly unacceptable approximations.

Estimates (2.1.16) and (2.1.17) show that the error between $u$ and its $N$-th order truncated Fourier series decays faster than algebraically in $1/N$, when $u$ is infinitely smooth and periodic with all its derivatives. As noted above, this property is commonly called *spectral accuracy*, or *infinite-order accuracy*, and we say that the series exhibits *infinite-order convergence*. (The term *exponential convergence* has also been used to characterize spectral methods. However, this term is no longer is common use as a descriptor of spectral accuracy for infinitely differentiable functions, since the error decay is only guaranteed to be exponential in $N$ if the function is also analytic. In this text we only use the term "exponential convergence" in the context of particular functions for which the convergence is actually exponentially fast.) However, in the analysis of spectral methods for PDEs, one is often interested in estimating global errors like (2.1.16) or (2.1.17) for those functions $u$ having finite regularity. In such cases, using (2.1.20) in (2.1.16) or (2.1.17) will result in a non-optimal rate of convergence of $P_N u$ to $u$. A different approach is then required, and it will be the subject of Sect. 5.1.2.

### 2.1.2 The Discrete Fourier Expansion

In many practical applications, numerical methods based upon Fourier series cannot be implemented in precisely the way suggested by the standard treatment of Fourier series that was reviewed in the previous subsection. Some of the difficulties are: The Fourier coefficients of an arbitrary function are not known in closed form and must therefore be approximated in some way; there needs to be an efficient way to recover in physical space the information that is calculated in transform space; and all but the simplest nonlinearities lead to extreme complications. The key to overcoming these difficulties is the use of the discrete Fourier transform and the related discrete Fourier series.

For any integer $N > 0$, consider the set of points

$$x_j = \frac{2\pi j}{N} , \qquad j = 0, \dots, N-1 , \tag{2.1.24}$$

referred to as nodes or grid points or knots. The *discrete Fourier coefficients* of a complex-valued function $u$ in $[0, 2\pi]$ with respect to these points are

$$\tilde{u}_k = \frac{1}{N} \sum_{j=0}^{N-1} u(x_j) e^{-ikx_j} , \qquad k = -N/2, \dots, N/2 - 1 . \tag{2.1.25}$$

Due to the orthogonality relation

$$\frac{1}{N} \sum_{j=0}^{N-1} e^{-ipx_j} = \begin{cases} 1 & \text{if } p = Nm, \ m = 0, \pm 1, \pm 2, \dots , \\ 0 & \text{otherwise} , \end{cases} \tag{2.1.26}$$

we have the inversion formula

$$u(x_j) = \sum_{k=-N/2}^{N/2-1} \tilde{u}_k e^{ikx_j} , \qquad j = 0, \dots, N-1 . \tag{2.1.27}$$

Consequently, the polynomial

$$I_N u(x) = \sum_{k=-N/2}^{N/2-1} \tilde{u}_k e^{ikx} \tag{2.1.28}$$

is the $N/2$-degree trigonometric interpolant of $u$ at the nodes (2.1.24), i.e., $I_N u(x_j) = u(x_j)$, $j = 0, \dots, N-1$. This polynomial is also known as the *discrete Fourier series* of $u$. Three examples of such series are provided in Fig. 2.1(d),(e),(f).

The $\tilde{u}_k$'s depend only on the $N$ values of $u$ at the nodes (2.1.24). The *discrete Fourier transform* (DFT) is the mapping between the $N$ complex numbers $u(x_j)$, $j = 0, \dots, N-1$, and the $N$ complex numbers $\tilde{u}_k$, $k = -N/2, \dots, N/2-1$. The two conventional forms for the DFT are given in (2.1.25) and (2.1.27), with the latter sometimes referred to as the inverse DFT. These equations show that the discrete Fourier transform is an orthogonal transformation in $\mathbb{C}^N$. From a computational point of view, it can be accomplished by the Fast Fourier Transform algorithm (Cooley and Tukey (1965)).

In this book we use the term *transform method* to refer to a computational procedure in a spectral method that employs the Fast Fourier Transform. This includes methods for transforming between physical space and transform space and methods for evaluating derivatives (as discussed above), as well as methods for evaluating convolution sums (as discussed in Sect. 3.4).

The simplest Fast Fourier Transform (FFT) requires $N$ to be a power of 2. If the data are fully complex it requires $5N \log_2 N - 6N$ real operations, where addition and multiplication are counted as separate operations. In most applications, $u$ is real and $\tilde{u}_{-k} = \overline{\tilde{u}_k}$. In this case the operation count is halved. Fast Fourier Transforms that allow factors of 2, 3, 4, 5 and 6 are widely available (Temperton (1983), Frigo and Johnson (2005)) and offer a 10–20% reduction in the operation count over the basic power-of-2 FFT. For simplicity, we shall often use just $5N \log_2 N$ as the operation count for a complex FFT. A more complete discussion of FFT's is contained in Appendix B.

Note that the continuous Fourier coefficients of the interpolant are precisely the values computed via the discrete Fourier transform (2.1.25). On the other hand, $\tilde{u}_k$ can be regarded as an approximation to $\hat{u}_k$ using the composite trapezoidal rule to evaluate the integral in (2.1.3). For infinitely differentiable, periodic functions the trapezoidal rule is the quadrature formula of Lagrange type with maximum precision.

Another form of the interpolant $I_N u$ that is of both theoretical and practical interest can be given. By substituting (2.1.25) into (2.1.28) and rearranging the sums, we obtain

$$I_N u(x) = \sum_{j=0}^{N-1} u(x_j) \psi_j(x) , \tag{2.1.29}$$

with

$$\psi_j(x) = \frac{1}{N} \sum_{k=-N/2}^{N/2-1} e^{ik(x-x_j)} . \tag{2.1.30}$$

The functions $\psi_j$ are the trigonometric polynomials in $S_N$ that satisfy

$$\psi_j(x_l) = \delta_{lj} , \qquad l, j = 0, \ldots, N-1 ; \tag{2.1.31}$$

this follows from (2.1.24) and (2.1.26). They are the *discrete delta-functions* at the nodes (2.1.24), also termed the *characteristic Lagrange trigonometric polynomials* at these nodes. The interpolant $I_N u$ is that particular linear combination of such functions whose coefficients are simply the values of $u$ at the grid points.

The interpolation operator $I_N$ can be regarded as an orthogonal projection upon the space $S_N$ of the trigonometric polynomials of degree $N/2$, with respect to the discrete approximation of the inner product (2.1.10). Actually, the bilinear form

$$(u, v)_N = \frac{2\pi}{N} \sum_{j=0}^{N-1} u(x_j) \overline{v(x_j)} \tag{2.1.32}$$

coincides with the inner product (2.1.10) if $u$ and $v$ are polynomials of degree $N/2$, due to (2.1.26):

$$(u, v)_N = (u, v) \quad \text{for all } u, v \in S_N . \tag{2.1.33}$$

As a consequence, (2.1.32) is an inner product on $S_N$, and

$$\|u\|_N = \sqrt{(u, u)_N} = \sqrt{(u, u)} = \|u\| \tag{2.1.34}$$

is the associated norm. The interpolant $I_N u$ of a continuous function $u$ satisfies trivially the identity

$$(I_N u, v)_N = (u, v)_N \quad \text{for all } v \in S_N \,. \tag{2.1.35}$$

The discrete Fourier coefficients can be expressed also in terms of the exact Fourier coefficients of $u$. If the Fourier series (2.1.6) converges to $u$ at every node (2.1.24), then by (2.1.25) one gets

$$\tilde{u}_k = \hat{u}_k + \sum_{\substack{m=-\infty \\ m \neq 0}}^{+\infty} \hat{u}_{k+Nm} \,, \qquad k = -N/2, \ldots, N/2-1 \,. \tag{2.1.36}$$

Formula (2.1.36) shows that the $k$-th mode of the trigonometric interpolant of $u$ depends not only on the $k$-th mode of $u$, but also on all the modes of $u$ that *alias* the $k$-th mode on the discrete grid. The $(k+Nm)$-th wavenumber aliases the $k$-th wavenumber on the grid; they are indistinguishable at the nodes since $\phi_{k+Nm}(x_j) = \phi_k(x_j)$. The phenomenon is illustrated in Fig. 2.2. Shown there are three sine waves with frequencies $k = 6$, $-2$, and $-10$. Superimposed upon each wave are the eight grid-point values of the function. In each case these grid-point values coincide with the $k = -2$ wave.

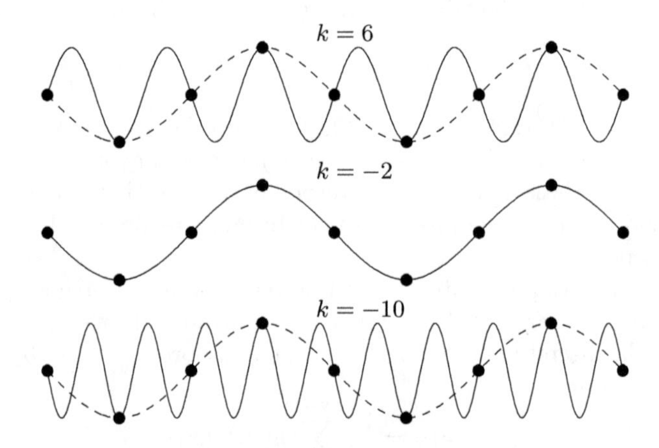


**Fig. 2.2.** Three sine waves that have the same $k = -2$ interpretation on an eight-point grid. The nodal values are denoted by the filled circles. The actual sine waves are denoted by the solid curves. Both the $k = 6$ and the $k = -10$ waves are misinterpreted as a $k = -2$ wave (dashed curves) on the coarse grid

An equivalent formulation of (2.1.36) is

$$I_N u = P_N u + R_N u \,, \tag{2.1.37}$$

with

$$R_N u = \sum_{k=-N/2}^{N/2-1} \left( \sum_{\substack{m=-\infty \\ m \neq 0}}^{\infty} \hat{u}_{k+Nm} \right) \phi_k \, . \tag{2.1.38}$$

The error $R_N u$ between the interpolating polynomial and the truncated Fourier series is called the *aliasing error*. It is orthogonal to the truncation error, $u - P_N u$, so that

$$\|u - I_N u\|^2 = \|u - P_N u\|^2 + \|R_N u\|^2 \, . \tag{2.1.39}$$

Hence, the error due to the interpolation is actually always larger than the error due to the truncation of the Fourier series.

Numerous papers have appeared over the years, especially in the early days of spectral methods, that have discussed the role of aliasing errors in spectral methods. The debate concerned the influence of these errors on both the stability and the accuracy of the methods. Clever methods were proposed to remove or control the aliasing effects on spectral calculations (Sect. 3.4 and CHQZ3, Sect. 3.3). Subsequently, it was proven that the influence of aliasing on the accuracy of spectral methods is asymptotically of the same order as the truncation error (Kreiss and Oliger (1979)). Indeed, error estimates (5.1.10) and (5.1.19) show that the truncation and interpolation errors decay at the same rate. This implies similar behavior of the approximation errors for a Galerkin and a collocation scheme. The influence of aliasing on the stability and accuracy of actual spectral solutions of PDEs will be discussed in Sect. 3.10 and in CHQZ3, Sects. 3.3.4 and 3.4.6. Rigorous analyses of aliasing errors in steady Navier-Stokes algorithms are given in CHQZ3, Sect. 3.7.

The sequence of interpolating polynomials exhibits convergence properties similar to those of the sequence of truncated Fourier series; furthermore, the continuous and the discrete Fourier coefficients share the same asymptotic behavior. More precisely, when $N \to \infty$, we have

(a) if $u$ is continuous, periodic and of bounded variation on $[0, 2\pi]$, then $I_N u$ converges to $u$ uniformly on $[0, 2\pi]$;
(b) if $u$ is of bounded variation on $[0, 2\pi]$, then $I_N u$ is uniformly bounded on $[0, 2\pi]$ and converges pointwise to $u$ at every continuity point for $u$;
(c) if $u$ is Riemann integrable, then $I_N u$ converges to $u$ in the mean.

Concerning the discrete Fourier coefficients, we have

(d) for any integer $k \neq 0$, and any positive $N$ such that $N/2 > |k|$, let $\tilde{u}_k = \tilde{u}_k^{(N)}$ be the $k$-th Fourier coefficient of $I_N u$. If $u$ is infinitely smooth and periodic with all its derivatives, formula (2.1.36) shows that $|\tilde{u}_k^{(N)}|$ decays faster than algebraically in $k^{-1}$, uniformly in $N$. More generally, if $u$ satisfies the hypotheses for which (2.1.20) holds, the same asymptotic behavior holds for $\tilde{u}_k^{(N)}$, uniformly in $N$.

### 2.1.3 Differentiation

The manner in which differentiation is accomplished in a spectral method depends upon whether one is working with a representation of the function in transform space or in physical space. Differentiation in transform space consists of simply multiplying each Fourier coefficient by the imaginary unit times the corresponding wavenumber. If $Su = \sum_{k=-\infty}^{\infty} \hat{u}_k \phi_k$ is the Fourier series of a function $u$, then

$$Su' = \sum_{k=-\infty}^{\infty} ik\hat{u}_k \phi_k \tag{2.1.40}$$

is the Fourier series of the derivative of $u$. Consequently,

$$(P_N u)' = P_N u' \; , \tag{2.1.41}$$

i.e., truncation and differentiation commute. The series (2.1.40) converges in $L^2$ provided that the derivative of $u$ (in the sense of distributions, see (A.10)) is a function in $L^2(0, 2\pi)$.

Differentiation in physical space is based upon the values of the function $u$ at the Fourier nodes (2.1.24). These are used in the evaluation of the discrete Fourier coefficients of $u$ according to (2.1.25), these coefficients are multiplied by $ik$, and the resulting Fourier coefficients are then transformed back to physical space according to (2.1.27). The values $(\mathcal{D}_N u)_j$ of the approximate derivative at the grid points $x_j$ are thus given by

$$(\mathcal{D}_N u)_j = \sum_{k=-N/2}^{N/2-1} \tilde{u}_k^{(1)} e^{2ikj\pi/N} \; , \qquad j = 0, 1, \dots, N-1 \; , \tag{2.1.42}$$

where

$$\tilde{u}_k^{(1)} = ik\tilde{u}_k = \frac{ik}{N} \sum_{l=0}^{N-1} u(x_l) e^{-2ikl\pi/N} \; , \qquad k = -N/2, \dots, N/2-1 \; . \tag{2.1.43}$$

(The use of the index $l$ in the latter sum, in lieu of $j$ as in (2.1.25), is motivated by the matrix formalism used in the sequel.)

The procedure (2.1.42)–(2.1.43) amounts to computing the grid-point values of the derivative of the discrete Fourier series of $u$, i.e.,

$$\mathcal{D}_N u = (I_N u)' \; , \tag{2.1.44}$$

where $I_N u$ is defined in (2.1.28). Since, in general,

$$\mathcal{D}_N u \neq P_N u' \; ,$$

the function $\mathcal{D}_N u$ is called the *Fourier interpolation derivative* of $u$, to distinguish it from the true spectral derivative of $u$ given by (2.1.41), which we refer to as the *Fourier projection derivative.*

Interpolation and differentiation do not commute, i.e.,

$$(I_N u)' \neq I_N(u') , \tag{2.1.45}$$

unless $u \in S_N$. However, we shall prove in Sect. 5.1.3 that the error,

$$(I_N u)' - I_N(u') ,$$

is of the same order as the truncation error for the derivative,

$$u' - P_N u' .$$

It follows that interpolation differentiation is spectrally accurate.

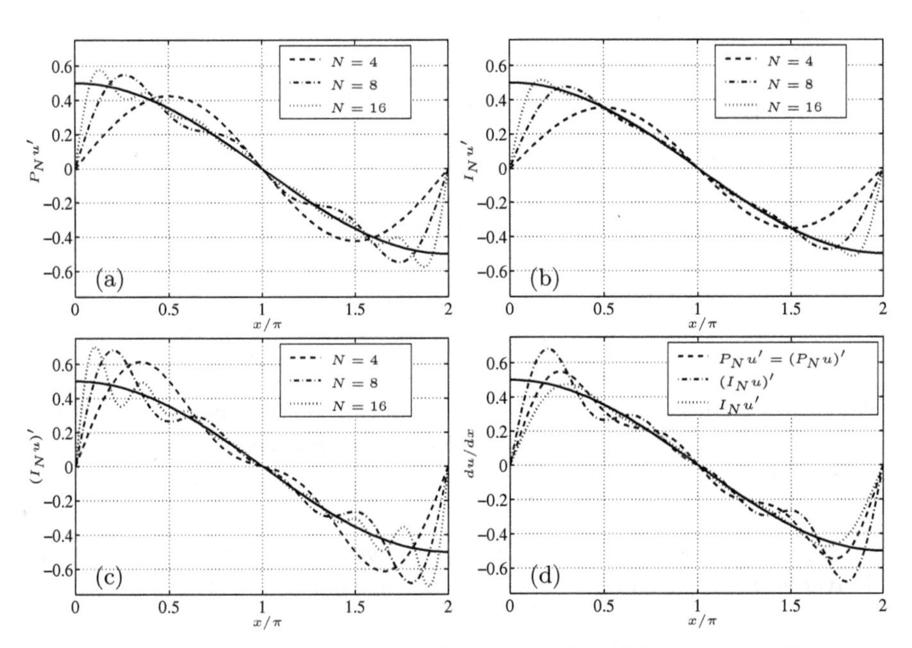


**Fig. 2.3.** Several versions of Fourier differentiation for $u(x) = \sin(x/2)$. The exact result is indicated by the solid curves and the approximate results for $N = 4$, 8 and 16 are indicated by the dashed curves. (a) $P_N u'$ and $(P_N u)'$; (b) $I_N u'$; (c) $(I_N u)'$. Part (d) shows all versions for $N = 8$

These various Fourier differentiation procedures are illustrated in Fig. 2.3 for the function $u(x) = \sin(x/2)$. Part (a) shows both $P_N u'$ and $(P_N u)'$, which are identical. Part (b) displays $I_N u'$, and part (c) shows $(I_N u)'$. The function $u'$ has a discontinuity of the same character as the square wave. The characteristic oscillations arising from a discontinuity, known as the Gibbs phenomenon, will be discussed at length in Sect. 2.1.4. The difference between $(I_N u')$ and $(I_N u)'$ is apparent in parts (b) and (c). Although the truncation errors of both have the same asymptotic behavior, in this example at least, the constant is much larger for $(I_N u)'$.

If $u \in S_N$, then $\mathcal{D}_N u = u'$. Thus, due to (2.1.33), $\mathcal{D}_N$ is a skew-symmetric operator on $S_N$:

$$(\mathcal{D}_N u, v)_N = -(u, \mathcal{D}_N v)_N \quad \text{for all } u, v \in S_N \,. \tag{2.1.46}$$

From a computational point of view, the Fourier interpolation derivative can be evaluated according to (2.1.43) and (2.1.42). These require $N$ multiplications and two discrete Fourier transforms. The total operation count is $(5\log_2 N - 5)N$ real multiplications or additions, provided that the discrete Fourier transforms are computed by an FFT that takes advantage of the reality of $u$, or that multiple derivatives are computed at once, as is the case for multidimensional problems.

Fourier interpolation differentiation can be represented by a matrix that will be called the Fourier interpolation derivative matrix. Equations (2.1.42) and (2.1.43) can be combined to yield

$$(\mathcal{D}_N u)_j = \sum_{l=0}^{N-1} (D_N)_{jl} u_l \,, \tag{2.1.47}$$

where

$$(D_N)_{jl} = \frac{1}{N} \sum_{k=-N/2}^{N/2-1} ik e^{2ik(j-l)\pi/N} \,. \tag{2.1.48}$$

We arrive at the same result by differentiating both sides of (2.1.29) and evaluating derivatives at the grid points (after exchanging the roles of $j$ and $l$). This shows that

$$(D_N)_{jl} = \psi_l'(x_j) \,, \tag{2.1.49}$$

i.e., the entries of the interpolation derivative matrix are the values of the derivative of the characteristic Lagrange polynomials (2.1.30) at the grid points.

Since the $k = -N/2$ term in the sum (2.1.48) makes a purely imaginary contribution if $u$ is a real function, its contribution effectively disappears. (See also the discussion of this point in Sect. 3.3.1.) Therefore, in practice (2.1.47) reduces to

$$(D_N)_{jl} = \frac{1}{N} \sum_{k=-N/2+1}^{N/2-1} ik e^{2ik(j-l)\pi/N} \,. \tag{2.1.50}$$

This sum may be evaluated in closed form:

$$(D_N)_{jl} = \begin{cases} \frac{1}{2}(-1)^{j+l} \cot\left[\frac{(j-l)\pi}{N}\right] \,, & j \neq l \,, \\ 0 \,, & j = l \,. \end{cases} \tag{2.1.51}$$

The skew symmetry of this real matrix is evident. Its eigenvalues are $ik$, $k = -N/2+1,\dots,N/2-1$. The eigenvalue 0 has double multiplicity. Its eigenvectors consist of the grid values of the functions 1 and $\cos(Nx/2)$. The latter function is associated with the $k = -N/2$ term in the sum (2.1.50). Note that central-difference operators for the first derivative also have a double zero eigenvalue.

Similarly, an explicit expression for the second-derivative matrix, again neglecting the $k = -N/2$ term, is

$$(D_N^{(2)})_{jl} = \begin{cases} \frac{1}{4}(-1)^{j+l}N + \dfrac{(-1)^{j+l+1}}{2\sin^2\left[\frac{(j-l)\pi}{N}\right]}, & j \neq l, \\ -\frac{(N-1)(N-2)}{12}, & j = l. \end{cases} \tag{2.1.52}$$

If a Fourier collocation method is based on an odd number of points rather than an even number, then the derivative matrix has a zero eigenvalue of single multiplicity. This alternative version of the Fourier method uses the collocation points

$$x_j = \frac{2j}{N+1}\pi, \qquad j = 0,\dots,N, \tag{2.1.53}$$

and keeps both the $\cos(Nx/2)$ and $\sin(Nx/2)$ terms in the discrete real Fourier series. The derivatives of these terms are both nonzero at the collocation points. Most applications use FFTs where $N$ is a multiple of 2. For this reason we have chosen to present Fourier methods here only for an even number of collocation points. Differentiation matrices for an odd number of collocation points can be found in Peyret (2002).

For an even number of collocation points there is a way to retain the information in the $\cos(Nx/2)$ mode for a diffusion operator of the form

$$\frac{\mathrm{d}}{\mathrm{d}x}\left(a(x)\frac{\mathrm{d}u}{\mathrm{d}x}\right).$$

The trick is to evaluate $\mathrm{d}u/\mathrm{d}x$ not at $x_j = 2\pi j/N$ but at $x_{j+1/2} = 2\pi\times(j+\frac{1}{2})/N$, to form the product $a(x_{j+1/2})\mathrm{d}u/\mathrm{d}x|_{j+1/2}$, and to evaluate the final result at $x_j$. This approach was suggested by Brandt, Fulton and Taylor (1985). They note that it can be implemented by standard FFTs, and that it does lead to more accurate approximations.

In principle, it is possible to compute Fourier interpolation differentiation by simply performing the matrix multiplication implied by (2.1.47) rather than resorting to Fourier transforms. This requires $2N^2$ operations. This operation count is lower than the operation count for transforms for $N \leq 8$. In practice, the exact crossover point will depend on the computer architecture and the programming details.

Figure 2.4 presents some timings on a desktop computer (with a clock speed of about a GigaHertz) for Fourier interpolation differentiation using

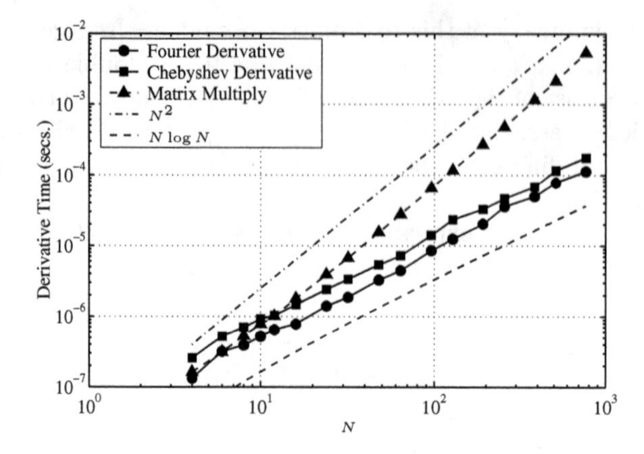


**Fig. 2.4.** Timing results for first-derivative computations using FFTs and derivative matrices

both the FFT (Fourier Derivative) and the derivative matrix (Matrix Multiply) approaches. (The Chebyshev derivative timings refer to material discussed in Sect. 2.4.) All routines were coded in Fortran and the FFTs did take advantage of the extra 10–15% efficiency available from using radix 4 rather than just radix 2 (see Appendix B). In the Fourier case the FFT method appears to be always faster than the derivative-matrix approach, and it is at least an order of magnitude faster for $N \geq 64$. The FFT method has the additional advantage of lesser contamination by round-off error. As noted in Sect. 8.3.1 of Deville, Fischer and Mund (2002), the performance of derivative matrix routines can be improved substantially, even in a high-level language such as Fortran, by hard-coding unrolled loops for each value of $N$.

### 2.1.4 The Gibbs Phenomenon

The Gibbs phenomenon describes the characteristic oscillatory behavior of the truncated Fourier series or the discrete Fourier series of a function of bounded variation in the neighborhood of a point of discontinuity. Figures 2.1(a), (b) and (c) furnish an interesting contrast. Each truncated Fourier series exhibits some oscillations about the exact function. However, the oscillations for the square wave example have some distinguishing features. The maximum amplitude of the oscillation nearest the discontinuity (the overshoot) tends to a finite limit, and the location of the overshoot tends toward the point of discontinuity as the number of retained wavenumbers is increased. The truncated series for the other two examples are uniformly convergent over $[0, 2\pi]$. They do not exhibit a finite limiting overshoot.

The behavior represented in Fig. 2.1(a) can be easily explained in terms of the singular integral representation of a truncated Fourier series. We assume here that the truncation is symmetric with respect to $N$, i.e., we set

$$P_N u = \sum_{|k| \leq N/2} \hat{u}_k \phi_k \ . \tag{2.1.54}$$

By (2.1.3) we have

$$\begin{aligned} P_N u(x) &= \sum_{|k| \leq N/2} \frac{1}{2\pi} \int_0^{2\pi} u(y) e^{-iky} \, dy \, e^{ikx} \\ &= \frac{1}{2\pi} \int_0^{2\pi} \left[ \sum_{|k| \leq N/2} e^{-ik(x-y)} \right] u(y) \, dy \ . \end{aligned}$$

The integral representation of $P_N u$ is therefore

$$P_N u(x) = \frac{1}{2\pi} \int_0^{2\pi} D_N(x-y) u(y) \, dy \ , \tag{2.1.55}$$

where $D_N(\xi)$ is the Dirichlet kernel (where in keeping with our notational convention for this part of the book we use $D_N$ for what is classically denoted by $D_{N/2}$)

$$\begin{aligned} D_N(\xi) &= 1 + 2 \sum_{k=1}^{N/2} \cos k\xi \\ &= \begin{cases} \dfrac{\sin((N+1)\xi/2)}{\sin(\xi/2)} \ , & \xi \neq 2j\pi \ , \\ N+1 \ , & \xi = 2j\pi \ , \end{cases} \qquad j \in \mathbb{Z} \ . \end{aligned} \tag{2.1.56}$$

It is illustrated in Fig. 2.5, where it is shown, for esthetic reasons, on the interval $[-\pi, \pi]$. The Dirichlet kernel can be considered as the orthogonal projection of the delta function upon the space of trigonometric polynomials of degree $N/2$, in the $L^2$-inner product. $D_N$ is an even function that changes sign at the points $\xi_j = 2j\pi(N+1)$ and that satisfies

$$\frac{1}{2\pi} \int_0^{2\pi} D_N(\xi) \mathrm{d}\xi = 1 \ , \tag{2.1.57}$$

as is evident from setting $u = 1$ in (2.1.55). Moreover, as $N \to \infty$, $D_N$ tends to zero uniformly on every closed interval excluding the singular points $\xi = 2j\pi$, $j \in \mathbb{Z}$. This means that for all $\delta > 0$ and all $\varepsilon > 0$ there exists an integer $N(\delta, \varepsilon) > 0$ such that

$$|D_N(\xi)| < \varepsilon \quad \text{if } N > N(\varepsilon, \delta) \text{ and } \delta \leq \xi \leq 2\pi - \delta \ . \tag{2.1.58}$$

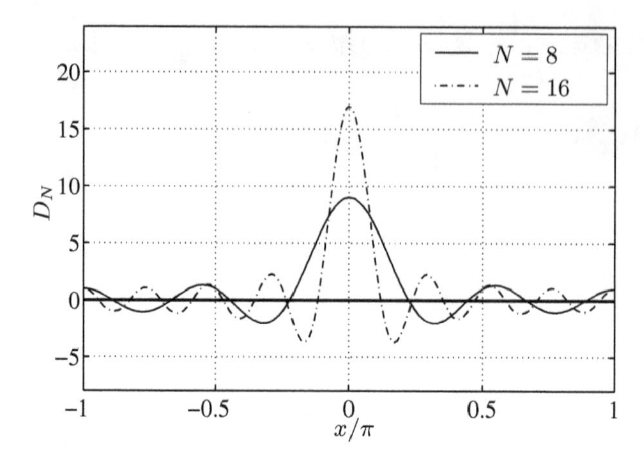


**Fig. 2.5.** The Dirichlet kernel for $N = 8$ and $N = 16$

We return now to the square wave represented in Fig. 2.1(a). For simplicity we shift the origin to the point of discontinuity, i.e., we consider the periodic function

$$\phi(x) = \begin{cases} 1\,, & 0 \le x < \pi\,, \\ 0\,, & \pi \le x < 2\pi\,, \end{cases} \tag{2.1.59}$$

whose truncated Fourier series is

$$\begin{aligned} P_N\phi(x) &= \frac{1}{2\pi}\int_{x-\pi}^{x} D_N(y)\mathrm{d}y \\ &= \frac{1}{2\pi}\left[\int_0^x D_N(y)\mathrm{d}y + \int_{-\pi}^{0} D_N(y)\mathrm{d}y + \int_{x-\pi}^{-\pi} D_N(y)\mathrm{d}y\right]\,. \end{aligned} \tag{2.1.60}$$

So long as $x$ is not close to $\pi$, the last integral on the right-hand side is arbitrarily small, provided $N$ is large enough, by (2.1.58). The middle integral equals $\pi$ by (2.1.57); hence,

$$P_N\phi(x) \simeq \frac{1}{2} + \frac{1}{2\pi}\int_0^x D_N(y)\mathrm{d}y \quad \text{as } N \to \infty\,. \tag{2.1.61}$$

This formula explains the Gibbs phenomenon for the square wave. If $x > 0$ is far enough from 0, then $1/2\pi\int_0^x D_N(y)\mathrm{d}y \simeq 1/2\pi\int_0^\pi D_N(y)\mathrm{d}y = 1/2$ by (2.1.57) and (2.1.58); hence, $P_N\phi(x)$ is close to 1. But the function $x \to 1/2\pi\int_0^x D_N(y)\mathrm{d}y$ has alternating maxima and minima at the points where $D_N$ vanishes, $\xi_j = 2j\pi/(N+1)$; this accounts for its oscillatory behavior. The absolute maximum occurs at $\xi_1 = 2\pi/(N+1)$, where for large enough $N$

$$\frac{1}{2\pi}\int_0^{2\pi/(N+1)} D_N(y)\mathrm{d}y \simeq \frac{1}{\pi}\int_0^\pi \frac{\sin t}{t} = 0.58949\ldots\,. \tag{2.1.62}$$

Thus, the sequence $\{(P_N\phi)[2\pi/(N+1)]\}$ tends to $1.08949\cdots > 1 = \phi(0^+)$ as $N \to \infty$. Equivalently,

$$\limsup_{\substack{N\to\infty \\ x\to 0^+}} (P_N\phi)(x) > \phi(0^+) . \tag{2.1.63}$$

Similarly, for $x$ negative one has

$$\liminf_{\substack{N\to\infty \\ x\to 0^-}} (P_N\phi)(x) < \phi(0^-) .$$

This is a mathematical characterization of the Gibbs phenomenon.

If now $u = u(x)$ is any function of bounded total variation (see Sect. A.8) in $[0, 2\pi]$ that has an isolated jump discontinuity at $x = x_0$, we can write

$$u(x) = \tilde{u}(x) + j(u; x_0)\phi(x - x_0) ,$$

where $j(u; x_0) = u(x_0^+) - u(x_0^-)$ is the jump of $u$ at $x_0$. The function $\tilde{u}(x) = u(x) - j(u; x_0)\phi(x - x_0)$ has at worst a removable singularity at $x = x_0$; hence, its Fourier series converges uniformly in a neighborhood of $x_0$. Thus, by (2.1.61),

$$\begin{aligned} P_N u(x) \simeq {} & \frac{1}{2}[u(x_0^+) + u(x_0^-)] \\ & + \frac{1}{2\pi}[u(x_0^+) - u(x_0^-)] \int_0^{x-x_0} D_N(y)\,\mathrm{d}y \quad \text{as } N \to \infty . \end{aligned} \tag{2.1.64}$$

This shows that the sequence $\{P_N u\}$ undergoes a Gibbs phenomenon at $x = x_0$ with the same structure as the Gibbs phenomenon for the square wave (2.1.59).

From a mathematical point of view it is worthwhile to observe that truncation does not preserve the boundedness of the total variation of a function. This means that even if the total variation of $u$ is finite, the total variation of $P_N u$ is not bounded independently of $N$. For the square wave (2.1.59), formula (2.1.61) shows that the total variation $V_N(\phi; a)$ of $P_N\phi$ in the neighborhood $[-a, a]$ of the origin is approximately

$$V_N(\phi; a) \simeq \frac{1}{\pi} \int_0^a |D_N(y)|\,\mathrm{d}y .$$

Since $D_N(y) = \sin(\frac{1}{2}(N+1)y)/y$ for $y$ close to 0, and $\int_0^{+\infty} |\sin t/t| \mathrm{d}t = \infty$, $V_N(\phi; a)$ diverges as $N \to \infty$.

The Gibbs phenomenon influences the behavior of the truncated Fourier series not only in the neighborhood of the point of singularity, but also over the entire interval $[0, 2\pi]$. The convergence rate of the truncated series is linear in $N^{-1}$ at a given nonsingular point. The point $x_0 = \pi/2$ is the farthest from all the singularity points. There one has

$$P_N\phi\left(\frac{\pi}{2}\right) = \frac{1}{2\pi}\int_{-\pi/2}^{\pi/2} D_N(y)\,dy\ ,$$

or

$$1 - P_N\phi\left(\frac{\pi}{2}\right) = \frac{1}{\pi}\int_{\pi/2}^{\pi} D_N(y)\,dy\ .$$

A primitive of the Dirichlet kernel is $(\int D_N)(x) = x + 2\sum_{k=1}^{N/2}(\sin kx)/k$; whence,

$$1 - P_N\phi\left(\frac{\pi}{2}\right) = \frac{2}{\pi}\sum_{p\geq N/4}\frac{(-1)^p}{2p+1} \simeq \frac{2}{N} \quad \text{as } N \to \infty\ .$$

This asymptotic behavior is evident in Fig. 2.1(a) for the square wave, for the corresponding point $x_0 = \pi$.

The Gibbs phenomenon also occurs for the sequence $\{I_N u\}$ of the trigonometric interpolating polynomials of $u$. If the points

$$x_l = \frac{2l\pi}{N+1}, \qquad l = 0, \ldots, N\ ,$$

already introduced in Sect. 2.1.3, are used in the interpolation process, then the interpolating polynomial has the following discrete integral representation:

$$I_N u(x) = \frac{1}{1+N}\sum_{l=0}^{N} D_N(x - x_l)u(x_l)\ . \tag{2.1.65}$$

Note that $D_N(x - x_j)/(N+1)$ is the characteristic Lagrange polynomial of degree $N/2$ at the nodes (2.1.53), i.e., the trigonometric polynomial of degree $N/2$ such that

$$\frac{1}{N+1}D_N(x_j - x_l) = \delta_{jl}, \qquad 0 \leq j,\ l \leq N\ .$$

The representation (2.1.65) for the discrete Fourier series can be related to the representation (2.1.55) for the truncated Fourier series via the use of the trapezoidal quadrature rule for evaluating the singular integral. This accounts, at least heuristically, for the similarity of the Gibbs phenomenon arising in the truncation and interpolation processes. Figure 2.1(d) shows the Gibbs phenomenon for the sequence of the discrete Fourier series of the square wave. The qualitative behavior is the same as for the truncated series, although quantitatively the oscillations appear here less pronounced. (Compare also Figs. 2.3(b), 2.3(d).)

We have seen so far how the Gibbs phenomenon occurs in the two most common trigonometric approximations of a discontinuous function: truncation and interpolation. The capability of constructing alternative trigonometric approximations that avoid or at least reduce the Gibbs phenomenon near the discontinuity points while producing a faithful representation of the function elsewhere in physical space is desirable both theoretically and practically. To be of any practical use this *smoothing* process (also referred to as a *filtering* process) ought to employ only such information that is available from a finite approximation to the function, namely a finite number of its Fourier coefficients or else its values at the grid points.

Since the Gibbs phenomenon is related to the slow decay of the Fourier coefficients of a discontinuous function (as seen in Sect. 2.1.1), it is natural to use smoothing procedures that attenuate the higher order coefficients. Thus, the oscillations associated with the higher modes in the trigonometric approximant are damped. On the other hand, the intrinsic structure of the coefficients carries information about the discontinuities, and this information should not be wasted. Too strong a smoothing procedure may result in excessively smeared approximations, which are again unfaithful representations of the true function. Therefore, the smoothing method has to be suitably tuned.

Let us now focus on smoothing for truncated Fourier series. A straightforward way to attenuate the higher order Fourier coefficients is to multiply each Fourier coefficient $\hat{u}_k$ by a factor $\sigma_k$. Thus, the truncated Fourier series $P_N u$ is replaced by the smoothed series

$$\mathcal{S}_N u = \sum_{k=-N/2}^{N/2} \sigma_k \hat{u}_k e^{ikx} . \tag{2.1.66}$$

Typically, the $\sigma_k$ are required to be real nonnegative numbers such that $\sigma_0 = 1$, $\sigma_k = \sigma_{-k}$ and $\sigma_{|k|}$ is a decreasing function of $|k|$.

The Cesáro sums are a classical way of smoothing the truncated Fourier series. They consist of taking the arithmetic means of the truncated series, i.e.,

$$\mathcal{S}_N u = \frac{1}{N/2+1} \sum_{k=0}^{N/2} P_k u = \sum_{-N/2}^{N/2} \left(1 - \frac{|k|}{N/2+1}\right) \hat{u}_k e^{ikx} . \tag{2.1.67}$$

In this case the smoothing factors are $\sigma_k = 1 - |k|/(N/2+1)$; they decay linearly in $|k|$.

Other simple smoothing methods are the Lanczos smoothing and the raised cosine smoothing. The factors that define the Lanczos smoothing are

$$\sigma_k = \frac{\sin(2k\pi/N)}{2k\pi/N} , \qquad k = -N/2, \dots, N/2 . \tag{2.1.68}$$

These are flat near $k = 0$ and approach 0 linearly as $k \to N/2$. The factors for the raised cosine smoothing are

$$\sigma_k = \frac{1 + \cos(2k\pi/N)}{2} , \qquad k = -N/2, \dots, N/2 . \tag{2.1.69}$$

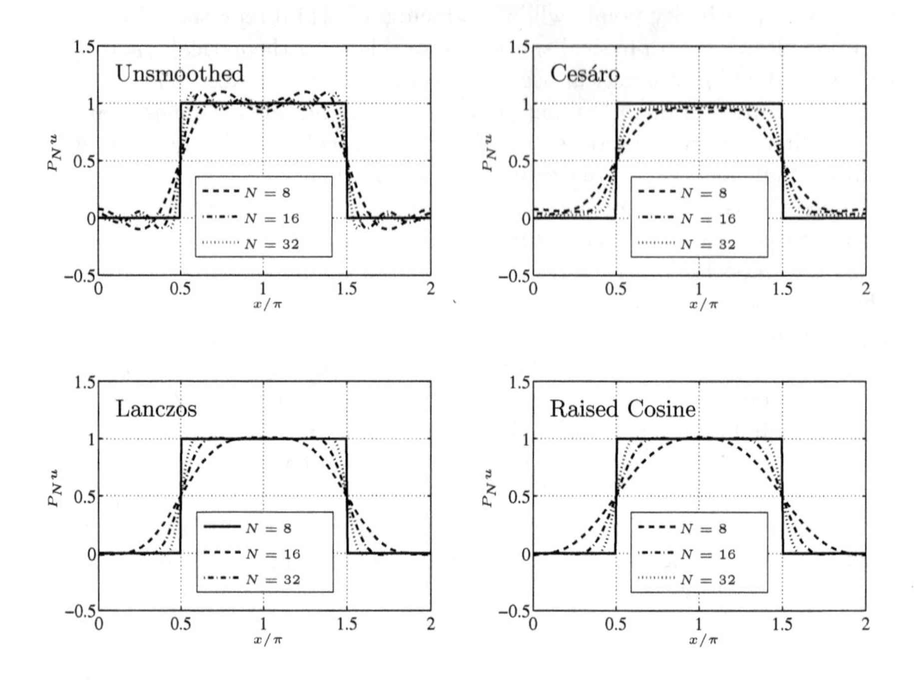


**Fig. 2.6.** Several smoothings for the square wave

These are flat at $k = N/2$ as well as at $k = 0$. The effect of each of these three smoothings upon the square wave is represented in Fig. 2.6.

The smoothed series (2.1.66) can be represented in terms of a singular integral as

$$\mathcal{S}_N u(x) = \frac{1}{2\pi} \int_0^{2\pi} K_N(x-y) u(y)\,\mathrm{d}y \,, \tag{2.1.70}$$

where the kernel $K_N(\xi)$ is given by

$$K_N(\xi) = 1 + 2 \sum_{k=1}^{N/2} \sigma_k \cos k\xi \,. \tag{2.1.71}$$

The representation (2.1.70) allows one to describe more general forms of smoothing than (2.1.66). The kernel $K_N(\xi)$ need not have the particular form (2.1.71). The only requirement is that $K_N$ be an approximate polynomial delta-function, i.e., a trigonometric polynomial of degree $N/2$ such that

$$\frac{1}{2\pi} \int_0^{2\pi} K_N(\xi)\mathrm{d}\xi = 1 \,, \tag{2.1.72}$$

and such that for $\delta > 0$ and all $\varepsilon > 0$ there exists an integer $N(\delta, \varepsilon) > 0$ for which

$$|K_N(\xi)| < \varepsilon \quad \text{if } N > N(\delta, \varepsilon) \text{ and } \delta \le \xi \le 2\pi - \delta \,. \tag{2.1.73}$$

Under these assumptions, one can repeat the arguments used in deriving (2.1.64) and obtain the asymptotic formula

$$\begin{aligned} \mathcal{S}_N u(x) \simeq & \frac{1}{2}[u(x_0^+) + u(x_0^-)] \\ & + \frac{1}{2\pi}[u(x_0^+) - u(x_0^-)] \int_0^{x-x_0} K_N(y)\,\mathrm{d}y \end{aligned} \tag{2.1.74}$$

near a point of discontinuity for $u$. Thus, the behavior of $\mathcal{S}_N u$ depends on the behavior of the function

$$\psi_N(z) = \frac{1}{2\pi}\int_0^z K_N(y)\,\mathrm{d}y \tag{2.1.75}$$

in a neighborhood of the origin. There will be a Gibbs phenomenon if there exists a sequence of points $z_N > 0$, with $z_N \to 0$ as $N \to \infty$, at which $\psi_N(z_N) \ge \alpha > \frac{1}{2}$ (for some $\alpha$ independent of $N$); in this case,

$$\lim_{N\to\infty} \mathcal{S}_N u(z_N) > u(x_0^+) \,.$$

The kernel $K_N^F$ generated by the Cesáro sums is known as the Fejér kernel. Its analytic expression is

$$\begin{aligned} K_N^F(\xi) &= 1 + 2\sum_{k=1}^{N/2}\left(1 - \frac{k}{N/2+1}\right)\cos k\xi \\ &= \begin{cases} \dfrac{1}{N/2+1}\left[\dfrac{\sin((N/2+1)\xi/2)}{\sin(\xi/2)}\right]^2, & \xi \ne 2j\pi\,, \\ N/2+1\,, & \xi = 2j\pi\,, \end{cases} \quad j \in \mathbb{Z}\,. \end{aligned} \tag{2.1.76}$$

This kernel is plotted in Fig. 2.7.

Since $K_N^F$ is nonnegative and $(1/2\pi)\int_{-\pi}^{\pi} K_N^F(y)\,\mathrm{d}y = 1$, the corresponding function $\psi_N(z)$ is monotonically increasing and satisfies $0 < \psi_N(z) < \frac{1}{2}$ in the interval $(0, \pi)$. It follows that the Cesáro sums do not exhibit the Gibbs phenomenon near a discontinuity point (see Fig. 2.6). The Cesáro sums have several useful theoretical properties of approximation: if $u$ is a continuous function in $[0, 2\pi]$, then the sequence $\mathcal{S}_N u$ converges to $u$ uniformly in the interval as $N \to \infty$. Moreover, the Cesáro sums preserve bounded variation in the sense that if $u$ is of bounded variation in $[0, 2\pi]$, then the total variation of $\mathcal{S}_N u$ can be bounded independently of $N$. However, as Fig. 2.6 shows, the Cesáro sums produce a heavy smearing of the function near a singularity point. In most applications it is desirable to have a sharper representation of the function, at the expense of retaining some oscillations. For this reason, other forms of smoothing, such as Lanczos' or the raised cosine, are preferred.

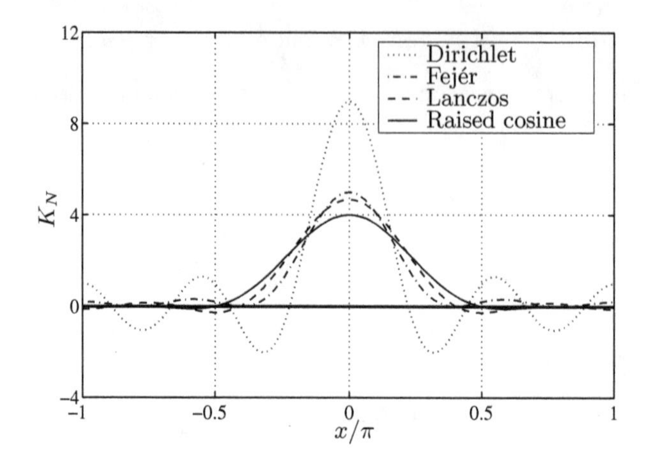


**Fig. 2.7.** Comparison of the Dirichlet kernel with the smoothed kernels for $N = 8$

The kernel $K_N^L$ corresponding to Lanczos' smoothing is given by

$$K_N^L(\xi) = 1 + \sum_{k=1}^{N/2} \frac{\sin\left(k\left(\xi + \frac{2\pi}{N}\right)\right) - \sin\left(k\left(\xi - \frac{2\pi}{N}\right)\right)}{2k\pi/N}, \tag{2.1.77}$$

while the kernel $K_N^R(\xi)$ associated with the raised cosine smoothing is

$$K_N^R(\xi) = \frac{1}{4}\left[D_N\left(\xi - \frac{2\pi}{N}\right) + 2D_N(\xi) + D_N\left(\xi + \frac{2\pi}{N}\right)\right] \tag{2.1.78}$$

(see again Fig. 2.7). Thus, the raised cosine kernel can be considered as a smoothing of the Dirichlet kernel by local averages. Both $K_N^L$ and $K_N^R$ change sign away from the origin. Thus, the associated functions $\psi_N$ defined in (2.1.75) exhibit an oscillatory behavior there. Since the first maximum value attained is larger than 1/2, both the Lanczos' and the raised cosine smoothing produce the Gibbs phenomenon near a discontinuity point. However,the oscillations of $K_N^L$ and $K_N^R$ away from the origin are considerably less pronounced than the oscillations of the Dirichlet kernel; hence the overshooting is dramatically reduced. Moreover, $K_N^R$ is better behaved than $K_N^L$ from the point of view of the oscillations. Consequently, the raised cosine smoothing is the most effective among those considered so far in this discussion.

A general strategy to design a smoothing operator of the form (2.1.66) consists of defining the smoothing factors $\sigma_k$ as

$$\sigma_k = \sigma(2k\pi/N), \qquad k = -N/2, \ldots, N/2, \tag{2.1.79}$$

where $\sigma = \sigma(\theta)$ is a real, even function that satisfies the following three conditions (Vandeven (1991)):

(i) $\sigma$ is $(p-1)$-times continuously differentiable in $\mathbb{R}$, for some $p \geq 1$;

(ii) $\sigma(\theta) = 0$ if $|\theta| \geq \pi$;

(iii) $\sigma(0) = 1, \ \sigma^{(j)}(0) = 0$ for $1 \leq j \leq p-1$.

Such a function is termed a *filtering function*, or simply a *filter*, of order $p$. (The terms *filtering* and *smoothing* are used interchangeably in this context.)

Condition (iii) guarantees that the zero mode of $u$ is kept unchanged, while the other low modes are only moderately damped (indeed, they are less and less damped as $p$ increases); thus, the smoothing procedure has little effect on a smooth function. On the contrary, condition (ii) and the smoothness of $\sigma$ imply

$$\sigma^{(j)}(\pi) = 0 \, , \qquad 0 \leq j \leq p-1 \, ;$$

this property induces a smooth and progressive damping of the higher order modes, an essential condition for properly curing the Gibbs phenomenon.

The Lanczos smoothing (2.1.68) corresponds to the filter $\sigma$ which in the interval $[0, \pi]$ is defined as

$$\sigma(0) = 1 \, , \qquad \sigma(\theta) = \frac{\sin\theta}{\theta} \quad \text{for } \theta \neq 0 \, . \tag{2.1.80}$$

Since $\sigma'(\pi) \neq 0$, this is a first-order filter, although $\sigma'(0) = \sigma''(0) = 0$ as for a second-order filter. The raised cosine smoothing (2.1.69) corresponds to

$$\sigma(\theta) = \frac{1+\cos\theta}{2} \, , \qquad \theta \in [0, \pi] \, , \tag{2.1.81}$$

and is second order. A modified form of the Cesáro smoothing (obtained by replacing $N/2+1$ by $N/2$ in the denominators of (2.1.67)) is given by the filter

$$\sigma(\theta) = 1 - \frac{\theta}{\pi} \, , \qquad \theta \in [0, \pi] \, , \tag{2.1.82}$$

which is first order. Higher order filters are

(a) the sharpened raised cosine filter, given by

$$\sigma(\theta) = \sigma_0(\theta)^4 [35 - 84\sigma_0(\theta) + 70\sigma_0(\theta)^2 - 20\sigma_0(\theta)^3] \tag{2.1.83}$$

(where $\sigma_0$ is the raised cosine filter (2.1.81)), which is eighth order;

(b) the Vandeven filter of order $p$ (Vandeven (1991)):

$$\sigma(\theta) = 1 - \frac{(2p-1)!}{(p-1)!} \int_0^{\theta/\pi} [t(1-t)]^{p-1} \mathrm{d}t \, , \tag{2.1.84}$$

which has optimal approximation properties; see (7.6.26) (interestingly, this filter essentially coincides with the Daubechies filter used in the construction of wavelets (Strang and Nguyen (1996));

(c) the exponential filter of order $p$, for $p$ even:

$$\sigma(\theta) = e^{-\alpha\theta^p}, \qquad \alpha > 0; \tag{2.1.85}$$

this filter does not satisfy condition (iii), however, the same effect is achieved in practice by choosing $\alpha$ so that $\sigma(\pi)$ is below machine accuracy.

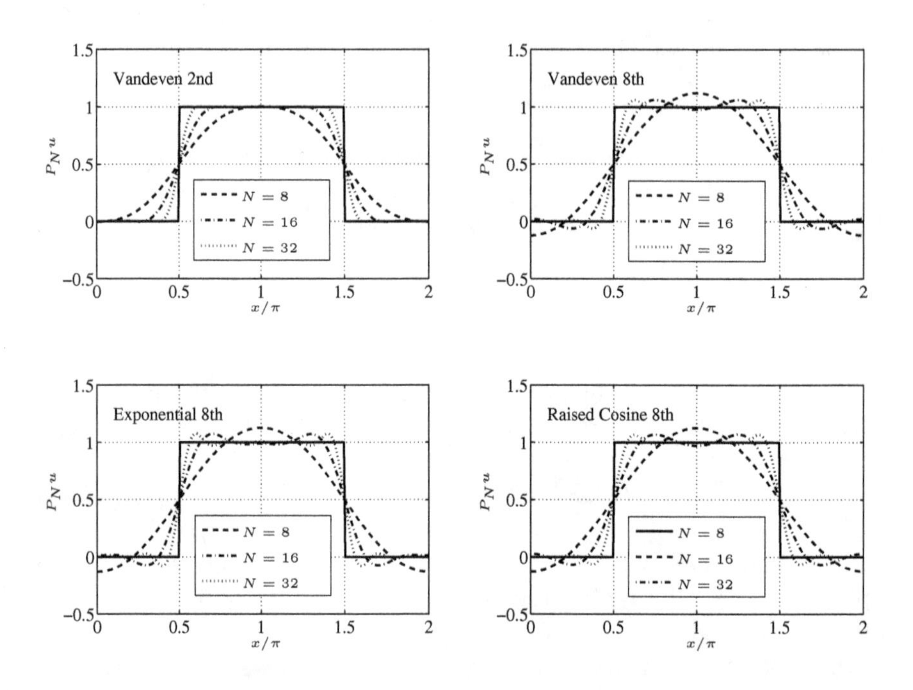


**Fig. 2.8.** Additional smoothings for the square wave

Figure 2.8 illustrates the effects of the Vandeven filter, for both $p = 2$ and $p = 8$, as well as the eighth-order raised cosine and exponential filters upon the square wave. The effect of the second-order Vandeven filter is similar to that of the second-order raised cosine filter shown in Fig. 2.6. Likewise, the effects of the various eighth-order filters in Fig. 2.8 are very similar to each other.

Figure 2.9 shows the effects of increasing order of filter (in this case the exponential filter) upon a Fourier truncation for a fixed value of $N$ (in this case $N = 128$). One clearly sees the progression from a heavily smoothed series with the low-order filter ($p = 2$) to a high-order filtering of the series ($p = 128$) that retains most of the oscillations of the original, unfiltered series.

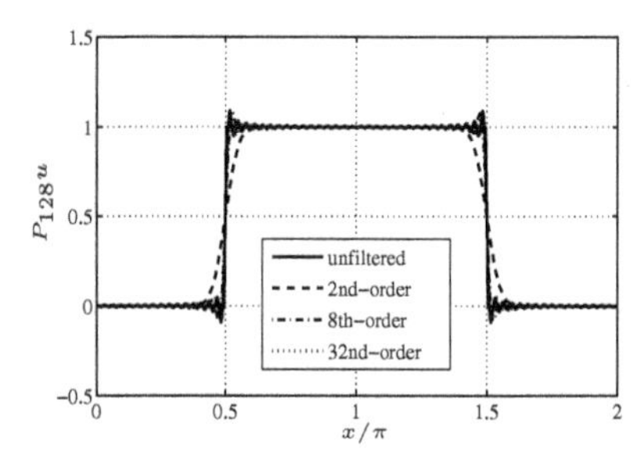


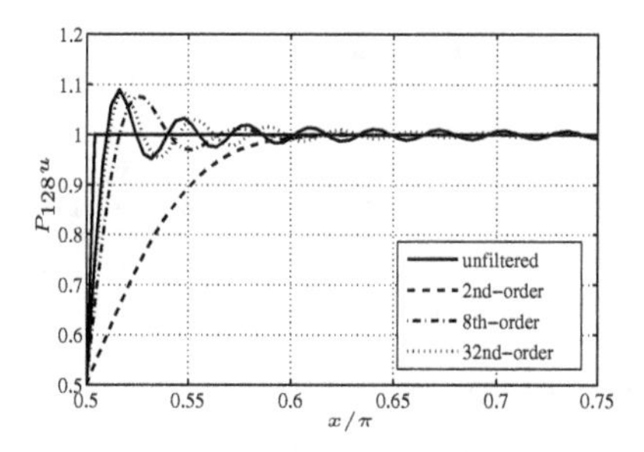


**Fig. 2.9.** Several orders of exponential smoothing for the square wave approximated by Fourier series with $N = 128$. Entire domain (*left*) and region of discontinuity (*right*)

The mathematical properties of the smoothing operators generated by filters of order $p$ are accounted for in Sect. 7.6.3.

Similar smoothing procedures can be implemented for discrete Fourier series by applying the smoothing factors to the discrete coefficients. Note however that care should be taken in operating on the discrete rather than exact expansion coefficients when the function to be smoothed is the approximate solution of a partial differential equation. The results by Majda, McDonough and Osher (1978) (see Sect. 7.6.4) are an example.

More sophisticated (although delicate) cures to the Gibbs phenomenon than simply smoothing have been proposed (see the references in Sects. 7.6.3–7.6.4). The idea underlying all of them is that, whenever the location of the singularities of a function is known, the unsmoothed coefficients contain enough information to allow for the reconstruction of an accurate, non-oscillatory approximation of the function in any interval between two consecutive singularities. This can be achieved either locally, at a point in which the function is smooth, by taking the convolutions in physical space of its truncated Fourier series with suitable smoothing kernels, or globally, in an interval between two consecutive singularities, by re-projecting the oscillating truncated Fourier series onto a sequence of orthogonal, nonperiodic basis functions defined in the interval. Techniques of singularity detection are then added in order to achieve a fully automated reconstruction procedure. Some details on these strategies are furnished in Sect. 7.6.3, as well as in Sect. 7.6.4 in the context of Fourier discretizations of hyperbolic differential equations.

The Gibbs phenomenon is not peculiar to Fourier expansions. All orthogonal polynomials introduced below, as well as more general orthogonal functions, yield Gibbs-like oscillations for truncated expansions or for interpolations of discontinuous functions. The cures indicated here and in Sect. 7.6.3

for the Fourier case extend to the nonperiodic cases, perhaps at the cost of some technical, although not conceptual, complications.

## 2.2 Orthogonal Polynomials in $(-1, 1)$

### 2.2.1 Sturm-Liouville Problems

The importance of Sturm-Liouville problems for spectral methods lies in the fact that the spectral approximation of the solution of a differential equation is usually regarded as a finite expansion of eigenfunctions of a suitable Sturm-Liouville problem. We recall that a Sturm-Liouville problem is an eigenvalue problem of the form

$$\begin{aligned} &-(pu')' + qu = \lambda wu \qquad \text{in the interval } (-1,1)\,, \\ &\text{suitable boundary conditions for } u\,. \end{aligned} \tag{2.2.1}$$

The coefficients $p, q$ and $w$ are three given, real-valued functions such that: $p$ is continuously differentiable, strictly positive in $(-1,1)$ and continuous at $x = \pm 1$; $q$ is continuous, nonnegative and bounded in $(-1,1)$; the weight function $w$ is continuous, nonnegative and integrable over $(-1,1)$.

The Sturm-Liouville problems of interest in spectral methods are those for which the expansion of an infinitely smooth function in terms of their eigenfunctions guarantees spectral accuracy. This means that the "Fourier" coefficients according to this basis decay faster than algebraically in the inverse of the eigenvalue. As pointed out in Gottlieb and Orszag (1977, Sect. 3) not all the Sturm-Liouville problems ensure this property. For instance, the Sturm-Liouville problem

$$\begin{aligned} &u'' + \lambda u = 0 \qquad \text{in } (-1,1)\,, \\ &u'(-1) = u'(1) = 0\,, \end{aligned}$$

has eigenvalues $\lambda_k = (\pi k)^2/2$ and corresponding eigenfunctions $\phi_k(x) = \cos(\pi/2)k(x+1)$. A smooth function can be approximated by the cosine series on $(-1,1)$ with spectral accuracy if and only if all its odd derivatives vanish at the boundary. This is due to the fact that the coefficient $p(x)$ in the operator does not vanish at the boundary in this case, i.e., the Sturm-Liouville problem is *regular*. Conversely, spectral accuracy is ensured if the problem is *singular*, i.e., if $p$ vanishes at the boundary. A mathematical proof of these facts is given in Sect. 5.2.

Among the singular Sturm-Liouville problems, particular importance rests with those problems whose eigenfunctions are algebraic polynomials because of the efficiency with which they can be evaluated and differentiated numerically. It is also proven in Sect. 5.2 that the Jacobi polynomials, whose properties are summarized in Sect. 2.5, are precisely the only polynomials arising as eigenfunctions of a singular Sturm-Liouville problem.

### 2.2.2 Orthogonal Systems of Polynomials

We shall consider here from a general point of view the problem of the expansion of a function in terms of a system of orthogonal polynomials. We denote by $\mathbb{P}_N$ the space of all polynomials of degree $\leq N$. Assume that $\{p_k\}_{k=0,1,\ldots}$ is a system of algebraic polynomials (with degree of $p_k = k$) that are mutually orthogonal over the interval $(-1, 1)$ with respect to a weight function $w$:

$$\int_{-1}^{1} p_k(x)p_m(x)w(x)\,\mathrm{d}x = 0 \quad \text{whenever } m \neq k\,. \tag{2.2.2}$$

The classical Weierstrass theorem implies that such a system is complete in the space $L^2_w(-1, 1)$. This is the space of functions $v$ such that the norm

$$\|v\|_w = \left(\int_{-1}^{1} |v(x)|^2 w(x)\,\mathrm{d}x\right)^{1/2} \tag{2.2.3}$$

is finite. The associated inner product is

$$(u, v)_w = \int_{-1}^{1} u(x)v(x)w(x)\,\mathrm{d}x\,. \tag{2.2.4}$$

When $w \equiv 1$ (Legendre weight), we will often use the simpler notation $L^2(-1, 1)$ instead of $L^2_w(-1, 1)$. The formal series of a function $u \in L^2_w(-1, 1)$ in terms of the system $\{p_k\}$ is

$$Su = \sum_{k=0}^{\infty} \hat{u}_k p_k\,,$$

where the expansion coefficients $\hat{u}_k$ are defined as

$$\hat{u}_k = \frac{1}{\|p_k\|^2_w} \int_{-1}^{1} u(x)p_k(x)w(x)\,\mathrm{d}x\,. \tag{2.2.5}$$

Equation (2.2.5) represents the *polynomial transform* of $u$. For an integer $N > 0$, the truncated series of $u$ of order $N$ is the polynomial

$$P_N u = \sum_{k=0}^{N} \hat{u}_k p_k\,. \tag{2.2.6}$$

Due to (2.2.2), $P_N u$ is the orthogonal projection of $u$ upon $\mathbb{P}_N$ in the inner product (2.2.4), i.e.,

$$(P_N u, v)_w = (u, v)_w \quad \text{for all } v \in \mathbb{P}_N\,. \tag{2.2.7}$$

The completeness of the system $\{p_k\}$ is equivalent to the property that, for all $u \in L^2_w(-1, 1)$,

$$\|u - P_N u\|_w \to 0 \quad \text{as } N \to \infty\,. \tag{2.2.8}$$

### 2.2.3 Gauss-Type Quadratures and Discrete Polynomial Transforms

We discuss here the close relation between orthogonal polynomials and Gauss-type integration formulas on the interval $[-1,1]$. The material of this subsection includes the interpolation formulas and discrete transforms pertinent to finite polynomial expansions.

First, we review Gaussian integration formulas, including those with some preassigned abscissas. The first result can be found in most textbooks on numerical analysis. For completeness we report the proofs concerning Gauss-Radau and Gauss-Lobatto formulas (see also Mercier (1981)).

***Gauss integration.*** *Let $x_0 < x_1 < \cdots < x_N$ be the roots of the $(N+1)$-th orthogonal polynomial $p_{N+1}$, and let $w_0, \ldots, w_N$ be the solution of the linear system*

$$\sum_{j=0}^{N} (x_j)^k w_j = \int_{-1}^{1} x^k w(x)\, \mathrm{d}x \,, \qquad 0 \le k \le N \,. \tag{2.2.9}$$

*Then*

*(i) $w_j > 0$ for $j = 0, \ldots, N$ and*

$$\sum_{j=0}^{N} p(x_j) w_j = \int_{-1}^{1} p(x) w(x)\, \mathrm{d}x \quad \text{for all } p \in \mathbb{P}_{2N+1} \,. \tag{2.2.10}$$

*The positive numbers $w_j$ are called* weights.

*(ii) It is not possible to find $x_j$, $w_j$, $j = 0, \ldots, N$, such that (2.2.10) holds for all polynomials $p \in \mathbb{P}_{2N+2}$.*

This version of Gauss integration is quite well known. However, the roots, which correspond to the collocation points, are all in the interior of $(-1,1)$. When boundary conditions are imposed strongly at one or both end points, one needs the generalized Gauss integration formulas that include these points.

To obtain the Gauss-Radau formula let us consider the polynomial

$$q(x) = p_{N+1}(x) + a p_N(x) \,, \tag{2.2.11}$$

where $a$ is chosen to produce $q(-1) = 0$ (hence, $a = -P_{N+1}(-1)/P_N(-1)$).

***Gauss-Radau integration.*** *Let $-1 = x_0 < x_1 < \cdots < x_N$ be the $N+1$ roots of the polynomial (2.2.11), and let $w_0, \ldots, w_N$ be the solution of the linear system*

$$\sum_{j=0}^{N} (x_j)^k w_j = \int_{-1}^{1} x^k w(x)\, \mathrm{d}x \,, \qquad 0 \le k \le N \,. \tag{2.2.12}$$

*Then*

$$\sum_{j=0}^{N} p(x_j)w_j = \int_{-1}^{1} p(x)w(x)\,\mathrm{d}x \quad \text{for all } p \in \mathbb{P}_{2N} \,. \tag{2.2.13}$$

The result can be established as follows. From the definition of $q$ and the orthogonality of the polynomials, it follows that

$$(q,\phi)_w = 0 \quad \text{for all } \phi \in \mathbb{P}_{N-1} \,. \tag{2.2.14}$$

For any $p \in \mathbb{P}_{2N}$ there exist $r \in \mathbb{P}_{N-1}$ and $s \in \mathbb{P}_N$ such that

$$p(x) = q(x)r(x) + s(x) \,.$$

Since $q(x_j) = 0$, $0 \le j \le N$, we have $p(x_j) = s(x_j)$, $0 \le j \le N$. It follows that

$$\begin{aligned}\sum_{j=0}^{N} p(x_j)w_j = \sum_{j=0}^{N} s(x_j)w_j &= \int_{-1}^{1} s(x)w(x)\,\mathrm{d}x \\ &= \int_{-1}^{1} p(x)w(x)\,\mathrm{d}x - \int_{-1}^{1} q(x)r(x)w(x)\,\mathrm{d}x \,.\end{aligned}$$

Now (2.2.13) is a consequence of (2.2.14).

In order to obtain the Gauss-Radau formula including the right-hand point $x = +1$, one has to take $a$ in (2.2.11) in such a way that $q(1) = 0$. If $x_0 < x_1 < \cdots < x_N = 1$ are the roots of $q(x)$, and $w_0, \ldots, w_N$ is the solution of the system (2.2.12) relative to these new points $x_j$, then (2.2.13) holds.

The Gauss-Lobatto formula is obtained in a similar way. We consider now

$$q(x) = p_{N+1}(x) + a p_N(x) + b p_{N-1}(x) \,, \tag{2.2.15}$$

where $a$ and $b$ are chosen so that $q(-1) = q(1) = 0$. Then we have

***Gauss-Lobatto integration.*** *Let $-1 = x_0 < x_1 < \cdots < x_N = 1$ be the $N+1$ roots of the polynomial (2.2.15), and let $w_0, \ldots, w_N$ be the solution of the linear system*

$$\sum_{j=0}^{N} (x_j)^k w_j = \int_{-1}^{1} x^k w(x)\,\mathrm{d}x \,, \qquad 0 \le k \le N \,. \tag{2.2.16}$$

*Then*

$$\sum_{j=0}^{N} p(x_j)w_j = \int_{-1}^{1} p(x)w(x)\,\mathrm{d}x \quad \text{for all } p \in \mathbb{P}_{2N-1} \,. \tag{2.2.17}$$

The proof of this result is similar to the previous one: here the decomposition $p = qr + s$ holds with $r \in \mathbb{P}_{N-2}$ and $s \in \mathbb{P}_N$.

In the important special case of a Jacobi weight (see Sect. 2.5), there is an alternative characterization of the Gauss-Lobatto points; namely they are the points $-1$, $+1$ and the roots of the polynomial

$$q(x) = p'_N(x) . \tag{2.2.18}$$

In fact, each $p \in \mathbb{P}_{2N-1}$ can be represented in the form

$$p(x) = (1 - x^2)p'_N(x)r(x) + s(x)$$

with $r \in \mathbb{P}_{N-2}$ and $s \in \mathbb{P}_N$. By partial integration we have

$$\begin{aligned}\int_{-1}^{1} p'_N(x)(1 - x^2)r(x)w(x)\,dx \\ = -\int_{-1}^{1} p_N(x)[(1 - x^2)r(x)]'w(x)\,dx \\ -\int_{-1}^{1} p_N(x)r(x)(1 - x^2)\frac{w'(x)}{w(x)}w(x)\,dx .\end{aligned}$$

If $w(x) = (1 - x)^\alpha(1 + x)^\beta$, with $\alpha$, $\beta > -1$, is a Jacobi weight, then the function $(1-x^2)[w'(x)/w(x)]$ is a polynomial of degree 1. It follows that $p'_N$ is orthogonal to $(1-x^2)r(x)$; hence, (2.2.17) holds when the interior quadrature nodes are the zeroes of $p'_N$, and the weights are defined by (2.2.16).

The Gauss-Lobatto points for the particular Jacobi polynomials corresponding to the weights $w(x) = (1 - x)^\alpha(1 + x)^\alpha$ are illustrated in Fig. 2.10

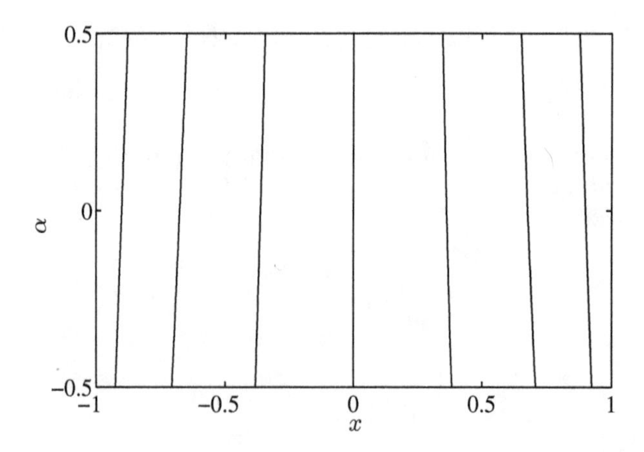


**Fig. 2.10.** The Gauss-Lobatto points for $N = 8$ for the Jacobi polynomials with the weight function $w(x) = (1 - x)^\alpha(1 + x)^\alpha$

for $N = 8$ and $-\frac{1}{2} \le \alpha \le \frac{1}{2}$. The plot shows a relevant monotonicity property of these nodes, and more generally of the zeros of ultraspherical Jacobi polynomials (see Sect. 2.5): they move toward the center of the interval $(-1, 1)$ as the parameter $\alpha$ increases (see, e.g., Szegö (1939), Chap. VI).

As observed at the beginning of this section, the nodes of the Gauss-type formulas play an important role in collocation approximations – they are precisely the collocation points at which the differential equations are enforced (see Sect. 5.4.3). We assume here that a weight function $w$ is given, together with the corresponding sequence of orthogonal polynomials $\{p_k\}$, $k = 0, 1, 2, \dots$ . For a given $N \ge 0$, we denote by $x_0, x_1, \dots, x_N$ the nodes of the $N+1$-point integration formula of Gauss, Gauss-Radau or Gauss-Lobatto type, and by $w_0, w_1, \dots, w_N$ the corresponding weights.

In a collocation method the fundamental representation of a smooth function $u$ on $(-1, 1)$ is in terms of its values at the discrete Gauss-type points. Derivatives of the function are approximated by analytic derivatives of the interpolating polynomial. The interpolating polynomial is denoted by $I_N u$. It is an element of $\mathbb{P}_N$ and satisfies

$$I_N u(x_j) = u(x_j) , \qquad 0 \le j \le N . \tag{2.2.19}$$

$I_N u$ is uniquely defined since the $x_j$'s are distinct. Since it is a polynomial of degree $N$, it admits an expression of the form

$$I_N u = \sum_{k=0}^{N} \tilde{u}_k p_k . \tag{2.2.20}$$

Obviously,

$$u(x_j) = \sum_{k=0}^{N} \tilde{u}_k p_k(x_j) , \qquad j = 0, \dots, N . \tag{2.2.21}$$

The $\tilde{u}_k$ are called the *discrete polynomial coefficients* of $u$. They are sometimes referred to as discrete expansion coefficients. The inverse relationship is

$$\tilde{u}_k = \frac{1}{\gamma_k} \sum_{j=0}^{N} u(x_j) p_k(x_j) w_j , \qquad k = 0, \dots, N , \tag{2.2.22}$$

where

$$\gamma_k = \sum_{j=0}^{N} p_k^2(x_j) w_j . \tag{2.2.23}$$

Equation (2.2.22) will be derived below. Explicit formulas for $\gamma_k$ for the more common orthogonal polynomials are supplied in Sects. 2.3 and 2.4.

Equations (2.2.21) and (2.2.22) enable one to transform freely between physical space $\{u(x_j)\}$ and transform space $\{\tilde{u}_k\}$. Such a transformation for orthogonal polynomials is the analogue of the transformation (2.1.27) and

(2.1.25) for trigonometric polynomials. We shall call it the *discrete polynomial transform* associated with the weight $w$ and the nodes $x_0, \ldots, x_N$.

For any $u, v$ continuous on $[-1, 1]$, we set

$$(u, v)_N = \sum_{j=0}^{N} u(x_j)v(x_j)w_j \ . \tag{2.2.24}$$

The Gauss integration formulas imply that

$$(u, v)_N = (u, v)_w \quad \text{if } uv \in \mathbb{P}_{2N+\delta} \ , \tag{2.2.25}$$

where $\delta = 1, 0, -1$ for Gauss, Gauss-Radau or Gauss-Lobatto integration, respectively. In particular, $(u, v)_N$ is an inner product on $\mathbb{P}_N$. The corresponding norm is

$$\|u\|_N = \sqrt{(u, u)_N} \ . \tag{2.2.26}$$

For any continuous $v$, (2.2.19) gives

$$(I_N u, v)_N = (u, v)_N \ . \tag{2.2.27}$$

This shows that, as for the trigonometric systems, the interpolant $I_N u$ is the orthogonal projection of $u$ upon $\mathbb{P}_N$ with respect to the discrete inner product (2.2.24).

The orthogonality of the $p_m$'s, together with (2.2.25) give

$$(p_m, p_k)_N = \gamma_k \delta_{km} \ , \qquad k, m = 0, \ldots, N, \tag{2.2.28}$$

where $\gamma_k$ is defined in (2.2.23). From (2.2.27) and (2.2.28) we obtain

$$(u, p_k)_N = (I_N u, p_k)_N = \sum_{m=0}^{N} \tilde{u}_m (p_m, p_k)_N = \gamma_k \tilde{u}_k \ , \qquad k = 0, \ldots, N \ ,$$

and (2.2.22) follows directly. In terms of the discrete inner product this is just

$$\tilde{u}_k = \frac{1}{\gamma_k}(u, p_k)_N \ , \qquad k = 0, \ldots, N \ . \tag{2.2.29}$$

The discrete polynomial coefficients $\tilde{u}_k$ can be expressed in terms of the continuous coefficients $\hat{u}_k$ as follows:

$$\tilde{u}_k = \hat{u}_k + \frac{1}{\gamma_k} \sum_{l>N} (p_l, p_k)_N \hat{u}_l \ , \qquad k = 0, \ldots, N \ . \tag{2.2.30}$$

This formula is an easy consequence of (2.2.29) and (2.2.28). Equivalently, one can write

$$I_N u = P_N u + R_N u \ , \tag{2.2.31}$$

where

$$R_N u = \sum_{k=0}^{N} \left( \frac{1}{\gamma_k} \sum_{l>N} (p_l, p_k)_N \hat{u}_l \right) p_k \tag{2.2.32}$$

can be viewed as the *aliasing error* due to interpolation (compare with (2.1.38)). The aliasing error is orthogonal to the truncation error $u - P_N u$ so that

$$\|u - I_N u\|_w^2 = \|u - P_N u\|_w^2 + \|R_N u\|_w^2 \, . \tag{2.2.33}$$

In general, $(p_l, p_k)_N \neq 0$ for all $l > N$. Thus the $k$-th mode of the algebraic interpolant of $u$ depends on the $k$-th mode of $u$ and *all* the modes whose wavenumber is larger than $N$. The aliasing error has a simpler expression for the Chebyshev interpolation points (see (2.4.20)).

## 2.3 Legendre Polynomials

### 2.3.1 Basic Formulas

We present here a collection of the essential formulas for Legendre polynomials. For proofs, the reader may refer to Szegö (1939). The Legendre polynomials $L_k(x)$, $k = 0, 1, \ldots$ , are the eigenfunctions of the singular Sturm-Liouville problem

$$((1 - x^2) L_k'(x))' + k(k+1) L_k(x) = 0 \, , \tag{2.3.1}$$

which is (2.2.1) with $p(x) = 1 - x^2$, $q(x) = 0$ and $w(x) = 1$. $L_k(x)$ is even if $k$ is even and odd if $k$ is odd. If $L_k(x)$ is normalized so that $L_k(1) = 1$, then for any $k$:

$$L_k(x) = \frac{1}{2^k} \sum_{l=0}^{[k/2]} (-1)^l \binom{k}{l} \binom{2k-2l}{k} x^{k-2l} \, , \tag{2.3.2}$$

where $[k/2]$ denotes the integral part of $k/2$. The Legendre polynomials satisfy the recursion relation

$$L_{k+1}(x) = \frac{2k+1}{k+1} x L_k(x) - \frac{k}{k+1} L_{k-1}(x) \, , \tag{2.3.3}$$

where $L_0(x) = 1$ and $L_1(x) = x$. Relevant properties are

$$|L_k(x)| \leq 1 \, , \quad -1 \leq x \leq 1 \, , \tag{2.3.4}$$

$$L_k(\pm 1) = (\pm 1)^k \, , \tag{2.3.5}$$

$$|L_k'(x)| \leq \tfrac{1}{2} k(k+1) \, , \quad -1 \leq x \leq 1 \, , \tag{2.3.6}$$

$$L_k'(\pm 1) = (\pm 1)^{k+1} \tfrac{1}{2} k(k+1) \, , \tag{2.3.7}$$

$$\int_{-1}^{1} L_k^2(x) \, \mathrm{d}x = (k + \tfrac{1}{2})^{-1} \, . \tag{2.3.8}$$

The expansion of any $u \in L^2(-1,1)$ in terms of the $L_k$'s is

$$u(x) = \sum_{k=0}^{\infty} \hat{u}_k L_k(x), \quad \hat{u}_k = (k+\tfrac{1}{2}) \int_{-1}^{1} u(x) L_k(x)\, dx \,. \tag{2.3.9}$$

We consider now discrete Legendre series. Since explicit formulas for the quadrature nodes are not known, such points have to be computed numerically as zeroes of appropriate polynomials. The quadrature weights can be expressed in closed form in terms of the nodes, as indicated in the following formulas (see, e.g., Davis and Rabinowitz (1984)):

*Legendre Gauss (LG).*

$$\begin{gathered} x_j \ (j=0,\dots,N) \text{ zeros of } L_{N+1}\,; \\ w_j = \frac{2}{(1-x_j^2)[L'_{N+1}(x_j)]^2}\,, \quad j=0,\dots,N\,. \end{gathered} \tag{2.3.10}$$

*Legendre Gauss-Radau (LGR).*

$$\begin{gathered} x_j \ (j=0,\dots,N) \text{ zeros of } L_N + L_{N+1}\,; \\ w_0 = \frac{2}{(N+1)^2}\,, \qquad w_j = \frac{1}{(N+1)^2}\frac{1-x_j}{[L_N(x_j)]^2}\,, \quad j=1,\dots,N\,. \end{gathered} \tag{2.3.11}$$

*Legendre Gauss-Lobatto (LGL).*

$$\begin{gathered} x_0=-1,\ x_N=1,\ x_j \ (j=1,\dots,N-1) \text{ zeros of } L'_N\,; \\ w_j = \frac{2}{N(N+1)}\frac{1}{[L_N(x_j)]^2}\,, \quad j=0,\dots,N\,. \end{gathered} \tag{2.3.12}$$

The normalization factors $\gamma_k$ introduced in (2.2.23) are given by

$$\begin{aligned} \gamma_k &= (k+\tfrac{1}{2})^{-1} && \text{for } k<N\,, \\ \gamma_N &= \begin{cases} (N+\tfrac{1}{2})^{-1} & \text{for Gauss and Gauss-Radau formulas}\,, \\ 2/N & \text{for the Gauss-Lobatto formula}\,. \end{cases} \end{aligned} \tag{2.3.13}$$

Certain bounds for the weights and nodes of these quadrature formulas are useful (see, e.g., Szegö (1939), Chap. VI). For the Gauss nodes, one has $x_j = -\cos\theta_j$ with

$$(j-\tfrac{1}{2})\frac{\pi}{N+1} < \theta_j < j\frac{\pi}{N+2}\,, \qquad j=1,2,\dots,\left[\tfrac{N+1}{2}\right]\,, \tag{2.3.14}$$

(nodes with higher values of $j$ are placed symmetrically with respect to the origin). The Gauss-Lobatto nodes $x_j = -\cos\eta_j$ are interlaced with the Gauss nodes corresponding to a polynomial of one degree smaller. On the other hand, in the interval $(-1,0)$, each one is placed left of the corresponding

Gauss-Lobatto node for the Chebyshev weight, given by (2.4.14); this follows from the monotonicity property with respect to $\alpha$ shown in Fig. 2.10. Hence,

$$j\frac{\pi}{N} < \eta_j < (j+1)\frac{\pi}{N+1}, \qquad j = 1, 2, \ldots, \left[\tfrac{N}{2}\right]. \tag{2.3.15}$$

For both families, each weight $w_j$ can be estimated in terms of the node $x_j$ as follows:

$$cN^{-1}(1-x_j^2)^{1/2} \le w_j \le c'N^{-1}(1-x_j^2)^{1/2}, \tag{2.3.16}$$

for suitable constants $0 < c < c'$ independent of $j$ and $N$.

### 2.3.2 Differentiation

As for the Fourier expansion, differentiation can be accomplished in transform space or in physical space, according to the representation of the function.

Differentiation in transform space consists of computing the Legendre expansion of the derivative of a function in terms of the Legendre expansion of the function itself. If $u = \sum_{k=0}^{\infty} \hat{u}_k L_k$, $u'$ can be (formally) represented as

$$u' = \sum_{k=0}^{\infty} \hat{u}_k^{(1)} L_k, \tag{2.3.17}$$

where

$$\hat{u}_k^{(1)} = (2k+1) \sum_{\substack{p=k+1 \\ p+k \text{ odd}}}^{\infty} \hat{u}_p, \qquad k \ge 0. \tag{2.3.18}$$

The key to proving this formula is the relation

$$(2k+1)L_k(x) = L'_{k+1}(x) - L'_{k-1}(x), \qquad k \ge 0. \tag{2.3.19}$$

This, in turn, is an easy consequence of the identity (see, e.g., Abramowitz and Stegun (1972, Chapter 22)),

$$(1-x^2)L'_k(x) = kL_{k-1}(x) - kxL_k(x) \tag{2.3.20}$$

and the recursion relation (2.3.3). By (2.3.19),

$$\begin{aligned}
u'(x) &= \sum_{k=0}^{\infty} \frac{\hat{u}_k^{(1)}}{2k+1} L'_{k+1}(x) - \sum_{k=0}^{\infty} \frac{\hat{u}_k^{(1)}}{2k+1} L'_{k-1}(x) \\
&= \sum_{k=1}^{\infty} \frac{\hat{u}_{k-1}^{(1)}}{2k-1} L'_k(x) - \sum_{k=-1}^{\infty} \frac{\hat{u}_{k+1}^{(1)}}{2k+3} L'_k(x) \\
&= \sum_{k=1}^{\infty} \left[ \frac{\hat{u}_{k-1}^{(1)}}{2k-1} - \frac{\hat{u}_{k+1}^{(1)}}{2k+3} \right] L'_k(x).
\end{aligned}$$

On the other hand,

$$u'(x) = \sum_{k=0}^{\infty} \hat{u}_k L'_k(x) ,$$

and since the $L'_k$ are linearly independent,

$$\hat{u}_k = \frac{\hat{u}^{(1)}_{k-1}}{2k-1} - \frac{\hat{u}^{(1)}_{k+1}}{2k+3} , \qquad k \geq 1 , \tag{2.3.21}$$

which imply (2.3.18). The previous identity generalizes, with obvious notation, to

$$\hat{u}^{(q-1)}_k = \frac{\hat{u}^{(q)}_{k-1}}{2k-1} - \frac{\hat{u}^{(q)}_{k+1}}{2k+3} , \qquad k \geq 1 , \tag{2.3.22}$$

from which it is possible to get explicit expressions for the Legendre coefficients of higher derivatives. For the second derivative we have

$$\hat{u}^{(2)}_k = (k + \tfrac{1}{2}) \sum_{\substack{p=k+2 \\ p+k \text{ even}}} [p(p+1) - k(k+1)]\hat{u}_p , \qquad k \geq 0 . \tag{2.3.23}$$

The previous expansions are not merely formal provided $u$ is smooth enough. For instance, the series (2.3.17) is convergent in the mean if the derivative of $u$ (in the sense of distributions) is a function in $L^2(-1,1)$.

Unlike for the Fourier system, differentiation and Legendre truncation do not commute, i.e., in general,

$$(P_N u)' \neq P_{N-1} u' . \tag{2.3.24}$$

This is an immediate consequence of (2.3.18). It is the quantity on the left that is referred to as the *Legendre projection derivative*. The error $(P_N u)' - P_{N-1} u'$ decays spectrally for infinitely smooth solutions. However, if $u$ has finite regularity then this difference decays at a rate slower than the truncation error for the derivative $u' - P_{N-1} u'$. This means that $(P_N u)'$ is asymptotically a worse approximation to $u'$ than $P_{N-1} u'$. This topic is discussed in Sect. 5.4.2.

The function $u(x) = |x|^{3/2}$ will serve as an illustration of the results produced by Legendre differentiation procedures. It has the Legendre coefficients

$$\hat{u}_k = \begin{cases} 0 , & k \text{ odd} , \\ 1/(a+1) , & k = 0 , \\ \dfrac{(2k+1)a(a-2)\dots(a-k+2)}{(a+1)(a+3)\dots(a+k+1)} & \text{otherwise} , \end{cases}$$

where $a = 3/2$. A comparison between $P_{N-1} u'$ and $(P_N u)'$ is furnished in Figs. 2.11(a) and (b). (Only the right half of the approximation interval $[-1,1]$

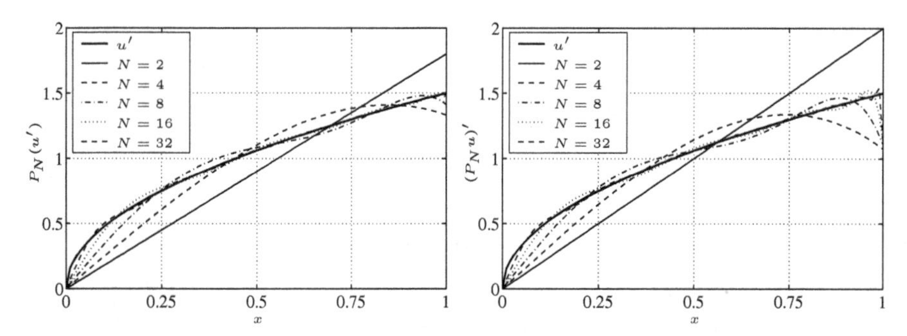


**Fig. 2.11.** Several versions of Legendre differentiation for $u(x) = |x|^{3/2}$ on $[-1, 1]$. The exact result is indicated by the solid curves and the approximate results for $N = 2$, 4, 8, 16, and 32 are indicated by the dashed curves. Only the right half of the interval is shown. ($P_N(u')$ *(left)*; $(P_N u)'$ *(right)*)

is displayed.) Both approximations yield the expected slow convergence near the singularity at $x = 0$. The global nature of the approximation leads to additional problems caused by the singularity that are most apparent at $x = \pm 1$. Further discussion of this behavior will be given in Sect. 5.4.2, after we have presented the general results on the error between $u'$ and $(P_N u)'$ in terms of $N$ and the regularity of $u$.

Let us consider now differentiation in physical space. If the function $u$ is known at one set of quadrature points (2.3.10), (2.3.11) or (2.3.12), one can compute an approximate derivative of $u$ by differentiating the interpolant $I_N u$ (as defined in (2.2.20)) and evaluating it at the same nodes. The polynomial of degree $N - 1$

$$\mathcal{D}_N u = (I_N u)' \tag{2.3.25}$$

is called the *Legendre interpolation derivative of* $u$ relative to the chosen set of quadrature nodes, since in general, it is different from the projection derivative $(P_N u)'$.

The error between $u'$ and the Legendre interpolation derivative of $u$ can be estimated in terms of $N$ and the regularity of $u$. This is done in Sect. 5.4.3 (see (5.4.36)).

In order to compute the values $(\mathcal{D}_N u)(x_j)$ , $j = 0, \dots, N$, from the values $u(x_l)$ , $l = 0, \dots, N$, one could use formula (2.2.22) to get the discrete Legendre coefficients of $u$, then use (2.3.18) to differentiate in transform space and finally compute $(\mathcal{D}_N u)_j$ through (2.2.21). However, this procedure is not efficient for $N$ of practical interest in the absence of a (fast) transform method for the Legendre expansion. Therefore, it is preferable to obtain the interpolation derivative at the nodes through matrix multiplication, namely,

$$(\mathcal{D}_N u)(x_j) = \sum_{l=0}^{N} (D_N)_{jl} u(x_l) \, , \qquad j = 0, \dots, N \, . \tag{2.3.26}$$

The entries $(D_N)_{jl}$ can be computed by differentiating the characteristic Lagrange polynomials $\psi_l$ of degree $N$, which are 1 at $x_l$ and 0 at all the other collocation points; the general expression for such polynomials is

$$\psi_l(x) = \prod_{\substack{k \neq l \\ 0 \leq k,l \leq N}} \frac{(x - x_k)}{(x_l - x_k)} . \tag{2.3.27}$$

In the Legendre case, these polynomials have been introduced in Chap. 1 (see (1.2.55)). For the commonly used Gauss-Lobatto points (2.3.12), the closed forms for the first-derivative and second-derivative matrices, respectively, are (see Gottlieb, Hussaini and Orszag (1984))

$$(D_N)_{jl} = \begin{cases} \dfrac{L_N(x_j)}{L_N(x_l)} \dfrac{1}{x_j - x_l} , & j \neq l , \\ -\dfrac{(N+1)N}{4} , & j = l = 0 , \\ \dfrac{(N+1)N}{4} , & j = l = N , \\ 0 & \text{otherwise} , \end{cases} \tag{2.3.28}$$

and

$$(D_N^{(2)})_{jl} = \begin{cases} -2\dfrac{L_N(x_j)}{L_N(x_l)} \dfrac{1}{(x_j - x_l)^2} , & \begin{array}{l} 1 \leq j \leq N-1 , \\ 0 \leq l \leq N , \quad j \neq l , \end{array} \\ \dfrac{L_N''(x_l)}{3L_N(x_l)} , & 1 \leq j = l \leq N-1 , \\ \dfrac{(-1)^N}{L_N(x_l)} \dfrac{N(N+1)(1+x_l) - 4}{2(1+x_l)^2} , & j = 0 , \quad 1 \leq l \leq N , \\ \dfrac{1}{L_N(x_l)} \dfrac{N(N+1)(1-x_l) - 4}{2(1-x_l)^2} , & j = N , \quad 0 \leq l \leq N-1 , \\ \dfrac{N(N+1)(N^2+N-2)}{24} , & j = l = 0 , \quad j = l = N . \end{cases} \tag{2.3.29}$$

(See the discussion at the end of Sect. 2.4 for alternative expressions that have more favorable round-off error properties.)

The matrix of the interpolation derivative can be obtained by a similarity transformation from the matrix of the projection derivative, which is associated with the linear transformation (2.3.18) with the summation truncated to $p \leq N$. Thus they both have 0 as generalized eigenvalue of order $N+1$; the only eigenvector is $L_0(x)$, while each $L_k(x), k = 1, \ldots, N$, is a generalized eigenvector, i.e., a function $f$ for which $f^{(k)}$ is 0.

In spectral methods of Legendre type, differentiation is usually associated with suitable boundary conditions. In this case, the spectra of the related operators may exhibit different behavior. This topic is discussed in Sects. 4.3 and 7.3.

### 2.3.3 Orthogonality, Diagonalization and Localization

In the discretization of boundary-value problems set in the interval $(-1,1)$, we will be interested in describing a polynomial of degree at most $N$ by the coefficients of its expansion upon a basis of $\mathbb{P}_N$. The Legendre basis $L_k(x)$, $k=0,\dots,N$, given by the first $N+1$ Legendre polynomials is an example of a *modal* basis (sometimes called a *hierarchical* basis), so termed because each basis function is associated with one particular wavenumber in the expansion. On the contrary, the Lagrange basis $\psi_j(x)$, $j=0,\dots,N$, given by the characteristic polynomials at the Gauss-Lobatto points (see (1.2.55)) is an example of a *nodal* basis, since each basis function is responsible for reproducing the value of the polynomial at one particular node in the domain.

The orthogonality of Legendre polynomials implies that the *mass matrix* $M=(M_{hk})$ (with $M_{hk}=(L_h,L_k)$, $0\le h,k\le N$) associated with the Legendre modal basis is diagonal. The mass matrix allows one to express the $L^2$-inner product of two polynomials in $\mathbb{P}_N$ in terms of their expansion coefficients, as

$$(u,v)=\sum_{h,k=0}^{N}\hat{u}_h M_{hk}\hat{v}_k=\mathbf{u}^T M\mathbf{v}\ ,$$

where $\mathbf{u}$, $\mathbf{v}$ are the vectors of the expansion coefficients of $u$, $v$ along the basis. Having a diagonal mass matrix may help in certain applications (such as the discontinuous Galerkin method for time-dependent problems with explicit time-advancing schemes – see CHQZ3, Sect. 5.3.3). The Lagrange nodal basis leads to a diagonal mass matrix as well, provided the exact inner product is replaced by the discrete one (2.2.24).

In applications to second-order boundary-value problems, even more crucial for the efficiency of the discretization is to have a diagonal *stiffness matrix*, which is the matrix $K$ expressing the $L^2$-inner product of the first derivatives of two polynomials in terms of their expansion coefficients, i.e.,

$$(u',v')=\mathbf{u}^T K\mathbf{v}\qquad \text{for all } u,v\in\mathbb{P}_N$$

(see Sect. 3.8 for a thorough discussion of mass and stiffness matrices in one dimension and Sect. 4.2.2 for the multidimensional case). None of the bases considered so far leads to a diagonal stiffness matrix. However, if we suitably integrate the Legendre basis, we obtain a new *modal* basis that does fulfill such a property. In particular, its elements are defined as

$$\begin{aligned}
\eta_0(x) &= \frac{1}{2}(L_0(x) - L_1(x)) = \frac{1-x}{2}\,, \\
\eta_1(x) &= \frac{1}{2}(L_0(x) + L_1(x)) = \frac{1+x}{2}\,, \\
\eta_k(x) &= \sqrt{\frac{2k-1}{2}}\int_x^1 L_{k-1}(s)\,\mathrm{d}s, \qquad 2 \le k \le N\,.
\end{aligned} \tag{2.3.30}$$

Recalling (2.3.19), one easily gets

$$\eta_k(x) = \frac{1}{\sqrt{2(2k-1)}}(L_{k-2}(x) - L_k(x))\,, \qquad 2 \le k \le N\,; \tag{2.3.31}$$

another useful expression for $\eta_k$ is

$$\eta_k(x) = \frac{\sqrt{2(2k-1)}}{k}\left(\frac{1-x}{2}\right)\left(\frac{1+x}{2}\right)P_{k-2}^{(1,1)}(x)\,, \qquad 2 \le k \le N\,, \tag{2.3.32}$$

where $P_k^{(1,1)}$ is the $k$-th Jacobi polynomial, orthogonal with respect to the weight $w(x) = 1 - x^2$ (see Sect. 2.5). A comparison of the behavior of the members of the three bases mentioned above is given in Fig. 2.12 for $N = 4$.

In the efficient design of multidomain spectral methods it is important that local bases within each subdomain can be easily matched to form global bases that enjoy as much localization as possible. If we think of the interval $(-1, 1)$ as a subdomain of a wider interval $(a, b)$, then the Legendre basis is clearly inappropriate to produce a global basis made up of *continuous* functions. Indeed, each Legendre polynomial is nonzero at both endpoints of the interval; hence, glueing together such functions would lead to globally continuous functions that are supported (i.e., not identically zero) over the whole domain. On the contrary, both the Lagrange nodal basis and the modal basis (2.3.30) lead to well-localized global bases. Indeed, each basis contains two functions – which we term *vertex basis functions* – that are nonzero at precisely one endpoint of the interval (these are the functions $\psi_0$ and $\psi_N$ in the Lagrange basis, and the functions $\eta_0$ and $\eta_1$ in the modal basis); all other basis functions – which we term *basis functions*, or *internal basis functions* – vanish at both endpoints. Each local bubble function, extended by zero outside the subdomain, generates a global continuous basis function supported over that subdomain; on the other hand, each local vertex basis function (not vanishing at $a$ or $b$) can be matched to the parent one living on the contiguous subdomain to form a global continuous basis function supported over the two subdomains (see the presentation in CHQZ3, Sect. 5.1 and, in particular, Fig. 5.2).

A basis in $\mathbb{P}_N(-1, 1)$ is termed *boundary-adapted* if it is composed of two vertex functions plus bubble functions. In addition to the Lagrange nodal basis at the Gauss-Lobatto points and the modal basis (2.3.30), another example of boundary-adapted basis is given by the set of functions $\eta_0$ and $\eta_1$ as in (2.3.30) and

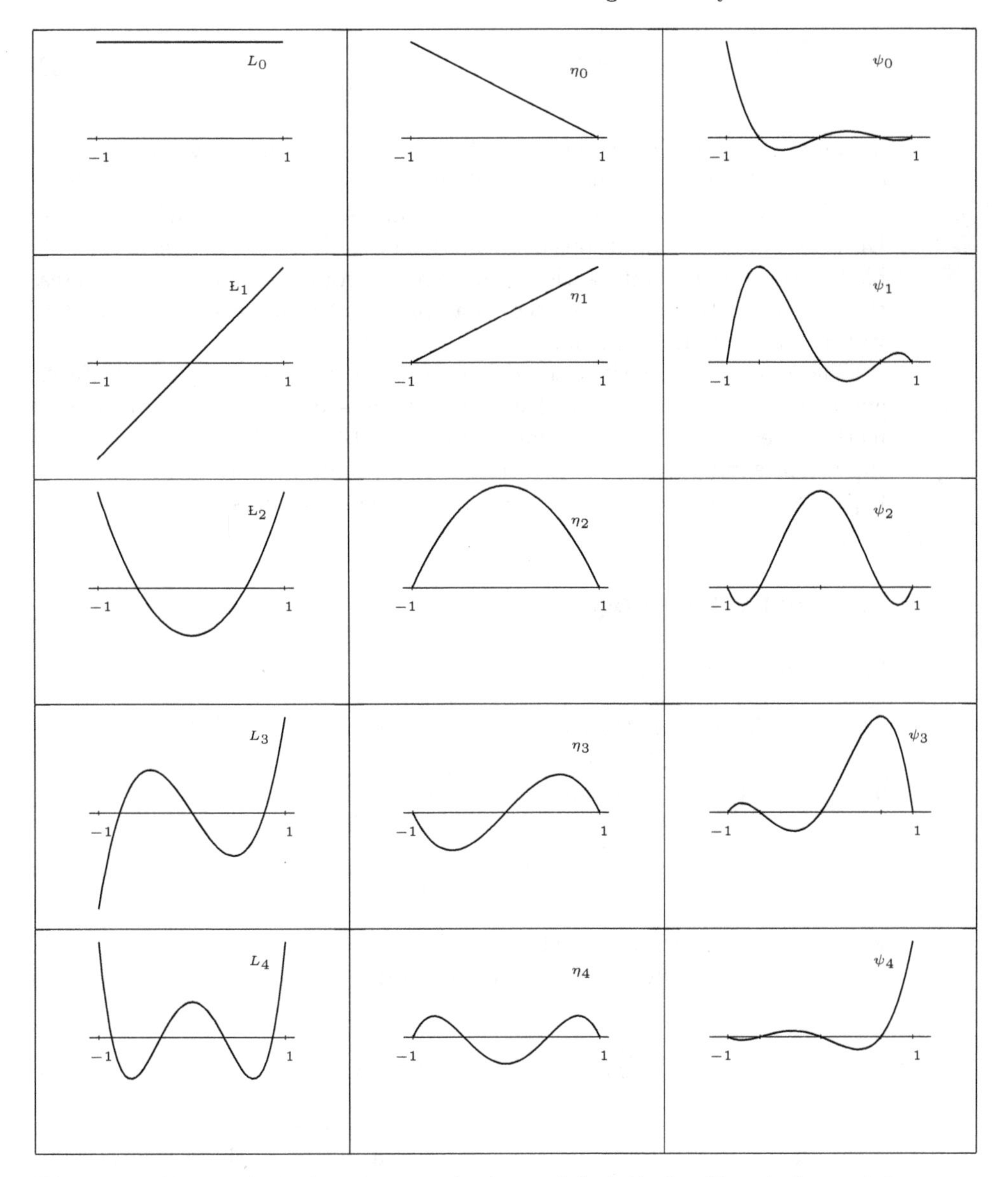


**Fig. 2.12.** Various basis functions on the interval $(-1, 1)$, for $N = 4$: the modal orthogonal basis $\{L_k\}$ (*left*), the modal boundary-adapted basis $\{\eta_k\}$ (*center*), the nodal basis at the Gauss-Lobatto points $\{\psi_k\}$ (*right*)

$$\tilde{\eta}_k(x) = \begin{cases} L_0(x) - L_k(x), & k \text{ even } \geq 2 \,, \\ L_1(x) - L_k(x), & k \text{ odd } \geq 3 \,; \end{cases} \tag{2.3.33}$$

however, since neither these basis functions nor their derivatives are $L^2$-orthogonal, their practical use is limited.

Obviously, a boundary-adapted basis allows for an easy enforcement of Dirichlet boundary conditions in a Galerkin or G-NI method. Indeed, the bubble functions individually satisfy homogeneous boundary conditions, whereas the two vertex functions are used to accommodate the prescribed boundary values, whenever they are nonzero.

The modal boundary-adapted basis (2.3.30) was developed by Babuška and co-workers in the late 1970's for use in the $p$-version of the finite-element method (see, e.g., Babuška, Szabó and Katz (1981)). Shen (1994) proposed the use of similar bases in spectral methods, built on both Chebyshev and Legendre polynomials. See Sect. 4.1.3 for some algorithms using such bases.

## 2.4 Chebyshev Polynomials

### 2.4.1 Basic Formulas

Classical references on the Chebyshev polynomials are Fox and Parker (1968) and Rivlin (1974). The Chebyshev polynomials of the first kind, $T_k(x)$, $k = 0, 1, \ldots$, are the eigenfunctions of the singular Sturm-Liouville problem

$$\left(\sqrt{1-x^2}\,T_k'(x)\right)' + \frac{k^2}{\sqrt{1-x^2}} T_k(x) = 0 \,, \tag{2.4.1}$$

which is (2.2.1) with $p(x) = (1-x^2)^{1/2}$, $q(x) = 0$ and $w(x) = (1-x^2)^{-1/2}$. For any $k$, $T_k(x)$ is even if $k$ is even, and odd if $k$ is odd. If $T_k$ is normalized so that $T_k(1) = 1$, then

$$T_k(x) = \cos k\theta \,, \quad \theta = \arccos x \,. \tag{2.4.2}$$

Thus, the Chebyshev polynomials are nothing but cosine functions after a change of independent variable. This property is the origin of their widespread popularity in the numerical approximation of nonperiodic boundary-value problems. The transformation $x = \cos\theta$ enables many mathematical relations as well as theoretical results concerning the Fourier system to be adapted readily to the Chebyshev system.

The Chebyshev polynomials can be expanded in power series as

$$T_k(x) = \frac{k}{2} \sum_{l=0}^{[k/2]} (-1)^k \frac{(k-l-1)!}{l!(k-2l)!} (2x)^{k-2l} \,, \tag{2.4.3}$$

where $[k/2]$ denotes again the integral part of $k/2$. Moreover, the trigonometric relation $\cos(k+1)\theta + \cos(k-1)\theta = 2\cos\theta\cos k\theta$ gives the recursion relation

$$T_{k+1}(x) = 2xT_k(x) - T_{k-1}(x) , \tag{2.4.4}$$

with $T_0(x) \equiv 1$ and $T_1(x) \equiv x$.

Some properties of the Chebyshev polynomials are

$$|T_k(x)| \le 1 , \quad -1 \le x \le 1 , \tag{2.4.5}$$

$$T_k(\pm 1) = (\pm 1)^k , \tag{2.4.6}$$

$$|T_k'(x)| \le k^2 , \quad -1 \le x \le 1 , \tag{2.4.7}$$

$$T_k'(\pm 1) = (\pm 1)^{k+1} k^2 , \tag{2.4.8}$$

$$\int_{-1}^{1} T_k^2(x) \frac{\mathrm{d}x}{\sqrt{1-x^2}} = c_k \frac{\pi}{2} , \tag{2.4.9}$$

where

$$c_k = \begin{cases} 2 , & k = 0 , \\ 1 , & k \ge 1 . \end{cases} \tag{2.4.10}$$

The Chebyshev expansion of a function $u \in L^2_w(-1,1)$ is

$$u(x) = \sum_{k=0}^{\infty} \hat{u}_k T_k(x) , \quad \hat{u}_k = \frac{2}{\pi c_k} \int_{-1}^{1} u(x) T_k(x) w(x) \, \mathrm{d}x . \tag{2.4.11}$$

If we define the even periodic function $\bar{u}$ by $\bar{u}(\theta) = u(\cos\theta)$, then

$$\bar{u}(\theta) = \sum_{k=0}^{\infty} \hat{u}_k \cos k\theta ;$$

hence, the Chebyshev series for $u$ corresponds to a cosine series for $\bar{u}$. It is easy to verify that if $u(x)$ is infinitely differentiable on $[-1,1]$, then $\bar{u}(\theta)$ is infinitely differentiable and periodic with all its derivatives on $[0, 2\pi]$. According to the integration-by-parts argument for Fourier series developed in Sect. 2.2.1, the Chebyshev coefficients of a sufficiently smooth function are guaranteed to decay faster than algebraically.

Turning now to relations of interest for discrete Chebyshev series, explicit formulas for the quadrature points and weights are

*Chebyshev Gauss (CG).*

$$x_j = \cos\frac{(2j+1)\pi}{2N+2} , \quad w_j = \frac{\pi}{N+1} , \qquad j = 0, \dots, N . \tag{2.4.12}$$

*Chebyshev Gauss-Radau (CGR).*

$$x_j = \cos \frac{2\pi j}{2N+1}, \quad w_j = \begin{cases} \dfrac{\pi}{2N+1}, & j = 0, \\ \dfrac{2\pi}{2N+2}, & j = 1, \dots, N. \end{cases} \tag{2.4.13}$$

*Chebyshev Gauss-Lobatto (CGL).*

$$x_j = \cos \frac{\pi j}{N}, \quad w_j = \begin{cases} \dfrac{\pi}{2N}, & j = 0, N, \\ \dfrac{\pi}{N}, & j = 1, \dots, N-1. \end{cases} \tag{2.4.14}$$

Note that the Chebyshev quadrature points as just defined are ordered from right to left. This violates our general convention that quadrature points are ordered from left to right (see Sect. 2.2.3). Virtually all of the classical literature on Chebyshev spectral methods uses this reversed order. Therefore, in the special case of the Chebyshev quadrature points we shall adhere to the ordering convention that is widely used in the literature (and implemented in the available software). We realize that our resolution of this dilemma imposes upon the reader the task of mentally reversing the ordering of the Chebyshev nodes whenever they are used in general formulas for orthogonal polynomials.

The most commonly used points are those for the Gauss-Lobatto case, which we consider in detail hereafter. The matrix representing the transformation from physical space to Chebyshev transform space (see (2.2.22)) is available in the simple form

$$C_{kj} = \frac{2}{N\bar{c}_j\bar{c}_k} \cos \frac{\pi jk}{N}, \tag{2.4.15}$$

where

$$\bar{c}_j = \begin{cases} 2, & j = 0, N, \\ 1, & j = 1, \dots, N-1. \end{cases} \tag{2.4.16}$$

Likewise, the inverse transformation (see (2.2.21)) is represented by

$$(C^{-1})_{jk} = \cos \frac{\pi jk}{N}. \tag{2.4.17}$$

Both transforms may be evaluated by the Fast Fourier Transform (Appendix B), i.e., by a transform method.

The normalization factors $\gamma_k$ introduced in (2.2.23) are here given by

$$\begin{aligned} \gamma_k &= \frac{\pi}{2} c_k \qquad \text{for } k < N, \\ \gamma_N &= \begin{cases} \dfrac{\pi}{2} & \text{for Gauss and Gauss-Radau formulas}, \\ \pi & \text{for the Gauss-Lobatto formula}. \end{cases} \end{aligned} \tag{2.4.18}$$

The structure of the aliasing error (2.2.32) due to interpolation takes a very simple form for the Chebyshev Gauss-Lobatto points. Recalling (2.4.2) and using the identity (2.1.26) with $N$ replaced by $2N$, one gets, for $k = 0, \ldots, N$,

$$(T_k, T_l)_N = \begin{cases} (T_k, T_k)_N & \text{if } l = 2mN \pm k,\ m \geq 0\ , \\ 0 & \text{otherwise}\ ; \end{cases} \tag{2.4.19}$$

hence, (2.2.30) becomes

$$\tilde{u}_k = \hat{u}_k + \sum_{\substack{j=2mN\pm k \\ j>N}} \hat{u}_j\ . \tag{2.4.20}$$

As for the Fourier points, the $k$-th Chebyshev mode of the interpolant polynomial depends upon all the Chebyshev modes that alias $T_k(x)$ on the grid.

### 2.4.2 Differentiation

The derivative of a function $u$ expanded in Chebyshev polynomials according to (2.4.11) can be represented formally as

$$u' = \sum_{k=0}^{\infty} \hat{u}_k^{(1)} T_k\ , \tag{2.4.21}$$

where

$$\hat{u}_k^{(1)} = \frac{2}{c_k} + \sum_{\substack{p=k+1 \\ p+k \text{ odd}}} p\hat{u}_p\ , \qquad k \geq 0\ . \tag{2.4.22}$$

This expression is a consequence of the relation

$$2T_k(x) = \frac{1}{k+1} T'_{k+1}(x) - \frac{1}{k-1} T'_{k-1}(x)\ , \qquad k \geq 1\ , \tag{2.4.23}$$

which, due to (2.4.2), is a different form of the trigonometric identity

$$2 \sin\theta \cos k\theta = \sin(k+1)\theta - \sin(k-1)\theta\ .$$

From (2.4.23) one has

$$2k\hat{u}_k = c_{k-1}\hat{u}_{k-1}^{(1)} - \hat{u}_{k+1}^{(1)}\ , \qquad k \geq 1\ ; \tag{2.4.24}$$

whence, (2.4.22) follows. Note that the last relation suggests an efficient way of differentiating a polynomial of degree $N$ in Chebyshev space. Since $\hat{u}_k^{(1)} = 0$ for $k \geq N$, the nonzero coefficients are computed in decreasing order by the recursion relation

$$c_k \hat{u}_k^{(1)} = \hat{u}_{k+2}^{(1)} + 2(k+1)\hat{u}_{k+1} , \qquad 0 \le k \le N-1 , \tag{2.4.25}$$

in $2N$ multiplications or additions. The generalization of this relation is

$$c_k \hat{u}_k^{(q)} = \hat{u}_{k+2}^{(q)} + 2(k+1)\hat{u}_{k+1}^{(q-1)} , \qquad k \ge 0 . \tag{2.4.26}$$

The coefficients of the second derivative are

$$\hat{u}_k^{(2)} = \frac{1}{c_k} \sum_{\substack{p=k+2 \\ p+k \text{ even}}}^{\infty} p(p^2-k^2)\hat{u}_p , \qquad k \ge 0 . \tag{2.4.27}$$

The *Chebyshev projection derivative* is just $(P_N u)'$. The *Chebyshev interpolation derivative* of a function $u$ known at one set of quadrature nodes – (2.4.12), (2.4.13) or (2.4.14) – is defined as the derivative of the discrete Chebyshev series of $u$ at the same nodes,

$$\mathcal{D}_N u = (I_N u)' . \tag{2.4.28}$$

As for Legendre polynomials, Chebyshev truncation and interpolation do not commute with differentiation. $(P_N u)'$ or $(I_N u)'$ are asymptotically worse approximations of $u'$ than $P_{N-1}u'$ and $I_{N-1}u'$, respectively, for functions with finite regularity. These results are made more precise in Sect. 5.5.2.

Chebyshev collocation differentiation can be accomplished efficiently by means of a transform method. The discrete Chebyshev coefficients of $u$ are computed according to (2.2.22), then (2.4.25) is used to differentiate in transform space, and finally the values of $\mathcal{D}_N u$ at the grid points are obtained by transforming back to physical space. If the discrete Chebyshev transforms are computed by an FFT algorithm that takes advantage of the reality and the parity of the function $\tilde{u}(\theta) = u(\cos\theta)$, the total number of operations required to differentiate in physical space is $(5\log_2 N + 8 + 2q)N$, where $q$ is the order of the derivative. The algorithmic details are furnished in Appendix B. The Chebyshev interpolation derivative can also be represented in matrix form as

$$(\mathcal{D}_N u)(x_j) = \sum_{l=0}^{N} (D_N)_{jl} u(x_l) , \qquad j = 0, \dots, N . \tag{2.4.29}$$

The entries $(D_N)_{jl}$ can be computed by differentiating the characteristic Lagrange polynomials $\psi_l$ of degree $N$, which are 1 at $x_l$ and 0 at all the other collocation points (see (2.3.27)).

For the popular Gauss-Lobatto points (2.4.14), these polynomials can be expressed as

$$\psi_l(x) = \frac{(-1)^{l+1}(1-x^2)T_N'(x)}{\bar{c}_l N^2 (x-x_l)} . \tag{2.4.30}$$

The first derivative matrix (Gottlieb, Hussaini and Orszag (1984)) is

$$(D_N)_{jl} = \begin{cases} \dfrac{\bar{c}_j}{\bar{c}_l} \dfrac{(-1)^{j+l}}{x_j - x_l}, & j \neq l, \\ -\dfrac{x_l}{2(1-x_l^2)}, & 1 \leq j = l \leq N-1, \\ \dfrac{2N^2+1}{6}, & j = l = 0, \\ -\dfrac{2N^2+1}{6}, & j = l = N, \end{cases} \quad (2.4.31)$$

and the second derivative matrix (Peyret (1986); see also Ehrenstein and Peyret (1989)) is

$$(D_N^{(2)})_{jl} = \begin{cases} \dfrac{(-1)^{j+l}}{\bar{c}_l} \dfrac{x_j^2 + x_j x_l - 2}{(1-x_j^2)(x_j - x_l)^2}, & 1 \leq j \leq N-1, \\ & 0 \leq l \leq N,\ j \neq l, \\ -\dfrac{(N^2-1)(1-x_j^2)+3}{3(1-x_j^2)^2}, & 1 \leq j = l \leq N-1, \\ \dfrac{2}{3} \dfrac{(-1)^l}{\bar{c}_l} \dfrac{(2N^2+1)(1-x_l)-6}{(1-x_l)^2}, & j = 0,\ 1 \leq l \leq N, \\ \dfrac{2}{3} \dfrac{(-1)^{(l+N)}}{\bar{c}_l} \dfrac{(2N^2+1)(1+x_l)-6}{(1+x_l)^2}, & j = N,\ 0 \leq l \leq N-1, \\ \dfrac{N^4-1}{15}, & j = l = 0, j = l = N. \end{cases} \quad (2.4.32)$$

However, alternative expressions that reduce the impact of the round-off errors resulting from subtraction of nearly equal quantities are preferred. With this aim, the most obvious approach to reducing the impact of subtracting nearly equal numbers for the Chebyshev derivative matrices is to use trigonometric identities, e.g.,

$$(D_N)_{jl} = \begin{cases} -\dfrac{\bar{c}_j}{2\bar{c}_l} \dfrac{(-1)^{j+l}}{\sin[(j+l)\pi/2N] \sin[(j-l)\pi/2N]}, & j \neq l, \\ -\dfrac{x_j}{2\sin^2(j\pi/N)}, & 1 \leq j = l \leq N-1, \\ \dfrac{2N^2+1}{6}, & j = l = 0, \\ -\dfrac{2N^2+1}{6}, & j = l = N, \end{cases} \quad (2.4.33)$$

for (2.4.31). (The expression (2.4.33) was used already within the computer programs included in Canuto et al. (1988).)

The matrix (2.4.31) is not skew symmetric, as opposed to the matrix (2.1.51) of the Fourier differentiation. Since it is obtained by a similarity transformation from the matrix of differentiation in transform space (see (2.4.22)), it is immediate that the only eigenvalue is 0 with algebraic multiplicity $N+1$. Clearly, introducing boundary conditions results in a different structure of the spectrum, as discussed in Sects. 4.3 and 7.3.

If the interpolation derivative is computed by matrix multiplication, the total number of operations is $2N^2$. Figure 2.4 (Sect. 2.1.3) also provides a timing comparison of matrix-multiply and transform-based Chebyshev derivatives. The operation counts are $2N^2$, $5N(\log_2 N+2)$ and $5N(\log_2 N-1)$ for the matrix-multiply, Chebyshev and Fourier first derivatives, respectively. The figure reflects the greater cost of the Chebyshev derivative compared with the Fourier derivative, by as much as a factor of 2 for small $N$. The figure also indicates that for $N$ small, say less than 12, the matrix-multiply derivative is actually faster than the Chebyshev derivative. But for $N \geq 128$, the Chebyshev derivative is at least an order of magnitude faster. (Naturally, the specific results quoted here for the crossover points depend very strongly on the computer architecture and the efficiency of the implementation. (Again, see Deville, Fischer and Mund (2002) for techniques to speed up the matrix-multiply derivatives.))

The use of trigonometric identities in computations of the elements of the the derivative matrices is confined to Chebyshev polynomials, and even in this special case further refinements have been discussed by several authors (e.g., Breuer and Everson (1992), Don and Solomonoff (1995)). Many authors have analyzed the sources of these errors as well as the extent to which they affect the spatial/temporal stability of discrete solutions of boundary-value/initial-value problems based on Chebyshev collocation methods. We refer the interested reader to, for example, Funaro (1988), Trefethen and Trummer (1987), Reddy and Trefethen (1992), Tang and Trummer (1996).

Alternatives applicable to general orthogonal polynomials have been provided by Welfert (1997), Schneider and Werner (1986), Baltensperger and Berrut (1999, 2001). Often in these approaches the Lagrangian functions (2.3.27) are reformulated in *barycentric form* as

$$
\begin{aligned}
\psi_l(x) &= \frac{\dfrac{\lambda_l}{x-x_l}}{\displaystyle\sum_{k=0}^{N} \frac{\lambda_k}{x-x_k}}\,, \qquad l=0,\dots,N\,, \\
\lambda_l &= \frac{1}{\displaystyle\prod_{k\neq l}(x_l-x_k)} = (-1)^l \begin{cases} 1/2\,, & l=0,N\,, \\ 1 & \text{otherwise}\,. \end{cases}
\end{aligned}
\tag{2.4.34}
$$

As pointed out in Baltenserger and Berrut (1999), it is desirable to satisfy the consistency condition that every diagonal element of the differentiation matrix equals the negative sum of all other elements on its row.

For the resulting representation of $D_N$ for general orthogonal polynomials we refer to Schneider and Werner (1986):

$$(D_N)_{jl} = \begin{cases} \dfrac{\delta_l}{\delta_j} \dfrac{(-1)^{j+l}}{x_j - x_l} , & j \neq l , \\ -\displaystyle\sum_{i=0, i \neq j}^{N} \dfrac{\delta_i}{\delta_j} \dfrac{(-1)^{i+j}}{x_j - x_i} , & j = l , \end{cases} \tag{2.4.35}$$

where $\delta_l = 1/2$ if $l = 0$ or $N$, $\delta_l = 1$ otherwise. Baltensperger and Berrut (1999) report that (2.4.35) even reduces the round-off errors for Chebyshev polynomials compared with (2.4.33).

Baltensperger and Berrut (1999) further recommend the use of

$$(D_N^{(2)})_{jl} = \begin{cases} (D_N)_{jl} \left( (D_N)_{jj} - \dfrac{1}{x_j - x_l} \right) , & j \neq l , \\ 2(D_N)_{jl}^2 + 2 \displaystyle\sum_{k=0,\ k \neq j}^{N} (D_N)_{jl} \dfrac{1}{x_j - x_k} , & j = l . \end{cases} \tag{2.4.36}$$

Modification of the barycentric form of the Lagrange functions (2.4.34) based on the replacement of $\{\lambda_l\}, l = 0, \ldots, N$, by alternative coefficients $\{\beta_l\}, l = 0, \ldots, N$, yields the rational functions $\{\psi_l^{(\beta)}(x)\}$, which still enjoy the Lagrangian property $\psi_l^{(\beta)}(x_k) = \delta_{kl}$, for $k, l = 0, \ldots, N$.

The interpolation formula based on the rational Lagrange functions, $\{\psi_l^{(\beta)}(x)\}$, underlies the so-called *linear-rational collocation method* (see Berrut and Baltensperger (2001), Berrut and Mittelmann (2001), Baltensperger, Berrut and Dubey (2003)). When accompanied with a new set of shifted nodes, $\tilde{x}_j = g(x_j)$, with $g$ being a suitable map so that $\{\tilde{x}_j\}$ are more uniformly distributed than the original Gauss-Lobatto nodes $\{x_j\}$, the corresponding collocation method can enjoy better temporal stability properties (Kosloff and Tal-Ezer (1993)).

## 2.5 Jacobi Polynomials

As noted in Sect. 2.2.1, the class of Jacobi polynomials comprises all the polynomial solutions to singular Sturm-Liouville problems on $(-1, 1)$. The Jacobi polynomials $P_k^{(\alpha,\beta)}(x)$ of indices $\alpha$, $\beta > -1$ and degree $k$ are the solutions to (2.2.1) with $p(x) = (1-x)^{1+\alpha}(1+x)^{1+\beta}$, $q(x) = 0$ and $w(x) = (1-x)^\alpha(1+x)^\beta$. The corresponding eigenfunctions are $\lambda_k = k(k+\alpha+\beta+1)$.

In this section we collect some useful formulas for these polynomials (for more details, see, e.g., Abramowitz and Stegun (1972, Chapter 22)).

Under the normalization $P_k^{(\alpha,\beta)}(1) = \binom{k+\alpha}{k}$, one has the expression

$$P_k^{(\alpha,\beta)}(x) = \frac{1}{2^k} \sum_{l=0}^{k} \binom{k+\alpha}{l} \binom{k+\beta}{k-l} (x-1)^l (x+1)^{k-l} . \quad (2.5.1)$$

The *Rodriguez formula* provides an alternative representation, namely,

$$P_k^{(\alpha,\beta)}(x) = \frac{(-1)^k}{2^k k!} (1-x)^{-\alpha} (1+x)^{-\beta} \frac{\mathrm{d}^k}{\mathrm{d}x^k} \left( (1-x)^{\alpha+k} (1+x)^{\beta+k} \right). \quad (2.5.2)$$

Jacobi polynomials satisfy the two recursion relations:

$$\begin{gathered} P_0^{(\alpha,\beta)}(x) = 1 , \quad P_1^{(\alpha,\beta)}(x) = \frac{1}{2}[(\alpha-\beta) + (\alpha+\beta+2)x] , \\ a_{1,k} P_{k+1}^{(\alpha,\beta)}(x) = a_{2,k} P_k^{(\alpha,\beta)}(x) - a_{3,k} P_{k-1}^{(\alpha,\beta)}(x) , \end{gathered} \quad (2.5.3)$$

where

$$\begin{gathered} a_{1,k} = 2(k+1)(k+\alpha+\beta+1)(2k+\alpha+\beta) , \\ a_{2,k} = (2k+\alpha+\beta+1)(\alpha^2-\beta^2) + x\Gamma(2k+\alpha+\beta+3)/\Gamma(2k+\alpha+\beta) , \\ a_{3,k} = 2(k+\alpha)(k+\beta)(2k+\alpha+\beta+2) ; \end{gathered}$$

and

$$b_{1,k}(x) \frac{\mathrm{d}}{\mathrm{d}x} P_k^{(\alpha,\beta)}(x) = b_{2,k}(x) P_k^{(\alpha,\beta)}(x) + b_{3,k}(x) P_{k-1}^{(\alpha,\beta)}(x) , \quad (2.5.4)$$

where

$$\begin{gathered} b_{1,k}(x) = (2k+\alpha+\beta)(1-x^2) , \qquad b_{2,k}(x) = k(\alpha-\beta-(2k+\alpha+\beta)x) , \\ b_{3,k}(x) = 2(k+\alpha)(k+\beta) . \end{gathered}$$

A useful formula that relates Jacobi polynomials and their derivatives is

$$\frac{\mathrm{d}^m}{\mathrm{d}x^m} P_k^{(\alpha,\beta)}(x) = 2^{-m} \frac{\Gamma(k+m+\alpha+\beta+1)}{\Gamma(k+\alpha+\beta+1)} P_{k-m}^{(\alpha+m,\beta+m)}(x) ; \quad (2.5.5)$$

in particular, one has

$$\frac{\mathrm{d}}{\mathrm{d}x} P_k^{(\alpha,\beta)}(x) = \tfrac{1}{2}(k+1+\alpha+\beta) P_{k-1}^{(\alpha+1,\beta+1)}(x) . \quad (2.5.6)$$

This shows that the internal Legendre Gauss-Lobatto nodes (2.3.12) are indeed the zeroes of the Jacobi polynomial $P_{N-1}^{(1,1)}$, i.e., they are Gauss nodes for the weight $w(x) = 1 - x^2$. A similar result holds for the Chebyshev Gauss-Lobatto nodes. The discussion in Sect. 2.2.3 contains the general formulas for the Jacobi nodes and their discrete quadrature weights.

Jacobi series are given by

$$\begin{aligned} u(x) &= \sum_{k=0}^{\infty} \hat{u}_k P_k^{(\alpha,\beta)}(x) , \\ \hat{u}_k &= \frac{2k+\alpha+\beta+1}{2^{\alpha+\beta+1}} \frac{k!\Gamma(k+\alpha+\beta+1)}{\Gamma(k+\alpha+1)\Gamma(k+\beta+1)} \\ &\quad \times \int_{-1}^{1} u(x) P_k^{(\alpha,\beta)}(x)(1-x)^{\alpha}(1+x)^{\beta}\, dx . \end{aligned} \tag{2.5.7}$$

Jacobi polynomials for which $\alpha = \beta$ are called *ultraspherical polynomials* and are denoted simply by $P_k^{(\alpha)}(x)$. They are related to the Legendre polynomials via

$$L_k(x) = P_k^{(0)}(x) \tag{2.5.8}$$

and to the Chebyshev polynomials via

$$T_k(x) = \frac{2^{2k}(k!)^2}{(2k)!} P_k^{(-1/2)}(x) . \tag{2.5.9}$$

A different normalization of the ultraspherical polynomials leads to the *Gegenbauer polynomials* $C_k^{\nu}$, which are defined as

$$C_k^{\nu}(x) = \frac{\Gamma(\nu+\frac{1}{2})\Gamma(2\nu+k)}{\Gamma(\nu+k+\frac{1}{2})\Gamma(2\nu)} P_k^{(\nu-1/2)}(x) . \tag{2.5.10}$$

Spectral methods based on Jacobi polynomials distinct from Chebyshev and Legendre polynomials have been developed. For instance, they are essential in the construction of warped tensor-product expansions in non-Cartesian domains (see Sect. 2.9). Gegenbauer polynomials appear in the spectrally accurate reconstruction of discontinuous functions (see Sect. 7.6.3). Jacobi polynomials are also used in some special Galerkin methods for wall-bounded incompressible flows (see CHQZ3, Sect. 3.4.3).

## 2.6 Approximation in Unbounded Domains

There are three basic ways to construct global approximations to functions defined on unbounded intervals, e.g., $[0,\infty)$ and $(-\infty,\infty)$: (1) expand in Laguerre or Hermite functions; (2) map the unbounded interval into a finite one and then expand in a set of Jacobi polynomials; and (3) truncate the domain to $[0, x_{\max}]$ or $[x_{\min}, x_{\max}]$ and use a Jacobi expansion. See Boyd (2001) for a detailed discussion of all three options.

We recall here the definitions and the most significant properties of Laguerre and Hermite expansions, leaving the two other strategies for the next section.

### 2.6.1 Laguerre Polynomials and Laguerre Functions

For any $\alpha > -1$, the *Laguerre polynomials* $l_k^{(\alpha)}(x)$, $k \geq 0$, are the eigenfunctions of the singular Sturm-Liouville problem in $(0, +\infty)$:

$$\left( x^{\alpha+1} e^{-x} \left( l_k^{(\alpha)} \right)' (x) \right)' + k x^{\alpha} e^{-x} l_k^{(\alpha)} = 0 . \tag{2.6.1}$$

They are orthogonal in $(0, +\infty)$ with respect to the weight $w(x) = x^{\alpha} e^{-x}$; precisely, assuming the normalization $l_k^{(\alpha)}(0) = \binom{k+\alpha}{k}$, one has

$$\int_0^{+\infty} l_k^{(\alpha)}(x) l_m^{(\alpha)}(x) x^{\alpha} e^{-x} \, \mathrm{d}x = \Gamma(\alpha+1) \binom{k+\alpha}{k} \delta_{km} , \quad k, m \geq 0 . \tag{2.6.2}$$

In the particular case $\alpha = 0$, the polynomials $l_k(x) = l_k^{(0)}(x)$ satisfy $l_k(0) = 1$ and are orthonormal in $(0, +\infty)$.

The analogue of the Rodriguez formula is

$$l_k^{(\alpha)}(x) = \frac{1}{k!} x^{-\alpha} e^{x} \frac{\mathrm{d}^k}{\mathrm{d}x^k} (x^{k+\alpha} e^{-x}) . \tag{2.6.3}$$

The Laguerre polynomials satisfy the recursion relation

$$l_{k+1}^{(\alpha)}(x) = (2k + \alpha + 1 - x) l_k^{(\alpha)}(x) - (k + \alpha) l_{k-1}^{(\alpha)}(x) , \tag{2.6.4}$$

where $l_0^{(\alpha)}(x) = 1$ and $l_1^{(\alpha)}(x) = \alpha + 1 - x$. The derivative of a Laguerre polynomial satisfies the relations

$$\frac{\mathrm{d}}{\mathrm{d}x} l_k^{(\alpha)}(x) = -l_{k-1}^{(\alpha+1)}(x) \tag{2.6.5}$$

and

$$x \frac{\mathrm{d}}{\mathrm{d}x} l_k^{(\alpha)}(x) = k l_k^{(\alpha)}(x) - l_{k-1}^{(\alpha)}(x) . \tag{2.6.6}$$

Any function $v \in L_w^2(0, +\infty)$ can be expanded in a Laguerre series as $v = \sum_k \hat{v}_k^{(\alpha)} l_k^{(\alpha)}$. The convergence of the series (in weighted square mean) is faster than algebraic, provided all the derivatives of the function belong to $L_w^2(0, +\infty)$. No boundary condition need be satisfied, since Laguerre polynomials are eigenfunctions of a Sturm-Liouville problem that is singular at both endpoints. On the other hand, convergence is in the mean, with a weight vanishing exponentially fast at infinity, where the expansion polynomials become unbounded. Thus, the quality of the approximation for a fixed truncation at $k = N$ may deteriorate as $x$ tends to $+\infty$, e.g., with oscillations that are unbounded as $x \to +\infty$.

In order to avoid such a problem for approximating functions that vanish at $+\infty$, it may be more appropriate to expand in the *Laguerre functions* defined as $\mathcal{L}_k(x) = e^{-x/2} l_k^{(0)}(x)$. Thanks to (2.6.2), they satisfy

$$\int_0^{+\infty} \mathcal{L}_k(x)\mathcal{L}_m(x)\,\mathrm{d}x = \delta_{km}\ , \qquad k,m \geq 0\ , \tag{2.6.7}$$

and thus form an orthonormal basis in $L^2(0,+\infty)$. Note however that for an infinitely smooth function $v \in L^2(0,+\infty)$, the spectral convergence of the truncated series in Laguerre functions occurs only if $v$ decays exponentially fast at $+\infty$.

### 2.6.2 Hermite Polynomials and Hermite Functions

The *Hermite polynomials* $H_k(x)$, $k \geq 0$, are the eigenfunctions of the singular Sturm-Liouville problem in $(-\infty,+\infty)$

$$\left(e^{-x^2}H_k'(x)\right)' + 2ke^{-x^2}H_k(x) = 0\ . \tag{2.6.8}$$

They are orthogonal in $(-\infty,+\infty)$ with respect to the weight $w(x) = e^{-x^2}$; precisely, they satisfy

$$\int_{-\infty}^{+\infty} H_k(x)H_m(x)e^{-x^2}\,\mathrm{d}x = \sqrt{\pi}\,2^k\,k!\,\delta_{km}\ , \qquad k,m \geq 0\ . \tag{2.6.9}$$

The analogue of the Rodriguez formula is

$$H_k(x) = (-1)^k e^{x^2}\frac{\mathrm{d}^k}{\mathrm{d}x^k}e^{-x^2}\ . \tag{2.6.10}$$

The Hermite polynomials satisfy the recursion relation

$$H_{k+1}(x) = 2xH_k(x) - 2kH_{k-1}(x)\ , \qquad k \geq 1\ . \tag{2.6.11}$$

where $H_0(x) = 1$ and $H_1(x) = 2x$. The derivative of a Hermite polynomial satisfies the relation

$$\frac{\mathrm{d}}{\mathrm{d}x}H_k(x) = 2kH_{k-1}(x)\ . \tag{2.6.12}$$

A related family of Hermite polynomials is given by

$$He_k(x) = (1/\sqrt{2^k})H_k(x/\sqrt{2})\ , \qquad k \geq 0\ . \tag{2.6.13}$$

Such polynomials are orthogonal with respect to the weight $\tilde{w}(x) = e^{-x^2/2}$.

The *Hermite functions* are defined as $\mathcal{H}_k(x) = e^{-x^2/2}H_k(x)$. Thanks to (2.6.9) they are orthogonal in $L^2(-\infty,+\infty)$:

$$\int_{-\infty}^{+\infty} \mathcal{H}_k(x)\mathcal{H}_m(x)\,\mathrm{d}x = \delta_{km}\ , \qquad k,m \geq 0\ , \tag{2.6.14}$$

and form an orthonomal basis of this space.

Considerations about the spectral convergence of the expansion of a function in Hermite polynomials or functions are similar to those described above for the Laguerre case.

## 2.7 Mappings for Unbounded Domains

We focus here on some fundamentals for the mapping approach, with a particular emphasis on Chebyshev expansions.

### 2.7.1 Semi-Infinite Intervals

In the present subsection we shall present some guidelines for selecting global approximations in $[0, +\infty)$ that yield faster than algebraic decay of the maximum error.

The combination of the mapping $x = \phi(\xi)$, for $\xi \in [-1, 1]$, with a Chebyshev polynomial expansion in $\xi$ is appealing because it allows the FFT to be employed for many of the requisite series manipulations. The convergence properties of the approximation to $u(x)$ can be determined from the behavior of the function $v(\xi) = u(\phi(\xi))$. Infinite-order accuracy is expected when $v(\xi)$ is infinitely differentiable on $[-1, 1]$. Assuming that $u(x)$ itself is infinitely differentiable on $[0, \infty)$, the critical issue is the behavior of the derivatives of $v(\xi)$ at $\xi = \pm 1$. Loosely put, uniform spectral accuracy can be achieved provided the derivatives of $u(x)$ decay fast enough and oscillate slowly enough as $x \to \infty$.

The most frequently used mappings are algebraic, exponential and logarithmic, given by the following formulas, in which the constant $L$ sets the length scale of the mappings:

*(Semi-Infinite) Algebraic Mapping.*

$$x = L\frac{1+\xi}{1-\xi}\,, \qquad \xi = \frac{x-L}{x+L}\,, \tag{2.7.1}$$

*(Semi-Infinite) Exponential Mapping.*

$$x = -L \ln\left(\frac{1-\xi}{2}\right)\,, \qquad \xi = 1 - 2e^{-x/L}\,, \tag{2.7.2}$$

*(Semi-Infinite) Logarithmic Mapping.*

$$x = \frac{L}{2}\ln\left(\frac{3+\xi}{1-\xi}\right)\,, \qquad \xi = -1 + 2\tanh(x/L)\,. \tag{2.7.3}$$

The algebraic mapping places the most collocation points at larger values of $x$ and the logarithmic mapping the fewest. Thus, the algebraic mapping is best suited to approximation of functions that decay relatively slowly, e.g., algebraically in $1/x$ as $x \to \infty$, whereas the exponential and logarithmic mappings are more appropriate for more rapidly decaying functions, e.g., decaying exponentially in $x$. Unlike expansions on a finite domain, spectral approximations on a semi-infinite domain have two discretization parameters – the length scale $L$ in addition to the usual series truncation parameter $N$. As a general rule, the length scale $L$ needs to be increased with $N$ in order to have spectral accuracy (see Boyd (2001)).

Numerous authors (Grosch and Orszag (1977), Boyd (1982), Herbert (1984)) have found that, in practice, algebraic mappings are more accurate and more robust (less sensitive to the scale factor $L$) than exponential ones. For functions that decay only algebraically the logarithmic map is the most robust.

Spalart (1984) observed that the use of the exponential mapping (2.7.2) for a function that decays faster than exponentially (as a Gaussian, for example) results in an inefficient distribution of grid points. Because of the clustering of nodes at $\xi = -1$ and $\xi = 1$, there will be more nodes for large $x$ than are required to resolve the function. Spalart proposed replacing (2.7.2) with

$$x = -L \ln \xi \,, \qquad \xi = e^{-x/L} \tag{2.7.4}$$

for $\xi \in [0, 1]$. Then the function $v(\xi)$ and all of its derivatives are zero at $\xi = 0$. Hence, $v(\xi)$ may be extended smoothly to a function on $[-1, 1]$. (In some cases, just the odd or just the even Chebyshev polynomials are appropriate expansion functions.) The grid points are clustered near $\xi = 1$ $(x = 0)$ and are coarsely distributed near $\xi = 0$ $(x = \infty)$. Likewise, for an exponentially decaying function, Chebyshev expansions may be combined with the map

$$x = L \frac{\xi}{1 - \xi} \,, \qquad \xi = \frac{x}{x + L} \,. \tag{2.7.5}$$

When the infinite interval is handled by truncating the domain to $[0, x_{\max}]$, infinite-order accuracy can only be achieved by increasing $x_{\max}$ as the number of terms in the series is increased. Boyd (1982) provides some guidance on how $x_{\max}$ should increase with $N$.

### 2.7.2 The Real Line

Similar considerations apply to expansions on $(-\infty, +\infty)$ as on semi-infinite intervals. The classical preference is for expansions in Hermite functions. However, there is no fast transform for them, and infinite-order accuracy requires that the function decay at least exponentially fast as $|x| \to \infty$ (Boyd (1984)).

Cain, Ferziger, and Reynolds (1984) suggested the use of the mapping

$$x = -L \cot(\xi/2) \,, \qquad \xi \in [0, 2\pi] \tag{2.7.6}$$

in conjunction with Fourier series. Infinite-order accuracy is only achieved if the function $u(x)$ and all of its derivatives exist and match at $x = -\infty$ and $x = +\infty$. The reason is that the function $v(\xi) = u(\phi(\xi))$ is implicitly extended periodically by the use of Fourier series, and continuity of $v(\xi)$ and all its derivatives is required for spectral accuracy.

For functions that approach different limits (but exponentially fast) at $x = \pm\infty$, such as $u(x) = \tanh x$, Cain et al. proposed the mapping

$$x = -L \cot \xi \,, \qquad \xi \in [0, \pi] \,, \tag{2.7.7}$$

with $v(\xi)$ extended to $\xi \in [\pi, 2\pi]$ by reflection. When coupled with Fourier series, this yields infinite-order accuracy.

Boyd (1987) has discussed the use of the mapping (2.7.7) on just $[0, \pi]$ in conjunction with a sine and cosine expansion (as opposed to the complex Fourier series on $[0, 2\pi]$). He noted that if just the cosine expansion is used, then $u(x)$ must at least have exponential decay (or special symmetries). If the decay is only algebraic and no special symmetries are present, then only algebraic convergence is possible with the cosine expansion.

An alternative approximation couples either the algebraic map,

$$x = L\frac{\xi}{\sqrt{1-\xi^2}}\,, \qquad \xi \in [-1, 1]\,, \tag{2.7.8}$$

or else the exponential map

$$x = L \tanh^{-1} \xi\,, \qquad \xi \in [-1, 1]\,, \tag{2.7.9}$$

with an expansion in Chebyshev polynomials. One expects infinite-order accuracy, even if $u(-\infty) \neq u(+\infty)$, provided that the derivatives of $u$ decay sufficiently fast, i.e., algebraic decay with (2.7.8) and exponential decay with (2.7.9), and of course, provided that $u(x)$ is analytic at $x = \pm\infty$.

## 2.8 Tensor-Product Expansions

In the previous sections we have introduced several one-dimensional expansions, and we have studied their orthogonality, localization, and differentiation properties. The most natural way to build a multidimensional expansion, exploiting all the one-dimensional features, is to take tensor products of one-dimensional expansions; the resulting functions are defined on the Cartesian product of intervals. Precisely, given $d$ families $\{\phi^{(l)}_{k_l}\}_{k_l}$ of one-dimensional basis functions on intervals $(a_l, b_l)$, then the family $\{\phi_{\mathbf{k}}(\mathbf{x})\}_{\mathbf{k}}$ defined as

$$\phi_{\mathbf{k}}(\mathbf{x}) = \prod_{l=1}^{d} \phi^{(l)}_{k_l}(x_l), \quad \mathbf{k} = (k_1, \ldots, k_d), \;\; \mathbf{x} = (x_1, \ldots, x_d), \tag{2.8.1}$$

is a multidimensional basis on the domain $\Omega = \prod_{l=1}^{d}(a_l, b_l)$.

The most familiar example is the multidimensional Fourier basis

$$\phi_{\mathbf{k}}(\mathbf{x}) = \prod_{l=1}^{d} \mathrm{e}^{ik_l x_l} = \mathrm{e}^{i\mathbf{k}\cdot\mathbf{x}}$$

defined on the periodic box $\Omega = (0, 2\pi)^d$. Another common example is the three-dimensional Fourier-Chebyshev basis

$$\phi_{\mathbf{k}}(\mathbf{x}) = \mathrm{e}^{i(k_1 x_1 + k_2 x_2)}\, T_{k_3}(x_3)$$

defined on $\Omega = (0, 2\pi)^2 \times (-1, 1)$, which is used, e.g., in Fourier-Chebyshev spectral simulations of plane channel flow (see CHQZ3, Sect. 3.4).

Orthogonality of each one-dimensional family with respect to a weight $w_l(x_l)$ implies orthogonality of the tensor-product family with respect to the weight $w(\mathbf{x}) = \prod_{l=1}^{d} w_l(x_l)$. On the other hand, if each individual factor $\phi_{k_l}^{(l)}$ is a characteristic Lagrange polynomial relative to a family of quadrature points in $[a_l, b_l]$, then $\phi_{\mathbf{k}}$ is a characteristic Lagrange polynomial relative to the family of tensorized quadrature points in $\overline{\Omega}$. For instance, the expression

$$\phi_{\mathbf{k}}(\mathbf{x}) = \prod_{l=1}^{d} \psi_{k_l}(x_l) ,$$

where $\psi_k$ is one of the $N$-degree characteristic Lagrange polynomials introduced in (1.2.56), defines a characteristic Lagrange polynomial relative to the $N$-degree tensorized Legendre-Gauss-Lobatto points in $\overline{\Omega} = [-1, 1]^d$. Such a basis is commonly used in multidimensional G-NI methods (see Sect. 6.4.3) in the reference domain $\Omega$.

The one-dimensional results on the precision of the quadrature rules and the decay rates of the coefficients extend to the tensor-product case as well.

First-order partial differentiation in wavenumber or in physical space can be accomplished by applying one-dimensional differentiation matrices to the coefficient vector in standard tensor-product fashion.

Boundary-adapted bases in each spatial direction tensorize to produce a boundary-adapted basis in $\Omega$; this is formed by bubble functions vanishing on the boundary $\partial\Omega$ and by vertex functions not vanishing at precisely one vertex, edge functions not vanishing at precisely one edge, face functions not vanishing at precisely one face, and so on. For instance, the tensor product $\{\eta_{\mathbf{k}}(\mathbf{x}) = \eta_{k_1}(x^{(1)})\eta_{k_2}(x^{(2)})\}$ of two modal bases (2.3.30) on the interval $(-1, 1)$ contains $(N-1)^2$ bubble functions (for $2 \le k_1, k_2 \le N$), 4 vertex functions (for $k_1, k_2 \in \{0, 1\}$) and $4(N-1)$ edge functions (for $k_1 \in \{0, 1\}, 2 \le k_2 \le N$, and $k_2 \in \{0, 1\}, 2 \le k_1 \le N$). Some of these functions are represented in the left half of Fig. 2.13; the right part shows the corresponding functions of the LGL nodal basis. As is typical, nodal basis functions are more localized than modal basis functions, but are more oscillatory.

### 2.8.1 Multidimensional Mapping

If one wishes to solve a two-dimensional problem by spectral methods, and the geometry is not directly conducive to the use of a tensor-product expansion, then one might be able to map the domain of interest onto a more standard computational domain, such as a square or a circle. (This might not always be possible or even desirable. One must then resort to the multidomain spectral methods discussed in CHQZ3, Chap. 5.)

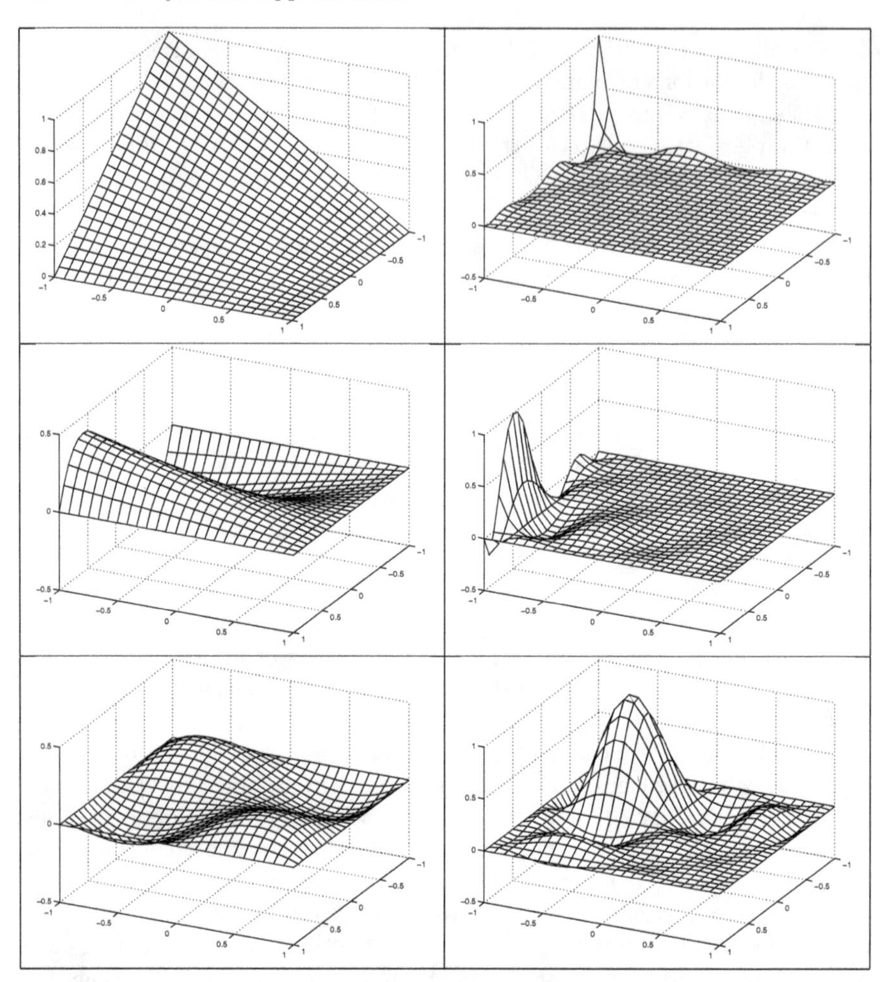

**Fig. 2.13.** Examples of boundary-adapted tensor-product basis functions on the square $(-1,1)^2$, for $N=4$: modal (*left*), nodal (*right*); vertex (*top*), edge (*center*), bubble (*bottom*). See also Fig. 2.12

One of the standard mapping techniques is based on conformal transformations. These are discussed in most elementary texts on complex variables (e.g., Carrier, Krook and Pearson (1966), Ahlfors (1979)). Among their advantages are the preservation of orthogonality and of simple operators such as divergence and gradient. Conformal mappings are widely used in two-dimensional fluid dynamical problems. The book by Milne-Thomson (1966) contains an extensive discussion. Several numerical methods have been devised for generating conformal mappings; see, for example, Meiron, Orszag and Israeli (1981) and Trefethen (1980).

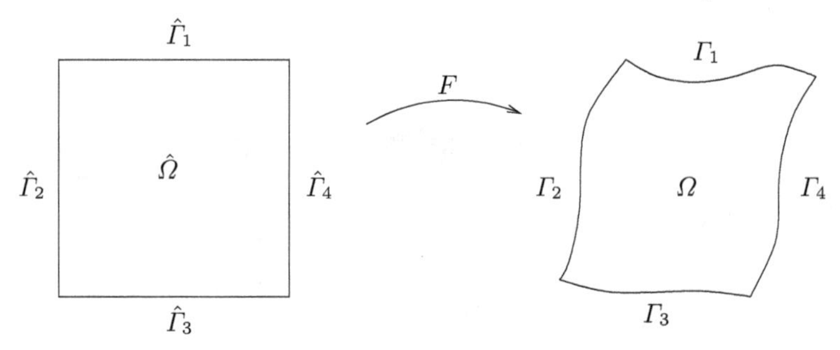


**Fig. 2.14.** Mapping of the unit square $\hat{\Omega} = [-1,1]^2$ into a quadrilateral $\Omega$ with curved boundaries

A fairly simple procedure exists for mapping a square $\hat{\Omega}$ into a quadrilateral $\Omega$ with curved boundaries. The basic geometry is illustrated in Fig. 2.14. Let the four sides of the quadrilateral be denoted by $\Gamma_i$, for $i = 1,2,3,4$, and those of the square by $\hat{\Gamma}_i$. One uses parametrizations $\pi_i$ from the interval $[0,1]$ to $\Gamma_i$ to construct the mapping $F$ from $\hat{\Omega}$ to $\Omega$, such that $F(\hat{\Gamma}_i) = \Gamma_i$ for all $i$. Gordon and Hall (1973a, 1973b) described a variety of mappings. The simplest is a linear blending mapping, for which $F$ can be expressed in terms of the $\pi_i$ as

$$\begin{aligned} F(\xi,\eta) = {} & \frac{1-\eta}{2}\pi_3(\xi) + \frac{1+\eta}{2}\pi_1(\xi) \\ & + \frac{1-\xi}{2}\left[\pi_2(\eta) - \frac{1+\eta}{2}\pi_2(1) - \frac{1-\eta}{2}\pi_2(-1)\right] \\ & + \frac{1+\xi}{2}\left[\pi_4(\eta) - \frac{1+\eta}{2}\pi_4(1) - \frac{1-\eta}{2}\pi_4(-1)\right] . \end{aligned} \tag{2.8.2}$$

(We assume that the arcs $\Gamma_1$ and $\Gamma_3$ are oriented from left to right and the arcs $\Gamma_2$ and $\Gamma_4$ from bottom to top.)

The Gordon-Hall transformation can be easily extended to three dimensions. A straightforward implementation is as follows: Let $\hat{\Omega} = [-1,1]^3$ be the reference cube with coordinates $(\xi,\eta,\zeta)$, and let $\hat{a}_i$ $(i = 1,\dots,8)$ and $\hat{\Sigma}_i$ $(i = 1,\dots,6)$ denote its vertices and faces, respectively, numbered as shown in Fig. 2.15. Let $\Omega \subset \mathbb{R}^3$ be the hexahedron, with faces $\Sigma_i$ $(i = 1,\dots,6)$, that is the image of $\hat{\Omega}$ under a smooth transformation $F$. We assume that we know each mapping $\pi_i : [-1,1]^2 \to \mathbb{R}^3$ from the reference square to the face $\Sigma_i$, which is the image of the face $\hat{\Sigma}_i$ under the transformation; with obvious notation, we have $\pi_1 = \pi_1(\xi,\zeta)$, $\pi_2 = \pi_2(\eta,\zeta)$, $\pi_3 = \pi_3(\xi,\zeta)$, $\pi_4 = \pi_4(\eta,\zeta)$, $\pi_5 = \pi_5(\xi,\eta)$, $\pi_6 = \pi_6(\xi,\eta)$. The vertices of $\Omega$ can be obtained as $a_1 = \pi_1(-1,-1)$, $a_2 = \pi_1(1,-1)$, $a_3 = \pi_3(1,-1)$, $a_4 = \pi_3(-1,-1)$, $a_5 = \pi_1(-1,1)$, $a_6 = \pi_1(1,1)$, $a_7 = \pi_3(1,1)$, and $a_8 = \pi_3(-1,1)$. Then, we can define $F$ as follows:

$$
\begin{aligned}
F(\xi,\eta,\zeta) = &\frac{1-\xi}{2}\pi_4(\eta,\zeta) + \frac{1+\xi}{2}\pi_2(\eta,\zeta) + \frac{1-\eta}{2}\pi_1(\xi,\zeta) \\
&+ \frac{1+\eta}{2}\pi_3(\xi,\zeta) + \frac{1-\zeta}{2}\pi_5(\xi,\eta) + \frac{1+\zeta}{2}\pi_6(\xi,\eta) \\
&- \frac{1-\xi}{2}\frac{1-\eta}{2}\frac{1-\zeta}{2}\pi_1(-1,-1) - \frac{1+\xi}{2}\frac{1-\eta}{2}\frac{1-\zeta}{2}\pi_1(1,-1) \\
&- \frac{1+\xi}{2}\frac{1+\eta}{2}\frac{1-\zeta}{2}\pi_3(1,-1) - \frac{1-\xi}{2}\frac{1+\eta}{2}\frac{1-\zeta}{2}\pi_3(-1,-1) \\
&- \frac{1-\xi}{2}\frac{1-\eta}{2}\frac{1+\zeta}{2}\pi_1(-1,1) - \frac{1+\xi}{2}\frac{1-\eta}{2}\frac{1+\zeta}{2}\pi_1(1,1) \\
&- \frac{1+\xi}{2}\frac{1+\eta}{2}\frac{1+\zeta}{2}\pi_3(1,1) - \frac{1-\xi}{2}\frac{1+\eta}{2}\frac{1+\zeta}{2}\pi_3(-1,1).
\end{aligned}
\tag{2.8.3}
$$

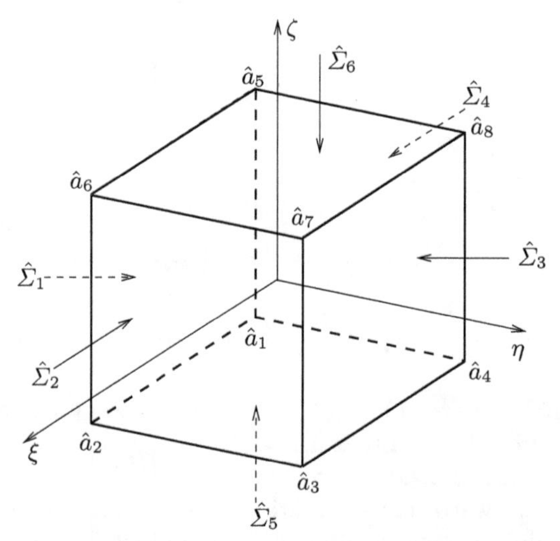


**Fig. 2.15.** Vertices and faces of the unit reference cube $\hat{\Omega} = [-1,1]^3$

More efficient implementations, in which the vertices, edges and faces are accounted for hierarchically, are available (see, e.g., Deville, Fischer and Mund (2002)).

In the event that the domain $\Omega$ is actually a subdomain in a multidomain spectral method (see CHQZ3, Chap. 5), the use of an isoparametric description of the curves $\Gamma_i$ may be desirable. Here one chooses the curves $\Gamma_i$ so that they are exactly parametrizable by polynomials of the same order as the discretization within $\Omega$. This approach is common in finite-element methods (see, e.g., Ciarlet (2002)) and was used first in spectral-element methods (see CHQZ3, Sect. 5.1) by Korczak and Patera (1986).

## 2.9 Expansions on Triangles and Related Domains

In this section we review some of the constructions of spectral polynomial approximations on elementary domains, such as triangles or tetrahedra, that are not Cartesian products of intervals. We distinguish between two different strategies, which are described in the following two subsections.

### 2.9.1 Collapsed Coordinates and Warped Tensor-Product Expansions

Simplicial domains, i.e., such non-tensor-product domains as triangles, tetrahedra, prisms and pyramids, are by far more flexible than Cartesian products of intervals, such as squares or cubes, in handling complex geometries by partitioning methods. (See Ciarlet (2002) and Hughes (2000) for complete descriptions of simplicial domains.) On the other hand, spectral methods prove themselves extremely efficient on tensor-product domains, due to the structure of the expansions employed therein. Therefore, it is tempting to try to marry the efficiency of tensor products with the flexibility of triangular geometries.

A successful realization of this marriage is provided, after Dubiner (1991b), by the concept of warped tensor-product expansion. Although similar ideas had appeared earlier in the literature (Proriol (1957), Koornwinder (1975)), Dubiner's paper was highly influential in the spectral methods community, as he introduced bases on a triangle geared towards the discretization of partial differential equations. Sherwin and Karniadakis (1995) extended the construction to the three-dimensional case. Warped tensor-product expansions exploit collapsed Cartesian coordinate systems in the simplices (see, e.g., Stroud (1971)).

We describe this approach in two dimensions. Let us introduce the reference triangle $\mathcal{T} = \{(x_1, x_2) \in \mathbb{R}^2 \ : \ -1 < x_1, x_2 \ ; \ x_1 + x_2 < 0\}$ as well as the reference square $\mathcal{Q} = \{(\xi_1, \xi_2) \in \mathbb{R}^2 \ : \ -1 < \xi_1, \xi_2 < 1\}$. The mapping

$$(x_1, x_2) \mapsto (\xi_1, \xi_2), \qquad \xi_1 = 2\frac{1 + x_1}{1 - x_2} - 1, \quad \xi_2 = x_2, \tag{2.9.1}$$

is a bijection between $\mathcal{T}$ and $\mathcal{Q}$. Its inverse is given by

$$(\xi_1, \xi_2) \mapsto (x_1, x_2), \qquad x_1 = \frac{1}{2}(1 + \xi_1)(1 - \xi_2) - 1, \quad x_2 = \xi_2. \tag{2.9.2}$$

Note that the mapping $(x_1, x_2) \mapsto (\xi_1, \xi_2)$ sends the ray in $\mathcal{T}$ issuing from the upper vertex $(-1, 1)$ and passing through the point $(x_1, -1)$ into the vertical segment in $\mathcal{Q}$ of the equation $\xi_1 = x_1$ (see Fig. 2.16). Consequently, the transformation becomes singular at the upper vertex of the triangle, although it stays bounded as one approaches the vertex. The determinant of the Jacobian of the inverse transformation is given by

$$\left| \frac{\partial(x_1, x_2)}{\partial(\xi_1, \xi_2)} \right| = \frac{1-\xi_2}{2} . \tag{2.9.3}$$

We term $(\xi_1, \xi_2)$ the *collapsed Cartesian coordinates* of the point on the triangle whose regular Cartesian coordinates are $(x_1, x_2)$.

Recall that $\{P_k^{(\alpha,\beta)}(\xi)\}$, $k \geq 0$, denotes the family of Jacobi polynomials that forms an orthogonal system with respect to the weight $(1-\xi)^\alpha(1+\xi)^\beta$ in $(-1, 1)$; see Sect. 2.5 (note that $P_k^{(0,0)}(\xi)$ is the Legendre polynomial $L_k(\xi)$ introduced in Sect. 2.3). For $\mathbf{k} = (k_1, k_2)$, define the *warped tensor-product basis* function on $\mathcal{Q}$:

$$\Phi_{\mathbf{k}}(\xi_1, \xi_2) = \Psi_{k_1}(\xi_1)\Psi_{k_1,k_2}(\xi_2), \tag{2.9.4}$$

where

$$\Psi_{k_1}(\xi_1) = P_{k_1}^{(0,0)}(\xi_1), \quad \Psi_{k_1,k_2}(\xi_2) = (1-\xi_2)^{k_1} P_{k_2}^{(2k_1+1,0)}(\xi_2), \tag{2.9.5}$$

which is a polynomial of degree $k_1$ in $\xi_1$ and $k_1 + k_2$ in $\xi_2$. By applying the mapping (2.9.1) one obtains the function defined on $\mathcal{T}$:

$$\begin{aligned} \varphi_{\mathbf{k}}(x_1, x_2) &= \Phi_{\mathbf{k}}(\xi_1, \xi_2) \\ &= P_{k_1}^{(0,0)}\left(2\frac{1+x_1}{1-x_2} - 1\right)(1-x_2)^{k_1} P_{k_2}^{(2k_1+1,0)}(x_2). \end{aligned} \tag{2.9.6}$$

It is easily seen that $\varphi_{\mathbf{k}}$ is a polynomial of global degree $k_1 + k_2$ in the variables $x_1, x_2$. Furthermore, thanks to the orthogonality of Jacobi polynomials,

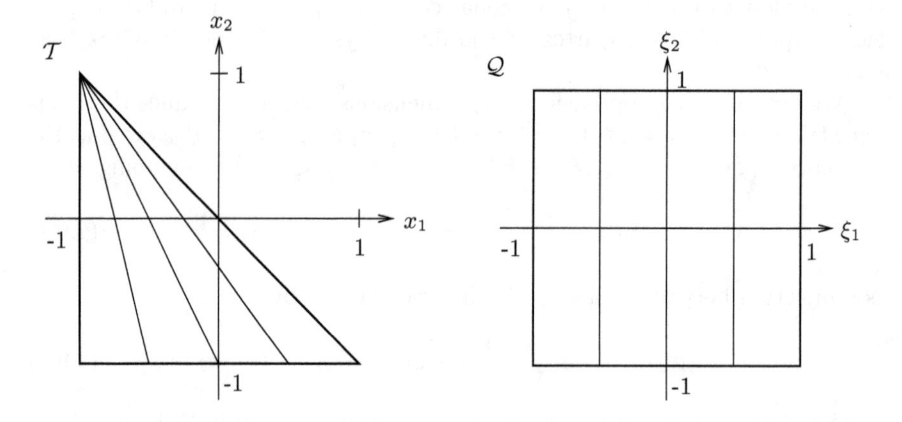


**Fig. 2.16.** The reference triangle $\mathcal{T}$ is mapped onto the reference square $\mathcal{Q}$. Oblique segments are transformed into vertical segments

one has, for $\mathbf{k} \neq \mathbf{h}$,

$$\int_T \varphi_{\mathbf{k}}(x_1, x_2)\varphi_{\mathbf{h}}(x_1, x_2)\, \mathrm{d}x_1 \mathrm{d}x_2 = \frac{1}{2}\int_{-1}^{1} P_{k_1}^{(0,0)}(\xi_1) P_{h_1}^{(0,0)}(\xi_1)\, \mathrm{d}\xi_1$$

$$\times \int_{-1}^{1} P_{k_2}^{(2k_1+1,0)}(\xi_2) P_{h_2}^{(2h_1+1,0)}(\xi_2)(1-\xi_2)^{k_1+h_1+1}\, \mathrm{d}\xi_2 = 0.$$

We conclude that the set $\{\varphi_{\mathbf{k}} \,:\, 0 \leq k_1, k_2 \text{ and } k_1+k_2 \leq N\}$ is an orthogonal modal basis of the space

$$\mathcal{P}_N(\mathcal{T}) = \operatorname{span}\{x_1^i x_2^j \,:\, 0 \leq i, j \text{ and } i+j \leq N\} \tag{2.9.7}$$

of the polynomials of global degree $\leq N$ in the variables $x_1, x_2$. The dimension of this space, i.e., the number of basis functions, is $\frac{1}{2}(N+1)(N+2)$. (Interestingly, Owens (1998) obtains an orthogonal basis in $\mathcal{P}_N(\mathcal{T})$ whose elements are the eigenfunctions of a singular Sturm-Liouville problem in $\mathcal{T}$. His construction extends to triangles the approach followed in one dimension to generate orthogonal polynomials; see Sect. 2.2.1.)

While orthogonality simplifies the structure of the mass matrix, it complicates the enforcement of boundary conditions, or of matching conditions between subdomains. This difficulty can be surmounted by building a new modal basis, say $\{\varphi_{\mathbf{k}}^{\mathrm{ba}}\}$, where "ba" stands for boundary adapted; it consists of boundary functions (3 vertex functions plus $3(N-1)$ edge functions) and internal functions ($\frac{1}{2}(N-2)(N-1)$ bubbles). Each basis function retains the same "warped tensor-product" structure as above. Indeed, it is enough to replace the one-dimensional Jacobi basis $P_k^{(\alpha,0)}(\xi)$ (with $\alpha = 0$ or $2k+1$) by the boundary-adapted basis given by the two boundary functions $\dfrac{1+\xi}{2}$ and $\dfrac{1-\xi}{2}$ and by the $N-1$ bubbles $\left(\dfrac{1+\xi}{2}\right)\left(\dfrac{1-\xi}{2}\right)P_{k-2}^{(\alpha,\beta)}(\xi)$, $k = 2, \ldots, N$, (for suitable $\alpha, \beta \geq 1$ fixed). Note that the choice $\alpha = \beta = 1$ yields the boundary-adapted basis $\eta_k$, $k = 0, \ldots, N$, defined in (2.3.30)–(2.3.32) (up to a normalization factor).

These univariate functions are then combined as in (2.9.4) to form the two-dimensional basis. To be precise, the vertex functions, expressed in the $(\xi_1, \xi_2)$-coordinates, are

$$\Phi^{V_1}(\xi_1, \xi_2) = \left(\frac{1-\xi_1}{2}\right)\left(\frac{1-\xi_2}{2}\right) \qquad (\text{vertex } V_1 = (-1,-1))\,,$$

$$\Phi^{V_2}(\xi_1, \xi_2) = \left(\frac{1+\xi_1}{2}\right)\left(\frac{1-\xi_2}{2}\right) \qquad (\text{vertex } V_2 = (+1,-1))\,,$$

$$\Phi^{V_3}(\xi_1, \xi_2) = \frac{1+\xi_2}{2} \qquad (\text{vertex } V_3 = (-1,+1))\,;$$

the edge functions are defined as

$$\Phi_{k_1}^{V_1V_2}(\xi_1,\xi_2) = \left(\frac{1-\xi_1}{2}\right)\left(\frac{1+\xi_1}{2}\right) P_{k_1-2}^{(\beta,\beta)}(\xi_1)\left(\frac{1-\xi_2}{2}\right)^{k_1}, \quad 2 \le k_1 \le N\,,$$

$$\Phi_{k_2}^{V_1V_3}(\xi_1,\xi_2) = \left(\frac{1-\xi_1}{2}\right)\left(\frac{1-\xi_2}{2}\right)\left(\frac{1+\xi_2}{2}\right) P_{k_2-2}^{(\beta,\beta)}(\xi_2)\,, \qquad 2 \le k_2 \le N\,,$$

$$\Phi_{k_2}^{V_2V_3}(\xi_1,\xi_2) = \left(\frac{1+\xi_1}{2}\right)\left(\frac{1-\xi_2}{2}\right)\left(\frac{1+\xi_2}{2}\right) P_{k_2-2}^{(\beta,\beta)}(\xi_2)\,; \qquad 2 \le k_2 \le N\,;$$

finally, the bubble functions are defined for $k_1, k_2 \ge 2$ and $k_1 + k_2 \le N$, as

$$\Phi_{k_1,k_2}^{B}(\xi_1,\xi_2) = \left(\frac{1-\xi_1}{2}\right)\left(\frac{1+\xi_1}{2}\right) P_{k_1-2}^{(\beta,\beta)}(\xi_1)$$
$$\times\left(\frac{1-\xi_2}{2}\right)^{k_1}\left(\frac{1+\xi_2}{2}\right) P_{k_2-2}^{(2k_1-1+\delta,\beta)}(\xi_2)\,.$$

The choice $\beta = \delta = 2$ yields orthogonality among the bubble functions (and certain boundary functions). However, usually the choice $\beta = 1$, $\delta = 0$ is preferred. Indeed, thanks to property (2.5.6)), it guarantees a good compromise in the sparsity pattern of both mass and stiffness matrices; furthermore, it leads to a more favorable conditioning of the stiffness matrix associated with a second-order operator.

It is conceptually important to notice that the vertex function $\Phi^{V_3}$ can be written as

$$\Phi^{V_3}(\xi_1,\xi_2) = \frac{1-\xi_1}{2}\frac{1+\xi_2}{2} + \frac{1+\xi_1}{2}\frac{1+\xi_2}{2}\,;$$

in other words, it is the sum of the two vertex functions associated with the vertices $(-1,+1)$ and $(+1,+1)$ of the square $\mathcal{Q}$. These vertices collapse into the vertex $V_3$ of the triangle $\mathcal{T}$ under the mapping (2.9.2).

With such bases in hand, one can discretize a boundary-value problem by the Galerkin with numerical integration (G-NI) method. To this end, one needs a high precision quadrature formula on $\mathcal{T}$. Since

$$\int_{\mathcal{T}} f(x_1,x_2)\,\mathrm{d}x_1\mathrm{d}x_2 = \frac{1}{2}\int_{-1}^{1}\mathrm{d}\xi_1\int_{-1}^{1}F(\xi_1,\xi_2)(1-\xi_2)\,\mathrm{d}\xi_2 \tag{2.9.8}$$

(where $f$ and $F$ are related by the change of variables (2.9.1)), it is natural to use a tensor-product Gaussian formula in $\mathcal{Q}$ for the weight $(1-\xi_2)$. This can be obtained by tensorizing the $(N+1)$-point Gauss-Lobatto formula for the weight 1 with the $(N+1)$-point Gauss-Lobatto formula for the weight $(1-\xi_2)$. Often, in the $\xi_2$-direction, the $N$-point Gauss-Radau formula for the weight $(1-\xi_2)$ with $\xi_2 = -1$ as integration node is preferred, since excluding the singular point $\xi_2 = 1$ from the integration nodes makes life easier in the construction of stiffness matrices (derivatives need not be computed therein) and improves the condition number of the matrices. The resulting formula is exact for all polynomials in $\mathcal{Q}$ of degree $\leq 2N-1$ in each variable, $\xi_1, \xi_2$; in particular, it is exact for all polynomials in $\mathcal{T}$ of global degree $\leq 2N-1$ in the variables $x_1, x_2$. Note, however, that the number of quadrature nodes in $\mathcal{T}$ is $N(N+1)$, nearly the double of the dimension of $\mathcal{P}_N(\mathcal{T})$, $\frac{1}{2}(N+1)(N+2)$; thus, no basis in $\mathcal{P}_N(\mathcal{T})$ can be the Lagrange basis associated with such quadrature nodes. This means that a G-NI method, based on the quadrature formula described above, cannot be equivalent to a collocation method at the quadrature points (as may occur on a simple, Cartesian domain (see, e.g., Sects. 3.5 and 3.8.2)).

Finally, we observe that the G-NI mass and stiffness matrices on $\mathcal{T}$ can be built efficiently by exploiting the tensor-product structure of both the basis functions and the quadrature points in $\mathcal{Q}$. We refer to Sect. 4.2.2 for more details.

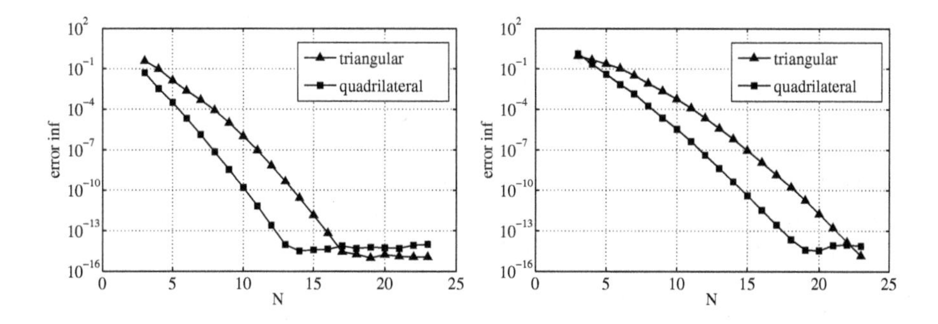


**Fig. 2.17.** Accuracy of the spectral discretization of a Poisson problem on the reference triangle $\mathcal{T}$ and on the reference square $\mathcal{Q}$. In both cases, the exact solution is $u(x,y) = (1+x)(1+y)(x+y)\exp(a(x+y))$, with $a = -1$ (*left*) and $a = -3$ (*right*)

In order to give a flavor of the behavior of a spectral method on a triangle, the Poisson problem

$$\begin{aligned} -\Delta u &= f \qquad \text{in } \mathcal{T}\,, \\ u &= g \qquad \text{on } \partial\mathcal{T} \end{aligned}$$

has been discretized by the Galerkin method, using the boundary-adapted basis described above (with the choice $\beta = 1$, $\delta = 0$) and mapped LGL numerical integration on the right-hand side. The data $f$ and $g$ have been chosen to produce the function $u(x,y) = (1+x)(1+y)(x+y)\exp(a(x+y))$, where $a < 0$ is a parameter, as the exact solution (note that $u$ vanishes on the boundary of the triangle, i.e., $g = 0$). Fig. 2.17 reports the errors $e^N = \max|u - u^N|$ vs. $N$ for two different values of $a$, where $u^N$ is the discrete solution and the maximum is taken over the LGL quadrature grid. For comparison, the same problem has been solved on the square $\mathcal{Q}$ using the standard tensor-product LGL nodal basis for the same $N$ and the same exact solution (which indeed is defined on the whole of $\mathcal{Q}$). Spectral accuracy is clearly documented for both methods. The Galerkin projection method on the triangle yields slightly less accurate results than on the square. However, the rates of decay of the error, indicated by the slopes of the curves, appear to approach each other as $N$ increases. This indicates that the loss of accuracy due to the geometry can be easily compensated for by increasing the polynomial degree of the triangular basis by a fixed (small) amount.

For expansions in three-dimensional simplicial domains, we furnish here just the basic principles; Karniadakis and Sherwin (1999) provide extensive coverage. Spectral expansions have been developed for three collapsed coordinate systems: prisms, pyramids and tetrahedra (Fig. 2.18). These are obtained by successively collapsing the cube $\mathcal{Q} = \{(\xi_1, \xi_2, \xi_3) \in \mathbb{R}^3 : -1 < \xi_1, \xi_2, \xi_3 < 1\}$. First, one applies an inverse collapsed transformation in the $\xi_2$ variable with respect to the $\xi_3$ variable, leaving the $\xi_1$ variable unchanged, i.e.,

$$\begin{aligned} &(\xi_1, \xi_2, \xi_3) \mapsto (x_1, x_2, x_3), \\ &\quad x_1 = \xi_1, \quad x_2 = \frac{1}{2}(1+\xi_2)(1-\xi_3) - 1, \quad x_3 = \xi_3. \end{aligned} \tag{2.9.9}$$

This transformation maps $\mathcal{Q}$ into the prism $\mathcal{T}_{23} = \{(x_1, x_2, x_3) \in \mathbb{R}^3 : -1 < x_1, x_2, x_3;\ x_1 < 1;\ x_2 + x_3 < 0\}$; see Fig. 2.18(a). The generic basis function in $\mathcal{T}_{23}$ associated with this transformation is

$$\varphi_{\mathbf{k}}(x_1, x_2, x_3) = P_{k_1}^{(0,0)}(\xi_1) P_{k_2}^{(0,0)}(\xi_2)(1-\xi_3)^{k_2} P_{k_3}^{(2k_2+1,0)}(\xi_3) \tag{2.9.10}$$

(with $\xi_1 = x_1,\ \xi_2 = 2\frac{1+x_2}{1-x_3} - 1,\ \xi_3 = x_3$), which is a polynomial of degree $k_1$ in the variable $x_1$ and of global degree $k_2 + k_3$ in the variables $x_2$ and $x_3$.

Next, starting from the prism, one applies an inverse collapsed transformation in the $x_1$ variable with respect to the $x_3$ variable, leaving the $x_2$ variable unchanged; in terms of the original variables $(\xi_1, \xi_2, \xi_3)$ in the cube, we now have

$$\begin{aligned} &(\xi_1, \xi_2, \xi_3) \mapsto (x_1, x_2, x_3), \\ &\quad x_1 = \frac{1}{2}(1+\xi_1)(1-\xi_3) - 1, \quad x_2 = \frac{1}{2}(1+\xi_2)(1-\xi_3) - 1, \quad x_3 = \xi_3. \end{aligned} \tag{2.9.11}$$

This transformation maps $\mathcal{Q}$ into the pyramid $\mathcal{T}_{123} = \{(x_1, x_2, x_3) \in \mathbb{R}^3 : -1 < x_1, x_2, x_3;\ x_1 + x_3 < 0;\ x_2 + x_3 < 0\}$; see Fig. 2.18(b). The generic basis function in $\mathcal{T}_{123}$ associated with this transformation is

$$\varphi_{\mathbf{k}}(x_1, x_2, x_3) = P_{k_1}^{(0,0)}(\xi_1) P_{k_2}^{(0,0)}(\xi_2)(1-\xi_3)^{k_1+k_2} P_{k_3}^{(2k_1+2k_2+2,0)}(\xi_3) \tag{2.9.12}$$

(with $\xi_1 = 2\frac{1+x_1}{1-x_3} - 1$, $\xi_2 = 2\frac{1+x_2}{1-x_3} - 1$, $\xi_3 = x_3$), which is a polynomial of global degree $k_1 + k_3$ in the variables $x_1$ and $x_3$ and of global degree $k_2 + k_3$ in the variables $x_2$ and $x_3$.

Finally, one cuts the pyramid by the planes $x_3 =$ constant and, in each such plane, applies an inverse collapsed transformation in the $x_1$ variable with respect to the $x_2$ variable; in terms of the original variables in the cube, $(\xi_1, \xi_2, \xi_3)$, we now get

$$\begin{aligned}
&(\xi_1, \xi_2, \xi_3) \mapsto (x_1, x_2, x_3), \\
&\quad x_1 = \frac{1}{2}(1 + \bar{\xi}_1)(1 - \xi_3) - 1 \quad \text{with} \quad \bar{\xi}_1 = \frac{1}{2}(1 + \xi_1)(1 - \xi_3) - 1, \\
&\quad x_2 = \frac{1}{2}(1 + \xi_2)(1 - \xi_3) - 1, \quad x_3 = \xi_3.
\end{aligned} \tag{2.9.13}$$

This transformation maps $\mathcal{Q}$ into the hexahedron $\mathcal{T} = \{(x_1, x_2, x_3) \in \mathbb{R}^3 : -1 < x_1, x_2, x_3;\ x_1 + x_2 + x_3 < 0\}$; see Fig. 2.18(c). The generic basis function

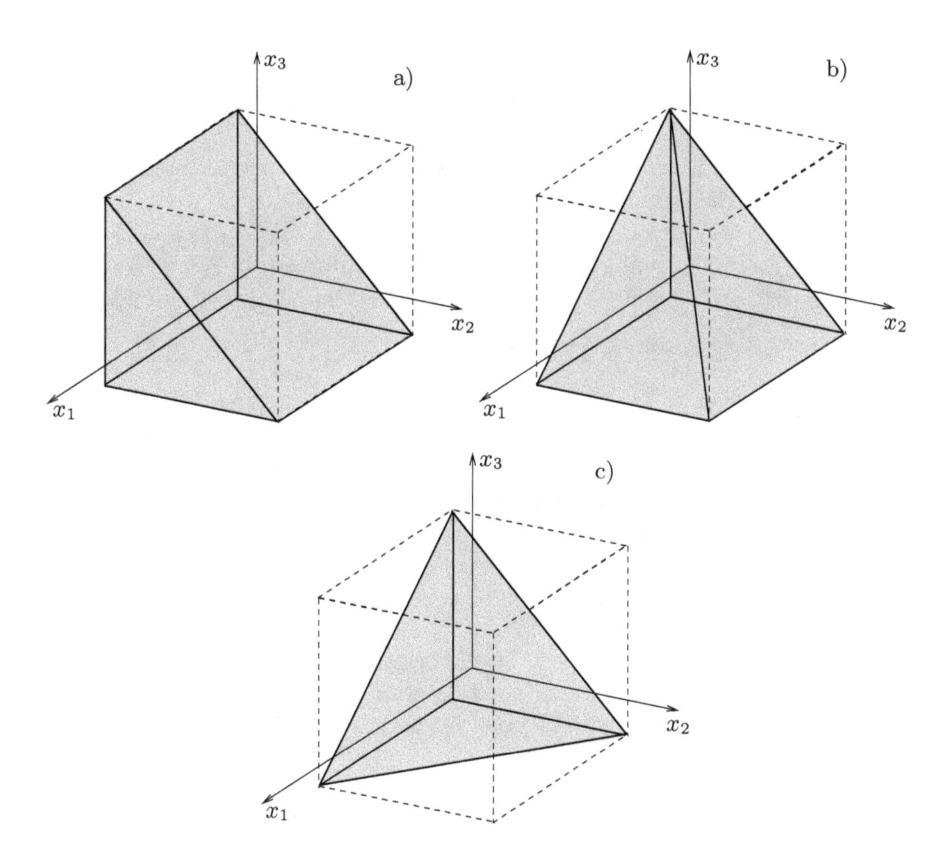


**Fig. 2.18.** The reference cube $(-1, 1)^3$ is sequentially collapsed into (**a**) a prism, (**b**) a pyramid and (**c**) a tetrahedron

in $\mathcal{T}$ associated with this transformation is

$$\begin{aligned}\varphi_{\mathbf{k}}(x_1,x_2,x_3) = P_{k_1}^{(0,0)}(\xi_1)(1-\xi_2)^{k_1} P_{k_2}^{(2k_1+1,0)}(\xi_2)\times \\ \times (1-\xi_3)^{k_1+k_2} P_{k_3}^{(2k_1+2k_2+2,0)}(\xi_3)\end{aligned} \tag{2.9.14}$$

(with $\xi_1 = 2\frac{1+x_1}{-x_2-x_3} - 1$, $\xi_2 = 2\frac{1+x_2}{1-x_3} - 1$, $\xi_3 = x_3$), which is a polynomial of global degree $k_1 + k_2 + k_3$ in the variables $x_1$, $x_2$ and $x_3$.

As for the two-dimensional case, the construction just sketched can be suitably modified to produce boundary-adapted modal bases on prisms, pyramids and tetrahedra. They consist of vertex functions, edge functions, face functions and bubble functions.

### 2.9.2 Non-Tensor-Product Expansions

We now describe several strategies to define, in a triangle $\mathcal{T}$, nodal bases that do not have any (warped) tensor-product structure. Each strategy relies upon the construction of an interpolation operator in the triangle; this is accomplished by looking for a set of points

$$\{\mathbf{x}_j\}\,, \qquad j = 1,\ldots,J_N = \tfrac{1}{2}(N+1)(N+2)\,,$$

in $\overline{\mathcal{T}} = \mathcal{T} \cup \partial\mathcal{T}$ having the following features: (i) the set is *unisolvent* for $\mathcal{P}_N(\mathcal{T})$, i.e., given an arbitrary distribution of values $f_j$ at the points $\mathbf{x}_j$, there exists a unique polynomial $p_N \in \mathcal{P}_N(\mathcal{T})$ such that $p_N(\mathbf{x}_j) = f_j$, $j = 1,\ldots,J_N$; (ii) the distribution of points in $\mathcal{T}$ fulfills certain symmetries and, possibly, certain boundary constraints; and (iii) the interpolation operator

$$I_N : C^0(\overline{\mathcal{T}}) \to \mathcal{P}_N(\mathcal{T})\,, \qquad I_N f(\mathbf{x}_j) = f(\mathbf{x}_j)\,, \quad j = 1,\ldots,J_N\,,$$

which exists by (i), has "good" approximation properties. Condition (i) above immediately yields the existence of the *nodal* basis in $\mathcal{P}_N(\mathcal{T})$ associated with the points $\{\mathbf{x}_j\}$ given by the characteristic Lagrange polynomials $\psi_j$ at these points. Condition (ii) enhances efficiency in the interpolation process. Classically, the third condition can be expressed by requiring a moderate growth with $N$ of the so-called *Lebesgue constant* $\Lambda_N$ of the set $\{\mathbf{x}_j\}$. This is the norm $\|I_N\|$ of the interpolation operator $I_N$ (see (A.3)), i.e., the smallest constant $C$ for which $\|I_N f\|_{\infty,\mathcal{T}} \le C\|f\|_{\infty,\mathcal{T}}$ for all $f \in C^0(\overline{\mathcal{T}})$, where $\|f\|_{\infty,\mathcal{T}} = \max_{\mathbf{x}\in\overline{\mathcal{T}}} |f(\mathbf{x})|$. The Lebesgue constant is significant, since through it we can relate the interpolation error in the maximum norm to the best approximation error in the same norm; indeed, the following Lebesgue inequality holds:

$$\|f - I_N f\|_{\infty,\mathcal{T}} \le (1+\Lambda_N) \inf_{p_N\in\mathcal{P}_N(\mathcal{T})} \|f - p_N\|_{\infty,\mathcal{T}} \qquad \text{for all } f \in C^0(\overline{\mathcal{T}}).$$

Thus, the smaller $\Lambda_N$, the closer the interpolation error to the smallest admissible value. In one dimension, the Lebesgue constant for interpolation on equally spaced points blows up exponentially with $N$ (as made apparent by the classical Runge phenomenon). On the contrary, the best possible Lebesgue constant among all distributions of $N$ points exhibits only a logarithmic growth with $N$ (see Erdös (1961), Natanson (1965); see also Hesthaven (1998) for a general overview in the context of spectral approximations); however, the corresponding points (the so-called *Lebesgue points*) are not known in a constructive form, although they are uniquely defined. Fortunately, the main families of Gaussian points (Gauss, Gauss-Radau, Gauss-Lobatto points) for the Legendre or Chebyshev weights have Lebesgue constants that grow logarithmically or sublinearly with $N$. In particular, Legendre or Chebyshev Gauss-Lobatto points have Lebesgue constants that are asymptotically close to the optimal one. In several dimensions, interpolation at equally-spaced points behaves just as unsatisfactorily as in one dimension; on the other hand, in domains that are not Cartesian products of intervals, there is no equivalent of Gaussian points defined as zeroes or extrema of suitable orthogonal polynomials. Therefore, the approach has been to select desirable properties satisfied by the Gaussian points in one dimension and extend them to higher dimension. In the sequel, we provide some examples of families of points fulfilling the three conditions above.

Stieltjes (1885) established that the Gauss quadrature points of the classical orthogonal polynomials can be determined as the steady-state, minimum energy solution to a problem of electrostatics; Szegö (1939) proved that such a minimum is unique. For instance, the internal Legendre Gauss-Lobatto points (2.3.12) minimize the electrostatic energy

$$E(x_1,\dots,x_{N-1}) = \\ = -\sum_{j=1}^{N-1}\left(\log|x_j+1| + \log|x_j-1| + \frac{1}{2}\sum_{i=1,i\neq j}^{N-1}\log|x_j-x_i|\right) .$$

This remarkable property led Hesthaven (1998) to define sets of points in the triangle via the minimization of the following electrostatic energy:

$$E(\mathbf{x}_1,\dots,\mathbf{x}_{J_N^*}) = \sum_{j=1}^{J_N^*}\left(\sum_{l=1}^{3}\sigma_l(\mathbf{x}_j) + \frac{1}{2}\sum_{i=1,i\neq j}^{J_N^*}\frac{1}{|\mathbf{x}_j-\mathbf{x}_i|}\right) ,$$

where $\mathbf{x}_j$ is a point internal to the triangle and $\sigma_l(\mathbf{x}) = \rho\int_0^1 \frac{1}{|\mathbf{x}-\mathbf{v}_t|}dt$ (for $\mathbf{v}_t = \mathbf{v}_a + t(\mathbf{v}_b - \mathbf{v}_a)$) is the potential at $\mathbf{x}$ generated by a continuous distribution of charges on the $l$-th side, $[\mathbf{v}_a,\mathbf{v}_b]$, of the triangle with a given line charge density $\rho > 0$, assumed to be constant. The $J_N^*$ internal nodes determined by the minimization process are augmented by $3N$ boundary nodes that are chosen as the (mapped) Gauss-Lobatto points on each side in order to simplify the matching between contiguous (triangular or quadrilateral) elements (see CHQZ3, Chap. 5). With the aim of defining a unisolvent set for

$\mathcal{P}_N(\mathcal{T})$, the number of internal nodes is chosen as $J_N^* = \frac{1}{2}(N+1)(N+2)-3N$. Several symmetries in the distribution of these nodes are imposed to facilitate the minimization process; the minimization is accomplished by driving (numerically) to steady-state a dynamical system of the $N$-body type. The constant density $\rho$ is used as a parameter to optimize the Lebesgue constant. Numerical computations indicate that the resulting constants are in the order of 2.6, 5.9 and 42.0 for $N = 4$, $N = 8$ and $N = 16$, respectively. The construction is extended to tetrahedra in Hesthaven and Teng (2000). They also exploit the symmetry of the nodes to derive efficient algorithms to compute derivatives at the same nodes.

Another approach to the construction of a unisolvent set for $\mathcal{P}_N(\mathcal{T})$ is based on the minimization or maximization of quantities related to the Lebesgue constant $\Lambda_N$. If $\psi_j$ are the characteristic Lagrange polynomials at the points $\mathbf{x}_j$, $j = 1, \ldots, J_N$, one has $\Lambda_N = \max_{\mathbf{x}\in\mathcal{T}} \sum_{j=1}^{J_N} |\psi_j(\mathbf{x})|$ . Chen and Babuška (1995, 1996) propose an algorithm to minimize the $L^2$-average $\left(\int_{\mathcal{T}} \sum_{j=1}^{J_N} |\psi_j(\mathbf{x})|^2 \, d\mathbf{x}\right)^{1/2}$ ; this is an easier task than minimizing the electrostatic energy. A closely related strategy considers the generalized Vandermonde matrix $V(\mathbf{x}_1, \ldots, \mathbf{x}_{J_N}) = (\,\varphi_i(\mathbf{x}_j)\,)$, where $\{\varphi_i\}$ is any basis in $\mathcal{P}_N(\mathcal{T})$. By Cramer's rule, one has

$$\psi_j(\mathbf{x}) = \frac{|V(\mathbf{x}_1, \ldots, \mathbf{x}_{j-1}, \mathbf{x}, \mathbf{x}_{j+1}, \ldots, \mathbf{x}_{J_N})|}{|V(\mathbf{x}_1, \ldots, \mathbf{x}_{J_N})|} \,,$$

which suggests the maximization of the denominator, leading to the so-called *Fekete points* (after the Hungarian mathematician M. Fekete). In this way, one is guaranteed to obtain those $|\psi_j(\mathbf{x})|$ for all $j$ and $\mathbf{x}$ that yield the upper bound for the Lebesgue constant $\Lambda_N \leq J_N$. Note that Fekete points are independent of the chosen basis, as a change of basis only results in the multiplication of $|V(\mathbf{x}_1, \ldots, \mathbf{x}_{J_N})|$ by a constant. Fejér (1932) proved that Fekete points on the interval are Legendre Gauss-Lobatto points; the same property holds in Cartesian-product domains for the tensorized Legendre Gauss-Lobatto points (Bos, Taylor and Wingate (2001)). Taylor, Wingate and Vincent (2000) have developed an algorithm to compute in an approximate way sets of points that locally maximize the Vandermonde determinants. The boundary points they utilize coincide with the (mapped) Legendre Gauss-Lobatto points on each side, as is customary when the interior points are chosen by the electrostatic analogy. The resulting Lebesgue constants are smaller than Hestaven's and Chen and Babuška's (for instance, they get $\Lambda_N \sim 12.1$ for $N = 16$). On the other hand, by increasing $N$ one may obtain undesired negative weights in the quadrature formulas constructed on these nodes. Fekete points are shown, together with the mapped LGL of the previous subsection, in Fig. 2.19.

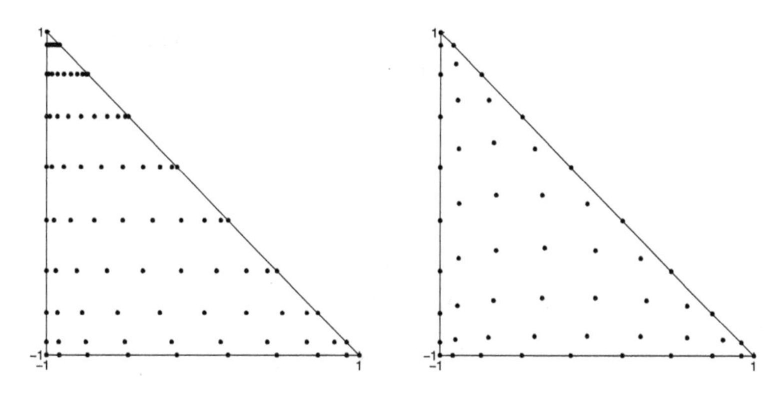

**Fig. 2.19.** Mapped LGL nodes (*left*) and Fekete points (*right*) for $N = 9$

In the context of the discretization of partial differential equations, the families of points mentioned above are appropriate for defining collocation methods in a triangle or a tetrahedron $\mathcal{T}$ since they are unisolvent for $\mathcal{P}_N(\mathcal{T})$ and have moderately growing Lebesgue constants. Taylor and Wingate (2000) (whose method, although formulated in an integral manner, can be viewed as a collocation scheme) and Pasquetti and Rapetti (2006) document the spectral accuracy of collocation methods based on Fekete points.

Another way to enforce boundary or interface conditions in a spectral method is through the use of a weak, or integral, formulation (such as in a Galerkin method). Such formulations call for appropriate quadrature rules to compute integrals on $\mathcal{T}$ or $\partial\mathcal{T}$ efficiently. The Galerkin with numerical integration (G-NI) approach precisely consists of replacing integrals by quadrature formulas in a Galerkin scheme. Unfortunately, unlike the tensor-product case, the interpolation points mentioned above *are not* Gaussian points: the quadrature formulae based on such $\mathcal{P}_N(\mathcal{T})$-unisolvent families of nodes are only exact for polynomials in $\mathcal{P}_N(\mathcal{T})$. This implies a poor approximation of the $L^2$-inner product in $\mathcal{T}$, which prevents the G-NI scheme from yielding spectral accuracy on smooth solutions, as clearly documented, e.g., in Pasquetti and Rapetti (2006).

The natural remedy consists of introducing a different quadrature formula on $\mathcal{T}$, of sufficiently high order to provide a good approximation of the $L^2$-inner product. Examples are given in Warburton, Pavarino and Hesthaven (2000), Hesthaven and Warburton (2002), Pasquetti and Rapetti (2004), Pasquetti, Pavarino, Rapetti and Zampieri (2006). Thus, two sets of points are involved: the set $X = \{\mathbf{x}_j\}$, $j = 1, \dots, J_N$, of the interpolation nodes, and the set $\tilde{X} = \{\tilde{\mathbf{x}}_l\}$, $l = 1, \dots, L_N$, of the quadrature nodes. The mapping $\mathbf{v} = (\,v(\mathbf{x}_j)\,) \mapsto \tilde{\mathbf{v}} = (\,v(\tilde{\mathbf{x}}_l)\,)$ between the values of a polynomial in $\mathcal{P}_N(\mathcal{T})$ at the two sets of nodes is a linear transformation; it can be accomplished as $\tilde{\mathbf{v}} = \tilde{V}V^{-1}\mathbf{v}$, where $V = (\,\varphi_k(\mathbf{x}_j)\,)$ is the (square) generalized Vandermonde matrix of the set $X$ with respect to any convenient basis in $\mathcal{P}_N(\mathcal{T})$ (such as, e.g., any of the modal bases introduced in Sect. 2.9.1), while $\tilde{V} = (\varphi_k(\tilde{\mathbf{x}}_l))$ is the (rectangular) Vandermonde matrix of the set $\tilde{X}$ for the same basis. An approximate $L^2$-inner product in $\mathcal{P}_N(\mathcal{T})$ is defined as

$$(u, v)_{\mathcal{T},N} = \sum_{l=1}^{L_N} u(\tilde{\mathbf{x}}_l) v(\tilde{\mathbf{x}}_l) \tilde{w}_l ,$$

where $\tilde{w}_l$ are the quadrature weights; in terms of grid-point values at the interpolation set $X$, it can be expressed as

$$(u, v)_{\mathcal{T},N} = \mathbf{v}^T M \mathbf{u} ,$$

where $M = (V^{-1})^T \tilde{V}^T \tilde{W} \tilde{V} V^{-1}$ and $\tilde{W} = \mathrm{diag}(\tilde{w}_l)$. Differentiation at the quadrature set $\tilde{X}$ (which is the basic ingredient for computing approximate stiffness matrices) can be accomplished as $\tilde{\mathbf{v}}^{\mathcal{D}} = \tilde{V} D V^{-1} \mathbf{v}$, where $\tilde{\mathbf{v}}^{\mathcal{D}} = (\mathcal{D}v(\tilde{\mathbf{x}}_l))$ is the vector of the grid-point values of a partial derivative $\mathcal{D}v$ at $\tilde{X}$ and $D$ is the matrix that expresses the partial differentiation of the basis $\{\varphi_k\}$ in transform space. (For the usual bases, its entries can be computed analytically quite cheaply.)

We are left with the problem of choosing a high-precision quadrature formula on $\mathcal{T}$. We refer to Cools (2003) for an overview of the state of the art on numerical integration on simplicial domains. A natural requirement is that the $L^2$-inner product be approximated by a formula that is exact for polynomials in $\mathcal{P}_{2N}(\mathcal{T})$ (actually, $\mathcal{P}_{2N-1}(\mathcal{T})$ may suffice). Quadrature formulas with minimal number of nodes for a prescribed degree of precision $p$ are known for several values of $p$, although not for all. Unfortunately, in addition to the drawback that negative weights, or nodes outside $\mathcal{T}$, may appear, the number of corresponding nodes is not significantly smaller than the number of nodes of the formula of equal precision obtained by mapping onto $\mathcal{T}$ a Gaussian tensor-product formula on $\mathcal{Q}$ (as those mentioned in Sect. 2.9.1). Therefore, to date, quadrature formulas of the latter type remain the preferred choice.

As far as the theoretical analysis is concerned, very little is known to date about the approximation properties of the interpolation operators at such points as the electrostatic or Fekete points in $\mathcal{T}$ in the (Sobolev) norms that will be introduced in Chap. 5 and which are appropriate for the study of differential problems.

### 2.9.3 Mappings

The idea underlying the Gordon-Hall transformations described in Sect. 2.8.1 provides the guidance needed to define a transformation between a reference triangle and a triangular domain with possibly curved sides, or between a non-tensorial reference domain such as a prism, a pyramid or a tetrahedron, and a similar domain with possibly curved faces and edges.

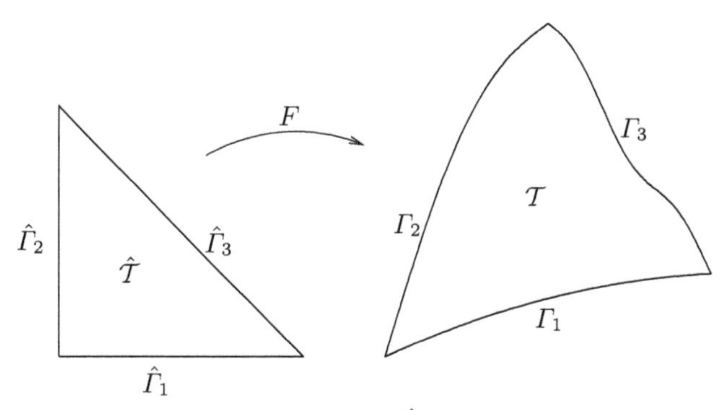


**Fig. 2.20.** Mapping of a reference triangle $\hat{T}$ into a triangle $T$ with curved boundaries

Let us consider the two-dimensional situation. At first, we choose the triangle $\{(\xi,\eta) \in \mathbb{R}^2 \ : \ 0 < \xi, \eta \ ; \ \xi + \eta < 1\}$ as the reference triangle $\hat{T}$; let $\hat{\Gamma}_1$ ($\hat{\Gamma}_2$, $\hat{\Gamma}_3$, resp.) denote the side whose equation is $\eta = 0$ ($\xi = 0$, $\xi + \eta = 1$, resp.) of $\hat{T}$ (see Fig. 2.20). Let $T$ be a triangular domain in the plane, with possibly curved sides, $\Gamma_1$, $\Gamma_2$, $\Gamma_3$, such that parametrizations $\pi_i : [0,1] \to \Gamma_i$ ($i = 1,2,3$) of the sides are known. We assume that the three vertices of $T$ are described by $\pi_1(0) = \pi_2(0)$, $\pi_1(1) = \pi_3(1)$ and $\pi_2(1) = \pi_3(0)$.

A mapping $F : \hat{T} \to T$ that extends smoothly the boundary mappings can be constructed in the form

$$F(\xi,\eta) = F_{\text{aff}}(\xi,\eta) + F_1(\xi,\eta) + F_2(\xi,\eta) + F_3(\xi,\eta) \ .$$

Here,

$$F_{\text{aff}}(\xi,\eta) = \xi\pi_1(1) + \eta\pi_2(1) + (1-\xi-\eta)\pi_1(0)$$

is the affine transformation that maps each vertex of $\hat{T}$ into the corresponding vertex of $T$. On the other hand, each $F_i$ provides the appropriate correction on $\hat{\Gamma}_i$ in the case of a curved side; the correction is extended to the whole of $\hat{T}$ in such a way that it vanishes on the sides $\hat{\Gamma}_k$ with $k \neq i$. For instance, setting

$$\tilde{\pi}_1(\xi) = \pi_1(\xi) - (1-\xi)\pi_1(0) - \xi\pi_1(1) \ ,$$

we define

$$F_1(\xi,\eta) = (1-\eta)\tilde{\pi}_1(\xi) - \xi\tilde{\pi}_1(1-\eta) \ ,$$

and we easily check that $F_1(\xi,0) = \tilde{\pi}_1(\xi)$, $F_1(0,\eta) = 0$ and $F_1(\xi,1-\xi) = 0$ for $0 \le \xi,\eta \le 1$.

The final expression for $F(\xi,\eta)$ is as follows:

$$\begin{aligned} F(\xi,\eta) = (1-\eta)\pi_1(\xi) - \xi\pi_1(1-\eta) + (1-\xi)\pi_2(\eta) - \eta\pi_2(1-\xi) \\ + (\xi+\eta)\pi_3(\xi) - \xi\pi_3(\xi+\eta) + \xi\pi_1(1) - (1-\xi-\eta)\pi_1(0) \ . \end{aligned} \tag{2.9.15}$$

If a different reference triangle $\hat{T}$ is preferred, namely, the triangle $\{(\xi,\eta) \in \mathbb{R}^2 \ : \ -1 < \xi,\eta \ ; \ \xi+\eta < 0\}$ already considered in Sect. 2.9.1

(in which case the parametrizations $\pi_i$ are defined on the interval $[-1,1]$), the expression for $F(\xi,\eta)$ is given by

$$\begin{aligned} F(\xi,\eta) = & \frac{1-\eta}{2}\pi_1(\xi) - \frac{1+\xi}{2}\pi_1(-\eta) + \frac{1-\xi}{2}\pi_2(\eta) - \frac{1+\eta}{2}\pi_2(-\xi) \\ & + \left(1+\frac{\xi+\eta}{2}\right)\pi_3(\xi) - \frac{1+\xi}{2}\pi_3(1+\xi+\eta) \\ & + \frac{1+\xi}{2}\pi_1(1) + \frac{\xi+\eta}{2}\pi_1(-1) . \end{aligned} \tag{2.9.16}$$

As for the tensorial case, a common practice for parametrizing the sides of $\mathcal{T}$ is to use isoparametric interpolation, i.e., polynomials of the same order as the basis chosen in $\hat{\mathcal{T}}$.

# 3. Basic Approaches to Constructing Spectral Methods

For the remainder of this book we shall be concerned with the use of spectral methods to obtain approximate solutions to ordinary differential equations (ODEs) and, especially, partial differential equations (PDEs). With very few exceptions spectral methods have only been applied to the approximation of spatial, and not temporal, derivatives. Our focus in this book is on the spatial approximations. The reader is referred to Appendix D for a review of time-discretization methods, including some brief comments on spectral approximations to time derivatives. When all or part of the time discretization is implicit, then the solution of implicit equations is required to advance in time. This topic is covered in the following chapter.

Our particular concern in this chapter is to illustrate how spectral approximations are actually constructed for the solutions to ODEs and PDEs. In the first part of this chapter we illustrate spectral methods on the Burgers equation. Although this is a relatively simple PDE, its discretization by spectral methods illuminates many points that occur for much more complicated problems. We begin, in Sect. 3.1, with some historical background and with a description of some exact solutions that are used in numerical examples. In Sect. 3.3 we derive the semi-discrete (discrete in space, continuous in time) ordinary differential equations which are satisfied by various spectral approximations to the Burgers equation. This involves a discussion of nonlinear terms, boundary conditions, projection operators, and different spectral discretizations. Section 3.4 provides a detailed discussion of transform methods for evaluating convolution sums. Sect. 3.5 closes this illustration of spectral discretizations for the Burgers equation with summary comments on the analogies and contrasts of the various spectral approaches.

The second part of this chapter covers some additional topics, again in the context of simple one-dimensional problems, that are essential for constructing spectral discretizations of PDEs. Some general remarks on conservation properties of spectral approximations are provided in Sect. 3.6. We then turn to scalar hyperbolic problems, for which the numerical boundary conditions are more subtle than for elliptic or parabolic problems, and examine in Sect. 3.7 various ways of enforcing the boundary conditions. In Sect. 3.8 we illustrate how the matrices associated with the different kinds of spectral methods are constructed. In Sect. 3.9 we then make some brief remarks on the treatment of coordinate singularities.

Finally, Sect. 3.10 is devoted to a brief summary of the salient theoretical results (provided in detail later in the text) on the effects of aliasing errors as a convenience for those readers especially interested in this always controversial subject but not particularly disposed to read the theoretical material in detail.

## 3.1 Burgers Equation

The nonlinear Burgers equation (in strong form)

$$\frac{\partial u}{\partial t} + u\frac{\partial u}{\partial x} - \nu\frac{\partial^2 u}{\partial x^2} = 0 \qquad \text{in } \Omega\,, \qquad \forall t > 0\,, \tag{3.1.1}$$

where $\nu$ is a positive constant and $\Omega$ is the spatial domain, provides a paradigm for more complex fluid-dynamics problems such as those described by the Navier-Stokes equations. It can be also written in conservation form as

$$\frac{\partial u}{\partial t} + \frac{\partial \mathcal{F}(u)}{\partial x} = 0 \qquad \text{in } \Omega\,, \qquad \forall t > 0\,, \tag{3.1.2}$$

where the flux $\mathcal{F}$ is given by

$$\mathcal{F}(u) = \frac{1}{2}u^2 - \nu\frac{\partial u}{\partial x}. \tag{3.1.3}$$

Of course, (3.1.1) or (3.1.2) must be supplemented with an initial condition,

$$u(x,0) = u_0(x) \qquad \text{in } \Omega\,, \tag{3.1.4}$$

and appropriate boundary conditions.

Burgers (1948) proposed this equation, later named after him, as a simplified model of full Navier-Stokes turbulence. The Burgers equation successfully models certain gas dynamic (Lighthill (1956)), acoustic (Blackstock (1966)) and turbulence phenomena (Burgers (1948)). Solutions to (3.1.1) exhibit a delicate balance between (nonlinear) advection and diffusion. It became a subject of extensive studies in the 1960s (Burgers (1974)) to investigate in isolation the specific feature of turbulence that balances generation of smaller scales by nonlinear advection with their dissipation by diffusion. The Burgers equation has also served as a benchmark for field-theoretic techniques such as direct interaction theory and renormalization group methods (Bouchaud, Mezard and Parisi (1995), Gurarie and Migdal (1996), Polyakov (1995)). As a simple model of nonequilibrium statistical mechanics, it has been employed as a qualitative model of a wide variety of physical phenomena including charge density waves (Feigelman (1980)), vortex lines in high-temperature superconductors (Blatter et al. (1994)), kinetic roughening of interfaces in epitaxial growth (Krug and Sophn (1992)), formation of large-scale cosmological structures (Shandarin and Zeldovich (1989), Vergassola et al. (1994)).

Since the Burgers equation is one of the few nonlinear PDEs for which exact and complete solutions are known in terms of the initial values (Hopf (1950), Cole (1951)), it remains a useful model problem for evaluating numerical algorithms (e.g., Berger and Colella (1989), Karniadakis et al. (1991), Dietachmayer and Droegameier (1992), Grauer and Marliani (1995), Mavriplis (1994), Huang and Russell (1997), Wei and Gu (2002)). Hopf (1950) and Cole (1951) showed that the transformation

$$u = -2\nu \frac{\phi_x}{\phi} \tag{3.1.5}$$

reduces the Burgers equation (3.1.1) for $u$ to a heat equation for $\phi$:

$$\frac{\partial \phi}{\partial t} - \nu \frac{\partial^2 \phi}{\partial x^2} = 0 \,. \tag{3.1.6}$$

Observe that if $u_b(x,t)$ is a solution of (3.1.1) and $c$ and $t_0$ are constants, then

$$u(x,t) = c + u_b(x - ct, t + t_0) \tag{3.1.7}$$

is also a solution. For the numerical examples that follow in this chapter, we will use two solutions for $u$ based on solutions for $u_b$ derived from the Hopf-Cole transformation. The solution that we will use for nonperiodic problems is based on the isolated N-wave solution (so called because of its shape) that is derived from

$$\phi_b(x,t) = \frac{x}{t} \frac{\sqrt{a/t}\, e^{-x^2/(4\nu t)}}{1 + \sqrt{a/t}\, e^{-x^2/(4\nu t)}} \,, \tag{3.1.8}$$

where $a$ is a constant. The subscript on $\phi$ emphasizes that this $\phi_b$ corresponds to the $u_b$ in (3.1.7). The solution that we will use for periodic problems is a sum of an infinite number of N-wave solutions spaced a distance $2\pi$ apart:

$$\phi_b(x,t) = \frac{1}{\sqrt{4\pi\nu t}} \sum_{n=-\infty}^{\infty} e^{-(x-2\pi n)^2/4\nu t} \,. \tag{3.1.9}$$

## 3.2 Strong and Weak Formulations of Differential Equations

In Chap. 1 we referred to weak and strong formulations of differential problems in a somewhat informal fashion. Here we make those concepts more precise, using the Burgers equation as a focus for the discussion. While these distinctions are well known in some circles, this material is provided as a service to the general reader.

Both (3.1.1) and (3.1.2) are in *strong form*, i.e., the PDE is required to be satisfied at each point in its domain and for each time. A *weak form* of the PDE is obtained by requiring that the integral of the PDE against all functions in an appropriate space $X$ of test functions be satisfied; precisely, we multiply both sides of (3.1.1) by each test function and integrate in space (for each time), to obtain

$$\int_a^b \frac{\partial u}{\partial t} v \, dx + \int_a^b u \frac{\partial u}{\partial x} v \, dx - \int_a^b \nu \frac{\partial^2 u}{\partial x^2} v \, dx = 0 \qquad \forall v \in X \,, \quad \forall t > 0 \,, \tag{3.2.1}$$

when $\Omega = (a, b)$. This is often referred to as an *integral form* of the PDE.

Another weak form of the PDE, which is often used, is obtained by performing an integration-by-parts on (3.2.1), yielding

$$\begin{aligned} &\int_a^b \frac{\partial u}{\partial t} v \, dx - \frac{1}{2} \int_a^b u^2 \frac{\partial v}{\partial x} \, dx + \nu \int_a^b \frac{\partial u}{\partial x} \frac{\partial v}{\partial x} \, dx \\ &+ (\frac{1}{2} u^2 v - \nu \frac{\partial u}{\partial x} v)\Big|_{x=b} - (\frac{1}{2} u^2 v - \nu \frac{\partial u}{\partial x} v)\Big|_{x=a} = 0 \qquad \forall v \in V \,, \quad \forall t > 0 \,. \end{aligned} \tag{3.2.2}$$

Equation (3.2.1) is meaningful if $u$ is twice differentiable, whereas the test functions need not be differentiable. In contrast, (3.2.2) requires less regularity on the solution, at the expense of increasing the regularity requirement on the test functions. This is reflected by restricting the test functions to lie in a subspace $V$ of the original space $X$. All three formulations are equivalent if the solution is smooth enough. The weak formulations, however, can accommodate less regular solutions. As a matter of fact, the solution to (3.2.2) is called the *distributional solution* to the original equation (3.1.1), since it can be shown that it satisfies (3.1.1) in the sense of distributions (Schwartz (1966), Lions and Magenes (1972), Renardy and Rogers (1993)). It is worth pointing out that for time-independent problems with a symmetric spatial operator, the weak formulation is also called the *variational formulation*, since it can be shown that its solution satisfies an extremal problem; for instance, the weak solution of the Dirichlet problem for the Poisson equation (1.2.67)–(1.2.68) minimizes the energy integral $\frac{1}{2} \int_\Omega |\nabla v|^2 - \int fv$.

Boundary conditions that should be satisfied by $u$ are incorporated in the boundary terms in (3.2.2) or are taken into account in the choice of test functions. For instance, if the flux is required to vanish at one boundary, then the corresponding boundary term drops out of (3.2.2). On the other hand, if the value of $u$ is prescribed at a boundary point, then all test functions are required to vanish at that point, and, consequently, the boundary term is zero.

Any discretization method considered in this book is derived from the strong or a weak formulation of the problem. Spectral collocation methods use the strong form of the PDE, as do finite-difference methods. For spectral

Galerkin, Galerkin with numerical integration (G-NI) and tau methods, as for finite-element methods, it is preferable to use the PDE in a weak form. While all the formulations of the differential problem are equivalent (provided that the solution is sufficiently smooth), this is not the case in general for the various discrete formulations derived from alternative formulations of the PDE. For example, the discrete solution based on a Galerkin method need not coincide with the discrete solution based on a collocation method. Moreover, Galerkin methods based on (3.2.1) are not necessarily equivalent to Galerkin methods based on (3.2.2).

In rough terms, strong and weak formulations are equivalent at the continuous (i.e., nondiscretized) level essentially because there are infinitely many test functions at our disposal. Their clever use allows one to recover the strong form from the weak form. This is not possible at the discrete (i.e., finite-dimensional) level, where only finitely many independent test functions are available. Hence, an appropriate way to design a numerical method is to first pick one of the formulations satisfied by the exact solution, then restrict the choice of test functions to a finite-dimensional space, to replace $u$ by the discrete solution $u^N$, and possibly to replace exact integration by quadrature rules.

The strong form (3.1.1) can be written compactly as

$$u_t + \mathcal{G}(u) + \mathcal{L}u = 0 \qquad \text{in } \Omega\,, \quad \forall t > 0\,, \tag{3.2.3}$$

where the nonlinear operator $\mathcal{G}$ is defined by $\mathcal{G}(u) = u(\partial u/\partial x)$, and the linear operator $\mathcal{L}$ is just $-\nu(\partial^2/\partial x^2)$. The corresponding compact version of the weak form (3.2.1) is

$$(u_t + \mathcal{G}(u) + \mathcal{L}u, v) = 0 \qquad \forall v \in X\,, \quad \forall t > 0\,, \tag{3.2.4}$$

where $(u, v)$ denotes the inner product in $X$. Likewise, the compact version of the second weak form (3.2.2) is

$$(u_t, v) - (\mathcal{F}(u), v_x) + [\mathcal{F}(u)v]_a^b = 0 \qquad \forall v \in V\,, \quad \forall t > 0\,. \tag{3.2.5}$$

## 3.3 Spectral Approximation of the Burgers Equation

This section will illustrate discretization processes for several spectral approximations to the Burgers equation. We consider here different treatments of the nonlinear and linear terms as well as different treatments of the boundary conditions. The rigorous discussion of these discretization processes is given in Chap. 6.

Each discretization that we present stems from one of the three formulations of the PDE. The solution is looked for in the space $X_N$ of trial functions. The weak formulations involve a finite-dimensional space of test functions that will be denoted by $Y_N \subset X$ if the weak formulation (3.2.1) is used as the starting point, or that will be denoted by $Y_N \subset V$ if the weak formulation (3.2.2) is used instead.

### 3.3.1 Fourier Galerkin

We look for a solution that is periodic in space on the interval $(0, 2\pi)$. The trial space $X_N$ is $S_N$, the set of all trigonometric polynomials of degree $\leq N/2$ (see (2.1.12)). The approximate function $u^N$ is represented as the truncated Fourier series

$$u^N(x,t) = \sum_{k=-N/2}^{N/2-1} \hat{u}_k(t) e^{ikx} . \tag{3.3.1}$$

In this method the fundamental unknowns are the coefficients $\hat{u}_k(t)$, $k = -N/2, \ldots, N/2-1$. Enforcement of the weak form (3.2.1) yields

$$\begin{gathered} \int_0^{2\pi} \left( \frac{\partial u^N}{\partial t} + u^N \frac{\partial u^N}{\partial x} - \nu \frac{\partial^2 u^N}{\partial x^2} \right) e^{-ikx} \, \mathrm{d}x = 0 , \\ k = -\frac{N}{2}, \ldots, \frac{N}{2} - 1 , \end{gathered} \tag{3.3.2}$$

which amounts to requiring that the residual of (3.1.1) be orthogonal to all the test functions in $Y_N = S_N$.

Due to the orthogonality property of the test and trial functions, we obtain a set of ODEs for the $\hat{u}_k$:

$$\frac{\mathrm{d}\hat{u}_k}{\mathrm{d}t} + \left( u^N \frac{\partial u^N}{\partial x} \right)_k^{\wedge} + k^2 \nu \hat{u}_k = 0 , \qquad k = -\frac{N}{2}, \ldots, \frac{N}{2} - 1 , \tag{3.3.3}$$

where

$$\left( u^N \frac{\partial u^N}{\partial x} \right)_k^{\wedge} = \frac{1}{2\pi} \int_0^{2\pi} u^N \frac{\partial u^N}{\partial x} e^{-ikx} \, \mathrm{d}x . \tag{3.3.4}$$

The initial conditions are clearly

$$\hat{u}_k(0) = \frac{1}{2\pi} \int_0^{2\pi} u(x,0) e^{-ikx} \, \mathrm{d}x . \tag{3.3.5}$$

The ODE initial-value problem (3.3.3)–(3.3.5) produced by the Fourier Galerkin spatial discretization is typically integrated in time by a method which treats the nonlinear, advection term explicitly and the linear, diffusion term either implicitly or else by an integrating-factor technique (see Sect. D.3).

The operator $\mathcal{L}_N$ is defined by $\mathcal{L}_N u^N = -\nu(\partial^2 u^N/\partial x^2)$, whereas the discrete nonlinear operator $\mathcal{G}_N$ is defined by $\mathcal{G}_N(u^N) = u^N(\partial u^N/\partial x)$.

The wavenumber $k = -N/2$ appears unsymmetrically in this approximation. If $\hat{u}_{-N/2}$ has a nonzero imaginary part, then the function $u^N(t)$ is not a real-valued function. This can lead to a number of difficulties, and it is advisable in practice simply to enforce the condition that $\hat{u}_{-N/2}$ is zero. This nuisance would, of course, be avoided if the approximation contained an odd rather than an even number of modes. However, the most widely used FFTs require an even number of modes.

Our objective is to describe spectral methods in a way that corresponds directly to the way they are implemented. This problem of the $k = -N/2$ mode arises, in practice, for all Fourier spectral methods using an even value of $N$. For all of the numerical examples and practical applications discussed in this text, $N$ is even and the $\hat{u}_{-N/2}$ coefficient is set to zero. (In more than one dimension, all Fourier coefficients with one or more indices equal to $-N/2$ are set to zero.)

The reader is advised that to apply the theoretical results in Chaps. 5–7 to the Fourier methods discussed here and in Chap. 4, and in CHQZ3, Chaps. 2–4 (where the $-N/2$ mode has been dropped), one needs to replace the $N$ in the theoretical chapters with $N/2 - 1$. (Recall, as discussed in Sect. 2.1.1, that the change from $N$ to $N/2$ comes from the truncation convention that is customary for the theory.)

The advection term (3.3.4) is a particular case of the general quadratic nonlinear term

$$(uv)^{\wedge}_k = \frac{1}{2\pi} \int_0^{2\pi} uv e^{-ikx} \, \mathrm{d}x \; , \tag{3.3.6}$$

where $u$ and $v$ denote generic trigonometric polynomials of degree $\leq N/2$, i.e., elements of $S_N$ (see (2.1.12)). They have expansions similar to (3.3.1). When these are inserted into (3.3.6) and the orthogonality property (2.1.2) is invoked, the expression

$$(uv)^{\wedge}_k = \sum_{p+q=k} \hat{u}_p \hat{v}_q \tag{3.3.7}$$

results. This is a convolution sum. The straightforward evaluation of (3.3.7) requires $O(N^2)$ operations. Fortunately, transform methods allow this term to be evaluated in only $O(N \log_2 N)$ operations (see Sect. 3.4). Integration of the implicitly treated diffusion terms takes only $O(N)$ operations, whether one uses an integrating-factor technique (see Sect. D.3) or a conventional time discretization, such as an Adams-Moulton method (see Sect. D.2.3). Hence, a single time-step for this Fourier Galerkin method takes only $O(N \log_2 N)$ operations.

### 3.3.2 Fourier Collocation

We again presume periodicity on $(0, 2\pi)$ and take $X_N = S_N$, but now think of the approximate solution $u^N$ as represented by its values at the grid points,

$x_j = 2\pi j/N$, $j = 0, \dots, N-1$. Recall that the grid-point values of $u^N$ are related to its discrete Fourier coefficients by (2.1.25) and (2.1.27). For the collocation method we require that the strong form (3.1.1) be satisfied at these points, i.e.,

$$\left. \frac{\partial u^N}{\partial t} + u^N \frac{\partial u^N}{\partial x} - \nu \frac{\partial^2 u^N}{\partial x^2} \right|_{x=x_j} = 0 , \qquad j = 0, 1, \dots, N-1 . \tag{3.3.8}$$

Initial conditions here are obviously

$$u^N(x_j, 0) = u_0(x_j) . \tag{3.3.9}$$

In vector form, with $\mathbf{u}(t) = (u^N(x_0,t), u^N(x_1,t), \dots, u^N(x_{N-1},t))^T$, (3.3.8) is

$$\frac{\mathrm{d}\mathbf{u}}{\mathrm{d}t} + \mathbf{u} \boxtimes D_N \mathbf{u} - \nu D_N^2 \mathbf{u} = \mathbf{0} , \tag{3.3.10}$$

where $D_N$ is the matrix, given by (2.1.41), that represents Fourier interpolation differentiation, and $\mathbf{u} \boxtimes \mathbf{v}$ is the component-wise product of two vectors $\mathbf{u}$ and $\mathbf{v}$.

Suppose that an explicit-advection/implicit-diffusion time discretization is employed. The derivative $\partial u^N/\partial x$ is most efficiently evaluated by the transform differentiation procedure described in Sect. 2.1.3. An efficient solution procedure for the implicit term is discussed in Sect. 4.1.1. It, too, resorts to transform methods. For a fully explicit time discretization, the diffusion term is also evaluated by a transform differentiation procedure. A single time-step thus can be performed in $O(N \log_2 N)$ operations for both mixed explicit/implicit and fully explicit time discretizations.

For the conservation form (3.1.2) of the Burgers equation, we approximate the nonlinear operator as $\mathcal{G}_N(u^N) = (1/2)\mathcal{D}_N[(u^N)^2]$. The collocation discretization of (3.1.2) is

$$\frac{\mathrm{d}\mathbf{u}}{\mathrm{d}t} + \frac{1}{2} D_N (\mathbf{u} \boxtimes \mathbf{u}) - \nu D_N^2 \mathbf{u} = \mathbf{0} . \tag{3.3.11}$$

Note that the nonlinear term is evaluated by first taking the pointwise square of $\mathbf{u}$ and then differentiating. The set of equations (3.3.11) is *not* equivalent to (3.3.10). In contrast, the Galerkin method produces the same discrete equations regardless of the precise form used for the PDE.

**Periodic Numerical Examples** The exact periodic solution $u$ corresponding to (3.1.9), (3.1.7) and (3.1.5) for $\nu = 0.2$, $c = 4$ and $t_0 = 1$ is shown in the upper left frame of Fig. 3.1 at $t = 0$ and $t = \pi/8$. The solution is nearly linear except for a "transition zone", which is the slowly diffusing (and advecting with speed $c = 4$) result of an initial discontinuity (for $t = -1$). This will be solved on the interval $[0, 2\pi]$ with initial data taken from this exact solution.

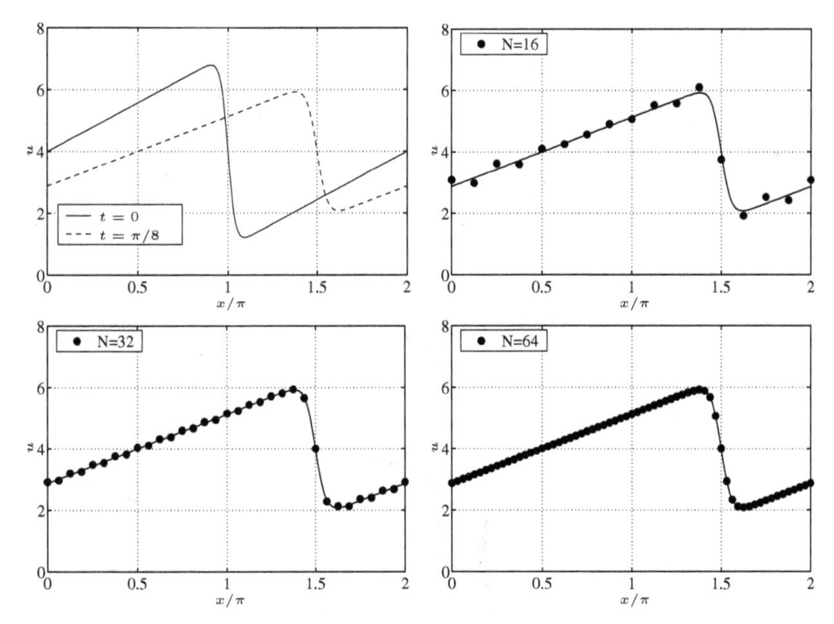


**Fig. 3.1.** The exact solution for the periodic Burgers equation problems (*top left*) and Fourier collocation solutions at $t = \pi/8$ for $N = 16$ (*top right*), $N = 32$ (*bottom left*), and $N = 64$ (*bottom right*)

We choose an explicit fourth-order Runge-Kutta method in time (RK4 – see (D.2.17)) to integrate the Fourier collocation spatial discretization given by (3.3.8). (The Runge-Kutta scheme provides the high temporal accuracy needed to demonstrate spectral accuracy in space, which is the objective of this example; for more challenging computations, such as for multidimensional Navier-Stokes computations, one would indeed treat the viscous term implicitly.) Figure 3.1 presents the computed solutions at $t = \pi/8$ for $N = 16, 32$ and 64. The approximation with only sixteen collocation points is unable to resolve the transition zone, and noticeable oscillations ensue. Once the transition zone has been well resolved these oscillations disappear, as illustrated by the $N = 32$ and $N = 64$ results in the bottom row of Fig. 3.1.

In Sect. 2.1.4 we discussed the Gibbs phenomenon, which arises in approximations to functions with discontinuities. The present example illustrates that similar oscillations arise whenever the solution contains gradients that are too steep for the trial functions to resolve. In principle, oscillations arising from solutions with finite gradients can always be avoided by increasing the spatial resolution. The theoretical discussion of spectral approximations to partial differential equations with discontinuous solutions is contained in Sect. 7.6. A summary of the various approaches to handling discontinuous solutions in fluid dynamics applications with spectral methods is provided in CHQZ3, Sect. 4.5.

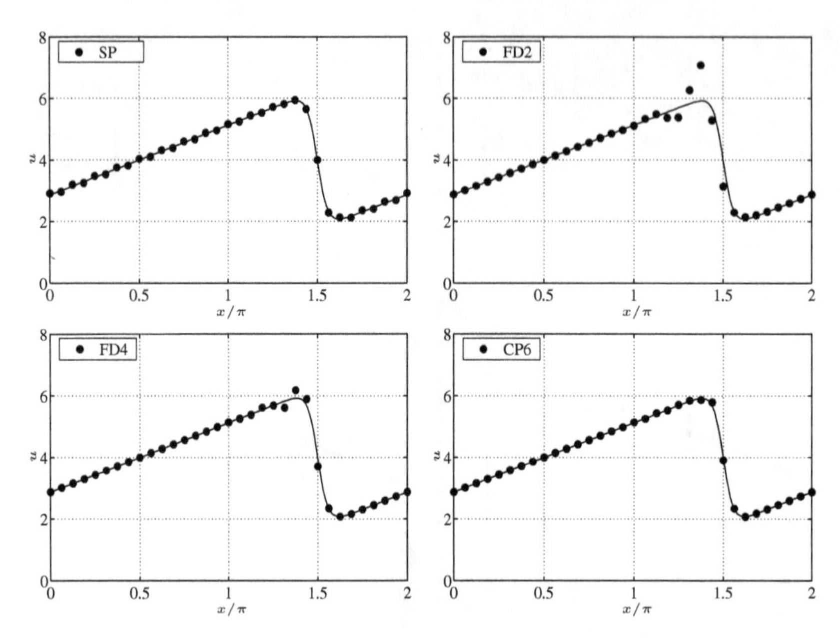


**Fig. 3.2.** Solutions to the periodic Burgers equation problem at $t = \pi/8$: comparison between Fourier collocation solution and finite-difference solutions of order 2, 4 and 6

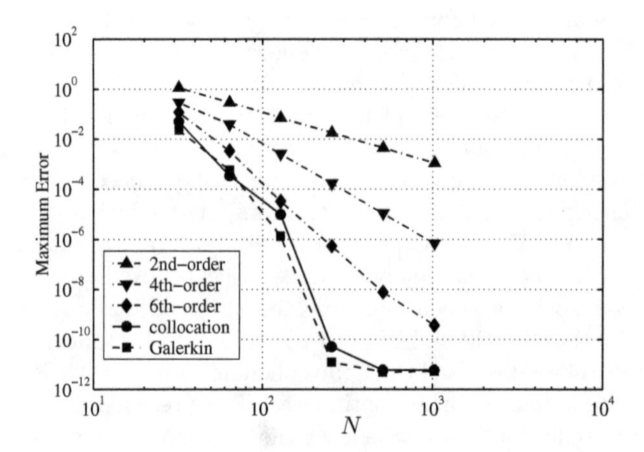


**Fig. 3.3.** Maximum errors for the periodic Burgers equation problem at $t = \pi/8$

A comparison with finite-difference schemes is instructive. Figure 3.2 displays second-order central-difference (FD2), fourth-order central-difference (FD4), and sixth-order compact (CP6), along with the Fourier collocation (SP) solutions, using the same number of gridpoints. The sixth-order compact scheme uses (1.2.19) for the first derivative and the periodic Padé approximation (see, for example, Collatz (1966) and Lele (1992))

$$\begin{aligned} \frac{2}{11}u''_{i-1} + u''_i + \frac{2}{11}u''_{i+1} = & \frac{12}{11(\Delta x)^2}(u_{i-1} - 2u_i + u_{i+1}) \\ & + \frac{3}{44(\Delta x)^2}(u_{i-2} - 2u_i + u_{i+2}) \,. \end{aligned} \tag{3.3.12}$$

for the second derivative. Figure 3.3 displays the maximum errors for these approximations as a function of $N$. The spectral scheme is already superior to the second-order method for $N = 16$, becomes superior to the fourth-order method for $N = 32$, and to the sixth-order method for $N = 128$. Notice that, as expected, the FD4 solution exhibits fourth-order asymptotic error decay, and the CP6 solutions exhibits sixth-order error decay because of the absence of special boundary stencils.

This is a fairly easy problem for a finite-difference method since the solution is essentially linear (and thus represented almost exactly even by a second-order finite-difference method) over all but the transition region. The real superiority of spectral methods emerges for problems with more structure in the solution – see the examples in Sect. 1.2 and CHQZ3, Sect. 4.3.

The above examples were geared towards illustrating the spatial accuracy of the method. The time-steps were typically well below the stability limit of the RK4 method. For the $N = 128$ spectral case, $\Delta t = .0005$ was needed in order to push the temporal errors below the spatial ones.

### 3.3.3 Chebyshev Tau

We now seek a solution to (3.1.1) on $(-1, 1)$ that satisfies the Dirichlet boundary conditions

$$u(-1, t) = u_L(t) \,, \qquad u(1, t) = u_R(t) \,, \tag{3.3.13}$$

where $u_L$ and $u_R$ are the prescribed Dirichlet boundary data. The trial space $X_N$ consists of all the members of $\mathbb{P}_N$ (the set of algebraic polynomials of degree $\leq N$). The discrete solution is expressed as the truncated Chebyshev series

$$u^N(x, t) = \sum_{k=0}^{N} \hat{u}_k(t) T_k(x) \,, \tag{3.3.14}$$

with the Chebyshev coefficients comprising the fundamental representation of the approximation. The equation (3.1.1) is enforced through its weak form (3.2.1), i.e., by insisting that the residual be orthogonal to the test functions in $Y_N = \mathbb{P}_{N-2}$:

$$\int_{-1}^{1}\left(\frac{\partial u^N}{\partial t}+u^N\frac{\partial u^N}{\partial x}-\nu\frac{\partial^2 u^N}{\partial x^2}\right)(x)\,T_k(x)(1-x^2)^{-1/2}\,\mathrm{d}x=0\,,\qquad k=0,\dots,N-2\,. \tag{3.3.15}$$

Note that the weight function, $w(x) = (1-x^2)^{-1/2}$, appropriate to the Chebyshev polynomials is used in the orthogonality condition. The boundary conditions (3.3.13) impose the additional constraints

$$u^N(-1,t)=u_L(t)\qquad\text{and}\qquad u^N(1,t)=u_R(t)\,. \tag{3.3.16}$$

Equation (3.3.15) reduces to

$$\frac{\partial \hat{u}_k}{\partial t}+\left(u^N\frac{\partial u^N}{\partial x}\right)^{\wedge}_k-\nu\hat{u}_k^{(2)}=0\,,\qquad k=0,1,\dots,N-2\,, \tag{3.3.17}$$

where $\hat{u}_k^{(2)}$ is given by (2.4.27) and

$$\left(u^N\frac{\partial u^N}{\partial x}\right)^{\wedge}_k=\frac{2}{\pi c_k}\int_{-1}^{1}\left(u^N\frac{\partial u^N}{\partial x}\right)(x)\,T_k(x)(1-x^2)^{-1/2}\,\mathrm{d}x\,, \tag{3.3.18}$$

where the $c_k$ are given by (2.4.10). In terms of the Chebyshev coefficients, the boundary conditions (3.3.16) become, through the use of (2.4.6),

$$\sum_{k=0}^{N}\hat{u}_k=u_R\,,\qquad \sum_{k=0}^{N}(-1)^k\hat{u}_k=u_L\,. \tag{3.3.19}$$

The initial conditions are

$$\hat{u}_k(0)=\frac{2}{\pi c_k}\int_{-1}^{1}u_0(x)T_k(x)(1-x^2)^{-1/2}\,\mathrm{d}x\,,\qquad k=0,1,\dots,N\,. \tag{3.3.20}$$

Equations (3.3.17), (3.3.19) and (3.3.20) form a complete set of ODEs for this approximation.

The expression in (3.3.18) is a special case of

$$(uv)^{\wedge}_k=\frac{2}{\pi c_k}\int_{-1}^{1}u(x)v(x)\,T_k(x)(1-x^2)^{-1/2}\,\mathrm{d}x\,, \tag{3.3.21}$$

which is equal to the following expression involving the convolution sums:

$$(uv)^{\wedge}_k=\frac{1}{2}\sum_{p+q=k}\hat{u}_p\hat{v}_q+\sum_{|p-q|=k}\hat{u}_p\hat{v}_q\,. \tag{3.3.22}$$

A typical time discretization is explicit for the nonlinear term and implicit for the linear one. Transform methods (see Sect. 3.4) are an efficient means of evaluating the nonlinear term. The implicit terms (including the boundary conditions) can be solved in $O(N)$ operations by the method described in Sect. 4.1.2.

If the boundary conditions are of Neumann type, $u'(-1,t) = 0$, $u'(1,t) = 0$, then conditions (3.3.19) are replaced by

$$\sum_{k=1}^{N} k^2 \hat{u}_k = 0\,, \qquad \sum_{k=1}^{N} (-1)^k k^2 \hat{u}_k = 0\,. \tag{3.3.23}$$

### 3.3.4 Chebyshev Collocation

For a collocation approximation to the Dirichlet problem the trial space $X_N$ is the same as for the previous example and the solution $u^N$ is represented by its values at the grid points $x_j = \cos \pi j/N$, $j = 0, 1, \ldots, N$. The grid-point values of $u^N$ are related to the discrete Chebyshev coefficients by (2.4.15) and (2.4.17). The discretization of the PDE in strong form is

$$\left. \frac{\partial u^N}{\partial t} + u^N \frac{\partial u^N}{\partial x} - \nu \frac{\partial^2 u^N}{\partial x^2} \right|_{x=x_j} = 0\,, \qquad j = 1, \ldots, N-1\,, \tag{3.3.24}$$

with

$$u^N(-1,t) = u_L(t)\,, \qquad u^N(1,t) = u_R(t)\,, \tag{3.3.25}$$

$$u^N(x_j, 0) = u_0(x_j)\,, \qquad j = 0, \ldots, N. \tag{3.3.26}$$

Let $\mathbf{u}(t) = (u^N(x_0,t), \ldots, u^N(x_N,t))^T$. Then, (3.3.24) can be written as, for all $t > 0$,

$$Z_N \left( \frac{\mathrm{d}\mathbf{u}}{\mathrm{d}t} + \mathbf{u} \boxtimes D_N \mathbf{u} - \nu D_N^2 \mathbf{u} \right) = 0\,, \tag{3.3.27}$$

where $D_N$ is the Chebyshev interpolation differentiation matrix given by (2.4.31), and $Z_N$ is the matrix that represents setting the first and last points of a vector to zero. The boundary conditions (3.3.25) are enforced by directly setting the first and last entries of $\mathbf{u}(t)$ to $\mathbf{u}_L(t)$ and $\mathbf{u}_R(t)$, respectively, for all $t > 0$. The numerical analysis of the Chebyshev collocation method is reviewed in Example 3 of Sect. 6.5.1 for the heat equation, and in Sect. 7.8 for the steady Burgers equation.

The nonlinear term can be evaluated efficiently by transform methods. The best direct solution method for an implicitly treated linear term (see Sect. 4.1.4), however, takes $O(N^2)$ operations. Iterative solution methods (see Sects. 4.5–4.6) are more efficient in some circumstances than direct methods, especially for multidimensional problems.

If a Chebyshev pseudospectral transform method (see Sect. 3.4) is used for the nonlinear term and a Chebyshev tau method for an implicitly treated linear term (see Sect. 4.1.2), then a single time-step takes only $O(N \log_2 N)$ operations. Such a mixed discretization scheme is, in fact, typical of most large-scale algorithms.

### 3.3.5 Legendre G-NI

As our concluding example, we seek a solution to the Burgers equation that satisfies the no-flux boundary conditions

$$\mathcal{F}(u) = \frac{1}{2}u^2 - \nu \frac{\partial u}{\partial x} = 0 \qquad \text{at } x = \pm 1 \quad \forall t > 0\,, \tag{3.3.28}$$

along with a specified initial condition at $t = 0$. In this case the preferred strong form of the Burgers equation is the conservation form (3.1.2). The weak form which is most convenient is (3.2.2); it becomes

$$\int_{-1}^{1} \frac{\partial u}{\partial t} v \, dx - \frac{1}{2} \int_{-1}^{1} u^2 \frac{\partial v}{\partial x} \, dx + \nu \int_{-1}^{1} \frac{\partial u}{\partial x} \frac{\partial v}{\partial x} \, dx = 0 \qquad \forall v \in V\,, \quad \forall t > 0\,, \tag{3.3.29}$$

after applying the boundary conditions (3.3.28). The trial and test function space $V$, technically indicated by $H^1(-1,1)$, collects all continuous functions in $[-1,1]$ having a square-integrable first derivative therein (see Appendix A); for functions in $V$, all integrals in the previous expression are meaningful. The boundary conditions are thus accounted for *naturally* in the weak formulation (3.3.29). Functions in $V$ need not satisfy them, but the solution $u$ does.

For the Legendre G-NI (Galerkin with numerical integration) method the trial function space $X_N = V_N$ is the whole of $\mathbb{P}_N$, and it coincides with the test function space $V_N$. With the discrete delta-functions $\psi_j$, $j = 0, 1, \ldots N$, defined as in (1.2.55), $u^N$ is represented through its grid-point values as

$$u^N(x,t) = \sum_{l=0}^{N} u_l^N(t) \psi_l(x).$$

The discrete weak formulation is obtained from (3.2.2) using the Gauss-Lobatto quadrature formula (2.2.17) with the Legendre weight, $w(x) = 1$, to approximate the integrals that appear therein. This results in the G-NI method

$$\left(\frac{\partial u^N}{\partial t}, v\right)_N - \frac{1}{2}\left((u^N)^2, \frac{\partial v}{\partial x}\right)_N + \nu \left(\frac{\partial u^N}{\partial x}, \frac{\partial v}{\partial x}\right)_N = 0 \qquad \forall v \in \mathbb{P}_N\,, \tag{3.3.30}$$

where the inner product is the discrete LGL inner product introduced in (2.2.24).

Using the discrete delta-functions as test functions, we obtain the equivalent form

$$\sum_{k=0}^{N} \left(\frac{\partial u^N}{\partial t} \psi_j\right)(x_k)\, w_k - \frac{1}{2} \sum_{k=0}^{N} \left((u^N)^2 \frac{\partial \psi_j}{\partial x}\right)(x_k)\, w_k + \\ + \nu \sum_{k=0}^{N} \left(\frac{\partial u^N}{\partial x} \frac{d\psi_j}{dx}\right)(x_k)\, w_k = 0\,, \qquad j = 0, \ldots, N. \tag{3.3.31}$$

As before, we set $\mathbf{u}(t) = (u_0^N(t), u_1^N(t), \ldots, u_N^N(t))^T$. We now indicate by $K_N^{(2)}$ the symmetric and positive-semi-definite matrix whose entries are

$$(K_N^{(2)})_{jl} = \sum_{k=0}^{N} \left( \frac{\mathrm{d}\psi_l}{\mathrm{d}x} \frac{\mathrm{d}\psi_j}{\mathrm{d}x} \right)(x_k)\, w_k \, .$$

We also introduce the matrix

$$(C_N)_{jl} = -(K_N^{(1)})_{lj} = -\sum_{k=0}^{N} \left( \psi_l \frac{\mathrm{d}\psi_j}{\mathrm{d}x} \right)(x_k)\, w_k = -\frac{\mathrm{d}\psi_j}{\mathrm{d}x}(x_l)\, w_l \, ,$$

as well as the diagonal mass matrix $M_N = K_N^{(0)} = \mathrm{diag}(w_0, w_1, \ldots, w_N)$ (see Sect. 3.8 for more details on the matrices $K_N^{(r)}$). Then (3.3.31) can be written as the system of ODEs

$$M_N \frac{\mathrm{d}\mathbf{u}}{\mathrm{d}t} + \frac{1}{2} C_N (\mathbf{u} \boxtimes \mathbf{u}) + \nu K_N^{(2)} \mathbf{u} = \mathbf{0} \, , \qquad t > 0, \tag{3.3.32}$$

which can be integrated, for instance, by an implicit method for the linear term coupled with an explicit method for the nonlinear term.

As we have already seen in Section 1.2.3, the G-NI method collocates at the internal nodes the approximation of the equation obtained by interpolating the flux $\mathcal{F}(u^N)$; for the Burgers equation, we have

$$\left. \frac{\partial u^N}{\partial t} + \frac{1}{2} \frac{\partial}{\partial x} I_N \left( (u^N)^2 \right) - \nu \frac{\partial^2 u^N}{\partial x^2} \right|_{x=x_j} = 0, \qquad j = 1, \ldots, N-1 \, . \tag{3.3.33}$$

On the other hand, the G-NI method enforces at the boundary points a particular linear combination of the approximate equation and the boundary condition; in particular, we have

$$\left. \frac{\partial u^N}{\partial t} + \frac{1}{2} \frac{\partial}{\partial x} I_N \left( (u^N)^2 \right) - \nu \frac{\partial^2 u^N}{\partial x^2} \right|_{x=\pm 1} - \alpha \, \mathcal{F}(u^N)\big|_{x=\pm 1} = 0, \tag{3.3.34}$$

with $\alpha = \pm \frac{1}{2} N(N+1)$ for $x = \pm 1$.

Note that the second terms in (3.3.33) and (3.3.34) are just the Legendre interpolation derivative of $(u^N)^2$ at the grid points. If one were using a Legendre collocation method for this problem in the traditional way, by enforcing the boundary condition explicitly, then one would obtain (3.3.33) plus

$$\mathcal{F}(u^N)\big|_{x=\pm 1} = 0 \tag{3.3.35}$$

instead of (3.3.34). Both theoretical arguments and computational experience suggest that (3.3.34) is better than (3.3.35) from stability as well as accuracy considerations. We recommend the use of (3.3.34) even in the context of a traditional collocation method.

**Nonperiodic Numerical Example** The nonperiodic exact solution $u$ corresponding to (3.1.8), (3.1.7) and (3.1.5) for $\nu = 0.01$, $c = 1$, $a = 16$ and $t_0 = 1$ is shown in Fig. 3.4 (left) at $t = 0$ and $t = 1$. The Burgers equation is solved on the interval $(-1, 1)$ with initial and boundary data taken from this exact solution.

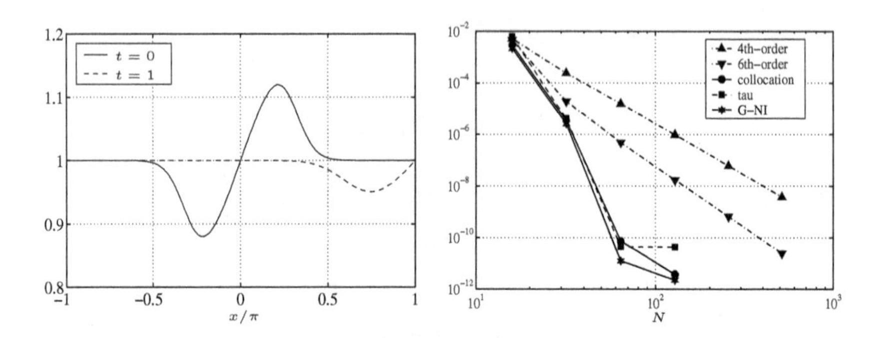


**Fig. 3.4.** The exact solution for the nonperiodic Burgers equation problems (*left*) and computed maximum errors at $t = 1$ (*right*)

Figure 3.4 (right) illustrates the errors from the Chebyshev tau, Chebyshev collocation and G-NI numerical schemes on this problem, integrated in time with the RK4 method (see (D.2.17)). Also included for comparison are solutions for fourth-order and sixth-order compact differences. Compact-difference approximations to the first and second derivatives require special one-sided stencils for the points at and adjacent to the boundaries. For the fourth-order scheme, the stencils used here are taken from Lele (1992); they are third-order accurate at the boundaries and fourth-order accurate for all the interior points. The asymptotic decay rate of the fourth-order solutions shown in Fig. 3.4 is fourth order. The stencils for the sixth-order scheme are third order at the boundary points, fourth order at the points adjacent to the boundaries and sixth order everywhere else. (See Sect. 3.7 and CHQZ3, Sect. 4.2 for further discussion of the challenges of appropriate boundary stencils for compact schemes. As illustrated in CHQZ3, Fig. 4.2, higher order stencils near the boundaries for this class of sixth-order schemes are temporally unstable.) The asymptotic decay rate of the sixth-order results is less than fifth order. All the spectral results decay faster than algebraically without requiring any special treatment at the boundaries.

## 3.4 Convolution Sums

A principal algorithmic component of efficient Galerkin methods for nonlinear or variable-coefficient problems is the evaluation of convolution sums.

Consider, however, the Fourier Galerkin treatment of the product

$$s(x) = u(x)v(x) \ . \tag{3.4.1}$$

In the case of an infinite series expansion, we have the familiar convolution sum

$$\hat{s}_k = \sum_{m+n=k} \hat{u}_m \hat{v}_n \ , \tag{3.4.2}$$

where

$$u(x) = \sum_{m=-\infty}^{\infty} \hat{u}_m e^{imx}, \qquad v(x) = \sum_{n=-\infty}^{\infty} \hat{v}_n e^{inx} \ , \tag{3.4.3}$$

and

$$\hat{s}_k = \frac{1}{2\pi} \int_0^{2\pi} s(x) e^{-ikx} \, dx \ . \tag{3.4.4}$$

In the present context $u$ and $v$ are finite Fourier series of degree $\leq N/2$, i.e., trigonometric polynomials belonging to $S_N$, whereas $s \in S_{2N}$. The values of $\hat{s}_k$, though, are only of interest for $|k| \leq N/2$. So, we truncate the product (3.4.1) at degree $N/2$ (i.e., taking $P_N(uv)$). Then (3.3.2) becomes

$$\hat{s}_k = \sum_{\substack{m+n=k \\ |m|,|n| \leq N/2}} \hat{u}_m \hat{v}_n \ , \qquad |k| \leq N/2 \ , \tag{3.4.5}$$

which amounts to requiring (3.4.4) for $|k| \leq N/2$. The direct summation implied by (3.4.5) takes $O(N^2)$ operations. (In three dimensions, the cost is $O(N^4)$, provided, as discussed in Orszag (1980), that one utilizes the tensor-product nature of multidimensional spectral approximations.) This is prohibitively expensive, especially when one considers that for a nonlinear term a finite-difference algorithm takes $O(N)$ operations in one dimension (and $O(N^3)$ in three). However, the use of transform methods enables (3.4.5) to be evaluated in $O(N \log_2 N)$ operations (and the three-dimensional generalization in $O(N^3 \log_2 N)$ operations). This technique was developed independently by Orszag (1969, 1970) and Eliasen, Machenhauer and Rasmussen (1970). It was the single most important development that made spectral Galerkin methods practical for large-scale computations.

### 3.4.1 Transform Methods and Pseudospectral Methods

The approach taken in the transform method for evaluating (3.4.5) for $u, v$ in $S_N$ is to use the inverse discrete Fourier transform (DFT) to transform $\hat{u}_m$ and $\hat{v}_n$ to physical space, to perform there a multiplication similar to (3.4.1), and then to use the DFT to determine $\hat{s}_k$. This must be done carefully, however. To illustrate the subtle point involved, we introduce the discrete transforms (Sect. 2.1.2):

$$u_j = \sum_{k=-N/2}^{N/2-1} \hat{u}_k e^{ikx_j} , \qquad j = 0, 1, \ldots, N-1 , \tag{3.4.6}$$
$$v_j = \sum_{k=-N/2}^{N/2-1} \hat{v}_k e^{ikx_j} ,$$

and define

$$s_j = u_j v_j , \qquad j = 0, 1, \ldots, N-1 , \tag{3.4.7}$$

and

$$\tilde{s}_k = \frac{1}{N} \sum_{j=0}^{N-1} s_j e^{-ikx_j} , \qquad k = -\frac{N}{2}, \ldots, \frac{N}{2} - 1 , \tag{3.4.8}$$

where

$$x_j = 2\pi j/N .$$

Note that the $\tilde{s}_k$ are the discrete Fourier coefficients of the function $s$ (see (2.1.25)). Use of the discrete transform orthogonality relation (2.1.26) leads to

$$\tilde{s}_k = \sum_{m+n=k} \hat{u}_m \hat{v}_n + \sum_{m+n=k\pm N} \hat{u}_m \hat{v}_n = \hat{s}_k + \sum_{m+n=k\pm N} \hat{u}_m \hat{v}_n . \tag{3.4.9}$$

The second term on the right-hand side is the aliasing error. If the convolution sums are evaluated as described above, then the differential equation is not approximated by a true spectral Galerkin method. Orszag (1971a) termed the resulting scheme a *pseudospectral method*. The convolution sum (3.4.5) in the pseudospectral method is evaluated at the cost of 3 FFTs and $N$ multiplications. The total operation count is $(15/2)N \log_2 N$ multiplications. The generalization of the pseudospectral evaluation of convolution sums to more than one dimension is straightforward.

There are two basic techniques for removing the aliasing error from (3.4.9). They are discussed in the following two subsections.

### 3.4.2 Aliasing Removal by Padding or Truncation

The key to this *de-aliasing* technique is the use of a discrete transform with $M$ rather than $N$ points, where $M \geq 3N/2$. Let

$$y_j = 2\pi j/M, \quad \bar{u}_j = \sum_{k=-M/2}^{M/2-1} \breve{u}_k e^{iky_j}, \quad \bar{v}_j = \sum_{k=-M/2}^{M/2-1} \breve{v}_k e^{iky_j}, \tag{3.4.10}$$

$$\bar{s}_j = u_j v_j , \tag{3.4.11}$$

for $j = 0, 1, \ldots, M-1$, where

$$\breve{u}_k = \begin{cases} \hat{u}_k , & |k| \lesssim N/2 \\ 0 & \text{otherwise} \end{cases} . \tag{3.4.12}$$

(Note that the $\bar{u}_j$ (and $\bar{v}_j$ and $\bar{s}_j$) are the values of $u$ at $y_j = 2\pi j/M$, whereas the $u_j$ defined in the previous section are the values of $u$ at $x_j = 2\pi j/N$.) Thus, the $\breve{u}_k$ coefficients are the $\hat{u}_k$ coefficients padded with zeros for the additional wavenumbers. Similarly, let

$$\breve{s}_k = \frac{1}{M} \sum_{j=0}^{M-1} \bar{s}_j e^{-iky_j} , \qquad k = -\frac{M}{2}, \ldots, \frac{M}{2} - 1 . \tag{3.4.13}$$

Then

$$\breve{s}_k = \sum_{m+n=k} \breve{u}_m \breve{v}_n + \sum_{m+n=k\pm M} \breve{u}_m \breve{v}_n . \tag{3.4.14}$$

We are only interested in $\breve{s}_k$ for $|k| \le N/2$, and choose $M$ so that the second term on the right-hand side vanishes for these $k$. Since $\breve{u}_m$ and $\breve{v}_m$ are zero for $|m| > N/2$, the worst-case condition is

$$-\frac{N}{2} - \frac{N}{2} \le \frac{N}{2} - 1 - M ,$$

or

$$M \ge \frac{3N}{2} - 1 . \tag{3.4.15}$$

With $M$ so chosen we have obtained the de-aliased coefficients

$$\hat{s}_k = \breve{s}_k , \qquad k = -\frac{N}{2}, \ldots, \frac{N}{2} - 1 . \tag{3.4.16}$$

The operation count for this transform method is $(45/4) N \log_2(\frac{3}{2}N)$, which is roughly 50% larger than the simpler, but aliased, method discussed earlier. For obvious reasons this technique is sometimes referred to as the *3/2-rule*. As described here it requires an FFT that can handle prime factors of 3. If only a prime factor 2 FFT is available, then this de-aliasing technique can be implemented by choosing $M$ as the smallest power of 2 that satisfies (3.4.15). This de-aliasing technique is also termed truncation and is sometimes referred to as the *2/3-rule*.

### 3.4.3 Aliasing Removal by Phase Shifts

A second method to remove the aliasing terms, due to Patterson and Orszag (1971), employs phase shifts. In this case (3.4.6) is replaced with

$$u_j^{\Delta} = \sum_{k=-N/2}^{N/2-1} \hat{u}_k e^{ik(x_j+\Delta)}, \quad v_j^{\Delta} = \sum_{k=-N/2}^{N/2-1} \hat{v}_k e^{ik(x_j+\Delta)} , \qquad j = 0, 1, \ldots, N-1 , \tag{3.4.17}$$

which are just the transforms on a grid shifted by the factor $\Delta$ in physical space. One then computes

$$s_j^\Delta = u_j^\Delta v_j^\Delta \,, \qquad j = 0, 1, \ldots, N-1, \tag{3.4.18}$$

and

$$\hat{s}_k^\Delta = \frac{1}{N} \sum_{j=0}^{N-1} s_j^\Delta e^{-ik(x_j+\Delta)} \,, \qquad k = -\frac{N}{2}, \ldots, \frac{N}{2} - 1 \,. \tag{3.4.19}$$

This last quantity is just

$$\hat{s}_k^\Delta = \sum_{m+n=k} \hat{u}_m \hat{v}_n + e^{\pm iN\Delta} \left( \sum_{m+n=k\pm N} \hat{u}_m \hat{v}_n \right) \,. \tag{3.4.20}$$

If one chooses $\Delta = \pi/N$, i.e., one shifts by half a grid cell, then

$$\hat{s}_k = \tfrac{1}{2} \left[ \tilde{s}_k + \hat{s}_k^\Delta \right] \,. \tag{3.4.21}$$

Thus, the aliasing contributions to the nonlinear term can be eliminated completely at the cost of two evaluations of the convolution sum. The cost here is $15N \log_2 N$. This is greater than the cost of the padding technique. However, if only a power of 2 FFT is available, then the padding technique requires the use of $M = 2N$ points rather than $(3/2)N$. Its cost then increases to $15N \log_2 N$.

The phase-shift technique and the padding method can both be extended to two and three dimensions. This discussion is postponed until Sect. 3.3, where it is given in the context of applications to simulations of incompressible, homogeneous turbulence.

Rogallo (1977) observed how the phase-shifting strategy can be incorporated at no extra cost into an otherwise pseudospectral algorithm to produce a method that has greatly reduced aliasing errors. Suppose that the time-differencing scheme is second-order Runge-Kutta (see (D.2.15)). At the first stage, the convolution sum is evaluated by the pseudospectral transform method described in Sect. 3.4.1 except that $\bar{u}_j$ and $\bar{v}_j$ are computed not by (3.4.6) but rather by (3.4.17), where $\Delta$ is a random number in $(0, 2\pi/N)$. In the second stage, (3.4.17) is again used for $\bar{u}_j$ and $\bar{v}_j$, but now with $\Delta$ replaced by $\Delta + \pi/N$. As a result the aliasing errors at the end of the full Runge-Kutta step are reduced to $O(\Delta t^2)$ times the pure pseudospectral aliasing errors, where $\Delta t$ is the size of the time-step. The use of a random shift $\Delta$ ensures that the remaining aliasing errors are uncorrelated from step to step.

### 3.4.4 Aliasing Removal for Orthogonal Polynomials

Quadratic nonlinearities also produce convolution-type sums in Chebyshev Galerkin and tau methods. A typical sum is given in (3.3.22). The simplest approach is to examine the nonlinear term from the perspective of quadrature.

Consider the product

$$s(x) = u(x)v(x) \,, \tag{3.4.22}$$

where $u$ and $v$ are in $\mathbb{P}_N$, i.e.,

$$u(x) = \sum_{k=0}^{N} \hat{u}_k T_k(x) \quad \text{and} \quad v(x) = \sum_{k=0}^{N} \hat{v}_k T_k(x) \,. \tag{3.4.23}$$

Then,

$$\hat{s}_k = \frac{2}{\pi c_k} \int_{-1}^{1} u(x)v(x)T_k(x)w(x)\,\mathrm{d}x \,, \qquad k = 0, 1, \ldots, N \,, \tag{3.4.24}$$

where $w(x)$ is the Chebyshev weight. The term $u(x)v(x)T_k(x)$ is a polynomial of degree $\leq 3N$. These coefficients $\hat{s}_k$ can be evaluated exactly by a Chebyshev Gauss-Lobatto quadrature using the points $y_j = \cos(\pi j/M)$, $j = 0, 1, \ldots, M$, provided that $2M - 1 \geq 3N$, or $M \geq 3N/2 + 1/2$. (See Sect. 2.2.3.)

Transform methods can be used to produce de-aliased representations of a quadratic product by choosing $M \geq 3N/2 + 1/2$ and then (1) padding $\hat{u}_k$ (and $\hat{v}_k$) as

$$\breve{u}_k = \begin{cases} \hat{u}_k \,, & k = 0, 1, \ldots, N \\ 0 \,, & k = N+1, N+2, \ldots, M \end{cases} \,; \tag{3.4.25}$$

(2) performing inverse discrete Chebyshev transforms on the Gauss-Lobatto points to obtain $u_j$, $v_j$, $j = 0, 1, \ldots, M$; (3) multiplying the physical space results to obtain $s_j = u_j v_j$, $j = 0, 1, \ldots, M$; (4) performing a discrete Chebyshev transform on $s_j$ to obtain $\breve{s}_k$, $k = 0, 1, \ldots, M$; and finally, (5) extracting $\hat{s}_k = \breve{s}_k$, $k = 0, 1, \ldots, N$.

Unlike the Fourier de-aliasing procedure discussed in Sect. 3.4.2, the choice here of $M = 3N/2$, which is desirable from the standpoint of efficient FFTs, does not produce a fully de-aliased set of coefficients. However, only the $\hat{s}_N$ term is not fully de-aliased; this can be handed separately at a relatively small cost. Alternatively, choosing $M = 2N$ produces a fully de-aliased set of coefficients, albeit at greater computational cost. A similar nuisance arises with the use of the Gauss and Gauss-Radau points. In the case of the tau method, this does not matter because the $N$-th coefficient of the nonlinear term is not used.

Of course, quadratic terms can also be evaluated pseudospectrally by transforming $\hat{u}_k$ and $\hat{v}_k$ to physical space at the points $x_j = 2\pi j/N$ with an $N$-mode Chebyshev transform, forming the product $u_j v_j$ there and then transforming back. This, of course, introduces aliasing errors. This modification to the algorithm discussed in Sect. 3.3.3 produces a pseudospectral Chebyshev tau method.

The approach of resorting to quadrature rules for de-aliasing procedures can be readily extended to cubic and high-order products. For example, a cubic product (such as the convective terms in compressible flow momentum equation) can be de-aliased by choosing $M \geq 2N$, again with special treatment of the $N$-th coefficient. This approach also applies to expansions in other sets of orthogonal polynomials in $(-1,1)$, although a fast transform is not available.

In a slightly different context, Debusschere et al. (2004) have recommended the solution of a system of equations to treat nonlinear terms. For example, to evaluate the expansion coefficients of $s = u/v$, one would write this as

$$\left(\sum_{j=0}^{N} \hat{v}_j T_j(x)\right)\left(\sum_{k=0}^{N} \hat{s}_k T_k(x)\right) = \left(\sum_{l=0}^{N} \hat{u}_l T_l(x)\right) . \tag{3.4.26}$$

After multiplying both sides of this equation by $T_n(x)w(x)$ and integrating over $(-1,1)$, one obtains the following linear system for the expansion coefficients of $s$:

$$\sum_{k=0}^{N}\left(\sum_{j=0}^{N} \frac{2}{\pi c_n} C_{jkn} \hat{v}_j\right) \hat{s}_k = \hat{u}_n , \tag{3.4.27}$$

where

$$C_{jkn} = \int_{-1}^{1} T_j(x) T_k(x) T_n(x) w(x) \mathrm{d}x . \tag{3.4.28}$$

The solution of the linear system (3.4.27) for $\hat{s}_k$ yields the de-aliased expansion coefficients of $s$.

## 3.5 Relation Between Collocation, G-NI and Pseudospectral Methods

In most cases Fourier pseudospectral methods are algebraically equivalent to collocation methods. Consider again the simple Burgers equation (3.1.1), periodic on $(0, 2\pi)$. The Galerkin approximation is

$$\frac{\mathrm{d}\hat{u}_k}{\mathrm{d}t} + \sum_{m+n=k} \hat{u}_m \hat{v}_n + \nu k^2 \hat{u}_k = 0 , \qquad k = -\frac{N}{2}, \ldots, \frac{N}{2} - 1 , \tag{3.5.1}$$

where $\hat{v}_k = ik\hat{u}_k$.

The pseudospectral approximation uses a fully aliased transform method to evaluate the convolution sum. Equation (3.5.1) is, in effect, replaced by

$$\frac{\mathrm{d}\hat{u}_k}{\mathrm{d}t} + \sum_{m+n=k} \hat{u}_m \hat{v}_n + \sum_{m+n=k\pm N} \hat{u}_m \hat{v}_n + \nu k^2 \hat{u}_k = 0 \tag{3.5.2}$$

(see (3.4.9)).

The collocation approximation may be written

$$\left.\frac{\partial u^N}{\partial t} + u^N v^N - \nu \frac{\partial^2 u^N}{\partial x^2}\right|_{x=x_j} = 0\,, \qquad j = 0, \dots, N-1\,, \tag{3.5.3}$$

where $v^N = \partial u^N / \partial x$. Resorting to the discrete Fourier series representations of $u$ and $v$ at the grid points, we have that (3.5.3) is

$$\begin{aligned}\sum_{l=-N/2}^{N/2-1} \frac{\mathrm{d}\tilde{u}_l}{\mathrm{d}t} e^{ilx_j} + \left(\sum_{m=-N/2}^{N/2-1} \tilde{u}_m e^{imx_j}\right)\left(\sum_{n=-N/2}^{N/2-1} \tilde{v}_n e^{inx_j}\right) \\ +\nu \sum_{l=-N/2}^{N/2-1} l^2 \tilde{u}_l e^{ilx_j} = 0\,, \qquad j = 0, 1, \dots, N-1\,.\end{aligned} \tag{3.5.4}$$

Applying the DFT to (3.5.4) and using the orthogonality relation (2.1.26), we find

$$\begin{aligned}\frac{\mathrm{d}\tilde{u}_k}{\mathrm{d}t} + \sum_{m+n=k} \tilde{u}_m \tilde{v}_n + \sum_{m+n=k\pm N} \tilde{u}_m \tilde{v}_n + \nu k^2 \tilde{u}_k = 0\,, \\ k = -\frac{N}{2}, \dots, \frac{N}{2} - 1\,.\end{aligned} \tag{3.5.5}$$

This is identical to (3.5.2). Thus, except for round-off error and provided the initial condition (and the right-hand side, if nonvanishing) is approximated in the same way, the pseudospectral and collocation discretizations of (3.1.1) are equivalent in the sense that they yield the same solution. So are the pseudospectral and collocation discretizations of the Burgers equation in the form (3.1.2). The same equivalence occurs for more complicated systems of equations such as incompressible Navier-Stokes (see CHQZ3, Sect. 3.3.4).

A scheme for the Burgers equation implemented as a standard collocation method can be de-aliased, if desired, by a truncation method. If at every time-step one sets to zero the discrete Fourier coefficients for which $|k| \geq (1/3)N$, the aliasing term in (3.5.2) vanishes. The collocation scheme then becomes algebraically equivalent to a Galerkin method. In this context the truncation method is known as the 2/3-*rule*. For the Burgers equation, this truncation can be accomplished as part of the solution of the implicit part of the equation. This is solved in transform space (see Sect. 4.1.1), and the unwanted Fourier coefficients are easily discarded.

The Chebyshev collocation method is not equivalent to the pseudospectral Chebyshev tau method mentioned in Sect. 3.4.4. As a matter of fact, at the operator level, in the latter method the quadratic term $u^N \dfrac{\partial u^N}{\partial x}$ is approximated by

$$P_{N-2}\left(I_N(u^N \frac{\partial u^N}{\partial x})\right),$$

whereas the Chebyshev collocation method uses

$$\tilde{I}_{N-2}\left(I_N(u^N\frac{\partial u^N}{\partial x})\right) = \tilde{I}_{N-2}\left(u^N\frac{\partial u^N}{\partial x}\right),$$

where $\tilde{I}_{N-2}v$ denotes the algebraic polynomial of degree $\leq N-2$ interpolating $v$ at the internal nodes $x_j,\ j = 1, \ldots, N-1$.

We might add that in some quarters the term pseudospectral method is used to refer to what we call in this book a collocation method. We use the adjective pseudospectral solely in terms of otherwise Galerkin or tau methods in which the nonlinear terms are subjected to a pseudospectral evaluation.

Finally, the relations between the Legendre collocation method and the G-NI method on an interval have already been pointed out in Sects. 1.2.3 and 3.3.5. Both methods enforce the same approximation of the differential equation at the internal quadrature points, whereas they may enforce the boundary conditions differently. In particular, for a second-order problem, Dirichlet boundary conditions are treated in the same way, whereas Neumann or Robin (flux) conditions are treated differently. This situation holds in multiple dimensions as well, if the methods are set on domains which are Cartesian products of intervals (possibly after a mapping) and use a tensor-product Gaussian grid. On simplicial domains, such as triangles and tetrahedra, collocation and G-NI methods may differ substantially, as mentioned in Sect. 2.9.2.

## 3.6 Conservation Forms

In many applications to hyperbolic problems, e.g., the inviscid Burgers equation or the Euler equations of fluid dynamics, the exact solution satisfies one or more conservation properties. Replicating some conservation properties in the approximate solution may be necessary for a physically meaningful result or for a numerically stable one. Even though strict conservation does not apply to advection-diffusion problems or to the Navier-Stokes equations of fluid dynamics, it is usually advisable to require some level of conservation for the hyperbolic part of the problem.

We begin this section with an illustration on the Burgers equation of some of the basic principles of assessing numerical conservation properties, and then turn to more general equations. The inviscid, periodic Burgers equation

$$\frac{\partial u}{\partial t} + u\frac{\partial u}{\partial x} = 0, \qquad 0 < x < 2\pi,\ t > 0\,, \tag{3.6.1}$$

satisfies an infinite set of conservation properties (for real-valued solutions)

$$\frac{\mathrm{d}}{\mathrm{d}t}\int_0^{2\pi} u^k \mathrm{d}t = 0\,, \qquad k = 1, 2, \ldots\,, \tag{3.6.2}$$

as can be seen by multiplying (3.6.1) by $u^{k-1}$ and integrating the second term by parts. As noted above, it is desirable for the discrete solution to satisfy analogous conservation laws. Both the spatial and temporal discretizations affect the conservation properties. We focus on the spatial discretization and consider the semi-discrete evolution equation. We assume here that both the solution and its approximation are real-valued functions.

Semi-discrete Fourier approximations to the inviscid Burgers equation satisfy only a small number of conservation properties. Consider first the Fourier Galerkin approximation. The Fourier Galerkin equations (3.3.2) with $\nu = 0$ are equivalent to

$$\int_0^{2\pi} \left( \frac{\partial u^N}{\partial t} + u^N \frac{\partial u^N}{\partial x} \right) v \, \mathrm{d}x = 0 \qquad \forall v \in S_N \, . \tag{3.6.3}$$

Taking $v \equiv 1$ yields

$$\frac{\mathrm{d}}{\mathrm{d}t} \int_0^{2\pi} u^N \mathrm{d}x = -\frac{1}{2} \int_0^{2\pi} \frac{\partial}{\partial x} \left( (u^N)^2 \right) \mathrm{d}x = -\frac{1}{2} (u^N)^2 \Big|_0^{2\pi} = 0 \, ,$$

and taking $v = u^N$ produces

$$\frac{\mathrm{d}}{\mathrm{d}t} \int_0^{2\pi} (u^N)^2 \mathrm{d}x = -\frac{1}{3} \int_0^{2\pi} \frac{\partial}{\partial x} \left( (u^N)^3 \right) \mathrm{d}x = -\frac{1}{3} (u^N)^3 \Big|_0^{2\pi} = 0 \, .$$

Hence, Fourier Galerkin approximations conserve $\int u^N$ and $\int (u^N)^2$. However, they do not necessarily conserve $\int (u^N)^k$ for $k \geq 3$. For example, the integral in (3.6.3) is not required to be satisfied for $v = (u^N)^2$, since $(u^N)^2$ is not guaranteed to be in $S_N$.

Fourier collocation approximations may conserve one or both of these two quantities, depending on precisely how the nonlinear term is approximated. On the space $S_N$ the bilinear form $(u, v)_N$, defined by (2.1.32), is an inner product. Moreover, the differentiation operator $\mathcal{D}_N$ is skew-symmetric with respect to this inner product when applied to functions in $S_N$; indeed, $\mathcal{D}_N v = \mathrm{d}v/\mathrm{d}x$ for such functions. The equations (3.3.11) with $\nu = 0$ are equivalent to

$$\frac{\partial u^N}{\partial t} + \frac{1}{2} \mathcal{D}_N \left( (u^N)^2 \right) = 0 \, . \tag{3.6.4}$$

Taking the discrete inner product of (3.6.4) with the function $v \in S_N$ produces

$$\begin{aligned} \frac{\mathrm{d}}{\mathrm{d}t} (u^N, v)_N &= -\frac{1}{2} (\mathcal{D}_N \left( (u^N)^2 \right), v)_N \\ &= \frac{1}{2} ((u^N)^2, \mathcal{D}_N v)_N = \frac{1}{2} ((u^N)^2, \frac{\mathrm{d}v}{\mathrm{d}x})_N \, . \end{aligned} \tag{3.6.5}$$

Taking $v \equiv 1$ and using the skew-symmetry of $\mathcal{D}_N$, this yields

$$\frac{\mathrm{d}}{\mathrm{d}t}\left(\frac{2\pi}{N}\sum_{j=0}^{N-1} u_j^N\right) = 0 \,,$$

the discrete analog of $\frac{\mathrm{d}}{\mathrm{d}t}\int_0^{2\pi} u^N \mathrm{d}x = 0$. (Actually, the two quantities coincide, due to the exactness of the quadrature formula.) However, $\frac{2\pi}{N}\sum(u_j^N)^2$ (which coincides with $\int_0^{2\pi}(u^N)^2\mathrm{d}x$ for the same reason) is not conserved, since the inner products on the right-hand side of (3.6.5) are not exact for $v = u^N$.

On the other hand, if the collocation method is applied in the form

$$\frac{\partial u^N}{\partial t} + \frac{1}{3}\mathcal{D}_N\left((u^N)^2\right) + \frac{1}{3}u^N\mathcal{D}_N u^N = 0 \,, \tag{3.6.6}$$

then taking the discrete scalar product with $u^N$, one has

$$\frac{\mathrm{d}}{\mathrm{d}t}(u^N, u^N)_N + \frac{1}{3}(\mathcal{D}_N\left((u^N)^2\right), u^N)_N + \frac{1}{3}(u^N\mathcal{D}_N u^N, u^N)_N = 0 \,.$$

Again, one has $(\mathcal{D}_N\left((u^N)^2\right), u^N)_N = -(u^N\mathcal{D}_N u^N, u^N)_N$, because $\mathcal{D}_N$ is skew-symmetric. Hence, the quadratic quantity $\frac{2\pi}{N}\sum(u_j^N)^2$ is conserved. Moreover, $\frac{2\pi}{N}\sum u_j^N$ is also conserved, as can be demonstrated by taking the inner product of (3.6.6) with $v \equiv 1$ and replacing the discrete inner product with the continuous inner product (permitted in this case by precision of the quadrature rule). These results are typical: collocation methods may or may not satisfy as many conservation properties as Galerkin ones.

For the inviscid, nonperiodic Burgers equation, (3.1.1) with $\nu = 0$, supplemented with the Dirichlet boundary condition $u(-1,t) = 0$ for all $t > 0$, the integrals $\int u^k$ are conserved up to a boundary term. For Legendre Galerkin approximations, conservation up to a boundary term holds for $\int u^N$ and $\int (u^N)^2$, by arguments analogous to those for the Fourier Galerkin case. (Integrals now are taken on $(-1,1)$.) The Legendre G-NI approximation in the form

$$\left(\frac{\partial u^N}{\partial t}, v\right)_N + \frac{1}{2}(\mathcal{D}_N\left((u^N)^2\right), v)_N = 0 \qquad \forall v \in X_N \,,$$

where $X_N = \mathbb{P}_N^{0-}(-1,1)$ is the space of all polynomials of degree $\leq N$ vanishing at $x = -1$, conserves $\int u^N$ but not $\int (u^N)^2$. (For $v = u^N$ the inner product generates a Legendre Gauss-Lobatto quadrature on a polynomial of degree $3N - 1$. Indeed,

$$\frac{1}{2}(\mathcal{D}_N((u^N)^2), u^N)_N = -\frac{1}{2}\left((u^N)^2, \frac{\partial u^N}{\partial x}\right)_N + \frac{1}{2}(u^N(1))^2.$$

This polynomial degree exceeds the precision of the quadrature formula.) However, the Legendre G-NI method in the form analogous to (3.6.6),

$$\left(\frac{\partial u^N}{\partial t}, v\right)_N + \frac{1}{3}\left(\mathcal{D}_N\left((u^N)^2\right), v\right)_N + \frac{1}{3}(u^N \mathcal{D}_N(u^N), v)_N = 0 \qquad \forall v \in X_N , \tag{3.6.7}$$

conserves both $\int u^N$ and $(u^N, u^N)_N = \sum_{j=0}^{N}(u_j^N)^2 w_j$ (which is equivalent to $\int (u^N)^2$, as we will see in Sect. 5.3). The former result follows from replacing the discrete inner product with the continuous inner product (permitted for $v \equiv 1$ by precision of the quadrature rule). The quadratic conservation property follows from choosing $v = u^N$, using $\mathcal{D}_N = \frac{\partial}{\partial x} I_N$, and noting that

$$\begin{aligned}
\frac{1}{3}(\mathcal{D}_N((u^N)^2), u^N)_N &= -\frac{1}{3}\left(I_N((u^N)^2), \frac{\partial u^N}{\partial x}\right) + \frac{1}{3}(u^N(1))^2 \\
&= -\frac{1}{3}\left((u^N)^2, \frac{\partial u^N}{\partial x}\right)_N + \frac{1}{3}(u^N(1))^2 \\
&= -\frac{1}{3}(u^N \mathcal{D}_N(u^N), u^N)_N + \frac{1}{3}(u^N(1))^2 .
\end{aligned}$$

Hence, $(u^N, u^N)_N$ is conserved up to the boundary term $\int_0^t (u^N(1,\tau))^2 \mathrm{d}\tau$. For Chebyshev approximations, the inner product of the approximation does not correspond to the physical inner product in which the conservation property holds.

Let us now consider more general problems, starting with the (possibly vector-valued) evolution equation

$$\frac{\partial \mathbf{u}}{\partial t} + \mathcal{M}(\mathbf{u}) = \mathbf{0} \quad \text{in } \Omega . \tag{3.6.8}$$

The independent variables themselves are conserved (except for boundary effects) if the spatial operator is in divergence form, i.e.,

$$\mathcal{M}(\mathbf{u}) = \nabla \cdot \mathcal{F}(\mathbf{u}) , \tag{3.6.9}$$

where the tensor $\mathcal{F}$ is called the flux function. Gauss' theorem implies that the solution to the evolution equation (3.6.8) satisfies

$$\frac{\mathrm{d}}{\mathrm{d}t}\int_\Omega \mathbf{u} = -\int_{\partial\Omega} \mathcal{F} \cdot \hat{\mathbf{n}} . \tag{3.6.10}$$

Hence, the only integral changes in $\mathbf{u}$ are those due to fluxes through the boundaries.

If the spatial operator is orthogonal to the solution, i.e.,

$$(\mathcal{M}(\mathbf{u}), \mathbf{u}) = 0 , \tag{3.6.11}$$

then the quadratic conservation law

$$\frac{\mathrm{d}}{\mathrm{d}t}(\mathbf{u}, \mathbf{u}) = \frac{\mathrm{d}}{\mathrm{d}t}\|\mathbf{u}\|^2 = 0 \tag{3.6.12}$$

holds. An important special case arises when the operator $\mathcal{M}$ is linear and skew-symmetric, i.e.,

$$\mathcal{M}(\mathbf{u}) = \mathcal{L}\mathbf{u} \,, \tag{3.6.13}$$

with

$$\mathcal{L}^* = -\mathcal{L} \tag{3.6.14}$$

(assuming real variables). In this case

$$\begin{aligned}(\mathcal{M}(\mathbf{u}), \mathbf{u}) &= (\mathcal{L}\mathbf{u}, \mathbf{u}) = \tfrac{1}{2}(\mathcal{L}\mathbf{u}, \mathbf{u}) + \tfrac{1}{2}(\mathbf{u}, \mathcal{L}^*\mathbf{u}) \\ &= \tfrac{1}{2}(\mathcal{L}\mathbf{u}, \mathbf{u}) - \tfrac{1}{2}(\mathbf{u}, \mathcal{L}\mathbf{u}) = 0 \,.\end{aligned}$$

Note that for a one-dimensional scalar problem with periodic boundary conditions, $\mathcal{M}(u) = \partial u/\partial x$ satisfies these conditions, but that $\mathcal{M}(u) = a(x)(\partial u/\partial x)$ does not unless $\mathrm{d}a/\mathrm{d}x = 0$. In more than one space dimension, $\mathcal{M}(u) = \mathbf{a} \cdot \nabla u$ with $\nabla \cdot \mathbf{a} \equiv 0$ satisfies (3.6.11), for then $\mathcal{M}(u) = \frac{1}{2}\mathbf{a} \cdot \nabla u + \frac{1}{2}\nabla \cdot (u\mathbf{a})$, which is skew-symmetric.

We can write the Galerkin approximation to (3.6.8) as

$$\left(\frac{\partial \mathbf{u}^N}{\partial t} + \mathcal{M}(\mathbf{u}^N) \,, \mathbf{v}\right) = 0 \qquad \forall \mathbf{v} \in X_N \,, \tag{3.6.15}$$

and the collocation and G-NI approximations as

$$\left(\frac{\partial \mathbf{u}^N}{\partial t} + \mathcal{M}_{\mathbf{N}}(\mathbf{u}^N) \,, \mathbf{v}\right)_N = 0 \qquad \forall \mathbf{v} \in X_N \,, \tag{3.6.16}$$

where $\mathcal{M}_N$ is a suitable discrete approximation of $\mathcal{M}$. Consider Fourier or Legendre approximations to periodic and nonperiodic problems, respectively. For spatial operators in the divergence form (3.6.2), the choice of $\mathbf{v}$ as the vector with each component identically equal to one, yields, as for the Burgers cases, that $\int \mathbf{u}^N$ is conserved (except for boundary terms in the nonperiodic case). For spatial operators of the form (3.6.11), by choosing $\mathbf{v} = \mathbf{u}^N$ we immediately obtain

$$\frac{1}{2}\frac{\mathrm{d}}{\mathrm{d}t}(\mathbf{u}^N, \mathbf{u}^N) = -(\mathcal{M}(\mathbf{u}^N), \mathbf{u}^N) = 0 \,, \tag{3.6.17}$$

which demonstrates a semi-discrete quadratic conservation property. For the linear, skew-symmetric problem, where (3.6.13) and (3.6.14) are satisfied, we can similarly show quadratic conservation for collocation and G-NI methods.

The semi-discrete conservation laws are not satisfied by the fully discrete solution unless the time discretization is symmetric (i.e. based on centered finite differences – see Appendix D). The leap frog and Crank-Nicolson methods are symmetric. However, the departure from conservation is small for unsymmetric time-discretization schemes, such as Adams-Bashforth and Runge-Kutta, and the departure decreases as the time-step is reduced.

Numerous spectral collocation computations have been presented in the literature in which the advantages of using a conservation form have been exhibited. Several of these demonstrations have indicated that collocation methods can be (temporally) stable if the discrete equations satisfy quadratic conservation, but unstable if a nonconservative form is utilized. An early, dramatic demonstration was provided by Dahlburg (1985) for ideal magnetohydrodynamics. One interpretation of this effect is that quadratic conservation forms of the discrete equations tend to reduce the effects of aliasing errors in collocation methods. Further discussion is provided in CHQZ3, Chap. 3 in the context of incompressible flow computations, and in CHQZ3, Chap. 4 for compressible flow simulations.

## 3.7 Scalar Hyperbolic Problems

The purpose of this section is to illustrate the essential features of the spectral boundary treatment for a scalar, one-dimensional, nonperiodic hyperbolic problem with an explicit time discretization. More complex situations, such as linear and nonlinear hyperbolic systems and implicit time discretizations, are discussed in CHQZ3, Sect. 4.2.

For an explicit time discretization, any errors produced in a finite-difference scheme, including those due to the boundary treatment, have a finite rate of propagation. Moreover, if the scheme is dissipative the growth of the errors will be retarded or perhaps even suppressed. However, spectral methods have little dissipation to slow the growth of the errors, and because of their global character the errors immediately affect the entire domain. Numerical experience has confirmed that spectral methods are far more sensitive than finite-difference methods to the boundary treatment. On the other hand, as we will see, spectral methods require no special formulas for derivatives at the boundary, whereas finite-difference methods typically do.

### 3.7.1 Enforcement of Boundary Conditions

Let us then consider the linear, scalar hyperbolic equation

$$\frac{\partial u}{\partial t} + \beta \frac{\partial u}{\partial x} = 0 , \qquad -1 < x < 1, \quad t > 0 . \tag{3.7.1a}$$

For simplicity, we assume that the wave speed $\beta$ is constant and strictly positive. Thus, the point $x = -1$ is the inflow boundary point, where the equation is supplemented with the inflow boundary condition

$$u(-1,t) = u_L(t) , \qquad t > 0 . \tag{3.7.1b}$$

The problem is completed by the initial condition

$$u(x,0) = u_0(x)\,, \qquad -1 < x < 1\,. \tag{3.7.1c}$$

An obvious approach is a *strong* imposition of the boundary condition, which is particularly straightforward to implement within a collocation scheme. Let the collocation points be the Legendre Gauss-Lobatto points $x_j$, $j = 0, \dots, N$, introduced in (2.3.12). The semi-discrete (in space) approximation, $u^N(t) \in \mathbb{P}_N(-1,1)$ for all $t > 0$, is defined by the conditions

$$\frac{\partial u^N}{\partial t}(x_j,t) + \beta\frac{\partial u^N}{\partial x}(x_j,t) = 0\,, \qquad j = 1, \dots, N, \quad t > 0\,, \tag{3.7.2a}$$

$$u^N(-1,t) = u_L(t)\,, \qquad t > 0\,, \tag{3.7.2b}$$

$$u^N(x_j,0) = u_0(x_j)\,, \qquad j = 0, \dots, N\,. \tag{3.7.2c}$$

The scheme can be interpreted as a G-NI scheme. Indeed, let $\mathbb{P}_N^{0-}(-1,1)$ denote the space, already introduced in the previous section, of the polynomials of degree $\le N$ vanishing at the left endpoint of the interval $(-1,1)$; multiplying (3.7.2a) by $v(x_j)w_j$, where $v \in \mathbb{P}_N^{0-}(-1,1)$ and $w_j$ is the Legendre Gauss-Lobatto weight associated with the point $x_j$, and summing up on $j$ we get

$$(u_t^N, v)_N + (\beta u_x^N, v)_N = 0 \qquad \text{for all } v \in \mathbb{P}_N^{0-}(-1,1), \quad t > 0\,, \tag{3.7.3}$$

where $(u,v)_N = \sum_{j=0}^N u(x_j)v(x_j)w_j$ is still the discrete $L^2$-inner product on $\mathbb{P}_N(-1,1)$ (see (2.2.24) and Sect. 5.3). Note that the trial function $u^N$ satisfies the inflow boundary condition at each time; correspondingly each test function vanishes at the inflow boundary point.

Since $\beta$ is constant and the Gauss-Lobatto quadrature is exact for polynomials of degree $\le 2N-1$, the spatial term in (3.7.3) is actually exact, i.e.,

$$(\beta u_x^N, v)_N = (\beta u_x^N, v) = \int_{-1}^1 \beta \frac{\partial u^N}{\partial x} v \,\mathrm{d}x\,.$$

This immediately yields a (uniform in $N$) bound for the spectral solution $u^N$ in the case of an homogeneous inflow condition, $u_L(t) = 0$ for all $t$. Indeed, taking $v = u^N$, we have

$$(\beta u_x^N, u^N)_N = \int_{-1}^1 \beta \frac{1}{2}\frac{\partial}{\partial x}(u^N)^2 \,\mathrm{d}x = \tfrac{1}{2}\beta[(u^N)^2]_{-1}^1 = \tfrac{1}{2}\beta(u^N)^2(1,t) \ge 0\,;$$

whence, from (3.7.3),

$$\frac{\mathrm{d}}{\mathrm{d}t}\|u^N\|_N^2 \le 0 \qquad \text{for all } t > 0\,. \tag{3.7.4}$$

Since the discrete and continuous $L^2$-norms are uniformly equivalent on $\mathbb{P}_N(-1,1)$ (see (5.3.2)), this implies that the $L^2$-norm of the spectral solution $u^N$ is uniformly bounded with respect to $N$ and $t$. As will be discussed in Chap. 6, this uniform bound establishes the $L^2$-stability of the approximation, and together with the consistency of the discretization, this implies convergence of $u^N$ to the exact solution $u$ as $N \to \infty$. The same result holds in the case of a nonhomogeneous inflow condition. Note that we have $\frac{\mathrm{d}}{\mathrm{d}t}\|u^N\|_N^2 = 0$, except possibly for the boundary terms, which is a type of conservation property discussed in the previous section.

A more flexible way to handle the boundary conditions, which turns out to be useful, e.g., in multidomain spectral methods (see CHQZ3, Sect. 5.3.3) or for systems of equations (see CHQZ3, Sect.4.2.2), is to enforce them in a *weak* sense. The rationale is that whenever stability holds, then accuracy is assured provided the boundary conditions are matched to within the same consistency error as for the equation in the interior. As already done for the Burgers equation at the beginning of Sect. 3.3.5, the starting point is integration-by-parts. Let $u$ be the solution of (3.7.1), and let $v = v(x)$ be any smooth function (not necessarily vanishing at $x = 1$). Then,

$$\begin{aligned}\int_{-1}^{1} \beta \frac{\partial u}{\partial x} v \,\mathrm{d}x &= -\int_{-1}^{1} \beta u \frac{\partial v}{\partial x} \,\mathrm{d}x + [\beta u v]_{-1}^{1} \\ &= -\int_{-1}^{1} \beta u \frac{\partial v}{\partial x} \,\mathrm{d}x + \beta u(1,t)v(1) - \beta u_L(t) v(-1) \,.\end{aligned}$$

This suggests the consideration of the following G-NI scheme with weak imposition of the boundary conditions: find $u^N(t) \in \mathbb{P}_N(-1,1)$ satisfying, for all $t > 0$ and all $v \in \mathbb{P}_N(-1,1)$,

$$(u_t^N, v)_N - (\beta u^N, v_x)_N + \beta u^N(1,t)v(1) = \beta u_L(t)v(-1) \,, \tag{3.7.5}$$

as well as the initial condition (3.7.1c). Note that in this case neither the trial function $u^N$ nor any test function $v$ is required to satisfy a boundary condition. Taking $v = u^N$, the same bound (3.7.4) is obtained as before, in the homogeneous case, $u_L = 0$.

An alternative, equivalent formulation is obtained by counter-integrating by parts in (3.7.5); precisely, $u^N(t) \in \mathbb{P}_N(-1,1)$ satisfies, for all $t > 0$ and all $v \in \mathbb{P}_N(-1,1)$,

$$(u_t^N + \beta u_x^N, v)_N + \beta\big(u^N(-1,t) - u_L(t)\big)v(-1) = 0 \,. \tag{3.7.6}$$

The approximate, or weak, way in which $u^N$ matches the inflow condition becomes apparent by taking as $v$ suitable discrete delta-functions, namely, the characteristic Lagrange polynomial function $\psi_j$ (see (1.2.55)) at each internal or outflow Gauss-Lobatto point $x_j$, $j = 1, \ldots, N$. We immediately see that $u^N$ still satisfies the collocation equations (3.7.2a). On the other hand, taking $v = \psi_0$ we obtain at the inflow point that

$$\left(\frac{\partial u^N}{\partial t}(-1,t) + \beta\frac{\partial u^N}{\partial x}(-1,t)\right) + \frac{1}{w_0}\beta(u^N(-1,t) - u_L(t)) = 0\,. \tag{3.7.7}$$

Recalling that $\frac{1}{w_0} = \frac{N(N+1)}{2}$, we see that the boundary condition is accounted for through a specific *penalty procedure*, and it is satisfied exactly only in the limit $N \to \infty$. The formula also demonstrates that if the differential equation is fulfilled to within spectral accuracy, so is the inflow boundary condition.

The G-NI scheme with the weak enforcement of the boundary condition just illustrated is but a particular case of the *penalty approach* to handle boundary conditions, whose use in spectral methods was first advocated by Funaro and Gottlieb (1988, 1991). The spectral approximation $u^N$ is defined as the solution of the polynomial equation

$$\left(\frac{\partial u^N}{\partial t} + \beta\frac{\partial u^N}{\partial x}\right)(x,t) + \tau\beta Q_N(x)\big(u^N(-1,t) - u_L(t)\big) = 0\,,$$
$$-1 \le x \le 1,\ t > 0\,, \tag{3.7.8}$$

where $\tau$ is the penalization parameter, and $Q_N$ is a fixed polynomial of degree $\le N$ which determines just how the equation is enforced. Choosing

$$Q_N(x) = \frac{(1-x)L'_N(x)}{2L'_N(-1)} = \begin{cases} 1 & \text{if } x = -1, \\ 0 & \text{if } x = x_j \text{ for } j = 1,\dots,N \end{cases}$$

(where $x_j$ are the Legendre Gauss-Lobatto points) yields again the collocation equations (3.7.2a) at the internal and outflow quadrature points, whereas at the inflow point one has

$$\left(\frac{\partial u^N}{\partial t}(-1,t) + \beta\frac{\partial u^N}{\partial x}(-1,t)\right) + \tau\beta\big(u^N(-1,t) - u_L(t)\big) = 0 \tag{3.7.9}$$

(compare with (3.7.7)). In order to establish the admissible values of $\tau$, let us evaluate (3.7.8) at $x = x_j$, multiply it by $u^N(x_j,t)w_j$ and sum up over $j$. Using the exactness of the quadrature rule to integrate by parts in space, we obtain, in the case of a homogeneous boundary condition, the following relation:

$$\frac{1}{2}\frac{\mathrm{d}}{\mathrm{d}t}\|u^N\|_N^2 + \frac{1}{2}\beta(u^N)^2(1,t) =$$
$$-\,(\tau w_0 - \frac{1}{2})\beta(u^N)^2(-1,t) \quad \text{for all } t > 0\,. \tag{3.7.10}$$

Thus, the penalty method solution is bounded (i.e., it satisfies (3.7.4)) provided the penalty parameter satisfies

$$\tau \ge \frac{1}{2w_0}\,.$$

Choosing $\tau$ smaller than the value $\tau = \frac{1}{w_0}$ which stems in a natural way from the G-NI approach, results in a loss of conservation (illustrated in the numerical example below); yet, in this way one may increase the allowable time-step in an explicit time-discretization scheme.

We refer to Hesthaven (2000) and to Gottlieb and Hesthaven (2001) for further details and generalizations of penalty methods.

So far, we have dealt with Legendre methods. Obviously, the weak imposition of the boundary condition (scheme (3.7.5) or (3.7.6)) requires the integration weight to be neither zero nor infinity at the boundary, thus confining the quadrature nodes to be of Legendre type. On the contrary, the strong imposition of the boundary condition (scheme (3.7.1)) is amenable to an implementation in terms of Chebyshev Gauss-Lobatto points as well.

Another scheme that can be implemented with nodes of either Legendre or Chebyshev type is the *staggered-grid method*, which we now briefly describe. It uses two families of interpolation/collocation nodes, the Gauss-Lobatto and the Gauss points. These two grids are staggered with respect to each other. In this method the solution $u$ is represented by a polynomial of degree $N-1$ using the Gauss points, whereas the "flux", $\mathcal{F}(u) = \beta u$, is represented by a polynomial of degree $N$ using the Gauss-Lobatto points. We denote their finite-dimensional approximations by $\bar{u}^N$ and $F^N$, respectively. Let $x_j$, $j = 0, \ldots, N$, denote the Gauss-Lobatto points and $\bar{x}_j$, $j = 1, \ldots, N$, the Gauss points. The boundary condition at $x = -1$ is enforced weakly in this method by first constructing the polynomial $\tilde{u}_N \in \mathbb{P}_N$ from the values

$$\tilde{u}^N(x_j, t) = \begin{cases} u_L(t) , & j = 0 , \\ \bar{u}^N(x_j, t) , & j = 1, \ldots, N , \end{cases} \tag{3.7.11}$$

at the Gauss-Lobatto points, then generating the flux, $\tilde{F}(x,t) = \mathcal{F}(\tilde{u}^N(x,t))$. Finally, the staggered-grid collocation conditions are

$$\frac{\partial \bar{u}^N}{\partial t}(\bar{x}_j, t) + \frac{\partial F^N}{\partial x}(\bar{x}_j, t) = 0 , \qquad j = 1, \ldots, N, \quad t > 0 , \tag{3.7.12a}$$

$$u^N(\bar{x}_j, 0) = u_0(\bar{x}_j) , \qquad j = 1, \ldots, N , \tag{3.7.12b}$$

with $F^N(x,t) = I_N^{GL}(\tilde{F}(x,t))$, where $I_N^{GL}$ denotes the interpolation operator at the Gauss-Lobatto nodes. Note that in general $F^N$ is a polynomial of degree $N$ even though $\bar{u}^N$ is a polynomial of degree $N-1$ because of the application of the boundary condition. However, $\partial F^N/\partial x$ is a polynomial of degree $N-1$ because of the differentiation.

This method requires interpolating $\bar{u}^N$ from the Gauss points to the Gauss-Lobatto points, and interpolating $\partial F^N/\partial x$ back from the Gauss-Lobatto points to the Gauss points. Procedures for this are described in CHQZ3, Sect. 3.4.2, where the staggered grid is discussed for incompressible flow computations. It requires twice as much work per step as the non-staggered-grid method since two matrix multiplies (or four FFTs if Chebyshev points are used) are needed per stage rather than a single matrix multiply (or two FFTs). This particular staggered-grid method for hyperbolic problems was introduced by Kopriva and Kolias (1996). Unlike the earlier work of Cai and Shu (1993), which defined cell-averaged values of the solution at the Gauss points and flux values at the Gauss-Lobatto points, the Kopriva and Kolias method uses simply the pointwise values of the solution and not the cell-averaged values at the Gauss points.

### 3.7.2 Numerical Examples

We now present several sets of numerical results to illustrate the behavior of the various strategies described above for enforcing the boundary conditions. We consider first the test problem

$$\begin{aligned} &\frac{\partial u}{\partial t} + \frac{3}{2}\frac{\partial u}{\partial x} = 0\,, && -1 < x < 1\,,\ t > 0\,, \\ &u(-1,t) = \sin(-2-3t)\,, && t > 0\,, \\ &u(x,0) = \sin 2x\,, && -1 < x < 1\,, \end{aligned} \tag{3.7.13}$$

whose solution is the right-moving wave $u(x,t) = \sin(2x - 3t)$.

The first set of experiments has been conducted with the Legendre quadrature/collocation points. In this way, we can compare all the formulations considered in this section: the collocation method (3.7.2) with strong imposition of the boundary conditions, the G-NI method (3.7.6) with weak enforcement of boundary conditions, the penalty method (3.7.8) for different values of $\tau$, and the staggered-grid method (3.7.12). Figure 3.5 shows the maximum error at $t = 4$ for each method as a function of the polynomial degree $N$. The time discretization has been conducted with $\Delta t = 10^{-4}$, using the RK4 scheme for all methods. The results show that the decay rate of the error is similar in all

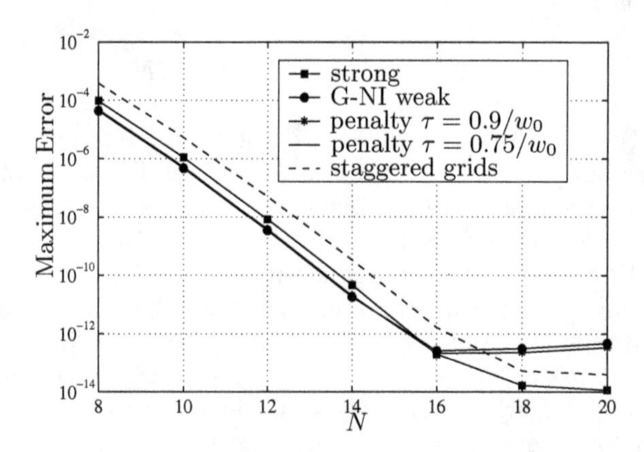


**Fig. 3.5.** Maximum error at $t = 4$ for the solution of problem (3.7.13) with different spectral schemes

cases. The staggered-grid method is the least accurate one; we refer to the discussion of the subsequent Fig. 3.10 for more comments on this method. The weak enforcement of the boundary condition (through G-NI or penalty) yields slightly better results than the strong enforcement, although a higher sensitivity to round-off errors appears as one approaches machine accuracy. The penalty scheme exhibits a quite moderate sensitivity to the parameter $\tau$, around the value corresponding to G-NI.

In order to assess the conservation properties of each scheme, we have monitored the evolution in time of the quantity

$$\begin{aligned}\Psi(t) = \left(\int_{-1}^{1} u^N(x,t)\,\mathrm{d}x + \beta \int_0^t u^N(1,s)\,\mathrm{d}s\right) \\ - \left(\int_{-1}^{1} u^N(x,0)\,\mathrm{d}x + \beta \int_0^t u_L(s)\,\mathrm{d}s\right),\end{aligned} \tag{3.7.14}$$

which is zero for the exact solution, as can be seen by integrating the equation in space and time. The results are shown in Fig. 3.6. All refer to the choice $N = 16$; Simpson's composite rule (which has the same accuracy as the RK4 time discretization) has been used to compute integrals in time. As expected, since the G-NI method is the one most consistent with the exact conservation form of the equation, this method yields the best results, although the penalty and staggered-grid methods are nearly comparable to the G-NI method in terms of conservation. All three of these methods achieve conservation to nearly the level of round-off error. The collocation method, however, results in about one significant digit loss of conservation. This suggests that strong imposition of boundary conditions should be avoided if conservation is a central issue of the numerical simulation.

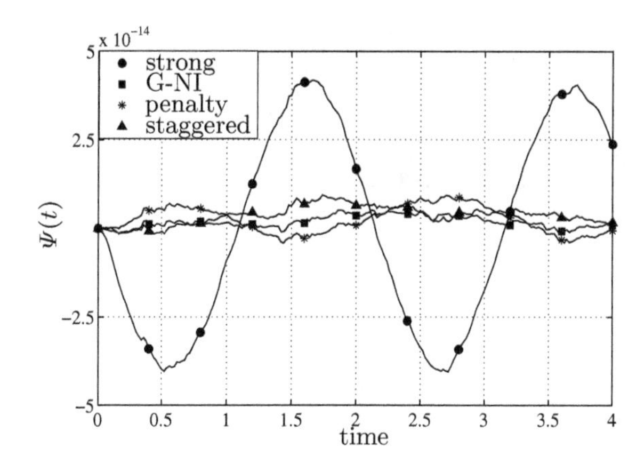


**Fig. 3.6.** Evolution in time of (the discretization of) the quantity $\Phi(t)$ defined in (3.7.14), for different spectral schemes

Next, we consider the stationary problem

$$\begin{aligned} &\frac{3}{2}\frac{\mathrm{d}u}{\mathrm{d}x} = 6\cos 6x \ , \qquad\qquad -1 < x < 1 \ , \\ &u(-1) = \sin(-6) \ , \end{aligned} \tag{3.7.15}$$

in order to investigate the spatial-discretization error alone. The exact solution is $u(x) = \sin 6x$. We have compared the collocation method (3.7.2) with strong imposition of the boundary conditions, the G-NI method (3.7.6) with weak enforcement of boundary conditions, and the penalty method (3.7.8) for two values of $\tau$. The corresponding results, reported in Fig. 3.7, show that all methods have the same convergence rate as the polynomial degree $N$ increases, and that the G-NI method is slightly more accurate than the other schemes.

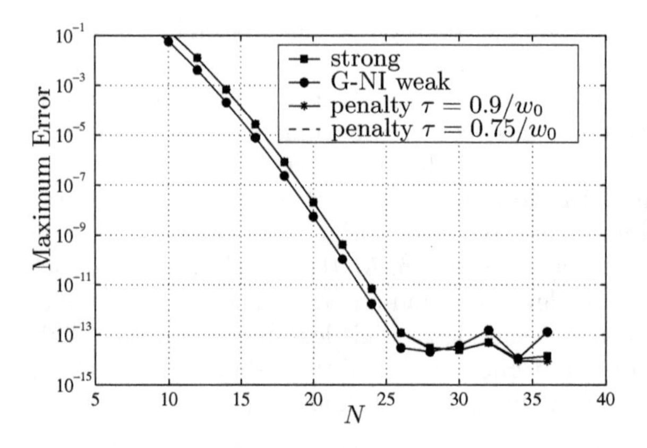


**Fig. 3.7.** Maximum error for the solution of problem (3.7.15), with different spectral schemes (note that the two penalty curves graphically coincide)

The third set of experiments compares the Chebyshev collocation methods (using only a non-staggered-grid) with high-order compact-difference schemes. Figure 3.8 illustrates the maximum errors in the discrete solution to (3.7.13) obtained with the strong scheme (3.7.2), as a function of $N$ at $t = 8$, and as a function of time for $N = 16$. (The time-step in the RK4 method was taken sufficiently small for the convergence results so that the time-discretization error was negligible; the time-step was fixed at 100 time-steps per period for the time-dependent results.) After each stage of the Runga-Kutta scheme, the boundary condition at $x = -1$ was enforced explicitly. The computed Chebyshev solution exhibits spectral accuracy as a function of $N$ and remains bounded for large $t$. Shown there for comparison are the results of several compact-difference schemes. These are (1) the classical fourth-order stencil

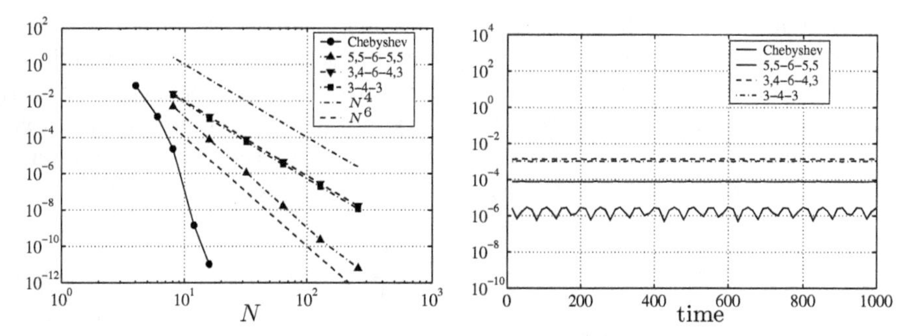


**Fig. 3.8.** Maximum error at $t = 8$ (*left*) and maximum error for $N = 16$ as a function of $t$ (*right*) for Chebyshev collocation and several compact-difference schemes for a scalar hyperbolic problem

(1.2.18) in the interior with third-order boundary stencils (3-4-3), (2) the classical sixth-order compact stencil (1.2.19) in the interior with third-order stencils at a boundary point and fourth-order stencils at a point adjacent to the boundary (3,4-6-4,3), and (3) the classical sixth-order compact stencil in the interior with fifth-order stencils at the boundary and points adjacent to the boundary (5,5-6-5,5). The stencils at and near the boundary for these schemes can be found in Carpenter, Gottlieb and Abarbanel (1993); the particular (5,5-6-5,5) stencil is given there on p. 293. The figure indicates that the global order of accuracy of these methods is one order greater than the order of the boundary stencil, and that all of these compact schemes are temporally stable, i.e., remain bounded in time (see Sect. D.1), for the scalar hyperbolic problem, as implied by the stability analysis of Carpenter, Gottlieb and Abarbanel (1993).

Figure 3.9 shows the corresponding results for some additional sixth-order compact-difference schemes. Again, all use the classical sixth-order compact stencil at interior points. The order of the stencil at a boundary point and

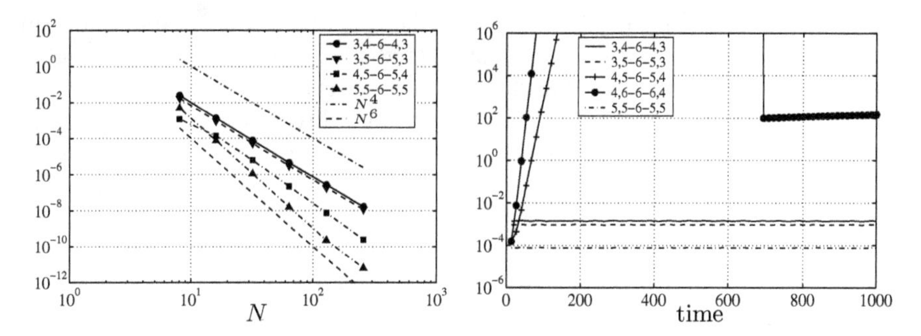


**Fig. 3.9.** Maximum error at $t = 8$ (*left*) and maximum error for $N = 16$ as a function of $t$ (*right*) for several sixth-order compact-difference schemes for a scalar hyperbolic problem

a point adjacent to the boundary point differ amongst the schemes, which are denoted by the same convention used above. The convergence rate in all cases is one order higher than the order of the boundary stencil. However, those schemes with fourth-order boundary closures are temporally unstable. (As discussed in CHQZ3, Sect. 4.2, however, all of the compact-difference methods used here are temporally unstable for a hyperbolic system.)

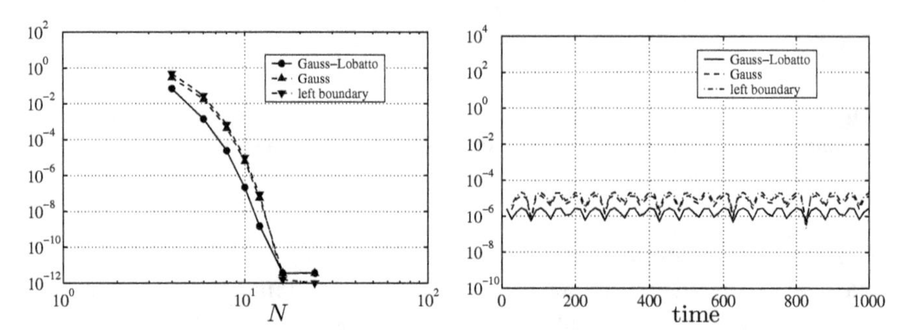


**Fig. 3.10.** Maximum error at $t = 8$ (*left*) and maximum error for $N = 16$ as a function of $t$ (*right*) for both non-staggered-grid and staggered-grid Chebyshev collocation schemes for a scalar hyperbolic problem

Finally, we compare the staggered-grid Chebyshev collocation method with its conventional, non-staggered-grid, counterpart. Figure 3.10 illustrates the convergence and temporal stability of the two Chebyshev collocation methods. Results for the staggered-grid collocation method include the maximum error at the left boundary. (In these calculations the maximum error of the staggered-grid solution interpolated to the Gauss-Lobatto points happened to always occur at the left boundary.) The staggered-grid method converges just as fast as the non-staggered-grid method, although with a slightly larger error at both the Gauss and Gauss-Lobatto points. Recall that for given $N$, the staggered-grid method has one fewer degree of freedom than the non-staggered-grid method. Even when compensation is made for this (not illustrated in the figure), the staggered-grid method still has a slightly larger error. The staggered-grid method is clearly also temporally stable. For problems with more than one nonperiodic direction, the staggered-grid method has several advantages over non-staggered-grid methods (see CHQZ3, Sect. 4.4.2).

## 3.8 Matrix Construction for Galerkin and G-NI Methods

When we apply any of the spectral techniques described so far to the spatial discretization of a linear boundary-value problem, we end up with a linear system of discrete equations (ordinary differential equations in the unsteady case, algebraic equations in the steady case). In the examples of Chap. 1, as well as in the discussion in the previous sections of the present chapter, we have written down the individual equations of the system, restraining ourselves from introducing a global matrix formalism. While this approach is often satisfactory for simple collocation and tau discretizations (for which explicit expressions for differentiation in physical or transform space may be available), it appears less appropriate for Galerkin and G-NI methods, particularly in several spatial dimensions and in complex geometries. Indeed, the modern efficient solution techniques for large algebraic systems require the access to the matrix entries, or at least the effect of applying the matrix to arbitrary vectors (see Chap. 4 for the details).

For these reasons, we present here the fundamentals of the construction of matrices arising from the Legendre Galerkin or G-NI discretization of a model second-order boundary-value problem in one spatial dimension, which incorporates variable coefficients and domain mapping. The discussion will continue in Sect. 4.2.2 for the multidimensional, single-domain case, and in CHQZ3, Chap. 5.1 for the multidomain case.

Let us assume that we want to solve the boundary-value problem

$$-\frac{\mathrm{d}}{\mathrm{d}x}\left(\alpha\frac{\mathrm{d}u}{\mathrm{d}x}\right)+\beta\frac{\mathrm{d}u}{\mathrm{d}x}+\gamma u=f \tag{3.8.1}$$

in a bounded interval, $I=(x_L,x_R)$, of the real line, supplemented by homogeneous Dirichlet or Neumann boundary conditions. The coefficients and the right-hand side are continuous functions defined in $\bar{I}=[x_L,x_R]$. The weak, or integral, formulation of the problem is

$$\int_I \alpha\frac{\mathrm{d}u}{\mathrm{d}x}\frac{\mathrm{d}v}{\mathrm{d}x}\,\mathrm{d}x+\int_I \beta\frac{\mathrm{d}u}{\mathrm{d}x}v\,\mathrm{d}x+\int_I \gamma uv\,\mathrm{d}x=\int_I fv\,\mathrm{d}x \qquad \text{for all } v\in V\,, \tag{3.8.2}$$

where the trial- and test-function space $V$ is composed of sufficiently smooth functions that satisfy homogeneous Dirichlet boundary conditions if $u$ is required to do so (technically, $V=H_0^1(-1,1)$ when Dirichlet conditions are applied; $V=H^1(-1,1)$ when Neumann conditions are applied; see Appendix A). A Galerkin method is obtained by restricting trial and test functions to a finite-dimensional space $V_N$; the G-NI version results from replacing exact integration by a high-precision quadrature formula.

In preparation for more complex situations, we assume that the interval $I$ is the image of the reference (or "parent") interval, $\hat{I}=(-1,1)$, under a smooth, invertible mapping $F$, i.e., $I=F(\hat{I})$. We denote by $\hat{x}$ the coordinate in $\hat{I}$, and by $x=F(\hat{x})$ its image in $I$. We assume that the transformation is nondegenerate, namely, $F'(\hat{x})\neq 0$ for all $\hat{x}\in[-1,1]$; we admit both the $F'>0$ and $F'<0$ cases. (An obvious instance is provided by the affine mapping $F(\hat{x})=\hat{x}(x_R-x_L)/2+(x_R+x_L)/2$.)

On the reference domain, we consider the space $\hat{V}_N$ defined as $\mathbb{P}_N^0(-1,1)=\{\hat{v}\in\mathbb{P}_N(-1,1)\,:\,\hat{v}(\pm1)=0\}$ in the case of Dirichlet boundary conditions, or

$\mathbb{P}_N(-1,1)$ in the case of Neumann boundary conditions. We use a boundary-adapted basis (see Sect. 2.3.3) for $\hat{V}_N$, which allows for an easy enforcement of the boundary/interface conditions. In particular, we choose either the modal basis defined in (2.3.30) or the Lagrange nodal basis defined in (1.2.55); Dirichlet boundary conditions are then simply enforced by dropping the first two elements from the modal basis and the first and last elements from the nodal basis. (This would not be the case if we were to choose the Legendre modal basis, $\{L_k\}_{k=0,\dots,N}$.) Compactly, we write

$$\hat{V}_N = \text{span}\,\{\hat{\phi}_k \,:\, \hat{\phi}_k \in \mathcal{B}\},$$

where $\mathcal{B}$ denotes either the modal basis or the nodal one. Trial and test functions on $I$ will be the images of the elements of $\hat{V}_N$ under the mapping $F$, i.e.,

$$V_N = \text{span}\,\{\phi_k(x) = \hat{\phi}_k(F^{-1}(x)) \,:\, \hat{\phi}_k \in \mathcal{B}\}.$$

Note that if $F$ is an affine mapping, then $V_N$ is just the space of polynomials of degree $\leq N$ on $I$, possibly vanishing on the boundary.

Setting $u^N = \sum_k u_k \phi_k \in V_N$ and choosing in (3.8.2) as $v$ any basis function $\phi_h \in V_N$, we obtain the Galerkin discretization

$$\int_I \alpha \frac{\mathrm{d}u^N}{\mathrm{d}x}\frac{\mathrm{d}\phi_h}{\mathrm{d}x}\,\mathrm{d}x + \int_I \beta \frac{\mathrm{d}u^N}{\mathrm{d}x}\phi_h\,\mathrm{d}x + \int_I \gamma u^N \phi_h\,\mathrm{d}x = \int_I f\phi_h\,\mathrm{d}x \qquad \text{for all } h\,, \tag{3.8.3}$$

which can be written in algebraic form as

$$L\mathbf{u} = \mathbf{b}\,, \tag{3.8.4}$$

where $\mathbf{u} = (u_k)$, $\mathbf{b} = \left(\int_I f\phi_h\,\mathrm{d}x\right)$ while $L = K$ is the *stiffness matrix* whose entries are

$$K_{hk} = \int_I \alpha \frac{\mathrm{d}\phi_k}{\mathrm{d}x}\frac{\mathrm{d}\phi_h}{\mathrm{d}x}\,\mathrm{d}x + \int_I \beta \frac{\mathrm{d}\phi_k}{\mathrm{d}x}\phi_h\,\mathrm{d}x + \int_I \gamma\phi_k\phi_h\,\mathrm{d}x \tag{3.8.5}$$

$$= K_{hk}^{(2)} + K_{hk}^{(1)} + K_{hk}^{(0)}\,. \tag{3.8.6}$$

The G-NI discretization is obtained by replacing each integral above by a quadrature formula, first defined on the reference interval and then transported on $I$ via the mapping $F$. In particular, suppose that the integral to be approximated is $\int_I g(x)\,\mathrm{d}x$; then, setting $\hat{g}(\hat{x}) = g(F(\hat{x})) = g(x)$, we have

$$\int_I g(x)\,\mathrm{d}x = \int_{\hat{I}} \hat{g}(\hat{x})F'(\hat{x})\,\mathrm{d}\hat{x} \simeq \sum_{j=0}^{\mathcal{N}} \hat{g}(\hat{x}_j)F'(\hat{x}_j)\hat{w}_j\,, \tag{3.8.7}$$

where $(\hat{x}_j, \hat{w}_j)$, $j = 0,\dots,\mathcal{N}$, are the nodes and weights of a suitable Gaussian quadrature formula on $\hat{I}$. In the subsequent discussion, we will invariably use the $N$-th order Legendre Gauss-Lobatto formula (i. e., we set $\mathcal{N} = N$); choosing $\mathcal{N} > N$ leads to a more accurate integration at some extra cost, which can be desirable in the presence of variable coefficients and variable Jacobian. (See Maday and Rønquist (1990) for a discussion of several quadrature strategies in the construction of stiffness matrices.) The resulting approximate stiffness matrix will be denoted by $L = K_{GNI}$.

### 3.8.1 Matrix Elements

We now detail the construction of the individual matrix elements. We treat the zeroth-, first- and second-order contributions separately, as well as the right-hand side. In the discussion, we assume Neumann boundary conditions, i.e., we include the vertex functions in the basis; in the case of homogeneous Dirichlet boundary conditions, these functions are not included, implying that the first and last rows and columns of the matrices below are deleted.

#### *Zeroth-order contributions*

We have

$$K_{hk}^{(0)} = \int_I \gamma(x)\phi_k(x)\phi_h(x)\,\mathrm{d}x = \int_{\hat{I}} \gamma^*(\hat{x})\hat{\phi}_k(\hat{x})\hat{\phi}_h(\hat{x})\,\mathrm{d}\hat{x}\,, \qquad (3.8.8)$$

with $\gamma^*(\hat{x}) = \gamma(F(\hat{x}))F'(\hat{x})$. Note that if $\gamma \equiv 1$, then $K^{(0)} = M$ coincides with the mass matrix of the chosen basis.

If $\gamma^*$ is constant, the use of the modal basis yields the sparsity pattern indicated in Fig. 3.11 (left). The pentadiagonal internal structure is easily derived from the expression (2.3.31) for the internal basis functions. Indeed, assuming $\gamma^* = 1$, the nonzero entries of the matrix, in its upper triangular part, are

$$K_{00}^{(0)} = K_{11}^{(0)} = \frac{2}{3}\,, \qquad K_{01}^{(0)} = \frac{1}{3}\,, \qquad (3.8.9a)$$

$$K_{02}^{(0)} = K_{12}^{(0)} = \frac{1}{\sqrt{6}}\,, \qquad K_{03}^{(0)} = -K_{13}^{(0)} = -\frac{1}{3\sqrt{10}}\,, \qquad (3.8.9b)$$

and, for $2 \le h \le k \le N$ ,

$$K_{hk}^{(0)} = \begin{cases} \dfrac{2}{(2h-3)(2h+1)}\,, & k = h\,, \\ -\dfrac{1}{(2h+1)\sqrt{(2h-1)(2h+3)}}\,, & k = h+2\,. \end{cases} \qquad (3.8.9c)$$

Since even and odd internal modes are decoupled, two tridiagonal matrices of half the size can be built instead. In all cases, the computational cost is $O(N)$ operations.

If $\gamma^*$ is a generic function, then $K^{(0)}$ is full. In this case, it is preferable to resort to the G-NI approximation

$$(K_{GNI}^{(0)})_{hk} = \sum_{j=0}^{N} \gamma^*(\hat{x}_j)\hat{\phi}_k(\hat{x}_j)\hat{\phi}_h(\hat{x}_j)\hat{w}_j \,. \qquad (3.8.10)$$

The use of the modal basis yields again a full matrix, which can be computed in $O(N^3)$ operations; note that the nodal values of the modal basis functions can be obtained using (2.3.31) and the recurrence relation (2.3.3). On the other hand, if the nodal basis is used, one has $\hat{\phi}_k(\hat{x}_j) = \delta_{kj}$ by definition; hence, the matrix is diagonal,

$$(K_{GNI}^{(0)})_{hk} = \gamma^*(\hat{x}_h)\hat{w}_h\delta_{hk} \qquad (3.8.11)$$

(realizing in this way the so-called *mass-lumping*), and obviously the cost of its construction is $O(N)$ operations.

### *First-order contributions*

Since $\dfrac{d\phi_k}{dx} = \dfrac{d\hat{\phi}_k}{d\hat{x}}\dfrac{d\hat{x}}{dx} = F'(\hat{x})^{-1}\dfrac{d\hat{\phi}_k}{d\hat{x}}$, we have

$$K_{hk}^{(1)} = \int_I \beta(x)\frac{d\phi_k}{dx}(x)\phi_h(x)\,dx = \int_{\hat{I}} \beta^*(\hat{x})\frac{d\hat{\phi}_k}{d\hat{x}}(\hat{x})\hat{\phi}_h(\hat{x})\,d\hat{x} \,, \qquad (3.8.12)$$

with $\beta^*(\hat{x}) = \beta(F(\hat{x}))$. If $\beta^*$ is constant and the modal basis is used, the resulting matrix has the tridiagonal internal structure indicated in Fig. 3.11 (*center*); its construction requires $O(N)$ operations. Indeed, assuming $\beta^* = 1$, the nonzero entries of the matrix are given by

$$K_{00}^{(1)} = -K_{11}^{(1)} = -\frac{1}{2}\,, \qquad K_{01}^{(1)} = -K_{10}^{(1)} = \frac{1}{2}\,, \qquad (3.8.13a)$$

$$K_{02}^{(1)} = -K_{20}^{(1)} = \frac{1}{\sqrt{6}}\,, \qquad K_{12}^{(1)} = -K_{21}^{(1)} = -\frac{1}{\sqrt{6}}\,, \qquad (3.8.13b)$$

and, for $2 \le h, k \le N$,

$$K_{h,h+1}^{(1)} = -K_{h+1,h}^{(1)} = \frac{1}{\sqrt{4h^2-1}}\,. \qquad (3.8.13c)$$

For a general $\beta^*$, the G-NI approximation yields

$$(K_{GNI}^{(1)})_{hk} = \sum_{j=0}^{N} \beta^*(\hat{x}_j)\frac{d\hat{\phi}_k}{d\hat{x}}(\hat{x}_j)\hat{\phi}_h(\hat{x}_j)\hat{w}_j \,, \qquad (3.8.14)$$

which simplifies to

$$(K_{GNI}^{(1)})_{hk} = \beta^*(\hat{x}_h)\hat{w}_h(D_N)_{hk} \,, \qquad (3.8.15)$$

where $(D_N)_{hk}$ is defined in (2.3.28), if the nodal basis is used. Note that with both bases the matrix has a full structure, and its construction requires $O(N^3)$ operations with the modal basis and $O(N^2)$ operations with the nodal basis.

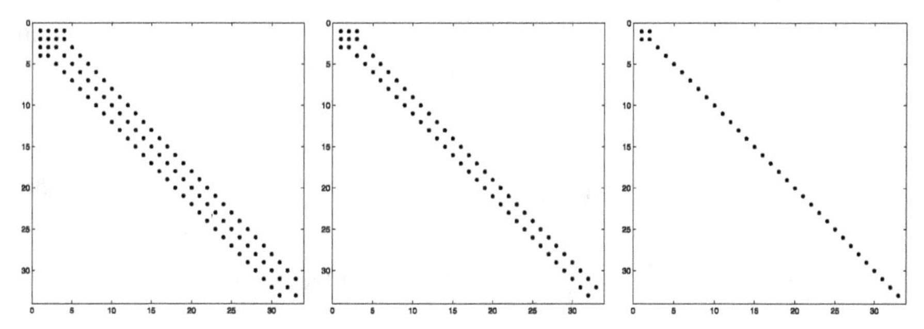

**Fig. 3.11.** Sparsity patterns of the mass and stiffness matrices for the modal basis (2.3.30), $N = 32$: mass matrix (*left*), first-derivative stiffness matrix (*center*), second-derivative stiffness matrix (*right*)

### *Second-order contributions*

We have

$$K_{hk}^{(2)} = \int_I \alpha(x) \frac{\mathrm{d}\phi_k}{\mathrm{d}x}(x) \frac{\mathrm{d}\phi_h}{\mathrm{d}x}(x)\,\mathrm{d}x = \int_{\hat{I}} \alpha^*(\hat{x}) \frac{\mathrm{d}\hat{\phi}_k}{\mathrm{d}\hat{x}}(\hat{x}) \frac{\mathrm{d}\hat{\phi}_k}{\mathrm{d}\hat{x}}(\hat{x})\,\mathrm{d}\hat{x}\ , \tag{3.8.16}$$

with $\alpha^*(\hat{x}) = \alpha(F(\hat{x}))F'(\hat{x})^{-1}$. If $\alpha^*$ is constant and the modal basis is used, the resulting matrix has the internal diagonal structure indicated in Fig. 3.11 (right). Precisely, assuming $\alpha^* = 1$, the nonzero entries of the matrix are

$$K_{01}^{(2)} = -\frac{1}{2}\ , \qquad K_{hh}^{(2)} = \begin{cases} \frac{1}{2}\ , & h = 0 \ \text{ or } \ h = 1\ , \\ 1\ , & 2 \le h \le N\ . \end{cases} \tag{3.8.17}$$

On the other hand, if $\alpha^*$ is variable and one resorts to the G-NI approximation, one has

$$(K_{GNI}^{(2)})_{hk} = \sum_{j=0}^{N} \alpha^*(\hat{x}_j) \frac{\mathrm{d}\hat{\phi}_k}{\mathrm{d}\hat{x}}(\hat{x}_j) \frac{\mathrm{d}\hat{\phi}_h}{\mathrm{d}\hat{x}}(\hat{x}_j)\hat{w}_j\ , \tag{3.8.18}$$

where the grid-point values of the derivatives are easily computed by (2.3.30) or by (2.3.28) according to the chosen basis. In both cases, the matrix is full, and its construction requires $O(N^3)$ operations.

### *Right-hand side*

We have

$$b_h = \int_I f(x)\phi_h(x)\,\mathrm{d}x = \int_{\hat{I}} f^*(\hat{x})\hat{\phi}_h(\hat{x})\,\mathrm{d}\hat{x}\ , \tag{3.8.19}$$

with $f^*(\hat{x}) = f(F(\hat{x}))F'(\hat{x})$. Unless $f^*$ has a particular polynomial expression (in which case the modal basis may give some advantage), it is preferable to approximate $b_h$ by the quantity $\tilde{b}_h$ defined via the quadrature formula

$$\tilde{b}_h = \sum_{j=0}^{N} f^*(\hat{x}_j)\hat{\phi}_h(\hat{x}_j)\hat{w}_j \ , \tag{3.8.20}$$

which simplifies into $\tilde{b}_h = f^*(\hat{x}_h)\hat{w}_h$ if the nodal basis is used.

### 3.8.2 An Example of Algebraic Equivalence between G-NI and Collocation Methods

Take $I = \hat{I}$ above and consider the homogeneous Dirichlet problem

$$\begin{aligned} -\frac{\mathrm{d}^2 u}{\mathrm{d}x^2} + \gamma u = f \qquad & \text{in } (-1,1) \ , \\ u(-1) = u(1) = 0 \ . & \end{aligned} \tag{3.8.21}$$

The standard Legendre G-NI method defines an approximation $u^N$ of $u$ as a polynomial in $\mathbb{P}_N^0(-1,1) = \{v \in \mathbb{P}_N \ : \ v(\pm 1) = 0\}$ of the form $u^N(x) = \sum_{k=1}^{N-1} u_k \psi_k(x)$, where the $\psi_k$'s are the characteristic Lagrange polynomials at the internal LGL nodes (2.3.12). The vector $\mathbf{u} = (u_k)$ is the solution of the algebraic system

$$K_{GNI}\mathbf{u} = \mathbf{b} \ , \tag{3.8.22}$$

which is obtained from (3.8.4) by applying the LGL quadrature formula to compute all integrals. Thus, $K_{GNI} = K_{GNI}^{(2)} + K_{GNI}^{(0)}$, with

$$(K_{GNI}^{(2)})_{hk} = \left(\frac{\mathrm{d}\psi_k}{\mathrm{d}x}, \frac{\mathrm{d}\psi_h}{\mathrm{d}x}\right)_N \quad \text{and} \quad (K_{GNI}^{(0)})_{hk} = (\gamma\psi_k, \psi_h)_N \ ,$$

where $(\cdot,\cdot)_N$ is the LGL discrete inner product and $1 \le h, k \le N-1$. On the other hand, $\mathbf{b} = (b_h)$ with $b_h = (f, \psi_h)_N = \sum_{k=1}^{N-1} (\psi_k, \psi_h)_N f(x_k)$ for all $h$; thus, $\mathbf{b} = M_{GNI}\mathbf{f}$, where $M_{GNI} = (\,(\psi_k, \psi_h)_N\,) = \mathrm{diag}(w_1, \dots, w_{N-1})$ is the lumped mass matrix and $\mathbf{f} = (f(x_k))$ is the vector of the nodal values of $f$. Consequently, (3.8.22) can be written as

$$K_{GNI}\mathbf{u} = M_{GNI}\mathbf{f} \ , \tag{3.8.23}$$

or, equivalently, as

$$M_{GNI}^{-1}K_{GNI}\mathbf{u} = \mathbf{f} \ .$$

The matrix on the left-hand side has a very precise meaning. To grasp it, observe that the exactness of the quadrature formula yields

$$
\begin{aligned}
(K_{GNI}^{(2)})_{hk} = \int_{-1}^{1} \frac{\mathrm{d}\psi_k}{\mathrm{d}x}\frac{\mathrm{d}\psi_h}{\mathrm{d}x}\,\mathrm{d}x &= -\int_{-1}^{1} \frac{\mathrm{d}^2\psi_k}{\mathrm{d}x^2}\psi_h\,\mathrm{d}x \\
&= -\left(\frac{\mathrm{d}^2\psi_k}{\mathrm{d}x^2}, \psi_h\right)_N = -w_h \frac{\mathrm{d}^2\psi_k}{\mathrm{d}x^2}(x_h)\,.
\end{aligned}
$$

Since $(K_{GNI}^{(0)})_{hk} = w_h\gamma(x_h)\delta_{hk}$, we obtain

$$
(K_{GNI})_{hk} = w_h\left(-\frac{\mathrm{d}^2\psi_k}{\mathrm{d}x^2}(x_h) + \gamma(x_h)\delta_{hk}\right) = w_h\left(-(D_N^{(2)})_{hk} + \gamma(x_h)\delta_{hk}\right)\,,
$$

where $D_N^{(2)}$ is the second-derivative matrix at the LGL-nodes, defined in (2.3.29). The term in brackets on the right-hand side is the entry $(L_{\mathrm{coll}})_{hk}$ of the matrix

$$
L_{\mathrm{coll}} = -\tilde{D}_N^{(2)} + \mathrm{diag}(\gamma(x_1), \dots, \gamma(x_{N-1}))
$$

(where $\tilde{D}_N^{(2)}$ is obtained from $D_N^{(2)}$ by deleting the first and last rows and columns, due to the boundary conditions), which corresponds to the *collocation* discretization of our problem. In other words, we have proven the relation

$$
K_{GNI} = M_{GNI}L_{\mathrm{coll}}\,, \qquad \text{i.e.,} \qquad L_{\mathrm{coll}} = M_{GNI}^{-1}K_{GNI}\,, \tag{3.8.24}
$$

which shows that the G-NI system (3.8.23) is equivalent to the collocation system

$$
L_{\mathrm{coll}}\mathbf{u} = \mathbf{f}\,. \tag{3.8.25}
$$

The same conclusions hold if we discretize the more general operator (3.8.1), again under Dirichlet boundary conditions.

The results just established are consistent with the fact, already observed in Sects. 1.2.3 (see (1.2.64)) and 3.3.5 (see (3.3.33)), that G-NI and collocation methods enforce the differential equation in the same (pointwise) manner at all *internal* LGL nodes; they may enforce different equations only at the boundary points (as for the weak or strong enforcement of a Neumann boundary condition). In the present case, however, both methods enforce the Dirichlet conditions exactly; consequently, they produce the *same* discrete solution.

The difference between (3.8.23) and (3.8.25) becomes apparent at the moment of solving the algebraic system: the matrix $K_{GNI}$ is symmetric and positive definite, whereas $L_{\mathrm{coll}}$ is not; in addition, the former matrix is better conditioned than the latter (see Sects. 4.3.1 and 7.3). These features have an impact on the solution techniques, as discussed in the next chapter.

## 3.9 Polar Coordinates

This section provides some basic material on spectral methods in polar coordinates. See Boyd (2001) and Fornberg (1996) for more comprehensive treatments and especially for their discussions on spectral methods in spherical coordinates. There are no particular subtleties for problems in an annulus. We focus here on Poisson's equation in a disk:

$$\begin{aligned} -\Delta u &= f\,, \qquad && 0 < r < 1\,,\ 0 \le \theta < 2\pi\,, \\ u &= 0\,, && r = 1\,, \end{aligned} \tag{3.9.1}$$

which presents the challenge of a coordinate singularity. A standard Fourier expansion in $\theta$, either Galerkin or collocation, is clearly in order. The numerical solution may be written

$$u(r,\theta) = \sum_{m=-M/2}^{M/2-1} \tilde{u}_m(r) e^{im\theta}\,. \tag{3.9.2}$$

There have been several proposals for Chebyshev expansions in radius. One of these is

$$\tilde{u}_m(r) = \sum_{\substack{n=0 \\ n+m \text{ even}}}^{N} a_{mn} T_n(r)\,. \tag{3.9.3}$$

Thus, the numerical solution to (3.9.1) will have the same parity, $\tilde{u}_m(-r) = (-1)^m \tilde{u}_m(r)$, as the analytic one. A further refinement (Orszag and Patera (1983)) is to incorporate the decay of $u(r,\theta)$ near the origin by using

$$\tilde{u}_m(r) = r^m \sum_{\substack{n=0 \\ n+m \text{ even}}}^{N} a_{mn} T_n(r)\,. \tag{3.9.4}$$

Both of these expansions have better resolution near the outer edge than near the origin, as is evident from the concentration of the zeroes of $T_n(r)$ near the edge. Improved center resolution can be achieved by expanding in

$$x = 2r - 1 \tag{3.9.5}$$

and using all of the Chebyshev polynomials.

These expansions must satisfy the condition

$$\frac{\partial u}{\partial \theta} = 0 \tag{3.9.6}$$

at the origin. This expresses the requirement that the solution be single-valued. Obviously, this requires

$$\tilde{u}_m(r) = 0 \qquad \text{at } r = 0 \text{ for } m \neq 0\,. \tag{3.9.7}$$

The appropriate condition on the remaining component is

$$\frac{d\tilde{u}_0}{dr} = 0 \qquad \text{at } r = 0 \,. \tag{3.9.8}$$

These latter two conditions are readily applied in a tau approach. Note that the expansion (3.9.4) automatically satisfies (3.9.7) and (3.9.8).

When $\mathbf{u}$ is a vector quantity, such as velocity, the necessary condition at the origin is

$$\frac{\partial \mathbf{u}}{\partial \theta} = 0 \,. \tag{3.9.9}$$

In polar coordinates, $\mathbf{u} = u_r\hat{\mathbf{r}} + u_\theta\hat{\boldsymbol{\theta}}$, where $\hat{\mathbf{r}}$ and $\hat{\boldsymbol{\theta}}$ are the unit vectors in the radial and azimuthal directions, and $u_r$ and $u_\theta$ are the respective velocity components. These unit vectors depend upon $\theta$, and this dependence must be included in applying (3.9.9). The result is

$$\begin{aligned} u_{r,m} = u_{\theta,m} &= 0 \qquad \text{for } |m| \neq 1 \\ u_{r,m} + imu_{\theta,m} &= 0 \qquad \text{for } |m| = 1 \,. \end{aligned} \tag{3.9.10}$$

These types of boundary conditions at the origin have been used (for mixed spectral/finite-difference calculations) by Schnack and Killeen (1980) and by Aydemir and Barnes (1984). They have been justified theoretically (for mixed spectral/finite-element calculations) by Mercier and Raugel (1982).

The expansions (3.9.3) and (3.9.4) are not well suited to pure collocation methods because there would need to be different collocation points in $r$ for the even $m$ and odd $m$ components. One needs a Fourier Galerkin-Chebyshev collocation method.

Suppose now that a standard Chebyshev expansion is combined with the mapping (3.9.5). If the Gauss-Lobatto points are used, then conditions such as (3.9.6) and (3.9.9) need to be imposed at $r = 0$ (or $x = -1$). Alternatively, one can use the Gauss-Radau points which include the point $r = 1$ (or $x = 1$) but exclude the origin. There is then no need to impose a boundary condition at $r = 0$.

## 3.10 Aliasing Effects

In Sect. 2.1.2 we noted that the discrete Fourier coefficients of a function are not identical to the continuous ones (see (2.1.36)). The difference is attributable to the aliasing phenomenon (see Fig. 2.2). Hence, the principal difference between Galerkin and collocation (or pseudospectral) methods is the presence of truncation error alone in the former versus the presence of both truncation and aliasing errors in the latter. The question of whether the additional aliasing errors in the collocation methods are indeed serious has been controversial, particularly in the early years of spectral methods.

The two most pertinent issues are the effects of aliasing upon the accuracy and, in evolution problems, the temporal stability of the calculation. The role of discrete conservation laws in assuring temporal stability was discussed in Sect. 3.6. Here we summarize the available theoretical results on the effects of aliasing upon accuracy, many of which are discussed at greater length elsewhere in this book.

Many approximation theory results are presented in Chap. 5 for Fourier, Legendre and Chebyshev series. Compare, for example, the Fourier Galerkin (truncation) and collocation (interpolation) approximation error bounds given in the $L^2$ norm by (5.1.9) and (5.1.16), respectively. These imply that although for fixed $N$ the collocation error will be larger than the Galerkin error, both errors exhibit the same asymptotic decay rate for large $N$. The Legendre Galerkin and collocation estimates are furnished in (5.4.11) and (5.4.33); the Chebyshev ones are (5.5.9) and (5.5.22). In the case of the Legendre polynomial approximation, the collocation approximation has an asymptotic error decay rate which is slower, by a factor of $\sqrt{N}$, than the rate of the Galerkin approximation. If the function has $m$ derivatives, then the Galerkin error decays as $N^{-m}$, whereas the collocation error decays as $N^{1/2-m}$. For smooth functions, this should be a very minor difference, once there are enough polynomials to resolve the essential structure. Nevertheless for marginally resolved cases we do anticipate more difficulty with aliasing in spectral approximations to nonperiodic problems than for periodic ones.

A number of theoretical results are available on the effect of aliasing upon solutions of differential equations by spectral methods. Kreiss and Oliger (1979) proved that the aliasing error decays at the same rate as the truncation error in Fourier approximations for the one-dimensional, linear wave equation. The spectral Galerkin (de-aliased) and collocation (aliased) approximations to the steady Burgers equation have the same asymptotic error decay rate, as discussed in Sect. 7.8. The theory of spectral approximations to the steady three-dimensional, Navier-Stokes equations states, too, that Galerkin (de-aliased) and collocation (aliased) approximations behave similarly in the asymptotic regime. This holds for Fourier, Chebyshev and Legendre approximations. The details of this analysis are supplied in CHQZ3, Sect. 3.7. Although the theoretical results of Sects. 7.8 and CHQZ3, Sect. 3.7 refer to steady cases only, the same conclusions can be rigorously drawn for the unsteady situations as well, since the source of the aliasing error is in the spatial terms of the equations.

On the other hand, Goodman, Hou and Tadmor (1994) considered the one-dimensional, variable-coefficient, linear wave equation in the form (3.6.8) with $\mathcal{M}(u) = \partial\,(a(x)u)\,/\partial x$. They proved that aliasing produces a slow, secular growth of the high-frequency modes if the coefficient $a(x)$ changes sign in the spatial domain. This can produce unacceptable errors unless the solution is well-resolved. Further details are provided in Sect. 7.6.1.

Thus, there is reasonable theoretical support for the claim that for any given problem, an aliased calculation will yield just as acceptable an answer as a de-aliased one, once sufficient resolution has been achieved. Moreover, the use of appropriate conservation forms for the discrete equations, as discussed in general in Sect. 3.6, ameliorates some temporal instabilities that may otherwise arise in aliased calculations. All the credible numerical evidence that we have seen supports this view.

However, in some applications of spectral methods, such as fully turbulent flow, the "sufficient resolution" threshold has been impractical due to computer resource limitations. Many turbulence flow computations have had only marginal resolution of the small scales. Some perspectives on the impact of aliasing in these circumstances is provided in CHQZ3, Chap. 3.

# 4. Algebraic Systems and Solution Techniques

The solution of implicit equations is an important component of many spectral algorithms. For steady problems this task is unavoidable, while spectral algorithms for many unsteady problems are only feasible if they incorporate implicit (or semi-implicit) time discretizations (see Appendix D for general information about time discretizations and Sect. 3.3 and CHQZ3, Chap. 3 for some uses of implicit time discretizations with spectral discretizations in space). We concentrate on linear systems, assuming that nonlinear ones are attacked by standard linearization techniques.

We focus primarily on problems involving the constant-coefficient Helmholtz equation

$$-\Delta u + \lambda u = f \tag{4.1}$$

on a $d$-dimensional ($d = 1, 2, 3$), tensor-product domain $\Omega \subset \mathbb{R}^d$, where $f$ is a function of $\mathbf{x}$, and $\lambda \geq 0$ is a constant. The simplest generalizations are to the self-adjoint, variable-coefficient form

$$-\nabla \cdot (a \nabla u) + \lambda u = f \ , \tag{4.2}$$

where $a > 0$ is a function of $\mathbf{x}$, and to the separable form

$$-\sum_{i=1}^{d} \frac{\partial}{\partial x_i} \left[ a_i(x_i) \frac{\partial u}{\partial x_i} \right] + \lambda u = f \ , \tag{4.3}$$

where now $a_i > 0$ is a function solely of $x_i$. More complex generalizations are

$$-\sum_{i=1}^{d} \frac{\partial}{\partial x_i} \left[ a_i \frac{\partial u}{\partial x_i} \right] + \lambda u = f \ , \tag{4.4}$$

where $a_i > 0$ is a function of $\mathbf{x}$, and

$$-\sum_{i=1}^{d} g_i \frac{\partial}{\partial x_i} \left[ a \ g_i \frac{\partial u}{\partial x_i} \right] + \lambda u = f \ , \tag{4.5}$$

where $a > 0$ is a function of $\mathbf{x}$, and $g_i > 0$ is a function of $x_i$. The form (4.4) is nonseparable. The case (4.5) arises, for example, when mappings are employed; $g_i$ then is is the inverse of the Jacobian of the mapping in the coordinate $x_i$, $i = 1, \ldots, d$.

We give occasional consideration to the advection-diffusion equation

$$-\nu \Delta u + \boldsymbol{\beta} \cdot \nabla u = f \tag{4.6}$$

and to the general (advection-diffusion-reaction) equation

$$-\sum_{i,j=1}^{d} \frac{\partial}{\partial x_i}\left(\alpha_{ij}\frac{\partial u}{\partial x_j}\right) + \sum_{i=1}^{d} \beta_i \frac{\partial u}{\partial x_i} + \gamma u = f \ . \tag{4.7}$$

All equations are, of course, subject to appropriate boundary conditions. Equation (4.1) contains as special cases the steady incompressible potential equation and implicit temporal discretizations of the heat equation.

All spectral discretization methods lead to a linear system of the form

$$L\mathbf{u} = \mathbf{b} \ . \tag{4.8}$$

For spectral collocation approximations and for most G-NI approximations the vector $\mathbf{u}$ consists of the grid-point values of $u^N$ (the discrete solution). The vector $\mathbf{b}$ collects the grid-point values of $f$ and all boundary data in a collocation approximation, whereas it is obtained from this vector upon multiplication by a suitable matrix in a G-NI approximation; the matrix $L$ is obtained from the nodal basis. For tau approximations and for most Galerkin approximations, $\mathbf{u}$ is the vector consisting of the expansion coefficients of $u^N$, whereas $\mathbf{b}$ collects the expansion coefficients of $f$ and the boundary data; $L$ is a matrix usually obtained from the modal basis.

The linear systems arising from (4.2)–(4.7) are usually full, albeit blockwise full and reasonably sparse in 2D and 3D. Gaussian elimination may, in principle, be applied. However, except for special cases for which efficient ad-hoc algorithms exist, solution of the linear systems requires $O(N^{3d})$ operations and $O(N^{2d})$ storage, where $d$ is the dimension of the problem. (We assume, for simplicity, that the number of degrees of freedom in *each* spatial dimension is $N$.)

In the first section of this chapter we discuss some direct techniques to very special problems, which in the case of Fourier and Chebyshev methods, where the FFT can be exploited, can yield the solution to (4.8) in $O(N^d)$, $O(N^d \log_2 N)$ or at worst $O(N^{d+1})$ operations with at most $O(N^d)$ additional storage. This is followed in Sect. 4.2 by a description of general-purpose direct methods; we describe the matrix structure produced by spectral methods owing to their tensor-product nature, and discuss Gaussian matrix factorization techniques.

Next, we briefly describe the eigenvalue structure of simple spectral operators, as they have important implications on the convergence properties of iterative methods and on the stability conditions of time-discretization schemes.

The three sections which follow are devoted to iterative techniques and to the critical issue of how to devise efficient low-cost preconditioners for spectral discretization matrices. We will review three families of iterative methods: descent methods, Krylov methods and spectral multigrid methods. The residual computation requires $O(N^d \log_2 N)$ operations per iteration for Chebyshev methods, $O(N^{d+1})$ for Legendre methods, and $O(N^d)$ additional storage. Finally, a comparison of the performance (accuracy, memory storage and CPU-time) of direct and iterative methods on some test problems is provided.

The discussion in the text of iterative methods presumes that the reader is familiar with the standard iterative schemes for linear systems such as minimum residual, steepest descent, conjugate gradient, generalized minimum residual and bi-conjugate gradient methods. Appendix C furnishes notation, algorithms and convergence properties for these schemes. The discussion in the text itself is confined to only those aspects of these iterative methods that are particularly relevant to linear systems arising from spectral discretizations (in space) of partial differential equations.

## 4.1 Ad-hoc Direct Methods

Our objectives in this section are to explain the principles underlying the basic direct techniques, to illustrate these on some specific problems that arise in practice, and to summarize the literature on more specialized applications.

We shall call a solution *efficient* if it enables the solution to (4.8) to be obtained in at most $O(N^d \log_2 N)$ operations. This makes the cost of solving (4.8) comparable, even for $N$ large, to the cost of typical explicit spectral operations such as differentiation and the evaluation of convolution sums. In many cases, a solution cost of $O(N^{d+1})$ is still acceptable in the sense that it only overwhelms the cost of other spectral operations for values of $N$ of 128 or so.

An important consideration is whether only a few or else a large number of solutions to (4.8) with different data $\mathbf{b}$ are sought. The latter case is typical of implicit or semi-implicit methods for unsteady problems: hundreds or even thousands of solutions to a linear system with the same left-hand-side but different right-hand-sides might be required. In such situations, it is reasonable to invest a substantial amount of calculations on a pre-processing stage that greatly reduces the subsequent cost of solving (4.8). The matrix-diagonalization techniques presented in Sect. 4.1.4 belong to this category. The discussion of Fourier, Chebyshev and Legendre methods in Sects. 4.1.1 and 4.1.3 is concerned with techniques for furnishing a solution to a single implicit equation. Naturally, they may also be employed in unsteady algorithms as well.

### 4.1.1 Fourier Approximations

The discussion will open with the simplest case – a one-dimensional, constant-coefficient, periodic problem:

$$\begin{aligned} -\frac{\mathrm{d}^2u}{\mathrm{d}x^2} + \lambda u = f \qquad & \text{in } (0, 2\pi) , \\ u \;\; 2\pi\text{-periodic} . & \end{aligned} \tag{4.1.1}$$

The Fourier Galerkin approximation takes the form

$$k^2 \hat{u}_k + \lambda \hat{u}_k = \hat{f}_k , \qquad k = -\frac{N}{2}, \dots, \frac{N}{2} - 1 , \tag{4.1.2}$$

where the Fourier coefficients $\hat{u}_k$ are defined by (2.1.3) and the corresponding truncated Fourier series by (2.1.7). The solution to (4.1.2) is trivially

$$\begin{aligned} \hat{u}_k = \hat{f}_k/(k^2 + \lambda) , \qquad & k = -\frac{N}{2}, \dots, \frac{N}{2} - 1 , \\ (\hat{u}_0 \text{ arbitrary for } \lambda = 0) & \end{aligned} \tag{4.1.3}$$

(where $\hat{f}_{-N/2} = 0$) with an operation count of $3N$, presuming $u$ is real, so that $\hat{u}_{-k} = \overline{\hat{u}_k}$.

A Fourier collocation approximation is (with $x_j$ given by (2.1.24))

$$-\frac{\mathrm{d}^2u}{\mathrm{d}x^2} + \lambda u - f \bigg|_{x=x_j} = 0 , \qquad j = 0, \dots, N-1 . \tag{4.1.4}$$

This may be solved by using the discrete Fourier transform (DFT) to diagonalize (4.1.4):

$$k^2 \tilde{u}_k + \lambda \tilde{u}_k = \tilde{f}_k , \qquad k = -\frac{N}{2}, \dots, \frac{N}{2} - 1 , \tag{4.1.5}$$

where the discrete Fourier coefficients $\tilde{u}_k$ and $\tilde{f}_k$ are defined by (2.1.25), then solving for $\tilde{u}_k$ as in (4.1.3), and finally reversing the discrete Fourier Transform to recover $u_j$ for $j = 0, 1, \dots, N-1$. The operation count for the direct solution of (4.1.4) is $5N \log_2 N$ real operations, with the FFT used to accomplish the discrete Fourier transform. (For those cases in which detailed operation counts are provided we count addition, subtraction, multiplication and division as separate operations. Lower order terms in the operation counts, such as those linear in $N$ in this case, are ignored unless they have especially large coefficients.)

Both the Galerkin and collocation approximations to the constant-coefficient Helmholtz problem (4.1) in more than one dimension are equally straightforward and efficient ($O(N^d \log_2 N)$ operations).

The problem

$$
\begin{aligned}
-\frac{\mathrm{d}}{\mathrm{d}x}\left[a(x)\frac{\mathrm{d}u}{\mathrm{d}x}\right] + \lambda u = f \qquad & \text{in } (0,2\pi)\,, \\
u \ \ 2\pi\text{-periodic}\,, &
\end{aligned}
\tag{4.1.6}
$$

represents the next level of complexity. The collocation approximation to (4.1.6) may be written in the form (4.8) with

$$
L = -D_N\, A\, D_N + \lambda I\,, \tag{4.1.7}
$$

where $D_N$ is the matrix given explicitly by (2.1.51), $A$ is the diagonal matrix representing multiplication by $a(x)$ in physical space, and $I$ is the identity matrix. An alternative expression to (2.1.51) for $D_N$ is

$$
D_N = C^{-1} K C\,, \tag{4.1.8}
$$

where

$$
C_{kj} = \frac{1}{N} e^{-ikx_j}\,, \qquad k = -\frac{N}{2},\ldots,\frac{N}{2}-1,\ j = 0,\ldots,N-1\,, \tag{4.1.9}
$$

represents the DFT and

$$
\begin{aligned}
K = \operatorname{diag}\{ik'\}\,, \qquad & k = -\frac{N}{2},\ldots,\frac{N}{2}-1\,, \\
k' = \begin{cases} k\,, & k = -\dfrac{N}{2}+1,\ldots,\dfrac{N}{2}-1\,, \\ 0\,, & k = -\dfrac{N}{2}\,, \end{cases} &
\end{aligned}
\tag{4.1.10}
$$

represents differentiation in transform space. Equation (4.1.6) admits an efficient direct solution only if $\lambda = 0$ or if the Fourier series of $a(x)$ contains just a few low-order terms.

In the former case we have for (4.8) that

$$
-C^{-1}KCAC^{-1}KC\mathbf{u} = \mathbf{b}\,, \tag{4.1.11}
$$

which is equivalent to

$$
\mathbf{u} = -C^{-1}K^{-1}CA^{-1}C^{-1}K^{-1}C\mathbf{b}\,. \tag{4.1.12}
$$

(Recall that $\mathbf{b}$ represents the grid-point values of $f$ for this collocation approximation.) Although $K$ is technically singular – because of the $k = 0$ and $k = -(N/2)$ components – this merely reflects the non-uniqueness of the problem. The offending Fourier components may be assigned arbitrary values. The solution procedure described by (4.1.12) involves four FFTs and three multiplications, for a total cost of $10N\log_2 N$.

In the latter case the condition on $a(x)$ is trivially satisfied by $a(x) \equiv 1$. A less trivial example is

$$a(x) = \sin^2(x/2) = \frac{1}{2} - \frac{1}{4}\left(e^{ix} + e^{-ix}\right) . \tag{4.1.13}$$

The collocation approximation to (4.1.6) can be expressed as

$$\begin{aligned} -\frac{1}{2} \sum_{k=-N/2+1}^{N/2-1} \Bigg[ k^2 \tilde{u}_k - \frac{k(k-1)}{2} \alpha_{k-1} \tilde{u}_{k-1} - \frac{k(k+1)}{2} \alpha_{k+1} \tilde{u}_{k+1} \\ + 2\lambda \tilde{u}_k \Bigg] e^{ikx_j} = f_j , \qquad j = 0, \ldots, N-1 , \end{aligned} \tag{4.1.14}$$

where we have ignored the contributions of the $\tilde{u}_{-(N/2)}$ term, and where

$$\alpha_k = \begin{cases} 1 , & |k| \le \dfrac{N}{2} - 1 , \\ 0 , & |k| > \dfrac{N}{2} - 1 . \end{cases} \tag{4.1.15}$$

The solution procedure clearly requires two FFTs and one tridiagonal solution. Since the cost of the latter is minor, the entire solution requires $5N \log_2 N$ operations.

A closely related system arises for the mapping (2.7.6) introduced by Cain, Ferziger, and Reynolds (1984) for problems on $(-\infty, \infty)$ with solutions which tend to the same constant at $\pm\infty$. In this case, the Poisson problem of interest is really not (4.1.6) but rather a one-dimensional version of (4.5) with $a(x) \equiv 1$:

$$\begin{aligned} -g(x) \frac{\mathrm{d}}{\mathrm{d}x} \left[ g(x) \frac{\mathrm{d}u}{\mathrm{d}x} \right] + \lambda u = f \qquad & \text{in } (0, 2\pi) , \\ u \;\; 2\pi\text{-periodic} , \qquad & \end{aligned} \tag{4.1.16}$$

where $g(x) = \sin^2(x/2)$ is the inverse of the Jacobian of the mapping $z = -\cot(x/2)$ which was discussed in Sect. 2.7.2. The relevant collocation approximation is now expressible as

$$\begin{aligned} \sum_{k=-N/2+1}^{N/2-1} \Bigg[ & \frac{1}{4}(k-1)(k-2)\alpha_{k-2}\tilde{u}_{k-2} - \frac{1}{2}(k-1)(2k-1)\alpha_{k-1}\tilde{u}_{k-1} \\ & + \left( \frac{3}{2}k^2 - \lambda \right) \tilde{u}_k - \frac{1}{2}(k+1)(2k+1)\alpha_{k+1}\tilde{u}_{k+1} \\ & + \frac{1}{4}(k+1)(k+2)\alpha_{k+2}\tilde{u}_{k+2} \Bigg] e^{ikx_j} = f_j , \qquad j = 0, 1, \ldots, N-1 , \end{aligned} \tag{4.1.17}$$

where $\alpha_k$ is again given by (4.1.15) and $\tilde{u}_k = 0$ for $|k| = N/2 - 1$, $N/2$ is assumed. The solution to (4.1.17) requires two FFTs and the solution of a pentadiagonal system. The linear system needs $19N$ operations, whereas the two FFTs together require $5N \log_2 N$ operations.

The mapping (2.7.7) leads to the following approximation to (4.1.16):

$$\frac{1}{4}\sum_{k=-N/2+1}^{N/2-1}\Bigg[\frac{1}{4}(k-2)(k-4)\alpha_{k-4}\tilde{u}_{k-4}-(k-1)(k-2)\alpha_{k-2}\tilde{u}_{k-2}$$
$$+\left(\frac{3}{2}k^2-4\lambda\right)\tilde{u}_k-(k+1)(k+2)\alpha_{k+2}\tilde{u}_{k+2} \tag{4.1.18}$$
$$+\frac{1}{4}(k+2)(k+4)\alpha_{k+4}\tilde{u}_{k+4}\Bigg]e^{ikx_j}=f_j\,,\quad j=0,1,\ldots,N-1\,.$$

This requires the same number of operations to solve as (4.1.17); since the odd modes decouple from the even ones, two pentadiagonal solutions of length $N/2$ suffice for the linear equations.

As a rule, the generality of efficient direct methods decreases as the dimensionality of the problem increases. Clearly, the generalization of techniques for (4.1.1) are straightforward. The operation count of a Galerkin solution to (4.1) is $(2d+1)N^d$ and that of a collocation approximation is $5dN^d\log_2 N$.

Two-dimensional versions of (4.2) and (4.5) are equally straightforward if, for (4.2) the coefficient $a$ depends only on $x$, for (4.5) the coefficient $g_1$ depends only on $x$, and $a$ and $g_2$ are constant. In this case, a Fourier transform in $y$ produces uncoupled sets of equations in $x$, which are of the form (4.1.6) and (4.1.16) with $\lambda$ replaced with $\lambda+k_y^2$. If, however, $a(\mathbf{x})$ in (4.1.6) contains a general dependence on $x$ and $y$, then even for $\lambda=0$ no efficient direct solution is available. The prospects for problems of the type (4.1.16) arising from the use of trigonometric mappings in two directions are almost as poor. In this case the matrix $L$ is banded, with half-bandwidth $O(N)$. Banded Gaussian elimination methods require $O(N^4)$ operations, which is quite expensive for a two-dimensional problem. Similar considerations apply to a third dimension.

### 4.1.2 Chebyshev Tau Approximations

Efficient solution processes are available for a limited class of Chebyshev and Legendre tau approximations to one-dimensional problems. An example of considerable importance is

$$\begin{aligned} -\frac{\mathrm{d}^2u}{\mathrm{d}x^2}+\lambda u&=f \qquad \text{in } (-1,1)\,,\\ u(-1)=u(1)&=0\,. \end{aligned} \tag{4.1.19}$$

We write the Chebyshev tau approximation as

$$-\hat{u}_k^{(2)}+\lambda\hat{u}_k=\hat{f}_k\,,\qquad k=0,1,\ldots,N-2\,, \tag{4.1.20}$$

$$\sum_{k=0}^{N}\hat{u}_k=0,\quad \sum_{k=0}^{N}(-1)^k\hat{u}_k=0\,. \tag{4.1.21a}$$

The boundary conditions may also be written as

$$\sum_{\substack{k=0 \\ k \text{ even}}}^{N} \hat{u}_k = 0, \quad \sum_{\substack{k=1 \\ k \text{ odd}}}^{N} \hat{u}_k = 0 \,. \tag{4.1.21b}$$

Equation (4.1.20) may be expressed as (see (2.4.27))

$$-\frac{1}{c_k} \sum_{\substack{p=k+2 \\ p+k \text{ even}}}^{N} p\left(p^2 - k^2\right) \hat{u}_p + \lambda \hat{u}_k = \hat{f}_k \,, \quad k = 0, 1, \ldots, N-2 \,. \tag{4.1.22}$$

Using (4.1.21b) and (4.1.22), we arrive at a linear system of the form (4.8) in which $L$ is upper triangular. The solution process requires $N^2$ operations. A far more efficient solution procedure is obtained by rearranging the equations. We invoke the recursion relation (2.4.26) with $q = 2$:

$$2k\hat{u}_k^{(1)} = c_{k-1}\hat{u}_{k-1}^{(2)} - \hat{u}_{k+1}^{(2)} \,,$$

and use (4.1.20) to obtain

$$2k\hat{u}_k^{(1)} = c_{k-1}\left(-\hat{f}_{k-1} + \lambda\hat{u}_{k-1}\right) - \left(-\hat{f}_{k+1} + \lambda\hat{u}_{k+1}\right) \,, \quad k = 1, \ldots, N-3 \,. \tag{4.1.23}$$

Next use (2.4.26) with $q = 1$ in combination with (4.1.23):

$$\begin{aligned} 2k\hat{u}_k &= \frac{c_{k-1}}{2(k-1)}[c_{k-2}(-\hat{f}_{k-2} + \lambda\hat{u}_{k-2}) - (-\hat{f}_k + \lambda\hat{u}_k)] \\ &\quad - \frac{1}{2(k+1)}[c_k(-\hat{f}_k + \lambda\hat{u}_k) - (-\hat{f}_{k+2} + \lambda\hat{u}_{k+2})] \,, \quad k = 2, \ldots, N-4 \,. \end{aligned}$$

This simplifies to

$$\begin{aligned} &\frac{c_{k-2}}{4k(k-1)}\lambda\hat{u}_{k-2} + \left(1 - \frac{\lambda}{2\left(k^2-1\right)}\right)\hat{u}_k + \frac{\lambda}{4k(k+1)}\hat{u}_{k+2} \\ &= -\frac{c_{k-2}}{4k(k-1)}\hat{f}_{k-2} + \frac{1}{2\left(k^2-1\right)}\hat{f}_k - \frac{1}{4k(k+1)}\hat{f}_{k+2} \,, \quad k = 2, \ldots, N-4 \,. \end{aligned} \tag{4.1.24}$$

By accounting carefully for the four equations which were dropped in going from (4.1.20) to (4.1.24), we can write (4.1.20) as

$$\begin{aligned} &\frac{c_{k-2}\lambda}{4k(k-1)}\hat{u}_{k-2} + \left[1 - \frac{\lambda\beta_k}{2\left(k^2-1\right)}\right]\hat{u}_k + \frac{\lambda\beta_{k+2}}{4k(k+1)}\hat{u}_{k+2} \\ &= -\frac{c_{k-2}}{4k(k-1)}\hat{f}_{k-2} + \frac{\beta_k}{2\left(k^2-1\right)}\hat{f}_k - \frac{\beta_{k+2}}{4k(k+1)}\hat{f}_{k+2} \,, \quad k = 2, \ldots, N \,, \end{aligned} \tag{4.1.25}$$

where

$$\beta_k = \begin{cases} 1, & 0 \le k \le N-2, \\ 0, & k > N-2. \end{cases} \tag{4.1.26}$$

Note that the even and odd coefficients are uncoupled in (4.1.25) and (4.1.21b). The structure of the linear system for the even coefficients is quasi-tridiagonal, namely,

$$\begin{pmatrix} 1\,1\,1 & \cdots & & 1 \\ *\,*\,* & & & \\ & *\,*\,* & & \\ & & *\,* & * \\ & & & \vdots \\ & & * & *\,* \\ & & & *\,* \\ & & & & *\,* \end{pmatrix} \begin{pmatrix} \hat{u}_0 \\ \hat{u}_2 \\ \hat{u}_4 \\ \\ \vdots \\ \hat{u}_{N-4} \\ \hat{u}_{k-2} \\ \hat{u}_N \end{pmatrix} = \begin{pmatrix} 0 \\ \hat{g}_0 \\ \hat{g}_2 \\ \\ \vdots \\ \hat{g}_{N-6} \\ \hat{g}_{N-4} \\ \hat{u}_{k-2} \end{pmatrix}, \tag{4.1.27}$$

where $*$'s denote the nonzero coefficients from (4.1.25), and $\hat{g}_k$ is the right-hand side of (4.1.25). This ordering has been chosen to minimize the round-off errors arising from a specially tailored Gauss elimination procedure for (4.1.27) which performs no pivoting (and works from the "bottom up" rather than the more customary "top down"). Assuming that the coefficients in (4.1.25) have already been calculated, the cost of solving for both the even and odd coefficients is $16N$. Note that if the boundary conditions were non-homogeneous, this would be reflected merely in an appropriate nonzero entry in the first component of the right-hand side of (4.1.27).

The coefficient of $\hat{u}_k$ in (4.1.25) is the largest coefficient, and it is desirable for it to be on the main diagonal. The system (4.1.27) is not diagonally dominant, and, in practice, round-off errors are a mild problem: typically four digits are lost for $N = 128$. The accuracy may be increased through iterative improvement (see Golub and Van Loan (1996), Chap. 3) or double-precision.

The solution process for a mixed collocation/tau approximation to (4.1.19) is: (1) perform a discrete Chebyshev transform on the grid-point values $f_j$; (2) solve the quasi-tridiagonal system (4.1.25), (4.1.21b); (3) perform an inverse Chebyshev transform on $\hat{u}_k$ to produce the $u_j$. Step (1) prevents this from being a pure tau method, since the Chebyshev coefficients are computed by quadrature rather than exact integration. This solution requires $5N \log_2 N + 24N$ operations, where we include the latter term because of its large coefficient.

The Neumann problem may be solved just as efficiently. In this case, (4.1.21) is replaced by

$$\sum_{k=1}^{N} k^2 \hat{u}_k = 0, \qquad \sum_{k=1}^{N} (-1)^k k^2 \hat{u}_k = 0, \tag{4.1.28a}$$

or equivalently,

$$\sum_{\substack{k=2 \\ k \text{ even}}}^{N} k^2 \hat{u}_k = 0, \qquad \sum_{\substack{k=1 \\ k \text{ odd}}}^{N} k^2 \hat{u}_k = 0 \,. \tag{4.1.28b}$$

The even and odd coefficients decouple, so that the cost is the same as that of the Dirichlet problem. If $\lambda = 0$, then the compatibility condition

$$\sum_{k=0}^{k-2} \frac{-2}{k^2-1} \hat{f}_k = 0$$

is required by the algebraic problem. This is the discrete analog of the compatibility condition

$$\int_{-1}^{1} f(x)\,\mathrm{d}x = 0$$

for the continuous problem.

Efficient tau approximations can be obtained for mild generalizations of the cases discussed above. For example, Haldenwang et al. (1984) have catalogued the relevant formulas for problems with nonhomogeneous boundary conditions of Robin type, and Dennis and Quartapelle (1985) have provided the formulas for including a constant-coefficient first-derivative term in (4.1.20). In both cases, however, the resulting systems are quasi-pentadiagonal, since the even and odd modes do not decouple.

Zebib (1984) introduced, albeit in a more general setting, the strategy of solving a differential equation in terms of the Chebyshev expansion coefficients of the highest derivative that appears in the equation rather than in terms of the coefficients of the function itself. There are several variations of this, including those of Greengard (1991) and Lundbladh, Henningson and Johansson (1992). The version described here is algebraically equivalent to the tau method described above. The starting point for this *integral Chebyshev tau* approximation to (4.1.19) is the discrete equations (4.1.20) and (4.1.21b). The Chebyshev expansions for $u$, $\mathrm{d}u/\mathrm{d}x$ and $\mathrm{d}^2u/\mathrm{d}x^2$ have degree $N$, $N-1$ and $N-2$, respectively. Instead of eliminating $\hat{u}_k^{(2)}$ in terms of $\hat{u}_k$ in (4.1.20) as before, we eliminate $\hat{u}_k$ in favor of $\hat{u}_k^{(2)}$ in both (4.1.20) and (4.1.21b). By invoking (2.4.26) first with $q=1$ and then with $q=2$, (4.1.20) becomes

$$\begin{aligned}
&-\hat{u}_0^{(2)} + \lambda \hat{u}_0 = \hat{f}_0 \,, \\
&-\hat{u}_1^{(2)} + \lambda \left( \hat{u}_0^{(1)} - \frac{1}{8}\hat{u}_1^{(2)} + \frac{1}{8}\hat{u}_3^{(2)} \right) = \hat{f}_1 \,, \\
&-\hat{u}_k^{(2)} + \lambda \frac{1}{4k} \left[ \frac{c_{k-2}}{k-1}\hat{u}_{k-2}^{(2)} - \left( \frac{1}{k-1} + \frac{1}{k+1} \right) \hat{u}_k^{(2)} + \frac{\beta_{k+2}}{k+1}\hat{u}_{k+2}^{(2)} \right] = \hat{f}_k \,, \\
&\quad k = 2, \dots, N-2 \,,
\end{aligned} \tag{4.1.29}$$

where $\beta_k$ is given by (4.1.26) and $\hat{u}_0^{(1)}$, $\hat{u}_0$ are integration constants. Similarly, the boundary conditions (4.1.21b) yield

$$\hat{u}_0 + \frac{1}{4}\hat{u}_0^{(2)} - \frac{7}{48}\hat{u}_2^{(2)} + \sum_{\substack{k=4 \\ k \text{ even}}}^{N-2} \frac{3}{(k-2)(k-1)(k+1)(k+2)}\hat{u}_k^{(2)} = 0 \,,$$

$$\hat{u}_0^{(1)} - \frac{1}{12}\hat{u}_1^{(2)} + \sum_{\substack{k=3 \\ k \text{ odd}}}^{N-3} \frac{3}{(k-2)(k-1)(k+1)(k+2)}\hat{u}_k^{(2)} = 0 \,. \tag{4.1.30}$$

Equations (4.1.29)–(4.1.30) decouple into 2 separate quasi-tridiagonal systems, with the boundary conditions filling the top row. For nonhomogeneous boundary conditions, the right-hand sides of (4.1.30) are nonzero.

The solution procedure is: (1) perform a discrete Chebyshev transform on the grid-point values $f_j$; (2) solve the quasi-tridiagonal system (4.1.29)–(4.1.30); (3) apply the recursion relation (2.4.26) twice to obtain the Chebyshev coefficients of the solution; and (4) perform an inverse Chebyshev transform on $\hat{u}_k$ to produce the $u_j$. If either $\mathrm{d}^2u/\mathrm{d}x^2$ or $\mathrm{d}u/\mathrm{d}x$ are desired in physical space, then the inverse Chebyshev transform is applied to their respective Chebyshev coefficients obtained in (2) and (3), respectively.

Greengard (1991) noted that the process of integrating a Chebyshev series (once) amplifies the errors by less than a factor of 2.4, whereas the process of differentiating a Chebyshev series amplifies errors by $O(N^2)$. He argued that one should expect greater accuracy in the first and second derivatives of the solution resulting from this method. Most of the applications of this integral method have used expansions with upper limits of $k = N, k = N+1$ and $k = N+2$ for the second derivative, first derivative and function itself. Naturally, this produces greater accuracy in the results than for a conventional tau method truncated at $k = N$.

Legendre tau methods are quite similar. Here, of course, one uses the recursion relation (2.3.22) in place of (2.4.26), and there is no fast transform.

### 4.1.3 Galerkin Approximations

For Legendre Galerkin approximations to (4.1.19) one simple choice of basis functions is

$$\phi_k(x) = \begin{cases} L_0(x) - L_k(x), & k \geq 2 \quad \text{even} \,, \\ L_1(x) - L_k(x), & k \geq 3 \quad \text{odd} \,, \end{cases} \tag{4.1.31}$$

that was already introduced in (2.3.33). The Legendre Galerkin approximation uses

$$u^N = \sum_{k=2}^{N} \check{u}_k \phi_k \,, \tag{4.1.32}$$

and requires that

$$-\left(\frac{\mathrm{d}^2u^N}{\mathrm{d}x^2},\phi_h\right)+\lambda(u^N,\phi_h)=(f,\phi_h)\equiv b_h\ ,\qquad h=2,\dots,N\ . \tag{4.1.33}$$

(We use $\check{u}_k$ to denote the expansion coefficients in the special basis $\phi_k$ to distinguish them from the expansion coefficients $\hat{u}_k$ in the standard Legendre basis $L_k$.) Using integration-by-parts, (4.1.33) can be rewritten

$$\left(\frac{\mathrm{d}u^N}{\mathrm{d}x},\frac{\mathrm{d}\phi_h}{\mathrm{d}x}\right)+\lambda(u^N,\phi_h)=b_h\ ,\qquad h=2,\dots,N\ . \tag{4.1.34}$$

This produces the linear system

$$K\check{\mathbf{u}}+\lambda M\check{\mathbf{u}}=\mathbf{b}\ , \tag{4.1.35}$$

where

$$\check{\mathbf{u}}=(\check{u}_2,\check{u}_3,\dots,\check{u}_N)^T\quad\text{and}\quad\mathbf{b}=(b_2,b_3,\dots,b_N)^T\ , \tag{4.1.36}$$

$$K_{hk}=\left(\frac{\mathrm{d}\phi_k}{\mathrm{d}x},\frac{\mathrm{d}\phi_h}{\mathrm{d}x}\right)\quad\text{and}\quad M_{hk}=(\phi_k,\phi_h)\ . \tag{4.1.37}$$

The matrices in the linear system (4.1.35) are full; a Chebyshev Galerkin approximation with an analogous choice of basis functions also leads to full matrices.

An alternative set of basis functions which produces a tridiagonal system for the coefficients in a Legendre Galerkin approximation to (4.1.19) is

$$\phi_k(x)=s_k\left(L_k(x)-L_{k+2}(x)\right)\ ,\qquad k\geq 0, \tag{4.1.38}$$

where

$$s_k=\frac{1}{\sqrt{4k+6}}\ . \tag{4.1.39}$$

Note that, up to a shift in the index, these are the bubble functions of the modal basis introduced in Sect. 2.3.3 (see (2.3.31)). Here we follow the notation of Shen (1994) in our review of the efficient solution schemes that he developed for Legendre and Chebyshev Galerkin approximations. The expansion is now

$$u^N=\sum_{k=0}^{N-2}\check{u}_k\phi_k\ , \tag{4.1.40}$$

and the Galerkin equations are

$$\left(\frac{\mathrm{d}u^N}{\mathrm{d}x},\frac{\mathrm{d}\phi_h}{\mathrm{d}x}\right)+\lambda(u^N,\phi_h)=(f,\phi_h)\equiv b_h\ ,\qquad h=0,\dots,N-2\ . \tag{4.1.41}$$

The linear system still has the form (4.1.35), but now

$$\check{\mathbf{u}}=(\check{u}_0,\check{u}_1,\dots,\check{u}_{N-2})^T\quad\text{and}\quad\mathbf{b}=(b_0,b_1,\dots,b_{N-2})^T\ . \tag{4.1.42}$$

The matrices $K$ and $M$ coincide, respectively, with the matrices $K^{(2)}$ and $K^{(0)}$ constructed in Sect. 3.8 for the modal basis, provided their first and last rows and columns are dropped. Indeed, (3.8.17) and (3.8.9c) yield, with the current notation,

$$K_{hk} = \begin{cases} 1, & k = h\,, \\ 0, & k \neq h\,, \end{cases} \qquad M_{hk} = M_{kh} = \begin{cases} s_h s_k \left(\frac{2}{2h+1} + \frac{2}{2h+5}\right), & k = h\,, \\ s_h s_k \frac{2}{2k+1}\,, & k = h+2\,, \\ 0 & \text{otherwise}\,. \end{cases} \tag{4.1.43}$$

Since the even and odd terms are decoupled, (4.1.35) reduces to two sets of tridiagonal equations; its solution therefore requires only $5N$ operations. We emphasize that although a tridiagonal system results from (4.1.38), the linear system for the basis (4.1.31) is full.

The right-hand side terms $b_h$ are related to the standard Legendre coefficients $\hat{f}_k$ by

$$b_h = s_h \left( \frac{2}{2h+1} \hat{f}_h - \frac{2}{2h+5} \hat{f}_{h+2} \right)\,, \qquad h = 0, \ldots, N-2\,. \tag{4.1.44}$$

The standard Legendre coefficients of the solution $u^N$ can be recovered from $\breve{u}_k$ via

$$\hat{u}_k = \begin{cases} s_k \breve{u}_k, & k = 0,\ 1\,, \\ s_k \breve{u}_k - s_{k-2} \breve{u}_{k-2}, & k = 2, \ldots, N\,. \end{cases} \tag{4.1.45}$$

Although only $O(N)$ operations are required in spectral space, transformations between spectral space and physical space, e.g., evaluation of the sum (4.1.40) at the Legendre Gauss-Lobatto quadrature points, take $O(N^2)$ operations.

Problems with the nonhomogeneous boundary conditions $u(-1) = u_-$ and $u(+1) = u_+$ are handled by a change of dependent variables to

$$\tilde{v}(x) = u(x) - (u_o + x u_e)\,, \tag{4.1.46}$$

where $u_e = \frac{1}{2}(u(1) + u(-1))$ and $u_o = \frac{1}{2}(u(1) - u(-1))$. Equivalently, this amounts to including the vertex basis functions $\eta_0$ and $\eta_N$ from (2.3.30) into the expansion of $u^N$. Shen (1994) discusses how to solve some other second-order problems and also demonstrates a Legendre basis that permits the fourth-order problem

$$\frac{\mathrm{d}^4 u}{\mathrm{d}x^4} - \nu \frac{\mathrm{d}^2 u}{\mathrm{d}x^2} + \lambda u = 0 \qquad \text{in } (-1,1)\,, \tag{4.1.47}$$

$$u(\pm 1) = \frac{\mathrm{d}u}{\mathrm{d}x}(\pm 1) = 0\,, \tag{4.1.48}$$

to be solved in $O(N)$ operations in Legendre space.

Shen (1995) also devised an efficient basis for a Chebyshev Galerkin approximation to (4.1.19), namely,

$$\phi_k(x) = T_k(x) - T_{k+2}(x) , \qquad k = 0, \dots, N-2 . \tag{4.1.49}$$

The presence of the Chebyshev weight leaves the Galerkin equations in the form

$$-\left(\frac{\mathrm{d}^2 u^N}{\mathrm{d}x^2}, \phi_h\right)_w + \lambda(u^N, \phi_h)_w = (f, \phi_h)_w \equiv b_h , \qquad h = 0, \dots, N-2 . \tag{4.1.50}$$

Shen shows that the basis (4.1.49) leads to the linear system (4.1.35) with $\tilde{\mathbf{u}}$ and $\mathbf{b}$ given by (4.1.42). Shen exploited (2.4.23) and (2.4.27) to show that

$$K_{hk} = \begin{cases} 2\pi(h+1)(h+2), & k = h , \\ 4\pi(h+1), & k = h+2, h+4, h+6, \dots , \\ 0 , & k < h \quad \text{or} \quad k+h \quad \text{odd} , \end{cases} \tag{4.1.51}$$

$$M_{hk} = M_{hk} = \begin{cases} \frac{c_h+1}{2}\pi , & k = h , \\ -\frac{\pi}{2} , & k = h-2 \quad \text{and} \quad k = h+2 , \\ 0 & \text{otherwise} , \end{cases} \tag{4.1.52}$$

where $c_k$ is given by (2.4.10). As before, the even and odd coefficients are decoupled. For both sets of equations, $M$ is tridiagonal and $K$ is an upper-triangular matrix, with the nonzero, off-diagonal elements in each row equal to a constant. The linear system for this Chebyshev Galerkin method is nonsymmetric, unlike the linear system for the Legendre Galerkin method, and this leads to slightly less favorable round-off error properties. Shen notes that a tailored solution procedure takes roughly $7N$ operations.

The right-hand side terms $b_h$ are related to the standard Chebyshev coefficients $\hat{f}_k$ by

$$b_h = \frac{\pi}{2}(c_h \hat{f}_h - c_{h+2}\hat{f}_{h+2}) , \qquad h = 0, \dots, N-2 , \tag{4.1.53}$$

and the standard Chebyshev coefficients of the solution $u^N$ can be recovered from the Galerkin expansion coefficients, $\breve{u}_k$, via

$$\hat{u}_k = \begin{cases} \breve{u}_k, & k = 0,\ 1 , \\ \breve{u}_k - \breve{u}_{k-2}, & k = 2, \dots, N . \end{cases} \tag{4.1.54}$$

Again, only $O(N)$ operations are needed in spectral space, but now the transformations between spectral space and physical space take just $O(N \log_2 N)$ operations because the FFT can be exploited. Shen (1995) discusses how to solve some other second-order problems and provides a Chebyshev basis for solving the fourth-order equation (4.1.47) in $O(N)$ operations in Chebyshev space.

**Numerical Example for Ad Hoc Methods in 1-D** Figure 4.1 shows the maximum error for the various methods for solving (4.1.19) for both $\lambda = 0$ and $\lambda = 1 \times 10^5$. The former choice of $\lambda$ yields a one-dimensional Poisson equation, and the latter is a representative value that occurs for the one-dimensional Helmholtz equation that arises in many algorithms for incompressible channel flow; in particular, for the numerical example in CHQZ3, Sect. 1.3 and for the algorithms discussed in CHQZ3, Sect. 3.4. The exact solution is taken to be $u(x) = \sin(4\pi x)$ and the right-hand side $f(x)$ is chosen accordingly. We see that the Galerkin methods are about an order of magnitude more accurate than the tau methods, and that roundoff errors are more of a concern for the Chebyshev methods, although not particularly significant below $N = 1024$. The results of the integral Chebyshev tau method are within $10^{-15}$ of those for the Chebyshev tau method for the function itself and are not shown here.

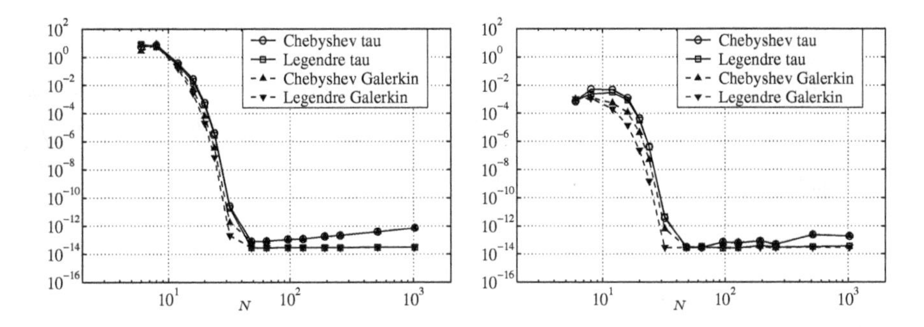


**Fig. 4.1.** Maximum error for various approximations to (4.1.19) with $\lambda = 0$ (*left*) and $\lambda = 1 \times 10^5$ (*right*)

### 4.1.4 Schur Decomposition and Matrix Diagonalization

Let us consider the Helmholtz equation in a square,

$$\begin{aligned} -\Delta u + \lambda u &= f \qquad \text{in } \Omega = (-1,1)^2 , \\ u &= 0 \qquad \text{on } \partial\Omega . \end{aligned} \tag{4.1.55}$$

The collocation approximation to this can be written

$$D_x U + U D_y^T + \lambda U = F , \tag{4.1.56}$$

where $U$ is the $(N_x - 1) \times (N_y - 1)$ matrix $(u_{ij})$ for $i = 1, \ldots, N_x - 1$, $j = 1, \ldots, N_y - 1$, $F$ is defined similarly, $D_x$ is the second-derivative operator (in $x$) in which the boundary conditions have been incorporated, and $D_y^T$ is the transpose of the second-derivative operator (in $y$). (In this subsection we do not assume $N_y = N_x$, as there are some important considerations for $N_y \neq N_x$.)

The Legendre G-NI approximation of problem (4.1.55) can be cast into the algebraic form (4.1.56) too, and solved by the techniques described below. Indeed, let $K^{(2)}$ be the stiffness matrix for the second-derivative operator in one space dimension with homogeneous Dirichlet boundary conditions, associated with the Lagrange nodal basis (see Sect. 3.8); let $M = K^{(0)}_{GNI}$ be the corresponding mass matrix, which is diagonal. The G-NI scheme for (4.1.55) can be written

$$K_x^{(2)} U M_y^T + M_x U K_y^{(2)^T} + \lambda M_x U M_y^T = M_x F M_y^T \,. \qquad (4.1.57)$$

(We refer to the subsequent Sect. 4.2.2 for a detailed description of the algebraic form of multidimensional Galerkin and G-NI schemes.) Then, we obtain a system of the form (4.1.56) with $\tilde{U} = M_x U M_y^T$ and $\tilde{F} = -M_x F M_y^T$ instead of $U$ and $F$, $D_x = K_x^{(2)} M_x^{-1}$ and $D_y^T = (K_y^{(2)} M_y^{-1})^T$.

Systems of the form (4.1.56) are solvable by Schur decomposition (Bartels and Stewart (1972)). An orthogonal transformation is used to reduce $D_x$ to block-lower-triangular form with blocks of size at most two. Similarly, $D_y^T$ is reduced to block-upper-triangular form. If $P$ and $Q$ denote the respective orthogonal transformations, then (4.1.56) is equivalent to

$$D_P U' + U' D_Q - \lambda U' = F' \,, \qquad (4.1.58)$$

where

$$D_P = P^T D_x P, \quad D_Q = Q^T D_y^T Q, \quad U' = P^T U Q, \quad F' = P^T F Q \,. \quad (4.1.59)$$

The solution process has four steps: (1) reduction of $D_x$ and $D_y^T$ to real Schur form (and determination of $P$ and $Q$); (2) construction of $F'$ via (4.1.59); (3) solution of (4.1.58) for $U'$; and (4) transformation of $U'$ to $U$ via (4.1.59).

The first step can be accomplished via the $QR$ algorithm (Wilkinson (1965)) in $(4 + 8\alpha)(N_x^3 + N_y^3)$ operations, where $\alpha$ is the average number of $QR$ steps. Step (3) requires $N_x N_y (N_x + N_y)$ operations and steps (2) and (4) take $2N_x N_y (N_x + N_y)$ operations apiece. Assuming $\alpha = 2$, a single solution requires $20(N_x^3 + N_y^3) + 5N_x N_y (N_x + N_y)$ operations. Hence, step (1) is the most time-consuming. When the same problem must be solved repeatedly, then step (1) need only be performed once, in a pre-processing stage. The matrices $D_P$, $D_Q$, $P$, and $Q$ may then be stored and used as needed. In this case a complete solution takes $5N_x N_y (N_x + N_y)$ operations, or $10N^3$ operations when $N_y = N_x = N$.

To date, however, this method has seen little use in spectral methods, in part because of the matrix-diagonalization technique described next. It would, however, be the method of choice for solving one equation of the form (4.1.56).

The matrix-diagonalization approach is similar to the Schur-decomposition method. The difference is that the matrices $D_x$ and $D_y^T$ in (4.1.56) are diagonalized rather than merely reduced to block-triangular form. An algebraic problem of the form (4.1.58) is obtained with (4.1.59) replaced by

$$\begin{aligned} &D_P = P^{-1} D_x P = \Lambda_{D_x}, \quad D_Q = Q^{-1} D_y^T Q = \Lambda_B, \\ &U' = P^{-1} U Q, \quad F' = P^{-1} F Q \,, \end{aligned} \tag{4.1.60}$$

where $\Lambda_{D_x}$ is the diagonal matrix with the eigenvalues of $D_x$ on the diagonal. Thus, we have

$$\Lambda_{D_x} U' + U' \Lambda_{D_y} - \lambda U' = F' \,. \tag{4.1.61}$$

The matrices $P$ and $Q$ are not necessarily orthogonal and their columns consist of the eigenvectors of $D_x$ and $D_y^T$, respectively.

The matrix-diagonalization scheme for (4.1.56) consists of the same four steps as the Schur-decomposition method except that the first, pre-processing stage also requires that the eigenvectors and the inverse transformations be computed. This takes an additional $4(N_x^3 + N_y^3)$ operations (Golub and Van Loan (1996), Algorithm 7.6-3). Step (3) takes only $3N_x N_y$ operations since the system is diagonal, and steps (2) and (4) require $2N_x N_y(N_x + N_y)$ operations apiece, as before. Software for performing the matrix transformations for the Schur-decomposition and matrix-diagonalization algorithms are readily available, e.g., in LAPACK (Anderson et al. (1999), Barker et al. (2001)).

For collocation problems requiring multiple solutions, the matrix-diagonalization method has the advantage of taking only 80% of the solution time of the Schur-decomposition method: $8N^3$ operations when $N_y = N_x = N$. Moreover, the entire solution process – steps 2, 3 and 4 – is extremely simple and can be optimized readily. The third stage of the Schur-decomposition method is more complicated.

This solution strategy is an application of the tensor-product approach devised by Lynch, Rice and Thomas (1964) for finite-difference approximations to Poisson's equation. For second-order approximations to (4.1.55) on a rectangular grid, the pre-processing stage can be performed analytically.

In the case of tau approximations to (4.1.55), further gains in efficiency are possible. The discrete problem may be written in the form (4.1.56) where $U$ is the $N_x - 1$ by $N_y - 1$ matrix $(\hat{u}_{nm})$ consisting of the Chebyshev coefficients of $u$ (minus those used to enforce the boundary conditions). $F$ is defined similarly, and $D_x$ and $D_y^T$ are the representations in transform space of the second-derivative operator (with the boundary conditions used to eliminate the two highest-order coefficients in each direction).

In the case of Dirichlet (or Neumann) boundary conditions, the even and odd modes decouple. Thus, $D_x$, $D_y^T$, $P$ and $Q$ contain alternating zero and nonzero elements. This property may be exploited to reduce the cost of both the pre-processing step (by a factor of 4) and the matrix multiplies (by a factor of 2). The cost of steps (2) through (4) is thus $2N_xN_y(N_x+N_y)$, or $4N^3$ when $N_y = N_x = N$.

The cost of the solution stages may be halved again by performing the diagonalization in only one direction and resorting to a standard tau solution in the other. Thus, (4.1.56) is reduced to

$$D_xU' + U'\Lambda_B + \lambda U' = F' , \tag{4.1.62}$$

where

$$U' = UQ, \qquad F' = FQ , \tag{4.1.63}$$

instead of to (4.1.61). The system (4.1.62) decouples into $N_y - 1$ systems of the form (4.1.20). Each of these may be reduced to a system like (4.1.25) and solved accordingly in $16N_x$ operations. The cost of the solution process is essentially halved, to $2N_xN_y(4+N_y)$ operations, since the number of matrix multiplies is cut in two. Note that if $N_x \neq N_y$, then it is preferable to apply diagonalization to $D_y^T$ if $N_y < N_x$ and to $D_x$ otherwise.

This particular algorithm has come to be known as the Haidvogel-Zang algorithm after the paper by Haidvogel and Zang (1979) in which the method was explained in detail and compared with finite-difference methods for the Poisson equation. The method had been used earlier by both Murdock (1977) and Haidvogel (1977) in computations of the Navier-Stokes equations with two nonperiodic directions.

In these algorithms, as indeed with matrix computations in general, the accumulation of round-off error is a concern. Haidvogel and Zang reported the loss of three to four digits (for $N$ between 16 and 64) with the Schur-decomposition method. These were recovered through iterative improvement. Since the computation of eigenvectors can be a sensitive process, double-precision is advisable for the pre-processing stage of the matrix-diagonalization method.

Both methods can be generalized. The use of Neumann or Robin boundary conditions is straightforward. However, with Robin boundary conditions the even and odd modes do not decouple, and hence, some of the economies of the tau method are lost. These methods can be applied to separable equations of the form (4.5). A third, periodic direction is trivial to include in (4.1.55) since, after Fourier transforming in this direction, one simply has an independent set of equations with different $\lambda$. The pre-processing is independent of $\lambda$ and hence of the third, periodic direction. A third, nonperiodic direction may be treated by diagonalizing in that direction and then using whichever of the preceding methods is most convenient. Haldenwang et al. (1984) discuss several alternatives. Of course, both algorithms may be applied to separable, variable-coefficient periodic problems.

In the case of Shen's Legendre Galerkin method (the Legendre Galerkin method which uses the modal basis (2.3.30)), the approximation to (4.1.55) can be written as

$$MU + UM + \lambda MUM = F , \tag{4.1.64}$$

where $U_{jl} = \tilde{u}_{jl}$, $F_{jl} = \bar{f}_{jl}$ and $M$ is given by (4.1.43), assuming that $N_x = N_y = N$, for simplicity. Shen (1994) describes the straightforward solution procedure using matrix diagonalization, including the three-dimensional case. Since $M$ is banded and symmetric, the eigenvalue decomposition is both significantly cheaper and better conditioned than for the tau method. Similarly, Shen (1995) describes how to treat two and three-dimensional equations with his Chebyshev Galerkin basis. The two-dimensional version of the fourth-order equation (4.1.47) can be readily handled with the Legendre Galerkin method but apparently not with the Chebyshev Galerkin method.

As it happens, these methods are *more* attractive in three-dimensional problems than in two-dimensional ones. Suppose that the number of degrees of freedom in each direction is $N$. The pre-processing cost is some large multiple of $N^3$. In two dimensions, the solution cost is a small multiple of $N^3$, and typical explicit spectral calculations take $O(N^2 \log_2 N)$ operations. Thus, the pre-processing cost is substantially larger than the cost of a single solution. In three dimensions, the solution cost is a small multiple of $N^4$ and typical explicit spectral calculations require $O(N^3 \log_2 N)$ operations. Thus, the pre-processing cost may even be smaller than the cost of the solution phase. Similarly, the extra memory required for $D_P$, $P$, and its inverse is proportionally smaller in three dimensions than in two.

**Numerical Example for Ad-hoc Methods in Two Dimensions** Figure 4.2 shows the maximum error for several approximations to (4.1.55) for $\lambda = 0$ and $f(x) = 32\pi^2 \sin(4\pi x)\sin(4\pi y)$, corresponding to $u(x) = \sin(4\pi x)\sin(4\pi y)$. Matrix diagonalization was used for the solution procedure. The Chebyshev tau results are taken from Haidvogel and Zang (1979), who performed their computations in 60-bit arithmetic (on a CDC 6600). The other data are 64-bit results taken from Shen (1994, 1995). The results are very similar to those shown in Fig. 4.1 for the one-dimensional case.

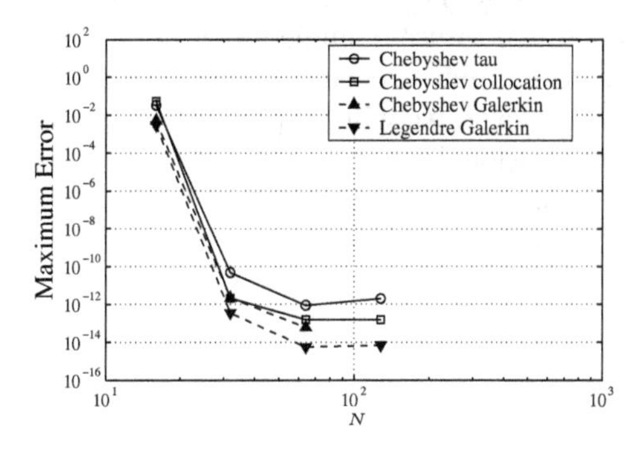


**Fig. 4.2.** Maximum error for various approximations to (4.1.19) (data from Haidvogel and Zang (1979) and Shen (1994, 1995))

## 4.2 Direct Methods

The *ad hoc* methods discussed in the previous section are very efficient when they are applicable, and many of the incompressible flow algorithms that we discuss in CHQZ3, Chap. 3 exploit these techniques. However, general purpose solution techniques are needed for application of spectral methods to a broader class of problems. This section describes the matrix structure produced by spectral Galerkin and G-NI methods. The focus is twofold: on how the tensor-product nature of the methods can be heavily exploited to build efficiently the matrices, and on how the sparseness of the matrices in 2D and especially in 3D can be accounted for in the direct techniques. The matrix structure and solution procedures are very similar for collocation methods. (We do not discuss tau methods and Galerkin methods with modal bases, as they are rarely used for general problems.)

### 4.2.1 Tensor Products of Matrices

In certain relevant circumstances, such as, e.g., constant-coefficient operators, the algebraic form of spectral discretizations to boundary-value problems takes advantage of the tensor-product structure of the expansion basis which is used to represent the discrete solution. In such cases, tensor-product matrices come into play. In view of the subsequent section, we recall the definition and some useful facts about this class of matrices. Let us start with some notation.

Let $\{\phi_{\mathbf{k}}\}$ be a tensor-product basis on the reference domain $\hat{\Omega} = (-1,1)^d$ (see (2.8.1)). For the sake of simplicity, we suppose that each entry $k_l$ of the multi-index $\mathbf{k} = (k_1, \dots, k_d)$ varies in the range $1, \dots, N$. (The use of different ranges in each coordinate direction adds no fundamental complication, just notational and mild implementation complexity.) A discrete function $v(\mathbf{x}) = \sum_{\mathbf{k}} v_{\mathbf{k}} \phi_{\mathbf{k}}(\mathbf{x})$ is identified by the vector $\mathbf{v} = (v_{\mathbf{k}})$ of the expansion coefficients. In principle, it is possible to use a single-index notation for the entries of $\mathbf{v}$, according to the lexicographic ordering $v_k = v_{\mathbf{k}}$ with $k = k_1 + (k_2 - 1)N + \cdots + (k_d - 1)N^{d-1}$; however, we will simply write $v_{\mathbf{k}} = v_{k_1 k_2 \dots k_d}$, i.e., we will consider $\mathbf{v}$ as a $d$-dimensional matrix.

Given $d$ 2-dimensional square matrices $A^{(l)} = (a_{hk}^{(l)})$ of order $N$, we can form their *tensor product*

$$A = \bigotimes_{l=1}^{d} A^{(l)} = A^{(1)} \otimes \cdots \otimes A^{(d)} , \tag{4.2.1}$$

which is a $2d$-dimensional square matrix of order $N$ whose entries are

$$A_{\mathbf{hk}} = \prod_{l=1}^{d} a_{h_l k_l}^{(l)} . \tag{4.2.2}$$

(assuming the same convention on the indices). The matrix $A$ inherits all relevant properties that are held by all the matrices $A^{(l)}$, $l = 1, \dots, d$ (e.g., the symmetry and possible diagonal or banded structure).

### 4.2.2 Multidimensional Stiffness and Mass Matrices

Sect. 3.8 is devoted to the study of stiffness and mass matrices in one space dimension. Here, we take again a second-order scalar equation, say,

$$-\sum_{i,j=1}^{d} \frac{\partial}{\partial x_i}\left(\alpha_{ij}\frac{\partial u}{\partial x_j}\right) + \sum_{i=1}^{d} \beta_i \frac{\partial u}{\partial x_i} + \gamma u = f \qquad \text{in } \Omega \subset \mathbb{R}^d , \tag{4.2.3}$$

supplemented with homogeneous Dirichlet or Neumann boundary conditions on $\partial\Omega$, as a model for our discussion on the algebraic aspects of Galerkin and G-NI methods in more than one space dimension. The integral formulation of the problem is

$$\sum_{i,j=1}^{d} \int_{\Omega} \alpha_{ij} \frac{\partial u}{\partial x_j}\frac{\partial v}{\partial x_i}\,\mathrm{d}\mathbf{x} + \sum_{i=1}^{d} \int_{\Omega} \beta_i \frac{\partial u}{\partial x_i} v\,\mathrm{d}\mathbf{x} + \int_{\Omega} \gamma u v\,\mathrm{d}\mathbf{x} = \int_{\Omega} f v\,\mathrm{d}\mathbf{x} \tag{4.2.4}$$

for all test functions $v$ which vanish on that part, $\partial\Omega_D$, of $\partial\Omega$ for which the homogeneous Dirichlet condition is imposed on $u$.

We assume that $\Omega = F(\hat{\Omega})$, where $\hat{\Omega} = (-1,1)^d$ and $F$ is a smooth invertible mapping, $F : \hat{\mathbf{x}} \to \mathbf{x} = F(\hat{\mathbf{x}})$, satisfying $|JF(\hat{\mathbf{x}})| = \det JF(\hat{\mathbf{x}}) > 0$ for all $\hat{\mathbf{x}} \in \hat{\Omega}$. $JF$ is the Jacobian matrix of the transformation $F$. We suppose that $\partial\Omega_D$ is the image through $F$ of a union $\partial\hat{\Omega}_D$ of "faces" (i.e., $(d-1)$-dimensional manifolds) contained in $\partial\hat{\Omega}$; thus we exclude boundary conditions that may be Dirichlet on part of a face and Neumann on another part of the same face.

Let $\{\hat{\phi}_{\mathbf{k}}\}$ be a finite tensor-product basis in $\hat{\Omega}$ (see (2.8.1)), built up by tensorizing copies of the univariate boundary-adapted modal or nodal bases already considered in Sect. 3.8; we assume that each $\hat{\phi}_{\mathbf{k}}$ vanishes on $\partial\hat{\Omega}_D$. We set $\hat{V}_N = \text{span}\,\{\hat{\phi}_{\mathbf{k}}\}$. The trial and test functions will be chosen in $V_N = F(\hat{V}_N) = \text{span}\,\{\phi_{\mathbf{k}}\}$, with $\phi_{\mathbf{k}}(\mathbf{x}) = \hat{\phi}_{\mathbf{k}}(F^{-1}(\mathbf{x}))$. Then, the Galerkin solution $u^N = \sum_{\mathbf{k}} u_{\mathbf{k}}\phi_{\mathbf{k}}$ is defined by the algebraic system (4.8), where $\mathbf{u} = (u_{\mathbf{k}})$, $\mathbf{b} = \left(\int_{\Omega} f\phi_{\mathbf{h}}\,\mathrm{d}\mathbf{x}\right)$ and $L$ is the stiffness matrix $K = (K_{\mathbf{hk}})$, which we decompose into its second, first and zeroth-order components as $K = K^{(2)} + K^{(1)} + K^{(0)}$. The entries of these matrices can be expressed in terms of integrals on the reference domain. To this end, set $G(\hat{\mathbf{x}}) = JF(\hat{\mathbf{x}})^{-1} = (g_{rs}(\hat{\mathbf{x}}))$ and observe that $\nabla_{\mathbf{x}}\phi(\mathbf{x}) = G^T(\hat{\mathbf{x}})\nabla_{\hat{\mathbf{x}}}\hat{\phi}(\hat{\mathbf{x}})$. Then, we have

$$K_{\mathbf{hk}}^{(2)} = \sum_{i,j=1}^{d} \int_{\Omega} \alpha_{ij} \frac{\partial \phi_{\mathbf{k}}}{\partial x_j}\frac{\partial \phi_{\mathbf{h}}}{\partial x_i}\,\mathrm{d}\mathbf{x} = \sum_{r,s=1}^{d} \int_{\hat{\Omega}} \alpha_{rs}^{*} \frac{\partial \hat{\phi}_{\mathbf{k}}}{\partial \hat{x}_r}\frac{\partial \hat{\phi}_{\mathbf{h}}}{\partial \hat{x}_s}\,\mathrm{d}\hat{\mathbf{x}} , \tag{4.2.5}$$

with $\alpha^*_{rs}(\hat{\mathbf{x}}) = \sum_{i,j=1}^d \alpha_{ij}(F(\hat{\mathbf{x}}))g_{ir}(\hat{\mathbf{x}})g_{js}(\hat{\mathbf{x}})|JF(\hat{\mathbf{x}})|$,

$$K^{(1)}_{\mathbf{hk}} = \sum_{i=1}^d \int_\Omega \beta_i \frac{\partial \phi_{\mathbf{k}}}{\partial x_i} \phi_{\mathbf{h}} \, d\mathbf{x} = \sum_{r=1}^d \int_{\hat{\Omega}} \beta^*_r \frac{\partial \hat{\phi}_{\mathbf{k}}}{\partial \hat{x}_r} \hat{\phi}_{\mathbf{h}} \, d\hat{\mathbf{x}} \,, \tag{4.2.6}$$

with $\beta^*_r(\hat{\mathbf{x}}) = \sum_{i=1}^d \beta_i(F(\hat{\mathbf{x}}))g_{ir}(\hat{\mathbf{x}})|JF(\hat{\mathbf{x}})|$,

$$K^{(0)}_{\mathbf{hk}} = \int_\Omega \gamma \phi_{\mathbf{k}} \phi_{\mathbf{h}} \, d\mathbf{x} = \int_{\hat{\Omega}} \gamma^* \hat{\phi}_{\mathbf{k}} \hat{\phi}_{\mathbf{h}} \, d\hat{\mathbf{x}} \,, \tag{4.2.7}$$

with $\gamma^*(\hat{\mathbf{x}}) = \gamma(F(\hat{\mathbf{x}}))|JF(\hat{\mathbf{x}})|$.

In the remainder of this section we provide the detailed expressions for the components of the stiffness matrices, not only for the Galerkin versions but also for their G-NI approximations. As we shall see, one benefit of the G-NI approximations is that for the case of arbitrary coefficients, the G-NI matrices have greater sparsity than their Galerkin counterparts. This has important implications for the efficiency of direct and iterative methods for the linear systems that these methods produce.

### *Zeroth-order contributions*

Let us first consider the (generalized) mass matrix $K^{(0)}$. If $\gamma^*$ has a tensor-product structure, i.e., $\gamma^*(\hat{\mathbf{x}}) = \prod_{l=1}^d \gamma^*_l(\hat{x}_l)$ (this is, e.g., the case if $\gamma^*$ is constant), then $K^{(0)}$ is a tensor-product matrix. Precisely, one has

$$K^{(0)} = \bigotimes_{l=1}^d K^{(0;\hat{x}_l)} \,,$$

where the matrices $K^{(0;\hat{x}_l)}$ are of the type (3.8.8), i.e., they are defined as

$$K^{(0;\hat{x}_l)}_{h_l k_l} = \int_{\hat{I}} \gamma^*_l(\hat{x}_l) \hat{\phi}^{(l)}_{k_l}(\hat{x}_l) \hat{\phi}^{(l)}_{h_l}(\hat{x}_l) \, d\hat{x}_l \tag{4.2.8}$$

(recall that $\{\hat{\phi}^{(l)}_k\}$ denotes the univariate basis used in the $l$-th direction, see (2.8.1)).

For a general $\gamma^*$, the usual practice is to resort to the G-NI approach and use the Lagrange nodal basis in each direction; this yields the diagonal matrix

$$(K^{(0)}_{GNI})_{\mathbf{hk}} = \gamma^*(\hat{\mathbf{x}}_{\mathbf{h}}) \hat{w}_{\mathbf{h}} \delta_{\mathbf{hk}} \,, \tag{4.2.9}$$

where $\hat{\mathbf{x}}_{\mathbf{h}} = (\hat{x}_{h_1}, \ldots, \hat{x}_{h_d})$ are the tensorized quadrature nodes, and $\hat{w}_{\mathbf{h}} = \hat{w}_{h_1} \cdots \hat{w}_{h_d}$ are the corresponding weights. If $\gamma \equiv 1$, we get the lumped mass matrix $M_{GNI}$.

The other matrices which contribute to the stiffness matrix $K$ can be analyzed in a similar manner.

***First-order contributions***

Consider one contribution,

$$K_{\mathbf{hk}}^{(1;r)} = \int_{\hat{\Omega}} \beta_r^* \frac{\partial \hat{\phi}_{\mathbf{k}}}{\partial \hat{x}_r} \hat{\phi}_{\mathbf{h}} \, d\hat{\mathbf{x}} \,,$$

to the first-order matrix $K^{(1)}$. If $\beta_r^*$ is a tensor-product function, then $K_r^{(1)}$ can be represented as

$$K^{(1;r)} = \bigotimes_{l=1}^{d} K^{(\delta_{lr};\hat{x}_l)} \,,$$

where $K^{(1;\hat{x}_r)}$ is a first-order univariate matrix like (3.8.12), whereas $K^{(0;\hat{x}_l)}$ for $l \neq r$ is a (generalized) mass matrix like (4.2.8).

For an arbitrary $\beta_r^*$, the use of a G-NI discretization with the Lagrange nodal basis leads to the sparse matrix

$$(K_{GNI}^{(1;r)})_{\mathbf{hk}} = \begin{cases} \beta_r^*(\hat{\mathbf{x}}_{\mathbf{h}})\hat{w}_{\mathbf{h}}(D_N)_{h_r k_r} & \text{if } h_l = k_l \text{ for all } l \neq r, \\ 0 & \text{otherwise,} \end{cases} \tag{4.2.10}$$

where $(D_N)_{hk}$ is defined in (2.3.28). A matrix-vector multiply with such a matrix requires $O(N^{d+1})$ operations. Summing up all first-order contributions, we obtain a sparse matrix $K^{(1)}$, which in 2D and 3D has the patterns shown in Fig. 4.3, where nonzero elements are noted by symbols or lines.

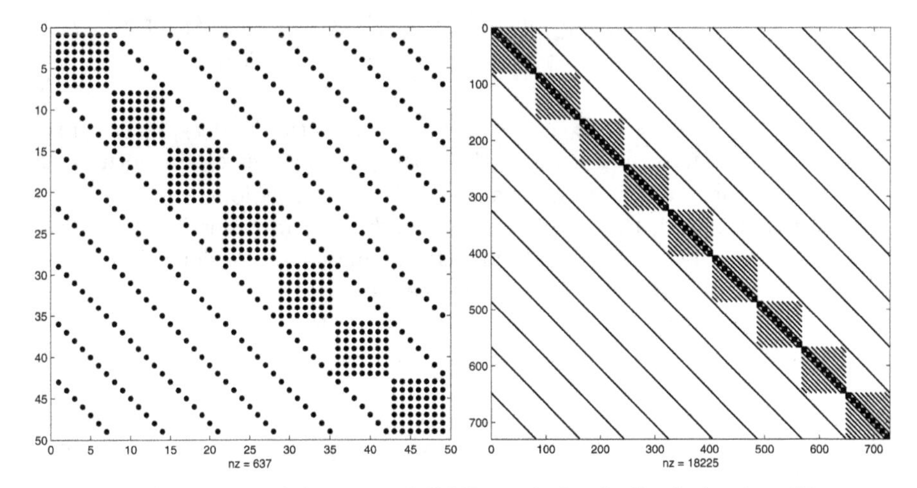


**Fig. 4.3.** The pattern of the spectral G-NI matrix for the Laplacian in a 2D square (*left*) and a 3D cube (*right*) with straight boundaries. The total number of nonzero elements $nz$ is $2N^3$ in 2D and $3N^4$ in 3D

***Second-order contributions***

Finally, let us consider one contribution,

$$K_{\mathbf{hk}}^{(2;r,s)} = \int_{\hat{\Omega}} \alpha_{rs}^* \frac{\partial \hat{\phi}_{\mathbf{k}}}{\partial \hat{x}_r} \frac{\partial \hat{\phi}_{\mathbf{h}}}{\partial \hat{x}_s} \, \mathrm{d}\hat{\mathbf{x}} \,,$$

to the second-order matrix $K^{(2)}$. In the case of tensorial $\alpha_{rs}^*$, the matrix $K^{(2;r,s)}$ has the structure

$$K^{(2;r,s)} = \bigotimes_{l=1}^{d} K^{(\delta_{lr}+\delta_{ls};\hat{x}_l)} \,,$$

i.e., it is a tensor product of zeroth and first-order univariate matrices if $r \neq s$, or of zeroth and second-order univariate matrices (the latter similar to (3.8.16)) if $r = s$.

For an arbitrary coefficient, if the Lagrange nodal basis is used within a G-NI scheme, we obtain a sparse approximate matrix; precisely, if $r \neq s$, we have

$$(K_{GNI}^{(2;r,s)})_{\mathbf{hk}} = \begin{cases} \alpha_{rs}^*(\hat{\mathbf{x}}_{\mathbf{n}})\hat{w}_{\mathbf{n}}(D_N)_{h_r k_r}(D_N)_{k_s h_s} & \text{if } h_l = k_l \text{ for all } l \neq r, s \,, \\ 0 & \text{otherwise,} \end{cases} \tag{4.2.11}$$

where $\hat{\mathbf{x}}_{\mathbf{n}}$ is the quadrature node whose components are $\hat{x}_{h_r}$ in the $r$-direction, $\hat{x}_{k_s}$ in the $s$-direction, and $\hat{x}_{h_l} = \hat{x}_{k_l}$ in the remaining directions, while $\hat{w}_{\mathbf{n}}$ is the corresponding weight; if $r = s$, we have

$$(K_{GNI}^{(2;s,s)})_{\mathbf{hk}} = \begin{cases} \sum_{j=0}^{N} \alpha_{ss}^*(\hat{\mathbf{x}}_{\mathbf{n}(j)})\hat{w}_{\mathbf{n}(j)}(D_N)_{jk_s}(D_N)_{jh_s} & \text{if } h_l = k_l \text{ for all } l \neq s \,, \\ 0 & \text{otherwise,} \end{cases} \tag{4.2.12}$$

where the components of $\hat{\mathbf{x}}_{\mathbf{n}(j)}$ are $\hat{x}_j$ in the $s$-direction and $\hat{x}_{h_l} = \hat{x}_{k_l}$ in the remaining directions, and $\hat{w}_{\mathbf{n}(j)}$ is the corresponding weight. In both cases, a matrix-vector multiply requires again $O(N^{d+1})$ operations.

Note that in 2D, the assembled matrix $K^{(2)}$ is in general full for arbitrary nonzero coefficients $\{\alpha_{rs}^*\}$; in 3D and higher dimensions, it recovers a sparse structure (for instance, all elements whose indices, $\mathbf{h}, \mathbf{k}$, differ for three or more components are zero). However, if the matrix $(\alpha_{rs}^*)$ is diagonal throughout $\hat{\Omega}$, then $K^{(2)}$ is also sparse in 2D and exhibits the same pattern as shown in the left half of Fig. 4.3.

It is worth observing that the algebraic form of a Legendre G-NI approximation to the Poisson problem (4.1.55) in the square, already considered in Sect. 4.1.4, can be written as

$$\left(K_x^{(2)} \otimes M_{GNI,y} + M_{GNI,x} \otimes K_y^{(2)} + \lambda M_{GNI,x} \otimes M_{GNI,y}\right) \mathbf{u} \\ = M_{GNI,x} \otimes M_{GNI,y} \mathbf{f} \,,$$

where $\mathbf{u}$ and $\mathbf{f}$, respectively, denote the vectors of the values of $u^N$ and $f$, respectively, at the internal LGL nodes. This is nothing but (4.1.57) under a different but equivalent notation. A more compact form of the system is

$$K_{GNI}\mathbf{u} = M_{GNI}\mathbf{f} \ ;$$

as in the one-dimensional case (see Sect. 3.8.2), we have

$$L_{\text{coll}} = M_{GNI}^{-1} K_{GNI} \ , \tag{4.2.13}$$

where $L_{\text{coll}}$ denotes the matrix of the collocation discretization of the problem at the internal LGL nodes.

### *Warped Tensor-Product Expansions*

We end this section by giving a short account of the structure of the matrices associated with the warped tensor-product expansions considered in Sect. 2.9.1. We confine ourselves to the two-dimensional case; we assume to be on the reference triangle $\mathcal{T} = \{(x_1, x_2) \in \mathbb{R}^2 \ : \ -1 < x_1, x_2 \ ; \ x_1 + x_2 < 0\}$. The basis functions have the general structure

$$\varphi_{\mathbf{k}}(x_1, x_2) = \psi_{k_1}(\xi_1)\psi_{k_1,k_2}(\xi_2) \ ,$$

with the transformation $(x_1, x_2) \mapsto (\xi_1, \xi_2)$ between $\mathcal{T}$ and the reference square $\mathcal{Q} = \{(\xi_1, \xi_2) \in \mathbb{R}^2 \ : \ -1 < \xi_1, \xi_2 < 1\}$ given by (2.9.1). The mass matrix $K^{(0)} = M$ associated with such an expansion has components

$$\begin{aligned}
(K^{(0)})_{\mathbf{hk}} &= \int_{\mathcal{T}} \varphi_{\mathbf{k}}(x_1, x_2)\varphi_{\mathbf{h}}(x_1, x_2)\, \mathrm{d}x_1 \mathrm{d}x_2 \\
&= \int_{\mathcal{Q}} \psi_{k_1}(\xi_1)\psi_{k_1,k_2}(\xi_2)\psi_{h_1}(\xi_1)\psi_{h_1,h_2}(\xi_2) \left(\frac{1-\xi_2}{2}\right) \mathrm{d}\xi_1 \mathrm{d}\xi_2 \\
&= \int_{-1}^{1} \psi_{k_1}(\xi_1)\psi_{h_1}(\xi_1)\, \mathrm{d}\xi_1 \int_{-1}^{1} \psi_{k_1,k_2}(\xi_2)\psi_{h_1,h_2}(\xi_2) \left(\frac{1-\xi_2}{2}\right) \mathrm{d}\xi_2 \\
&= m^{(1)}_{h_1 k_1} m^{(2)}_{h_1 k_1 h_2 k_2} \ .
\end{aligned}$$

Thus, as in the pure tensor-product case, the elements of $K^{(0)}$ are products of elements of suitable one-dimensional matrices. Note, however, that the second factor, $m^{(2)}_{h_1 k_1 h_2 k_2}$, also depends on the indices, $h_1, k_1$, of the first factor. This complication, inherent to the warped nature of the tensor-products considered here, makes the construction of the matrix, as well as the application of the matrix to a vector, less efficient than in the pure tensor-product case.

The stiffness matrix for a constant-coefficient operator is a sum of matrices having the same structure as $K^{(0)}$. This easily stems from the expressions of the partial derivatives of the basis functions $\varphi_{\mathbf{k}}$; indeed, the chain rule yields

$$\begin{aligned}
\frac{\partial \varphi_{\mathbf{k}}}{\partial x_1}(x_1,x_2) &= \left( \frac{\mathrm{d}\psi_{k_1}}{\mathrm{d}\xi_1}\frac{\partial \xi_1}{\partial x_1}\,\psi_{k_1,k_2} + \psi_{k_1}\frac{\mathrm{d}\psi_{k_1,k_2}}{\mathrm{d}\xi_2}\frac{\partial \xi_2}{\partial x_1}\right)(\xi_1,\xi_2) \\
&= \frac{\mathrm{d}\psi_{k_1}}{\mathrm{d}\xi_1}(\xi_1)\,\frac{2}{1-\xi_2}\,\psi_{k_1,k_2}(\xi_2)\,, \\
\frac{\partial \varphi_{\mathbf{k}}}{\partial x_2}(x_1,x_2) &= \left( \frac{\mathrm{d}\psi_{k_1}}{\mathrm{d}\xi_1}\frac{\partial \xi_1}{\partial x_2}\,\psi_{k_1,k_2} + \psi_{k_1}\frac{\mathrm{d}\psi_{k_1,k_2}}{\mathrm{d}\xi_2}\frac{\partial \xi_2}{\partial x_2}\right)(\xi_1,\xi_2) \\
&= (1+\xi_1)\,\frac{\mathrm{d}\psi_{k_1}}{\mathrm{d}\xi_1}(\xi_1)\,\frac{1}{1-\xi_2}\,\psi_{k_1,k_2}(\xi_2) + \psi_{k_1}(\xi_1)\,\frac{\mathrm{d}\psi_{k_1,k_2}}{\mathrm{d}\xi_2}(\xi_2)\,.
\end{aligned}$$

Each addend on the right-hand sides is a product of a function of $\xi_1$ alone times a function of $\xi_2$ alone. The same feature holds for all higher order partial derivatives.

Further details on the construction of stiffness and mass matrices can be found in the books by Karniadakis and Sherwin (1999) and Deville, Fischer and Mund (2002).

### 4.2.3 Gaussian Elimination Techniques

The classical direct methods for solving the linear system (4.8) are based on decomposing the system matrix $L$ into the product $L = RS$ of two factors, $R$ and $S$, the former being lower triangular, the latter upper triangular. (This is universally known as the *LU*-decomposition of a matrix $A$, where $L$ and $U$ denote the two triangular factors of $A$; we use the unconventional notation $R$ and $S$ in lieu of $L$ and $U$, as we have reserved the symbol $L$ for the matrix of the algebraic system to be solved.) Then, we obtain the solution of $L\mathbf{u} = \mathbf{b}$ by solving $S\hat{\mathbf{u}} = \mathbf{b}$ (backward substitution), followed by $R\mathbf{u} = \hat{\mathbf{u}}$ (forward elimination). Let $n$ be the dimension of the original system. Then, assuming that $R_{ii} = 1$ for $i = 1, \dots, n$ ($n = (N-1)^d$ in the case of Dirichlet boundary conditions, whereas $n = (N+1)^d$ for Neumann boundary conditions), if the first $n-1$ principal minors of $R$ are nonsingular, the algorithm reads

$$\begin{aligned}
&\mathit{for}\ k = 1,\dots,n-1 \\
&\quad \mathit{for}\ i = k+1,\dots,n \\
&\qquad L_{ik} = L_{ik}/L_{ii} \\
&\qquad \mathit{for}\ j = k+1,\dots,n \\
&\qquad\quad L_{ij} = L_{ij} - L_{ik}\,L_{kj}\,,
\end{aligned} \tag{4.2.14}$$

then set

$$S_{ij} = \begin{cases} L_{ij}\,, & i \le j\,, \\ 0\,, & i > j\,, \end{cases} \qquad R_{ij} = \begin{cases} L_{ij}\,, & i > j\,, \\ 1\,, & i = j\,, \\ 0\,, & i < j\,. \end{cases}$$

The Gauss decomposition requires $2n^3/3$ operations, and the solution phase takes $2n^2$ operations. These operation counts are $O(N^{3d})$ and $O(N^{2d})$, respectively. In two dimensions, the solution phase alone takes $O(N^4)$ operations, significantly more than the $O(N^3)$ operations required by the *ad hoc* methods discussed in Sect. 4.1.4, which, however, are applicable only in special cases.

It is well known that the Gauss decomposition can be successfully carried out (without resorting to a pivoting strategy) if the given matrix $L$ is diagonally dominant (either by rows or by columns) or if it is symmetric and positive definite. Unfortunately, spectral matrices arising from collocation methods are neither diagonally dominant nor symmetric. Consequently, the Gauss decomposition has to be modified to allow for pivoting. This yields the factorization $PL = RS$ (in the case of row permutation), where $R$ and $S$ are still lower and upper-triangular matrices, $P$ is the matrix which accounts for the permutation of rows yielding that pivot element $L_{ii}$ that is the largest (in modulus) of the elements on the $i$-th column (see, e.g., Quarteroni, Sacco and Saleri (2000) or Golub and Van Loan (1996)). When the matrix $L$ is symmetric and positive definite (e.g., this is the case for G-NI approximation of elliptic, self-adjoint operators), the more efficient Cholesky factorization is preferred. Here the factorization is $L = CC^T$, where $C$ is a lower-triangular matrix. This yields the two systems, $C\hat{\mathbf{u}} = \mathbf{b}$ and $C^T\mathbf{u} = \hat{\mathbf{u}}$. An algorithm for computing the entries of $C$ is

$$\begin{aligned}
&C_{11} = \sqrt{L_{11}} \\
&for\ i = 2, \dots, n \\
&\quad for\ j = 1, \dots, i-1 \\
&\qquad C_{ij} = (L_{ij} - \textstyle\sum_{k=1}^{j-1} C_{ik}C_{jk})/C_{jj} \\
&\quad C_{ii} = (L_{ii} - \textstyle\sum_{k=1}^{i-1} C_{ik}^2)^{1/2}
\end{aligned} \qquad (4.2.15)$$

Its computational cost is $n^3/3$ operations; an additional $2n^2$ operations are needed for the solution of the two corresponding triangular systems.

For a G-NI approximation on a rectangular domain $\Omega$ of a second-order elliptic operator that does not contain mixed derivatives, the lower (and upper) bandwidth of $L$ is $n-(N+1)^{d-1}$ in the Neumann case and $n-(N-1)^{d-1}$ in the Dirichlet case, and the number of nonzero entries is about $2n^{3/2}$ (see Fig. 4.3 for the sparsity pattern of $L$). For a general quadrilateral domain, $L$ is a full matrix. When the Gauss or Cholesky factorization is used, the matrix, $L$, is formed with the so-called "full" format (all entries are stored, including the zeros).

As the problem dimensionality $d$ increases, the cost in terms of time and storage for the Gauss and Cholesky decompositions quickly becomes prohibitive. Fortunately, the relative sparsity of the matrix $L$ increases with the dimensionality. However, this sparsity pattern is lost for the two factors, $R$ and $S$, of the Gauss decomposition (or $C$ and $C^T$ of the Cholesky decomposition) due to *fill-in*. Consequently, the sparsity results in little or no savings in the solution cost. In order to benefit from the sparsity a reordering of $L$ prior to factorization is required. In Fig. 4.4 we show two examples of reordering on the 3D matrix $L$: the Cuthill-McKee ordering and the minimum-degree ordering (see George and Liu (1981), Gilbert et al. (1992), Saad (1996)).

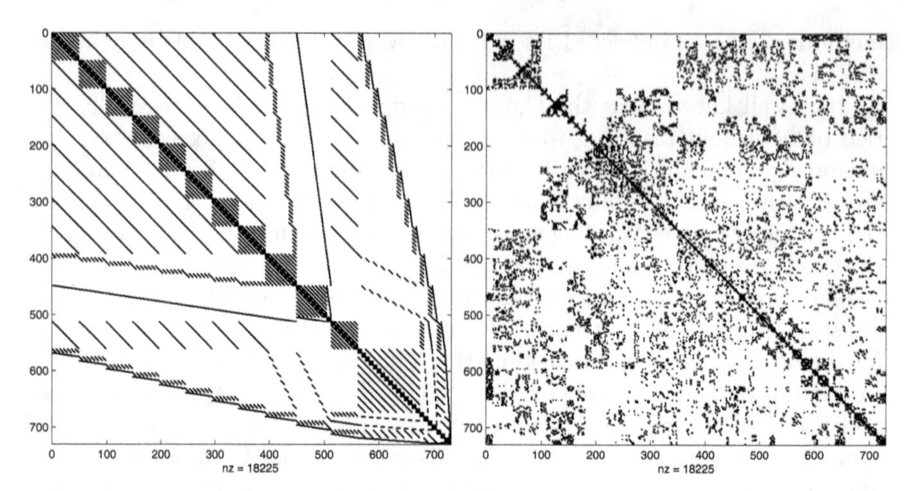


**Fig. 4.4.** The pattern of the 3D spectral G-NI matrix after reordering: Cuthill-McKee ordering (*left*), minimum-degree ordering (*right*)

A comparison of the CPU time needed to solve the Poisson equation in a cube (using the Cholesky factorization for the G-NI or, for that matter, the Galerkin matrix) with and without reordering is reported in Fig. 4.5. In all cases, the curves grow like $cN^9$, but the reorderings reduce the overall solution time by about a third.

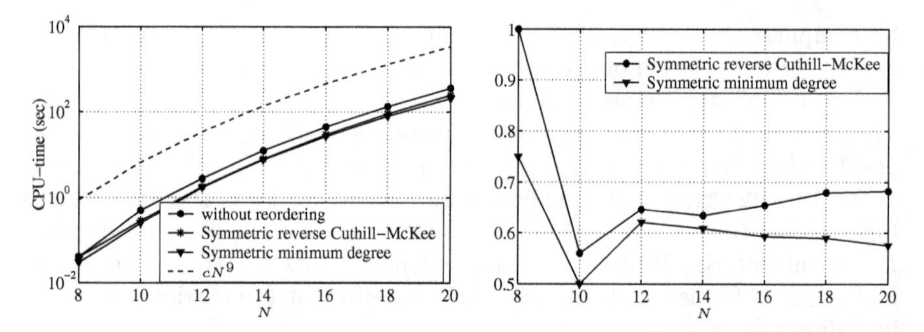


**Fig. 4.5.** At left the CPU-time needed to solve $-\Delta u = f$ with Dirichlet boundary conditions in a cube, with Cholesky factorization, without reordering or two types of reordering: with the symmetric reverse Cuthill-McKee ordering, and the symmetric minimum-degree ordering. At right the ratio between CPU-time of Cholesky factorization with and without reordering. The comparison curve $cN^9$ is drawn for $c = 2/3 \cdot 10^{-8}$

The frontal and multifrontal methods are attractive alternatives to the Gauss or Cholesky decompositions. The basic frontal method performs the Gauss factorization of a large sparse matrix by partly factorizing a sequence of small dense submatrices, called frontal matrices (see Davis and Duff (1999)). Frontal methods work well for matrices with small profile, so that they are usually preceded by an ordering technique to reduce the larger bandwidth of the matrix and to bound the fill-in. Nevertheless high fill-in occurs if the matrix cannot be ordered into a matrix with small bandwidth.

The multifrontal method is a generalization of the frontal method. As the name suggests, several frontal matrices are employed simultaneously during the factorization process; while one frontal matrix is being generated, the work on the other frontal matrices is suspended. The generation of the frontal matrices is suggested by a graph structure (a tree or a forest) with the aim of bounding fill-in as much as possible. Multifrontal methods, like frontal methods, are combined with ordering techniques, usually of "minimum degree" type. However, multifrontal methods do require extra work with respect to frontal methods (for the composition of the constituent frontal matrices).

More refined algorithms that exploit features of both frontal and multifrontal methods enable a general fill-in reduction ordering to be applied while avoiding the data movement of classical multifrontal approaches (see Davis (2004)).

A comparison between the storage needed to solve the linear system using the Cholesky decomposition and the multifrontal method for a two-dimensional Poisson problem with Dirichlet boundary conditions is reported in Table 4.1. Here, $N-1$ is the number of interior LGL points in every direction, $n=(N-1)^2$ is the number of rows (or columns) of $L$. This table indicates that, compared with the straightforward Cholesky scheme, the multifrontal approach requires more than twice as much memory, whereas, as is illustrated later in Fig. 4.46, it takes substantially less CPU time.

**Table 4.1.** Memory requirements for the Cholesky and multifrontal methods

| Method | Words (real) | Words (integer) |
|---|---|---|
| Cholesky | $n(n+1)/2$ | 0 |
| Multifrontal | $n^2+3n^{3/2}$ | $5n^{3/2}+36n$ |

## 4.3 Eigen-Analysis of Spectral Derivative Matrices

Before turning to a discussion of iterative methods for the solution of the implicit equations arising from spectral methods, we shall briefly discuss the eigenvalues and condition numbers of some of the matrices which arise from spectral discretizations of one-dimensional problems with first and/or second derivatives. These have important implications for the conditioning of the matrices in direct methods (Sect. 4.2), for the convergence rate of iterative methods (Sects. 4.4–4.7), and also for the stability restrictions of time discretizations (Appendix D).

As far as the solution of linear systems is concerned, the *condition number* $\kappa$ of a matrix $L$ in some norm $\|\cdot\|$ is given by

$$\kappa_{\|\cdot\|}(L) = \|L\| \left\|L^{-1}\right\| \ . \tag{4.3.1}$$

The condition number in the 2-norm, termed the *spectral condition number*, is given by

$$\kappa_2(L) = \left[\frac{\lambda_{\max}\left(L^T L\right)}{\lambda_{\min}\left(L^T L\right)}\right]^{1/2} \ . \tag{4.3.2}$$

In the case that $L$ is symmetric and positive definite, this becomes

$$\kappa_2(L) = \frac{\lambda_{\max}(L)}{\lambda_{\min}(L)} \ . \tag{4.3.3}$$

This ratio of the largest to the smallest eigenvalue can be an important parameter for symmetric and positive-definite matrices. This is discussed further in Sect. 4.4.1. Precisely, the larger the condition number, the greater the impact of round-off errors in direct methods, and the larger the required number of iterations for iterative techniques. For nonsymmetric matrices, (4.3.1) is still the right indicator for the sensitivity to round-off in direct methods, but not necessarily for the convergence rate of iterative methods.

With respect to explicit time discretizations, the subsequent analysis will demonstrate that, in most cases, spectral methods require a more restrictive time-step limit than standard low-order methods. This is due to the property that the eigenvalues of spectral spatial discretization operators that correspond to the high frequencies grow more rapidly with respect to the discretization parameter $N$ than those of low-order operators.

We note that in the analysis of temporal stability for time-discretization methods, it is rather the *generalized* eigenvalues that matter, i.e., the eigenvalues of the matrix $M^{-1}L$, where $L$ is the matrix associated with the given differential operator, and $M$ is the mass matrix (see Appendix D).

The spatial eigenfunctions of Fourier approximations to constant-coefficient problems are just $e^{ikx}$ for $-N/2 \le k < N/2-1$. The eigenvalues of such problems are apparent. We shall discuss here the behavior of the eigenvalues of various nonperiodic spectral approximations of the second-order diffusion operator $\mathcal{L}u = -\mathrm{d}^2u/\mathrm{d}x^2$, the first-order hyperbolic operator $\mathcal{L}u = \mathrm{d}u/\mathrm{d}x$, and the advection-diffusion operator, $\mathcal{L}u = -\nu\mathrm{d}^2u/\mathrm{d}x^2 + \mathrm{d}u/\mathrm{d}x$. The theoretical discussion of the spectra of these types of matrices, including rigorous bounds on the eigenvalues, is postponed until Sect. 7.3.

### 4.3.1 Second-Derivative Matrices

We consider here second-order eigenvalue problems

$$-\frac{\mathrm{d}^2 u}{\mathrm{d}x^2} = \lambda u \qquad \text{on } (-1,1) \tag{4.3.4}$$

subject to boundary conditions of either Dirichlet type

$$u(-1) = u(+1) = 0 \,, \tag{4.3.5a}$$

or of Neumann type

$$\frac{\mathrm{d}u}{\mathrm{d}x}(-1) = \frac{\mathrm{d}u}{\mathrm{d}x}(1) = 0 \,. \tag{4.3.5b}$$

Spectral discretizations of the above problem lead to algebraic (generalized) eigenvalue problems of the form

$$A\mathbf{u} = \lambda B\mathbf{u} \,, \tag{4.3.6}$$

where $\mathbf{u}$ represents either the expansion coefficients of the discrete solution or else its nodal values. In the case of a collocation approximation to the Dirichlet problem, $A$ is the $(N-1) \times (N-1)$ matrix $L_{\mathrm{coll}}$, constructed from the negative of the square of the interpolation differentiation matrix $D_N$ by deleting its first and last rows and columns, and $B$ is the identity matrix. For the Neumann problem, $A$ is the $(N+1) \times (N+1)$ matrix with its second through next-to-last rows the same as those of the negative of $D_N^2$ and its first and last rows the same as those of $D_N$, while $B$ is equal to the identity matrix except for having 0 rather than 1 on the diagonal of the first and last rows. For tau approximations, the first $N-1$ rows of $A$ are taken from the matrix representation of (the negative of) (2.4.27) for Chebyshev approximation, or (2.3.23) for Legendre, and its last two rows represent the boundary conditions, e.g. (4.1.21b) for Chebyshev approximation to the Dirichlet problem. Similarly, $B$ is the identity matrix, except for having 0 on the diagonal of the last two rows. For Chebyshev and Legendre Galerkin approximations using the bases (4.1.38) and (4.1.49), the matrix $A$ in (4.3.6) is the same as the matrix $K$ in (4.1.35), and $B$ is the matrix $M$ in (4.1.35). For Legendre G-NI approximations, $A$ is the stiffness matrix $K_{GNI}$ and $B$ is the mass matrix $M_{GNI}$ given by (3.8.18) with $\alpha \equiv 1$ and (3.8.11) with $\gamma^* \equiv 1$, respectively. (For the Dirichlet problem, we already noted in Sect. 3.8.2 that $L_{\mathrm{coll}} = M_{GNI}^{-1} K_{GNI}$.)

We show in Sect. 7.3.1 that, except for the zero eigenvalue of the Neumann problem, all the (generalized) eigenvalues of collocation, G-NI and tau approximations are real and positive, and there exist two positive constants, $c_1, c_2$, independent of $N$ such that

$$0 < c_1 \leq \lambda \leq c_2 N^4 \,. \tag{4.3.7}$$

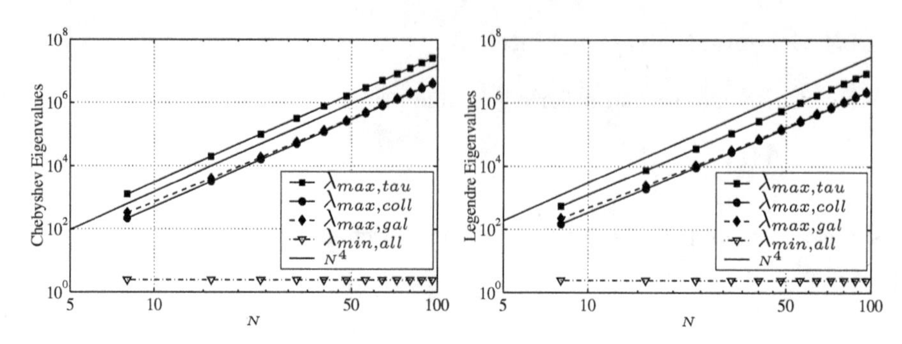


**Fig. 4.6.** Extreme eigenvalues of Chebyshev (*left*) and Legendre (*right*) approximations to the second-order derivative operator with Dirichlet boundary conditions

The extreme eigenvalues of the discrete algebraic eigenvalue problems produced by the second-order differentiation operator with Dirichlet boundary conditions are illustrated in Fig. 4.6 for Chebyshev and Legendre approximations using collocation (here equivalent to G-NI), tau and Galerkin methods. (The corresponding plots for Neumann boundary conditions are similar, differing only by a vertical offset.) Each part of the figure contains a solid line representing a constant times $N^4$, which the maximum eigenvalues track very well. The asymptotic constants for the maximum Dirichlet eigenvalues are given in Table 4.2, for all three methods, plus some results for the Neumann case. The maximum eigenvalue of the Legendre methods is typically only half as large as that of the corresponding Chebyshev method. The tau method has a maximum eigenvalue that can be as much as 6 times larger than the corresponding collocation or Galerkin result.

Of course, the smaller discrete eigenvalues are good approximations to the eigenvalues of the corresponding analytic problem. It is only the upper third of the discrete eigenvalue spectrum which differs from the analytic eigenvalues by more than 10% (see, e.g., Vandeven (1990)). The minimum eigenvalues for the three methods are indistinguishable graphically and are increasingly better approximations to $\pi^2/4$ as $N$ increases.

**Table 4.2.** Asymptotic growth of the largest second-derivative eigenvalues

| Approximation | Collocation | Galerkin | Tau |
|---|---|---|---|
| Chebyshev Dirichlet | $0.04737\ N^4$ | $0.04735\ N^4$ | $0.3028\ N^4$ |
| Legendre Dirichlet | $0.02532\ N^4$ | $0.02532\ N^4$ | $0.1013\ N^4$ |
| Chebyshev Neumann | $0.01418\ N^4$ | | $0.02531\ N^4$ |
| Legendre Neumann | $0.006332\ N^4$ | | $0.02531\ N^4$ |

**Table 4.3.** Asymptotic trends of the G-NI second-derivative eigenvalues

| Approximation | $\lambda_{\max}$ | $\lambda_{\min}$ |
|---|---|---|
| Legendre G-NI Dirichlet | $0.3624\ N^2$ | $7.0416\ N^{-1}$ |
| Legendre G-NI Neumann | $0.4629\ N^2$ | $3.6326\ N^{-1}$ |

The (generalized) algebraic eigenvalue problem (4.3.6) is the discrete counterpart of the differential eigenvalue problem (4.3.4). On the other hand, for the G-NI method another algebraic eigenvalue problem that matters is simply

$$K_{GNI}\mathbf{u} = \lambda\mathbf{u} \,. \tag{4.3.8}$$

Indeed, these are the eigenvalues that affect the direct or iterative solution of the system $K_{GNI}\mathbf{u} = \mathbf{b}$, which is nothing but (4.8). For the considered boundary conditions (again, aside from the zero eigenvalue of the Neumann problem), we have

$$0 < c_3 N^{-1} \leq \lambda \leq c_4 N^2$$

for suitable positive constants $c_3$ and $c_4$ independent of $N$ (see Sect. 7.3.1). Figure 4.7 illustrates the extreme eigenvalues of just the stiffness matrix for Legendre G-NI approximations to the Dirichlet and Neumann problems; they behave as predicted by the theory. Indeed, the maximum eigenvalues grow as $O(N^2)$, whereas the minimum eigenvalues decay as $O(N^{-1})$. Table 4.3 provides the asymptotic constants. Therefore, the spectral condition numbers grow like $O(N^3)$. This feature makes the sensitivity to round-off errors in direct methods less dramatic. Moreover, the different spectral properties of $K_{GNI}$ and $L_{\text{coll}}$ may suggest different preconditioning strategies; see Sect. 4.4.2.

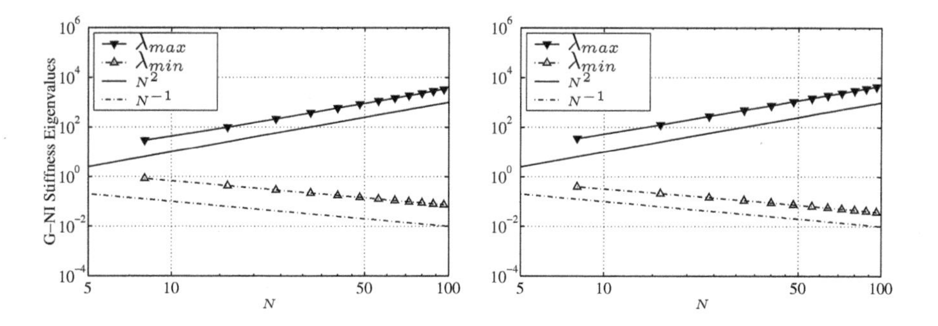


**Fig. 4.7.** Extreme eigenvalues of Legendre G-NI stiffness matrices for the second-order derivative operator with Dirichlet (*left*) and Neumann (*right*) boundary conditions

### 4.3.2 First-Derivative Matrices

We consider next the advection operator

$$\mathcal{L}u = \frac{\mathrm{d}u}{\mathrm{d}x} \qquad \text{on } (-1,1)\,, \tag{4.3.9}$$

subject to the boundary condition

$$u(1) = 0\,. \tag{4.3.10}$$

We confine ourselves in this subsection to just collocation and G-NI methods, since tau and Galerkin methods are rarely used on first-order problems. The eigenanalysis of spectral discretizations for first-order operators can certainly also be discussed in terms of an eigenvalue problem and written in the form (4.3.6). However, for the present discussion, which is restricted to fewer approximation approaches, we prefer to focus on the derivative matrix for collocation methods (with the boundary condition used to eliminate the variable $u_N$ from the matrix) and the stiffness matrix for the G-NI methods, which incorporates the boundary conditions. We denote the collocation first-derivative matrix by $L_{\mathrm{coll}}$ and the G-NI stiffness and mass matrices by $K_{GNI}$ and $M_{GNI}$, respectively. We also examine the matrix $M_{GNI}^{-1}K_{GNI}$, referred to as the generalized G-NI matrix.

Consider first the matrices for collocation differentiation. The boundary condition (4.3.10) implies that $L_{\mathrm{coll}}$ is an $N \times N$ matrix obtained from the interpolation differentiation matrix $D_N$, by deleting its last row, while using that last row to eliminate the last column. For Chebyshev and Legendre collocation, we show in Sect. 7.3.3 that the real parts of the eigenvalues $\lambda$ of $L_{\mathrm{coll}}$ are strictly negative, while their moduli satisfy a bound of the form

$$|\lambda| \leq O(N^2)\,. \tag{4.3.11}$$

Figure 4.8 illustrates the eigenvalues of $L_{\mathrm{coll}}$ computed in 64-bit arithmetic for a Chebyshev collocation method. These results indicate that the estimate (4.3.11) is sharp. However, as noted by Trefethen and Trummer (1987), round-off errors have a significant effect upon numerical computations of first-derivative eigenvalues. For the 64-bit computations illustrated in Fig. 4.8, these round-off error effects become apparent for $N > 32$. Trefethen and Trummer explain that the source of the problem is the exponentially (in $x$) decaying character of the eigenvectors: these behave roughly as $e^{x\mathrm{Re}\{\lambda\}}$. Once $e^{+2\mathrm{Re}\{\lambda\}}$ falls below the machine precision (the 2 comes from the length of the interval), the eigenfunctions cannot be approximated in any meaningful sense. Since the real part of $\lambda$ becomes increasingly negative as $N$ increases, there will be a value of $N$ beyond which the eigenvalues can no longer be computed reliably (with fixed-precision arithmetic).

At a more fundamental level, neither the Chebyshev collocation nor the Legendre collocation first-derivative matrices are normal matrices.

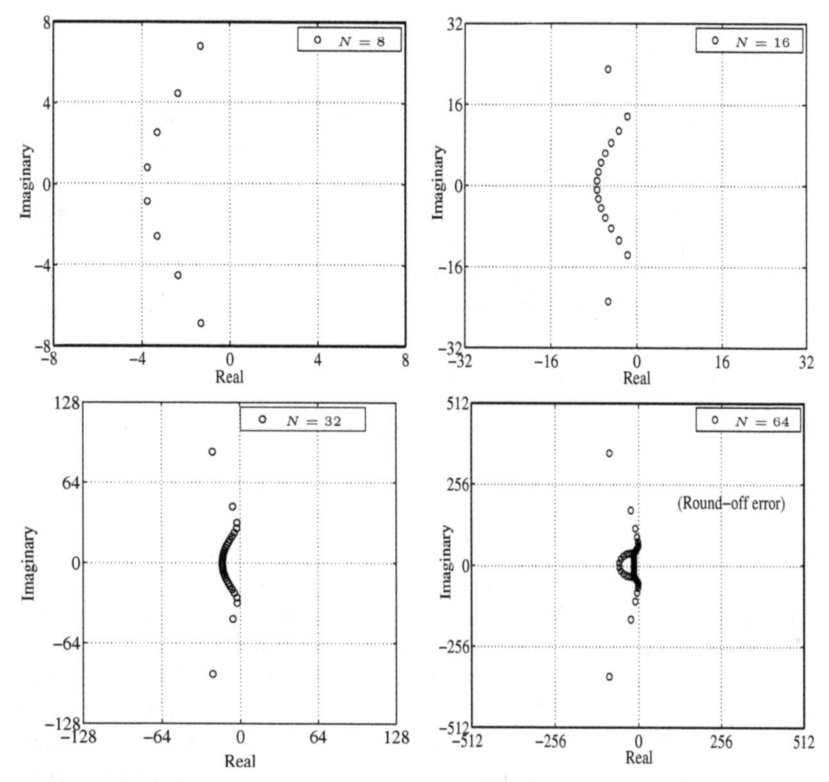


**Fig. 4.8.** Chebyshev collocation first-derivative eigenvalues computed with 64-bit precision. Results contaminated by round-off error are indicated

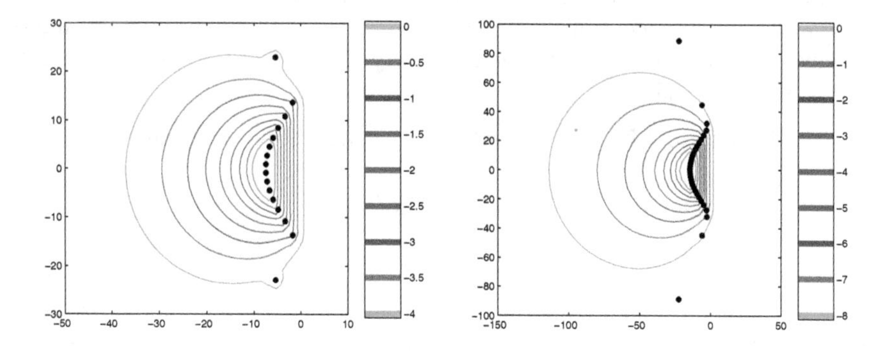

**Fig. 4.9.** $\epsilon$-pseudospectra, $\Lambda_\epsilon$, of Chebyshev collocation first-derivative matrix. $\Lambda_\epsilon$ is plotted for $\epsilon = 10^{-4}, 10^{-3.5}, \ldots, 10^0$ for $N = 16$ (*left*) and for $\epsilon = 10^{-8}, 10^{-7}, \ldots, 10^0$ for $N = 32$ (*right*). The innermost isoline corresponds to the minimum value of $\log_{10} \epsilon$, the outermost to the maximum value of $\log_{10} \epsilon$. These are $-4$ and $0$ in the left-hand figure, and $-8$ and $0$ in the right-hand figure

(A matrix $L$ is termed *normal* if $L^T L = LL^T$.) Normal matrices have a complete set of orthogonal eigenvectors. It is the lack of orthogonality of the eigenvectors of the first-derivative matrices that is responsible for the numerical difficulties in computing the eigenvalues. Even though the matrices for the collocation and tau methods for the second-derivative problem are not normal, they are not nearly as sensitive to round-off error.

Trefethen (1992) provided a more complete explanation of this sensitivity of the computed eigenvalues of the first-derivative matrices in terms of pseudospectra. (See Trefethen (1997) for a review of origins of the study of pseudospectra and Embree and Trefethen (2005) for comprehensive coverage of the subject.) For any $\epsilon > 0$, the *$\epsilon$-pseudospectrum* of a matrix $L$ is defined to be that subset of the complex plane defined by

$$\Lambda_\epsilon(L) = \{z \in \mathbb{C} : z \in \Lambda(L + E) \text{ for some } E \text{ with } \|E\| \le \epsilon\} , \qquad (4.3.12)$$

where $\Lambda(L)$ is the set of the eigenvalues of $L$. The usual spectrum is produced for $\epsilon = 0$. Loosely speaking, for $\epsilon > 0$, the $\epsilon$-pseudospectrum is the set of points which are elements of the spectrum of some matrix which differs from $L$ (in norm) by no more than $\epsilon$. For a normal matrix, $\Lambda_\epsilon(L)$ is the same as the set $\{z \in \mathbb{C} : |z - \Lambda(L)| \le \epsilon\}$. However, if $L$ is not normal, $\Lambda_\epsilon(L)$ can be a much larger set. This is precisely the situation for these first-derivative matrices. They are not normal, and the $\Lambda_\epsilon(L)$ sets have a radius much larger than $\epsilon$. This is illustrated in Fig. 4.9, which displays some pseudospectra for the Chebyshev collocation first-derivative matrix. (All pseudospectra figures in this section were generated using the Eigtool software of Wright; see Wright and Trefethen (2001) for the details of the algorithm.) Clearly, the size of the $\epsilon$-pseudospectra sets are orders of magnitude larger than $\epsilon$.

The eigenvalues and pseudospectra of the Legendre collocation first-derivative matrix have similar behavior (see Figs. 4.10 and 4.11). The largest eigenvalues are much closer to the imaginary axis, but still have negative real parts (see Sect. 7.3).

The Legendre G-NI approximation of the advection problem

$$\begin{aligned} \frac{du}{dx} &= f \qquad \text{in } (-1,1) , \\ u(1) &= 0 , \end{aligned}$$

with the weak imposition of the boundary conditions (by analogy to what was done in (3.7.5) for the time-dependent problem (3.7.1)) reads as follows: find $u^N \in \mathbb{P}_N$ such that

$$-(u^N, v_x^N)_N - u^N(-1)v^N(-1) = (f, v^N)_N \qquad \forall v^N \in \mathbb{P}_N , \qquad (4.3.13)$$

where $(\cdot,\cdot)_N$ is the LGL inner product. Note that, by analogy with what was already shown in Sect. 3.8 for the corresponding time-dependent problem, (4.3.13) can be equivalently written as

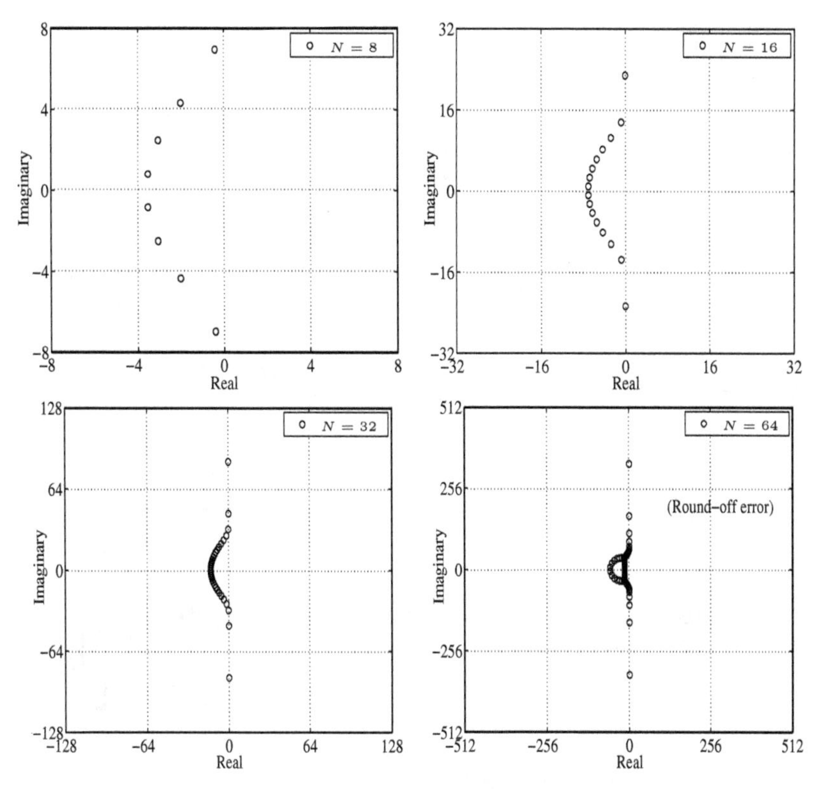


**Fig. 4.10.** Legendre collocation first-derivative eigenvalues computed with 64-bit precision. Results contaminated by round-off error are indicated

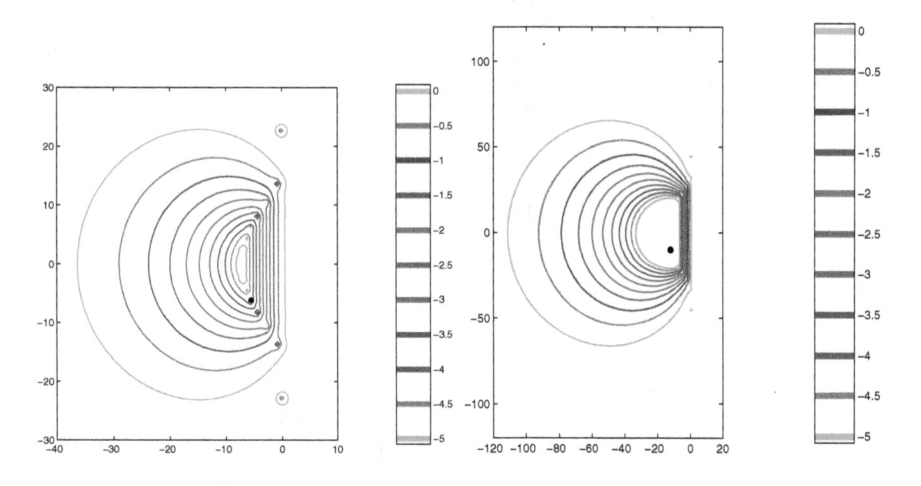

**Fig. 4.11.** $\epsilon$-pseudospectra, $\Lambda_\epsilon$, of Legendre collocation first-derivative matrix. $\Lambda_\epsilon$ is plotted for $\epsilon = 10^{-5}, 10^{-4.5}, \ldots, 10^{0}$ for both $N = 16$ (*left*) and $N = 32$ (*right*). The range of the isoline values are $[-5, 0]$ for both figures

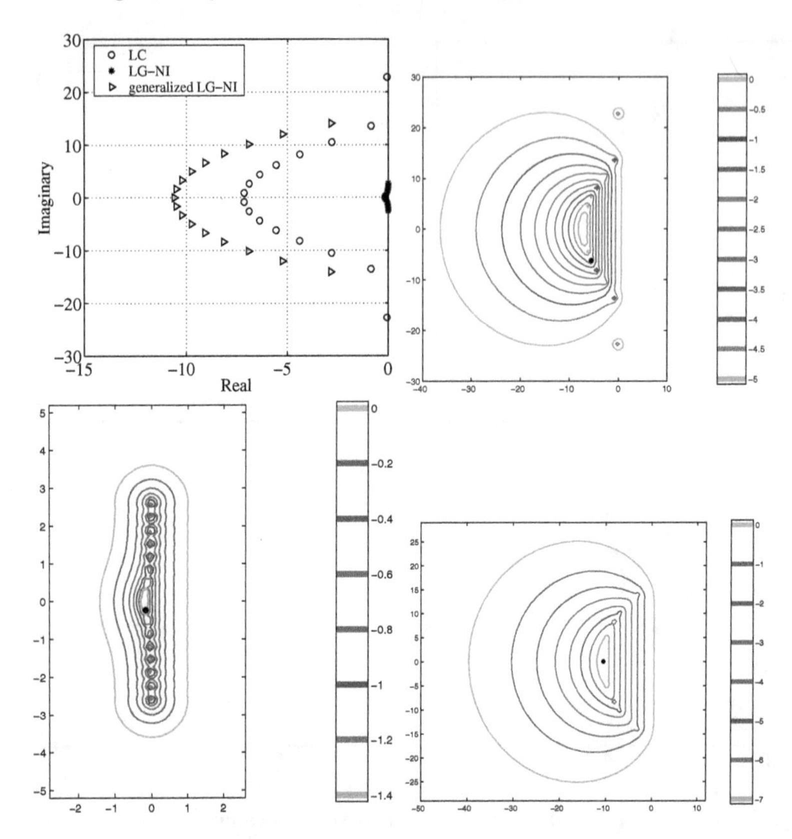


**Fig. 4.12.** Legendre first-derivative spectra and pseudospectra for $N = 16$. *Top-left*: spectra. *Top-right*: spectrum and pseudospectra of $L_{\text{coll}}$ (LC), *bottom-left*: spectrum and pseudospectra of $K_{GNI}$ (LG-NI), *bottom-right*: spectrum and pseudospectra of $M_{GNI}^{-1}K_{GNI}$ (generalized LG-NI). The range for isolines is $[-5, 0]$ for the upper right figure, $[-1.4, 0]$ for the lower left figure, $[-7, 0]$ for the lower right figure

$$(u_x^N, v_N)_N - u^N(1)v_N(1) = (f, v_N)_N \qquad \forall v^N \in \mathbb{P}_N \,. \tag{4.3.14}$$

The associated $(N+1) \times (N+1)$ matrix that represents the left-hand side of (4.3.13) for the nodal basis is

$$K_{GNI} = -D_N^T M_{GNI} - \text{diag}\{1, 0, \ldots, 0\} \,, \tag{4.3.15}$$

where $D_N$ is the first-derivative matrix (2.3.28) and $M_{GNI} = \text{diag}\{w_0, \ldots, w_N\}$ is the diagonal mass matrix of the LGL integration weights.

Figures 4.12 and 4.13 illustrate the spectra and pseudospectra for $N = 16$ and $N = 64$, respectively, of the matrices for Legendre collocation ($L_{\text{coll}}$), Legendre G-NI ($K_{GNI}$, see (4.3.15)), and generalized Legendre G-NI ($M_{GNI}^{-1}K_{GNI}$) approximations. The spectra for the G-NI matrix are relatively insensitive to round-off errors, unlike the spectra for the other two

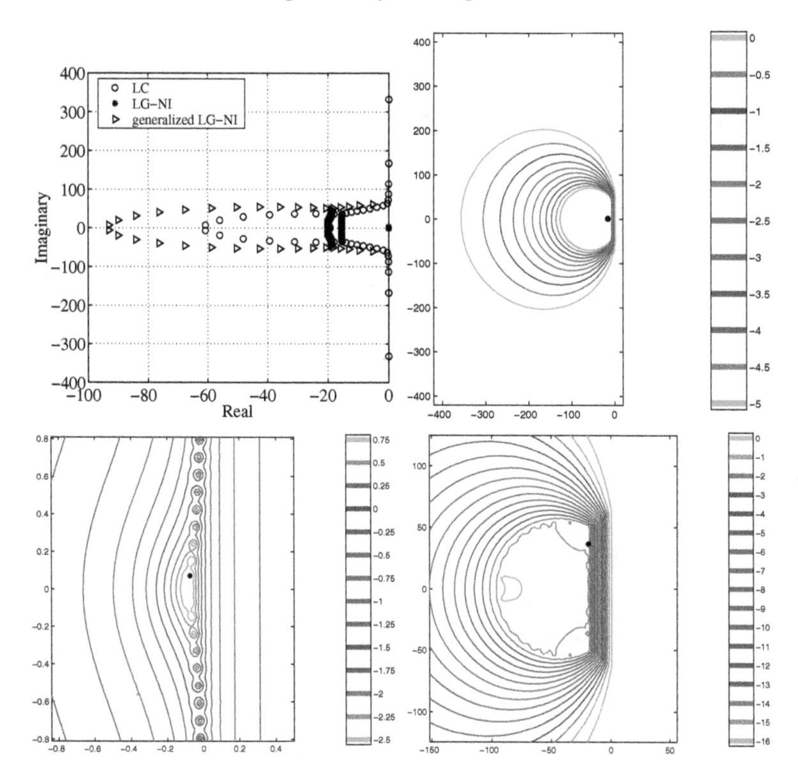


**Fig. 4.13.** Legendre first-derivative spectra and pseudospectra for $N = 64$. *Top-left*: spectra. *Top-right*: spectrum and pseudospectra of $L_{\rm coll}$ (LC), *bottom-left*: spectrum and pseudospectra of $K_{GNI}$ (LG-NI), *bottom-right*: spectrum and pseudospectra of $M_{GNI}^{-1}K_{GNI}$ (generalized LG-NI). The range for isolines is [-5,0] for the upper right figure, [-2.5,0.75] for the lower left figure, [-16,0] for the lower right figure

matrices. The generalized G-NI matrix is even more sensitive than the collocation matrix. The extreme eigenvalues for these matrices, as computed in 64-bit arithmetic, are displayed in the left half of Fig. 4.14. The abrupt slope changes in some of the curves for the extreme eigenvalues are produced by round-off error effects, as can be seen by careful comparison of Figs. 4.13 and 4.14.

The condition numbers $\kappa_2(L)$ in the 2-norm for these matrices, again as computed in 64-bit arithmetic, are displayed in the right half of Fig. 4.14. The condition numbers of both $L_{\rm coll}$ and $M_{GNI}^{-1}K_{GNI}$ scale as $O(N^2)$, whereas those of $K_{GNI}$ scale sublinearly with $N$.

The Fourier first-derivative matrix is skew-symmetric (see Sect. 2.1.3), and therefore is a normal matrix. Hence, the numerically computed eigenvalues of the Fourier collocation first-derivative matrix are not nearly so

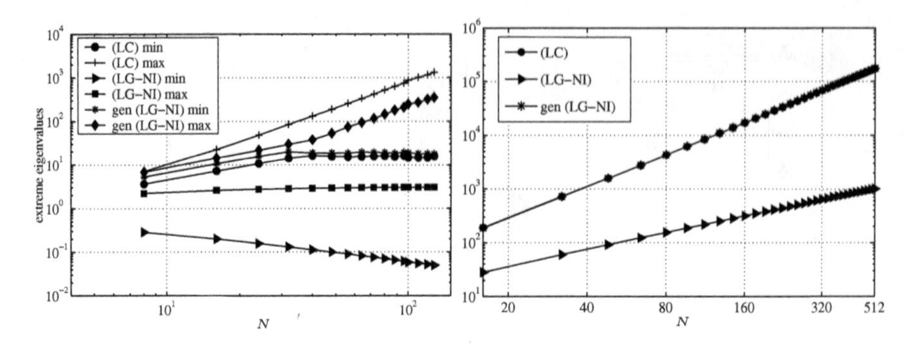


**Fig. 4.14.** Maximum and minimum moduli of Legendre first-derivative matrix eigenvalues (*left*) and the spectral condition numbers, $\kappa_2(\cdot)$ of these matrices (*right*)

susceptible to round-off errors. Moreover, the eigenvalues can be determined analytically.

### 4.3.3 Advection-Diffusion Matrices

The operator of interest here is the advection-diffusion operator

$$\mathcal{L}u = -\nu \frac{\mathrm{d}^2 u}{\mathrm{d}x^2} + \frac{\mathrm{d}u}{\mathrm{d}x} \qquad \text{on } (-1,1) . \tag{4.3.16}$$

Theoretical bounds on the eigenvalues of the matrices resulting from some spectral approximations to the advection-diffusion operator are discussed in Sect. 7.3.2. The eigenvalues are complex, but with real parts bounded from below. For Legendre Galerkin or G-NI (collocation) methods, they have indeed non-negative real parts. (The same behavior occurs if a variable coefficient $\beta$ multiplies the first derivative operator, provided $\nu$ is sufficiently large.)

Here, we shall illustrate only Dirichlet boundary conditions (see (4.3.5a)) and confine ourselves to Legendre G-NI approximations. Let $K_{GNI}$ and $M_{GNI}$ denote the stiffness and mass matrices for a G-NI approximation to this advection-diffusion problem. The matrix $M_{GNI}^{-1}K_{GNI}$ is the generalized G-NI matrix. We examine three cases: $\nu = 1$, $\nu = 10^{-2}$, and $\nu = 10^{-3}$. Figures 4.15 and 4.16 display the minimum and maximum moduli of the eigenvalues of $K_{GNI}$, and $M_{GNI}^{-1}K_{GNI}$, respectively. The lines on these figures represent the asymptotic trends of the eigenvalues. As one would expect, for $N$ large enough the extreme eigenvalues have the same asymptotic scaling as for the pure second-order problem, i.e., $|\lambda(K_{GNI})|_{\min} = O(N^{-1})$ and $|\lambda(K_{GNI})|_{\max} = O(N^2)$ for the stiffness matrix, and $|\lambda(M_{GNI}^{-1}K_{GNI})|_{\min} = O(1)$ and $|\lambda(M_{GNI}^{-1}K_{GNI})|_{\max} = O(N^4)$ for the generalized G-NI matrix.

The behavior is different when $N$ is insufficiently large with respect to $1/\nu$ to guarantee that $\nu N^2 \gtrsim 1$. For the model boundary-value problem

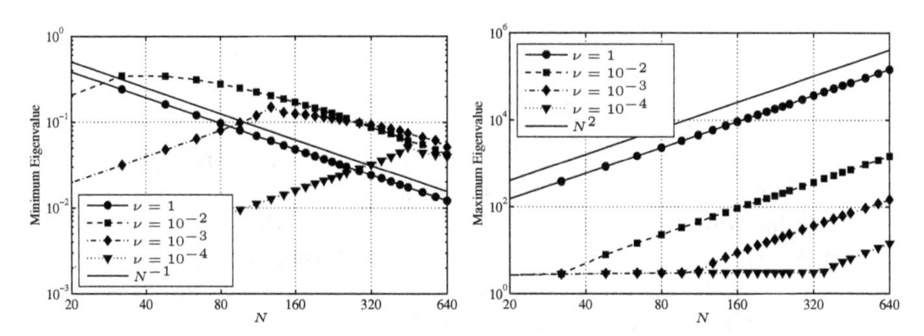


**Fig. 4.15.** Minimum (*left*) and maximum (*right*) moduli of the eigenvalues of Legendre G-NI advection-diffusion stiffness matrices ($K_{GNI}$) for Dirichlet boundary conditions

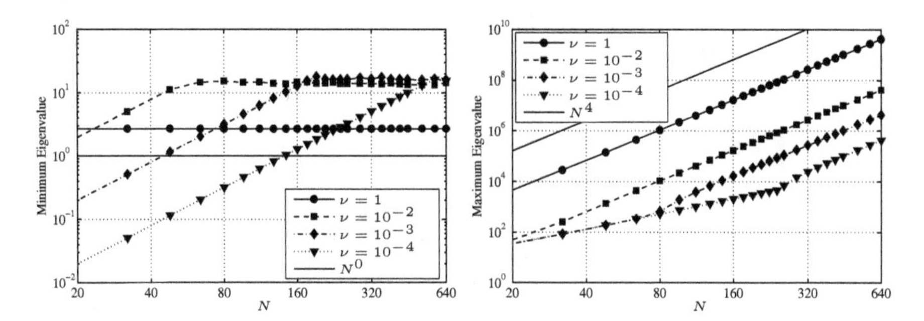


**Fig. 4.16.** Minimum (*left*) and maximum (*right*) moduli of the eigenvalues of Legendre generalized G-NI advection-diffusion matrices ($M_{GNI}^{-1}K_{GNI}$) for Dirichlet boundary conditions

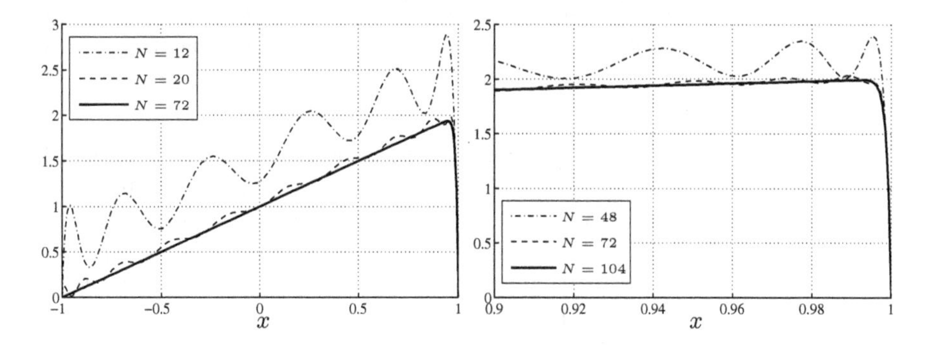


**Fig. 4.17.** Solution of the advection-diffusion problem (4.3.17) for different values of $N$ and for $\nu = 10^{-2}$ (*left*) and $\nu = 10^{-3}$ (*right*)

$$
\begin{aligned}
&-\nu \frac{\mathrm{d}^2 u}{\mathrm{d}x^2} + \frac{\mathrm{d}u}{\mathrm{d}x} = 1\,, \qquad\qquad -1 < x < 1\,, \\
&u(-1) = 0\,,\ u(1) = 0\,,
\end{aligned}
\tag{4.3.17}
$$

this situation corresponds to a numerically unresolved boundary layer. The numerical solution of such an unresolved problem contains spurious oscillations, as illustrated in Fig. 4.17. (See the theoretical discussion in Sect. 7.2; in particular, see (7.2.3) or (7.2.13), and the discussion after (7.2.16).) Then, the regime of behavior of extreme eigenvalues is that of the pure-convection, first-order GNI matrix; a numerical stabilization (see Sect. 7.2.1) should be used in order to get rid of potential instabilities.

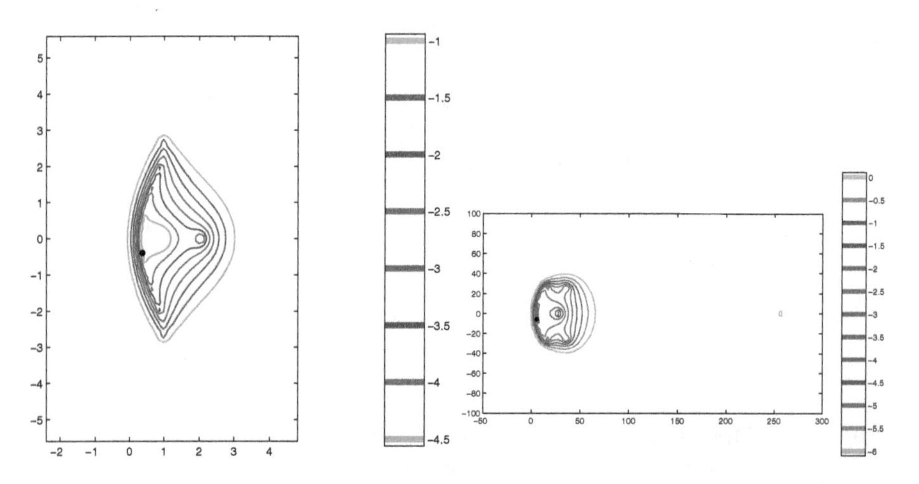

**Fig. 4.18.** Spectrum and pseudospectra of Legendre G-NI advection-diffusion matrices with $N = 32$ for $\nu = 10^{-2}$. Stiffness matrix $K_{GNI}$ (*left*) and generalized matrix $M_{GNI}^{-1} K_{GNI}$ (*right*). The range for isolines is $[-4.5, -1]$ on the left, and $[-6, 0]$ on the right

To illustrate the sensitivity of the spectra to round-off errors, we furnish the pseudospectra in Figs. 4.18 and 4.19 for $\nu = 10^{-2}$ and $\nu = 10^{-5}$, respectively. For the advection-diffusion problem, as for the pure first-order problem, the generalized G-NI matrix is more sensitive to round-off error than the stiffness matrix. Perhaps surprisingly, there is greater sensitivity to round-off error for the $\nu = 10^{-2}$ case than for the $\nu = 10^{-5}$ one.

## 4.4 Preconditioning

From the previous eigen-analysis it appears clear that spectral matrices ought to be preconditioned when solving the associated systems by iterative

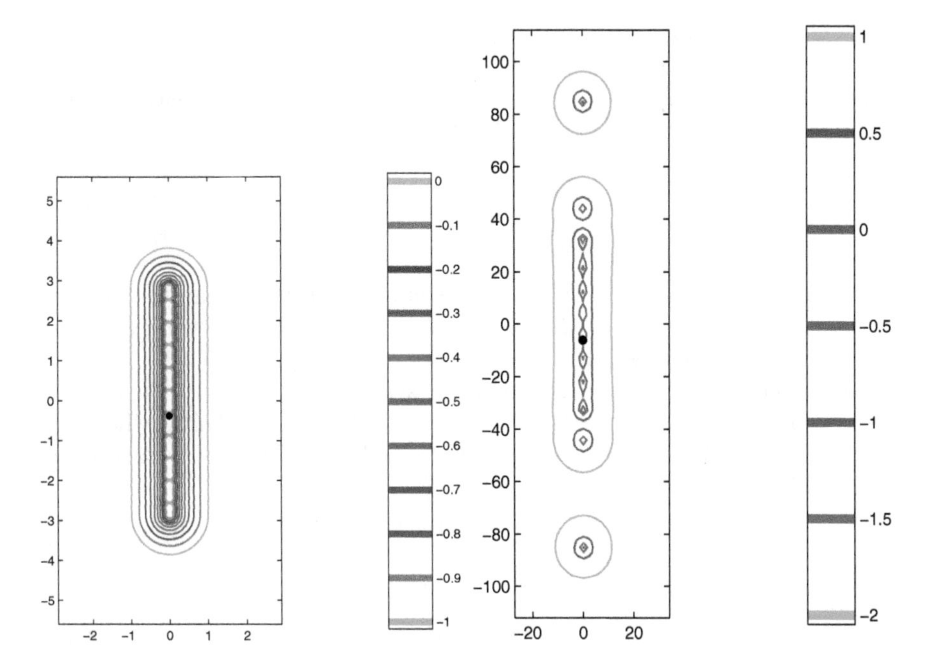


**Fig. 4.19.** Spectrum and pseudospectra of Legendre G-NI advection-diffusion matrices with $N = 32$ for $\nu = 10^{-5}$. Stiffness matrix $K_{GNI}$ (*left*) and generalized matrix $M_{GNI}^{-1}K_{GNI}$ (*right*). The range for isolines is $[-1, 0]$ on the left, and $[-2, 1]$ on the right

methods. We begin this section with an elementary discussion of iterative methods that serves to motivate the practical necessity for using preconditioning. Then we examine the basics of low-order finite-difference and finite-element preconditioning for spectral discretizations by considering several one-dimensional model problems. Next, we survey the alternatives for efficient preconditioning in several dimensions. Finally, we summarize the use of spectral discretizations of constant-coefficient operators as preconditioners for variable-coefficient operators.

### 4.4.1 Fundamentals of Iterative Methods for Spectral Discretizations

The fundamentals of iterative methods for spectral equations, as well as the effect of preconditioning, are perhaps easiest to grasp for the simple one-dimensional model problem

$$\begin{aligned} -\frac{\mathrm{d}^2 u}{\mathrm{d}x^2} &= f \qquad \text{in } (0, 2\pi)\,, \\ u & \quad 2\pi\text{-periodic}\,, \end{aligned} \tag{4.4.1}$$

even though the practical motivation for iterative methods in general and preconditioning in particular becomes obvious only for multidimensional problems. The Fourier approximation to (4.4.1) at the collocation points, $x_j = 2\pi j/N$ for $j = 0, \ldots, N-1$ (for $N$ even), is

$$\sum_{p=-N/2+1}^{N/2-1} p^2 \tilde{u}_p e^{ipx_j} = f_j \;, \tag{4.4.2}$$

where $\tilde{u}_p$ are the discrete Fourier coefficients of $u$. (Other than to note here that we disable the $p = -N/2$ mode for the usual reasons, we won't comment on this mode in this section.)

This may be represented by the linear system (4.8) with $\mathbf{u} = (u_0, u_1, \ldots, u_{N-1})$, $\mathbf{b} = \mathbf{f} = (f_0, f_1, \ldots, f_{N-1})$, and $L = -D_N^2 = -C^{-1}K^2C$ where $D_N$ is given by (2.1.51), $C$ by (4.1.9) and $K$ by (4.1.10). The eigenvectors of this approximation are

$$\xi_j(p) = e^{2\pi ijp/N} \;, \tag{4.4.3}$$

with the corresponding eigenvalues

$$\lambda(p) = p^2 \;, \tag{4.4.4}$$

where $j = 0, 1, \ldots, N-1$ and $p = -N/2+1, \ldots, N/2-1$. The index $p$ has a natural interpretation as the frequency of the eigenvector. The $p = 0$ eigenvector corresponds to the mean level of the solution. Since it is at one's disposal for this problem, it can essentially be ignored.

The conceptually simplest iterative method to solve the linear system obtained in this way is the Richardson method, which is reviewed in Sect. C.1 of Appendix C. Given an initial guess $\mathbf{v}^0$ to $\mathbf{u}$, subsequent approximations are obtained via

$$\mathbf{v}^{n+1} = \mathbf{v}^n + \omega \mathbf{r}^n \;, \tag{4.4.5}$$

where $\omega$ is a relaxation parameter and $\mathbf{r}^n = \mathbf{b} - L\mathbf{v}^n$ is the residual associated with $\mathbf{v}^n$. The Richardson method is applicable, since all the eigenvalues of the matrix $L$ are positive (ignoring the eigenvalue for $p = 0$) and lie in the interval $[\lambda_{\min}, \lambda_{\max}]$, where $\lambda_{\min} = 1$ and $\lambda_{\max} = N^2/4$. However, the iterative condition number $\mathcal{K} = \lambda_{\max}/\lambda_{\min}$ introduced in (C.1.10) is given by

$$\mathcal{K} = \frac{1}{4} N^2 \;. \tag{4.4.6}$$

Thus, even with the optimal choice of the relaxation parameter given by (C.1.8), the number $\mathcal{J}$ of iterations required to reduce the error by a factor of $e$ satisfies

$$\mathcal{J} \cong \frac{1}{8} N^2 \tag{4.4.7}$$

(see (C.1.13)), i.e., it is proportional to the square of the cut-off parameter $N$.

The major expense in Richardson iteration is the evaluation of $L\mathbf{v}^n$. For the problem at hand, this requires $5N\log_2 N$ operations via transform methods and $2N^2$ operations for matrix multiplies. To reduce the error by a single order of magnitude takes $1.4N^3\log_2 N$ operations, which is more than the cost of a direct solution in one dimension. This observation motivates the introduction of a preconditioner for the linear system, as discussed in Appendix C.

### 4.4.2 Low-Order Preconditioning of Model Spectral Operators in One Dimension

Preconditioning techniques have been investigated extensively for finite-difference and finite-element methods (see Evans (1983), Saad (1996)). The preconditioned version of Richardson's method is (see (C.1.16))

$$H\left(\mathbf{v}^{n+1}-\mathbf{v}^n\right)=\omega\mathbf{r}^n \tag{4.4.8}$$

instead of (4.4.5), where $H$ is the preconditioning matrix. Orszag (1980) proposed a preconditioning for spectral methods in physical space which amounts to using a low-order finite-difference approximation as $H$. The subsequent discussion will presume periodic boundary conditions, which lead us to analyze Fourier methods; later on in this section, we will consider nonperiodic conditions (hence, Chebyshev and Legendre methods).

#### *Fourier collocation operators*

Let $H^{(fd2)}$, $H^{(fd4)}$ and $L$ denote second-order finite-difference, fourth-order finite-difference and spectral collocation discretizations of the operator $\mathcal{L}=-\mathrm{d}^2/\mathrm{d}x^2$ with periodic boundary conditions in $(0,2\pi)$. For example, the second-order finite-difference approximation to (4.4.1) is given by

$$-\frac{u_{j+1}-2u_j+u_{j-1}}{(\Delta x)^2}=f_j\,,\qquad j=0,1,\dots,N-1\,, \tag{4.4.9}$$

where $\Delta x=2\pi/N$, $u_j\simeq u(x_j)$, $f_j=f(x_j)$, with $x_j=j\Delta x$. (We adopt the obvious convention that $u_{j\pm N}=u_j$ for all $j=0,\dots,N-1$.) The inversion of (4.4.9) requires the solution of a cyclic tridiagonal system. The fourth-order approximation is equally straightforward and requires the solution of a cyclic pentadiagonal system. Both types of systems can be inverted far more quickly than the computation of $L\mathbf{v}^n$. The eigenfunctions of these discretizations are all given by (4.4.3), and the eigenvalues of $H^{(fd2)}$ and $H^{(fd4)}$ are

$$\lambda_p^{(fd2)}=4\frac{\sin^2\left(\dfrac{p\Delta x}{2}\right)}{(\Delta x)^2}\,,\qquad \lambda_p^{(fd4)}=\frac{\cos(2p\Delta x)-16\cos(p\Delta x)+15}{6(\Delta x)^2}\,, \tag{4.4.10}$$

where $p = -N/2+1, \ldots, N/2-1$. Since the spectral operator and the finite-difference operator have the same eigenfunctions, it is clear that the effective eigenvalues of the preconditioned iterations based on $(H^{(fd2)})^{-1}L$ and $(H^{(fd4)})^{-1}L$ are then given by

$$\Lambda_p^{(fd2)} = (p^2)(\lambda_p^{(fd2)})^{-1} = \frac{(p\Delta x/2)^2}{\sin^2(p\Delta x/2)}, \tag{4.4.11}$$

$$\Lambda_p^{(fd4)} = (p^2)(\lambda_p^{(fd4)})^{-1} = \frac{6(p\Delta x)^2}{\cos(2p\Delta x) - 16\cos(p\Delta x) + 15}. \tag{4.4.12}$$

The argument $p\Delta x$ lies in $(-\pi, \pi)$ and, in fact, only $[0, \pi)$ need be considered due to symmetry. Similar results for even higher order finite-difference preconditionings are straightforward but, as we shall see, of dubious utility.

An alternative type of preconditioning is based on finite elements rather than finite differences. The use of linear finite elements to precondition Chebyshev collocation approximations was originally advocated by Canuto and Quarteroni (1985) and by Deville and Mund (1985); the latter authors suggested the form that provides better performance than finite-difference preconditioning. The preconditioned operator has a smaller spread of eigenvalues (when used in strong form – see below) and hence a reduced condition number. This feature is mainly due to the weighting of the spectral residuals that is produced when they are operated on by the mass matrix. Other advantages, which emerge for nonperiodic boundary conditions, will be mentioned later on.

The finite-element preconditioning is based on the matrices $K_{FE}$ and $M_{FE}$, which are the stiffness and mass matrices generated by linear finite elements on the equally-spaced grid $\{x_j\}$ $(j = 0, \ldots, N)$; in other words, denoting by $\varphi_j$ the periodic, piecewise linear characteristic Lagrange functions at these nodes, we have $(K_{FE})_{ij} = \int_0^{2\pi} \varphi_{j,x}\varphi_{i,x}\,dx$ and $(M_{FE})_{ij} = \int_0^{2\pi} \varphi_j\varphi_i\,dx$. The corresponding spectral matrices, $K$ and $M$, are defined by $K_{ij} = \int_0^{2\pi} \psi_{j,x}\overline{\psi_{i,x}}\,dx$ and $M_{ij} = \int_0^{2\pi} \psi_j\overline{\psi_i}\,dx$, where $\psi_j$ are the trigonometric polynomial characteristic Lagrange functions at the same nodes (see Sect. 2.1.2). Note that $K_{ij} = (\psi_{j,x}, \psi_{i,x})_N$ and $M_{ij} = (\psi_j, \psi_i)_N$, where $(\cdot,\cdot)_N$ is the discrete inner product defined in (2.1.32); indeed, (2.1.33) holds. Thus, $K = K_{GNI}$ and $M = M_{GNI}$ in the present constant-coefficient, periodic case. Furthermore, the collocation matrix $L_{\text{coll}}$ can be written as $L_{\text{coll}} = M^{-1}K$ (see Sect. 3.8.2).

The finite-element preconditioning can be utilized in either strong form or weak form. For the strong form, (4.4.8) is in effect replaced by

$$K_{FE}\left(\mathbf{v}^{n+1} - \mathbf{v}^n\right) = \omega M_{FE}\mathbf{r}_s^n, \tag{4.4.13}$$

where the strong form of the residual is given by

$$\mathbf{r}_s^n = \mathbf{f} - L_{\text{coll}}\mathbf{v}^n. \tag{4.4.14}$$

The matrix that governs convergence of the strong form of finite-element preconditioning is therefore $K_{FE}^{-1} M_{FE} L_{\text{coll}} = \left(M_{FE}^{-1} K_{FE}\right)^{-1} \left(M^{-1} K\right)$.

The weak form of finite-element preconditioning is

$$K_{FE}(\mathbf{v}^{n+1} - \mathbf{v}^{n}) = \omega \mathbf{r}_w^n \,, \tag{4.4.15}$$

where the weak form of the residual is given by

$$\mathbf{r}_w^n = M\mathbf{f} - K\mathbf{v}^n \,. \tag{4.4.16}$$

The matrix that governs convergence of the weak form of finite-element preconditioning is $K_{FE}^{-1} K$.

For the Fourier model problem, the finite-element stiffness and mass matrices are described by

$$(K_{FE}\mathbf{u})_j = -\frac{u_{j+1} - 2u_j + u_{j-1}}{\Delta x} \,, \tag{4.4.17}$$

$$(M_{FE}\mathbf{f})_j = \frac{1}{6} \Delta x \left(f_{j+1} + 4f_j + f_{j-1}\right) \,. \tag{4.4.18}$$

For the model problem with linear, finite-element preconditioning in strong form, the effective eigenvalues of the preconditioned matrix $K_{FE}^{-1} M_{FE} L_{\text{coll}}$ are

$$\Lambda_p^{(\text{fes})} = \frac{(p\Delta x/2)^2}{\sin^2(p\Delta x/2)} \frac{2 + \cos(p\Delta x)}{3} \,, \tag{4.4.19}$$

whereas for the weak-form preconditioning the eigenvalues $\Lambda_p^{(\text{few})}$ of the preconditioned matrix $K_{FE}^{-1} K$ are identical to those of second-order finite-difference preconditioning (see (4.4.11)).

Figure 4.20 illustrates the eigenvalues for these four preconditionings. Notice that the eigenvalues for the two finite-difference preconditionings and the weak finite-element preconditioning are monotonically increasing with the mode number, whereas the minimum eigenvalue for the model problem preconditioned by linear finite elements in weak form occurs for an interior mode. Table 4.4 summarizes the key properties of this class of preconditioning. Unlike the original system, which has a spectral condition number scaling as $N^2$, the preconditioned system for this model problem for both finite-difference and finite-element preconditioning has a spectral condition number which is independent of $N$. The relatively small reduction in spectral radius achieved by moving from second-order finite-difference preconditioning to fourth-order finite-difference preconditioning suggests that this higher order preconditioning is of doubtful utility. A single iteration with finite-element preconditioning in strong form produces the same reduction in the error that follows from two iterations with second-order finite-difference preconditioning, clearly off-setting the extra cost of applying the mass matrix.

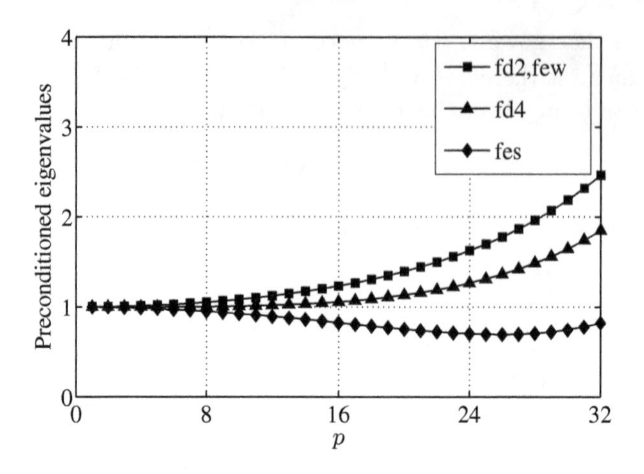


**Fig. 4.20.** Preconditioned eigenvalues for the model problem

**Table 4.4.** Properties of finite-difference and finite-element preconditionings for the model problem (4.4.1)

| Preconditioning | $\Lambda_{\min}$ | $\Lambda_{\max}$ | $\omega_{opt}$ | $\rho$ |
|---|---|---|---|---|
| fd2 | 1.000000 | 2.467401 | 0.5768009 | 0.4231991 |
| fd4 | 1.000000 | 1.850551 | 0.7016188 | 0.2983812 |
| few | 1.000000 | 2.467401 | 0.5768009 | 0.4231991 |
| fes | 0.6928333 | 1.000000 | 1.1814512 | 0.1814512 |

The weak form of finite-element preconditioning converges at the same rate as second-order finite-difference preconditioning.

Effective preconditioning of the first-order advection equation

$$\frac{\mathrm{d}u}{\mathrm{d}x} = f \tag{4.4.20}$$

is far more challenging than for the second-order equation (4.4.1). We continue to presume periodic boundary conditions. Using the second-order central-difference approximation

$$\frac{u_{j+1} - u_{j-1}}{2\Delta x} = f_j \ , \qquad j = 0, 1, \ldots, N-1 \ , \tag{4.4.21}$$

we arrive at the following eigenvalues for the preconditioned matrix:

$$\Lambda_p^{(2)} = \frac{p\Delta x}{\sin(p\Delta x)} \tag{4.4.22}$$

for $|p\Delta x| \in [0, \pi)$. The obvious difficulty is that $\Lambda_{\max}^{(2)}$ is unbounded. No iterative scheme can overcome this property. (Since finite-element preconditioning does not overcome these problems, our discussion will be couched in terms of the simpler finite-difference preconditioning.)

Orszag (1980) suggested one way around this difficulty: in the Fourier collocation evaluation of $du/dx$, simply set the upper third or so of the frequency spectrum to zero. The prescription for this is first to compute

$$\tilde{u}_k = \frac{1}{N}\sum_{j=0}^{N-1} u_j e^{-ikx_j} , \qquad k = -\frac{N}{2}, -\frac{N}{2}+1, \ldots, \frac{N}{2}-1 , \quad (4.4.23)$$

as usual; then to apply a high-mode cut-off, for example,

$$\tilde{u}_k^{(1)} = \begin{cases} ik\tilde{u}_k , & |k| \leq \dfrac{N}{3} , \\ 0 , & \dfrac{N}{3} < |k| \leq \dfrac{N}{2} ; \end{cases} \qquad (4.4.24)$$

and finally to use

$$\left.\frac{du}{dx}\right|_j = \sum_{k=-N/2+1}^{N/2-1} \tilde{u}_k^{(1)} e^{ikx_j} , \qquad j = 0, 1, \ldots, N-1 . \qquad (4.4.25)$$

The relevant range of $|p\Delta x|$ is $[0, 2\pi/3]$. The upper bound on $\Lambda_p^{(2)}$ is 2.42; the lower bound is still 1. In addition to the *loss of accuracy* of the resulting eigenvalues which this method produces, there is also the need to remove the upper third of the spectrum of $f$ so that the residual may be used to monitor the convergence of the scheme.

Another approach is to use a first-order, one-sided finite-difference approximation such as

$$\frac{u_{j+1} - u_j}{\Delta x} = f_j , \qquad j = 0, 1, \ldots, N-1 . \qquad (4.4.26)$$

The eigenvalues resulting from this preconditioning are

$$\Lambda_p^{(1)} = \frac{\dfrac{p\Delta x}{2}}{\sin\left(\dfrac{p\Delta x}{2}\right)} e^{-i(p\Delta x/2)} . \qquad (4.4.27)$$

These eigenvalues are bounded in absolute value but are complex. Since the entire frequency spectrum has been retained, there is no loss of accuracy. However, the iterative scheme must be able to handle complex eigenvalues.

Yet another alternative is to shift, or stagger, the grid on which the derivative is evaluated with respect to the grid on which the function itself is defined. This is illustrated in Fig. 4.21. Fourier derivative evaluations are performed by computing $\tilde{u}_k$ as usual and then using

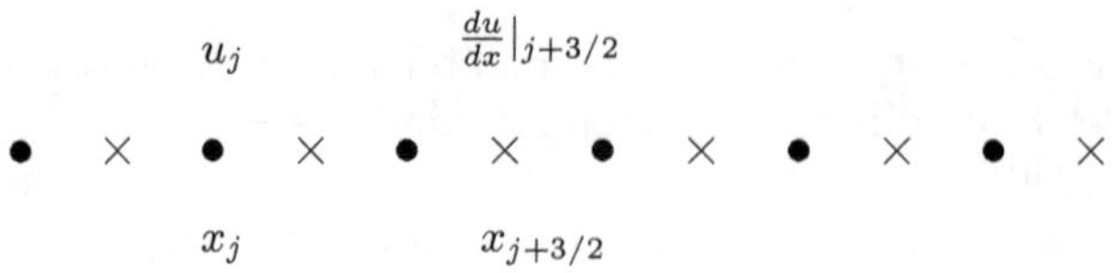

$u_j$ $\frac{du}{dx}|_{j+3/2}$

• × • × • × • × • × • ×

$x_j$ $x_{j+3/2}$

**Fig. 4.21.** The staggered Fourier grid. The standard collocation points are denoted by the circles and the shifted points by the $x$'s

**Table 4.5.** Preconditioned eigenvalues for a one-dimensional first-derivative model problem

| Preconditioning | Eigenvalues |
|---|---|
| Central differences | $\frac{p\Delta x}{\sin(p\Delta x)}$ |
| One-sided differences | $e^{-i(p\Delta x/2)}\frac{p\Delta x/2}{\sin(p\Delta x/2)}$ |
| High-mode cut-off | $\begin{cases} \frac{p\Delta x}{\sin(p\Delta x)} & 0 \le \lvert p\Delta x\rvert \le (2\pi/3) \\ 0 & (2\pi/3) < \lvert p\Delta x\rvert \le \pi \end{cases}$ |
| Staggered grid | $\frac{p\Delta x/2}{\sin(p\Delta x/2)}$ |

$$\left.\frac{\mathrm{d}u}{\mathrm{d}x}\right|_{j+1/2} = \sum_{k=-N/2+1}^{N/2-1} ik\tilde{u}_k e^{ik(x_j+(\pi/N))} \ . \tag{4.4.28}$$

The finite-difference eigenvalues on this staggered grid are

$$\lambda_p^{(s)} = ik\frac{\sin\left(\frac{p\Delta x}{2}\right)}{\frac{p\Delta x}{2}}e^{i(p\Delta x/2)} \ , \tag{4.4.29}$$

but the spectral eigenvalues have a similar complex phase shift. Thus, the preconditioned eigenvalues are

$$\Lambda_p^{(s)} = \frac{p\Delta x}{2} \Big/ \sin\left(\frac{p\Delta x}{2}\right) \ . \tag{4.4.30}$$

These are real and confined to the narrow interval $[1, \pi/2)$. Even the simple Richardson method will perform well with the staggered-grid preconditioning. These alternative first-order preconditionings are summarized in Table 4.5.

Another difficulty is posed by the traditional Helmholtz equation

$$-\frac{\mathrm{d}^2u}{\mathrm{d}x^2} - \lambda u = f \ , \tag{4.4.31}$$

where $\lambda > 0$. The Helmholtz problem with periodic boundary conditions is indefinite with eigenvalues

$$\lambda_p = p^2 - \lambda \,, \tag{4.4.32}$$

but has a well-defined solution so long as $\lambda \neq p^2$ for any integer $p$. Second-order finite-difference preconditioning leads to

$$\begin{aligned} \lambda_p^{(2)} &= 4\sin^2\left(\frac{p\Delta x}{2}\right) \Big/ \Delta x^2 - \lambda, \\ \Lambda_p^{(2)} &= (p^2 - \lambda) / \left( \frac{p^2 \sin^2\left(\frac{p\Delta x}{2}\right)}{\left(\frac{p\Delta x}{2}\right)^2} - \lambda \right) . \end{aligned} \tag{4.4.33}$$

There is likely to be a range of $p$ for which $\Lambda_p^{(2)} < 0$. Thus, a preconditioned version of this Helmholtz problem will have both positive and negative eigenvalues.

As a final example of the complications which can arise in practice, let us consider the advection-diffusion problem

$$-\nu \frac{\mathrm{d}^2 u}{\mathrm{d}x^2} + \frac{\mathrm{d}u}{\mathrm{d}x} = f \,, \tag{4.4.34}$$

still with periodic conditions. Second-order finite-difference preconditioning leads to

$$\begin{aligned} \Lambda_p^{(2)} &= \frac{\nu p^2 + ip}{\nu p^2 \dfrac{\sin^2(p\Delta x/2)}{(p\Delta x/2)^2} + ip \dfrac{\sin(p\Delta x)}{(p\Delta x)}} \\ &= \frac{\nu^4 p^4 \dfrac{\sin^2(p\Delta x/2)}{(p\Delta x/2)^2} + p^2 \sin(p\Delta x)/(p\Delta x)}{\left( \nu^2 p^4 \dfrac{\sin^4(p\Delta x/2)}{(p\Delta x/2)^4} + p^2 \dfrac{\sin^2(p\Delta x)}{(p\Delta x)^2} \right)} \\ &\quad + i \frac{\nu p^3 \sin^2(p\Delta x/2)/(p\Delta x/2)^2 - \nu p^3 \sin(p\Delta x)/(p\Delta x)}{\left( \nu^2 p^4 \dfrac{\sin^4(p\Delta x/2)}{(p\Delta x/2)^4} + p^2 \dfrac{\sin^2(p\Delta x)}{(p\Delta x)^2} \right)} . \end{aligned} \tag{4.4.35}$$

The eigenvalues are complex and although the real parts are positive, there are some real parts which are close to zero for small $\nu$. The staggered-grid preconditioning produces complex eigenvalues as well, but their real parts are safely bounded greater than zero.

***Chebyshev collocation operators***

The eigenvalue ranges of the preconditioned Fourier operator are a good guide to the range of the preconditioned Chebyshev one as well. Chebyshev polynomials would be employed in place of trigonometric functions for problems with Dirichlet or Neumann boundary conditions if a fast transform were desired. The appropriate preconditioning is a second-order finite-difference or a linear finite-element approximation on the non-uniform, Chebyshev grid. For (4.1.19) with $\lambda = 0$, the finite-difference preconditioning is

$$\frac{-2}{h_{j-1}\left(h_j + h_{j-1}\right)} u_{j-1} + \frac{2}{h_j h_{j-1}} u_j + \frac{-2}{h_j\left(h_j + h_{j-1}\right)} u_{j+1} = f_j \qquad (4.4.36)$$

for $j = 1, \ldots, N-1$, with $u_0 = 0$ and $u_N = 0$, where $h_j = x_j - x_{j+1}$ with $x_j = \cos \pi j/N$. Haldenwang et al. (1984) have shown analytically that the eigenvalues of the preconditioned matrix $H^{-1}L_{\text{coll}}$ are given exactly by

$$\Lambda_p^{(2)} = \frac{p(p-1)\sin^2\dfrac{\pi}{2N}\cos\dfrac{\pi}{2N}}{\sin\dfrac{(p-1)\pi}{2N}\sin\dfrac{\pi}{2N}}, \qquad p = 2, 3, \ldots, N. \qquad (4.4.37)$$

Hence,

$$\Lambda_{\min}^{(2)} = 1, \qquad \Lambda_{\max}^{(2)} = N(N-1)\sin^2\frac{\pi}{2N}. \qquad (4.4.38)$$

Note that $\Lambda_{\max}^{(2)} \le \pi^2/4$, which is the same upper bound that applies to the second-order preconditioned Fourier operator. Francken, Deville and Mund (1990) derived an analytical approximation to the eigenvalues for linear, finite-element preconditioning in strong form of this same one-dimensional problem:

$$\Lambda_p^{(2)} \simeq \frac{p(p-1)\sin^2\dfrac{\pi}{2N}\cos\dfrac{\pi}{2N}}{\sin\dfrac{(p-1)\pi}{2N}\sin\dfrac{\pi}{2N}} \times\left[2 + \cos\frac{(p-2)\pi}{N} - \tan\frac{\pi}{2N}\sin\frac{(p-2)\pi}{N}\right], \qquad p = 2, 3, \ldots, N. \qquad (4.4.39)$$

They show reasonable agreement between their estimate of the eigenvalues and numerically computed ones. However, there are a small number of eigenvalues with imaginary parts as large as $\pm 0.1$.

The finite-difference preconditioning matrix for a Chebyshev collocation approximation to the second-derivative operator is not symmetric. However, Heinrichs (1988) noted that it can be symmetrized by scaling the i-th row by $\sin(i\pi/N)$.

Funaro (1987) has analyzed the staggered grid preconditioning for the nonperiodic first-order problem (4.4.20) with Dirichlet boundary conditions at $x = +1$ using Chebyshev collocation. He has shown that the preconditioned eigenvalues are

$$\Lambda_p^{(s)} = p \sin \frac{\pi}{2N} \Big/ \sin \frac{p\pi}{N} , \qquad p = 1, \ldots, N . \tag{4.4.40}$$

These are confined to the interval $[1, \pi/2]$, just as they are for the periodic problem. Funaro also presents some theoretical and numerical results for preconditioned, one-dimensional first-order systems.

Numerical eigenvalue calculations by Phillips, Zang and Hussaini (1986) indicate that the largest eigenvalue for the fourth-order finite-difference preconditioning of the Chebyshev second-derivative operator is bounded by 1.85. Once again, the estimate from the preconditioned Fourier operator is reliable for the more complicated Chebyshev case. Even for the periodic problem, fourth-order preconditioning seemed not worthwhile. The case is even more compelling for nonperiodic problems since (1) special difference formulas are needed at points adjacent to a boundary, and (2) stable fourth-order finite-difference approximations on a non-uniform grid to variable-coefficient problems can be tedious to obtain.

***Legendre G-NI operators***

In the remaining part of this subsection, we focus on linear finite-element preconditioning for Legendre G-NI approximations. Such an approach was first devised by Quarteroni and Zampieri (1992). We consider the Dirichlet boundary-value problem,

$$\begin{aligned} -\frac{\mathrm{d}^2 u}{\mathrm{d}x^2} &= f, \qquad && -1 < x < 1 , \\ u(-1) &= u(1) = 0 . \end{aligned} \tag{4.4.41}$$

Its Legendre G-NI approximation consists of finding $u^N \in \mathbb{P}_N^0(-1,1)$ (the space of the algebraic polynomials of degree $\leq N$ that vanish at the endpoints $x = \pm 1$) satisfying

$$(u_x^N, v_x^N)_N = (f, v^N)_N \qquad \text{for all } v^N \in \mathbb{P}_N^0(-1,1) , \tag{4.4.42}$$

where $(\cdot,\cdot)_N$ is the LGL discrete inner product defined in (2.2.24) and (2.3.12). The corresponding algebraic system reads

$$K_{GNI}\mathbf{u} = M_{GNI}\mathbf{f} , \tag{4.4.43}$$

where $\mathbf{u}$ and $\mathbf{f}$ are the vectors whose components are $u^N(x_j)$, $j = 1, \ldots, N-1$, and $f(x_j)$, $j = 1, \ldots, N-1$, respectively, whereas $K_{GNI}$ and $M_{GNI}$ are the Legendre G-NI stiffness and mass matrices, respectively, already introduced in Sect. 3.8. Precisely, denoting by $\psi_j$ the characteristic Lagrange polynomials at the LGL nodes $x_j$ (see Sect. 2.3.2), we have $(K_{GNI})_{ij} = (\psi_{j,x}, \psi_{i,x})_N$ and

$(M_{GNI})_{ij} = (\psi_j, \psi_i)_N = w_i \delta_{ij}$. We recall that the LGL collocation matrix for the problem at hand is $L_{\text{coll}} = M_{GNI}^{-1} K_{GNI}$ (see Sect. 3.8.2).

For the finite-element approximations used as preconditioners, we denote by $K_{FE}$ and $M_{FE}$ the stiffness and mass matrices associated with linear finite elements built on the LGL grid; precisely, denoting by $\varphi_j$ the piecewise linear, characteristic Lagrange functions at the nodes $x_j$, we have $(K_{FE})_{ij} = (\varphi_{j,x}, \varphi_{i,x})$ and $(M_{FE})_{ij} = (\varphi_j, \varphi_i)$, where $(u,v) = \int_{-1}^{1} u(x)v(x)\,dx$. For some versions of the finite-element preconditioner, we employ the *lumped* mass matrix $M_{FE,d}$, which is the diagonal matrix obtained from the mass matrix by using the composite trapezoidal numerical integration formula in evaluating the integrals; precisely, setting $h_k = x_{k+1} - x_k$, we have

$$
\begin{aligned}
(M_{FE,d})_{ij} &= \sum_{k=0}^{N-1} [\varphi_j(x_k)\varphi_i(x_k) + \varphi_j(x_{k+1})\varphi_i(x_{k+1})] h_k \\
&= \left(\frac{1}{2}h_i + \frac{1}{2}h_{i-1}\right)\delta_{ij}.
\end{aligned}
\tag{4.4.44}
$$

Finally, as discussed in Sect. C.1, $\mathcal{K}(B)$ denotes the iterative condition number of a matrix $B$ whose eigenvalues are all real and strictly positive, i.e., $\mathcal{K}(B) = \lambda_{\max}(B)/\lambda_{\min}(B)$.

We investigate several ways of preconditioning the linear system (4.4.43). They are defined by the preconditioned matrices and associated transformed linear systems reported in Table 4.6.

The algebraic system (4.4.45) corresponds to the *weak* form of finite-element preconditioning (according to the distinction introduced at the beginning of the present subsection in the Fourier case), whereas (4.4.46) corresponds to the *strong* form. The system (4.4.47) is obtained by merely replacing the exact (nondiagonal) finite-element mass matrix by its lumped, diagonal approximation. Since the preconditioning matrix in (4.4.45) is symmetric and positive definite, this system can be solved by the preconditioned conjugate gradient (PCG) method; see Sect. 4.5 and Appendix C. This is not the case for (4.4.46) and (4.4.47); hence, for the solution of these systems one can resort, e.g., to the preconditioned Bi-CGStab or GMRES iterative methods.

One PCG iteration costs 1 matrix-vector product plus 1 solution of the linear system on the preconditioner, whereas one PBi-CGStab iteration costs 2 matrix-vector products plus 2 solutions of the linear system on the preconditioner. The linear system on the preconditioner is solved by either Cholesky factorization (in the symmetric case) or LU factorization (in the nonsymmetric case).

The systems (4.4.48) and (4.4.49) are symmetrized versions of (4.4.46) and (4.4.47), respectively, motivated by the desire to exploit the PCG method, which is more robust and usually more efficient than PBi-CGStab. These two systems involve square roots of matrices. We recall that for any given symmetric and positive definite matrix $B$, $B^{1/2}$ denotes its *square root*, i.e.,

**Table 4.6.** Preconditioned matrices and associated transformed linear systems for (4.4.43)

| Preconditioned matrix | Preconditioned system | |
|---|---|---|
| $P_1 = K_{FE}^{-1} K_{GNI}$ | $P_1 \mathbf{u} = \widetilde{\mathbf{f}}$, with $\widetilde{\mathbf{f}} = K_{FE}^{-1} M_{GNI} \mathbf{f}$ , | (4.4.45) |
| $P_2 = (M_{FE}^{-1} K_{FE})^{-1} M_{GNI}^{-1} K_{GNI}$ | $P_2 \mathbf{u} = \widetilde{\mathbf{f}}$, with $\widetilde{\mathbf{f}} = (M_{FE}^{-1} K_{FE})^{-1} \mathbf{f}$ , | (4.4.46) |
| $P_3 = (M_{FE,d}^{-1} K_{FE})^{-1} M_{GNI}^{-1} K_{GNI}$ | $P_3 \mathbf{u} = \widetilde{\mathbf{f}}$, with $\widetilde{\mathbf{f}} = (M_{FE,d}^{-1} K_{FE})^{-1} \mathbf{f}$ , | (4.4.47) |
| $P_4 = (M_{FE}^{-1/2} K_{FE} M_{FE}^{-1/2})^{-1} M_{GNI}^{-1/2} K_{GNI} M_{GNI}^{-1/2}$ | $P_4 \widetilde{\mathbf{u}} = (M_{FE}^{-1/2} K_{FE} M_{FE}^{-1/2})^{-1} \widetilde{\mathbf{f}}$ , with $\widetilde{\mathbf{v}} = M_{GNI}^{1/2} \mathbf{v}$ , | (4.4.48) |
| $P_5 = (M_{FE,d}^{-1/2} K_{FE} M_{FE,d}^{-1/2})^{-1} M_{GNI}^{-1/2} K_{GNI} M_{GNI}^{-1/2}$ | $P_5 \widetilde{\mathbf{u}} = (M_{FE,d}^{-1/2} K_{FE} M_{FE,d}^{-1/2})^{-1} \widetilde{\mathbf{f}}$ , with $\widetilde{\mathbf{v}} = M_{GNI}^{1/2} \mathbf{v}$. | (4.4.49) |

the matrix such that $B^{1/2}B^{1/2} = B$; now let $B^{-1/2}$ be a short-hand notation for $(B^{1/2})^{-1}$. We note that if $A$ and $B$ are two symmetric and positive-definite matrices, then the two matrices $B^{-1}A$ and $B^{-1/2}A\, B^{-1/2}$ are similar, and therefore

$$\mathcal{K}(B^{-1/2}A\, B^{-1/2}) = \mathcal{K}(B^{-1}A) .$$

The final detail concerns the computation of the the square root of a matrix $B$. When $B$ is diagonal, such as, e.g., $M_{FE,d}$, $B^{1/2}$ is simply given by the square root of its diagonal elements. A nondiagonal (but symmetric and positive-definite) matrix $B$ can be diagonalized by $W^T BW = \Lambda$ (where $\Lambda$ is the diagonal matrix of the eigenvalues of $B$, and $W$ is the matrix of the corresponding orthogonal eigenvectors). We then have that $B^{1/2} = W\Lambda^{1/2}W^T$. However, this procedure requires the computation of the eigenvalues and eigenvectors of the matrix; furthermore, in general it leads to a full matrix even if the original matrix is sparse or banded.

As an alternative to diagonalization, one can employ the Cholesky decomposition of $B$, namely $B = B_{Ch}B_{Ch}^T$, with $B_{Ch}$ lower triangular. Then, setting $B_{Ch}^{-T} = (B_{Ch}^{-1})^T = (B_{Ch}^T)^{-1}$, $B_{Ch}^{-1}A(B_{Ch}^{-T})$ is still symmetric and positive definite and is similar to $B^{-1}A$. Since

$$\mathcal{K}(B_{Ch}^{-1}A\,(B_{Ch}^{-T})) = \mathcal{K}(B^{-1}A) ,$$

instead of $M_{FE}^{-1/2}K_{FE}M_{FE}^{-1/2}$ we can use $M_{FE,Ch}^{-1}K_{FE}M_{FE,Ch}^{-T}$ in (4.4.48).

**Table 4.7.** Iterative condition numbers of the preconditioned matrices $P_1, \ldots, P_5$ associated with problem (4.4.42)

| $N$ | $\mathcal{K}(P_1)$ | $\mathcal{K}^*(P_2)$ | $\mathcal{K}(P_3)$ | $\mathcal{K}(P_4)$ | $\mathcal{K}(P_5)$ |
|---|---|---|---|---|---|
| 16 | 2.18516 | 1.35975 | 2.18512 | 1.60205 | 2.18512 |
| 32 | 2.32011 | 1.38172 | 2.32010 | 1.59526 | 2.32010 |
| 48 | 2.36773 | 1.40196 | 2.36772 | 1.59491 | 2.36772 |
| 64 | 2.39207 | 1.41180 | 2.39207 | 1.59483 | 2.39207 |
| 80 | 2.40686 | 1.41813 | 2.40686 | 1.59479 | 2.40686 |
| 96 | 2.41680 | 1.42170 | 2.41680 | 1.59477 | 2.41680 |
| 112 | 2.42393 | 1.42507 | 2.42393 | 1.59476 | 2.42393 |
| 128 | 2.42930 | 1.42703 | 2.42930 | 1.59475 | 2.42930 |

In Table 4.7 we report the iterative condition numbers of the preconditioned matrices $P_1, \ldots, P_5$, whereas Fig. 4.22 collects their extreme eigenvalues. All the condition numbers are small and uniformly bounded with respect to $N$; whence all of the preconditioners are optimal. $P_3$ and $P_5$, which are almost similar to each other (their eigenvalues coincide up to 7 significant digits), are also almost similar to $P_1$; in fact, we can write

$P_3 = K_{FE}^{-1}(M_{FE,d}M_{GNI}^{-1})K_{GNI}$, and since we have that $(M_{FE,d}M_{GNI}^{-1})_{ii} = \left(\frac{1}{2}h_{i-1} + \frac{1}{2}h_i\right)/w_i \sim 1$ for all $i$ (see Sect. 7.4), we conclude that $P_3 \sim P_1$. Their extreme eigenvalues coincide up to the 4-th significant digit.

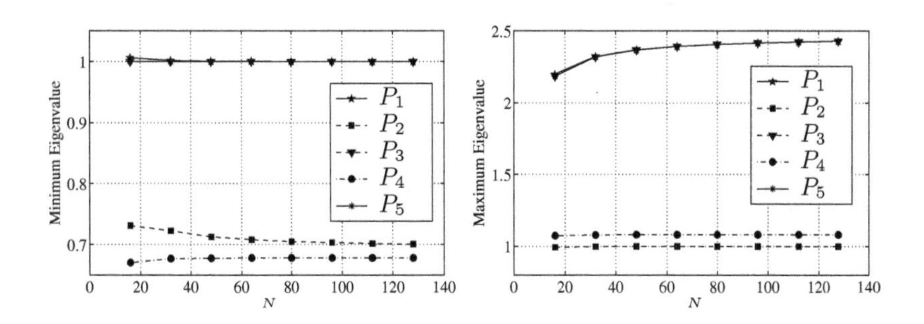


**Fig. 4.22.** Extreme eigenvalues of the preconditioned matrices considered in Table 4.7

In Fig. 4.23 we report the number of iterations that are required for convergence of two iterative methods, CG and Bi-CGStab, on a one-dimensional problem. A thorough discussion of the performance of iterative methods on more challenging, two-dimensional problems is provided in Sect. 4.7.

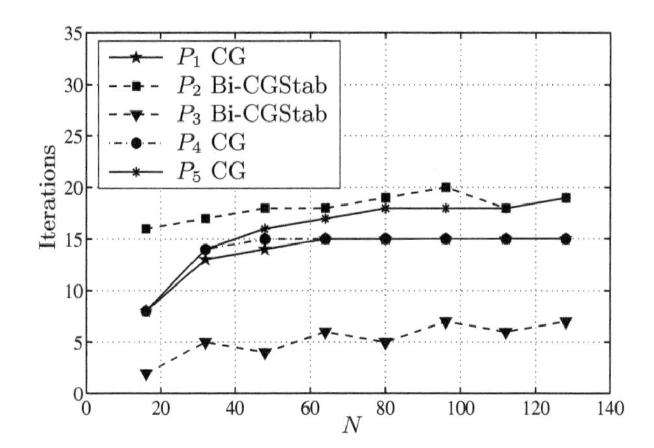


**Fig. 4.23.** Number of PCG or PBi-CGStab iterations needed to solve problem (4.4.42) with $f = 1$ and $u(-1) = u(1) = 0$. Stopping criterion is $||\mathbf{r}^{(k)}||_2/||\mathbf{r}^{(0)}||_2 < 10^{-14}$. The initial vector is $\mathbf{u}^0 = \mathbf{0}$

The preconditioned matrices $P_1$, $P_4$ and $P_5$ have real, positive eigenvalues since they are the product of two symmetric and positive-definite matrices. The theoretical analysis given in Sect. 7.4 guarantees that the eigenvalues of

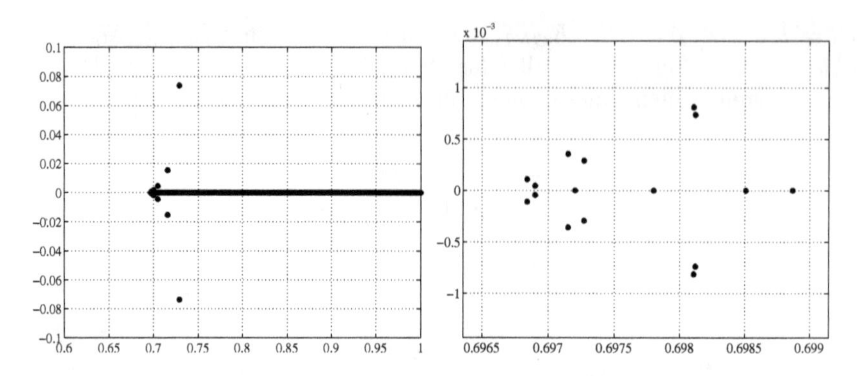


**Fig. 4.24.** The eigenvalues of $P_2$ for $N = 256$ (1D case). The picture on the right is a zoom of the one on the left

these matrices are uniformly bounded from above, and uniformly bounded away from zero (with respect to $N$). Numerical evidence indicates that $P_2$ has complex eigenvalues, whereas $P_3$ has real, positive eigenvalues. The eigenvalues of $P_2$ (for $N = 256$) are plotted in Fig. 4.24. As predicted by the theory (see Sect. 7.4), their real parts are positive and uniformly bounded away from 0, and their moduli are uniformly bounded. Their imaginary parts are bounded by roughly one-tenth of the corresponding moduli. For a matrix with this type of eigenstructure, the parameter

$$\mathcal{K}^* = \mathcal{K}^*(L) = \frac{\max_j |\lambda_j|}{\min_j |\lambda_j|} \simeq \mathcal{K}(L_S) \tag{4.4.50}$$

(where $L_S$ denotes the symmetric part of $L$) is an effective surrogate for $\mathcal{K}$ as an indicator of the convergence properties of the Richardson iterative scheme. (In the sequel, we will not usually comment on our use of this surrogate for $\mathcal{K}$ for those matrices for which the surrogate is more appropriate; however, the relevant figure labels and captions will reflect the use of the surrogate in those cases.)

In Fig. 4.25 we plot the iterative condition numbers of the preconditioned matrices $P_1 = K_{FE}^{-1} K_{GNI}$ and $P_2 = (M_{FE}^{-1} K_{FE})^{-1} M_{GNI}^{-1} K_{GNI}$ corresponding to the elliptic problem (4.1.19) with several values of $\lambda$, not only for Dirichlet but also for Neumann boundary conditions. In all cases the iterative condition numbers are bounded from above by a small number ($\pi^2/4$ or even less).

Preconditioners based on the piecewise linear finite elements are still optimal (with respect to $N$) for the G-NI approximation of the same problem with Robin conditions, say, $u_x(1) + \alpha u(1) = 0$ for $\alpha > 0$, and $u(-1) = 0$. In Fig. 4.26 we report the iterative condition numbers of the preconditioned matrices $P_1$ and $P_2$ for different values of $N$ and several values of $\alpha$. Note that they are uniformly bounded with respect to $N$, and that the condition numbers for different values of $\alpha$ are graphically indistinguishable.

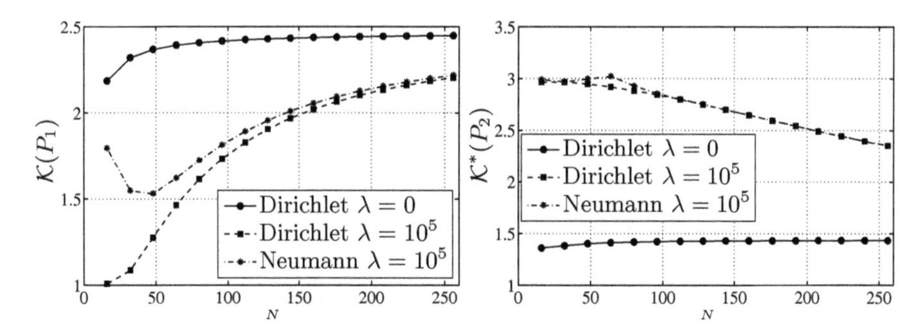


**Fig. 4.25.** The iterative condition numbers $\mathcal{K}(P_1)$ (*left*) and $\mathcal{K}^*(P_2)$ (*right*) for the problem $-u_{xx} + \lambda u = f$, $-1 < x < 1$, with either Dirichlet or Neumann boundary conditions and different values of $\lambda$

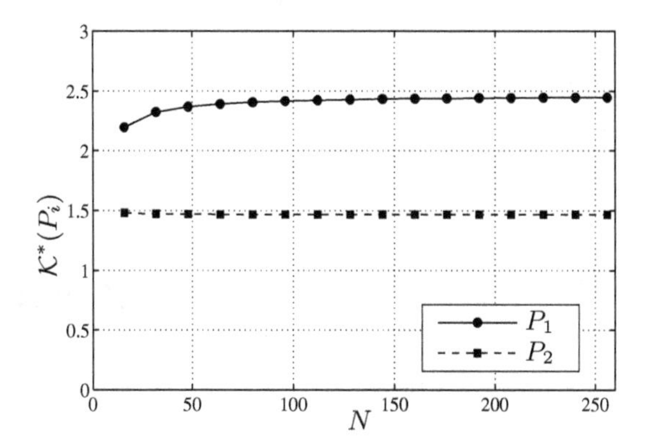


**Fig. 4.26.** The iterative condition numbers $\mathcal{K}^*(P_1)$ and $\mathcal{K}^*(P_2)$ for the problem $-u_{xx} = f$, $-1 < x < 1$, with Robin boundary condition $u_x + \alpha u = 0$ at $x = -1$ and $x = 1$, for $\alpha = 0.01, 1$ and $100$; the curves for different values of $\alpha$ are graphically indistinguishable. Similar results are obtained when a Dirichlet condition is enforced on $x = 1$

For advection-diffusion equations such as (4.4.34) or pure advection equations like (4.4.20) the situation is more varied. For values of $N$ large enough with respect to $1/\sqrt{\nu}$ (see the analysis on the stabilization of advection-diffusion equations in Sect. 7.2), the pure G-NI method provides stable and accurate solutions for (4.4.34) with, say, Dirichlet boundary conditions. In that case, the standard Galerkin piecewise-linear finite-element matrix can still be used to precondition the G-NI matrix, as the results of Fig. 4.27 show. However, in order to get condition numbers close to 2.5, smaller values of $\nu$ require larger values of $N$, or else suitable stabilization strategies for both the G-NI and the FEM approximation. In Sect. 7.2.1 we consider stabilization techniques for spectral Galerkin or G-NI discretizations of advection-diffusion operators, inspired by the popular SUPG stabilization used in finite-element

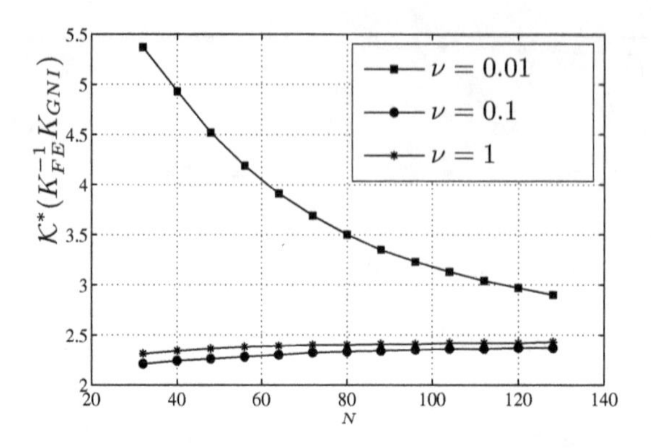


**Fig. 4.27.** The iterative condition number $\mathcal{K}^*(K_{FE}^{-1}K_{GNI})$ for the problem $-\nu u_{xx}+u_x = f$, $-1 < x < 1$, with homogeneous Dirichlet boundary conditions

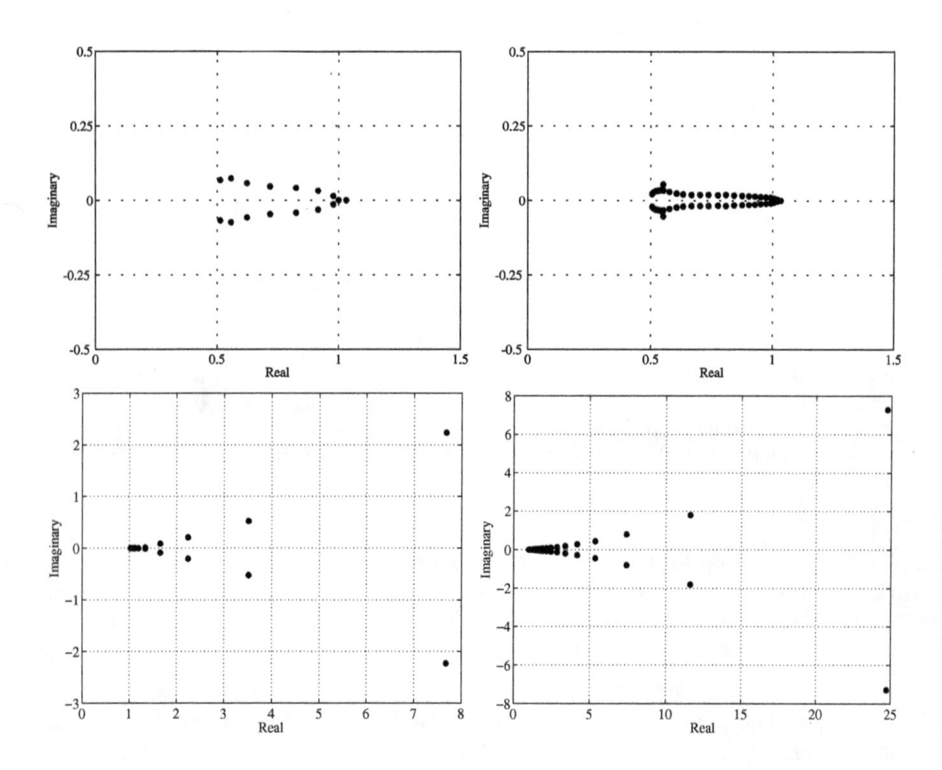


**Fig. 4.28.** Eigenvalues of the finite-element preconditioned, advection-diffusion operator $-\nu u_{xx} + u_x = f$, $-1 < x < 1$, for $\nu = 10^{-4}$:
(*left column*) $N = 16$, (*right column*) $N = 48$;
(*upper row*) with SUPG-stabilization in both spectral and finite-element scheme,
(*lower row*) without any stabilization in either scheme

methods (see (7.2.20)). In that case, the natural preconditioner is provided by the low-order finite-element scheme, stabilized by the same SUPG technique. Such a preconditioner is quite effective in all regimes. The upper row of Fig. 4.28 shows the spectra of the resulting preconditioned operator for two values of $N$. The eigenvalues, although complex, are close to a segment in the positive real semi-axis, and the resulting condition numbers are on the order of 2.5. For comparison, the lower row of the figure shows the spectra of the preconditioned operator when stabilization is switched-off from both the spectral scheme and the finite-element preconditioner.

### 4.4.3 Low-Order Preconditioning in Several Dimensions

As in the one dimensional case, preconditioning in several dimensions can be accomplished by either finite-difference or finite-element operators at the same nodal points used for the spectral discretization. This yields structured, sparse matrices. The solution of the associated systems can be achieved by direct or iterative algorithms. In the latter case, particularly when the system size is very large, the preconditioning matrix itself needs to be preconditioned, for instance by resorting to one of its inexact factorizations.

Although the separation is not at all sharp, we prefer to split our presentation into two logical parts for the sake of clarity. The first part will mostly deal with finite-difference preconditioners applied to Chebyshev collocation discretizations and to their inexact factorizations. The second part will be concerned with finite-element preconditioners for Legendre G-NI discretizations.

#### *Inexact Factorizations of Low-Order Preconditioners*

The structure of the preconditioning matrix is similar for second-order finite-difference preconditioning and linear finite-element preconditioning of a spectral collocation discretization. Much of the discussion in this section is common to both. Where this is the case, we will just use the term *low-order preconditioning* to refer to either of these cases. The available theoretical results for such preconditioning are summarized in Sect. 7.4. For one-dimensional problems, low-order preconditionings of the spectral operator (see Sect. 4.4.2) are a quite inexpensive part of the iterative scheme. The low-order inversion part of the algorithm (by that we mean the solution of a linear system whose matrix is the preconditioner) typically costs $O(N)$ operations, compared with the $O(N \log_2 N)$ or $O(N^2)$ cost for the application of the spectral operator to get the residual. In higher dimensions, however, the low-order inversion becomes relatively expensive and/or complicated. The best that one can do in terms of a direct solution is for separable problems, for which the cost of a low-order solver using cyclic reduction (see the review by Swarztrauber (1977)) is $O(N^d (\log_2 N)^{d-1})$. For nonseparable problems, direct solution of the low-order equations is still more expensive – scaling as $N^{3d-2}$ for a banded solver. For separable problems in more that two dimensions and for nonseparable problems already in two dimensions, the cost of a direct inversion of the low-order preconditioner is much larger than the cost of evaluating the residual – $O(N^d \log_2 N)$ when fast transforms are applicable and $O(N^{2d})$ otherwise. General banded solver software is available in LAPACK (Anderson et al (1999), Barker et al. (2001)) and general sparse solvers in (Davis (2004)).

Iterative methods, particularly conjugate gradient methods (and their generalizations to nonsymmetric problems), have been the preferred strategy for inverting the low-order preconditioner to the spectral operator. These are reviewed in Appendix C, and their use in spectral methods is discussed in Sects. 4.5.1 and 4.5.2. Multigrid methods, despite their asymptotically smaller cost of only $O(N^d)$ operations to invert a low-order approximation to an elliptic problem, have not seen much use for solving the preconditioned system. One exception is the work of Heinrichs (1993), who demonstrated multigrid solutions of a finite-element preconditioner for several two-dimensional Poisson examples. Perhaps it has been their greater complexity (compared with conjugate gradient methods) that has led to their lack of use for inverting low-order preconditioners to spectral methods. Some of the flavor of multigrid methods is conveyed in Sect. 4.6 in the context of spectral discretizations. Thorough discussions of multigrid methods for finite-difference methods are given by Stuben and Trottenberg (1982), Hackbusch (1985) and Wesseling (2004). General multigrid software that appears suitable for performing low-order multigrid on the non-uniform grids arising from Chebyshev (or Legendre) spectral methods is available in MADPACK (Douglas (1995)).

In practice, when the size of the algebraic system is very large, preconditioning of the *low-order* approximation is essential. Hence, the use of iterative methods for the solution of the preconditioned spectral equations requires an inner iteration (for the low-order equation) embedded within an outer iteration (for the spectral equation itself). Preconditioning even for low-order approximations continues to be an active field of research; see e.g. Evans (1983), Axelsson (1994), Saad (1996), Benzi (2002), van der Vorst (2003). The most commonly used low-order preconditioners of spectral methods, at least for conjugate gradient-type iterative schemes, are based on *incomplete-LU decompositions* (or *incomplete-Cholesky decompositions* for symmetric and positive-definite matrices). Alternating-line relaxation in 2 dimensions (or plane relaxation in 3 dimensions) has seen a more limited amount of use, primarily for spectral multigrid methods.

To illustrate the incomplete-LU decomposition, we consider the approximate inversion of a finite-difference preconditioner for Chebyshev collocation approximation to the two-dimensional Poisson equation (4.1.55). Let the matrix $H_{FD}$ represent the standard five-point second-order finite-difference approximation to the differential equation (4.1.55). The standard incomplete-LU decomposition (Meijerink and van der Vorst (1981), Axelsson (1994)) is given by

$$H_{IN} = R_{IN} S_{IN} \tag{4.4.51}$$

(using notation in keeping with that of Sect. 4.2.3), where $R_{IN}$ (apart from the diagonal) is identical to the lower-triangular portion of $H_{FD}$, and $S_{IN}$ is chosen so that the two super diagonals of $H_{IN}$ agree with those of $H_{FD}$. A modified type of incomplete-LU preconditioning – the so-called *row-sum-equivalence incomplete-LU decomposition* – is obtained similarly, but the diagonal elements of $R_{IN}$ are altered from those of $H_{FD}$ so as to ensure that the row sums of $H_{IN}$ and $H_{FD}$ are identical. We denote the resulting preconditioning matrix for the standard incomplete-LU decomposition by $H_{ILU}$, and the one for the row-sum-equivalence version by $H_{IRS}$.

A five-point approximation on a Chebyshev grid to (4.1.55) may be written as

$$\begin{aligned}(H_{FD}U)_{i,j} = {} & E_{i,j}U_{i,j} + D_{i,j}U_{i-1,j} + F_{i,j}U_{i+1,j} \\ & + H_{i,j}U_{i,j+1} + B_{i,j}U_{i,j-1} \,,\end{aligned} \tag{4.4.52}$$

where $U_{i,j}$ denotes the value of the spectral solution at the grid point $\mathbf{x}_{i,j}$. Figure 4.29 shows the structure of the matrix $H_{FD}$. A five-diagonal incomplete-LU factorization is given by (4.4.51) with

$$(R_{IN}U)_{i,j} = v_{i,j}U_{i,j} + t_{i,j}U_{i-1,j} + g_{i,j}U_{i,j-1} \tag{4.4.53}$$

and

$$(S_{IN}U)_{i,j} = U_{i,j} + e_{i,j}U_{i+1,j} + f_{i,j}U_{i,j+1} \,. \tag{4.4.54}$$

Figure 4.30 shows the structure of the factors $R_{IN}$ and $S_{IN}$. The coefficients in (4.4.53) and (4.4.54) are related to those in (4.4.51) by

$$\begin{aligned} t_{i,j} &= D_{i,j} \,, \quad g_{i,j} = B_{i,j} \,, \\ v_{i,j} &= E_{i,j} - t_{i,j}f_{i,j-1} - g_{i,j}e_{i-1,j} \\ &\quad - \alpha\left[t_{i,j}e_{i,j-1} + g_{i,j}f_{i-1,j}\right] \,, \\ e_{i,j} &= F_{i,j}/v_{i,j}, \quad f_{i,j} = H_{i,j}/v_{i,j} \,. \end{aligned} \tag{4.4.55}$$

The choice $\alpha = 0$ gives the standard incomplete-LU result ($H_{ILU5}$), and $\alpha = 1$ gives the row-sum-equivalence version ($H_{IRS5}$). Since neither version is an exact factorization of the original finite-difference matrix, some error is inevitable. Roughly speaking, the standard incomplete-LU decomposition ($H_{ILU5}$) does better on the high-frequency components and the row-sum-equivalence alternative ($H_{IRS5}$) is more accurate on the low-frequency end.

A more accurate factorization can be achieved by including one extra nonzero diagonal in $R_{IN}$ and $S_{IN}$ as indicated in Fig. 4.31. This seven-diagonal incomplete-LU factorization is a straightforward generalization (see Wong, Zang and Hussaini (1986) for details in the context of spectral

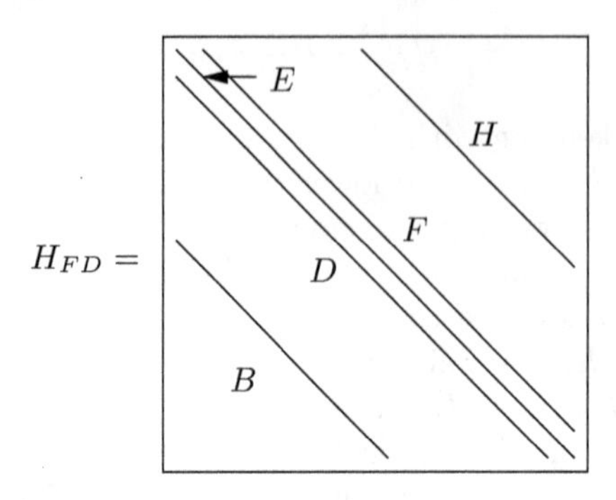


**Fig. 4.29.** Structure of the full finite-difference preconditioning for a two-dimensional problem

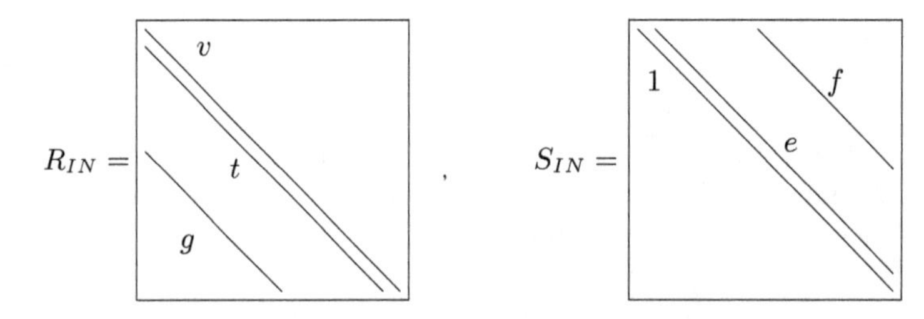


**Fig. 4.30.** Structure of the five-diagonal incomplete-LU preconditioning

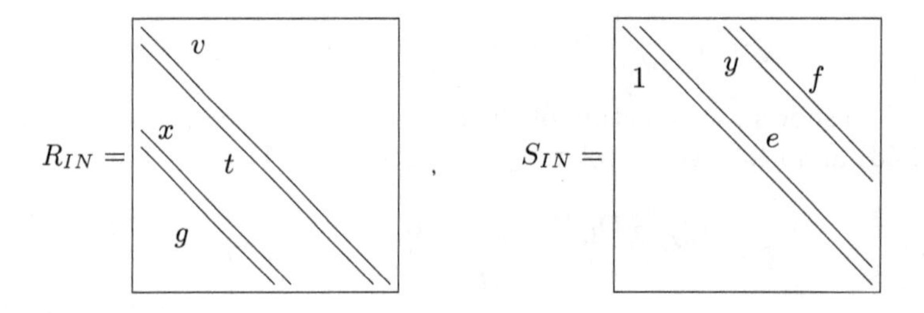


**Fig. 4.31.** Structure of the seven-diagonal incomplete-LU preconditioning

methods). Once again, there are both the standard, $H_{ILU7}$, and the row-sum equivalence, $H_{IRS7}$, versions.

A good indication of the effectiveness of these preconditionings is provided by their eigenvalue distribution. Let us consider the case of a $16 \times 16$ grid. The fully finite-difference preconditioning produces eigenvalues which are purely real and confined to the interval [1,2.31]. All but two of the eigenvalues resulting from the standard incomplete-LU preconditioning are real; the imaginary parts of the two complex eigenvalues are only of order $10^{-3}$. The real parts are in $[.22, 2.4]$. The row-sum-equivalence preconditioned eigenvalues have real parts in $[1, 2.7]$, and the imaginary parts of the only two complex ones are of order $10^{-3}$ as well.

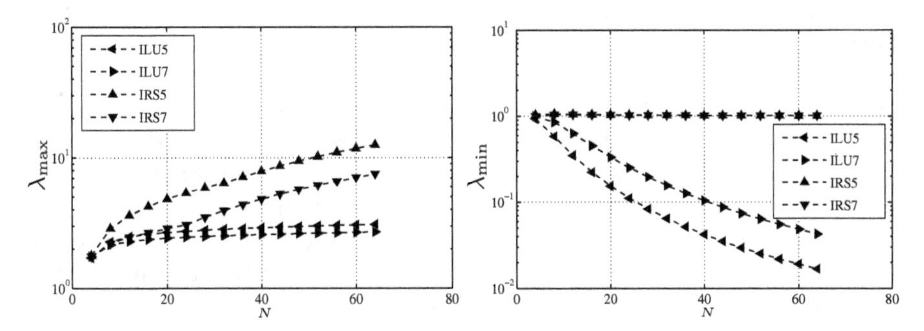


**Fig. 4.32.** Maximum (*left*) and minimum (*right*) eigenvalues for the preconditioned Chebyshev Laplace matrix in two dimensions using incomplete-LU decompositions with 5 and 7 nonzero diagonals, both without (ILU) and with (IRS) the row-sum-equivalence modification

As $N$ increases beyond 16, more complex eigenvalues arise for the factored preconditionings, but their imaginary parts are still very small. The complex eigenvalues remain small in number and are well removed from the extreme moduli of the spectra; in particular, the eigenvalues with the maximum and minimum moduli are purely real. Figure 4.32 summarizes how the extreme eigenvalues depend on the $N \times N$ grid. (The interval $[1.09, 2.60]$ encompasses the real parts of all the complex eigenvalues for the cases shown

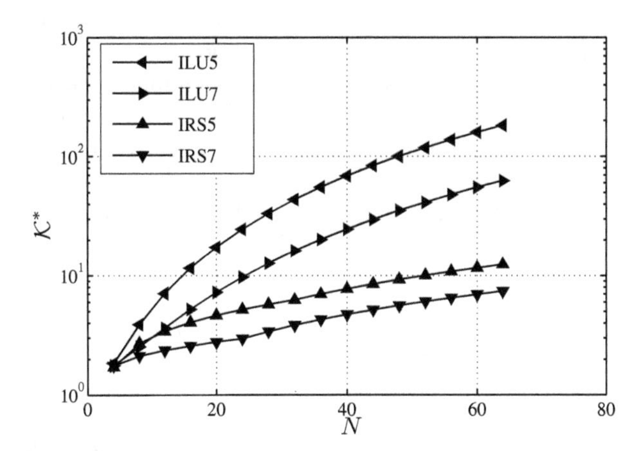


**Fig. 4.33.** Iterative condition numbers $\mathcal{K}^*$ for the preconditioned Chebyshev Laplace matrix using incomplete-LU decompositions, both without (ILU) and with (IRS) the row-sum-equivalence modification

in this figure; the imaginary parts are all less than $5 \times 10^{-3}$.) These empirical results indicate that the largest eigenvalue grows very slowly for the standard incomplete-LU decompositions (approximately as $N^{1/8}$ and $N^{1/10}$ for the 5-diagonal and 7-diagonal versions, respectively), but grows roughly as $N$ for both row-sum-equivalence versions. On the other hand, the smallest eigenvalue for the incomplete-LU decomposition decreases rapidly towards zero, approximately as $N^{-5}$ for the 5-diagonal preconditioning and approximately as $N^{-2}$ for the 7-diagonal preconditioning. Figure 4.33 displays the iterative condition numbers for these cases. Eigenvalue computations for the full $H_{FD}$ preconditioner indicate that its iterative condition number is bounded in this two-dimensional case by the one-dimensional bound of $\pi^2/4 \approx 2.47$. Since all the iterative methods discussed in Sect. 4.5 perform better for smaller $\mathcal{K}^*$, the choice there lies between $H_{FD}$ and $H_{IRS7}$. For small values of $N$, $H_{IRS7}$ is clearly preferable since, in practice, the inversion of $H_{IRS7}$ takes only a few percent of the time of the evaluation of $L\mathbf{v}$ and its iterative condition number is nearly as good as that of the much more expensive $H_{FD}$. Wong, Zang and Hussaini (1986) present several numerical examples of these incomplete-LU preconditionings used with iterative schemes. The preference eventually changes for large enough $N$, however, since the iterative condition number of $H_{IRS7}^{-1}L$ grows as $\sqrt{N}$ whereas that of $H_{FD}^{-1}L$ remains bounded by 2.47. In this case the best approach may be to use either a multigrid or a direct method for inverting $H_{FD}$.

Although the standard incomplete-LU decompositions are clearly outperformed by their row-sum-equivalence version for standard iterative methods, they do have distinct advantages in the spectral multigrid context, as will be discussed in Sect. 4.6.

In the case of bilinear finite-element preconditioning in two dimensions, the stencil contains 9 points rather than the 5 points for second-order finite-difference approximations. Hence, the corresponding class of incomplete-Cholesky or incomplete-LU decompositions (depending upon whether or not the preconditioning matrix is symmetric) require at least 9 diagonals.

Another class of preconditioners for the low-order approximation is based upon line relaxation. The simplest description of alternating line relaxation (ALR) uses the notation employed in Sect. 4.1.4. We write the spectral collocation discretization of (4.1.55) as (4.1.56) with $\lambda = 0$ and the corresponding full finite-difference preconditioned problem as

$$\begin{aligned} R^n &= (F - D_x V^n + V^n D_y^T) \, , \\ H_x(V^{n+1} - V^n) + (V^{n+1} - V^n) H_y^T &= \omega R^n \, , \end{aligned} \tag{4.4.56}$$

where $R$ denotes here the residual matrix, and $H_x$ and $H_y$ are the respective finite-difference approximations to $D_x$ and $D_y$, and $\omega$ is the relaxation parameter. The approximate finite-difference preconditioned problem is

$$\begin{aligned} H_x(V^{n+1/2} - V^n) &= \omega R^n - (V^{n+1/2} - V^n) H_y^T \, , \\ (V^{n+1} - V^{n+1/2}) H_y^T &= \omega R^n - H_x(V^{n+1/2} - V^n) \, . \end{aligned} \tag{4.4.57}$$

For second-order finite differences, the odd rows (or columns) are decoupled from the even rows (or columns). One can then solve for all the odd rows (columns) in parallel and then add the even rows (columns) in parallel. This refinement is referred to as alternating zebra line relaxation (AZLR). It was introduced as a preconditioner for approximating second-order finite-difference preconditioners for spectral discretizations by Brandt, Fulton and Taylor (1985). The ALR scheme is a relaxed line-Jacobi iteration, whereas the AZLR version is relaxed line-Gauss-Seidel.

The incomplete-LU preconditionings makes more poor use of parallel computers because of their recursive nature. The ALR and AZLR techniques, however, parallelize well. Their primary use for spectral methods has been in the context of spectral multigrid methods (see Sect. 4.6).

Yet another type of line relaxation that has been applied to spectral methods by Streett, Zang and Hussaini (1985) is based upon approximate factorization (AF) of the low-order preconditioner:

$$V^{n+1} = V^n + \omega \Delta V^n \,, \tag{4.4.58}$$

where $\Delta V^n$ is the solution to

$$[\alpha_n I - H_x V^n] \left[\alpha_n I - V^n H_y^T\right] \Delta V^n = \alpha_n R^n \,. \tag{4.4.59}$$

This is just the Douglas and Gunn (1964) version of alternating direction implicit (ADI) relaxation applied to the full finite-difference approximation. (The solution algorithm for these equations may include parallelization over the $y$-direction for the inversion of the matrix in the first brackets and over the $x$-direction for the other matrix.) An essential part of this type of preconditioning is the choice of the parameters $\alpha_n$ and $\omega_n$. A brief discussion is provided by Streett, Zang and Hussaini (1985). Trial and error is a major component of the selection process.

### *Finite Element Preconditioning of G-NI Operators*

We now consider the multidimensional counterpart of problem (4.4.41), namely,

$$\begin{aligned} -\Delta u &= f \qquad && \text{in } \Omega = (-1,1)^2 \,, \\ u &= 0 && \text{on } \partial\Omega \,. \end{aligned} \tag{4.4.60}$$

(Although we confine ourselves to homogeneous Dirichlet boundary conditions for the Laplacian operator, the extension of the subsequent arguments to the case of other boundary conditions and operators is straightforward.) The Legendre G-NI discretization of this problem consists of finding a polynomial $u^N$ in $\mathbb{P}_N^0(\Omega)$ (the space of the algebraic polynomials of degree $\leq N$ in each direction, vanishing on $\partial\Omega$) satisfying

$$(\nabla u^N, \nabla v^N)_N = (f, v^N)_N \qquad \text{for all } v^N \in \mathbb{P}_N^0(\Omega) \,, \tag{4.4.61}$$

where $(\cdot,\cdot)_N$ denotes the two-dimensional Legendre Gauss-Lobatto (LGL) discrete inner product in $\Omega$. The algebraic system corresponding to (4.4.61) reads again as (4.4.43), i.e.,

$$K_{GNI}\mathbf{u} = M_{GNI}\mathbf{f} \ , \tag{4.4.62}$$

where now $\mathbf{u}$ and $\mathbf{f}$ are the vectors whose components are the values of $u^N$ and $f$ at the $(N-1)\times(N-1)$ interior LGL nodes $\mathbf{x}_j$ (here numbered in lexicographical order). Correspondingly, $\psi_j$ will denote the characteristic Lagrange polynomial at $\mathbf{x}_j$, defined by the conditions $\psi_j \in \mathbb{P}_N^0(\Omega)$ and $\psi_j(\mathbf{x}_k) = \delta_{jk}$ for all $k = 1, \dots, (N-1)^2$. Thus, for $i, j = 1, \dots, (N-1)^2$,

$$(K_{GNI})_{ij} = (\nabla\psi_j, \nabla\psi_i)_N \qquad \text{and} \qquad (M_{GNI})_{ij} = (\psi_j, \psi_i)_N \ .$$

The finite-element preconditioner is built on the partition (or mesh) of $\overline{\Omega} = [-1,1]^2$ made of the rectangles, $R$, whose vertices are two consecutive LGL nodes in each direction (see Fig. 4.34).

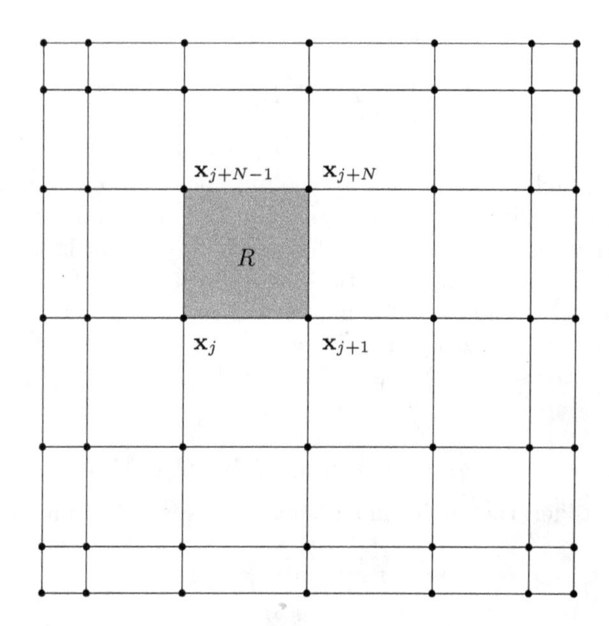


**Fig. 4.34.** The finite-element mesh in $\Omega$ induced by the two dimensional LGL grid. An internal rectangle is highlighted with its vertices

Let $\varphi_j$ denote the finite-element characteristic Lagrange function at $\mathbf{x}_j$, i.e., the globally continuous, piecewise bilinear function in each $R$, vanishing on $\partial\Omega$, such that $\varphi_j(\mathbf{x}_k) = \delta_{jk}$ for all $k = 1, \dots, (N-1)^2$. The associated finite-element stiffness matrix $K_{FE}$ is defined by

$$(K_{FE})_{ij} = (\nabla \varphi_j, \nabla \varphi_i) \,, \qquad i,j = 1, \ldots, (N-1)^2 \,, \tag{4.4.63}$$

where $(\cdot, \cdot)$ denotes the standard ($L^2$) inner product in $\Omega$. We will also consider its numerical approximation $K_{FE,app}$ , defined by

$$(K_{FE,app})_{ij} = \sum_R \int_R \Pi_{1,R}(\nabla \varphi_j^T \nabla \varphi_i) \, \mathrm{d}\mathbf{x} \,, \qquad i,j = 1, \ldots, (N-1)^2 \,,$$

where $\Pi_{1,R}(g)$ denotes the bilinear interpolant of a function $g$ at the four vertices of $R$. The finite-element mass matrix $M_{FE}$ is defined by

$$(M_{FE})_{ij} = (\varphi_j, \varphi_i) \,, \qquad i,j = 1, \ldots, (N-1)^2 \,.$$

Its diagonal approximation is the *lumped mass matrix* $M_{FE,d}$ , defined by

$$(M_{FE,d})_{jj} = \frac{1}{4}(h_{j_1-1} + h_{j_1})(h_{j_2-1} + h_{j_2}) \,, \qquad j = 1, \ldots, (N-1)^2 \,,$$

if $\mathbf{x}_j = (x_{j_1}, y_{j_2})$.

Similarly to what we have done in Sect. 4.4.2 for the one-dimensional case, we now introduce several preconditioned matrices, which lead to corresponding linear systems equivalent to (4.4.62). They are reported in Table 4.8.

Note that (4.4.64) and (4.4.65) are exactly the counterparts of the one-dimensional preconditioned systems (4.4.45) and (4.4.46), respectively. On the contrary, (4.4.66) and (4.4.67) are obtained from (4.4.64) and (4.4.65), respectively, using the approximate versions of both finite-element matrices.

As in the one-dimensional case, we can symmetrize the systems (4.4.65) and (4.4.67), resulting in the entries (4.4.68)–(4.4.70) in Table 4.8. In (4.4.69), $M_{FE,Ch}$ is the Cholesky factor of $M_{FE}$. This variant is motivated by the rapidly prohibitive (as $N$ increases) cost of computing the square root $M_{FE}^{1/2}$ in the present two dimensional geometry.

The matrices $P_1$, $P_3$, $P_5$, $P_6$ and $P_7$ have real positive eigenvalues, being products of two symmetric and positive-definite matrices; as predicted by the theory (see Sect. 7.4), they are uniformly bounded from above and from below with respect to $N$. As for the one-dimensional case, numerical evidence indicates that $P_4$ has real eigenvalues, $P_2$ has complex eigenvalues of bounded moduli and real parts positive and uniformly bounded away from 0 (see again Sect. 7.4), with imaginary parts hardly larger than one-tenth of the corresponding moduli. As in the one-dimensional case, we still look at $\mathcal{K}^*$ (see (4.4.50)) as a surrogate of the iterative condition number.

In Table 4.9 we report the iterative condition numbers of the preconditioned matrices $P_1, \ldots, P_7$. All of them are uniformly bounded with respect to $N$. The values for $P_2$ to $P_5$ are similar to those obtained in the one-dimensional case (see Table 4.7). On the contrary, the condition numbers of $P_1$ is significantly larger than the others; this behavior can be understood by carefully exploring the tensor-product structure of the preconditioning matrix – we refer to the last paragraph of Sect. 7.4 for the details.

**Table 4.8.** Preconditioned matrices and associated transformed linear systems for (4.4.62)

| Preconditioned matrix | Preconditioned system | |
|---|---|---|
| $P_1 = K_{FE}^{-1} K_{GNI}$ | $P_1 \mathbf{u} = \widetilde{\mathbf{f}}$, with $\widetilde{\mathbf{f}} = K_{FE}^{-1} M_{GNI} \mathbf{f}$ , | (4.4.64) |
| $P_2 = (M_{FE}^{-1} K_{FE})^{-1} M_{GNI}^{-1} K_{GNI}$ | $P_2 \mathbf{u} = \widetilde{\mathbf{f}}$, with $\widetilde{\mathbf{f}} = (M_{FE}^{-1} K_{FE})^{-1} \mathbf{f}$ , | (4.4.65) |
| $P_3 = K_{FE,app}^{-1} K_{GNI}$ | $P_3 \mathbf{u} = \widetilde{\mathbf{f}}$, with $\widetilde{\mathbf{f}} = K_{FE,app}^{-1} M_{GNI} \mathbf{f}$ , | (4.4.66) |
| $P_4 = (M_{FE,d}^{-1} K_{FE,app})^{-1} M_{GNI}^{-1} K_{GNI}$ | $P_4 \mathbf{u} = \widetilde{\mathbf{f}}$, with $\widetilde{\mathbf{f}} = (M_{FE,d}^{-1} K_{FE,app})^{-1} \mathbf{f}$ , | (4.4.67) |
| $P_5 = (M_{FE}^{-1/2} K_{FE} M_{FE}^{-1/2})^{-1} M_{GNI}^{-1/2} K_{GNI} M_{GNI}^{-1/2}$ | $P_5 \widetilde{\mathbf{u}} = (M_{FE}^{-1/2} K_{FE} M_{FE}^{-1/2})^{-1} \widetilde{\mathbf{f}}$ , with $\widetilde{\mathbf{v}} = M_{GNI}^{1/2} \mathbf{v}$ , | (4.4.68) |
| $P_6 = (M_{FE,Ch}^{-1} K_{FE} M_{FE,Ch}^{-T})^{-1} M_{GNI}^{-1/2} K_{GNI} M_{GNI}^{-1/2}$ | $P_6 \widetilde{\mathbf{u}} = (M_{FE,Ch}^{-1} K_{FE} M_{FE,Ch}^{-T})^{-1} \widetilde{\mathbf{f}}$ , where $B^{-T} = (B^T)^{-1}$ , | (4.4.69) |
| $P_7 = (M_{FE,d}^{-1/2} K_{FE,app} M_{FE,d}^{-1/2})^{-1} M_{GNI}^{-1/2} K_{GNI} M_{GNI}^{-1/2}$ | $P_7 \widetilde{\mathbf{u}} = (M_{FE,d}^{-1/2} K_{FE,app} M_{FE,d}^{-1/2})^{-1} \widetilde{\mathbf{f}}$ . | (4.4.70) |

**Table 4.9.** Iterative condition numbers of the preconditioned matrices $P_1, \ldots, P_7$ associated with problem (4.4.60)

| $N$ | $\mathcal{K}(P_1)$ | $\mathcal{K}^*(P_2)$ | $\mathcal{K}(P_3)$ | $\mathcal{K}(P_4)$ | $\mathcal{K}(P_5)$ | $\mathcal{K}(P_6)$ | $\mathcal{K}(P_7)$ |
|---|---|---|---|---|---|---|---|
| 8 | 5.4728 | 1.3525 | 1.9454 | 1.9451 | 1.6020 | 2.6893 | 1.9451 |
| 16 | 6.4402 | 1.3597 | 2.1852 | 2.1851 | 1.6020 | 2.9846 | 2.1851 |
| 24 | 6.7655 | 1.3720 | 2.2739 | 2.2738 | 1.5957 | 3.0511 | 2.2738 |
| 32 | 6.9273 | 1.3817 | 2.3201 | 2.3201 | 1.5953 | 3.0756 | 2.3201 |
| 40 | 7.0238 | 1.3978 | 2.3485 | 2.3485 | 1.5950 | 3.0872 | 2.3485 |
| 48 | 7.0878 | 1.4020 | 2.3677 | 2.3677 | 1.5949 | 3.0937 | 2.3677 |
| 56 | 7.1333 | 1.4077 | 2.3816 | 2.3816 | 1.5949 | 3.0976 | 2.3816 |
| 64 | 7.1674 | 1.4118 | 2.3921 | 2.3921 | 1.5949 | 3.1001 | 2.3921 |

The smallest condition number is obtained, as in the one-dimensional case, for $P_2$; this amounts to using the *strong* form of finite-element preconditioning, or, equivalently, to preconditioning the collocation matrix $L_{\text{coll}} = M_{GNI}^{-1} K_{GNI}$ by the corresponding finite-element matrix $M_{FE}^{-1} K_{FE}$ involving the consistent mass matrix (i.e., the mass matrix with no lumping). However, the size of the iterative condition number of the preconditioned matrix is but one element in the evaluation of the performance of an iterative method, the cost of the single iteration being another important element of analysis. We will report in Sect. 4.7 numerical results concerning number of iterations and CPU times for several iterative solution schemes applied to the preconditioned systems (4.4.64)–(4.4.70). From them it emerges that, unlike the one-dimensional case, the overall best performance is guaranteed by the matrix $P_3$ (corresponding to the *weak* form of finite-element preconditioning, but with an approximate stiffness matrix) within a conjugate gradient method.

The operation counts for the major steps in the iterative solution of the 2D Poisson problem with a Legendre G-NI method are supplied in Table 4.10. The cost of a matrix assembly scales linearly with the dimension $n = (N-1)^2$ of the matrix; precisely it takes about $50n$ operations to assemble $K_{FE}$ or $K_{FE,\text{app}}$, about $20n$ to assemble $M_{FE}$ and about $4n$ to assemble $M_{FE,d}$. The factorization of the preconditioning matrix takes $O(N^4)$ operations for each type of preconditioning listed in the table. Thus, the total pre-processing cost – assembly of the preconditioning mass and stiffness matrices plus factorization of the preconditioning stiffness matrix – is driven by the cost of the factorization. The residual computation is dominated by the matrix-vector product, which also takes $O(N^3)$ operations. The back-substitution and forward-substitution stages of the solution of the preconditioned system also take $O(N^3)$ operations. Thus, the total operation count per iteration scales as $O(N^3)$. As documented in Sect. 4.7, the number of iterations typi-

**Table 4.10.** Number of floating point operations for the major steps in the construction and application of the various preconditioners for the 2D Poisson problem. $N$ denotes the polynomial degree, $n = (N-1)^2$ the matrix dimension, $b$ the preconditioner bandwidth. Residual computation takes $2nz+n$ operations in all cases, where the number $nz$ of non-zero elements is $\simeq N^3$.

| | Preconditioner assembly | Bandwidth ($b$) of $K_{FE}$ or $K_{FE,\text{appr}}$ | Factorization of $K_{FE}$ or $K_{FE,\text{appr}}$ | Preconditioner solve |
|---|---|---|---|---|
| $P_1$ | $50n$ | $N+2$ | $n(b^2+3b)+n$ sqrt | $4n \cdot b$ |
| $P_2$ | $50n+20n$ | $N+2$ | $n(b^2+3b)+n$ sqrt | $8n \cdot b$ |
| $P_3$ | $50n$ | $N+1$ | $n(b^2+3b)+n$ sqrt | $4n \cdot b$ |
| $P_4$ | $50n+4n$ | $N+1$ | $n(b^2+3b)+n$ sqrt | $4n \cdot b+n$ |
| $P_5$ | $50n+20n$ | $N+2+2n^3$ $+n^2(1+8b/3)$ | $n(b^2+3b)+n$ sqrt | $4n \cdot b+4n^2$ |
| $P_6$ | $50n+20n$ | $N+2$ | $2n(b^2+3b)+2n$ sqrt | $8n \cdot b$ |
| $P_7$ | $50n+4n$ | $N+1$ | $n(b^2+3b)+2n$ sqrt | $4n \cdot b+2n$ |

cally required to solve the 2D Poisson problem with the preconditioners discussed in this subsection is $O(10)$. Thus, for small $N$, direct solution methods are the most efficient and for intermediate $N$ the preferred approach is iterative solution with full factorization of the preconditioner. For sufficiently large $N$, an inexact factorization of the preconditioner, similar to that discussed above in the context of Chebyshev collocation methods, may be attractive.

### 4.4.4 Spectral Preconditioning

The vast majority of the work on preconditioners for spectral methods has focused on the use of low-order preconditioners. Yet, for some problems, a spectral preconditioner can be competitive. One can precondition a variable-coefficient problem such as (4.4) by the spectral approximation to the corresponding equation in which $a_i$ are constants rather than functions of $\mathbf{x}$. The solution of the constant-coefficient preconditioner can be obtained quite efficiently by *ad hoc* methods (Sect. 4.1). Candidate iterative schemes for this type of preconditioned problem include not only the methods discussed in Sect. 4.5, but also the classic method of Concus and Golub (1973), which was developed for low-order discretizations. The robustness and efficiency of these methods are very problem dependent. Some examples of applications of spectral preconditioners can be found in Guillard and Desideri (1990), Zhao and Yedlin (1994), Strain (1994), and Dimitropoulos and Beris (1997).

## 4.5 Descent and Krylov Iterative Methods for Spectral Equations

The past four decades have witnessed extensive research into iterative schemes for linear equations. Some standard references include the books by Varga (1962), Young (1971), Hageman and Young (1981), Saad (1996) and van der Vorst (2003). Appendix C furnishes a summary in a generic context of many widely-used iterative methods. The present section is focused on the specialized context of iterative algorithms for spectral methods. In particular, this section directs the users of spectral methods towards those classes of iterative algorithms that have proven the most useful for solving the linear system (4.8) produced by spectral discretizations.

The most thorough analyses of iterative methods are available for symmetric and positive-definite systems. Descent methods are simple, robust and efficient schemes for such systems. Unfortunately, they are strictly applicable to a limited subset of spectral equations, e.g., for Fourier collocation or Legendre G-NI approximations to self-adjoint problems. Of course, a nonsymmetric system of the form (4.8) can always be transformed into a positive-definite system given by the normal equation

$$L^T L\mathbf{u} = L^T\mathbf{b} . \tag{4.5.1}$$

But the normal equation generally has a condition number that is the square of that for the original system, and the operator $L$ must be applied twice. In most cases, effective alternatives to the normal equation approach are available and our discussion is confined to these alternatives.

### 4.5.1 Multidimensional Matrix-Vector Multiplication

Before entering this discussion, it is worth pointing out that at every iteration, all iterative methods require the computation of a new residual as well as (possibly) the solution (exact or inexact) of a linear system governed by the preconditioning matrix. The latter issue has been extensively covered in the previous section.

The residual evaluation can be accomplished either by a direct computation of the derivatives involved in the underlying differential operator or by a matrix-vector multiplication. The former strategy can benefit from the use of the transform methods that were illustrated in the previous chapters. On the other hand, when using matrix-vector multiplication, the structure of the spectral matrix $L$ can be conveniently exploited. For tensor-product matrices (whose impact on the efficiency of spectral methods was first pointed out by Orszag (1980)), a matrix-vector product can be performed reasonably efficiently, i.e., at a cost of only $O(N^{d+1})$ operations, as opposed to the $O(N^{2d})$ operations of the standard algorithm. The implementation, which is described below, does not resort to any fast transform method and does not require the storage or even the formation of the $O(N^{2d})$ entries of the spectral matrix. Another common structure is the sparse pattern of a G-NI matrix which originates from the discretization of a separable operator (in the reference tensor-product domain). In this case, the number of non-zero elements in each row is $O(N)$, yielding again a cost of $O(N^{d+1})$ operations for a matrix-vector multiplication algorithm tailored to sparse matrices.

We now focus on the details of the algorithm for matrix-vector multiplication in the tensor-product case. Let us assume that the spectral matrix is a sum of matrices individually having the tensor-product structure illustrated in (4.2.1). Then, for any given vector, $\mathbf{v}^n$, the computation of the residual, $\mathbf{r}^n = \mathbf{b} - L\mathbf{v}^n$, can take advantage of the following algorithm for computing a product $\mathbf{z} = A\mathbf{v}^n$. Using the following factorization of the multidimensional sum:

$$\begin{aligned} z_{\mathbf{h}} &= \sum_{\mathbf{k}} a^{(d)}_{h_d k_d} \cdots a^{(2)}_{h_2 k_2} a^{(1)}_{h_1 k_1} v_{\mathbf{k}} \\ &= \sum_{k_d=1}^{N} a^{(d)}_{h_d k_d} \left( \cdots \sum_{k_2=1}^{N} a^{(2)}_{h_2 k_2} \left( \sum_{k_1=1}^{N} a^{(1)}_{h_1 k_1} v_{k_1 k_2 \dots k_d} \right) \cdots \right) , \end{aligned}$$

we obtain $\mathbf{z} = \mathbf{z}^{(d)}$ as the output of the recursion algorithm

$$\begin{aligned} &\mathbf{z}^{(0)} = \mathbf{v} \, ; \\ &for\ l = 1, \dots, d, \text{define } \mathbf{z}^{(l)} \text{ by} \\ &\qquad z^{(l)}_{h_1 \dots h_l k_{l+1} \dots k_d} = \sum_{k_l=1}^{N} a^{(l)}_{h_l k_l} z^{(l-1)}_{h_1 \dots h_{l-1} k_l \dots k_d} \, . \end{aligned}$$

Each recursion step requires $O(N^d)$ operations using only the action of one of the matrices $A^{(l)}$, whence the result. This algorithm is termed the *sum factorization* technique.

The interest of this procedure is obviously not restricted to the computation of residuals. It can be applied to any linear transformation which can be factorized into successive applications of one-dimensional operators (such as the transforms from coefficient space to physical space and back, or the numerical evaluation of partial derivatives or integrals of a function).

The sum factorization technique can be efficiently applied also to some linear transformations related to the warped tensor-product expansions considered in Sect. 2.9.1. Assume for instance that the vector $\mathbf{z} = A\mathbf{v}$ has to be computed, given a vector $\mathbf{v} = (v_{\mathbf{k}}) = (v_{k_1 k_2})$ and a matrix $A = (a_{\mathbf{hk}})$, whose entries can be factorized as

$$a_{\mathbf{hk}} = a^{(1)}_{h_1 k_1} a^{(2)}_{k_1 h_2 k_2} \, .$$

(This is the structure of the matrix which describes, e.g., the evaluation of a polynomial in $\mathcal{P}_N(\mathcal{T})$ at the $O(N^2)$ mapped LGL nodes, given the $O(N^2)$ coefficients of its warped tensor-product expansion.) Then, we can write

$$z_{\mathbf{h}} = \sum_{\mathbf{k}} a_{\mathbf{hk}} v_{\mathbf{k}} = \sum_{k_1} a^{(1)}_{h_1 k_1} \left( \sum_{k_2} a^{(2)}_{k_1 h_2 k_2} v_{k_1 k_2} \right) .$$

The recursive evaluation of the right-hand side yields $\mathbf{z}$ in $O(N^3)$ operations if $\mathbf{v}$ contains $O(N)$ entries in each of the two directions. Note that, unlike the pure tensor-product case considered above, here the order in which the factorization is performed is uniquely determined.

Unfortunately, the mass and stiffness matrices associated with warped tensor-product expansions are sums of matrices whose entries rather have the structure

$$a_{\mathbf{hk}} = a^{(1)}_{h_1 k_1} a^{(2)}_{h_1 k_1 h_2 k_2}$$

(see Sect. 4.2.2); in this case, no gain is obtained from the sum factorization technique, leaving at the standard $O(N^4)$ operations the computational cost of applying such matrices to a vector with $O(N^2)$ entries.

We refer to Karniadakis and Sherwin (1999) for further details on the latter topics.

### 4.5.2 Iterative Methods

Iterative algorithms of descent type include several variants of the minimum residual Richardson method, the steepest descent (or gradient) Richardson method, the conjugate gradient method, and the conjugate residual method; see Appendix C. We abbreviate these as MRR, SDR, CG and CR respectively, when used without preconditioning, and as PMRR, PSDR, PCG and PCR respectively, when used with preconditioning. They represent a natural choice for solving symmetric and positive-definite spectral equations, such as those generated by Fourier collocation approximations or Legendre G-NI approximations to second-order self-adjoint problems. Efficiency requires that the algorithms be applied to the ill-conditioned spectral matrix $L$ in their preconditioned version; the preconditioning matrix $H$ should be symmetric and positive definite as well. Sometimes, a gradient method is used on a non-symmetric problem with the same algorithm as used for the symmetric and positive-definite case; in such circumstances we insert the adjective *truncated* in the name to emphasize that the orthogonality properties of the method when applied to a symmetric and positive-definite system are lost. An example is the truncated conjugate residual (TCR) method, which we use in Sect. 4.7 in some numerical examples.

The description of several preconditioned descent methods, accompanied by pseudocodes, can be found in Sect. C.2 of Appendix C. Comparative numerical results on their use for solving spectral equations, concerning convergence histories and CPU times, are deferred to Sect. 4.7.

Standard Chebyshev collocation approximations do not yield symmetric discretizations, even for self-adjoint problems. However, it is possible to perform collocation on a symmetric, weak formulation of the problem (4.2) – see

Spalart (1986) for the weak formulation of the corresponding Fourier-Jacobi Galerkin approximation.

More generally, descent methods are applicable in certain situations even to nonsymmetric systems. A common situation is when the eigenvalues of the symmetric part of $H^{-1}L$, defined by

$$(H^{-1}L)_S = \frac{1}{2}\left[(H^{-1}L) + (H^{-1}L)^T\right] , \tag{4.5.2}$$

are positive. Malik, Zang and Hussaini (1985) and Zang, Wong and Hussaini (1986) provide one- and two-dimensional examples, respectively, of the use of the PMRR method for solving nonsymmetric systems resulting from Chebyshev collocation. See Sect. 4.7 for some two-dimensional numerical examples.

Scaling can be crucial for these descent methods. Suppose that the rows of $L$ are scaled by $Q_1$ and the columns by $Q_2$, and likewise, for $H$. Then we are interested in

$$\begin{aligned} L_Q &= Q_1 L Q_2 , \\ H_Q &= Q_1 H Q_2 . \end{aligned} \tag{4.5.3}$$

We have that

$$L_Q H_Q^{-1} = Q_1 L H^{-1} Q_1^{-1} .$$

Although the spectrum of $L_Q H_Q^{-1}$ corresponds to that of $LH^{-1}$, the same is not true of their symmetric parts. An example of the crucial role that scaling can play is furnished in Malik, Zang and Hussaini (1985).

For general nonsymmetric systems, as for symmetric systems which are preconditioned in a nonsymmetric way, iterative methods that usually work can be found in the family of Krylov methods; they include the (restarted) generalized minimum residual (GMRES) method, as well as the bi-conjugate gradient stabilized (Bi-CGStab) method. Again, their preconditioned versions should be used for spectral systems.

This family of iterative schemes are described in Sect. C.3 of Appendix C, where pseudocodes for the preconditioned GMRES and Bi-CGStab methods are given. Numerical results concerning their application to the solution of spectral equations are again reported in Sect. 4.7.

## 4.6 Spectral Multigrid Methods

For elliptic problems such as Poisson's equation some of the preconditioned iterative methods described above require an increasing number of iterations to achieve convergence as the size of the problem increases. Methods which use either a direct method or a multigrid technique to invert the full low-order (second-order finite-difference or linear finite-element) preconditionings are optimal, i.e., the iterative condition number is independent of $N$, the number of polynomials in one dimension, and so is the convergence rate of the overall iterative procedure. However, the cost of a direct inversion of the low-order preconditioner increases faster than the cost of the evaluation of the residual of the approximate solution to the discrete spectral equations. In the case of an $N \times N$ two-dimensional problem, the cost of the direct inversion of the preconditioning matrix is $O(N^4)$, which is large compared with the $O(N^3)$ or $O(N^2 \log_2 N)$ cost of the residual computation (depending upon whether matrix multiplies or fast transforms are used). Multigrid solution of the low-order preconditioning equations takes only $O(N^2)$ operations. As noted in Sect. 4.4.3 this approach has attracted little attention, perhaps undeservedly so.

Our focus in this section will be on yet another alternative – spectral multigrid (SMG) iterative methods – originally proposed by Zang, Wong, Hussaini (1982). These resort to relatively cheap preconditioning schemes such as incomplete-$LU$ factorization or line relaxation, but within an overall iterative scheme for which the number of iterations is independent of the number of unknowns.

### 4.6.1 One-Dimensional Fourier Multigrid Model Problem

We begin the discussion of spectral multigrid techniques by reverting to the non-preconditioned Richardson scheme with which Sect. 4.4.1 began. The condition number of this method increases as $N^2$. The resulting slow convergence was the outcome of balancing the damping of the lowest frequency eigenfunction with that of the highest frequency one in (C.1.7). The multigrid approach takes advantage of the fact that the low frequency modes ($|p| < N/4$) can be represented just as well on coarser grids. It settles for balancing the middle frequency one ($|p| = N/4$) with the highest frequency one ($|p| = N/2$), and hence damps effectively only those modes which cannot be resolved on coarser grids. In (C.1.8) and (C.1.9), $\lambda_{\text{min}}$ is replaced by $\lambda_{\text{mid}} = \lambda(N/4)$. The optimal relaxation parameter in this context is

$$\omega_{MG} = \frac{2}{\lambda_{\text{max}} + \lambda_{\text{mid}}} . \tag{4.6.1}$$

The multigrid smoothing factor

$$\mu_{MG} = \frac{\lambda_{\text{max}} - \lambda_{\text{mid}}}{\lambda_{\text{max}} + \lambda_{\text{mid}}} \tag{4.6.2}$$

measures the damping rate of the high-frequency modes. Alternatively, we may write

$$\mu_{MG} = \frac{\mathcal{K}_{MG} - 1}{\mathcal{K}_{MG} + 1} , \tag{4.6.3}$$

where

$$\mathcal{K}_{MG} = \lambda_{\max}/\lambda_{\text{mid}} \tag{4.6.4}$$

is known as the multigrid condition number. (In the case of complex eigenvalues, we use the a surrogate $\mathcal{K}^*_{MG}$, defined analogously to (4.4.50).) In this example, $\mu_{MG} = 0.60$, independent of $N$. Figure 4.35 illustrates the single grid and multigrid damping factors for the positive modes $p$ for $N = 64$. Although the high-frequency errors (for $p \in [N/4, N/2]$) overall are damped more effectively than the low-frequency errors, the low-frequency errors (for $p \in [1, N/4]$) are damped less effectively than they are in a conventional Richardson scheme. However, on a grid with $N/2$ collocation points, the modes for $|p| \in [N/8, N/4]$ are now the high-frequency ones. They get damped on this grid. Still coarser grids can be used until relaxations are so cheap that one can afford to damp all the remaining modes, or even to solve the discrete equations exactly. For the case illustrated in Fig. 4.35 the high frequency error reduction in the multigrid context is roughly 250 times as fast as the single grid reduction for $N = 64$ – $\mathcal{K}_{SG} = 1024$ (see (C.1.10)) whereas $\mathcal{K}MG = 4$ (see (4.6.4)).

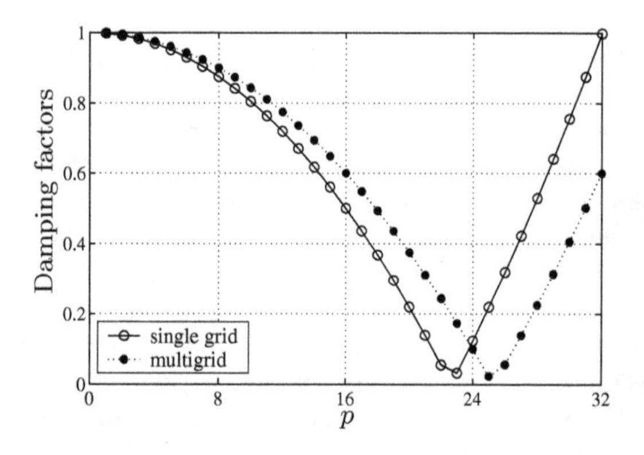


**Fig. 4.35.** Damping factors for the Fourier model problem for $N - 64$

We describe the multigrid process for solving a collocation problem by considering the interplay between two grids. The fine grid problem can be written in the form

$$L^f \mathbf{u}^f = \mathbf{f}^f \; . \tag{4.6.5}$$

The decision to switch to the coarse grid is made after the fine grid approximation $\mathbf{v}^f$ has been sufficiently smoothed by the relaxation process, i.e., after the high-frequency content of the error, $\mathbf{v}^f - \mathbf{u}^f$, has been sufficiently reduced. For the model problem, three relaxations on a grid reduce the error by a factor of $(.60)^3$, which is roughly an order of magnitude. The auxiliary equation on the coarse grid is

$$L^c \mathbf{u}^c = \mathbf{f}^c , \tag{4.6.6}$$

where

$$\mathbf{f}^c = R_{MG} \left[ \mathbf{f}^f - L^f \mathbf{v}^f \right] . \tag{4.6.7}$$

The restriction operator $R_{MG}$ interpolates a function from the fine grid to the coarse grid. The coarse grid operator and the correction are denoted by $L^c$ and $\mathbf{u}^c$, respectively. After an adequate approximation $\mathbf{v}^c$ to the coarse grid problem has been obtained, the fine grid approximation is updated using

$$\mathbf{v}^f \longleftarrow \mathbf{v}^f + P_{MG} \mathbf{v}^c . \tag{4.6.8}$$

The prolongation operator $P_{MG}$ interpolates a function from the coarse grid to the fine grid. Figure 4.36 shows one possible control structure. The symbols $N_d$ and $N_u$ denote the number of relaxations on each level after the restriction operation and after the prolongation operation, respectively. This particular fixed algorithm is known as a $V$-cycle.

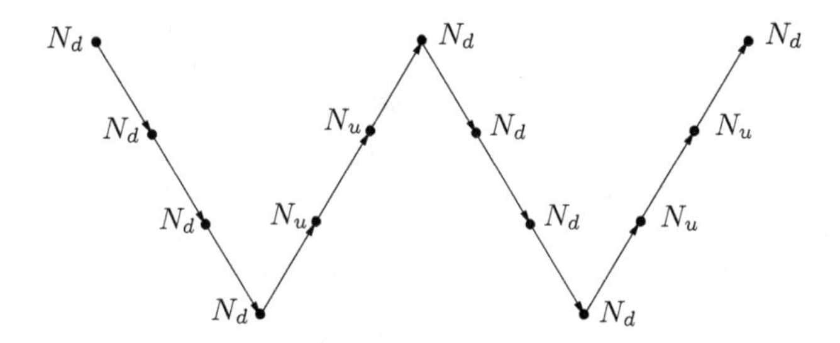


**Fig. 4.36.** Two multigrid $V$-cycles. The number of relaxations after restriction is denoted by $N_d$ and the number of relaxations before prolongation is denoted by $N_u$

For the model problem it is clear that the ideal interpolation operators – both restriction and prolongation – are those which transfer the eigenfunctions intact and without contamination. Trigonometric interpolation accomplishes precisely this and can be implemented efficiently by the FFT. Consider first the prolongation process: given a function on a coarse grid (with $N_c$ points), compute the discrete Fourier coefficients and then use the resulting discrete Fourier series to construct the interpolated function on the fine grid (with $N_f$ points). This may be accomplished by performing two FFTs.

On the coarse grid, the discrete Fourier coefficients of the corrections $u_j^c$ at the coarse-grid collocation points $x_j^c$, $j = 0, 1, \ldots, N_c - 1$, are computed using

$$\tilde{u}_p^c = \frac{1}{N_c} \sum_{j=0}^{N_c-1} u_j^c e^{-ipx_j^c} , \qquad p = -N_c/2, \ldots, N_c/2 - 1 . \tag{4.6.9}$$

The fine-grid approximation is then updated using

$$u_j^f \longleftarrow u_j^f + \sum_{p=-N_c/2}^{N_c/2-1} \tilde{u}_p^c e^{ipx_j^f} , \tag{4.6.10}$$

where $x_j^f$, $j = 0, 1, \ldots, N_f - 1$, are the fine-grid collocation points. Similarly, the restriction operation is given by

$$\tilde{r}_p^f = \frac{1}{N_f} \sum_{j=0}^{N_f-1} r_j^f e^{-ipx_j^f} , \qquad p = -N_f/2, \ldots, N_f/2 - 1 , \tag{4.6.11}$$

followed by

$$r_j^c = \sum_{p=-N_c/2}^{N_c/2-1} \tilde{r}_p^f e^{ipx_j^c} . \tag{4.6.12}$$

Except for a multiplicative factor of $(N_f/N_c)$, the restriction operator is the adjoint of the prolongation operator. Zang, Wong and Hussaini (1984) provide closed form representations of these operators. For most purposes it suffices to use as the coarse-grid correction operator $L_c$ just the standard collocation approximation on the coarse grid. See Zang, Wong and Hussaini (1984) for a discussion of more complex formulations of the coarse-grid correction operator.

Inversion of the low-order preconditioning matrix takes only $O(N)$ operations for the one-dimensional Fourier model problem, and yields a single-grid condition number which is independent of $N$. Recall that the spectral radii for finite-difference and finite-element preconditionings are 0.43 and 0.18, respectively, which are lower than the smoothing rate of 0.60 exhibited by the non-preconditioned multigrid scheme.

Preconditioning, though, improves the multigrid method as well. Figure 4.37 displays the damping achieved when low-order preconditioning is applied in this multigrid context. For second-order finite-difference preconditioning on this model problem, $\omega_{MG} = 16/(3\pi^2) = 0.5403796$ and $\mu_{MG} = 1/3$, and for linear (strong) finite-element preconditioning $\omega_{MG} = 1.1319870$ and $\mu_{MG} = 0.08554985$. These are better, but not dramatically better, than what can be achieved without multigrid. Even for the one-dimensional Dirichlet model problem multigrid offers only this same modest improvement in convergence rate over the corresponding single grid relaxation scheme. Thus, in one dimension spectral multigrid does not seem worth the trouble.

### 4.6.2 General Spectral Multigrid Methods

Suppose now that the one-dimensional periodic problem is of the self-adjoint, variable-coefficient form

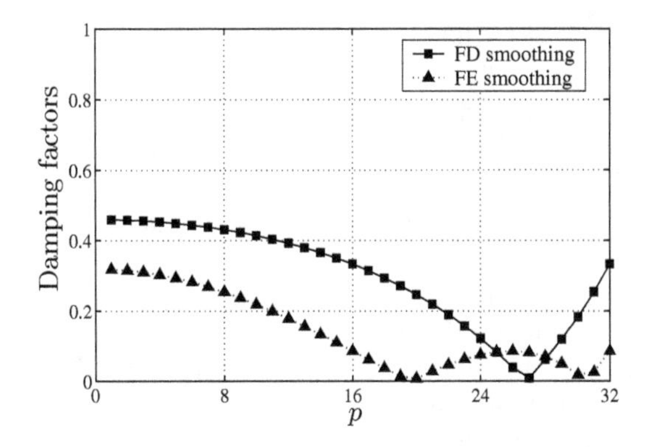


**Fig. 4.37.** Damping factors for the preconditioned Fourier model problem for $N = 64$

$$-\frac{\mathrm{d}}{\mathrm{d}x}\left[a(x)\frac{\mathrm{d}u}{\mathrm{d}x}\right] = f(x) \ . \tag{4.6.13}$$

Brandt, Fulton and Taylor (1985) recommended use of a relaxation parameter that depends on position according to

$$\omega(x) = \frac{\omega_{MG}}{a(x)} \ , \tag{4.6.14}$$

where $\omega_{MG}$ is the parameter appropriate for the $a(x) \equiv 1$ case. This maintains a smoothing rate of 0.60 for the variable-coefficient problem. Moreover, they noted that by weighting the residuals, one can reduce the smoothing rate. In the one-dimensional case, instead of using $r_i$, one can use

$$r_i \ \leftarrow \ \beta r_{i-1} + \alpha r_i + \beta r_{i+1} \ , \tag{4.6.15}$$

where $\alpha$ and $\beta$ are chosen to maximize the smoothing. This is called a *residual smoothing* method (RSM). Choosing $\alpha = 0.380125$ and $\beta = 0.138155$, the smoothing rate is reduced from 0.60 to 0.0620992. This is better than the smoothing rate for direct inversion of a strong finite-element preconditioner (0.18) or for Fourier multigrid using strong finite-element preconditioning (0.086) and doesn't require any matrix inversions.

Consider next a two-dimensional, variable-coefficient Poisson problem – (4.4) with $\lambda = 0$ on $[0, 2\pi]^2$ with periodic boundary conditions. A single SMG iteration takes only $O(N^2 \log_2 N)$ operations, and the number of Richardson iterations can be independent of $N$ even without preconditioning provided that the discrete problem is isotropic, i.e., that $a_1(\mathbf{x})$ is strictly proportional to $a_2(\mathbf{x})$ and that $(\Delta x)^2$ and $(\Delta y)^2$ are in the same proportion. Using the local relaxation parameter and the residual smoothing, we have

$$\begin{aligned} r_{ij} \ \leftarrow & \ \alpha r_{ij} + \beta(r_{i-1,j} + r_{i+1,j} + r_{i,j-1} + r_{i,j+1}) \\ & + \gamma(r_{i-1,j-1} + r_{i-1,j+1} + r_{i+1,j-1} + r_{i+1,j+1}) \ , \end{aligned} \tag{4.6.16}$$

with $\alpha = 0.2240$, $\beta = 0.07000$ and $\gamma = 0.28800$, Brandt et al. demonstrated that a smoothing rate of 0.1058 is obtained (compared with a smoothing rate of 0.78 on this two-dimensional problem using a local relaxation parameter but not the residual smoothing). In contrast, exact inversion of the full low-order preconditioner takes $O(N^4)$ operations, and yields worse smoothing rates – the results in Table 4.4 apply to the two-dimensional Poisson problem as well.

Erlebacher, Zang and Hussaini (1987) have examined the periodic, isotropic three-dimensional Poisson problem and demonstrated that residual smoothing reduces the smoothing rate for stationary Richardson iteration from 0.85 to 0.20. If the problem is not isotropic, however, these refinements are little help. Moreover, Erlebacher et al. also pointed out that residual weighting is not very effective for the Helmholtz problem (4.2) with $\lambda \neq 0$.

For the non-isotropic Fourier SMG problem in two dimensions, Brandt et al. resorted to finite-difference preconditioning. They used alternating line Gauss-Seidel relaxation (AZLR with $\omega_{AZLR} = 1.0$) for approximate inversion of the finite-difference preconditioner in the underlying Richardson scheme and achieved smoothing rates for the purely periodic, two-dimensional problem of roughly 0.4. With periodic boundary conditions, line-relaxation preconditioning seems preferable to incomplete-LU, because the enforcement of periodicity is much simpler. Moreover, this relaxation scheme is more economical of storage than incomplete-LU because one needs auxiliary storage for just a few one-dimensional vectors rather than for many two-dimensional vectors.

We turn now to nonperiodic boundary conditions, and hence to Chebyshev multigrid methods. Interpolation for nonperiodic coordinates in one dimension employs Chebyshev series in a fashion analogous to (4.6.9)–(4.6.12), which can be accomplished with fast transforms. The prolongation operation is accomplished by

$$\tilde{u}_p^c = \frac{2}{N_c \bar{c}_p} \sum_{j=0}^{N_c} \bar{c}_j^{-1} u_j^c \cos \frac{p\pi j}{N_c}, \qquad p = 0, 1, \ldots, N_c, \tag{4.6.17}$$

and

$$u_j^f \longleftarrow u_j^f + \sum_{p=0}^{N_c} \tilde{u}_p^c \cos \frac{p\pi j}{N_c}, \qquad j = 0, 1, \ldots, N_c, \tag{4.6.18}$$

where $\bar{c_k}$ is defined by (2.4.16) with $N = N_c$. The recommended restriction process is

$$\tilde{r}_p^f = \frac{2}{N_f \bar{c}_p} \sum_{j=0}^{N_f} \bar{c}_j^{-1} r_j^f \cos \frac{p\pi j}{N_f}, \qquad p = 0, 1, \ldots, N_c, \tag{4.6.19}$$

and

$$r_j^c = \sum_{p=0}^{N_c} \tilde{r}_p^f \cos \frac{p\pi j}{N_c}\,, \qquad j = 0, 1, \ldots, N_c\,. \tag{4.6.20}$$

In order for the restriction operator (4.6.19)–(4.6.20) to be the adjoint of the prolongation operator (4.6.17)–(4.6.18), as is the recommended practice in multigrid, one needs to use the identical definition of $\bar{c}_p$ in both cases, i.e., based on $N = N_c$ in (2.4.16). See Zang, Wong and Hussaini (1984) for more discussion on this point, as well as for a closed form solution of these interpolation operators and for a discussion of alternatives to using the standard collocation approximation on the coarse grid as the coarse-grid correction operator.

Preconditioning is essential for Chebyshev multigrid. Consider incomplete-LU decomposition applied in conjunction with second-order finite-differences. Recall that the relevant eigenvalues are the largest eigenvalue and the lowest high-frequency eigenvalue, which in this two-dimensional case is roughly the eigenvalue that separates the smallest 25% of the eigenvalues from the largest 75%. The left frame of Fig. 4.32 displays the largest eigenvalue and the left frame of Fig. 4.38 furnishes the lowest high-frequency one. The lowest high-frequency eigenvalue ($\lambda_{\text{mid}}$) turns out to be relatively insensitive to $N$. The multigrid condition numbers $\mathcal{K}^*_{MG}$ for these four preconditioners are shown in the right frame of Fig. 4.38. Note that the standard incomplete-LU factorization is superior to the row-sum-equivalence alternative in the multigrid context. The former evidently does a better job on the high-frequency components of the solution. Although it performs far worse on the low-frequency components, this is immaterial for a multigrid scheme.

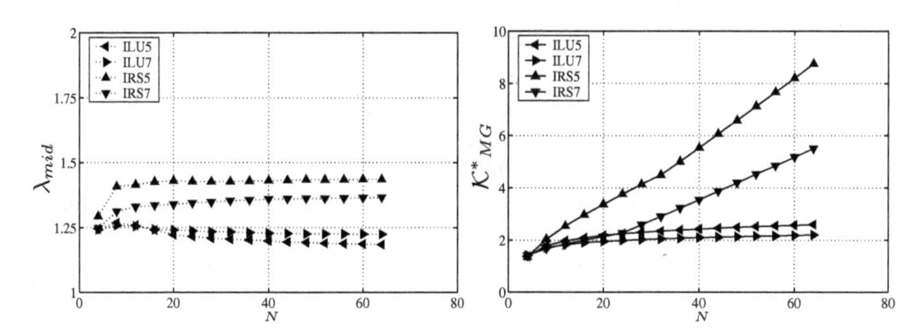


**Fig. 4.38.** Lowest high-frequency eigenvalue (*left*) and multigrid condition number (*right*) for the preconditioned Chebyshev Laplace operator in two dimensions using incomplete LU decompositions with 5 and 7 nonzero diagonals, both without (ILU) and with (IRS) the row-sum-equivalence modification

For the stationary Richardson iterative method the 7-diagonal incomplete-LU factorization on the constant-coefficient Poisson problem has a multigrid condition number not much larger than that of the full finite-difference preconditioning (for which $\mathcal{K}_{MG} = 2$). The incomplete-LU decomposition costs far less than the evaluation of the spectral residual for all but small values of $N$, whereas the inversion of the original finite-difference matrix is more expensive than the residual evaluation. The attraction of multigrid approach in two (or more) dimensions is that it offers the prospect of having the number of iterations (required for convergence) virtually independent of $N$ with a very inexpensive preconditioner.

Heinrichs (1988, 1993) has explored various preconditioners and relaxation schemes for Chebyshev multigrid applications to two-dimensional problems of the form (4.4) with Dirichlet boundary conditions. In particular, Heinrichs (1988) considered AZLR and incomplete-LU (ILU) approximations to second-order finite-differences as preconditioners, and, among others, stationary Richardson (SR) and minimum residual Richardson (MRR) as relaxation schemes. He considered several versions of ILU schemes which differed in whether the unknowns were ordered first in $x$ or first in $y$. He concluded that the best combination for the general non-isotropic problem used AZLR as preconditioner and MRR as the relaxation scheme. One can also find in this reference a discussion of the appropriate multigrid cycle and recommended numbers of iterations at each level for the various alternatives. Heinrichs (1991) extended these methods to the three-dimensional counterpart of (4.4). As in the case of standard finite-difference or finite-element multigrid methods, he showed that the use of alternating plane (as opposed to line) relaxation on the finite-difference preconditioner for the Chebyshev multigrid method did produce results that converged in a number of iterations that was virtually independent of $N$.

Relatively little attention has been paid to use of finite-element preconditioning in the context of spectral multigrid methods. The analysis of finite-element preconditioning for the 2D periodic Poisson equation is straightforward. Table 4.11 summarizes the key parameters for strong (bilinear) finite-element preconditioning of the 2D periodic Poisson equation, with second-order finite-difference preconditioning results included for comparison. The smoothing rates given there assume exact inversion of the preconditioners, and these rates are based on the maximum damping factor over the high-frequency range. In the case of the finite-difference preconditioning, the relevant eigenvalues are $\Lambda_{\rm mid}$ and $\Lambda_{\rm max}$, whereas for the finite-element precon-

**Table 4.11.** Properties of finite-difference and finite-element preconditionings for the periodic 2D Poisson problem

| Preconditioning | $\Lambda_{\rm min}$ | $\Lambda_{\rm mid}$ | $\Lambda_{\rm max}$ | $\omega_{MG}$ | $\mu_{MG}$ |
|---|---|---|---|---|---|
| fd2 | 1.000000 | 1.233701 | 2.467401 | 0.540380 | 0.333333 |
| fes | 0.6928333 | 0.848666 | 1.000000 | 1.297438 | 0.101092 |

ditioning they are $\Lambda_{\min}$ and $\Lambda_{\mathrm{mid}}$. Presumably, the ILU or AZLR strategies would produce nearly this rate of convergence, as they have done for the finite-difference preconditioning. Numerical computations of the eigenvalues for the nonperiodic counterpart, using Chebyshev collocation, indicate that these estimates for the periodic problem are reasonably reliable for the non-periodic problem.

## 4.7 Numerical Examples of Direct and Iterative Methods

In this section we illustrate the results obtained by direct and iterative methods for the solution of the linear systems associated with Fourier collocation, Chebyshev collocation and Legendre G-NI discretizations of two-dimensional problems. Numerical results are furnished here for just a small sample of the various algorithms discussed in Sects. 4.2–4.6. We begin with some examples for Fourier collocation methods, focusing on descent methods and spectral multigrid methods. We then proceed in roughly historical order, starting with solutions to Chebyshev discretizations using rather venerable preconditioned iterative schemes, continue to direct and iterative solutions of Legendre G-NI discretizations, and conclude with a detailed illustration of various preconditioners for Legendre G-NI methods.

### 4.7.1 Fourier Collocation Discretizations

This first set of numerical examples is for the simple, periodic Poisson problem

$$\begin{aligned} -\Delta u &= f && \text{in } \Omega = (0, 2\pi)^2 , \\ u & \ 2\pi\text{-periodic} && \text{in each direction,} \end{aligned} \tag{4.7.1}$$

and for its variable-coefficient, self-adjoint generalization

$$\begin{aligned} -\nabla \cdot (a \nabla u) &= f && \text{in } \Omega = (0, 2\pi)^2 , \\ u & \ 2\pi\text{-periodic} && \text{in each direction,} \end{aligned} \tag{4.7.2}$$

where $a(x, y) = 1 + \varepsilon e^{\cos(x+y)}$, for suitable $\varepsilon > 0$. The function $f$ is chosen in such a way that the exact solution is $u(x, y) = \sin(4\pi \cos(x)) \sin(4\pi \cos(y))$.

Figure 4.39 presents results using conventional iterative methods on a single grid (left) and spectral multigrid methods (right) for the Poisson problem (4.7.1). The schemes illustrated are (static) non-stationary Richardson with 3 parameters (NSR – see (C.1.1) and (C.1.14), for $k = 3$), minimum residual Richardson without preconditioning (MRR – see (C.2.2) and Table C.1 with the $PMRR_2$ choice), residual smoothing combined with stationary Richardson (RSM – see (4.6.16)), and non-stationary Richardson with 3 parameters preconditioned with second-order finite differences (PNSR). For all of these examples the convergence criterion is that $\|\mathbf{r}^n\|/\|\mathbf{f}\| < 10^{-14}$, and a random initial guess $\mathbf{u}^0$, i.e., with each component drawn from a uniform random distribution in $(0, 1)$.

The iterative condition number $\mathcal{K}$ of the linear system for this problem for the non-preconditioned schemes (NSR, MRR, RSM and CG) is $N^2/2$, and it is the usual $\pi^2/4$ for the preconditioned method (PNSR). The NSR, MRR and RSM methods are expected to require a number of iterations for convergence that scale as $N^2$ (see (C.1.13)), the CG method only requires $O(N)$ iterations (see (C.2.12)), and the number of iterations required for the PNSR method should be independent of $N$. The left part of Fig. 4.39 reports the number of iterations required to achieve convergence versus the number of collocation points in each direction. The results are all consistent with the expected growth with $N$. The NSR and MRR methods require nearly the same number of iterations, whereas the RSM scheme takes an order of magnitude fewer. The superiority of the conjugate gradient method over the simpler non-preconditioned schemes is evident in the orders of magnitude fewer iterations required at a relatively minor extra cost. The number of iterations required by the preconditioned scheme (PNSR) is virtually constant with $N$, again, as expected. Although each iteration of the preconditioned scheme is more expensive than the non-preconditioned schemes, the dramatic reduction in the required number of iterations yields a substantial net benefit.

Results for multigrid schemes are shown in the right part of Fig. 4.39 using the same underlying iterative methods as for the single-grid schemes with the exceptions that the CG method is not used and that for the PNSR scheme alternating line relaxation is applied with only 3 sweeps rather than resorting to an exact solution of the full finite-difference preconditioner. For multigrid methods there are numerous alternative strategies. For these examples we follow the recommendations in Brandt, Fulton and Taylor (1985) and used a simple V-cycle (Fig. 4.36) starting on the finest grid. For the NSR and MRR iterative methods we use $N_d = 3$ and $N_u = 0$, whereas for the RSM method

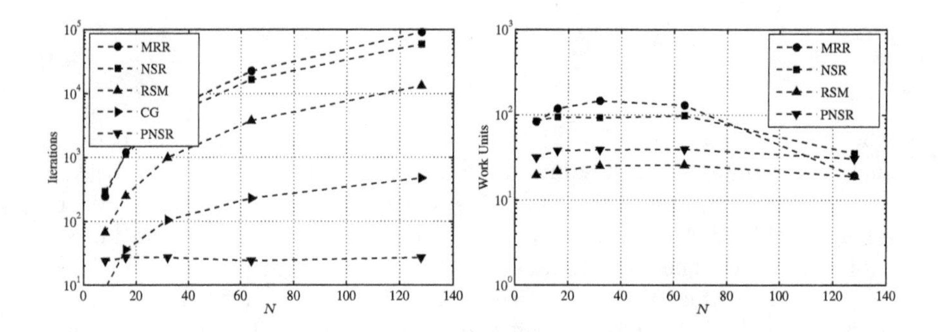


**Fig. 4.39.** Single-grid (*left*) and multigrid (*right*) iterations necessary to converge for the 2D periodic Poisson problem (4.7.1)

we use $N_d = 2$ and $N_u = 1$. The performance is reported in the standard multigrid measure of *work units* rather than iterations. A work unit is the time required for a single iteration on the finest grid. This measure includes the time required for iterations on the coarser grids, as well as for the inter-grid transfers. In all of the multigrid examples here, the actual number of iterations performed on the finest grid was very close to two-thirds of the work units. Note that for $16 \le N \le 64$ the number of work units required for convergence is nearly constant with $N$. There is a substantial decrease in work for $N = 128$, but the solution is already resolved to nearly machine precision by $N = 64$, so that there is very little smoothing needed on the finest grid. Note also how remarkably effective the residual smoothing is for this simple problem; it even outperforms the PNSR method, for which each iteration is much more expensive.

Observe that the best of the non-preconditioned multigrid methods (RSM) converges in less that 40 effective fine-grid iterations, whereas on the finer grids $10^3$–$10^5$ iterations are needed for the simpler non-preconditioned single-grid methods (NSR MRR and RSM), and $10^2$–$10^3$ iterations are needed for the non-preconditioned CG single-grid method. Moreover, although the preconditioned single-grid results (PNSR) take slightly fewer iterations, they come with the requirement to solve the full finite-difference approximation at each iteration, which makes them more costly than their multigrid counterpart.

Results for the variable-coefficient problem (4.7.2) are given in Fig. 4.40 for three values of the parameter $\varepsilon$ for the best of iterative schemes. In particular, results are given for the CG and PNSR methods on a single grid, and for the RSM (using NSR) and the PNSR (preconditioned with 3 sweeps of alternating line relaxation) multigrid methods. Note that the conjugate gradient method is still applicable because the discrete problem remains symmetric and positive definite for $\varepsilon > 0$. The methods converge in all cases, but more iterations are required as $\varepsilon$ increases, presumably because the increasing variation of the coefficient worsens the conditioning of the linear system. For this example the multigrid method using RSM is the most efficient.

### 4.7.2 Chebyshev Collocation Discretizations

We continue with some results for iterative solutions of Chebyshev collocation discretizations to the Poisson problem

$$\begin{aligned} -\Delta u &= f \qquad && \text{in } \Omega = (-1,1)^2 , \\ u &= 0 && \text{on } \partial\Omega , \end{aligned} \tag{4.7.3}$$

and to its variable-coefficient, self-adjoint generalization

$$\begin{aligned} -\nabla \cdot (a \nabla u) &= f \qquad && \text{in } \Omega = (-1,1)^2 , \\ u &= 0 && \text{on } \partial\Omega , \end{aligned} \tag{4.7.4}$$

where $a(x, y) = 1 + \varepsilon x^2 y^2$, for suitable $\varepsilon > 0$. The particular choice $f = 1$ is made.

As discussed in Haidvogel and Zang (1979), a spectral solution to this problem converges only algebraically (as $1/N^4$) because of the corner singularities. This problem was chosen because the relatively slow decay of the expansion coefficients ensures that the results on the performance of the iterative methods are representative of the most challenging problems arising in practice. Recall that examples of some Chebyshev and Legendre ad hoc methods for (4.7.3) have been provided in Fig. 4.2, albeit for a choice of $f$ and of boundary conditions that produces solutions with spectral accuracy.

In the Chebyshev collocation case the matrix $L$ is nonsymmetric. The preconditionings are the incomplete-LU factorizations of the second-order finite-difference approximation discussed in Sect. 4.4.3. The iterative schemes are the preconditioned minimum residual Richardson $\mathrm{PMRR}_2$ method described by (C.2.2) and Table C.1 and the (truncated) preconditioned conjugate residual method described by (C.2.21). (Since the linear system is nonsymmetric, the orthogonality conditions (C.2.13) of the conjugate residual method are not satisfied. Hence, we refer to this method as the preconditioned truncated conjugate residual (PTCR) method.) The convergence criterion is the same

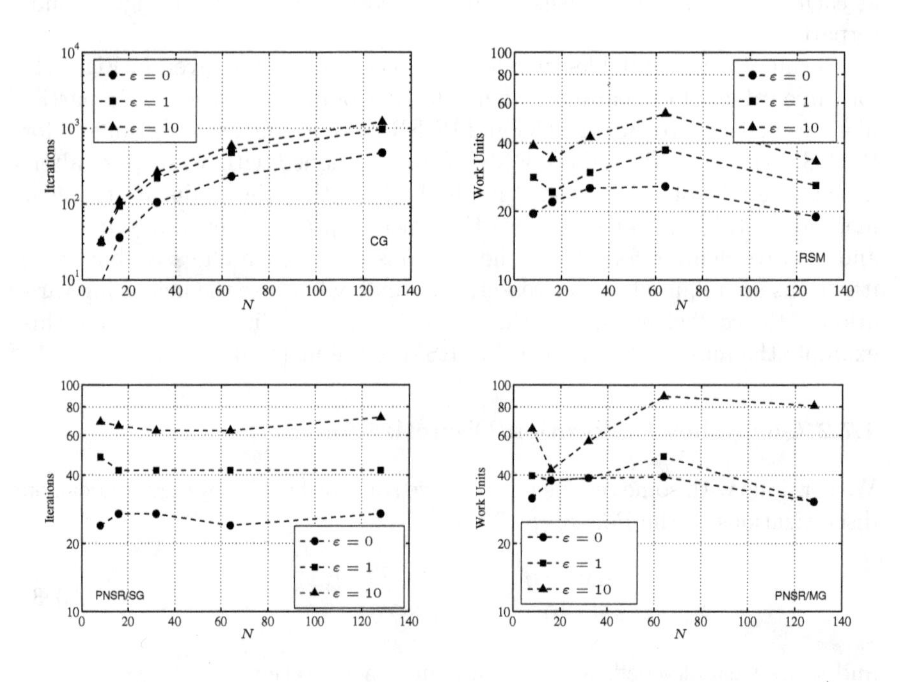


**Fig. 4.40.** Single-grid (*left*) and multigrid (*right*) iterations necessary to converge for the 2D periodic, variable-coefficient problem (4.7.2)

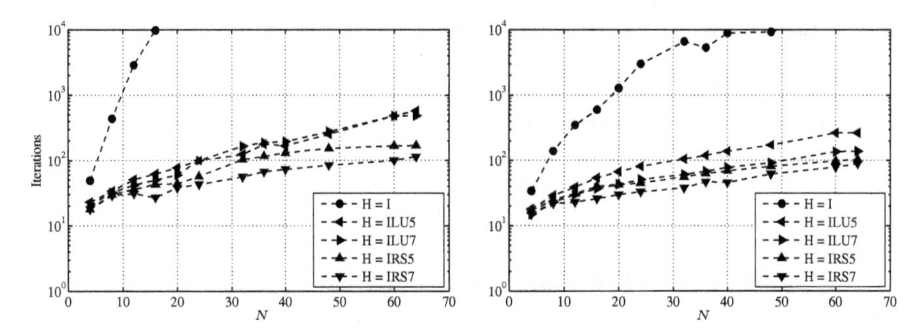


**Fig. 4.41.** PMRR$_2$ (*left*) and PTCR (*right*) iterations necessary to converge for the 2D Poisson problem (4.7.3) for non-preconditioned ($H = I$) and various incomplete-LU preconditionings with $\mathbf{u}^0 = \mathbf{0}$

as for the Fourier collocation example ($\|\mathbf{r}^n\|/\|\mathbf{f}\| < 10^{-14}$), but the initial guess is just $\mathbf{u}^0 = \mathbf{0}$.

Figure 4.41 reports the results for the Poisson problem (4.7.3). Even though $H^{-1}L$ is not symmetric and positive definite, these methods still converge for the Poisson problem since the eigenvalues of the symmetric part (4.5.2) of the linear system are positive. The iterative condition number of the non-preconditioned linear system scales as $N^4$, implying, according to (C.1.13), the same quartic growth with $N$ of the number of required iterations. Note that the performance of the various preconditioned iterative schemes follows the trends of the iterative condition numbers illustrated in Fig. 4.33, namely, the row-sum equivalence versions (IRS5 and IRS7) perform better than the straight incomplete-LU versions (ILU5 and ILU7), and the inclusion of the two extra nonzero diagonals in the incomplete decompositions is beneficial. The PMRR$_2$ performance is noticeably worse than that of PTCR for the plain incomplete-LU preconditionings (ILU5 and ILU7), but the row-sum equivalence results are comparable.

The performance of the best of the incomplete-LU preconditioners – the IRS7 version – with these same two iterative schemes on the variable-coefficient, self-adjoint problem (4.7.4) is reported in Fig. 4.42. For small and moderate values of $N$ the iterative schemes converge even for fairly large values of $\varepsilon$. However, for large $N$ the schemes eventually fail to converge as $\varepsilon$ increases. Hence, for Chebyshev collocation methods these older iterative methods (PMRR$_2$ and PTCR) must yield to the more modern Krylov methods discussed in Sect. C.3 of Appendix C or to the spectral multigrid method; see Zang, Wong and Hussaini (1985) and Heinrichs (1988, 1993) for some numerical results of the latter methods.

### 4.7.3 Legendre G-NI Discretizations

The next set of examples are again for the Poisson problem (4.7.3), but this time with Legendre G-NI discretizations. We assume that the data are such that the exact solution is $u(x,y) = \sin(4\pi x)\sin(4\pi y)$. Both direct and iterative methods are illustrated. The associated system $K_{GNI}\mathbf{u} = M_{GNI}\mathbf{f}$ has a matrix with $(N-1)^2$ rows and columns, where $N$ is the spectral polynomial degree; $\mathbf{u}$ is the vector of the $(N-1)^2$ values at the internal Gauss-Lobatto nodes. The stiffness matrix $K_{GNI}$ is symmetric and positive definite; its spectral condition number scales as $N^3$.

We begin with some comparisons between the behavior of direct and iterative methods of solution. Two different direct methods are used: the *Cholesky factorization* of $K_{GNI}$ as implemented in the library LAPACK (see Sect. 4.4.1) and the *multifrontal method* as implemented in the library UMFPACK (see Sect. 4.4.3). Iterative methods are based on the PCG method with four different kinds of preconditioners: $H = I$ (no preconditioning, i.e., a simple CG method); $H = \text{diag}(K_{GNI})$, i.e., the diagonal matrix whose entries are the diagonal elements of $K_{GNI}$; $H = ICHOL(K_{GNI})$, i.e., the incomplete-Cholesky factorization of $K_{GNI}$ with no fill-in (see Saad (1996), Chap. 10); $H = K_{FE}$, i.e., the finite-element stiffness matrix based on the use of bilinear elements on the two-dimensional mesh whose vertices are the LGL nodes, as defined in (4.4.63), or its approximation $K_{FE,\text{app}}$. Bear in mind that the incomplete-Cholesky factorization for the G-NI example was performed for the full spectral discretization, whereas in the Chebyshev collocation example above, the incomplete-LU factorizations were performed for the low-order preconditioning. On the other hand, the linear systems for the $K_{FE}$ preconditioner were invariably solved by a direct method, based on the Cholesky factorization of this matrix.

The first comparison is on accuracy with respect to round-off. In Fig. 4.43 we report for several values of $N$ the relative error in the discrete maximum norm,

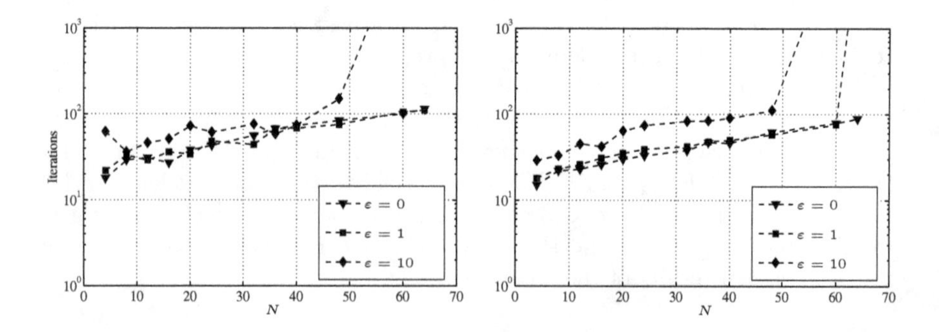


**Fig. 4.42.** PMRR$_2$ (*left*) and PTCR (*right*) iterations necessary to converge for the 2D self-adjoint problem (4.7.4) using the IRS7 preconditioning

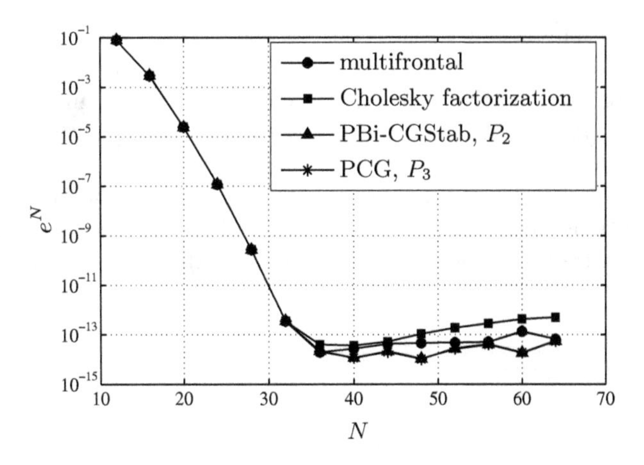


**Fig. 4.43.** Comparison of accuracy for direct and iterative solution of the 2D Poisson problem (4.7.3). The relative error $e^N$ versus $N$ obtained with direct and iterative methods. $f$ is chosen such that the exact solution is $u(x, y) = \sin(4\pi x)\sin(4\pi y)$ and the initial guess for iterative methods is $\mathbf{u}^0 = \mathbf{0}$

$$e^N = \max_{i,j} |u(x_i, y_j) - u^N(x_i, y_j)| / \max_{i,j} |u(x_i, y_j)| ,$$

where the maximum is taken over all internal LGL nodes, $u$ is the exact solution, $u^N$ is the G-NI solution obtained by two different direct methods (multifrontal and the Cholesky factorization) or by two different preconditioned iterative methods (the PCG, with preconditioned matrix $P_3$ given in (4.4.66), and PBi-CGStab with preconditioned matrix $P_2$ given in (4.4.65)). When $P_3$ is used, the stopping criterion is $\|\mathbf{r}^n\| / \|M_{GNI}\mathbf{f}\| < 10^{-14}$, with $\mathbf{r}^n = M_{GNI}\mathbf{f} - K_{GNI}\mathbf{u}$ (the same results are obtained using a more strict tolerance of $10^{-18}$). When $P_2$ is used instead, the stopping criterion is $||\mathbf{r}^n||_H / ||\mathbf{f}||_H < 10^{-14}$, with $\mathbf{r}^n = \mathbf{f} - M_{GNI}^{-1} K_{GNI} \mathbf{u}^n$. The iterative schemes have slightly more favorable round-off error behavior than the two direct methods.

We now focus on the iterative methods, starting, in Fig. 4.44, with the dependence upon $N$ of the iterative condition numbers of the preconditioned matrix $H^{-1}K_{GNI}$ for several choices of the preconditioner $H$. As expected, the iterative condition number for the finite-element-based preconditioners is independent of problem size, whereas the iterative condition number grows with $N$ for the other preconditioners. Note that the choice $H = K_{FE,\text{app}}$ yields the preconditioned matrix $P_3$ introduced in (4.4.66), while $H = (M_{FE}^{-1} K_{FE})^{-1} M_{GNI}^{-1}$ yields the matrix $P_2$ introduced in (4.4.65); the numerical values of the iterative condition numbers for these two preconditionings are reported in Table 4.9.

In cross-comparing these iterative methods we switch to the choices $f = 1$ and $\mathbf{u}^0 = \mathbf{0}$ for the reasons described in the previous subsection. In Fig. 4.45

we report the number of PCG iterations (versus $N$), while in Fig. 4.46 we report the CPU-time (versus $N$) that is needed to solve the linear system (4.4.62). (For the iterative methods we include both the time necessary for constructing (and factorizing) the preconditioner along with that required for the iterations.) The behavior of the curves is consistent with the behavior of the iterative condition numbers shown above. Note that the total number of iterations needed here for the incomplete-Cholesky preconditioning for the Legendre G-NI discretization is roughly the same as was needed be-

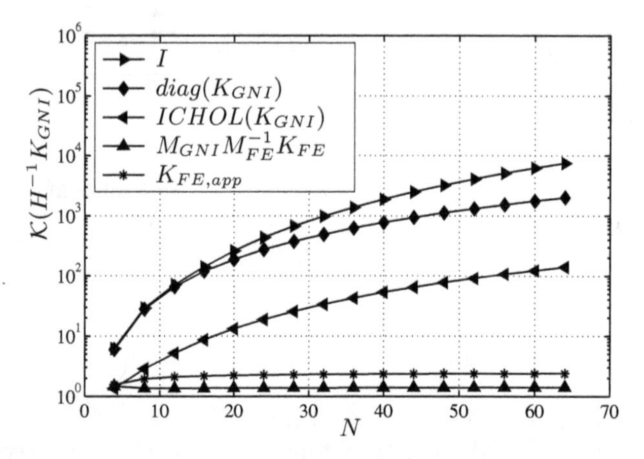


**Fig. 4.44.** Iterative condition number (versus $N$) of the preconditioned matrix $H^{-1}K_{GNI}$ for different choices of $H$ for the solution of the 2D Poisson problem (4.7.3)

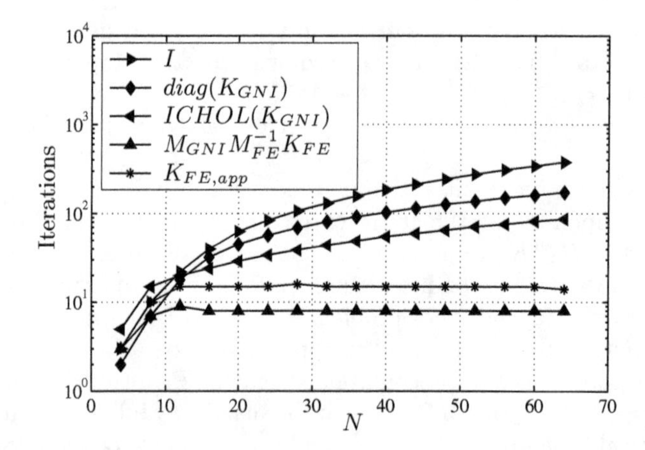


**Fig. 4.45.** Iterations necessary to converge for the 2D Poisson problem (4.7.3) with $f = 1$ and $\mathbf{u}^0 = \mathbf{0}$. PBi-CGStab has been used for $H = M_{GNI}M_{FE}^{-1}K_{FE}$, PCG for the other preconditioners

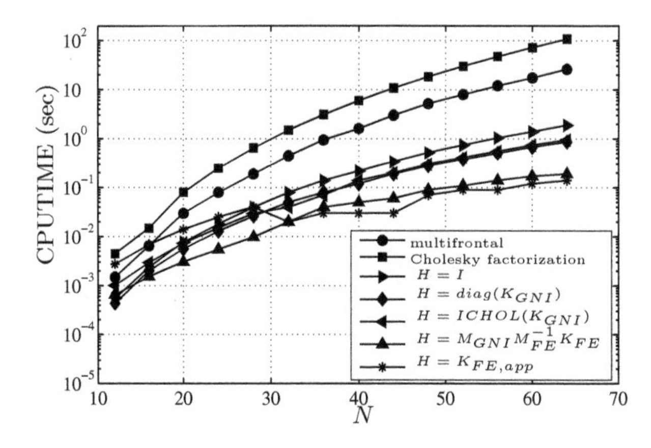


**Fig. 4.46.** CPU-time necessary to solve the linear system for the 2D Poisson problem (4.7.3) with $f = 1$. For iterative solution methods, $\mathbf{u}^0 = \mathbf{0}$, PBi-CGStab has been used for $H = M_{GNI}M_{FE}^{-1}K_{FE}$, PCG otherwise

fore for the row-sum equivalence incomplete factorizations (RS5 and RS7) of the low-order finite-difference preconditioning for the Chebyshev collocation discretization. The results on the CPU times neatly indicate that from moderate to large values of $N$, the PCG method that uses the (exactly factorized) finite-element matrices yields the best performance. The plot also clearly documents that the multifrontal strategy invariably outperforms the Cholesky factorization even for spectral matrices. However, for small values of $N$ (say, on the order of 10), typical of those used in each subdomain of a multidomain spectral method (see CHQZ3, Chaps. 5–6), the CPU times for factorization are negligible. In this case, the most convenient choice, in terms of easiness of programming and memory storage, remains the direct Cholesky factorization of the spectral matrix.

Table 4.12 compares the memory requirements of the different approaches on a 2D problem as a function of $n = (N-1)^2$, the number of rows (or columns) of the matrix $K_{GNI}$. The memory demands of the two direct methods scale as $n^2$, whereas the memory demands for the iterative methods scale only as $n^{3/2}$. In general, the multifrontal method is the most memory intensive, followed by the Cholesky factorization, and then by the PCG methods.

### 4.7.4 Preconditioners for Legendre G-NI Matrices

In Sect. 4.4.3, we introduced several preconditioned forms of the algebraic system $K_{GNI}\mathbf{u} = M_{GNI}\mathbf{f}$ (see (4.4.64)–(4.4.70)), and we documented the iterative condition numbers of the corresponding matrices $P_1, \ldots, P_7$ (see Table 4.9). Here we aim at investigating the performance of the various preconditioners when inserted within an appropriate iterative method to solve

**Table 4.12.** Comparison of memory requirements for the solution of 2D GNI systems

| Method | Words (real) | Words (integer) |
|---|---|---|
| Cholesky | $n(n+1)/2$ | 0 |
| Multifrontal | $n^2 + 3n^{3/2}$ | $5n^{3/2} + 36n$ |
| PCG $H = I$ or $H = \mathrm{diag}(K_{GNI})$ | $n^{3/2} + 5n$ | $n^{3/2} + n$ |
| PCG $H = ICHOL(K_{GNI})$ | $2n^{3/2} + 6n$ | $2n^{3/2} + 3n$ |
| PCG $H = K_{FE,\mathrm{app}}$ | $2n^{3/2} + 5n$ | $n^{3/2} + 2n$ |

the test problem considered in the previous subsection. We recall that the systems with matrices $P_1, P_3, P_5, P_6$ and $P_7$ can be solved by the PCG method, whereas those with matrices $P_2$ and $P_4$ require an iterative algorithm for nonsymmetric matrices; we focus on the PBi-CGStab method, but the GMRES method (Sect. C.3 of Appendix C) is a viable alternative. In all cases a direct Cholesky factorization of the finite-element matrix is performed in a pre-processing stage, and at each iteration only the forward elimination and back substitution are required.

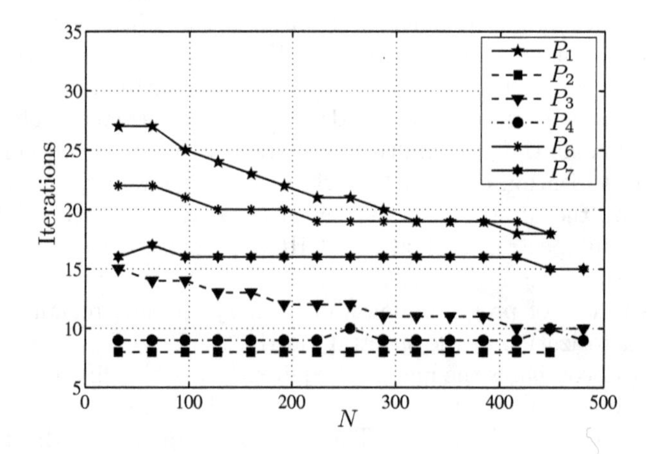


**Fig. 4.47.** Number of PCG and PBi-CGStab iterations to solve problem (4.7.3) with $\mathbf{f} = 1$ and $\mathbf{u}^0 = \mathbf{0}$, for the different preconditioners given by (4.4.64)–(4.4.70)

Figure 4.47 reports the number of iterations needed to meet the stopping criterion $||\mathbf{r}^n||_H/||\mathbf{r}^0||_H < 10^{-14}$ with the initial guess $\mathbf{u}^0 = \mathbf{0}$. Note that no results for $P_5$ are reported, as the cost of computing the square root of the finite-element matrix makes the method noncompetitive in practice even for moderate values of $N$. The corresponding CPU times (in seconds) are reported in Fig. 4.48. (The numerical results in this subsection were performed on a more powerful computer – in order to explore the regime of large $N$ – than those in the previous subsection. Hence, the CPU times are

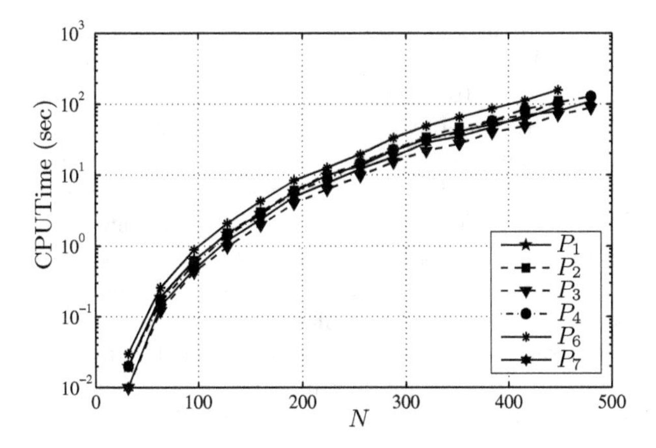


**Fig. 4.48.** Total CPUtime (sec) of PCG and PBi-CGStab iterations to solve problem (4.7.3) with $f = 1$ and $\mathbf{u}^0 = \mathbf{0}$, for the different preconditioners given by (4.4.64)–(4.4.70)

smaller for the present results.) Note that there is a factor of at most three difference between the fastest and the slowest methods. The precise ranking of the methods no doubt depends on the details of the implementation, the compiler, and the host computer. A comparison of Figs. 4.47 and 4.48 clearly indicates that the actual cost of a solution scheme cannot be inferred solely from the number of iterations – the PBi-CGStab methods are not the fastest ones despite requiring fewer iterations than the CG methods. In the two-dimensional case, the fastest solution was obtained from preconditioned matrix $P_3$. It corresponds to the weak form of finite-element preconditioning, but with a numerical approximation of the stiffness matrix $K_{FE}$ that is more consistent with the structure of the two-dimensional stiffness matrix $K_{GNI}$. Remarkably, this approximation is responsible by itself of producing the best results, without even involving the mass matrix. We recall that in one dimension the best results were guaranteed by the strong form of finite-element preconditioning, which does incorporate the mass matrix. Note however that $P_7$, which includes the mass matrix but in a symmetric way, yielded nearly as good results as $P_3$. The slowest solution was produced by the $P_6$ preconditioner.

The major components of the iterative methods are assembly and factorization of the preconditioning matrix, solution of the preconditioned system at each iteration from the forward elimination/back substitution algorithms, and residual computation. According to Table 4.10, the cost of factorizing the preconditioner scales as $N^4$; in contrast, the cost of assembly the preconditioner scales as $N^2$, while the cost of the other two components only scales as $N^3$. Hence, one expects that as $N$ increases, factorization will increasingly dominate the computational time. Numerical results indicate that the overall cost of the finite-element preconditioner can be confined between 40 and 60 percent of the total solution cost with a clever programming, provided $N$ stays significantly below 100. For large values of $N$ the matrix construction takes an increasing fraction of the computational time. In that range it seems essential to employ some form of inexact solution of the finite-element system, such as inexact factorization. That would reduce the operation count for factorization from $O(N^4)$ to $O(N^3)$, albeit, as we saw in the Chebyshev collocation examples of Sect. 4.7.2, at the price of a slow increase in the number of iterations required for convergence. A trade-off analysis depends on so many factors (such as the kind of differential problem at hand, its spatial dimension, the choice of the inexact factorization, the range of N and the implementation details) that we refrain from drawing any general conclusions here. See Canuto, Gervasio and Quarteroni (2006) for a comprehensive study.

At the end of Sect. 4.7.2 we saw that the PMRR and PTCR schemes failed to converge for Chebyshev collocation discretizations to the 2D self-adjoint problem (4.7.4) for sufficiently large $\varepsilon$. Recall that the Chebyshev collocation discretization is nonsymmetric, and that there is not a general convergence guarantee for the PMRR and PTCR schemes unless the eigenvalues of the symmetric part of the preconditioned operator are positive. The Legendre G-NI discretization to (4.7.4) is, of course, symmetric and positive definite. Moreover, this property is retained for most of the corresponding preconditioned systems that we have discussed. In these cases, there is a convergence guarantee for the various iterative methods. Figure 4.49 illustrates the performance of the PCG method using the $P_3$ preconditioner on the Legendre G-NI discretization of the variable-coefficient, self-adjoint problem (4.7.4). (The choice of the $P_3$ preconditioner was made because not only is it quite efficient in terms of overall computation time – see Fig. 4.48 – but also the $P_3$ preconditioner is much easier to construct than the alternatives for the variable-coefficient problem.) Note that convergence is always achieved, and that the required number of iterations is independent of $N$. As was observed for the corresponding Fourier collocation (Fig. 4.40) and Chebyshev collocation (Fig. 4.42) examples, the required number of iterations roughly doubled as $\varepsilon$ increased from 0 to 10.

As a final set of numerical examples, we consider again the Legendre G-NI method and we apply it to the solution of the advection-diffusion boundary-value problem

$$\begin{aligned} -\nu\Delta u + \boldsymbol{\beta}\cdot\nabla u = f \qquad & \text{in } \Omega = (-1,1)^2 , \\ u = g \qquad & \text{on } \partial\Omega_D , \\ \nu\frac{\partial u}{\partial n} = h \qquad & \text{on } \partial\Omega_N , \end{aligned} \tag{4.7.5}$$

with $\partial\Omega_D = \{(x,y)\in\partial\Omega \,:\, \boldsymbol{\beta}\cdot\mathbf{n} < 0\}$ and $\partial\Omega_N = \partial\Omega\setminus\partial\Omega_D$. We have set $\nu = 0.1$ and $\boldsymbol{\beta} = (1,1)^T$. The functions $g$, $h$ and $f$ are chosen in such a way that the exact solution is $u(x,y) = \sin(\pi x)\sin(\pi y)$.

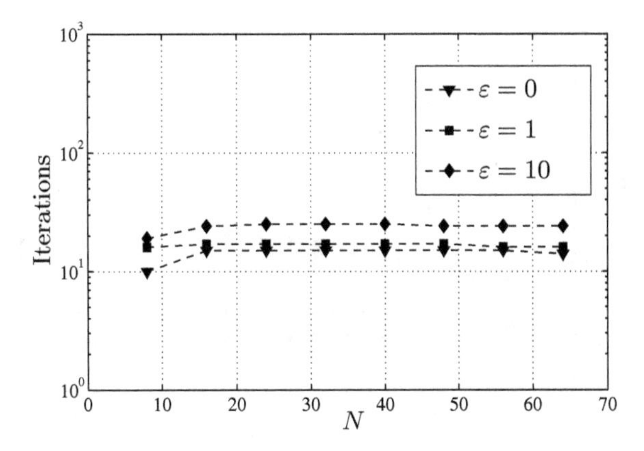


**Fig. 4.49.** CG iterations necessary to converge for the 2D self-adjoint problem (4.7.4) using the $P_3$ preconditioning

The associated system (whose matrix $K_{GNI}$ is still positive definite but no longer symmetric) has been solved by the multifrontal method, the Gauss-LU factorization (implemented in LAPACK), the PBi-CGStab method introduced in Appendix C. Several kinds of preconditioners $H$ have been used: $H = I$ (i.e., no preconditioning), $H = \text{diag}(K_{GNI})$, $H = ILU(K_{GNI})$ (i.e., the incomplete LU-factorization with no fill-in), $H = M_{GNI}M_{FE}^{-1}K_{FE}$, which yields the preconditioned matrix $P_2$ (see (4.4.65)), $H = K_{FE,\text{app}}$, which yields the matrix $P_3$. (In the last two cases the linear system associated with $H$ is solved by the banded-LU factorization of LAPACK.) In Figs. 4.50 we report the condition numbers with respect to the 2-norm and the iterative condition number of the preconditioned matrix $H^{-1}K_{GNI}$ versus $N$ for different choices of the preconditioner. As noted in Appendix C, it is the iterative condition number $\mathcal{K}^*$ which is the most useful indicator of the performance

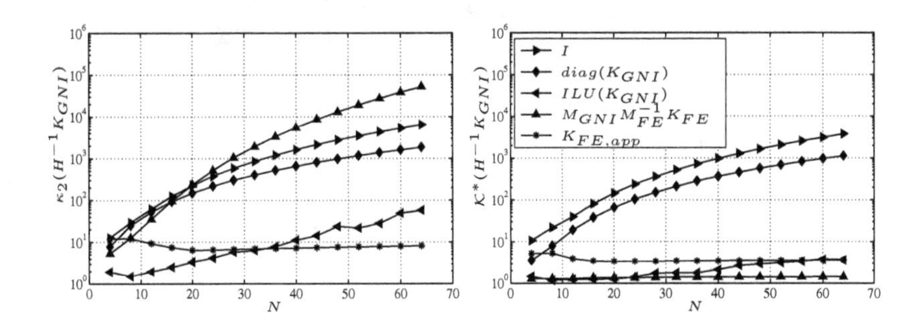


**Fig. 4.50.** Condition numbers for the advection-diffusion problem (4.7.5). 2-norm condition number (*left*) and iterative condition number $\mathcal{K}^*$ (*right*) of the preconditioned matrix, for several preconditioners

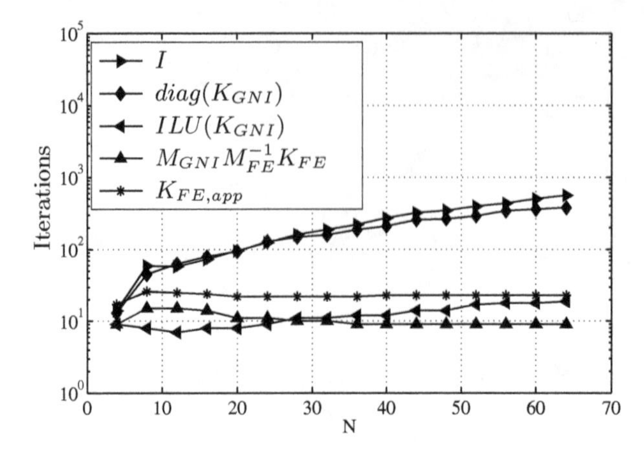


**Fig. 4.51.** PBi-CGStab iterations to solve problem (4.7.5), with $f = 1$, $\mathbf{u}_0 = \mathbf{0}$ and tolerance $10^{-14}$. Different curves refer to different choices of preconditioners

of the iterative methods. In Fig. 4.51 we report the number of iterations that are needed to fulfill the stopping criterion $\|\mathbf{r}^n\|/\|M_{GNI}\mathbf{f}\| < 10^{-14}$ when using the PBi-CGStab method. The corresponding CPU-times for the solution of the linear system by the various direct and iterative methods mentioned above are provided in Fig. 4.52.

There is a clear preference for iterative methods over direct methods for all but the smallest values of $N$, but the preference of iterative methods does not extend to as small a value of $N$ as it does for the Poisson example.

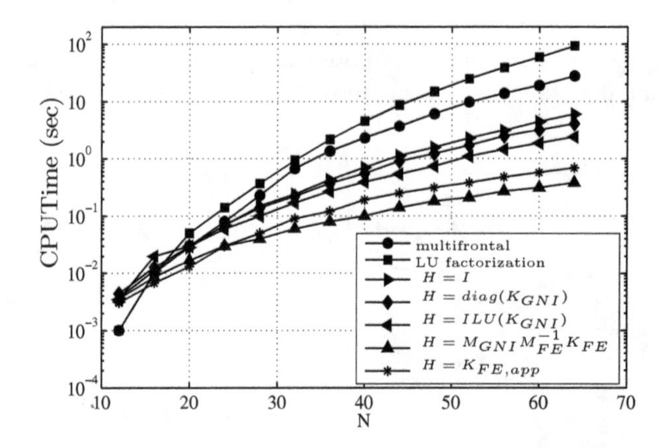


**Fig. 4.52.** CPU-time for the PBi-CGStab to solve problem (4.7.5), with $f = 1$, $\mathbf{u}_0 = \mathbf{0}$ and tolerance $10^{-14}$. Different curves refer to different choices of preconditioners

These results suggest that $P_2$ and $P_3$ are the methods of choice, and amongst these, $P_2$, which involves the finite-element mass matrix, provides a significant advantage.

## 4.8 Interlude

The emphasis in these past three chapters has been on providing a practical foundation for the application of spectral methods to differential equations in simple domains. We have laid out the basic elements of approximations of smooth functions by spectrally accurate expansions in orthogonal polynomials (trigonometric and algebraic), described how to construct a spectral approximation to a differential problem using Galerkin, Galerkin with numerical integration, collocation and tau methods, covered how to construct the matrices connected with the numerical approximation, presented an eigenanalysis of the matrices representing the spatial discretization, and surveyed the key aspects of approximating boundary conditions. Several critical issues of numerical efficiency were addressed, including transform methods and solution of implicit equations by direct and iterative methods.

The focus of this book now shifts to the theoretical analysis of spectral methods in simple domains. We cover in depth the relevant approximation theory, expound on a general theory of stability and convergence for spectral methods, and provide a number of key applications of the theory to model problems.

Our companion book (CHQZ3) furnishes extensive coverage of spectral algorithms for fluid dynamics applications in simple domains and then describes the evolution of spectral methods from the classical spectral methods covered in this book and the first part of CHQZ3 to the modern-day multidomain spectral methods capable of furnishing efficient, highly accurate approximations to differential equations in general domains.

# 5. Polynomial Approximation Theory

In the remainder of this book we concentrate on summarizing the fundamental spectral methods theory for approximation errors, stability and convergence, and apply this to the analysis of model equations. We will not present here all the details of all the proofs of the results that are cited. Rather, we illustrate the basic principles of the theory by presenting proofs for representative results. In many cases these proofs are delayed until later in the chapter (in the interests of having a coherent summary). For the same reason bibliographic references for the main contributions to the theory are likewise deferred to the end of the appropriate section.

More specifically, in this chapter we present error estimates for the approximation of functions by orthogonal polynomials. The results will cover the following topics:

(i) inverse inequalities for polynomials concerning summability and differentiability;
(ii) error estimates for the truncation error $u - P_N u$, where $P_N u$ denotes the truncated "Fourier" series of $u$;
(iii) existence, uniqueness and error estimates for the polynomials of best approximation in $L^p$ or Sobolev norms;
(iv) error estimates for the interpolation error $u - I_N u$, where $I_N u$ denotes the polynomial interpolating $u$ at a selected set of points in the domain.

Many of the results we present are taken from the general theory of approximation by polynomials. Their interest extends beyond the boundaries of approximation theory, since they are applied to the convergence analysis of spectral methods (see Chap. 6). We include proofs of those results that are most significant for the analysis of such methods.

In all the estimates contained in this chapter, $C$ will denote a positive constant that depends upon the type of norm involved in the estimate, but which is independent of the function $u$, the integer $N$, and the diameter of the domain.

## 5.1 Fourier Approximation

In this section, as well as throughout the remaining chapters, we will deal with trigonometric polynomials of degree up to $N$, rather than $N/2$ as in the previous chapters. This change is motivated by the desire for simplicity in the mathematical notation. Thus, we denote here by $S_N$ the space of the trigonometric polynomials of degree up to $N$:

$$S_N = \text{span}\left\{e^{ikx} \mid -N \leq k < N\right\} . \tag{5.1.1}$$

### 5.1.1 Inverse Inequalities for Trigonometric Polynomials

We consider the problem of the equivalence of the $L^p$-norms for trigonometric polynomials. We recall that the $L^p$-norm of a function $u$ over $(0, 2\pi)$ is defined as follows:

$$\|u\|_{L^p(0,2\pi)} = \left(\int_0^{2\pi} |u(x)|^p \, dx\right)^{1/p} , \qquad 1 \leq p < \infty , \tag{5.1.2}$$

and

$$\|u\|_{L^\infty(0,2\pi)} = \sup_{0 \leq x \leq 2\pi} |u(x)| , \qquad p = \infty \tag{5.1.3}$$

(rigorously speaking, the supremum in the latter norm should exclude subsets of $[0, 2\pi]$ of zero measure (see Sect. A.9)). The set of functions for which each particular norm is finite forms a Banach space denoted by $L^p(0, 2\pi)$ (see (A.9.f)). The following several inequalities enable one to relate the norms of a given polynomial in different $L^p$ spaces.

If $p$, $q$ are any real numbers such that $1 \leq p \leq q \leq \infty$, and if $u \in L^q(0, 2\pi)$, then $u \in L^p(0, 2\pi)$, and $\|u\|_{L^p(0,2\pi)} \leq C\|u\|_{L^q(0,2\pi)}$, where $C$ depends on $p$ and $q$. If $u$ is a periodic function with a finite expansion this inequality can be inverted. Indeed, the following *Nikolski's inequality* holds:

$$\|\phi\|_{L^q(0,2\pi)} \leq CN^{1/p-1/q}\|\phi\|_{L^p(0,2\pi)} \quad \text{for all } \phi \in S_N . \tag{5.1.4}$$

A different kind of inverse inequality, the *Bernstein inequality*, relates the norm of a function $u \in S_N$ to that of its derivatives. For all real $p$, $1 \leq p \leq \infty$, and for all integers $r \geq 1$,

$$\|\phi^{(r)}\|_{L^p(0,2\pi)} \leq N^r\|\phi\|_{L^p(0,2\pi)} \quad \text{for all } \phi \in S_N , \tag{5.1.5}$$

where $\phi^{(r)}$ denotes the derivative of order $r$ of $\phi$.

### 5.1.2 Estimates for the Truncation and Best Approximation Errors

Let $P_N : L^2(0,2\pi) \to S_N$ be the orthogonal projection upon $S_N$ in the inner product of $L^2(0,2\pi)$ (see (2.1.10)):

$$(u - P_N u, \phi) = 0 \quad \text{for all } \phi \in S_N \ .$$

With the present definition of $S_N$ (see (5.1.1)), $P_N u$ is the truncated Fourier series of $u$, i.e.,

$$P_N \left( \sum_{k=-\infty}^{\infty} \hat{u}_k \phi_k \right) = \sum_{k=-N}^{N-1} \hat{u}_k \phi_k \ ,$$

where $\phi_k(x) = e^{ikx}$.

A natural family of norms for the modern numerical analysis of differential equations is comprised of the Sobolev norms. Hence, we present approximation results with respect to these norms. We recall that the Sobolev norm of integer order $m \geq 0$ is given by

$$\|u\|_{H^m(0,2\pi)} = \left( \sum_{k=0}^{m} \int_0^{2\pi} |u^{(k)}(x)|^2 \, dx \right)^{1/2} . \tag{5.1.6}$$

The reader unfamiliar with Sobolev spaces can think of $u^{(k)}$ as the classical (continuous) derivative of $u$ of order $k$. However, this norm can be defined for a wider class of functions. These form a Hilbert space, called $H^m(0,2\pi)$, which is introduced in (A.11.a). We are concerned here with functions periodic in $(0,2\pi)$. We consider the subspace $H_p^m(0,2\pi)$ of $H^m(0,2\pi)$ that consists of functions whose first $m-1$ derivatives are periodic (see (A.11.d)). Since $(e^{ikx})' = ike^{ikx}$, it follows that for any $u = \sum_{k=-\infty}^{\infty} \hat{u}_k \phi_k \in H_p^m(0,2\pi)$, the norm $\|u\|_{H^m(0,2\pi)}$ is equivalent to

$$\|u\|_m = \left( \sum_{k=-\infty}^{\infty} \left(1 + |k|^{2m}\right) |\hat{u}_k|^2 \right)^{1/2} , \tag{5.1.7}$$

i.e, for some positive constants $C_1$ and $C_2$ that are independent of $u$,

$$C_1 \|u\|_{H^m(0,2\pi)} \leq \|u\|_m \leq C_2 \|u\|_{H^m(0,2\pi)} \ .$$

The spaces $H_p^m(0,2\pi)$ consist of functions for which it is permissible to differentiate termwise the Fourier series $m$ times, provided the convergence is in the square mean. For instance, $H_p^1(0,2\pi)$ is the space of all functions $u$ for which

$$u' = \sum_{k=-\infty}^{\infty} ik\hat{u}_k \phi_k \quad \text{in } L^2(0,2\pi) \ . \tag{5.1.8}$$

This means that the Fourier series of $u'$ converges in the squared mean to the derivative of $u$. Result (5.1.8) is a direct consequence of the commutability of the operators $\mathrm{d}/\mathrm{d}x$ and $P_N$ on $H_p^1(0,2\pi)$, i.e.,

$$(P_N u)' = P_N u' \quad \text{for all } u \in H_p^1(0,2\pi) \, .$$

This, in turn follows from the identity

$$2\pi (u')_k^\wedge = (u', \phi_k) = -\,(u, \phi_k') = ik(u, \phi_k) = ik\hat{u}_k \quad \text{for all } k \, .$$

Since $u$ is in $H_p^1(0,2\pi)$, the first inner product is well-defined. By the same arguments, a similar characterization can be given also for $H_p^m(0,2\pi)$. It is enough to replace the first derivative with the $m$-th order derivative in (5.1.8).

The first error estimate we present concerns the truncation error in the $L^2$-norm. We recall that, by definition, $P_N u$ is the best approximation of $u$ in the $L^2$-norm among all the functions in $S_N$. One has, for any $u \in H_p^m(0,2\pi)$ and $m \geq 0$,

$$\|u - P_N u\|_{L^2(0,2\pi)} \leq C N^{-m} \|u^{(m)}\|_{L^2(0,2\pi)} \, . \tag{5.1.9}$$

This follows from the Parseval identity (2.1.14). Indeed,

$$\begin{aligned}
\frac{1}{\sqrt{2\pi}} \|u - P_N u\|_{L^2(0,2\pi)} &= \left( \sum_{|k| \gtrsim N} |\hat{u}_k|^2 \right)^{1/2} = \left( \sum_{|k| \gtrsim N} \frac{1}{|k|^{2m}} |k|^{2m} \, |\hat{u}_k|^2 \right)^{1/2} \\
&\leq N^{-m} \left( \sum_{|k| \gtrsim N} |k|^{2m} \, |\hat{u}_k|^2 \right)^{1/2} ,
\end{aligned}$$

where the symbol $\sum_{|k| \gtrsim N}$ has been introduced in (2.1.16). The last bracket can be bounded by the $L^2$-norm of $u^{(m)}$; hence, (5.1.9) follows.

Moreover, we can estimate the truncation error in higher Sobolev norms as follows:

$$\|u - P_N u\|_{H^l(0,2\pi)} \leq C N^{l-m} \|u^{(m)}\|_{L^2(0,2\pi)} \tag{5.1.10}$$

for any $m \geq 0$ and any $0 \leq l \leq m$. The proof of (5.1.10) is very similar to the one of (5.1.9). Indeed,

$$\begin{aligned}
\|u - P_N u\|_{H^l(0,2\pi)} &= \left( \sum_{|k| \gtrsim N} (1 + |k|^{2l}) \, |\hat{u}_k|^2 \right)^{1/2} \\
&\leq 2 \left( \sum_{|k| \gtrsim N} |k|^{2m-2(m-l)} |\hat{u}_k|^2 \right)^{1/2} \leq C N^{l-m} \|u^{(m)}\|_{L^2(0,2\pi)} \, .
\end{aligned}$$

We have seen that truncation and differentiation commute. Hence, $P_N u$ is the best approximation of $u$ in $S_N$ for any Sobolev norm 5.1.6. However, it is not so if we consider the $L^p$-norms, $1 \leq p \leq \infty$. An estimate of $u - P_N u$ in these norms can be given as a consequence of a preliminary investigation of the best approximation error. Results of this kind are known as *Jackson's theorems*. We shall recall here those applied to the forthcoming convergence analysis.

The first result is concerned with the best approximation in $S_N$, relative to the maximum norm; it states that for any $m \geq 0$

$$\inf_{\phi \in S_N} \|u - \phi\|_{L^\infty(0,2\pi)} \leq \frac{\pi}{2} N^{-m} \|u^{(m)}\|_{L^\infty(0,2\pi)} \, . \tag{5.1.11}$$

This is a particular case of the following general result concerning best approximation errors in $L^p$ for the whole range $1 \leq p \leq \infty$:

$$\inf_{\phi \in S_N} \|u - \phi\|_{L^p(0,2\pi)} \leq C N^{-m} \|u^{(m)}\|_{L^p(0,2\pi)} \, . \tag{5.1.12}$$

In the two previous estimates we have assumed that the $m$-th order derivative of $u$ (in the sense of periodic distributions, see (A.10.c)) belongs to the space $L^p(0, 2\pi)$ for which the norm on the right-hand side is finite.

We deal now with the evaluation of the truncation error $u - P_N u$ in the $L^p$-norms, $1 \leq p \leq \infty$. We recall first that if $u \in L^p(0, 2\pi)$, with $1 < p < \infty$, then its Fourier series converges, i.e.,

$$\|u - P_N u\|_{L^p(0,2\pi)} \to 0 \quad \text{as } N \to \infty \, . \tag{5.1.13}$$

This result includes and generalizes the property (2.1.9) which corresponds to the case $p = 2$. Furthermore, if $1 < p < \infty$,

$$\|u - P_N u\|_{L^p(0,2\pi)} \leq C \inf_{\phi \in S_N} \|u - \phi\|_{L^p(0,2\pi)} \, . \tag{5.1.14}$$

Hence, $P_N u$ approximates $u$ in the $L^p$-norms with the same order as the best approximation. If $p = 1$ or $p = \infty$, inequality (5.1.14) still holds provided the constant $C$ is replaced by $C(1 + \log N)$. (Here and in the sequel, we do not specify the base of logarithms, since the choice of the particular base only influences the constant $C$.)

When the function $u$ is $2\pi$-periodic and analytic in a strip of the complex plane about the real axis, the error $u - P_N u$ decays exponentially in $N$. For instance, if $u$ belongs to the Gevrey space $G_{\eta,m}(0, 2\pi)$, i.e., if $\|u\|^2_{G_{\eta,m}(0,2\pi)} = \sum_{k \in \mathbb{Z}} e^{2\eta(1+|k|)} \left(1 + |k|^{2m}\right) |\hat{u}_k|^2 < \infty$ (which implies that $u$ is analytic in the strip $|\operatorname{Im} z| < \eta$), then the same arguments as above prove that

$$\|u - P_N u\|_{H^l(0,2\pi)} \leq C N^{l-m} e^{-\eta N} \|u\|_{G_{\eta,m}(0,2\pi)} \, , \tag{5.1.15}$$

for $0 \leq l \leq m$.

### 5.1.3 Estimates for the Interpolation Error

Let $I_N u \in S_N$ denote the trigonometric interpolant of the function $u$ at the nodes $x_j = \pi j/N$, $j = 0, \dots, 2N-1$ (see (2.1.28), where on the right-hand side $N/2$ must be replaced by $N$). We shall give some approximation results for the interpolation error $u - I_N u$. For the estimate in the $L^2$-norm we have

$$\|u - I_N u\|_{L^2(0,2\pi)} \le CN^{-m}\|u^{(m)}\|_{L^2(0,2\pi)} \quad \text{for all } u \in H_p^m(0,2\pi) \text{ with } m \ge 1\,. \tag{5.1.16}$$

A comparison of (5.1.9) and (5.1.16) reveals that the interpolation error behaves asymptotically like the truncation error. A proof of this estimate will be presented at the end of the section.

The following result provides an estimate of the interpolation error in the maximum norm:

$$\|u - I_N u\|_{L^\infty(0,2\pi)} \le C(\log N)N^{-m}\|u^{(m)}\|_{L^\infty(0,2\pi)}\,. \tag{5.1.17}$$

Result (5.1.16) allows one to estimate the *aliasing error* $R_N u = I_N u - P_N u$ (see (2.1.37)). Indeed, since by (2.1.39) $\|R_N u\|_{L^2(0,2\pi)} \le \|u - I_N u\|_{L^2(0,2\pi)}$, one gets

$$\|R_N u\|_{L^2(0,2\pi)} \le CN^{-m}\|u^{(m)}\|_{L^2(0,2\pi)} \tag{5.1.18}$$

under the same hypotheses as (5.1.16). *The important implication of this estimate is that the aliasing error is asymptotically no worse than the interpolation error in the $L^2$-norm.*

An evaluation of the interpolation error in all Sobolev norms is now possible, and it is given by the estimate

$$\|u - I_N u\|_{H^l(0,2\pi)} \le CN^{l-m}\|u^{(m)}\|_{L^2(0,2\pi)} \quad \text{for } 0 \le l \le m \text{ and } u \in H_p^m(0,2\pi)\,, \text{ with } m \ge 1\,. \tag{5.1.19}$$

This inequality follows directly from the preceding results. It is a consequence of (5.1.18), (5.1.10) and the Bernstein inequality (5.1.5) used with $p = 2$ and $r = l$. Indeed we get

$$\begin{aligned}\|u - I_N u\|_{H^l(0,2\pi)} &\le \|u - P_N u\|_{H^l(0,2\pi)} + \|R_N u\|_{H^l(0,2\pi)} \\ &\le CN^{l-m}\|u^{(m)}\|_{L^2(0,2\pi)} + CN^l\,\|R_N u\|_{L^2(0,2\pi)} \\ &\le CN^{l-m}\|u^{(m)}\|_{L^2(0,2\pi)}\,.\end{aligned}$$

As a particular relevant case of (5.1.19), one can estimate the error produced in evaluating the interpolation derivative of a function (see (2.1.44)):

$$\|u' - \mathcal{D}_N u\|_{L^2(0,2\pi)} \leq CN^{1-m}\|u^{(m)}\|_{L^2(0,2\pi)} \tag{5.1.20}$$

for all $u \in H_p^m(0,2\pi)$, $m \geq 1$. Equivalently, recalling the identity (2.1.33) and noting that $u'(x_j) = (I_N u')(x_j)$ for $j = 0,\dots,2N-1$, one has under the same hypotheses

$$\left(\frac{\pi}{N}\sum_{j=0}^{2N-1} |u'(x_j) - \mathcal{D}_N u(x_j)|^2\right)^{1/2} \leq CN^{1-m}\|u^{(m)}\|_{L^2(0,2\pi)} \,. \tag{5.1.21}$$

Finally, when the function $u$ is analytic, the error $u' - \mathcal{D}_N u$ decays exponentially in $N$. Precisely, if $u$ is a $2\pi$-periodic analytic function in the strip $|\operatorname{Im} z| < \eta_0$, then

$$\|u' - \mathcal{D}_N u\|_{L^2(0,2\pi)} \leq \frac{4}{\sinh(\eta)} N e^{-\eta N} M(u,\eta) \tag{5.1.22}$$

for all $\eta$, $0 < \eta < \eta_0$, where $M(u,\eta) = \max_{|\operatorname{Im} z| \leq \eta} |u(z)|$.

PROOF OF (5.1.16). For each function $u : (0,2\pi) \to \mathbb{C}$ we consider the function $\mathcal{F}u : (0,2\pi N) \to \mathbb{C}$ such that $\mathcal{F}u(x) = u(x/N)$ for all $x \in (0,2\pi N)$. Then we define

$$S_N^* = \{\mathcal{F}\phi \mid \phi \in S_N\} \,.$$

Let $x_j = j\pi/N$, $j = 0,\dots,2N-1$, be the interpolation points, and set $\theta_j = N x_j$ for $j = 0,\dots,2N-1$. We denote by $I_N^*$ the interpolation operator with respect to these points, i.e., for all $u \in C^0([0,2\pi N])$,

$$I_N^* u \in S_N^*, \quad I_N^* u(\theta_j) = u(\theta_j) \quad \text{for } j = 0,\dots,2N-1 \,. \tag{5.1.23}$$

The following three relations can be easily proved:

$$\mathcal{F}(I_N u) = I_N^*(\mathcal{F}u) \quad \text{for all } u \in C^0([0,2\pi]) \,; \tag{5.1.24}$$

$$I_N^* u = u \quad \text{for all } u \in S_N^* \,; \tag{5.1.25}$$

$$\|u^{(l)}\|_{L^2(0,2\pi)} = N^{l-1/2}\|(\mathcal{F}u)^{(l)}\|_{L^2(0,2\pi N)} \,, \quad l \geq 0 \,. \tag{5.1.26}$$

Then, if we denote by $I$ the identity operator (i.e., $I(u) = u$ for all $u$), it follows that

$$\begin{aligned} \|u - I_N u\|_{L^2(0,2\pi)} &= N^{-1/2}\|\mathcal{F}u - I_N^*(\mathcal{F}u)\|_{L^2(0,2\pi N)} \\ &= N^{-1/2}\|(I - I_N^*)(\mathcal{F}u - \mathcal{F}(P_N u))\|_{L^2(0,2\pi N)} \\ &\leq N^{-1/2}\|I - I_N^*\|_{\mathcal{L}_m}\|\mathcal{F}(u - P_N u)\|_{H^m(0,2\pi N)} \,. \end{aligned} \tag{5.1.27}$$

We have denoted by $\mathcal{L}_m = \mathcal{L}(H_p^m(0,2\pi N), L^2(0,2\pi N))$ the space of all linear and continuous applications from $H_p^m(0,2\pi N)$ into $L^2(0,2\pi N)$ (see (A.3)). Using (5.1.26) and (5.1.10) gives

$$\begin{aligned}\|\mathcal{F}(u - P_N u)\|^2_{H^m(0,2\pi N)} &= \sum_{l=0}^{m} N^{1-2l} \|(u - P_N u)^{(l)}\|^2_{L^2(0,2\pi)} \\ &\leq C N^{1-2m} \|u^{(m)}\|^2_{L^2(0,2\pi)} \,.\end{aligned}$$

Then, from (5.1.27) we obtain

$$\|u - I_N u\|_{L^2(0,2\pi)} \leq C N^{-m} \|u^{(m)}\|_{L^2(0,2\pi)} \|I - I_N^*\|_{\mathcal{L}_m} \,. \quad (5.1.28)$$

Since $\|I\|_{\mathcal{L}_m} = 1$, it remains to prove that there is a constant $C$ independent of $N$ such that

$$\|I_N^*\|_{\mathcal{L}_m} \leq C \,. \quad (5.1.29)$$

We note that (see (A.3))

$$\|I_N^*\|_{\mathcal{L}_m} = \sup\{\|I^* v\|_{L^2(0,2\pi N)} | v \in H_p^m(0,2\pi N), \|v\|_{H^m(0,2\pi N)} = 1\}. \quad (5.1.30)$$

Using (5.1.24) and (5.1.26) it follows that

$$\begin{aligned}\|I_N^* v\|_{L^2(0,2\pi N)} &= N^{1/2} \|I_N(\mathcal{F}^{-1} v)\|_{L^2(0,2\pi)} = N^{1/2} \left( \int_0^{2\pi} |I_N(\mathcal{F}^{-1} v)|^2 \, dx \right)^{1/2} \\ &= N^{1/2} \left( \frac{\pi}{N} \sum_{j=0}^{2N-1} |(\mathcal{F}^{-1} v)(x_j)|^2 \right)^{1/2} \\ &= \sqrt{\pi} \left( \sum_{j=0}^{2N-1} |v(\theta_j)|^2 \right)^{1/2} .\end{aligned} \quad (5.1.31)$$

We can write $[0, 2\pi N] = \bigcup_{j=0}^{2N-1} [\theta_j, \theta_{j+1}]$, and by the Sobolev inequality (see (A.12)) we get, for each $m \geq 1$,

$$|v(\theta_j)| \leq C \|v\|_{H^m(\theta_j, \theta_{j+1})} \quad \text{for } j = 0, \ldots, 2N-1 \,.$$

Thus,

$$\sum_{j=0}^{2N-1} |v(\theta_j)|^2 \leq C \|v\|^2_{H^m(0,2\pi N)} \,,$$

and (5.1.29) follows now from (5.1.30) and (5.1.31). □

***Bibliographical Notes***

Nikolskii's inequality has been proven in Nikolskii (1951). In Butzer and Nessel (1971) one can find proofs of the Bernstein inequality (Theorem 2.3.1 and Corollary 2.3.2), estimate (5.1.12) (Theorem 2.2.3), estimate (5.1.14) (Proposition 9.3.8), and the convergence result (5.1.13) (Theorem 9.3.6). Estimate (5.1.11) is proved in Cheney (1966, p. 145). Estimate (5.1.16) was first proved by Kreiss and Oliger (1979). The proof given here is due to Pasciak (1980), who actually proved (5.1.19). Estimate (5.1.17) was proven by Jackson (1930, p. 123). Finally, inequality (5.1.22) has been established by Tadmor (1986).

## 5.2 Sturm-Liouville Expansions

In this section we consider expansions with respect to eigenfunctions of Sturm-Liouville problems. We refer for notation to Sect. 2.2.1. We analyze the decay properties of the coefficients of a function with respect to such a basis, distinguishing between regular and singular Sturm-Liouville problems.

We assume that the coefficients $p$, $q$ and $w$ satisfy the assumptions made in Sect. 2.2.1. Moreover, we suppose that $\int_{-1}^{1} w(x)^{-1}\mathrm{d}x < +\infty$.

### 5.2.1 Regular Sturm-Liouville Problems

If the function $p$ is bounded from below by a positive constant, say $p(x) \geq p_0 > 0$, then the two boundary conditions to be specified in (2.2.1) assume the form

$$\begin{aligned} \alpha_1 u(-1) + \beta_1 u'(-1) = 0 \,, \qquad & \alpha_1^2 + \beta_1^2 \neq 0 \,, \\ \alpha_2 u(1) + \beta_2 u'(1) = 0 \,, \qquad & \alpha_2^2 + \beta_2^2 \neq 0 \,, \end{aligned} \tag{5.2.1}$$

for suitable $\alpha_1$, $\beta_1$, $\alpha_2$, $\beta_2$. In this case we are speaking of a *regular* Sturm-Liouville boundary-value problem.

Under the assumptions that $\alpha_1\beta_1 \leq 0$ and $\alpha_2\beta_2 \geq 0$, it is known (see, e.g., Courant and Hilbert (1953, vol. I)), that the eigenvalues of the regular Sturm-Liouville problem (2.2.1), (5.2.1) form an infinite, unbounded sequence of nonnegative numbers, $0 \leq \lambda_0 < \cdots < \lambda_k < \lambda_{k+1} < \cdots$, and have multiplicity 1. The corresponding eigenfunctions $\phi_k$, determined up to a constant, have exactly $k$ zeroes in the open interval $(-1,1)$. The asymptotic behavior of the eigenvalues as $k \to \infty$ is given by the formula

$$\lim_{k\to\infty} \frac{k^2}{\lambda_k} = \frac{\pi^2}{4} \int_{-1}^{1} \sqrt{w/p}\,\mathrm{d}x \,. \tag{5.2.2}$$

The asymptotic behavior of the eigenfunctions depends on the type of boundary conditions. For instance, for the Neumann boundary conditions $u'(-1) = u'(1) = 0$, one has

$$\phi_k(x) = A_k \cos \frac{\pi}{2} k(x+1) + \frac{O(1)}{k} , \qquad k \to \infty .$$

Eigenfunctions are mutually orthogonal with respect to the weighted inner product

$$(u, v)_w = \int_{-1}^{1} u(x)v(x)w(x)\,\mathrm{d}x , \tag{5.2.3}$$

namely,

$$(\phi_k, \phi_m)_w = 0 \quad \text{if } k \neq m . \tag{5.2.4}$$

Moreover, the system $\{\phi_k,\ k = 0, 1, \ldots,\}$ is complete in the weighted $L^2_w(-1,1)$ space (see (A.9.g)). This means that if we define the sequence of the "Fourier" coefficients of a function $u \in L^2_w(-1,1)$ as

$$\hat{u}_k = (u, \phi_k)_w , \qquad k = 0, 1, \ldots$$

($\phi_k$ is assumed to be normalized by $\|\phi_k\|_{L^2_w(-1,1)} = 1$), and we set

$$P_N u = \sum_{k=0}^{N} \hat{u}_k \phi_k \quad \text{for integer } N > 0 ,$$

then

$$\|u - P_N u\|_{L^2_w(-1,1)} \to 0 \quad \text{as } N \to +\infty .$$

In other words, the "Fourier" series $\sum_{k=0}^{\infty} \hat{u}_k \phi_k$ of $u$ is convergent to $u$ in the weighted squared mean for any $u \in L^2_w(-1,1)$.

Local convergence properties require more regularity on $u$. For instance, as in the case of the Fourier expansion, if $u$ is of bounded variation on $[-1,+1]$ (see (A.8)), $P_N u(x)$ converges pointwise to $[u(x^+) + u(x^-)]/2$ for any $x \in [-1,1]$ (see, e.g., Titchmarsh (1962)).

The rate of decay of the coefficients of a function $u \in L^2_w(-1,1)$ depends not only on its regularity but also on the fulfillment of a suitable set of boundary conditions. This can be seen as follows. Equation (2.2.1) and integration-by-parts yield

$$\begin{aligned} \hat{u}_k = (u, \phi_k)_w &= \frac{1}{\lambda_k} \int_{-1}^{1} u[-(p\phi_k')' + q\phi_k]\,\mathrm{d}x \\ &= \frac{1}{\lambda_k} \int_{-1}^{1} [-(pu')' + qu]\phi_k\,\mathrm{d}x - \frac{1}{\lambda_k}[p(\phi_k' u - \phi_k u')]_{-1}^{1} \\ &= \frac{1}{\lambda_k} \left( \frac{1}{w} \mathcal{L}u, \phi_k \right)_w - \frac{1}{\lambda_k}[p(\phi_k' u - \phi_k u')]_{-1}^{1} . \end{aligned} \tag{5.2.5}$$

This deduction is rigorous under the assumption that the function $u_{(1)} = \frac{1}{w}\mathcal{L}u$ satisfy

$$u_{(1)} \in L^2_w(-1,1) \, . \tag{5.2.6}$$

Due to the regularity of the elliptic operator $\mathcal{L}$, this means that the second derivative of $u$ must be square integrable with respect to the weight $1/w$. Under this hypothesis, $u$ and $u'$ are continuous up to the boundary.

Now, *if $u$ satisfies the boundary conditions* (5.2.1), the boundary term in (5.2.5) vanishes, so that

$$\hat{u}_k = \frac{1}{\lambda_k}(u_{(1)}, \phi_k)_w \, .$$

The iteration of this argument yields $\hat{u}_k = 1/(\lambda_k)^m (u_{(m)}, \phi_k)_w$, for $m \geq 2$, provided $u_{(m)} = (1/w)\mathcal{L}u_{(m-1)} \in L^2_w(-1,1)$ *and* $u_{(m-1)}$ satisfies the boundary conditions (5.2.1). We deduce the asymptotic decay estimate

$$|\hat{u}_k| \leq \frac{C}{k^{2m}} \|u_{(m)}\|_{L^2_w(-1,1)} \, .$$

If for some $m$, $u_{(m)}$ does not satisfy (5.2.1), then $\hat{u}_k$ decays no faster than $1/k^{2m}$, even if $u$ is infinitely smooth. In this case $u$ cannot be approximated with spectral accuracy by the system of the $\phi_k$'s.

### 5.2.2 Singular Sturm-Liouville Problems

A singular Sturm-Liouville problem occurs when $p$ vanishes for at least one point on the boundary. We will consider here only the case $p(-1) = p(1) = 0$. The boundary conditions (5.2.1) are replaced by conditions on the type of singularities allowed on the boundary. Precisely, one requires the solution to satisfy

$$p(x)u'(x) \to 0 \quad \text{as } x \to \pm 1 \, . \tag{5.2.7}$$

Let us assume that $u$ is square integrable with respect to both the weights $q$ and $w$, and that $u'$ is square integrable with respect to the weight $p$, i.e., let us assume that $u \in X$, where

$$X = \left\{ v \in L^2_w(-1,1) \cap L^2_q(-1,1) \mid v' \in L^2_p(-1,1) \right\} \, .$$

($X$ is a Hilbert space for the norm $\|v\|^2 = \int_{-1}^1 v^2 w \, \mathrm{d}x + \int_{-1}^1 v^2 q \, \mathrm{d}x + \int_{-1}^1 (v')^2 p \, \mathrm{d}x$.) Then, it is possible to give the following variational formulation of (2.2.1):

$$\int_{-1}^1 (pu'v' + quv) \, \mathrm{d}x = \lambda \int_{-1}^1 uvw \, \mathrm{d}x \quad \text{for all } v \in X \, . \tag{5.2.8}$$

This takes into account the new boundary conditions in a natural way. As for the regular Sturm-Liouville problem, the eigenvalues of (5.2.8) form an unbounded sequence of nonnegative real numbers $0 \le \lambda_0 \le \cdots \lambda_k \le \cdots$; each of them has finite multiplicity. The system of corresponding eigenfunctions $\phi_k$ is orthogonal and complete in $L^2_w(-1,1)$. In order to prove these results, let us consider the following problem:

$$
\begin{aligned}
&u \in X\,,\\
&\int_{-1}^{1} (pu'v' + quv + uvw)\,\mathrm{d}x = \int_{-1}^{1} fvw\,\mathrm{d}x \quad \text{for all } v \in X\,.
\end{aligned}
\tag{5.2.9}
$$

For each $f \in L^2_w(-1,1)$, there exists a unique solution to this problem. This follows from the Riesz representation theorem (see (A.1.d)), since the left-hand side of (5.2.9) is precisely the inner product in $X$. Let $\mathcal{T} : L^2_w(-1,1) \to L^2_w(-1,1)$ be the linear operator that maps $f$ into $u$. The eigenvalues $\lambda$ of (5.2.8) are obtained from the eigenvalues $\mu$ of $\mathcal{T}$ by the relation $\lambda + 1 = \mu^{-1}$. The eigenfunctions are the same. It is immediate that $\mathcal{T}$ is a symmetric, positive operator in the inner product of $L^2_w(-1,1)$ (i.e., it satisfies $(\mathcal{T}f, f) > 0$ for any $f \neq 0$), and that each eigenvalue of $\mathcal{T}$ is $\le 1$. Moreover, one can prove that $\mathcal{T}$ is compact (see (A.3)). The proof of this property is based on the observation that if $u$ is the solution of (5.2.9), then $(pu')' \in L^1(-1,1)$ and $pu'$ is continuous on $[-1,1]$; thus, one can apply Ascoli's Theorem (see, e.g., Taylor (1958), Sect. 5.5). At this point one can invoke a fundamental result of spectral analysis in Hilbert spaces (see, e.g., Taylor (1958), Theorem 6.4-D) that states that the eigenvalues of $\mathcal{T}$ form an infinite sequence of positive numbers that converges to 0. The corresponding eigenfunctions form a complete orthogonal basis in $L^2_w(-1,1)$. This yields the desired properties for the eigenvalues of (5.2.8).

In order to investigate the behavior of the expansion coefficients $\hat{u}_k = (u, \phi_k)_w$ of a function $u \in L^2_w(-1,1)$ with respect to the system of eigenfunctions of a singular Sturm-Liouville problem, we proceed as in (5.2.5):

$$
\begin{aligned}
\hat{u}_k &= \frac{1}{\lambda_k}\int_{-1}^{1} (p\phi_k' u' + q\phi_k u)\,\mathrm{d}x \quad \text{(by (5.2.8))}\\
&= \frac{1}{\lambda_k}\int_{-1}^{1} [-(pu')' + qu]\phi_k\,\mathrm{d}x + \frac{1}{\lambda_k}[pu'\phi_k]_{-1}^{1}\\
&= \frac{1}{\lambda_k}\left(\frac{1}{w}\,\mathcal{L}u, \phi_k\right)_w + \frac{1}{\lambda_k}[pu'\phi_k]_{-1}^{1}\,.
\end{aligned}
\tag{5.2.10}
$$

Again, this holds provided (5.2.6) is satisfied. Note that under this assumption, $pu'$ is continuous up to the boundary, since

$$
\begin{aligned}
|(pu')(x_1) - (pu')(x_2)| &= \left|\int_{x_1}^{x_2} (pu')'\,\mathrm{d}x\right|\\
&\le \left(\int_{x_1}^{x_2} \frac{1}{w}|(pu')'|^2\right)^{1/2} \left(\int_{x_1}^{x_2} w\right)^{1/2}.
\end{aligned}
$$

Thus, condition (5.2.7) makes sense, and it implies that the boundary term in (5.2.8) vanishes. We stress that, unlike the case of regular Sturm-Liouville boundary-value problems, (5.2.7) is just a *regularity* assumption on $u$ over the closed interval $[-1,1]$, i.e., $u$ is not required to satisfy specific boundary conditions. One can easily check that (5.2.7) is satisfied if, for instance, $(p/w)u'' \in L^2_w(-1,1)$. Again, one can iterate the argument and get the representation $\hat{u}_k = 1/(\lambda_k)^m (u_{(m)}, \phi_k)_w$ provided $u_{(m)} = (1/w)\mathcal{L}u_{(m-1)} \in L^2_w(-1,1)$ and $u_{(m-1)}$ satisfies (5.2.7) for $m \geq 2$. In the cases of interest (see Sects. 2.3.1 and 2.4.1), $\lambda_k = O(k^2)$ as $k \to \infty$. Hence, the expansion coefficients of $u$ decay faster than algebraically under the sole assumption that $u$ be infinitely differentiable.

This result does not necessarily hold if $q$ is unbounded in $[-1,1]$. For instance, let us consider the singular Sturm-Liouville boundary-value problem (Bessel equation) after changing the interval to $[0,2]$:

$$-(xu')' + \frac{n^2}{x}u = \lambda x u\,, \qquad 0 < x < 2\,,$$
$$u(2) = 0\,,\ u \text{ bounded near } 0\,.$$

For $n \neq 0$, the condition $u_{(m)} \in L^2_w(-1,1)$ forces $u_{(m)}$ to vanish at $x = 0$, since $q^2/w$ is not integrable. In order to achieve spectral accuracy in this case, an infinite number of boundary conditions must be satisfied even though the operator is singular.

We conclude this section by showing that the only polynomial eigenfunctions of a singular Sturm-Liouville problem are the Jacobi polynomials. Actually, if $\phi_k = (1/(\lambda_k w_k))\mathcal{L}\phi_k$ is a polynomial of degree $k$ for $k = 0,1,2,\ldots$, it is readily seen by taking $k = 0,1,2$ that $q/w$ is a polynomial of degree zero (i.e., $q(x) = q_0 w(x)$) and $p/w$ and $p'/w$ are, respectively, polynomials of degree two and one. Since $p$ must vanish at the boundary, necessarily one has $w(x) = c_1(1-x)^\alpha(1+x)^\beta$ and $p(x) = c_2(1-x)^{\alpha+1}(1+x)^{\beta+1}$. Finally, the integrability of $w$ in $(-1,1)$ implies $\alpha, \beta > -1$.

## 5.3 Discrete Norms

Before stating the approximation results for the Legendre and the Chebyshev polynomials, we give here some general theoretical results concerning the discrete inner product $(u,v)_N$ defined in (2.2.24). This bilinear form is a high-precision approximation of the inner product $(u,v)_w$, with respect to which the polynomials $p_k$ introduced in Sect. 2.2.2 are orthogonal. The quantity

$$\|v\|_N = (v,v)_N^{1/2}\,, \tag{5.3.1}$$

which is meaningful for all continuous functions $v$ in $[-1,1]$, defines a norm for the polynomials of $\mathbb{P}_N$ associated with the discrete inner product. If the quadrature points $x_j$ are of Gauss or Gauss-Radau type, then $\|\phi\|_N = \|v\|_{L^2_w(-1,1)}$ for all $\phi \in \mathbb{P}_N$. If the points $x_j$ are of Gauss-Lobatto type, this equality holds for $\phi \in \mathbb{P}_{N-1}$, but in general $\|p_N\|_N \neq \|p_N\|_{L^2_w(-1,1)}$. However, for the polynomials of $\mathbb{P}_N$, the discrete norm $\|\phi\|_N$ is uniformly equivalent to the norm $\|\phi\|_{L^2_w(-1,1)}$ in the more important cases, such as Legendre, Chebyshev or other Jacobi polynomials. This means that there exist positive constants $C_1$ and $C_2$, independent of $N$, such that

$$C_1\|\phi\|_{L^2_w(-1,1)} \leq \|\phi\|_N \leq C_2\|\phi\|_{L^2_w(-1,1)} \quad \text{for all } \phi \in \mathbb{P}_N \,. \tag{5.3.2}$$

This result has been established by Canuto and Quarteroni (1982a). For the Legendre and Chebyshev polynomials, one has

$$1 \leq \frac{\|p_N\|_N}{\|p_N\|_{L^2_w(-1,1)}} = \begin{cases} \sqrt{2} & \text{(Chebyshev)}\,, \\ \sqrt{2+\dfrac{1}{N}} & \text{(Legendre)}\,, \end{cases}$$

as a consequence of (2.2.23), (2.3.13) and (2.4.18). Thus, (5.3.2) holds with $C_1 = 1$ and $C_2 = \sqrt{3}$, thanks to the orthogonality of the polynomials $p_k$.

The uniform equivalence of the discrete and continuous norms on $\mathbb{P}_N$ is used in a variety of ways in the analysis of stability and convergence, as will be seen in Chaps. 6 and 7. For instance, at each stage of the analysis one may use whichever of the two norms is more convenient, and, if desired, convert to the other norm by the uniform equivalence property. Moreover, error estimates obtained for the continuous norm can be readily converted to error estimates in the discrete norm, and conversely.

A trivial application of (5.3.2) is the estimate

$$\|v\|_N \leq C_2\|I_N v\|_{L^2_w(-1,1)} \,, \tag{5.3.3}$$

which holds for all the continuous functions on $[-1, 1]$.

The difference between the $L^2_w$-inner product $(u, v)_w$ and the discrete inner product $(u, v)_N$ can be bounded in terms of truncation and interpolation errors. Such estimates will be used in the convergence analysis of the subsequent chapters. Hereafter we denote by $u$ any continuous function on $[-1, 1]$, and by $\phi$ any polynomial of $\mathbb{P}_N$.

For the *Gauss* and *Gauss-Radau integration*, we have

$$|(u,\phi)_w - (u,\phi)_N| \leq \|u - I_N u\|_{L^2_w(-1,1)}\|\phi\|_{L^2_w(-1,1)} \,. \tag{5.3.4a}$$

Indeed, from (2.2.25) and (2.2.27) we get

$$(u,\phi)_w - (u,\phi)_N = (u,\phi)_w - (I_N u,\phi)_w \,;$$

hence, (5.3.4a) follows from the Cauchy-Schwarz inequality.

For the *Gauss-Lobatto integration*, if (5.3.2) holds, then there exists a positive constant $C$ independent of $N$ such that

$$\begin{aligned}|(u,\phi)_w - (u,\phi)_N| \le C(&\|u - P_{N-1}u\|_{L^2_w(-1,1)} \\ &+ \|u - I_N u\|_{L^2_w(-1,1)})\|\phi\|_{L^2_w(-1,1)} \,.\end{aligned} \tag{5.3.4b}$$

Actually we have

$$\begin{aligned}
&|(u,\phi)_w - (u,\phi)_N| \\
&\quad = |(u,\phi)_w - (P_{N-1}u,\phi)_w + (P_{N-1}u,\phi)_w - (I_N u,\phi)_N| \\
&\quad \le |(u - P_{N-1}u,\phi)_w| + |(P_{N-1}u - I_N u,\phi)_N| \quad \text{(by (2.2.25))} \\
&\quad \le C\left(\|u - P_{N-1}u\|_{L^2_w(-1,1)} + \|P_{N-1}u - I_N u\|_N\right)\|\phi\|_{L^2_w(-1,1)} \\
&\qquad\qquad\qquad \text{(by the Cauchy-Schwarz inequality and (5.3.2))} \\
&\quad \le C\left(2\|u - P_{N-1}u\|_{L^2_w(-1,1)} + \|u - I_N u\|_{L^2_w(-1,1)}\right)\|\phi\|_{L^2_w(-1,1)} \\
&\qquad\qquad\qquad \text{(by (5.3.2))}\,;
\end{aligned}$$

whence, (5.3.4b) follows.

## 5.4 Legendre Approximations

We present in this section various results concerning polynomial approximations in $L^p$-spaces or in Sobolev spaces, in which integration is performed with respect to the Legendre weight $w(x) \equiv 1$. Additional results can be found, e.g., in Bernardi and Maday (1997) and in Schwab (1998).

### 5.4.1 Inverse Inequalities for Algebraic Polynomials

We recall here the inverse inequalities concerning summability and differentiability for algebraic polynomials on the interval $(-1,1)$. These results are expressed in terms of $L^p$-norms, which are defined as follows:

$$\|u\|_{L^p(-1,1)} = \left(\int_{-1}^{1} |u(x)|^p \,\mathrm{d}x\right)^{1/p}, \qquad 1 \le p < \infty\,, \tag{5.4.1}$$

and

$$\|u\|_{L^\infty(-1,1)} = \sup_{-1\le x\le 1} |u(x)|\,, \qquad p = \infty\,. \tag{5.4.2}$$

These are the norms of the Banach spaces $L^p(-1,1)$ defined in (A.9.f).

The inverse inequality concerning summability states that for any real $p$ and $q$ with $1 \le p \le q \le \infty$, there exists a positive constant $C$ independent of $N$ such that

$$\|\phi\|_{L^q(-1,1)} \le CN^{2(1/p-1/q)} \|\phi\|_{L^p(-1,1)} \quad \text{for all } \phi \in \mathbb{P}_N . \qquad (5.4.3)$$

The following inequality relates the $L^2$-norm to a weaker weighted $L^2$-norm:

$$\|\phi\|_{L^2(-1,1)} \le CN^{\alpha} \|\phi\|_{L^2_{\eta^\alpha}(-1,1)} \quad \text{for all } \phi \in \mathbb{P}_N , \qquad (5.4.4)$$

where the weight on the right-hand side is $\eta^\alpha(x) = (1-x^2)^\alpha$, with $\alpha \ge 0$, and $C$ is a positive constant independent of $N$.

On the other hand, the inverse inequality concerning differentiation states that for any $p$ with $2 \le p \le \infty$, and for all integers $r \ge 1$, there exists a positive constant $C$ independent of $N$ such that

$$\|\phi^{(r)}\|_{L^p(-1,1)} \le CN^{2r} \|\phi\|_{L^p(-1,1)} \quad \text{for all } \phi \in \mathbb{P}_N . \qquad (5.4.5)$$

The exponent of $N$ in both (5.4.3) and (5.4.5) is the smallest possible. However, it is exactly twice the exponent in the Fourier inverse inequalities (5.1.4) and (5.1.5), or in the corresponding uniform-grid finite-element inequalities. This has some important consequences for the stability and convergence analysis of orthogonal polynomial spectral methods. Result (5.4.5) is also used in Sect. 7.3 to discuss the growth with $N$ of the eigenvalues of the discrete first- and second-derivative operators. With one exception (the Legendre tau first derivative operator), these eigenvalues grow twice as fast as those of the corresponding matrices generated by, say, finite-difference or finite-element methods on uniform grids with the same number of unknowns. The implication is that, for evolution equations, explicit time-advancing schemes applied with spectral methods in space have a more restrictive time-step limitation than standard low-order methods.

Inverse inequalities with smaller powers of $N$ than in (5.4.5) can be obtained, at the expense of inserting a weaker weight in the left-hand side norm or a stronger weight in the right-hand side norm. For instance, setting $\eta(x) = (1-x^2)$, the following inequality holds:

$$\| \phi' \sqrt{\eta} \|_{L^2(-1,1)} \le \sqrt{2} N \|\phi\|_{L^2(-1,1)} \quad \text{for all } \phi \in \mathbb{P}_N(-1,1) . \qquad (5.4.6)$$

If $\phi$ vanishes at the endpoints of the interval, then $\phi^2 \eta^{-1}$ is integrable and we have the bound

$$\|\phi'\|_{L^2(-1,1)} \le \sqrt{2} N \| \phi / \sqrt{\eta} \|_{L^2(-1,1)} \quad \text{for all } \phi \in \mathbb{P}^0_N(-1,1) . \qquad (5.4.7)$$

The latter estimate is used in Sect. 7.3.1 to study the growth of the largest eigenvalue of the stiffness matrix generated by a G-NI method.

Finally, we mention another inequality that allows one to bound the maximum norm of a polynomial by its norm in the Sobolev space of fractional order $H^{1/2}(-1,1)$ (see (A.11.e)). Precisely, there exists a positive constant $C$ independent of $N$ such that

$$\|\phi\|_{L^\infty(-1,1)} \le C\sqrt{\log(1+N)}\, \|\phi\|_{H^{1/2}(-1,1)} \quad \text{for all } \phi \in \mathbb{P}_N . \qquad (5.4.8)$$

### 5.4.2 Estimates for the Truncation and Best Approximation Errors

As for the Fourier system, we will measure several approximation errors for the Legendre system in terms of Sobolev norms. The most commonly used Sobolev norm of order $m \geq 0$ is given by

$$\|u\|_{H^m(-1,1)} = \left( \sum_{k=0}^{m} \|u^{(k)}\|^2_{L^2(-1,1)} \right)^{1/2} . \tag{5.4.9}$$

Again, one can consider $u^{(k)}$ to be the classical continuous derivative of $u$ of order $k$. These norms can actually be defined for less regular functions, which form a Hilbert space called $H^m(-1,1)$. This space is introduced in (A.11.a).

In bounding from above the approximation error, only some of the $L^2$-norms appearing on the right-hand side of (5.4.9) enter into play. Thus, it is convenient to introduce the seminorms

$$|u|_{H^{m;N}(-1,1)} = \left( \sum_{k=\min(m,N+1)}^{m} \|u^{(k)}\|^2_{L^2(-1,1)} \right)^{1/2} ; \tag{5.4.10}$$

note that whenever $N \geq m-1$, one has

$$|u|_{H^{m;N}(-1,1)} = \|u^{(m)}\|_{L^2(-1,1)} = |u|_{H^m(-1,1)} .$$

The truncation error $u - P_N u$, where $P_N u = \sum_{k=0}^{N} \hat{u}_k L_k$ is the truncated Legendre series of $u$, can be estimated as follows: for all $u \in H^m(-1,1)$, $m \geq 0$, one has

$$\|u - P_N u\|_{L^2(-1,1)} \leq C N^{-m} |u|_{H^{m;N}(-1,1)} \tag{5.4.11}$$

where $C$ depends on $m$.

A brief comment on the right-hand side of this inequality is in order. Obviously, we have $|u|_{H^{m;N}(\Omega)} \leq \|u\|_{H^m(\Omega)}$; hence, (5.4.11) implies the estimate

$$\|u - P_N u\|_{L^2(-1,1)} \leq C N^{-m} \|u\|_{H^m(-1,1)} . \tag{5.4.12}$$

Not only is (5.4.11) sharper than the latter estimate, but the presence of the seminorm, rather than the norm, on its right-hand side expresses the fact that the projection operator $P_N$ is exact for all polynomials in $\mathbb{P}_N$. Indeed, take $m = N+1$ in (5.4.11) and observe that the condition $|u|_{H^{N+1;N}(-1,1)} = \|u^{(N+1)}\|_{L^2(-1,1)} = 0$ is equivalent to $u^{(N+1)}$ vanishing identically in $(-1,1)$, which in turn is equivalent to $u$ being a polynomial of degree $\leq N$. Thus, if $|u|_{H^{N+1;N}(-1,1)} = 0$, (5.4.11) implies $u - P_N u = 0$, i.e., $P_N u = u$.

Sharper estimates than (5.4.11) can be obtained. One of them is given by the first inequality in the subsequent estimate (5.6.1); this surfaces in a natural way in Sect. 5.6 in the proof of (5.4.11). Another one is the bound

$$\|u - P_N u\|_{L^2(-1,1)} \leq \left( \frac{(N+1-s)!}{(N+1+s)!} \right)^{1/2} \|u^{(s)}\|_{L^2_s(-1,1)} , \qquad (5.4.13)$$

which holds for all $0 \leq s \leq \min(m, N+1)$, with

$$\|u^{(s)}\|_{L^2_s(-1,1)} = \left( \int_{-1}^{1} |u^{(s)}(x)|^2 (1-x^2)^s \, dx \right)^{1/2} . \qquad (5.4.14)$$

In the limit $N \to \infty$, $m$ fixed, the preceding estimate takes the form of (5.4.11) with $|u|_{H^{m;N}(-1,1)} = \|u^{(m)}\|_{L^2(-1,1)}$ replaced by $\|u^{(m)}\|_{L^2_m(-1,1)}$. This sharper estimate arises because the allowable growth of the derivative at the endpoints of the interval is damped there by the vanishing weight. Although we will not explicitly mention it in the sequel, we remark that such an improvement applies to all the error estimates given throughout this section.

The truncated Legendre series $P_N u$ is the polynomial of best approximation of $u$ in the $L^2$-norm. One can consider the problem of the best approximation polynomial of $u$ with respect to a general norm. For any normed linear space $X$ and any $u \in X$, it is known that there exists a polynomial $\phi^* \in \mathbb{P}_N$ such that

$$\|u - \phi^*\|_X = \inf_{\phi \in \mathbb{P}_N} \|u - \phi\|_X ; \qquad (5.4.15)$$

$\phi^*$ is called a best approximation polynomial of $u$ in the norm of $X$. We are interested in the case where $X = L^p(-1,1)$ for $1 \leq p \leq \infty$. For these norms $\phi^*$ is unique.

The best approximation error in any $L^p$-norm with $2 < p \leq \infty$ decays as the truncation error in the $L^2$-norm, i.e.,

$$\inf_{\phi \in \mathbb{P}_N} \|u - \phi\|_{L^p(-1,1)} \leq C N^{-m} \left( \sum_{k=\min(m,N+1)}^{m} \|u^{(k)}\|^p_{L^p(-1,1)} \right)^{1/p} . \qquad (5.4.16)$$

This estimate holds for all the functions $u$ whose (distributional) derivatives of order up to $m$ belong to $L^p(-1,1)$.

The rate of convergence of the truncation error in $L^p$-norms, $p > 2$, is not as fast as the rate of convergence of the best approximation. For instance, for any function $u$ with an $m$-th derivative of bounded variation (see (A.8)), one has

$$\|u - P_N u\|_{L^\infty(-1,1)} \leq C N^{1/2-m} V(u^{(m)}) , \qquad (5.4.17)$$

where $V(u^{(m)})$ is the total variation of $u^{(m)}$. Comparing this result with (5.4.16) for $p = \infty$, and noting that a function of bounded variation is certainly bounded, we see that the rate of convergence of the truncation error is slower by at least a factor of $\sqrt{N}$.

In those cases for which the truncation error of the derivatives is relevant, the following estimate extends (5.4.11) to higher order Sobolev norms:

$$\|u - P_N u\|_{H^l(-1,1)} \leq C N^{2l-1/2-m} |u|_{H^{m;N}(-1,1)} , \tag{5.4.18}$$

for $u \in H^m(-1,1)$ with $m \geq 1$ and for any $l$ such that $1 \leq l \leq m$. Note that in the important case $l = m = 1$, this inequality does not imply convergence of the derivative of the truncated series. Indeed, it is possible to construct a function $u$ such that the truncated Legendre series converges in $L^2(-1,1)$ but not in $H^1(-1,1)$. Thus, the derivative of the series does not converge.

A simple manifestation of this phenomenon is provided by considering a sequence of functions rather than a series. In particular, let

$$u^{(N)} = \frac{1}{N+1} L_{N+1} - \frac{1}{N-1} L_{N-1} .$$

The seminorm $|u^{(N)}|_{H^1(-1,1)}$ is bounded, as can be verified by using the Parseval equality to evaluate the norm of the first derivative and then using (2.3.18), which expresses the coefficients of the derivative in terms of the coefficients of the function. Nevertheless, in a similar fashion one obtains

$$\|u^{(N)} - P_N u^{(N)}\|_{H^1(-1,1)} \sim \sqrt{N} .$$

Fourier series are better behaved in this regard. If $u$ itself is in $H_p^1(0, 2\pi)$, then the $L^2$-norm of the derivative of the truncated series of $u$ is at least bounded. The analogous example is

$$u^{(N)}(x) = \frac{1}{N+1} e^{i(N+1)x} - \frac{1}{N-1} e^{i(N-1)x} .$$

Clearly,

$$\|u^{(N)} - P_N u^{(N)}\|_{H^1(0,2\pi)} = \sqrt{2\pi \left(1 + \frac{1}{(N+1)^2}\right)} .$$

The difference between the two types of expansions can be attributed to the loss of two powers of $N$ in (5.4.5) for every derivative as opposed to only one power of $N$ in the Fourier case.

The function $u(x) = |x|^{3/2}$ displayed in Fig. 5.1 is almost in $H^2(-1,1)$, i.e., for all real $p < 2$,

$$\int_{-1}^{1} |u''(x)|^p \, dx < \infty .$$

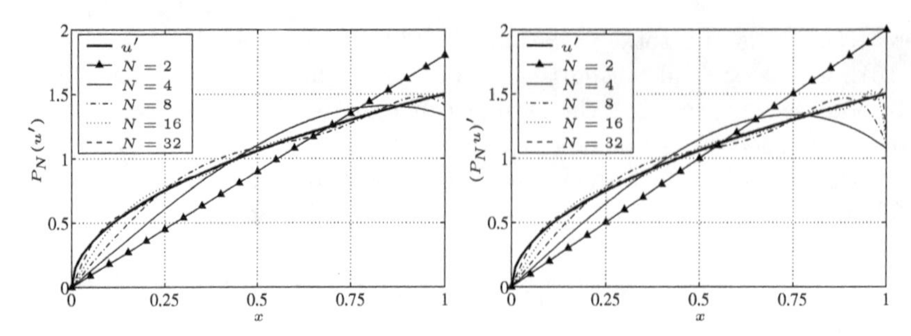


**Fig. 5.1.** Several versions of Legendre differentiation for $u(x) = |x|^{3/2}$ on $[-1, 1]$. The exact result is indicated by the solid, thick curve, the approximate results are obtained for $N = 2, 4, 8, 16$ and 32. Only the right half of the interval is shown, *(left)* $P_N u'$; *(right)* $(P_N u)'$

Result (5.4.18) then implies that $(P_N u)'$ converges to $u'$ in the $L^2$-norm. But it does not imply convergence in the $L^\infty$-norm, as is evident from the figure. Indeed, a sharp upper bound in the maximum norm for all functions in $H^2(-1, 1)$ can be obtained from the Sobolev inequality (A.12) and the estimate (5.4.18):

$$\begin{aligned}\|u' - (P_N u)'\|_{L^\infty(-1,1)} &\leq C\|u - P_N u\|^{1/2}_{H^1(-1,1)}\|u - P_N u\|^{1/2}_{H^2(-1,1)} \\ &\leq CN^{1/2}|u|_{H^{2;N}(-1,1)} \,.\end{aligned}$$

On the other hand, Fig. 5.1 suggests that $P_N u'$ does converge to $u'$ in the $L^\infty$-norm. This is true for all functions in $H^2(-1, 1)$, as follows from the estimate (5.4.17) applied with $u'$ replacing $u$ and with $m = 1$.

The rate of decay in (5.4.18) is not optimal in the sense that the best approximation error has a faster rate of convergence in the same norms. We will confine the discussion here to the $H^1(-1, 1)$ norm. Since $H^1(-1, 1)$ is a Hilbert space, the best approximation polynomial for $u$ is the orthogonal projection of $u$ upon $\mathbb{P}_N$ in the scalar product that induces the norm of $H^1(-1, 1)$. This is defined as

$$((u, v)) = \int_{-1}^{1} (u'v' + uv)\, \mathrm{d}x \quad \text{for all } u, v \in H^1(-1, 1)\,. \tag{5.4.19}$$

Then, the polynomial $P^1_N u \in \mathbb{P}_N$ such that

$$((P^1_N u, \phi)) = ((u, \phi)) \quad \text{for all } \phi \in \mathbb{P}_N \tag{5.4.20}$$

satisfies the identity

$$\|u - P^1_N u\|_{H^1(-1,1)} = \inf_{\phi \in \mathbb{P}_N} \|u - \phi\|_{H^1(-1,1)} \,. \tag{5.4.21}$$

The approximation error (5.4.21) satisfies, for all $u \in H^m(-1, 1)$, with $m \geq 1$, the estimate

$$\|u - P^1_N u\|_{H^1(-1,1)} \leq CN^{1-m}|u|_{H^{m;N}(-1,1)} \,. \tag{5.4.22}$$

On the other hand, the error $u - P_N^1 u$ in the $L^2$-norm satisfies

$$\|u - P_N^1 u\|_{L^2(-1,1)} \leq CN^{-m} |u|_{H^{m;N}(-1,1)} \, . \tag{5.4.23}$$

The exponent of $N$ is the same here as it is for the best approximation error in the $L^2$-norm.

An illustration of both the $L^2(-1,1)$ and $H^1(-1,1)$-projections is provided in Fig. 5.2, again for the function $u(x) = |x|^{3/2}$. The maximum pointwise error for the $H^1$-projection appears to decay slightly faster than the corresponding error for the $L^2$-projection (see Figs. 5.2(c) and (a)). In fact, for all functions $u \in H^m(-1,1)$, $m \geq 1$, one has

$$\|u - P_N u\|_{L^\infty(-1,1)} \leq CN^{3/4-m} |u|_{H^{m;N}(-1,1)} \tag{5.4.24}$$

and

$$\|u - P_N^1 u\|_{L^\infty(-1,1)} \leq CN^{1/2-m} |u|_{H^{m;N}(-1,1)} \, . \tag{5.4.25}$$

These estimates follow from the Sobolev inequality (A.12) together with the previous estimates in the Sobolev norms: (5.4.24) is obtained using (5.4.11) and (5.4.18) with $l = 1$; (5.4.25) is a consequence of (5.4.22) and (5.4.23). On the other hand, it is evident in Figs. 5.2(d) and (b) that the $H^1$-projection is definitely superior to the $L^2$-projection in the approximation of the first derivative of $u$.

The approximation results in the Sobolev norms are of importance for the analysis of spectral approximations of boundary-value problems. In this case it may be more appropriate to project not just onto the space of polynomials, but onto the space of polynomials satisfying the boundary data. Result (5.4.22) holds for this projection as well (provided, of course, that $u$ satisfies the same boundary data). Let us consider, for instance, homogeneous Dirichlet conditions at both endpoints of the interval $(-1,1)$. The functions of $H^1(-1,1)$ that satisfy such conditions form a subspace that is usually denoted by $H_0^1(-1,1)$ (see (A.11.c)), i.e.,

$$H_0^1(-1,1) = \left\{ v \in H^1(-1,1) \mid v(-1) = v(1) = 0 \right\} \, . \tag{5.4.26}$$

Similarly, the polynomials of degree $N$ that vanish at the endpoints form a subspace $\mathbb{P}_N^0$ of $\mathbb{P}_N$:

$$\mathbb{P}_N^0 = \{ v \in \mathbb{P}_N \mid v(-1) = v(1) = 0 \} \, . \tag{5.4.27}$$

The inner product that is most commonly used for functions in $H_0^1(-1,1)$ is defined by

$$[u, v] = \int_{-1}^{1} u'(x) v'(x) \, dx \quad \text{for } u, v \in H_0^1(-1,1) \, . \tag{5.4.28}$$

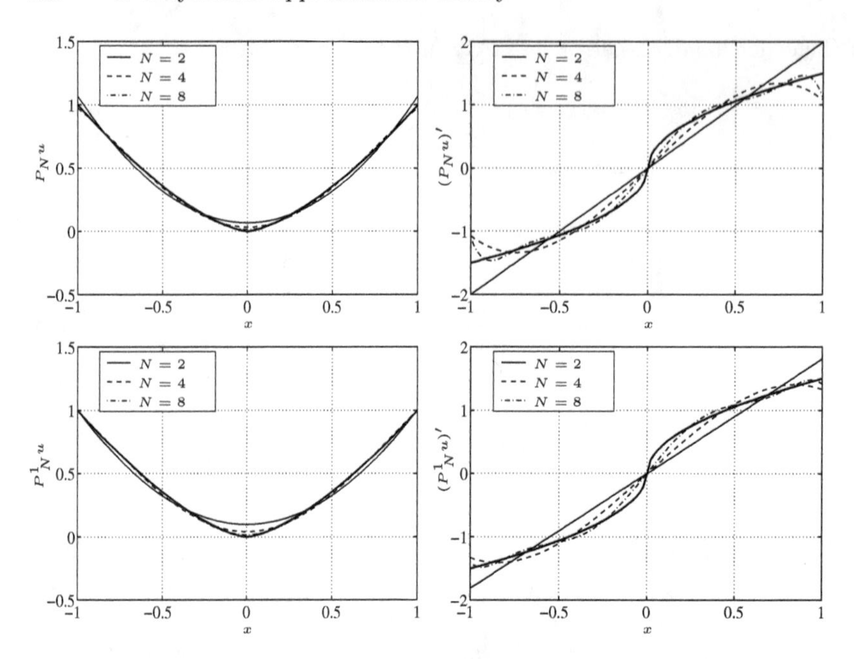


**Fig. 5.2.** $L^2(-1,1)$- and $H^1(-1,1)$-Legendre projections for $u(x) = |x|^{3/2}$. The exact result is indicated by the solid, thick curve, and the approximate results for $N = 2, 4$ and 8 by the thin curves:
$u$ and its $L^2(-1,1)$-projections (*upper left*);
$u'$ and the first derivative of the $L^2(-1,1)$-projections of $u$ (*upper right*);
$u$ and its $H^1(-1,1)$-projections (*lower left*);
$u'$ and the first derivative of the $H^1(-1,1)$-projections of $u$ (*lower right*)

It induces a norm on $H_0^1(-1,1)$ which is equivalent to the $H^1$-norm, due to the Poincaré inequality (A.13) (see also (A.11.c)). The $H_0^1$-projection of a function $u \in H_0^1(-1,1)$ upon $\mathbb{P}_N^0$ is the polynomial $P_N^{1,0}u \in \mathbb{P}_N^0$ such that

$$[P_N^{1,0}u, \phi] = [u, \phi] \quad \text{for all } \phi \in \mathbb{P}_N^0 \, . \tag{5.4.29}$$

We have the error estimate

$$\|u - P_N^{1,0}u\|_{H^k(-1,1)} \leq CN^{k-m}|u|_{H^{m;N}(-1,1)} \tag{5.4.30}$$

for all $u \in H^m(-1,1)$ vanishing at the boundary, with $m \geq 1$ and $k = 0, 1$.

More generally, for any function $u \in H^1(-1,1)$, we can introduce the affine polynomial $r(x) = u(-1)\dfrac{1-x}{2} + u(1)\dfrac{1+x}{2}$ in $(-1,1)$ and then construct the polynomial

$$P_N^{1,b}u = r + P_N^{1,0}(u - r) \, . \tag{5.4.31}$$

Note that $P_N^{1,b}u(\pm 1) = u(\pm 1)$, i.e., $P_N^{1,b}u$ matches the boundary values of $u$. The difference $u - P_N^{1,b}u$ can be estimated exactly as in (5.4.30), provided $u \in H^m(-1,1)$ for some $m \geq 1$.

The error bound (5.4.22) extends to higher order Sobolev norms as follows: Let $P_N^l u$ be the orthogonal projection of $u$ onto $\mathbb{P}_N$, under the inner product of $H^l(-1,1)$ that induces the norm (5.4.9) (with $m = l$). Then

$$\|u - P_N^l u\|_{H^k(-1,1)} \leq CN^{k-m}|u|_{H^{m;N}(-1,1)} \tag{5.4.32}$$

for $m \geq l$, $0 \leq k \leq l$, provided $u \in H^m(-1,1)$. The same estimate holds if we replace $P_N^l$ by $P_N^{l,\lambda}$ $(0 \leq \lambda \leq l-1)$, which is the orthogonal projection operator from the subspace of $H^l(-1,1)$ of the functions vanishing at the boundary with their derivatives of order up to $\lambda$, upon the subspace of $\mathbb{P}_N$ of the polynomials satisfying the same boundary conditions. In case of nonhomogeneous boundary conditions, one can construct as in (5.4.31) a polynomial $P_N^{l,\lambda,b}u$ matching the boundary values of $u$ and its first $\lambda$ derivatives, for which an error estimate similar to (5.4.32) holds.

Finally, if $k > l$, i.e., if the norm in which the error is measured is stronger than the norm for which the error is minimal, then the exponent of $N$ in all the previous estimates is $2k - l - \frac{1}{2} - m$.

### 5.4.3 Estimates for the Interpolation Error

We consider now the interpolation error. Let $x_j$, $0 \leq j \leq N$, be the Gauss, or the Gauss-Radau, or the Gauss-Lobatto points relative to the Legendre weight $w(x) \equiv 1$, considered in Sect. 2.3.1. Let $I_N u$ denote the polynomial of degree $N$ that interpolates $u$ at one of these sets of points. We give some estimates for the interpolation error $u - I_N u$ in the norms of the Sobolev spaces $H^l(-1,1)$.

In the familiar $L^2(-1,1)$-norm, whenever $u \in H^m(-1,1)$ with $m \geq 1$, one has

$$\|u - I_N u\|_{L^2(-1,1)} \leq CN^{-m}|u|_{H^{m;N}(-1,1)} \,, \tag{5.4.33}$$

i.e., the interpolation error behaves asymptotically as the truncation error in the $L^2$-norm. The generalization of this formula for $1 \leq l \leq m$ is

$$\|u - I_N u\|_{H^l(-1,1)} \leq CN^{2l-1/2-m}|u|_{H^{m;N}(-1,1)} \,, \tag{5.4.34}$$

exactly as for the truncation error (recall (5.4.18)). For instance, if $I_N$ denotes the interpolation operator at the Gauss points and if $u = u^{(N)} = L_{N+1} - L_{N-1}$, one has $\|u^{(N)} - I_N u^{(N)}\|_{H^1(-1,1)} \sim cN^{1/2}|u^{(N)}|_{H^1(-1,1)}$ as $N \to \infty$. However, for the Gauss-Lobatto interpolation, one has the following optimal error estimate in $H^1(-1,1)$

$$\|u - I_N u\|_{H^1(-1,1)} \leq CN^{1-m}|u|_{H^{m;N}(-1,1)} \,. \tag{5.4.35}$$

Comparing this estimate to (5.4.22) and to (5.4.30), we see that the polynomial $I_N u$ behaves asymptotically both as the best $N$-degree polynomial approximation of $u$ in the $H^1$-norm, and as the best $N$-degree polynomial approximation of $u$ (again in the $H^1$-norm) which matches the boundary values of $u$.

The last inequality includes the following bound on the error between the exact derivative $u'$ and the Legendre interpolation derivative $\mathcal{D}_N u = (I_N u)'$ (see (2.3.25)):

$$\|u' - \mathcal{D}_N u\|_{L^2(-1,1)} \leq C N^{1-m} |u|_{H^{m;N}(-1,1)} . \tag{5.4.36}$$

According to (5.3.3) and (5.4.33), the same estimate holds if the continuous $L^2$-norm of the error is replaced by the discrete $L^2$-norm at the interpolation points. Furthermore, (5.4.35) easily implies that Gauss-Lobatto interpolation is *stable* in the $H^1$-norm (which, in many applications, is the natural energy norm for the problem; see Chap. 6); indeed, one has

$$\|I_N u\|_{H^1(-1,1)} \leq C \|u\|_{H^1(-1,1)} , \tag{5.4.37}$$

with $C$ independent of $N$.

We conclude this section by providing an estimate for the integration error arising from the use of Gauss quadrature formulae relative to the Legendre weight. Assume that a $(N+1)$-point Gauss, or Gauss-Radau, or Gauss-Lobatto quadrature formula relative to the Legendre weight is used to integrate the product $u\phi$, where $u \in H^m(-1,1)$ for some $m \geq 1$ and $\phi \in \mathbb{P}_N$. Then combining (5.3.4a) or (5.3.4b) with (5.4.33) and (5.4.11), one can show that

$$\left| \int_{-1}^{1} u(x)\phi(x)\, dx - (u,\phi)_N \right| \leq C N^{-m} |u|_{H^{m;N-1}(-1,1)} \|\phi\|_{L^2(-1,1)} . \tag{5.4.38}$$

### 5.4.4 Scaled Estimates

In view of the multidomain spectral approximations given in Chap. 5 of the companion book CHQZ3, it is useful to consider polynomial approximations of a function $u$ defined not on the standard interval $(-1,1)$ but on a generic interval $I = (x_L, x_R)$ of length $h = x_R - x_L$. The orthogonal projections $P_N u$, $P_N^1 u$ and $P_N^l u$ onto $\mathbb{P}_N$ with respect to the $L^2$-, the $H^1$- and the $H^l$-inner product, respectively, are constructed as above, simply by replacing integrals over $(-1,1)$ by integrals over $I$. Boundary conditions are accounted for as above, with the obvious modifications. Each interpolating polynomial $I_N u$ is based on the Gaussian points $x_j = F(\hat{x}_j)$, $0 \leq j \leq N$, that are the images of the corresponding Gaussian points $\hat{x}_j$ on the *reference interval* $\hat{I} = (-1,1)$ under the affine mapping $x = F(\hat{x}) = x_L + \frac{1}{2}h(\hat{x}+1)$.

On the interval $I$, the error estimates given above are modified by the presence of a power of the size $h$ of the interval. We report hereafter the most significant ones; the constants $C$ are now independent of both $N$ and $h$, although they depend as above on $m$.

Estimate (5.4.11) for the Legendre truncation error in $I$ reads as follows: for all $u \in H^m(I)$, $m \geq 0$,

$$\|u - P_N u\|_{L^2(I)} \leq C h^{\min(m,N)} N^{-m} |u|_{H^{m;N}(I)} . \tag{5.4.39}$$

Estimates (5.4.32) for the orthogonal projection errors in $H^l(I)$, $l \geq 1$, become: for all $u \in H^m(I)$, $m \geq l$,

$$\|u - P_N^l u\|_{H^k(I)} \leq C h^{k-\min(m,N)} N^{k-m} |u|_{H^{m;N}(I)} , \tag{5.4.40}$$

for all $0 \leq k \leq l$. Similar estimates hold if we replace $P_N^l$ by $P_N^{l,\lambda,b}$ in order to match the values of $u$ and certain derivatives of $u$ at the endpoints of the interval. In particular, for all $u \in H^m(I)$, $m \geq 1$, one has

$$\|u - P_N^{1,b} u\|_{H^1(I)} \leq C h^{1-\min(m,N)} N^{1-m} |u|_{H^{m;N}(I)} . \tag{5.4.41}$$

The interpolation error at the Gauss-Lobatto points is estimated as follows: for all $u \in H^m(I)$, $m \geq 1$, one has, for $k = 0, 1$,

$$\|u - I_N u\|_{H^k(I)} \leq C h^{k-\min(m,N)} N^{k-m} |u|_{H^{m;N}(I)} . \tag{5.4.42}$$

Finally, we notice that the inverse inequality (5.4.3) becomes

$$\|\phi\|_{L^q(I)} \leq C h^{1/q-1/p} N^{2(1/p-1/q)} \|\phi\|_{L^p(I)} \qquad \text{for all } \phi \in \mathbb{P}_N , \tag{5.4.43}$$

whereas (5.4.5) becomes

$$\|\phi^{(r)}\|_{L^p(I)} \leq C h^{-r} N^{2r} \|\phi\|_{L^p(I)} \qquad \text{for all } \phi \in \mathbb{P}_N . \tag{5.4.44}$$

## *Bibliographical Notes*

The inverse inequality (5.4.3) is proven, e.g., in Timan (1963, p. 236). Inequality (5.4.5) for $p = \infty$ is the classical Markov inequality (see, e.g., Timan (1963, p. 218)); for $p = 2$ we refer to Babuška, Szabó, and Katz (1981) or Canuto and Quarteroni (1982a), where different proofs are given; for $2 < p < \infty$, it can be obtained by interpolation of spaces (see Quarteroni (1984)). The inverse inequalities (5.4.4), (5.4.6) and (5.4.7) can be found in Bernardi and Maday (1992a). Estimates (5.4.11) and (5.4.18) have been obtained by Canuto and Quarteroni (1982a) with the full norm (5.4.9) on the right-hand side; here, we have refined them by introducing the seminorm (5.4.10), as is common in the error estimates for finite-element approximations. Estimate (5.4.13) can be found in Schwab (1998), Thm. 3.11. The discussion on the optimality of the truncation error in higher Sobolev norms is also based on results from Canuto and Quarteroni (1982a). For the existence and uniqueness of the polynomials of best approximation in the $L^p$-norms we refer to Nikolskii (1975), Theorem 1.3.6, and Timan (1963), pp. 35–40. Estimate (5.4.16) is proven in Quarteroni (1984), while estimate (5.4.17) is due to Jackson (1930), Theorem XV. Estimates (5.4.22), (5.4.23) and (5.4.30) for the $H^1$- and $H_0^1$-projection operators are due to Maday and Quarteroni (1981), while their extension to higher order projections (5.4.32) has been carried out by Maday (1990). The results of Sect. 5.4.3 have been established by Bernardi and Maday (1992a). Finally, the scaled estimates of Sect. 5.4.4 are typical of the analysis of the *hp*-version of the finite-element method (see, e.g., Schwab (1998) and the references therein).

## 5.5 Chebyshev Approximations

This section will be dedicated to Chebyshev approximation and will be similar in spirit to the section on Legendre approximation. Since the Chebyshev polynomials are orthogonal with respect to the nonconstant weight $w(x) = (1-x^2)^{-1/2}$, it is natural to frame the results in terms of weighted $L^p$ and Sobolev spaces. For additional results we refer to Bernardi and Maday (1997).

### 5.5.1 Inverse Inequalities for Polynomials

We define weighted $L^p$-norms as follows:

$$\|u\|_{L^p_w(-1,1)} = \left( \int_{-1}^{1} |u(x)|^p w(x)\, \mathrm{d}x \right)^{1/p} \quad \text{for } 1 \le p < \infty\,, \tag{5.5.1}$$

and we again set

$$\|u\|_{L^\infty_w(-1,1)} = \sup_{-1\le x\le 1} |u(x)| = \|u\|_{L^\infty(-1,1)}\,. \tag{5.5.2}$$

The space of functions for which a particular norm is finite forms a Banach space, indicated by $L^p_w(-1,1)$ (see (A.9.g)).

The inverse inequality concerning the summability in the Chebyshev $L^p$-norm for polynomials states that for any $p$ and $q$, $1 \le p \le q \le \infty$, there exists a positive constant $C$ such that, for each $\phi \in \mathbb{P}_N$,

$$\|\phi\|_{L^q_w(-1,1)} \le (2N)^{(1/p-1/q)} \|\phi\|_{L^p_w(-1,1)}\,. \tag{5.5.3}$$

Note that the power of $N$ is half the corresponding power in the Legendre estimate (5.4.3).

The inverse inequality concerning differentiation states that for any $p$, $2 \le p \le \infty$, and any integer $r \ge 1$, there exists a positive constant $C$ such that, for any $\phi \in \mathbb{P}_N$,

$$\|\phi^{(r)}\|_{L^p_w(-1,1)} \le CN^{2r}\|\phi\|_{L^p_w(-1,1)} \,. \tag{5.5.4}$$

Note that this estimate shares with the Legendre estimate (5.4.5) the double power of $N$ on the right-hand side.

Estimates (5.4.6) and (5.4.7) have their Chebyshev counteparts, obtained by inserting the Chebyshev weight on both sides of the integrals that define the norms. Precisely, setting again $\eta(x) = 1 - x^2$, one has

$$\|\,\phi'\sqrt{\eta}\,\|_{L^2_w(-1,1)} \le CN\|\phi\|_{L^2_w(-1,1)} \qquad \text{for all } \phi \in \mathbb{P}_N \,, \tag{5.5.5}$$

and

$$\|\phi'\|_{L^2_w(-1,1)} \le CN\|\,\phi/\sqrt{\eta}\,\|_{L^2_w(-1,1)} \qquad \text{for all } \phi \in \mathbb{P}^0_N(-1,1) \,, \tag{5.5.6}$$

where $C$ is a positive constant independent of $N$.

### 5.5.2 Estimates for the Truncation and Best Approximation Errors

The natural Sobolev norms in which to measure approximation errors for the Chebyshev system involve the Chebyshev weight in the quadratic averages of the error and its derivatives over the interval $(-1,1)$. Thus, we set

$$\|u\|_{H^m_w(-1,1)} = \left(\sum_{k=0}^{m} \|u^{(k)}\|^2_{L^2_w(-1,1)}\right)^{1/2} . \tag{5.5.7}$$

The Hilbert space associated to this norm is denoted by $H^m_w(-1,1)$ and is introduced in (A.11.b). Similarly to (5.4.10), we also define the seminorms

$$|u|_{H^{m;N}_w(-1,1)} = \left(\sum_{k=\min(m,N+1)}^{m} \|u^{(k)}\|^2_{L^2_w(-1,1)}\right)^{1/2} . \tag{5.5.8}$$

The truncation error $u - P_N u$, where now $P_N u = \sum_{k=0}^{N} \hat{u}_k T_k$ is the truncated Chebyshev series of $u$, satisfies the inequality

$$\|u - P_N u\|_{L^2_w(-1,1)} \le CN^{-m}|u|_{H^{m;N}_w(-1,1)} \,, \tag{5.5.9}$$

for all $u \in H^m_w(-1,1)$, with $m \ge 0$. This is a particular case of the estimate for the truncation error in the weighted $L^p$-norms, which reads as follows:

$$\|u - P_N u\|_{L^p_w(-1,1)} \le C\sigma_p(N)N^{-m} \sum_{k=\min(m,N+1)}^{m} \|u^{(k)}\|_{L^p_w(-1,1)} \,, \tag{5.5.10}$$

for all functions $u$ whose distributional derivatives of order up to $m$ belong to $L^p_w(-1,1)$. The constant $\sigma_p(N)$ equals 1 for $1 < p < \infty$, and $1 + \log N$ for $p = 1$ or $p = \infty$. As a consequence of this result, one gets an optimal estimate for the error of best approximation in the $L^p_w$-norms for $1 < p < \infty$. (Note that this error in the norm of $L^\infty_w(-1,1) = L^\infty(-1,1)$ is estimated in (5.4.16).)

As for the Legendre case, the seminorm on the right-hand side of (5.5.9) can be replaced by a weaker seminorm, which is defined as in (5.4.14) with the measure $\mathrm{d}x$ replaced by $w(x)\mathrm{d}x$. Thus, the error decay rate predicted by (5.5.9) is realized also for functions $u$ that are more singular at the boundary points than functions in $H^m_w(-1,1)$. This observation applies to all the subsequent estimates as well.

The truncation error in higher order Sobolev norms is estimated by the inequality

$$\|u - P_N u\|_{H^l_w(-1,1)} \leq C N^{2l-1/2-m} |u|_{H^{m;N}_w(-1,1)} , \qquad (5.5.11)$$

for $u \in H^m_w(-1,1)$, with $m \geq 1$ and $1 \leq l \leq m$. Thus, the asymptotic behavior of the Chebyshev truncation error is the same as for Legendre polynomials; hence, it is non-optimal with respect to the exponent of $N$.

In order to define the polynomial of best approximation in $H^1_w(-1,1)$, we introduce the inner product

$$((u,v))_w = \int_{-1}^{1} (u'v' + uv) w \, \mathrm{d}x \quad \text{for all } u, v \in H^1_w(-1,1) , \qquad (5.5.12)$$

and we define the related orthogonal projection on $\mathbb{P}_N$ as the polynomial $P^1_N u \in \mathbb{P}_N$ such that

$$((P^1_N u, \phi))_w = ((u, \phi))_w \quad \text{for all } \phi \in \mathbb{P}_N . \qquad (5.5.13)$$

The corresponding general error estimate is

$$\|u - P^1_N u\|_{H^k_w(-1,1)} \leq C N^{k-m} |u|_{H^{m;N}_w(-1,1)} , \qquad (5.5.14)$$

for all $u \in H^m_w(-1,1)$ with $m \geq 1$, and $k = 0, 1$. Fig. 5.3 provides an example of the different behavior of the $L^2_w(-1,1)$ and $H^1_w(-1,1)$ projections. In higher order Sobolev norms one can prove the following result. For all integer $l$ such that $0 \leq l \leq m$, and for every function $u \in H^m_w(-1,1)$, there exists a polynomial $u^N \in \mathbb{P}_N$ such that

$$\|u - u^N\|_{H^k_w(-1,1)} \leq C N^{k-m} |u|_{H^{m;N}_w(-1,1)} , \qquad (5.5.15)$$

for $0 \leq k \leq l$. The polynomial $u^N$ can be defined as the orthogonal projection of $u$ upon $\mathbb{P}_N$ in an inner product on $H^l_w(-1,1)$ that induces a norm equivalent to $\|u\|_{H^l_w(-1,1)}$.

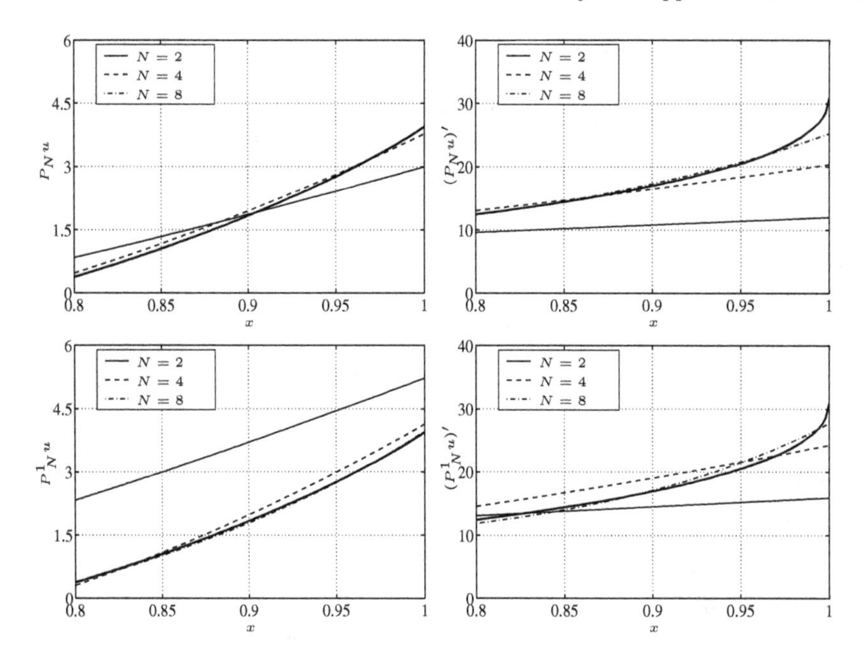


**Fig. 5.3.** $L^2_w(-1,1)$- and $H^1_w(-1,1)$-Chebyshev projections for the function $u(x) = \frac{1}{48}\left[2\pi^2(\theta-\pi)^2 - (\theta-\pi)^4\right] - x$ where $\theta = \cos^{-1} x$. The exact result is indicated by the solid, thick curve, and the approximate results for $N = 2, 4$ and $8$ by the thin curves: $u$ and its $L^2(-1,1)$-projections (*upper left*);
$u'$ and the first derivative of the $L^2(-1,1)$-projections of $u$ (*upper right*);
$u$ and its $H^1(-1,1)$-projections (*lower left*);
$u'$ and the first derivative of the $H^1(-1,1)$-projections of $u$ (*lower right*)

These estimates extend to functions satisfying prescribed boundary data in the same way that the Legendre estimates did. For instance, assume that $u$ is a function in $H^1_w(-1,1)$ that vanishes at $x = \pm 1$, i.e., $u$ belongs to the subspace of $H^1_w(-1,1)$ defined as

$$H^1_{w,0}(-1,1) = \left\{v \in H^1_w(-1,1) \mid v(-1) = v(1) = 0\right\} \tag{5.5.16}$$

(see (A.11.c)). The projection of $u$ upon $\mathbb{P}^0_N$ (see (5.4.27)) in the norm of this space is the polynomial $P^{1,0}_N u \in \mathbb{P}^0_N$ such that

$$\left[P^{1,0}_N u, \phi\right]_w = [\,u, \phi\,]_w \quad \text{for all } \phi \in \mathbb{P}^0_N\,. \tag{5.5.17}$$

Here we use the natural inner product in $H^1_{w,0}(-1,1)$:

$$[\,u, v\,]_w = \int_{-1}^{1} u'v'w \,\mathrm{d}x \quad \text{for } u, v \in H^1_{w,0}(-1,1) \tag{5.5.18}$$

(see (A.11.c)). For the projector $P^{1,0}_N$ we have the following estimate:

$$\|u - P_N^{1,0} u\|_{H_w^1(-1,1)} \leq C N^{1-m} |u|_{H_w^{m;N}(-1,1)} , \tag{5.5.19}$$

for all $u \in H_w^m(-1,1)$, $m \geq 1$, vanishing at the boundary points.

Furthermore, one can find a polynomial $u^N \in \mathbb{P}_N^0$ whose distance from $u$ decays in an optimal way both in the $H_w^1$-norm and in the $L_w^2$-norm, i.e.,

$$\|u - u^N\|_{H_w^k(-1,1)} \leq C N^{k-m} |u|_{H_w^{m;N}(-1,1)} , \tag{5.5.20}$$

for $k = 0$ and $k = 1$. For instance, $u^N$ can be defined as the solution of the Galerkin problem

$$\int_{-1}^{1} (u - u^N)' (\phi w)' \, dx = 0 \quad \text{for all } \phi \in \mathbb{P}_N^0 \tag{5.5.21}$$

(see Sect. 7.1).

Finally, we mention that, as for the Legendre approximation, if $u$ belongs to $H_w^l(-1,1)$ and vanishes at the boundary with the derivatives of order up to $\lambda$ for an integer $\lambda \leq l-1$, then one can find a polynomial $u^N$ satisfying the same boundary conditions as $u$ such that an estimate like (5.5.15) holds. A similar conclusion holds in the nonhomogeneous case, i.e., when the derivatives of $u$ of order up to $\lambda$ are not necessarily zero at the endpoints.

### 5.5.3 Estimates for the Interpolation Error

We consider now the interpolation error. Let $I_N u \in \mathbb{P}_N$ denote the interpolant of $u$ at any of the three families of Chebyshev Gauss points (2.4.12) or (2.4.13) or (2.4.14). Then the following estimate holds:

$$\|u - I_N u\|_{L_w^2(-1,1)} \leq C N^{-m} |u|_{H_w^{m;N}(-1,1)} , \tag{5.5.22}$$

if $u \in H_w^m(-1,1)$ for some $m \geq 1$. In higher order Sobolev norms, as for the Legendre case, optimal error estimates do not hold. For instance, a standard argument that uses (5.5.9), (5.5.22) and the inverse inequality (5.5.4) yields

$$\|u - I_N u\|_{H_w^l(-1,1)} \leq C N^{2l-m} |u|_{H_w^{m;N}(-1,1)} , \tag{5.5.23}$$

for $0 \leq l \leq m$. As for the Legendre case, this estimate can be improved for the Gauss-Lobatto interpolation. Indeed, in the case $l = 1$, the power $N^{2-m}$ can be replaced by the optimal one $N^{1-m}$. As a consequence, Gauss-Lobatto interpolation is stable in the $H_w^1$-norm, i.e.,

$$\|I_N u\|_{H_w^1(-1,1)} \leq C \|u\|_{H_w^1(-1,1)} , \tag{5.5.24}$$

for all $u \in H_w^m(-1,1)$, with $m \geq 1$. The optimal error estimate can be phrased in terms of the error between the exact derivative $u'$ and the Chebyshev interpolation derivative $\mathcal{D}_N u = (I_N u)'$ (see (2.4.28)) as follows:

$$\|u' - \mathcal{D}_N u\|_{L^2_w(-1,1)} \leq CN^{1-m} |u|_{H^{m;N}_w(-1,1)} \, . \tag{5.5.25}$$

The same estimate holds in the discrete $L^2_w$-norm at the interpolation points. In higher order Sobolev norms, the interpolation error at these points is again suboptimal; precisely, one has, for $2 \leq l \leq m$ ,

$$\|u - I_N u\|_{H^l_w(-1,1)} \leq CN^{2l-1-m} |u|_{H^{m;N}_w(-1,1)} \, . \tag{5.5.26}$$

When the function $u$ is analytic, the error $u^{(l)} - (I_N u)^{(l)}$ decays exponentially in $N$ for all $l \geq 0$. Precisely, if $u$ is analytic in $[-1,1]$ and has an analytic extension to the ellipse $E_\eta$ with foci in $z = \pm 1$ and sum of semi-axes equal to $e^\eta > 1$ for some $\eta > 0$, then

$$\|u^{(l)} - (I_N u)^{(l)}\|_{L^2_w(-1,1)} \leq \frac{C(l)}{\sinh \eta} N^{2l} e^{-\eta N} M(u,\eta) \, , \tag{5.5.27}$$

where $M(u,\eta) = \max_{z \in E_\eta} |u(z)|$.

The interpolation error in the maximum norm is also of interest. An estimate of it is given by

$$\|u - I_N u\|_{L^\infty(-1,1)} \leq CN^{1/2-m} |u|_{H^{m;N}_w(-1,1)} \tag{5.5.28}$$

under the same assumptions as for (5.5.22).

By (5.5.9) and (5.5.22) we can obtain an estimate for the integration error produced by a Gauss-type quadrature formula relative to the Chebyshev weight. If $u \in H^m_w(-1,1)$ for some $m \geq 1$ and $\phi \in \mathbb{P}_N$, then using (5.3.4a) and (5.3.4b) we get the following result:

$$\left| \int_{-1}^{1} u(x)\phi(x)w(x)\mathrm{d}x - (u,\phi)_N \right| \leq CN^{-m} |u|_{H^{m;N}_w(-1,1)} \|\phi\|_{L^2_w(-1,1)} \, . \tag{5.5.29}$$

### *Bibliographical Notes*

A proof of the inverse inequality (5.5.3) is given in Quarteroni (1984). Inequality (5.5.4) has been established by Canuto and Quarteroni (1982a) for $p = 2$ and extended to arbitrary $p$ by Quarteroni (1984). Estimates (5.5.9) and (5.5.11) are proven in Canuto and Quarteroni (1982a), while estimate (5.5.10) has been obtained by Quarteroni (1984); as for the Legendre case, the presence of the seminorm (5.5.8) rather than the norm (5.5.7) on the right-hand side of the estimates is new. Inequalities (5.5.14), (5.5.19) and (5.5.20) are due to Maday and Quarteroni (1981); here, we give a different proof of (5.5.14). The extension of these results to higher order norms has been carried out by Maday (1990). Finally, the results given in Sect. 5.5.3 are due to Bernardi and Maday (1992a), except for (5.5.27), which is due to Tadmor (1986).

## 5.6 Proofs of Some Approximation Results

We present in this section the proofs of some of the most relevant approximation error estimates given in the two previous sections. We confine ourselves to estimates in Hilbert norms of the truncation, interpolation and projection operators. Indeed, these are precisely the error estimates that most frequently occur in this book for the convergence analysis of spectral methods.

PROOF OF (5.4.5) AND (5.5.4). Let us begin with (5.5.4); we confine ourselves to the case $p = 2$ and $r = 1$. Let $\phi = \sum_{k=0}^{N} \hat{\phi}_k T_k$. By (2.4.22) we obtain $\phi' = \sum_{k=0}^{N-1} \hat{\phi}_k^{(1)} T_k$, with

$$c_k \hat{\phi}_k^{(1)} = 2 \sum_{\substack{\ell=k+1 \\ \ell+k \text{ odd}}}^{N} \ell \hat{\phi}_\ell ,$$

where the coefficients $c_k$ are defined in (2.4.10). The Cauchy-Schwarz inequality and the identity $\sum_{m=1}^{N} m^2 = N(N+1)(2N+1)/6$ give

$$\left(c_k \hat{\phi}_k^{(1)}\right)^2 \leq 4 \left( \sum_{\substack{\ell=k+1 \\ \ell+k \text{ odd}}}^{N} \ell^2 \right) \left( \sum_{\substack{\ell=k+1 \\ \ell+k \text{ odd}}}^{N} (\hat{\phi}_\ell)^2 \right) \leq \frac{2}{3} N(N+1)(2N+1) \sum_{\ell=0}^{N} (\hat{\phi}_\ell)^2 .$$

On the other hand, from (2.4.9) we have

$$\begin{aligned} \|\phi'\|^2_{L^2_w(-1,1)} &= \sum_{k=0}^{N-1} \frac{\pi c_k}{2} \left(\hat{\phi}_k^{(1)}\right)^2 \\ &\leq \frac{\pi}{3} N(N+1)(2N+1) \sum_{k=0}^{N-1} \frac{1}{c_k} \sum_{\ell=0}^{N} (\hat{\phi}_\ell)^2 \leq C N^4 \|\phi\|^2_{L^2_w(-1,1)} . \end{aligned}$$

Although the proof of (5.5.4) may seem very crude, the exponent of $N$ in (5.5.4) cannot be reduced. To convince oneself, it is sufficient to consider the function $\phi = \sum_{\substack{k=0 \\ k+N \text{ odd}}}^{N} T_k$, for which one has $\|\phi'\|_{L^2_w(-1,1)} \simeq N^2 \|\phi\|_{L^2_w(-1,1)}$.

The proof of (5.4.5), again in the case $p = 2$ and $r = 1$, follows the same guidelines as above, using now (2.3.18) and (2.3.9). □

A USEFUL LEMMA. The following results, which are an elementary version of the so called Deny-Lions Lemma (see, e.g., Quarteroni and Valli (1994), Proposition 3.4.4), will be used in some of the subsequent proofs. Let $w$ denote the Legendre or the Chebyshev weight (or, more generally, any Jacobi weight).

**Lemma 5.1.** *Let $m \geq 1$ and $0 \leq r \leq m-1$ be integers. There exists a constant $C$ (depending on $r$) such that for all $v \in H_w^m(-1,1)$ one has*

$$\inf_{\phi \in \mathbb{P}_r} \|v - \phi\|_{H_w^r(-1,1)} \leq C \|v^{(r+1)}\|_{L_w^2(-1,1)}.$$

PROOF. We recall that any function $\psi \in H_w^1(-1,1)$ satisfying $\int_{-1}^1 \psi(x)\,\mathrm{d}x = 0$ necessarily vanishes for at least one point in $(-1,1)$; hence, the Poincaré inequality $\|\psi\|_{L_w^2(-1,1)} \leq C\|\psi'\|_{L_w^2(-1,1)}$ holds true (see Sect. A.13).

Given $v \in H_w^m(-1,1)$, let $\psi \in \mathbb{P}_r$ be the unique polynomial defined by the relations

$$\int_{-1}^1 \psi^{(k)}(x)\,\mathrm{d}x = \int_{-1}^1 v^{(k)}(x)\,\mathrm{d}x, \quad 0 \leq k \leq r.$$

By repeated application of the Poincaré inequality to $v-\psi$ and its derivatives, we get

$$\begin{aligned} \|v-\psi\|_{L_w^2(-1,1)} &\leq C\|v'-\psi'\|_{L_w^2(-1,1)} \leq \cdots \leq \\ &\leq C^r \|v^{(r)} - \psi^{(r)}\|_{L_w^2(-1,1)} \leq C^{r+1}\|v^{(r+1)} - \psi^{(r+1)}\|_{L_w^2(-1,1)}. \end{aligned}$$

Since $\psi^{(r+1)} \equiv 0$, we obtain the result. □

**Corollary 5.6.1.** *Let $m \geq 1$ and $N \geq 0$ be integers. There exists a constant $C$ depending only on $m$ such that for all $v \in H_w^m(-1,1)$ one has*

$$\inf_{\phi \in \mathbb{P}_N} \|v - \phi\|_{H_w^m(-1,1)} \leq C |v|_{H_w^{m;N}(-1,1)}$$

*(where the seminorm on the right-hand side is defined in (5.5.8)).*

PROOF. It is enough to set $r = \min(m-1, N)$ in the previous Lemma. □

PROOF OF (5.4.11). We give a proof that only exploits the fact that Legendre polynomials are the eigenfunctions of the singular Sturm-Liouville operator $\mathcal{L}\phi = -((1-x^2)\phi')'$. Precisely, denoting by $\phi_k$ the $k$-th Legendre polynomial (normalized in $L^2$, i.e., divided by the square root of the right-hand side of (2.3.8)), we have $\mathcal{L}\phi_k = \lambda_k \phi_k$; thus,

$$\hat{u}_k = (u, \phi_k) = \frac{1}{\lambda_k}(u, \mathcal{L}\phi_k) = \frac{1}{\lambda_k}(\mathcal{L}u, \phi_k)\,,$$

provided $\mathcal{L}u \in L^2(-1,1)$, which is implied by the condition $u \in H^2(-1,1)$. Iterating the argument $\mu$ times, we get

$$\hat{u}_k = \frac{1}{\lambda_k^\mu}(\mathcal{L}^\mu u, \phi_k)\,,$$

provided $\mathcal{L}^\mu u \in L^2(-1,1)$, which is certainly true if $u \in H^m(-1,1)$ with $m = 2\mu$. Thus, we have

$$\begin{aligned}\|u-P_Nu\|^2{}_{L^2(-1,1)} &= \sum_{k>N}|\hat{u}_k|^2 = \sum_{k>N}\frac{1}{\lambda_k^m}|(\mathcal{L}^\mu u,\phi_k)|^2\\ &\le \frac{1}{\lambda_{N+1}^m}\sum_{k>N}|(\mathcal{L}^\mu u,\phi_k)|^2 \le \frac{1}{\lambda_{N+1}^m}\|\mathcal{L}^\mu u\|^2_{L^2(-1,1)}\\ &\le \frac{C_m}{\lambda_{N+1}^m}\|u\|^2_{H^m(-1,1)}.\end{aligned}$$

Recalling that $\lambda_k \sim k^2$, we conclude that the bound

$$\|u-P_Nu\|_{L^2(-1,1)} \le c_m N^{-m}\|\mathcal{L}^{m/2}u\|_{L^2(-1,1)} \le C_m N^{-m}\|u\|_{H^m(-1,1)} \tag{5.6.1}$$

holds for all $u \in H^m(-1,1)$ and all even integers $m \ge 0$. By space interpolation between two consecutive even integers, we extend the bound to all odd integers. (This is not a simple argument. For the interested reader, interpolation between Hilbert spaces is discussed, e.g., in Bergh and Löfström (1976).) Finally, we observe that replacing $u$ by $u-\phi$, where $\phi \in \mathbb{P}_N$ is arbitrary, leaves the left-hand side unchanged since $P_N\phi = \phi$. Hence,

$$\|u-P_Nu\|_{L^2(-1,1)} \le C_m N^{-m} \inf_{\phi\in\mathbb{P}_N}\|u-\phi\|_{H^m(-1,1)} \quad \forall u \in H^m(-1,1),$$

and Corollary 5.6.1 concludes the proof. □

The first inequality in (5.6.1) shows that the $O(N^{-m})$ rate of decay of the error is achieved not only for those functions having their $m$-th derivative in $L^2(-1,1)$, but also for functions whose $m$-th derivative is more singular at the boundary points; indeed, such derivatives appear in $\mathcal{L}^{m/2}u$ multiplied by $1-x^2$. This property is expressed, in equivalent form, by estimate (5.4.13).

PROOF OF (5.4.13). We first recall that the Legendre polynomials $L_k(x) = P_k^{(0,0)}(x)$ satisfy, by (2.5.5),

$$\frac{\mathrm{d}^s}{\mathrm{d}x^s}L_k(x) = 2^{-s}\frac{(k+s)!}{k!}P_{k-s}^{(s,s)}(x)\ ;$$

hence, their $s$-derivatives are orthogonal in the inner product associated with the norm (5.4.14), i.e., using also (2.5.7),

$$\int_{-1}^{1} L_h^{(s)}(x)L_k^{(s)}(x)\left(1-x^2\right)^s \mathrm{d}x = \frac{2}{2k+1}\frac{(k+s)!}{(k-s)!}\delta_{hk}\,, \qquad h,k = 0,1,\ldots$$

It follows that if $u = \sum_{k=0}^{\infty}\hat{u}_k c_k L_k$, where $c_k = \sqrt{2k+\frac{1}{2}}$ are the normalization factors, then one has

$$\int_{-1}^{1}|u^{(s)}(x)|^2\left(1-x^2\right)^s \mathrm{d}x = \sum_{k=0}^{\infty}\frac{(k+s)!}{(k-s)!}|\hat{u}_k|^2\,.$$

On the other hand, as in the previous proof, $\|u - P_N u\|^2_{L^2(-1,1)} = \sum_{k>N} |\hat{u}_k|^2$, whence the result. □

PROOF OF (5.5.9). We shall make use of the transformation

$$x \in (-1,1)\,, \quad u(x) \mapsto u^*(\theta) = u(\cos\theta)\,, \quad \theta \in (0, 2\pi)\,. \tag{5.6.2}$$

Since $\theta = \arccos x$, we have $\mathrm{d}\theta/\mathrm{d}x = -w(x)$ (the Chebyshev weight); thus,

$$\|u\|^2_{L^2_w(-1,1)} = \frac{1}{2}\|u^*\|^2_{L^2(0,2\pi)}\,. \tag{5.6.3}$$

It follows that the map $u \mapsto u^*$ is an isomorphism between $L^2_w(-1,1)$ and the subspace of $L^2(0,2\pi)$ of the even real functions. Moreover, it maps $H^m_w(-1,1)$ into the space of periodic functions $H^m_p(0,2\pi)$ (see (A.11.d)). Indeed, since $u \in C^{m-1}([-1,1])$, then $u^* \in C^{m-1}(-\infty,+\infty)$ and is $2\pi$-periodic with all the derivatives of order up to $m-1$; whence, $u^* \in H^m_p(0,2\pi)$. Finally, since $|\mathrm{d}x/\mathrm{d}\theta| = |-\sin\theta| \le 1$, we also have

$$\|u^*\|_{H^m(0,2\pi)} \le C\|u\|_{H^m_w(-1,1)} \quad \text{for } m \ge 1\,. \tag{5.6.4}$$

Let $P^*_N$ denote the symmetric truncation of the Fourier series up to degree $N$, i.e.,

$$P^*_N\left(\sum_{k=-\infty}^{\infty} \hat{v}_k e^{ik\theta}\right) = \sum_{k=-N}^{N} \hat{v}_k e^{ik\theta}\,.$$

It is easily seen that

$$(P_N u)^* = P^*_N u^* \quad \text{for all } u \in L^2_w(-1,1)\,. \tag{5.6.5}$$

Indeed, since $u(x) = \sum_{k=0}^{\infty} \hat{u}_k T_k(x)$, $u^*(\theta) = \sum_{k=0}^{\infty} \hat{u}_k \cos k\theta = \sum_{k=0}^{\infty} \hat{u}_k \times (e^{ik\theta} + e^{-ik\theta})/2$; whence, (5.6.5). Now, from (5.6.3) and (5.1.9) one gets

$$\|u - P_N u\|_{L^2_w(-1,1)} = \frac{1}{\sqrt{2}}\|u^* - P^*_N u^*\|_{L^2(0,2\pi)} \le CN^{-m}\|u^{*(m)}\|_{L^2(0,2\pi)}\,;$$

whence, by (5.6.4),

$$\|u - P_N u\|_{L^2_w(-1,1)} \le C_m N^{-m}\|u\|_{H^m_w(-1,1)}, \quad \forall u \in H^m_w(-1,1).$$

As in the proof of (5.4.11), we conclude by invoking Corollary 5.6.1. □

We remark that a sharper estimate in which a weaker semi-norm appears on the right-hand side of (5.5.9) can be obtained. Indeed, the bound (5.6.4) is rather crude, as it neglects the information that the Jacobian of the mapping, $|\mathrm{d}x/\mathrm{d}\theta| = \sqrt{1-x^2}$, vanishes at the endpoints of the interval. Taking this into account, one can allow the $m$-th derivative of $u$ to have a stronger singularity at the endpoints than permitted by being in $L^2_w(-1,1)$. An alternative argument to arrive at the same conclusions consists of adapting the proof of (5.4.13) to the Chebyshev case.

PROOF OF (5.4.18) AND (5.5.11). Let us first deal with (5.5.11). We consider the case $l = 1$ only; the result corresponding to $l > 1$ follows by an inductive procedure. Using the triangle inequality and the estimate (5.5.9), we obtain

$$\begin{aligned}\|u - P_N u\|_{H^1_w(-1,1)} &\le \|u - P_N u\|_{L^2_w(-1,1)} + \|u' - P_N u'\|_{L^2_w(-1,1)} \\ &\quad + \|P_N u' - (P_N u)'\|_{L^2_w(-1,1)} \\ &\le C N^{1-m} |u|_{H^{m;N}_w(-1,1)} + \|P_N u' - (P_N u)'\|_{L^2_w(-1,1)} \,.\end{aligned} \tag{5.6.6}$$

In order to bound the last term let us expand $u$ and $u'$ in Chebyshev polynomials as

$$u = \sum_{k=0}^{\infty} \hat{u}_k T_k \,, \qquad u' = \sum_{k=0}^{\infty} \hat{u}_k^{(1)} T_k \,.$$

Let us show that the polynomial $q_N = P_N u' - (P_N u)'$ has the form

$$q_N = \begin{cases} \hat{u}_N^{(1)} \phi_0^N + \hat{u}_{N+1}^{(1)} \phi_1^N & \text{if } N \text{ is even} \,, \\ \hat{u}_{N+1}^{(1)} \phi_0^N + \hat{u}_N^{(1)} \phi_1^N & \text{if } N \text{ is odd} \,, \end{cases} \tag{5.6.7}$$

where $\phi_0^N = \sum_{\substack{k=0 \\ k \text{ even}}}^{N} (1/c_k) T_k$ and $\phi_1^N = \sum_{\substack{k=1 \\ k \text{ odd}}}^{N} T_k$. We can assume first that $u$ is continuous with all its derivatives in $[-1,1]$, so that (see (2.4.22))

$$c_k \hat{u}_k^{(1)} = 2 \sum_{\substack{p=k+1 \\ p+k \text{ odd}}}^{\infty} p \hat{u}_p \,, \quad k = 0, 1, 2, \dots \,.$$

The series is absolutely convergent, since each $\hat{u}_p$ decays faster than any power of $1/p$ (this follows from (5.5.9)). Still using (2.4.22) we get

$$(P_N u)' = \sum_{k=0}^{N-1} \hat{v}_k T_k \quad \text{with} \quad c_k \hat{v}_k = 2 \sum_{\substack{p=k+1 \\ p+k \text{ odd}}}^{N} p \hat{u}_p \,;$$

thus,

$$c_k (\hat{u}_k^{(1)} - \hat{v}_k) = \begin{cases} 2 \displaystyle\sum_{\substack{p=N+2 \\ p+N \text{ even}}}^{\infty} p \hat{u}_p = \hat{u}_{N+1}^{(1)} & \text{if } k+N \text{ is odd} \,, \\ 2 \displaystyle\sum_{\substack{p=N+1 \\ p+N \text{ odd}}}^{\infty} p \hat{u}_p = \hat{u}_N^{(1)} & \text{if } k+N \text{ is even} \,; \end{cases}$$

whence, the result (5.6.7) if $u$ is smooth. Next, we remove this assumption. If $u$ is just in $H^1_w(-1,1)$, it can be approximated by a sequence of infinitely differentiable functions $u_n$ (see (A.11.b)), for which (5.6.7) holds. Then we can pass to the limit as $n \to \infty$, since both sides of (5.6.7) are continuous in the norm of $H^1_w(-1,1)$.

From estimate (5.5.9) it follows that

$$|\hat{u}^{(1)}_{N+1}| \leq \|u' - P_N u'\|_{L^2_w(-1,1)} \leq C N^{1-m} |u|_{H^{m;N}_w(-1,1)} ,$$

and similarly for $\hat{u}^{(1)}_N$. On the other hand,

$$\|\phi_0^N\|^2_{L^2_w(-1,1)} = \sum_{\substack{k=0 \\ k \text{ even}}}^{N} \frac{2}{c_k \pi} \simeq N , \qquad \|\phi_1^N\|^2_{L^2_w(-1,1)} = \sum_{\substack{k=1 \\ k \text{ odd}}}^{N} \frac{2}{\pi c_k} \simeq N .$$

Thus, noting that $\phi_0^N$ and $\phi_1^N$ are orthogonal, we have

$$\|P_N u' - (P_N u)'\|_{L^2_w(-1,1)} \leq C N^{(3/2)-m} |u|_{H^{m;N}_w(-1,1)} ; \tag{5.6.8}$$

whence, (5.5.11) follows by (5.6.6).

One can check that the exponent of $N$ in (5.5.11) is optimal, in the sense that one cannot expect a faster decay of the error for all $u \in H^m_w(-1,1)$.

The proof of (5.4.18) is similar, using now the expression (2.3.18) for the Legendre coefficients of the first derivative. □

PROOF OF (5.4.22), PROOF OF (5.5.14) IN THE CASE $k = 1$. Let us first deal with (5.5.14) for $k = 1$. Let us set

$$V = \left\{ v \in H^1_w(-1,1) \mid \hat{v}_0 = \frac{1}{\pi} \int_{-1}^{1} v T_0 w \, \mathrm{d}x = 0 \right\} . \tag{5.6.9}$$

$V$ is a Hilbert space for the inner product $[u,v]_w$ defined in (5.5.18). Actually, if $u \in V$, there exists at least one point $\xi \in (-1,1)$ where $v(\xi) = 0$. Hence, the Poincaré inequality (A.13) holds, and $\|v\|_V = [v,v]_w^{1/2} = \|v'\|_{L^2_w(-1,1)}$ is a norm equivalent to the standard norm $\|v\|_{H^1_w(-1,1)}$. For any $u \in H^1_w(-1,1)$, let us define the polynomial

$$u^N(x) = \alpha + \int_{-1}^{x} (P_{N-1} u')(s) \, \mathrm{d}s . \tag{5.6.10}$$

As usual, $P_{N-1} v$ is the truncation of degree $N-1$ of the Chebyshev series of $v$. The constant $\alpha$ is chosen in such a way that $\left(u^N\right)^{\wedge}_0 = \hat{u}_0$. Then by (5.5.9) it follows that

$$\|u - u^N\|_V = \|u' - P_{N-1} u'\|_{L^2_w(-1,1)} \leq C N^{1-m} |u|_{H^{m;N}_w(-1,1)} . \tag{5.6.11}$$

The result (5.5.14) for $k = 1$ follows, noting that

$$\|u - P_N^1 u\|_{H_w^1(-1,1)} \leq \|u - v\|_{H_w^1(-1,1)} \quad \text{for all } v \in \mathbb{P}_N \, .$$

The proof of (5.4.22) follows the same guidelines. □

In order to prove (5.4.23) and (5.5.14) for $k = 0$, we need the following regularity result. Let $w$ denote the Legendre weight or the Chebyshev weight.

**Lemma 5.2.** *For each $g \in L_w^2(-1,1)$, there exists a unique $\psi \in H_w^1(-1,1)$ such that*

$$\int_{-1}^{1} (\psi' v' + \psi v) w \, \mathrm{d}x = \int_{-1}^{1} g v w \, \mathrm{d}x \qquad \text{for all } v \in H_w^1(-1,1) \, . \quad (5.6.12)$$

*Moreover, $\psi \in H_w^2(-1,1)$, and there is a constant $C > 0$ such that*

$$\|\psi\|_{H_w^2(-1,1)} \leq C \|g\|_{L_w^2(-1,1)} \, . \quad (5.6.13)$$

PROOF. Since the left-hand side of (5.6.12) is the inner product of $H_w^1(-1,1)$, the existence and uniqueness of $\psi$ follows from the Riesz representation theorem (see (A.1.d)). Choosing $v = \psi$ in (5.6.12), we get

$$\|\psi\|_{H_w^1(-1,1)} \leq \|g\|_{L_w^2(-1,1)} \, . \quad (5.6.14)$$

Letting $v$ vary in $\mathscr{D}(-1,1)$ (this space is defined in (A.10)), we obtain from (5.6.12) that

$$-(\psi' w)' = (g - \psi) w \quad \text{in the sense of distributions} \quad (5.6.15)$$

(see (A.10.a)). In the Legendre case, this identity together with (5.6.14) immediately implies (5.6.13), and the proof is finished.

So, from now on, we assume that $w$ is the Chebyshev weight. At first, we show that $\psi' w$ is continuous in $[-1,1]$. Indeed, for any $x_1, x_2 \in (-1,1)$ it follows by (5.6.15) and the Cauchy-Schwarz inequality (see (A.2)) that

$$\begin{aligned} |(\psi' w)(x_1) - (\psi' w)(x_2)| &= \left| \int_{x_1}^{x_2} (g - \psi) w \, \mathrm{d}x \right| \\ &\leq \|g - \psi\|_{L_w^2(-1,1)} |\arccos x_2 - \arccos x_1|^{1/2} \, . \end{aligned}$$

Hence, $(\psi' w)(\pm 1)$ makes sense. Multiplying (5.6.15) by $v \in H_w^1(-1,1)$ and integrating by parts yields

$$[\psi' w v]_{-1}^{1} = \int_{-1}^{1} \psi' v' w \, \mathrm{d}x - \int_{-1}^{1} (g - \psi) v w \, \mathrm{d}x \quad \text{for all } v \in H_w^1(-1,1) \, .$$

Hence, $\psi' w(-1) = \psi' w(1)$ by (5.6.12). By (5.6.15), $-\psi'' = (g-\psi) - \psi'(w'/w)$. Thus, it remains to prove that $\psi'(w'/w) \in L^2_w(-1,1)$. Since $w'/w = xw^2$, we have

$$\int_{-1}^{1} (\psi' w'/w)^2 w \, dx \le \int_{-1}^{1} (\psi')^2 w^5 \, dx \,.$$

Moreover,

$$\begin{aligned}
\int_{-1}^{0} (\psi')^2 w^5 \, dx &= \int_{-1}^{0} \left[ \int_{-1}^{x} (\psi' w)' \, d\xi \right]^2 w^3 \, dx \\
&= \int_{-1}^{0} \left[ w^2 \int_{-1}^{x} (\psi - g) w \, d\xi \right]^2 w^{-1} \, dx \\
&\le C \int_{-1}^{0} \left[ \frac{1}{1+x} \int_{-1}^{x} (\psi - g) w \, d\xi \right]^2 \sqrt{1+x} \, dx \,.
\end{aligned}$$

Using the Hardy inequality (A.14) with $\alpha = 1/2$, $a = -1$ and $b = 0$, we obtain

$$\int_{-1}^{0} (\psi')^2 w^5 \, dx \le C \int_{-1}^{0} (\psi - g)^2 w \, dx \,.$$

Similarly, we can prove that $\int_0^1 (\psi')^2 w^5 \, dx \le C \int_0^1 (\psi - g)^2 w \, dx$. Therefore, we conclude that $\psi'' \in L^2_w(-1,1)$, with

$$\|\psi''\|_{L^2_w(-1,1)} \le C \left( \|\psi\|_{L^2_w(-1,1)} + \|g\|_{L^2_w(-1,1)} \right) \,.$$

This, using (5.6.14), gives (5.6.13). □

Proof of (5.4.23), proof of (5.5.14) in the case $k = 0$. Let $w$ denote again the Legendre weight or the Chebyshev weight. We use a well-known duality argument, the so-called Aubin-Nitsche trick, based on the identity

$$\|u - P^1_N u\|_{L^2_w(-1,1)} = \sup_{\substack{g \in L^2_w(-1,1) \\ g \ne 0}} \frac{\int_{-1}^{1} (u - P^1_N u) g w \, dx}{\|g\|_{L^2_w(-1,1)}} \,.$$

Let $\psi$ be the solution of (5.6.12) corresponding to a given $g$. Then, choosing $v = u - P^1_N u$ in (5.6.12) and recalling the definition of $P^1_N$, we get

$$\begin{aligned}
\int_{-1}^{1} (u - P^1_N u) g w \, dx &= \int_{-1}^{1} \left[ \psi' (u - P^1_N u)' + \psi (u - P^1_N u) \right] w \, dx \\
&= \int_{-1}^{1} \left[ (\psi - P^1_N \psi)' (u - P^1_N u)' + (\psi - P^1_N \psi)(u - P^1_N u) \right] w \, dx \,.
\end{aligned}$$

The Cauchy-Schwarz inequality, estimates (5.4.23) or (5.5.14) with $k = 1$, and (5.6.13) yield

$$\left|\int_{-1}^{1}(uP_N^1 u)gw\,\mathrm{d}x\right| \le \|\psi - P_N^1\psi\|_{H_w^1(-1,1)}\|u - P_N^1 u\|_{H_w^1(-1,1)}$$
$$\le CN^{-1}\|\psi\|_{H_w^2(-1,1)}\|u - P_N^1 u\|_{H_w^1(-1,1)}$$
$$\le CN^{-1}\|g\|_{L_w^2(-1,1)}\|u - P_N^1 u\|_{H_w^1(-1,1)}\,.$$

Then,

$$\|u - P_N^1 u\|_{L_w^2(-1,1)} \le CN^{-1}\|u - P_N^1 u\|_{H_w^1(-1,1)}\,.$$

Hence, the desired result follows again using (5.4.23) or (5.5.14) with $k = 1$. □

PROOF OF (5.4.30) WITH $k = 1$, PROOF OF (5.5.19). As above, let $w$ denote the Legendre weight or the Chebyshev weight. Let us define $u^N$ as in (5.6.10), now with $\alpha = 0$. Next, define

$$R_N u(\xi) = \int_{-1}^{\xi}\left(P_{N-1}u' - \frac{1}{2}u^N(1)\right)\mathrm{d}x\,,$$

so that $R_N u \in \mathbb{P}_N^0$. We have, by the triangle inequality,

$$\|u' - (R_N u)'\|_{L_w^2(-1,1)} \le \|u' - P_{N-1}u'\|_{L_w^2(-1,1)} + \frac{1}{2}\left(\int_{-1}^{1} w\,\mathrm{d}x\right)^{1/2}|u^N(1)|\,.$$

On the other hand, by the Cauchy-Schwarz inequality one has

$$|u^N(1)| = |u(1) - u^N(1)| = \left|\int_{-1}^{1}(u' - P_{N-1}u')\,\mathrm{d}x\right|$$
$$\le \left(\int_{-1}^{1} w^{-1}\,\mathrm{d}x\right)^{1/2}\|u' - P_{N-1}u'\|_{L_w^2(-1,1)}\,.$$

Using (5.4.11) in the Legendre case or (5.5.9) in the Chebyshev case, and the two previous inequalities, we obtain

$$\|u' - (R_N u)'\|_{L_w^2(-1,1)} \le CN^{1-m}|u|_{H_w^{m;N}(-1,1)}\,.$$

Finally, estimates (5.4.30) with $k = 1$ or (5.5.19) follow since $P_N^{1,0}u$ is the polynomial of best approximation of $u$ in the norm associated to the $H_0^1$-inner product (5.4.28) in the Legendre case, or the $H_{w,0}^1$-inner product (5.5.18) in the Chebyshev case. □

PROOF OF (5.4.30) WITH $k = 0$, PROOF OF (5.5.20). As above, let $w$ denote the Legendre weight or the Chebyshev weight. We define $u^N \in \mathbb{P}_N^0$ to be the solution of the problem

$$a(u - u^N, v) = 0 \quad \text{for all } v \in \mathbb{P}_N^0 \,, \tag{5.6.16}$$

where $a(\phi, \psi) = \int_{-1}^{1} \phi'(\psi w)' \, dx$ (see (7.1.13)). This is precisely the polynomial $P_N^{1,0} u$ defined in (5.4.29) in the Legendre case, or the polynomial defined in (5.5.21) in the Chebyshev case. It is shown in Sect. 7.1 that the bilinear form $a(\phi, \psi)$ defined on $H^1_{w,0}(-1,1) \times H^1_{w,0}(-1,1)$ satisfies the hypotheses of the Lax-Milgram Theorem (A.5) (see (7.1.11) and (7.1.12)). Then the existence and uniqueness of $u^N$ is assured. Moreover, by the coercivity and the continuity of $a$ we get

$$\begin{aligned} \|u - u^N\|^2_{H^1_w(-1,1)} &\le C_1 a(u - u^N, u - u^N) \\ &= C_1 a(u - u^N, u - v) \qquad \text{(by (5.6.16))} \\ &\le C_2 \|u - u^N\|_{H^1_w(-1,1)} \|u - v\|_{H^1_w(-1,1)} \end{aligned}$$

for all $v \in \mathbb{P}_N^0$. Thus,

$$\|u - u^N\|_{H^1_w(-1,1)} \le C_2 \inf_{v \in \mathbb{P}_N} \|u - v\|_{H^1_w(-1,1)} \,. \tag{5.6.17}$$

Estimate (5.5.20) for $k = 1$ follows now from (5.5.19).

In order to prove both estimates (5.4.30) and (5.5.20) for $k = 0$, we use an Aubin-Nitsche duality argument similar to the one we have used to prove (5.5.14). We have

$$\|u - u^N\|_{L^2_w(-1,1)} = \sup_{\substack{g \in L^2_w(-1,1) \\ g \neq 0}} \frac{\int_{-1}^{1} (u - u^N) g w \, dx}{\|g\|_{L^2_w(-1,1)}} \,. \tag{5.6.18}$$

For each fixed $g \in L^2_w(-1,1), g \neq 0$, let $\psi = \psi(g) \in H^1_{w,0}(-1,1)$ be the solution of the problem

$$a(v, \psi) = \int_{-1}^{1} g v w \, dx \quad \text{for all } v \in H^1_{w,0}(-1,1) \,, \tag{5.6.19}$$

which is uniquely defined since the form $a(u, v)$ is symmetric in the Legendre case, whereas in the Chebyshev case the transpose form $a^T(u, v) = a(v, u)$ satisfies again the hypotheses of the Lax-Milgram theorem. A very technical argument (in the Chebyshev case) allows us to prove that $\psi \in H^2_w(-1,1)$ and

$$\|\psi\|_{H^2_w(-1,1)} \le C \|g\|_{L^2_w(-1,1)} \,. \tag{5.6.20}$$

Then, using (5.6.19) and the definitions (5.4.29) or (5.5.21), we obtain, for each $\psi^N \in \mathbb{P}_N^0$,

$$\begin{aligned} \left| \int_{-1}^{1} (u - u^N) g w \, dx \right| &= |a(u - u^N, \psi)| = |a(u - u^N, \psi - \psi^N)| \\ &\le C \|u - u^N\|_{H^1_w(-1,1)} \|\psi - \psi^N\|_{H^1_w(-1,1)} \,. \end{aligned}$$

Using (5.4.30) or (5.5.20) with $k = 1$ for both $u$ and $\psi$ yields

$$\left| \int_{-1}^{1} (u - u_N) g w \, \mathrm{d}x \right| \leq C N^{-m} \|\psi\|_{H^2_w(-1,1)} |u|_{H^{m;N}_w(-1,1)} .$$

Now estimates (5.4.30) and (5.5.20) with $k = 0$ follow using (5.6.18) and (5.6.20). □

PROOF OF (5.4.33) AND (5.5.22). Let $w$ denote the Legendre or Chebyshev weight. By a rather technical argument, which can be found in Bernardi and Maday (1997), Chap. 3, one first proves the bound

$$\|I_N v\|_{L^2_w(-1,1)} \leq C \left( \|v\|_{L^2_w(-1,1)} + N^{-1} \|v'\|_{L^2_w(-1,1)} \right) \tag{5.6.21}$$

for all $v \in H^1_{0,w}(-1,1)$, where $I_N$ is the interpolation operator at any of the families of Gaussian points. Next, one applies this bound to the function $v = u - u^N \in H^1_{0,w}(-1,1)$, where $u^N$ is any polynomial matching the boundary values of $u$ and providing an optimal approximation to $u$ in the $H^1_w(-1,1)$-norm, i.e.,

$$\|u - u^N\|_{H^k_w(-1,1)} \leq C N^{k-m} |u|_{H^{m;N}_w(-1,1)} , \qquad 0 \leq k \leq 1 , \quad m \geq 1 .$$

Such a polynomial exists, as indicated in Sects. 5.4.2 and 5.5.2. Noting that $I_N u^N = u^N$, we obtain the desired estimates via the triangle inequality

$$\|u - I_N u\|_{L^2_w(-1,1)} \leq \|u - u^N\|_{L^2_w(-1,1)} + \|I_N(u - u^N)\|_{L^2_w(-1,1)} .$$

□

PROOF OF (5.4.35) AND (5.5.25). Invoking again the triangle inequality as in the above proof, it is enough to estimate $\|(I_N u)' - (u^N)'\|_{L^2_w(-1,1)}$. At first, one applies the inverse inequality (5.4.7) or (5.5.6) to the polynomial $I_N u - u^N \in \mathbb{P}^0_N$. Next, one uses, with $v = I_N u - u^N$, the bound

$$\| (I_N v)/\sqrt{\eta} \|_{L^2_w(-1,1)} \leq C \left( \| v/\sqrt{\eta} \|_{L^2_w(-1,1)} + N^{-1} \|v'\|_{L^2_w(-1,1)} \right)$$

for all $v \in H^1_{0,w}(-1,1)$, whose proof can be found again in Bernardi and Maday (1997), Chapter 3. Finally, the resulting right-hand side is bounded in the desired way, thanks to the approximation results for $u^N$. □

PROOF OF (5.5.22) AND (5.5.23). We give here a self-contained proof of these results. We consider the Gauss-Lobatto interpolation points $x_j = \cos(\pi j/N)$, for $j = 0, \ldots, N$. The proof for the other two sets of points (Gauss and Gauss-Radau one) is similar. We still make use of the mapping (5.6.2). We define

$$\tilde{S}_N = \left\{ v : (0, 2\pi) \longrightarrow \mathbb{C} \,\middle|\, v(\theta) = \sum_{k=-N}^{N} \hat{v}_k e^{ik\theta}, \hat{v}_N = \hat{v}_{-N} \right\} ,$$

and, for every $v \in C^0([0, 2\pi])$, we denote by $I_N^* v$ the unique function of $\tilde{S}_N$ that interpolates $v$ at the points $\theta_j = \pi j/N$, for $j = 0, \ldots, 2N$. Note that these points are symmetrically distributed around the point $\theta = \pi$. Moreover, for each continuous function $u : [-1, 1] \to \mathbb{R}$, both $u^*$ and $(I_N u)^*$ are even functions with respect to the point $\theta = \pi$. Therefore,

$$(I_N u)^* = I_N^* u^* \in \tilde{S}_N . \tag{5.6.22}$$

Now we use the error estimate (5.1.16) for the Fourier interpolation and we obtain, by (5.6.3) and (5.6.4),

$$\|u - I_N u\|_{L^2_w(-1,1)} = \frac{1}{\sqrt{2}} \|u^* - I_N^* u^*\|_{L^2(0,2\pi)} \leq C N^{-m} \|u\|_{H^m_w(-1,1)} , \tag{5.6.23}$$

i.e., (5.5.22). For $m \geq 1$, the inverse inequality (5.5.4) yields

$$\|u - I_N u\|_{H^l_w(-1,1)} \leq \|u - P_N u\|_{H^l_w(-1,1)} + C N^{2l} \|P_N u - I_N u\|_{L^2_w(-1,1)} .$$

Now (5.5.23) follows using (5.5.9) and (5.6.23). □

## 5.7 Other Polynomial Approximations

The orthogonal systems described so far have been the ones most commonly used in building up spectral approximations to partial differential equations. However, other relevant sets of orthogonal polynomials guarantee spectral accuracy as well.

### 5.7.1 Jacobi Polynomials

The Jacobi polynomials $\{P_k^{\alpha,\beta}(x),\ k = 0, 1, 2, \ldots\}$ have been introduced in Sect. 2.5. They are the eigenfunctions of the singular Sturm-Liouville problem (2.2.1) where $p(x) = (1-x)^{\alpha+1}(1+x)^{\beta+1}$ ($\alpha$ and $\beta > -1$), $q(x) \equiv 0$ and $w(x) = (1-x)^\alpha(1+x)^\beta$. The eigenvalue whose eigenfunction is $P_k^{(\alpha,\beta)}$ is $\lambda_k = k(k + \alpha + \beta + 1)$. The Legendre polynomials correspond to the choice $\alpha = \beta = 0$, while the Chebyshev polynomials of the second kind correspond to $\alpha = \beta = -1/2$.

Jacobi polynomials for other choices of $\alpha$ and $\beta$ have asymptotic approximation properties similar to those of Legendre or Chebyshev polynomials. Although a fast transform is not available for them, they can lead to small matrix bandwidths in Galerkin methods (see Sect. 3.4.3). The mathematical difficulties in the analysis of the Chebyshev methods that arise from the singularity of the Chebyshev weight are shared by the Jacobi methods.

We anticipate that some Chebyshev approximations to hyperbolic problems (see Sect. 7.6.2) are stable in some weighted norms corresponding to a Jacobi weight. Furthermore, Jacobi polynomials enter into the construction of polynomial bases on triangles and related domains (see Sect. 2.9).

The best approximation and interpolation error estimates given in Sect. 5.5 for the Chebyshev case hold unchanged when $w$ denotes a Jacobi weight. We refer to Bernardi and Maday (1997) (see also Bernardi and Maday (1992a) for the details of proofs in the ultraspherical case). Most of the results hold for $-1 < \alpha, \beta < 1$, as their proofs require the integrability of both $w$ and $w^{-1}$. However, the inverse inequalities (5.5.9) or (5.5.14) with $k = 1$ hold for all $\alpha, \beta > -1$. On the other hand, (5.5.25) holds under the stronger assumption $m \geq 2$ if $\alpha$ or $\beta$ are strictly positive.

Sharper estimates that involve sophisticated weighted Sobolev spaces and that exhibit the explicit dependence on the Jacobi parameters $\alpha$ and $\beta$ have been given by Guo and Wang (2004).

### 5.7.2 Laguerre and Hermite Polynomials

The Laguerre and Hermite polynomials have been introduced in Sect. 2.6. The Laguerre polynomials $\{l_k(x) = l_k^{(0)}(x),\ k = 0, 1, \dots\}$ are the eigenfunctions of the singular Sturm-Liouville problem (2.2.1) on the semi-infinite interval $\mathbb{R}_+ = (0, +\infty)$ with $p(x) = xe^{-x}$, $q(x) \equiv 0$ and $w(x) = e^{-x}$. The eigenvalue corresponding to $l_k$ is $\lambda_k = k$. The Hermite polynomials $\{H_k(x),\ k = 0, 1, \dots\}$ are the eigenfunctions of (2.2.1) on the real line $\mathbb{R} = (-\infty, +\infty)$ with $p(x) = e^{-x^2}$, $q(x) \equiv 0$ and $w(x) = e^{-x^2}$. The eigenvalue corresponding to $H_k$ is $\lambda_k = 2k$.

By adapting the arguments of Sect. 5.2 to the case of an unbounded interval, one can prove that the coefficients of the Laguerre (or Hermite, resp.) expansion of a smooth function defined over $\mathbb{R}_+$ (over $\mathbb{R}$, resp.) decay faster than algebraically, provided all the derivatives of the function are square-integrable with respect to the weight $w$.

Laguerre approximations in weighted Sobolev spaces on the half-line $\mathbb{R}_+$ were first investigated by Maday, Pernaud-Thomas and Vandeven (1985), Coulaud, Funaro and Kavian (1990), Funaro (1991). The basic weighted space is

$$L^2_w(\mathbb{R}_+) = \left\{v : \mathbb{R}_+ \to \mathbb{R} \text{ measurable} \mid \|v\|_{L^2_w(\mathbb{R}_+)} < +\infty\right\} , \qquad (5.7.1)$$

where

$$\|v\|_{L^2_w(\mathbb{R}_+)} = \left(\int_{\mathbb{R}_+} v^2(x) e^{-x} \mathrm{d}x\right)^{1/2} .$$

It generates the family of weighted Sobolev spaces

$$H^m_w(\mathbb{R}_+) = \left\{v \in L^2_w(\mathbb{R}_+) \mid \|v\|_{H^m_w(\mathbb{R}_+)} < +\infty\right\} , \qquad m \geq 0 , \qquad (5.7.2)$$

where

$$\|v\|_{H^m_w(\mathbb{R}_+)} = \left( \sum_{j=0}^{m} \|v^{(j)}\|^2_{L^2_w(\mathbb{R}_+)} \right)^{1/2} .$$

A related family of weighted Sobolev spaces is useful, namely,

$$H^m_{w;\alpha}(\mathbb{R}_+) = \left\{ v \in L^2_w(\mathbb{R}_+) \mid (1+x)^{\alpha/2} \in H^m_w(\mathbb{R}_+) \right\} , \qquad m \geq 0 , \quad (5.7.3)$$

equipped with the natural norm $\|v\|_{H^m_{w;\alpha}(\mathbb{R}_+)} = \|(1+x)^{\alpha/2} v\|_{H^m_w(\mathbb{R}_+)}$.

For each $u \in L^2_w(\mathbb{R}_+)$, let $P_N u \in \mathbb{P}_N$ be the truncation of its Laguerre series, i.e., the orthogonal projection of $u$ upon $\mathbb{P}_N$ with respect to the inner product of $L^2_w(\mathbb{R}_+)$:

$$\int_{\mathbb{R}_+} (u - P_N u)\, \phi\, e^{-x}\, dx = 0 \quad \text{for all } \phi \in \mathbb{P}_N .$$

The following error estimate holds for any $m \geq 0$ and $0 \leq k \leq m$:

$$\|u - P_N u\|_{H^k_w(\mathbb{R}_+)} \leq C N^{k - \frac{m}{2}} \|u\|_{H^m_{w;m}(\mathbb{R}_+)} . \qquad (5.7.4)$$

For the orthogonal projection $P^1_N$ upon $\mathbb{P}_N$ in the norm of $H^1_w(\mathbb{R}_+)$, the following estimate holds for $m \geq 1$:

$$\|u - P^1_N u\|_{H^1_w(\mathbb{R}_+)} \leq C N^{\frac{1}{2} - \frac{m}{2}} \|u\|_{H^m_{w;m-1}(\mathbb{R}_+)} ; \qquad (5.7.5)$$

the same result holds for the projection $P^{1,0}_N$ upon $\mathbb{P}^0_N$ (Guo and Shen (2000)).

Concerning interpolation, let us consider the $N+1$ Gauss-Radau points $x_j$, $j = 0, \dots, N$, where $x_0 = 0$ and $x_j$, for $j = 1, \dots, N$, are the zeros of $l'_{N+1}(x)$, the derivative of the $(N+1)$-th Laguerre polynomial. For each continuous function $u$ on $\mathbb{R}_+$, let $I_N u \in \mathbb{P}_N$ be the interpolant of $u$ at the points $x_j$. In order to estimate the interpolation error, let us introduce the the space $H^m_{w;*}(\mathbb{R}_+)$ of the functions such that

$$\|v\|_{H^m_{w;*}(\mathbb{R}_+)} = \left( \sum_{j=0}^{m} \|x^{j/2} v^{(j)}\|^2_{L^2_w(\mathbb{R}_+)} \right)^{1/2}$$

(see Monegato and Mastroianni (1997), who give an equivalent definition, meaningful also for non-integer $m$). Then, for any $m \geq 1$, $0 \leq k \leq m$ and $0 < \epsilon < 1$, one has

$$\|u - I_N u\|_{H^k_w(\mathbb{R}_+)} \leq C_\epsilon N^{k + \frac{1}{2} + \epsilon - \frac{m}{2}} \left( \|u\|^2_{H^m_{w;m}(\mathbb{R}_+)} + \|u\|^2_{H^m_{w;*}(\mathbb{R}_+)} \right)^{1/2} \quad (5.7.6)$$

for all functions $u \in H^m_{w;m}(\mathbb{R}_+) \cap H^m_{w;*}(\mathbb{R}_+)$ (see Xu and Guo (2002), where additional approximation results can be found). Examples of applications to spectral Laguerre discretizations of boundary-value problems in $\mathbb{R}_+$ are provided in the above references. Usually, an appropriate change of unknown function is needed to cast the differential problem into the correct functional setting based on Laguerre-weighted Sobolev spaces.

Hermite approximations can be studied in a similar manner. The basic weighted space $L^2_w(\mathbb{R})$ involves the norm

$$\|v\|_{L^2_w(\mathbb{R})} = \left( \int_{\mathbb{R}} v^2(x) e^{-x^2} \mathrm{d}x \right)^{1/2} .$$

The Sobolev spaces $H^m_w(\mathbb{R})$ are defined as above, with respect to this norm. The $L^2_w$-orthogonal projection operator $P_N$ upon $\mathbb{P}_N$ satisfies the estimate

$$\|u - P_N u\|_{H^k_w(\mathbb{R})} \le C N^{\frac{k}{2} - \frac{m}{2}} \|u\|_{H^m_w(\mathbb{R})} \tag{5.7.7}$$

for all $m \ge 0$ and $0 \le k \le m$ (Guo (1999)). Interestingly, all $H^\ell_w$-orthogonal projection operators $P^\ell_N$ upon $\mathbb{P}_N$, for $\ell \ge 0$, coincide with $P_N$, due to property (2.6.12) of Hermite polynomials. For the interpolation operator $I_N$ at the Hermite-Gauss nodes in $\mathbb{R}$, Guo and Xu (2000) proved the estimate

$$\|u - I_N u\|_{H^k_w(\mathbb{R})} \le C_\epsilon N^{\frac{1}{3} + k - \frac{m}{2}} \|u\|_{H^m(\mathbb{R})} , \tag{5.7.8}$$

for $m \ge 1$ and $0 \le k \le m$.

When dealing with the unbounded intervals $\mathbb{R}_+$ and $\mathbb{R}$, an alternative to polynomials as approximating functions is given by functions that are the product of a polynomial times the natural weight for the interval. Thus, one uses the Laguerre functions $\psi(x) = \phi(x)e^{-x}$ in $\mathbb{R}_+$ or the Hermite functions $\psi(x) = \phi(x)e^{-x^2}$ in $\mathbb{R}$, where $\phi$ is any polynomial in $\mathbb{P}_N$. The behavior at infinity of the function to be approximated may suggest such a choice. We refer, e.g., to Funaro and Kavian (1990) and to Guo and Shen (2003) for the corresponding approximation results and for applications.

## 5.8 Approximation in Cartesian-Product Domains

We shall now extend to several space dimensions some of the approximation results we presented in the previous sections for a single spatial variable. The three expansions of Fourier, Legendre and Chebyshev will be considered. However, we will only be concerned with those Sobolev-type norms that are most frequently applied to the convergence analysis of spectral methods.

### 5.8.1 Fourier Approximations

Let us consider the domain $\Omega = (0, 2\pi)^d$ in $\mathbb{R}^d$, for $d = 2$ or 3, and denote an element of $\mathbb{R}^d$ by $\mathbf{x} = (x_1, \dots, x_d)$. The space $L^2(\Omega)$, as well as the Sobolev spaces $H^m_p(\Omega)$ of periodic functions, are defined in Appendix A (see (A.9.h) and (A.11.d)). Since $\Omega$ is the Cartesian product of $d$ copies of the interval $(0, 2\pi)$, it is natural to use, as an orthogonal system in $L^2(\Omega)$, the tensor product of the trigonometric system in $L^2(0, 2\pi)$. Thus, we set

$$\phi_{\mathbf{k}}(\mathbf{x}) = e^{i\,\mathbf{k}\cdot\mathbf{x}} , \quad \text{with } \mathbf{k}\cdot\mathbf{x} = k_1x_1 + \cdots + k_dx_d , \tag{5.8.1}$$

by analogy with (2.1.1), and

$$S_N = \text{span}\{\phi_{\mathbf{k}}(\mathbf{x}) \mid -N \leq k_j \leq N-1 \text{ for } j = 1, \ldots, d\} . \tag{5.8.2}$$

Moreover, we still denote by $P_N$ the orthogonal projection operator from $L^2(\Omega)$ upon $S_N$. Then, for any $u \in L^2(\Omega)$ we have

$$P_N u = \sum_{\|\mathbf{k}\| \lesssim N} \hat{u}_{\mathbf{k}} \phi_{\mathbf{k}} , \quad \hat{u}_{\mathbf{k}} = \left(\frac{1}{2\pi}\right)^d \int_\Omega u(\mathbf{x})\overline{\phi_{\mathbf{k}}(\mathbf{x})}\, \mathrm{d}x , \tag{5.8.3}$$

where the above summation is extended to all $\mathbf{k} \in \mathbb{Z}^d$ such that $-N \leq k_j \leq N-1$, for $j = 1, \ldots, d$. The following result provides an estimate in all Sobolev norms for the remainder of the Fourier series of $u$:

$$\|u - P_N u\|_{H^l(\Omega)} \leq CN^{l-m}|u|_{H^m(\Omega)} \quad \text{for } 0 \leq l \leq m , \tag{5.8.4}$$

where $|u|_{H^m(\Omega)} = \left(\sum_{j=1}^d \|D_j^m u\|^2_{L^2(\Omega)}\right)^{1/2}$. It can be obtained for all $u \in H_p^m(\Omega)$ by a proof that mimics the one of (5.1.10).

Concerning interpolation, let us introduce the $(2N)^d$ points

$$\mathbf{x} = (x_{j_1}, \ldots, x_{j_d}) \qquad \text{where } x_j = \frac{\pi}{N} j , \tag{5.8.5}$$

with $0 \leq j_m \leq 2N-1$ for $m = 1, \ldots, d$. For every function $u$ continuous in the closure of $\Omega$, we denote by $I_N u$ the function of $S_N$ interpolating $u$ at the points (5.8.5). By analogy with the one-dimensional case (cf. (2.1.28) and (2.1.32)) one has

$$I_N u = \sum_{\|\mathbf{k}\| \lesssim N} \tilde{u}_{\mathbf{k}} \phi_{\mathbf{k}} , \quad \tilde{u}_{\mathbf{k}} = \left(\frac{1}{2N}\right)^d \sum_{\mathbf{j}} u(x_{\mathbf{j}})\overline{\phi_{\mathbf{k}}(x_{\mathbf{j}})} , \tag{5.8.6}$$

where $\tilde{u}_{\mathbf{k}}$ is the $k$-th discrete Fourier coefficient of $u$. The error estimate for this interpolation is

$$\|u - I_N u\|_{H^l(\Omega)} \leq CN^{l-m}|u|_{H^m(\Omega)} \quad \text{for } 0 \leq l \leq m . \tag{5.8.7}$$

It holds for all $u \in H_p^m(\Omega)$ with $m > d/2$. For $l = 0$ the proof can be done as for (5.1.16) by mapping $\Omega$ onto the reference domain $\Omega_N = (0, 2\pi N)^d$. For $l > 0$ the estimate (5.8.7) is obtained using the corresponding one for $l = 0$, the estimate (5.8.4), and the following inverse inequality: for $0 \leq k \leq m$,

$$\|\phi\|_{H^m(\Omega)} \leq CN^{m-k}\|\phi\|_{H^k(\Omega)} \qquad \text{for all } \phi \in S_N , \tag{5.8.8}$$

which extends (5.1.5) for $p = 2$.

### 5.8.2 Legendre Approximations

We consider now the domain $\Omega = (-1,1)^d$ in $\mathbb{R}^d$ with $d = 2$ or 3, and we still denote an element of $\mathbb{R}^d$ by $\mathbf{x} = (x_1, \ldots, x_d)$. We denote by $L^2(\Omega)$ the space of square-integrable functions in $\Omega$ and by $H^m(\Omega)$ the corresponding Sobolev space of order $m$ (see (A.9.h) and (A.11.a)). The tensor products of the Legendre polynomials,

$$\phi_{\mathbf{k}}(\mathbf{x}) = L_{k_1}(x_1) \cdots L_{k_d}(x_d) \qquad \text{for } \mathbf{k} = (k_1, \ldots, k_d) \in \mathbb{N}^d\,, \tag{5.8.9}$$

form an orthogonal basis for $L^2(\Omega)$. Let $\mathbb{P}_N = \mathbb{P}_N(\Omega)$ be the space of all algebraic polynomials of degree up to $N$ in each variable $x_i$ for $i = 1, \ldots, d$. Denote by $P_N$ the orthogonal projection operator from $L^2(\Omega)$ upon $\mathbb{P}_N$, so that (see (2.3.9))

$$P_N u = \sum_{\|\mathbf{k}\| \le N} \hat{u}_{\mathbf{k}} \phi_{\mathbf{k}}, \quad \hat{u}_{\mathbf{k}} = \prod_{i=1}^{d} \left(k_i + \frac{1}{2}\right) \cdot \int_\Omega u(\mathbf{x}) \phi_{\mathbf{k}}(\mathbf{x})\, d\mathbf{x} \tag{5.8.10}$$

for all $u \in L^2(\Omega)$. We set here and in the sequel $\|\mathbf{k}\| = \max_{1 \le i \le d} k_i$ .

Concerning the truncation error, the following estimate holds for all $u \in H^m(\Omega), m \ge 0$:

$$\|u - P_N u\|_{H^l(\Omega)} \le C N^{\sigma(l) - m} |u|_{H^{m;N}(\Omega)} \quad 0 \le l \le m\,, \tag{5.8.11}$$

where $\sigma(l) = 0$ if $l = 0$ and $\sigma(l) = 2l - \frac{1}{2}$ for $l > 0$. The seminorm on the right-hand side is defined as

$$|u|_{H^{m;N}(\Omega)} = \left( \sum_{k=\min(m,N+1)}^{m} \sum_{i=1}^{d} \|D_i^k u\|^2_{L^2(\Omega)} \right)^{1/2} \tag{5.8.12}$$

(compare with (5.4.10)). Note that only pure derivatives in each spatial direction appear in this expression.

We consider now the operators of orthogonal projection for the inner product of the Sobolev spaces $H^1(\Omega)$ and $H_0^1(\Omega)$ (the latter space being defined in (A.11.c)). By analogy with (5.4.20), we set

$$\begin{aligned} &P_N^1 : H^1(\Omega) \longrightarrow \mathbb{P}_N \quad \text{such that} \\ &((P_N^1 u, \phi)) = ((u, \phi)) \quad \text{for all } \phi \in \mathbb{P}_N\,, \end{aligned} \tag{5.8.13}$$

where $((u,v)) = \int_\Omega (\nabla u \cdot \nabla v + uv)\, dx$ is the inner product of $H^1(\Omega)$. Moreover, denoting by $\mathbb{P}_N^0$ the subspace of $\mathbb{P}_N$ of those polynomials vanishing at the boundary of $\Omega$, we set, by analogy with (5.4.29),

$$\begin{aligned} &P_N^{1,0} : H_0^1(\Omega) \longrightarrow \mathbb{P}_N^0 \quad \text{such that} \\ &[P_N^{1,0} u, \phi] = [u, \phi] \quad \text{for all } \phi \in \mathbb{P}_N^0\,, \end{aligned} \tag{5.8.14}$$

where $[u,v] = \int_\Omega \nabla u \cdot \nabla v \, d\mathbf{x}$ is the inner product of $H_0^1(\Omega)$ (see (A.11.c)). For all $u \in H^m(\Omega)$ (respectively, $H^m(\Omega) \cap H_0^1(\Omega)$) with $m \geq 1$, set $u^N = P_N^1 u$ (respectively $P_N^{1,0} u$). Then the following estimates hold:

$$\left\| u - u^N \right\|_{H^k(\Omega)} \leq C N^{k-m} |u|_{H^{m;N}(\Omega)} , \qquad 0 \leq k \leq 1 . \tag{5.8.15}$$

These estimates are optimal and generalize to more dimensions those given in Sect. 5.4.2 for a single spatial variable. They can be extended to higher order Sobolev norms and to cover different kinds of boundary behavior of $u$. This very general result reads as follows:

Let $l$ and $m$ be two integers such that $0 \leq l \leq m$, and let $\lambda$ be another integer so that $0 \leq \lambda \leq l$. Let $u$ be a function of $H^m(\Omega)$ such that, if $\lambda \geq 1$, $u$ vanishes at the boundary together with its derivatives of order up to $\lambda - 1$. Then there exists a polynomial $u^N \in \mathbb{P}_N$ having the same boundary behavior as $u$ such that

$$\|u - u^N\|_{H^k(\Omega)} \leq C N^{k-m} |u|_{H^{m;N}(\Omega)} \quad \text{for } 0 \leq k \leq l . \tag{5.8.16}$$

We refer to Bernardi, Dauge and Maday (1992) for the case of non-homogeneous boundary conditions.

Finally, we consider multidimensional Legendre interpolation. Let $\{x_j, 0 \leq j \leq N\}$ denote one of the Gauss Legendre quadrature families (2.3.10), (2.3.11), or (2.3.12) on the interval $(-1,1)$. Let us introduce the points in $\bar{\Omega}$:

$$\mathbf{x_j} = (x_{j_1}, \dots, x_{j_d}) \quad \text{for } \mathbf{j} = (j_1, \dots, j_d) \in \mathbb{N}^d, \quad \|\mathbf{j}\| \leq N , \tag{5.8.17}$$

and let us denote by $I_N$ the interpolation operator at these points, i.e., for each continuous function $u$, $I_N u \in \mathbb{P}_N$ satisfies

$$(I_N u)(\mathbf{x_j}) = u(\mathbf{x_j}) \qquad \text{for all } \mathbf{j} \in \mathbb{N}^d, \ \|\mathbf{j}\| \leq N . \tag{5.8.18}$$

We can represent $I_N u$ as follows:

$$I_N u = \sum_{\|\mathbf{k}\| \leq N} \tilde{u}_{\mathbf{k}} \phi_{\mathbf{k}} , \tag{5.8.19}$$

with

$$\tilde{u}_{\mathbf{k}} = (\gamma_{k_1} \cdots \gamma_{k_d})^{-1} \sum_{\mathbf{j}} u(\mathbf{x_j}) \phi_{\mathbf{k}}(\mathbf{x_j}) w_{j_1} \cdots w_{j_d} ,$$

where the $\gamma_k$'s are defined in (2.3.13), and the $w_m$'s are one of the weights (2.3.10)–(2.3.12), according to the choice of the interpolation points. The interpolation error estimate in the $L^2(\Omega)$-norm is

$$\|u - I_N u\|_{L^2(\Omega)} \leq C N^{-m} |u|_{H^{m;N}(\Omega)} , \tag{5.8.20}$$

for all $u \in H^m(\Omega)$, with $m > d/2$. The same result holds if we blend different Gauss Legendre quadrature families along different space directions (e.g., Gauss points in the $x_1$-direction but Gauss-Radau points in the $x_2$-direction).

Estimate (5.8.20) is optimal, since the interpolation error behaves as the best approximatoin error in $\mathbb{P}_N$ for the $L^2(\Omega)$-norm. However, similarly to (5.8.11), the interpolation error is suboptimal in higher Sobolev norms, with one significant exception that occurs when Gauss-Lobatto points are used in all space directions. In this case, we have the optimal estimate

$$\|u - I_N u\|_{H^1(\Omega)} \leq CN^{1-m}|u|_{H^{m;N}(\Omega)} \tag{5.8.21}$$

for all $u \in H^m(\Omega)$ with $m \geq (d+1)/2$. This result is quite important, particularly for the multidomain spectral setting discussed in CHQZ3, Chaps. 5 and 6. Indeed, it allows one to construct a global approximation operator that is optimal in the $H^1(\Omega)$-norm and that appears naturally in second-order boundary-value problems, by simply glueing together local interpolation operators defined on the subdomains.

Finally, let us introduce the bilinear form

$$(f, g)_{N,\Omega} = \sum_{\mathbf{j}} f(\mathbf{x_j})g(\mathbf{x_j})w_{j_1} \cdots w_{j_d} ,$$

which approximates the $L^2(\Omega)$-inner product, $(f, g)_\Omega = \int_\Omega fg$, and which is indeed a discrete inner product in $\mathbb{P}_N$. The one-dimensional estimates (5.3.4a)-(5.3.4b) imply the following general estimate: for all $u \in C^0(\bar{\Omega})$ and $\phi \in \mathbb{P}_N$, one has

$$\begin{aligned} |(u, \phi)_\Omega - (u, \phi)_{N,\Omega}| \leq C(&\|u - P_{N-1}u\|_{L^2(\Omega)} \\ &+ \|u - I_N u\|_{L^2(\Omega)})\|\phi\|_{L^2(\Omega)} . \end{aligned} \tag{5.8.22}$$

### 5.8.3 Mapped Operators and Scaled Estimates

The approximation operators considered above act on functions defined in the Cartesian-product domain $[-1, 1]^d$, $d = 2, 3$; we will refer to this domain as to the *reference domain* and we will indicate it by $\hat{\Omega}$. A smooth, invertible mapping $F$ between $\hat{\Omega}$ and a bounded domain $\Omega \subset \mathbb{R}^d$ (see Sect. 2.8.1 for examples of such transformations) induces a mapping between functions defined in $\hat{\Omega}$ and functions defined in $\Omega$; consequently, approximation operators can be defined in $\Omega$ as images of those defined in $\hat{\Omega}$.

To be precise, let $F : \hat{\Omega} \to \Omega$ be an $m$-times differentiable, invertible mapping such that the determinant $|JF|$ of its Jacobian matrix $JF$ satisfies $|JF| \geq \gamma$ in $\hat{T}$, for some constant $\gamma > 0$. This implies that the inverse mapping $F^{-1}$ is $m$-times differentiable as well. Any real-valued function $\hat{v}$ defined in $\hat{\Omega}$ gives rise to a function $v$ defined in $\Omega$ by setting $v(\mathbf{x}) = \hat{v}(F^{-1}(\mathbf{x}))$ for all $\mathbf{x} \in \Omega$; the inverse transformation is $\hat{v}(\hat{\mathbf{x}}) = v(F(\hat{\mathbf{x}}))$ for all $\hat{\mathbf{x}} \in \hat{\Omega}$. The assumptions on $F$ imply that $v$ is just as differentiable as $\hat{v}$, up to derivatives of order $m$. Now, let $\hat{\mathcal{P}}_N : \mathcal{W}(\hat{\Omega}) \to X_N(\hat{\Omega})$ be one of the approximation operators introduced in the previous subsection; here, $\mathcal{W}(\hat{\Omega})$ is a space of sufficiently smooth functions defined in $\hat{\Omega}$, whereas $X_N(\hat{\Omega})$ is the subspace of the polynomial space $\mathbb{P}_N(\hat{\Omega})$ defined by the boundary conditions enforced by $\hat{\mathcal{P}}_N$. We set $\mathcal{W}(\Omega) = \{v : \hat{v} \in \mathcal{W}(\hat{\Omega})\}$ and $X_N(\Omega) = \{v : \hat{v} \in X_N(\hat{\Omega})\}$, and we define the approximation operator $\mathcal{P}_N : \mathcal{W}(\Omega) \to X_N(\Omega)$ by setting, for all $v \in \mathcal{W}(\Omega)$,

$$(\mathcal{P}_N v)(\mathbf{x}) = (\hat{\mathcal{P}}_N \hat{v})(\hat{\mathbf{x}}) , \quad \hat{\mathbf{x}} = F^{-1}(\mathbf{x}) , \qquad \text{for all } \mathbf{x} \in \Omega .$$

The simplest situation occurs when $F$ is an affine mapping, $F(\hat{\mathbf{x}}) = B\hat{\mathbf{x}} + \mathbf{b}$, with $|B| \neq 0$. In this case, $X_N(\Omega)$ is a subspace of the space $\mathbb{P}_N(\Omega)$ of all polynomials of degree at most $N$ in each variable, restricted to $\Omega$. It is easily seen that if $\hat{\mathcal{P}}_N = \hat{P}_N$ is the $L^2(\hat{\Omega})$-orthogonal projection operator upon $\mathbb{P}_N(\hat{\Omega})$, then $\mathcal{P}_N = P_N$ is the $L^2(\Omega)$-orthogonal projection operator upon $\mathbb{P}_N(\Omega)$. Similarly, if $\hat{\mathcal{P}}_N = \hat{I}_N$ is the interpolation operator at a set of tensor-product Gaussian nodes in $\Omega$, then $\mathcal{P}_N = I_N$ is the interpolation operator at the set of mapped nodes in $\Omega$.

Let us assume that, for all $\hat{u} \in H^m(\hat{\Omega}) \cap \mathcal{W}(\hat{\Omega})$, an error bound for $\hat{\mathcal{P}}_N$ holds in the following form:

$$|\hat{u} - \hat{\mathcal{P}}_N \hat{u}|_{H^k(\hat{\Omega})} \leq \hat{C} N^{\sigma(k)-m} |\hat{u}|_{H^m(\hat{\Omega})} , \tag{5.8.23}$$

for some $k \leq m$ and under the condition $N > m$. Here, $\hat{C}$ is a constant depending on $k$ and $m$ but independent of $N$, $\sigma(k)$ is a given function of $k$, and the seminorms are defined as $|\hat{v}|_{H^r(\hat{\Omega})} = \left(\sum_{j=1}^d \|\hat{D}_j^r\|^2_{L^2(\hat{\Omega})}\right)^{1/2}$, with $\hat{D}_i = \partial/\partial\hat{x}_i$. All estimates presented in the previous subsection have this form (if $N > m$), with $\sigma(k) = k$ in the optimal cases.

We aim at deriving from (5.8.23) an error bound for $\mathcal{P}_N$. Let us start with the case in which $F$ performs a dilation in each direction, with arbitrary dilation factors $h_i > 0$; thus, $B = \mathrm{diag}(h_i)$. Setting $D_i = \partial/\partial x_i$, we have $\hat{D}_i = h_i D_i$ for all $i$. Using these relations in both sides of (5.8.23) and noting that $(u - \mathcal{P}_N u)^\wedge = \hat{u} - \hat{\mathcal{P}}_N \hat{u}$, we obtain

$$\left(\sum_{i=1}^d h_i^{2k} \|D_i^k (u - \mathcal{P}_N u)\|^2_{L^2(\Omega)}\right)^{1/2} \leq \hat{C} N^{\sigma(k)-m} \left(\sum_{i=1}^d h_i^{2m} \|D_i^m u\|^2_{L^2(\Omega)}\right)^{1/2} \tag{5.8.24}$$

for all $u \in H^m(\Omega) \cap \mathcal{W}(\Omega)$. This bound accounts for the possible anisotropy of $u$ (i.e., a significant variation in the order of magnitude of the partial derivatives of $u$ in the different directions) in the sharpest way. On the other hand, if all dilation factors satisfy $ch \leq h_i \leq c'h$ for a suitable $h > 0$ and constants $c, c' > 0$ of order of magniture one, then the previous bound simplifies to

$$|u - \mathcal{P}_N u|_{H^k(\Omega)} \le C h^{m-k} N^{\sigma(k)-m} |u|_{H^m(\Omega)} \; ; \tag{5.8.25}$$

the constant $C$ depends only on $\hat{C}, c$ and $c'$. Such a bound is more appropriate for an isotropic situation. If $m \le N$, then the seminorm $|\hat{u}|_{H^m(\hat{\Omega})}$ on the right-hand side of (5.8.23) has to be replaced by the seminorm $|\hat{u}|_{H^{m;N}(\hat{\Omega})}$ (see (5.8.12)). In this case, assuming without loss of generality that $h \le 1$, (5.8.25) becomes

$$|u - \mathcal{P}_N u|_{H^k(\Omega)} \le C h^{\min(m,N+1)-k} N^{\sigma(k)-m} |u|_{H^{m;N}(\Omega)} \; . \tag{5.8.26}$$

The isotropic condition on the dilation factors corresponds to the geometric requirement that the aspect ratio(s) of the Cartesian-product domain $\Omega$ be of order of magnitude one. More generally, this is equivalent to the condition that the ratio between the diameter of $\Omega$ and the diameter of the largest ball contained in $\Omega$ be, again, of order of magnitude one. (A general mapping $F$ for which $\Omega = F(\hat{\Omega})$ satisfies this condition is termed a *regular* mapping.) The error bound (5.8.26) can be derived from (5.8.23) for any domain which is the image of $\hat{\Omega}$ under a regular affine mapping; in this case, $h$ denotes the diameter of $\Omega$. The proof, which is essentially based on sharply estimating how seminorms in the mapped domain vary with $h$, is classical in the analysis of the $h$-version of finite-element methods (see, e.g, Quarteroni and Valli (1994), Chapter 3, for the details).

If the mapping $F$ is not affine, the transformation of a seminorm induced by $F$ involves partial derivatives of order lower than the order of the seminorm. Therefore, for a regular mapping (5.8.26) takes the form

$$|u - \mathcal{P}_N u|_{H^k(\Omega)} \le C h^{\min(m,N+1)-k} N^{\sigma(k)-m} \|u\|_{H^m(\Omega)} \; . \tag{5.8.27}$$

Again, the constant $C$ is independent of $h$ and $N$.

Summarizing, if $\Omega$ is a domain obtained from $\hat{\Omega}$ by a smooth, invertible and regular mapping, then an optimal (i.e., with $\sigma(k) = k$) error bound of the form (5.8.26) or (5.8.27) holds for the $L^2$-orthogonal projection upon $\mathbb{P}(\Omega)$ (for $k = 0$, $m \ge 0$), for the $H^1$- and $H^1_0$-orthogonal projection (for $k = 0, 1$ and $m \ge 1$) and for the interpolation operator at the mapped Gauss-Lobatto nodes (for $k = 0, 1$ and $m \ge (d+1)/2$).

### 5.8.4 Chebyshev and Other Jacobi Approximations

Unless otherwise specified, we keep the notation of the previous section. Instead of (5.8.9) we set now

$$\phi_{\mathbf{k}}(\mathbf{x}) = T_{k_1}(x_1) \cdots T_{k_d}(x_d) \qquad \text{for } \mathbf{k} = (k_1, \ldots, k_d) \in \mathbb{N}^d \; .$$

This is an orthogonal basis of $L^2_w(\Omega)$, the space of the measurable functions on $\Omega$ that are square integrable for the multidimensional Chebyshev weight

$w(\mathbf{x}) = \left(\prod_{i=1}^d (1-x_i^2)\right)^{-1/2}$ (see (A.9.h)). For each $u \in L^2_w(\Omega)$, the truncation of its Chebyshev series is given by

$$P_N u = \sum_{\|\mathbf{k}\| \le N} \hat{u}_{\mathbf{k}} \phi_{\mathbf{k}} , \qquad (5.8.28)$$

with

$$\hat{u}_{\mathbf{k}} = \left(\frac{2}{\pi}\right)^d \left(\prod_{i=1}^d \frac{1}{c_{k_i}}\right) \int_\Omega u(\mathbf{x}) \phi_{\mathbf{k}}(\mathbf{x}) w(\mathbf{x})\, d\mathbf{x} ,$$

(see (2.4.10) and (2.4.11)). Denoting by $H^m_w(\Omega)$ the weighted Sobolev spaces relative to the Chebyshev weight (see (A.11.b)), the remainder of the Chebyshev series of a function $u \in H^m_w(-1,1)$, $m \ge 0$, can be bounded as follows:

$$\|u - P_N u\|_{H^l_w(\Omega)} \le C N^{\sigma(l)-m} |u|_{H^{m;N}_w(\Omega)} , \qquad 0 \le l \le m , \quad (5.8.29)$$

where $\sigma(l) = 0$ if $l = 0$, and $\sigma(l) = 2l - \frac{1}{2}$ if $l > 0$. The seminorm on the right-hand side is defined as in (5.8.12), using now weighted $L^2$-norms.

Concerning the projection errors in the higher order Sobolev norms, we have essentially the same kind of results as for the Legendre expansion. For instance, let us define the operator

$$\begin{aligned} &P^1_N : H^1_w(\Omega) \longrightarrow \mathbb{P}_N \quad \text{such that} \\ &((P^1_N u, \phi))_w = ((u,\phi))_w \quad \text{for all } \phi \in \mathbb{P}_N , \end{aligned} \qquad (5.8.30)$$

where $((u,v))_w = \int_\Omega (\nabla u \cdot \nabla v + uv) w\, dx$ is the inner product of $H^1_w(\Omega)$. Moreover, we define the operator

$$\begin{aligned} &P^{1,0}_N : H^1_{w,0}(\Omega) \longrightarrow \mathbb{P}^0_N \quad \text{such that} \\ &[P^{1,0}_N u, \phi]_w = [u,\phi]_w \quad \text{for all } \phi \in \mathbb{P}^0_N . \end{aligned} \qquad (5.8.31)$$

Here, $[u,v]_w = \int_\Omega (\nabla u \cdot \nabla v) w\, dx$ is the inner product of $H^1_{w,0}(\Omega)$ (see (A.11.c)).

For each $u \in H^m_w(\Omega)$ ($u \in H^1_{w,0}(\Omega) \cap H^m_w(\Omega)$, resp.), with $m \ge 1$, we set $u^N = P^1_N u$ ($u^N = P^{1,0}_N u$, resp.). Then we have the estimate

$$\|u - u^N\|_{H^1_w(\Omega)} \le C N^{1-m} |u|_{H^{m;N}_w(\Omega)} . \qquad (5.8.32)$$

Optimal approximation estimates that extend (5.8.16) to the Chebyshev case are also available (see Bernardi and Maday (1997), Chapter 4).

Let us now deal with Chebyshev interpolation in $\Omega$. Let the interpolation points be defined as in (5.8.17), where now the $x_{j_i}$ belong to any of the families (2.4.12), (2.4.13), or (2.4.14). The Chebyshev interpolation at these points is defined as in (5.8.18) or (5.8.19), where the $\gamma_k$'s are defined in (2.4.18) and the $w_j$'s are defined in (2.4.12), (2.4.13) or (2.4.14). The interpolation error estimate is

$$\|u - I_N u\|_{L^2_w(\Omega)} \leq CN^{-m}|u|_{H^{m;N}_w(\Omega)} \tag{5.8.33}$$

for all $u \in H^m_w(\Omega)$ with $m > d/2$. If interpolation is taken at the Chebyshev Gauss-Lobatto points (2.4.14) in all space directions, we also have the optimal estimate in the $H^1_w(\Omega)$-norm:

$$\|u - I_N u\|_{H^1_w(\Omega)} \leq CN^{1-m}|u|_{H^{m;N}_w(\Omega)} \tag{5.8.34}$$

for all $u \in H^m_w(\Omega)$ with $m > (3d+2)/4$.

All the results concerning Chebyshev approximations mentioned in this section hold unchanged for any Jacobi approximation, as defined in Sect. 2.5, for the full range $-1 < \alpha, \beta < 1$. Obviously, $w$ is now the Jacobi weight, $w(x) = (1-x)^\alpha(1+x)^\beta$. Again, we refer to Bernardi and Maday (1997) for the details.

### 5.8.5 Blended Trigonometric and Algebraic Approximations

Several spectral approximations provide a numerical solution which is a finite expansion in terms of trigonometric (Fourier) polynomials in some Cartesian directions and of algebraic (Jacobi) polynomials in the others.

This is typically the case of those problems set in Cartesian geometry, whose physical solution is periodic with respect to one (or more) variables, and submitted to Dirichlet or Neumann boundary conditions in the direction of the remaining variables.

We consider here for the sake of simplicity a two-dimensional domain, say $\Omega = (-1,1) \times (0, 2\pi)$, but what we are going to present is extendable in an obvious manner to a domain of the form $\Omega = (-1,1)^{d_1} \times (0,2\pi)^{d_2}$ for $d_1, d_2 \geq 1$. We introduce first some notation. For each integer $M$ we denote by $\mathbb{P}_M$ the space of algebraic polynomials in the variable $x$ of degree up to $M$. Moreover, for each integer $N$ we denote by $S_N$ the space

$$S_N = \text{span}\left\{e^{iky} \mid -N \leq k \leq N-1\right\} .$$

Then we define the space $V_{M,N}$ as the tensor product of $\mathbb{P}_M$ and $S_N$, i.e.,

$$V_{M,N} = \left\{\phi(x,y) = \sum_{m=0}^{M}\sum_{n=-N}^{N-1} a_{mn}p_m(x)e^{iny}, \; a_{mn} \in \mathbb{C}\right\} ,$$

where we use the notation $p_m(x)$ to indicate the $m$-th Jacobi polynomial with respect to a given Jacobi weight $w(x) = (1-x)^\alpha(1+x)^\beta$, with $-1 < \alpha, \beta < 1$. Let us denote by $L^2_y(H^k_{w,x})$ the space of the measurable functions $u : \Omega \to \mathbb{R}$ such that

$$\|u\|_{k,0} = \left( \int_0^{2\pi} \|u(\cdot, y)\|^2_{H^k_w(-1,1)} \, dy \right)^{1/2} < +\infty \,. \tag{5.8.35}$$

For $k = 0$, this norm will be denoted briefly by

$$\|u\|_0 = \left( \int_0^{2\pi} dy \int_{-1}^{1} |u(x, y)|^2 w(x) \, dx \right)^{1/2} . \tag{5.8.36}$$

Moreover, for any positive integer $h$ we define

$$H^\ell_y(L^2_{w,x}) = \left\{ u \in L^2(\Omega) \;\middle|\; \frac{\partial^j u}{\partial y^j} \in L^2_y(L^2_{w,x}),\ 0 \le j \le \ell \right\};$$

the norm is given by

$$\|u\|_{0,\ell} = \left( \sum_{j=0}^{\ell} \left\| \frac{\partial^j u}{\partial y^j} \right\|_0^2 \right)^{1/2} . \tag{5.8.37}$$

The space $H^\ell_{p,y}(L^2_{w,x})$ is the closure in $H^\ell_y(L^2_{w,x})$ of $C^\infty_p(\Omega)$, which is the set of all functions that are continuous with all their derivatives up to the boundary of $\Omega$ and $2\pi$-periodic with all their derivatives with respect to the $y$-direction.

For any function $u \in L^2_y(L^2_{w,x})$, let $P_{M,N}u$ denote the projection of $u$ upon $V_{M,N}$, i.e.,

$$P_{M,N}u = \sum_{m=0}^{M} \sum_{n=-N}^{N-1} \hat{u}_{mn} p_m(x) e^{iny} \,, \tag{5.8.38}$$

where

$$\hat{u}_{mn} = \frac{1}{c_k \pi^2} \int_0^{2\pi} \int_{-1}^{1} u(x, y) p_m(x) e^{-iny} w(x) \, dx \, dy \,.$$

The $c_k$'s are given in (2.4.10). Then, for all $k, \ell \ge 0$, we have

$$\|u - P_{M,N}u\|_0 \le C_1 M^{-k} \|u\|_{k,0} + C_2 N^{-\ell} \|u\|_{0,\ell} \,, \tag{5.8.39}$$

for all $u$ for which the norms on the right-hand side are finite. The proof of this result can be done as follows. Denote by $P^J_M$ and $P^F_N$ the $L^2$-orthogonal projection operators upon $\mathbb{P}_M$ and $S_N$ in the Jacobi and Fourier expansions, respectively. Then,

$$u - P^F_N P^J_M u = (u - P^F_N u) + P^F_N (u - P^J_M u) \,. \tag{5.8.40}$$

Now (5.8.39) follows, noting that $\|u - P_{M,N}u\|_0 \le \|u - P^F_N P^J_M u\|_0$ and using (5.1.9) and (5.5.9), which, as mentioned in Sect. 5.7.1, holds for all Jacobi weights.

In higher order norms, the best approximation error can be estimated by a splitting technique similar to the one used in (5.8.40). For instance, using instead of $P_M^J$ the $H_w^1(-1,1)$-orthogonal projector $P_M^{J,1}$ defined in (5.5.13), it follows that, for all $k,\ell \geq 1$, we have

$$\begin{aligned} \|u - P_N^F P_M^{J,1} u\|_1 \leq C_1 M^{1-k} & \left( \|u\|_{k,0} + \left\| \frac{\partial u}{\partial y} \right\|_{k-1,0} \right) \\ & + C_2 N^{1-\ell} \left( \|u\|_{0,\ell} + \left\| \frac{\partial u}{\partial x} \right\|_{0,\ell-1} \right) \end{aligned} \tag{5.8.41}$$

for all $u$ for which the norms on the right-hand side are finite, where we have used (5.1.10) and (5.5.14). Obviously, a similar estimate holds if $u$ and $v$ are assumed to vanish on the sides $x=-1$ and $x=1$ of the boundary of $\Omega$. It is enough to take the operator defined in (5.5.17) and to use (5.5.19) instead of (5.5.14). Best approximation error estimates in higher norms can be proven similarly.

Concerning interpolation, let us consider the points

$$\xi_{ij} = \left( \xi_i, \frac{\pi j}{N} \right), \quad 0 \leq i \leq M,\ 0 \leq j \leq 2N-1\,, \tag{5.8.42}$$

where $\xi_i$ denote the GL points for the Jacobi weight $\omega(x)$.

Then, denote by $I_M^J$ the Jacobi interpolation operator with respect to the points $\xi_i$ and by $I_N^F$ the Fourier interpolation operator relative to the points $\pi j/N$. Of course, $I_{M,N} = I_M^J I_N^F = I_N^F I_M^J$ is the interpolation operator relative to the points $\xi_{ij}$, i.e., for all $u \in C^0(\overline{\Omega})$,

$$I_{M,N} u \in V_{M,N} \quad \text{satisfies} \quad I_{M,N} u(\xi_{ij}) = u(\xi_{ij})\,, \tag{5.8.43}$$

for $0 \leq i \leq M$ and $0 \leq j \leq 2N-1$.

Using the error bound (5.1.19) for Fourier interpolation and the bounds (5.5.22) and (5.5.25) as well as the stability estimate (5.5.24) for Jacobi interpolation, one proves, with the same arguments as above, that for all $k,\ell \geq 2$ and $0 \leq r \leq 1$, we have

$$\begin{aligned} \|u - I_N^F I_M^J u\|_r \leq C_1 M^{r-k} & \left( \|u\|_{k,0} + \left\| \frac{\partial u}{\partial y} \right\|_{k-1,0} \right) \\ & + C_2 N^{r-\ell} \left( \|u\|_{0,\ell} + \left\| \frac{\partial u}{\partial x} \right\|_{0,\ell-1} \right) \end{aligned} \tag{5.8.44}$$

for all $u$ for which the norms on the right-hand side are finite.

### *Bibliographical Notes*

The estimate (5.8.7) is due to Pasciak (1980). The results on the truncation operators for both the Legendre and the Chebyshev systems can be found in

Canuto and Quarteroni (1982a). Maday (1981) proved the estimates on the higher order projection operators. The results for the interpolation operators are due to Bernardi and Maday (1992a).

Blended Fourier-Jacobi expansions were first studied by Quarteroni (1987) in the Chebyshev case and by Bernardi, Maday and Métivet (1987) in the Legendre case.

The coupling of Fourier and finite-element approximations, for instance in a domain of the form $\Omega \times (0, 2\pi)$, where $\Omega$ is a polygonal region of $\mathbb{R}^d$ $(d \geq 2)$ and the solution is periodic with respect to the last variable, is of interest as well. Early studies of this kind were made by Canuto, Maday and Quarteroni (1982), and Mercier and Raugel (1982). More recent results can be found in Belhachmi et al. (2006).

## 5.9 Approximation in Triangles and Related Domains

In this section we present a few approximation results for a function defined in a simplicial domain by means of suitable algebraic polynomials. As usual, we confine ourselves to those results that are useful in the analysis of spectral methods.

Let us first consider the simplest situation of the reference triangle $\mathcal{T} = \{(x_1, x_2) \in \mathbb{R}^2 : -1 < x_1, x_2 \text{ and } x_1 + x_2 < 0\}$, which is contained in the reference square $\mathcal{Q} = \{(\xi_1, \xi_2) \in \mathbb{R}^2 : -1 < \xi_1, \xi_2 < 1\}$. For clarity, we shall denote by $\mathcal{P}_N(\mathcal{T})$ the space of polynomials of total degree $\leq N$ in the variables $x_1, x_2$, and by $\mathbb{P}_N(\mathcal{Q})$ the space of polynomials of degree $\leq N$ in each of the variables $\xi_1, \xi_2$. A first set of results is as follows (as usual, $C$ denotes a constant independent of $u$ and $N$, but depending on $m$):

*$L^2$-approximation.* The orthogonal projection operator $P_N : L^2(\mathcal{T}) \to \mathcal{P}_N(\mathcal{T})$ in the inner product $(u, v) = \int_{\mathcal{T}} uv \, d\mathbf{x}$ satisfies

$$\|u - P_N u\|_{L^2(\mathcal{T})} \leq C N^{-m} |u|_{H^{m;N}(\mathcal{T})} \tag{5.9.1}$$

for all $u \in H^m(\mathcal{T})$, $m \geq 0$.

*$H^1$-approximation.* The orthogonal projection operator $P_N^1 : H^1(\mathcal{T}) \to \mathcal{P}_N(\mathcal{T})$ in the inner product $((u, v)) = \int_{\mathcal{T}} (\nabla u \cdot \nabla v + uv) \, d\mathbf{x}$ satisfies

$$\|u - P_N^1 u\|_{H^k(\mathcal{T})} \leq C N^{k-m} |u|_{H^{m;N}(\mathcal{T})} \,, \qquad k = 0, 1 \,, \tag{5.9.2}$$

for all $u \in H^m(\mathcal{T})$, $m \geq 1$.

*$H_0^1$-approximation.* The orthogonal projection operator $P_N^{1,0} : H_0^1(\mathcal{T}) \to \mathcal{P}_N^0(\mathcal{T}) = \mathcal{P}_N(\mathcal{T}) \cap H_0^1(\mathcal{T})$ in the inner product $[u, v] = \int_{\mathcal{T}} \nabla u \cdot \nabla v \, d\mathbf{x}$ satisfies

$$\|u - P_N^{1,0} u\|_{H^k(\mathcal{T})} \leq C N^{k-m} |u|_{H^{m;N}(\mathcal{T})} \,, \qquad k = 0, 1 \,, \tag{5.9.3}$$

for all $u \in H^m(\mathcal{T}) \cap H_0^1(\mathcal{T})$, $m \geq 1$.

We sketch proofs of these results. Following Schwab (1998), we first recall that, given any $m \geq 0$ and $u \in H^m(\mathcal{T})$, there exists a function $\tilde{u} \in H^m(\mathcal{Q})$ that extends $u$ to the whole square $\mathcal{Q}$ in a continuous way, i.e., $\tilde{u}$ is such that $\tilde{u}_{|\mathcal{T}} = u$ and $\|\tilde{u}\|_{H^m(\mathcal{Q})} \leq C\|u\|_{H^m(\mathcal{T})}$. Denote by $\tilde{u}_{N/2} = \tilde{P}_{N/2}\tilde{u}$ the $L^2$-orthogonal projection of $\tilde{u}$ upon $\mathbb{P}_{N/2}(\mathcal{Q})$, which by (5.8.11) satisfies $\|\tilde{u} - \tilde{u}_{N/2}\|_{L^2(\mathcal{Q})} \leq CN^{-m}\|\tilde{u}\|_{H^m(\mathcal{Q})}$; set $u_N = (\tilde{u}_{N/2})_{|\mathcal{T}}$ and note that $u_N \in \mathcal{P}_N(\mathcal{T})$. Then we have

$$\begin{aligned}\|u - P_N u\|_{L^2(\mathcal{T})} = & \inf_{v_N \in \mathcal{P}_N(\mathcal{T})} \|u - v_N\|_{L^2(\mathcal{T})} \leq \|u - u_N\|_{L^2(\mathcal{T})} \\ & \leq \|\tilde{u} - \tilde{u}_{N/2}\|_{L^2(\mathcal{Q})} \leq CN^{-m}\|\tilde{u}\|_{H^m(\mathcal{Q})} \\ & \leq C'N^{-m}\|u\|_{H^m(\mathcal{T})} \, .\end{aligned}$$

This estimate is refined further into (5.9.1) by an argument, already used in the previous sections, based on the fact that $P_N$ is exact on $\mathcal{P}_N(\mathcal{T})$, i. e., $P_N v = v$ whenever $v \in \mathcal{P}_N(\mathcal{T})$.

In order to establish (5.9.2), we rather choose $\tilde{u}_{N/2} = \tilde{P}^1_{N/2}\tilde{u}$ as defined in (5.8.13), for which (5.8.15) holds. Proceeding as above, we arrive at $\|u - P^1_N u\|_{H^1(\mathcal{T})} \leq \|u - u_N\|_{H^1(\mathcal{T})} \leq CN^{1-m}\|u\|_{H^m(\mathcal{T})}$. The Aubin-Nitsche duality argument, already used in Sect. 5.6, applies since the solution of the boundary-value problem $-\Delta\psi + \psi = g$ in $\mathcal{T}$, $\partial u/\partial n = 0$ on $\partial\mathcal{T}$, satisfies $\|\psi\|_{H^2(\mathcal{T})} \leq C\|g\|_{L^2(\mathcal{T})}$, as $\mathcal{T}$ is convex. This yields $\|u - P^1_N u\|_{L^2(\mathcal{T})} \leq CN^{-1}\|u - P^1_N u\|_{H^1(\mathcal{T})}$; whence, (5.9.2) follows.

At last, we consider (5.9.3). We choose again $\tilde{u}_{N/2} = \tilde{P}^1_{N/2}\tilde{u}$, but we have to correct it on the boundary of $\mathcal{T}$. To this end, we observe that $g_N = (\tilde{u}_{N/2})_{|\partial\mathcal{T}}$ is a continuous function that is a polynomial of degree $\leq N$ on each side. Furthermore, we recall that the fractional $H^{1/2}$-norm of a function on the boundary of a domain can be bounded in terms of its $H^1$-norm in the domain (see (A.11.e)). Since $u$ vanishes on the boundary of $\mathcal{T}$, we have $\|g_N\|_{H^{1/2}(\partial\mathcal{T})} = \|g_N - u_{|\partial\mathcal{T}}\|_{H^{1/2}(\partial\mathcal{T})} \leq C\|(\tilde{u}_{N/2})_{|\mathcal{T}} - u\|_{H^1(\mathcal{T})} \leq C\|\tilde{u}_{N/2} - \tilde{u}\|_{H^1(\mathcal{Q})} \leq CN^{1-m}\|u\|_{H^m(\mathcal{T})}$. We now use a polynomial lifting result (see, e.g., Schwab (1998), Theorem 4.84) that ensures that there exists a polynomial $G_N \in \mathcal{P}_N(\mathcal{T})$ such that $(G_N)_{|\partial\mathcal{T}} = g_N$ and $\|G_N\|_{H^1(\mathcal{T})} \leq C\|g_N\|_{H^{1/2}(\partial\mathcal{T})}$. Setting $u_N = (\tilde{u}_{N/2})_{|\mathcal{T}} - G_N \in \mathcal{P}_N(\mathcal{T})$, we have $u_N = 0$ on $\partial\mathcal{T}$ and $\|u - u_N\|_{H^1(\mathcal{T})} \leq C(\|\tilde{u} - \tilde{u}_{N/2}\|_{H^1(\mathcal{Q})} + \|G_N\|_{H^1(\mathcal{T})}) \leq CN^{1-m}\|u\|_{H^m(\mathcal{T})}$. Then, we proceed as above. This concludes the proofs of estimates (5.9.1)-(5.9.3).

Sharper estimates than those stated above could have been provided, but only at the price of introducing a significantly heavier notation. As for the one-dimensional case, a weaker seminorm can be put on the right-hand side of (5.9.1)–(5.9.3), which makes the error bound true even for functions that are less regular near the boundary. Owens (1998) derives an $L^2$-error estimate of this type using his orthogonal basis of $\mathcal{P}_N(\mathcal{T})$ that is formed by eigenfunctions of a singular Sturm-Liouville problem in $\mathcal{T}$. The estimate stems from this property, following the same arguments given in Sect. 5.6

for proving (5.4.11). Guo and Wang (2006) get $L^2$- and $H^1$-error estimates, in which norms of sophisticated anisotropic weighted Sobolev spaces appear on the right-hand side. They use the warped tensor-product basis (2.9.6), allowing for a different maximal degree in each variable. For the proof, they first map $\mathcal{T}$ onto the reference square $\mathcal{Q}$ via the singular transformation (2.9.1); then, they combine suitable one-dimensional projection operators of Jacobi type in a warped manner. Error estimates for such operators (Guo and Wang (2004)), in which the dependence on the Jacobi indices $\alpha$ and $\beta$ is made explicit, are essential in their analysis. Indeed, some of the Jacobi indices are functions on the polynomial degree, see (2.9.6), hence, they grow unboundedly with $N$.

In view of the numerical analysis of G-NI and collocation methods, it is of interest to estimate the error incurred by using the quadrature formula in $\mathcal{T}$ based on the collapsed coordinates introduced in Sect. 2.9.1. Precisely, recall (2.9.8) and denote by $\boldsymbol{\xi}_j$, with $j$ belonging to a suitable index set $J_N$, the tensorized Gaussian nodes in the square $\mathcal{Q}$ for the measure $(1-\xi_2)\,d\xi_1\,d\xi_2$ mentioned therein; let $w_j'$ denote the corresponding weights. If $\mathbf{x}_j$ are the mapped nodes in the triangle $\mathcal{T}$, and if $w_j = \frac{1}{2}w_j'$, we have

$$\int_{\mathcal{T}} f(\mathbf{x})\,d\mathbf{x} = \frac{1}{2}\int_{\mathcal{Q}} F(\boldsymbol{\xi})(1-\xi_2)\,d\boldsymbol{\xi} \simeq \frac{1}{2}\sum_{j\in J_N} F(\boldsymbol{\xi}_j)w_j' = \sum_{j\in J_N} f(\mathbf{x}_j)w_j \,.$$

This formula is exact for all polynomials in $\mathcal{P}_{2N-1}(\mathcal{T})$. Furthermore, the bilinear form $(f,g)_{N,\mathcal{T}} = \sum_{j\in J_N} f(\mathbf{x}_j)g(\mathbf{x}_j)w_j$ is a discrete inner product in $\mathcal{P}_N(\mathcal{T})$ that approximates the $L^2$-inner product $(f,g)_{\mathcal{T}} = \int_{\mathcal{T}} f(\mathbf{x})g(\mathbf{x})\,d\mathbf{x}$. By mapping $\mathcal{T}$ to $\mathcal{Q}$ and by tensorizing the results of Sect. 5.3 about the uniform equivalence of discrete and continuous norms on the interval $(-1,1)$, one easily checks that the same results hold for the discrete and continuous $L^2$-norms on $\mathcal{T}$: there exist constants $c_1, c_2 > 0$ independent of $N$ such that

$$c_1\|\phi\|_{L^2(\mathcal{T})} \le \|\phi\|_{N,\mathcal{T}} \le c_2\|\phi\|_{L^2(\mathcal{T})} \qquad \text{for all } \phi\in\mathcal{P}_N(\mathcal{T})\,. \tag{5.9.4}$$

Furthermore, the following estimate, which extends (5.4.38), holds: for all $u\in H^m(\mathcal{T})$ with $m>1$, and for all $\phi\in\mathcal{P}_N(\mathcal{T})$,

$$|(u,\phi)_{\mathcal{T}} - (u,\phi)_{N,\mathcal{T}}| \le CN^{-m}|u|_{H^{m;N-1}(\mathcal{T})}\|\phi\|_{L^2(\mathcal{T})}\,. \tag{5.9.5}$$

For the proof, we apply the transformation $(x_1,x_2)\mapsto(\xi_1,\xi_2)$ and we introduce the weight $w(\boldsymbol{\xi}) = (1-\xi_2)$. Proceeding as in the proofs of (5.3.4a)–(5.3.4b) and using the results of Sect. 5.8, we get, with obvious notation,

$$\begin{aligned}
|(u,\phi)_{\mathcal{T}} - (u,\phi)_{N,\mathcal{T}}| &= \frac{1}{2}\left|\int_{\mathcal{Q}} U(\boldsymbol{\xi})\Phi(\boldsymbol{\xi})(1-\xi_2)\,d\boldsymbol{\xi} - \sum_{j\in J_N} U(\boldsymbol{\xi}_j)\Phi(\boldsymbol{\xi})w_j'\right| \\
&\le C(\|U-\tilde{P}_{N-1}U\|_{L^2_w(\mathcal{Q})} + \|U-\tilde{I}_N U\|_{L^2_w(\mathcal{Q})})\,\|\Phi\|_{L^2_w(\mathcal{Q})} \\
&\le CN^{-m}\|U\|_{H^m_w(\mathcal{Q})}\,\|\Phi\|_{L^2_w(\mathcal{Q})}\,.
\end{aligned}$$

Next, we observe that $\dfrac{\partial(x_1,x_2)}{\partial(\xi_1,\xi_2)} = \begin{pmatrix} \frac{1}{2}(1-\xi_2) & \frac{1}{2}(1+\xi_1) \\ 0 & 1 \end{pmatrix}$, which easily implies $\|U\|_{H^m_w(\mathcal{Q})} \leq C\|u\|_{H^m(\mathcal{T})}$; furthermore, $\|\Phi\|_{L^2_w(\mathcal{Q})} = \|\phi\|_{L^2(\mathcal{T})}$. Finally, we refine the resulting estimate using the property that the error vanishes if $u \in \mathcal{P}_{N-1}(\mathcal{T})$.

All previous results can be extended, with similar proofs, to three-dimensional reference domains $\mathcal{T}$, such as the tetrahedron, the pyramid or the prism, already considered in Sect. 2.9.1.

Scaled error estimates are easily obtained from the previous ones if $\mathcal{T}$ is now a two- or three-dimensional element that is the image of one of the reference simplicial domains $\hat{\mathcal{T}}$ under a smooth, invertible mapping $F : \hat{\mathcal{T}} \to \mathcal{T}$. Precisely, let $h_{\mathcal{T}}$ denote the diameter of $\mathcal{T}$ and let us assume that the mapping is *regular*, in the sense that the diameter $\varrho_{\mathcal{T}}$ of the ball inscribed into $\mathcal{T}$ satisfies $\varrho_{\mathcal{T}} \geq c h_{\mathcal{T}}$ for a constant $c > 0$ of order of magniture one. If the mapping $F$ is affine, then $\mathcal{T}$ is still a simplicial domain and the approximating functions in $\mathcal{T}$ are still polynomials of total degree at most $N$. Then the estimates given above for the reference domain hold as well for $\mathcal{T}$, with the appearence of a power of $h$ that accounts for the geometric scaling (as in the tensor-product case; see Sect. 5.8.3). For instance, estimate (5.9.2) gives rise to the following one:

$$\|u - P^1_N u\|_{H^k(\mathcal{T})} \leq C h_{\mathcal{T}}^{\min(m,N+1)-k} N^{k-m} |u|_{H^{m;N}(\mathcal{T})}\,, \quad k = 0,1\,, \qquad (5.9.6)$$

for all $u \in H^m(\mathcal{T})$, $m \geq 1$; similarly, the counterpart of (5.9.5) is as follows: for all $u \in H^m(\mathcal{T})$ with $m > 1$, and for all $\phi \in \mathcal{P}_N(\mathcal{T})$,

$$|(u,\phi)_{\mathcal{T}} - (u,\phi)_{N,\mathcal{T}}| \leq C h_{\mathcal{T}}^{\min(m,N)} N^{-m} |u|_{H^{m;N-1}(\mathcal{T})} \|\phi\|_{L^2(\mathcal{T})}\,. \qquad (5.9.7)$$

Beyond the affine case, we may assume that $F$ is $m$-times continuously differentiable in $\hat{\mathcal{T}}$ and that the determinant $|JF|$ of its Jacobian matrix $JF$ satisfies $|JF| \geq \gamma$ in $\hat{\mathcal{T}}$, for some constant $\gamma > 0$. Then the estimates above hold as well, provided the seminorm $|u|_{H^{m;N-1}(\mathcal{T})}$ is replaced by the full norm $\|u\|_{H^m(\mathcal{T})}$.

The methods of proof for obtaining all these results are similar to those for the mapped tensor-product domain (see again Sect. 5.8.3).

# 6. Theory of Stability and Convergence

In this chapter we present a fairly general approach to the stability and convergence analysis of spectral methods. We confine ourselves to linear problems. Analysis of several nonlinear problems is presented in Chap. 7 and in CHQZ3, Chap. 3. For time-dependent problems, only the discretizations in space are considered. Stability for fully discretized time-dependent problems is discussed in Appendix D by a classical eigenvalue analysis and in Chap. 7 by variational methods.

It may be worthwhile to specify precisely what is meant here by stability of a spatial approximation based on a spectral method. A scheme will be called *stable* if it is possible to control the discrete solution by the data in a way independent of the discretization parameter $N$ (the degree of the polynomials used). This means that a suitable norm of the solution is bounded by a constant multiple of a suitable norm of the data, and all the norms involved, as well as the constant, do not depend on $N$. In other words, for a fixed data, all the discrete solutions produced by the spectral scheme, as $N$ tends to infinity, lie in a bounded subset of a normed linear space.

The most representative methods of spectral type, i.e., Galerkin (with or without numerical integration), collocation and tau, are considered. We begin with a reexamination of some of the examples of Chap. 1. The aim here is to introduce the salient aspects of the different methods of analysis. We then proceed to the general theory with the objective of achieving a unified methodology. Time-independent problems are considered first, and then both parabolic and hyperbolic equations are analyzed. All spectral schemes can be obtained from some weak (or variational) formulation of the differential problem by restricting the function spaces to polynomials and possibly by introducing some further approximation (e.g., numerical integration); most of them can indeed be interpreted as projection methods over a finite-dimensional space of polynomials with respect to a certain inner product. The stability is proved either by the energy method or by a generalized variational principle. The convergence analysis uses stability results and the results of approximation theory given in Chap. 5 for several projection operators. Applications of these general results to the analysis of many pertinent examples are given.

## 6.1 Three Elementary Examples Revisited

Some basic aspects of the analysis of stability and convergence for spectral methods can be illustrated by considering three of the examples already discussed in Sect. 1.2. Other aspects of the analysis will be highlighted in Sect. 6.4.3 by considering a multidimensional version of the remaining example, presented in Sect. 1.2.4. The nature of the theory presented in this section is deliberately pedestrian, since the purpose is to introduce the reader to the more sophisticated and abstract mathematics in the remaining sections of this chapter.

### 6.1.1 A Fourier Galerkin Method for the Wave Equation

The linear hyperbolic problem

$$\begin{aligned} &\frac{\partial u}{\partial t} - \frac{\partial u}{\partial x} = 0\,, && 0 < x < 2\pi\,,\ t > 0\,, \\ &u(x,t)\ 2\pi\text{-periodic in } x, && t > 0\,, \\ &u(x,0) = u_0(x)\,, && 0 < x < 2\pi\,, \end{aligned}$$

was approximated in Sect. 2.1 by the Galerkin scheme (1.2.3). For any $t \geq 0$, $u^N(x,t)$ is a trigonometric polynomial of degree $N$ in $x$, i.e., $u^N(t) \in S_N$ where

$$S_N = \text{span}\{e^{ikx} \mid -N \leq k \leq N-1\}\,,$$

(see (5.1.1)). (Note that in Chaps. 5–7 we are following the convention that Fourier series are truncated at degree $N$ rather than degree $N/2$, as this is more convenient for the theoretical discussion.) The solution $u^N$ satisfies the integral relation

$$\int_0^{2\pi} \left( \frac{\partial u^N}{\partial t}(x,t) - \frac{\partial u^N}{\partial x}(x,t) \right) \overline{v(x)} \mathrm{d}x = 0 \quad \text{for all } v \in S_N\,,\ t > 0, \quad (6.1.1)$$

which is equivalent to (1.2.3) since the $\psi_k$'s are a basis in $S_N$, and, by the initial condition (1.2.11),

$$u^N(0) = P_N u_0 = \sum_{k=-N}^{N-1} \hat{u}_{0,k} e^{ikx}\,.$$

For any $t > 0$, let us set $v(x) = u^N(x,t)$ in (6.1.1). An integration-by-parts yields

$$\text{Re} \int_0^{2\pi} \frac{\partial u^N}{\partial x}(x,t) \overline{u}^N(x,t) \mathrm{d}x = \frac{1}{2} \{ |u^N(2\pi,t)|^2 - |u^N(0,t)|^2 \} = 0$$

by the periodicity condition. It follows that

$$\frac{1}{2}\frac{\mathrm{d}}{\mathrm{d}t}\int_0^{2\pi}|u^N(x,t)|^2\mathrm{d}x = \mathrm{Re}\int_0^{2\pi}\frac{\partial u^N}{\partial t}(x,t)\overline{u}^N(x,t)\mathrm{d}x = 0\,,$$

i.e., the $L^2$-norm (in space) of the spectral solution is constant in time. Therefore, for any $t > 0$,

$$\int_0^{2\pi}|u^N(x,t)|^2\mathrm{d}x = \int_0^{2\pi}|P_N u_0(x)|^2\mathrm{d}x \le \int_0^{2\pi}|u_0(x)|^2\mathrm{d}x\,.$$

Since the right-hand side is a constant, the Galerkin scheme (1.2.3) is stable in the $L^2$-norm.

On the other hand, projecting the equation $(\partial u/\partial t)-(\partial u/\partial x) = 0$ on $S_N$ yields the result that the truncated Fourier series $P_N u$ of the exact solution $u$ satisfies, at any $t > 0$,

$$\int_0^{2\pi}\left(\frac{\partial}{\partial t}P_N u - \frac{\partial}{\partial x}P_N u\right)(x,t)\overline{v(x)}\mathrm{d}x = 0 \quad \text{for all } v \in S_N\,.$$

This is the same variational relation that defines $u^N$. Since $u^N = P_N u$ at time $t = 0$, it follows that

$$u^N = P_N u \quad \text{for all } t \ge 0\,.$$

Since $P_N u$ converges to $u$ as $N$ tends to infinity, the approximation is convergent. Moreover, (5.1.9) provides an estimate of the error between the exact and the spectral solution. For all $t > 0$ we have

$$\int_0^{2\pi}|u(x,t)-u^N(x,t)|^2\mathrm{d}x \le CN^{-2m}\int_0^{2\pi}\left|\frac{\partial^m u}{\partial x^m}(x,t)\right|^2\mathrm{d}x\,.$$

### 6.1.2 A Chebyshev Collocation Method for the Heat Equation

Consider now the linear heat equation

$$\frac{\partial u}{\partial t}-\frac{\partial^2 u}{\partial x^2} = 0\,, \qquad -1 < x < 1\,,\; t > 0\,, \tag{6.1.2}$$

with homogeneous Dirichlet conditions

$$u(-1,t) = u(1,t) = 0\,, \qquad t > 0\,, \tag{6.1.3}$$

and initial condition

$$u(x,0) = u_0(x)\,, \qquad -1 < x < 1\,. \tag{6.1.4}$$

A Chebyshev collocation scheme was considered for this problem in Sect. 1.2.2 (see (1.2.26)–(1.2.28)). For any $t > 0$, the spectral solution $u^N$ is an algebraic polynomial of degree $N$ on the interval $(-1, 1)$, vanishing at the endpoints. It is defined through the collocation equations

$$\frac{\partial u^N}{\partial t}(x_k, t) - \frac{\partial^2 u^N}{\partial x^2}(x_k, t) = 0 \,, \qquad k = 1, \ldots, N-1 \,, \tag{6.1.5}$$

and the initial condition

$$u^N(x_k, 0) = u_0(x_k) \,, \qquad k = 0, \ldots, N \,.$$

The collocation points are given by $x_k = \cos(k\pi/N)$ (see (1.2.31) or (2.4.14)). They are the nodes of the Gauss-Lobatto quadrature formula relative to the Chebyshev weight, $w(x) = 1/\sqrt{1-x^2}$, whose weights are given by $w_0 = w_N = \pi/2N$ and $w_k = \pi/N$ if $k = 1, \ldots, N-1$ (see (2.4.14)). This property will be constantly used in the subsequent analysis of Chebyshev collocation methods. Its relevance in the theory of spectral methods was first pointed out by Gottlieb (1981).

Let us multiply the $k$-th equation of (6.1.5) by $u^N(x_k, t)w_k$ and sum over $k$. We get

$$\frac{1}{2}\frac{\mathrm{d}}{\mathrm{d}t}\sum_{k=0}^{N}\left[u^N(x_k, t)\right]^2 w_k - \sum_{k=0}^{N}\frac{\partial^2 u^N}{\partial x^2}(x_k, t)u^N(x_k, t)w_k = 0 \,, \tag{6.1.6}$$

where we are allowed to include the boundary points in the sum since $u^N$ vanishes there. The product $(\partial^2 u^N/\partial x^2)u^N$ is a polynomial of degree $2N-2$; hence, by the exactness of the quadrature formula (see (2.2.17)),

$$-\sum_{k=0}^{N}\frac{\partial^2 u^N}{\partial x^2}(x_k, t)u^N(x_k, t)w_k = -\int_{-1}^{1}\frac{\partial^2 u^N}{\partial x^2}(x, t)u^N(x, t)w(x)\mathrm{d}x \,.$$

In Sect. 7.1.2 it is proved, as a part of a general result, that the right-hand side is positive and actually dominates a weighted "energy" of the solution, i.e.,

$$-\int_{-1}^{1}\frac{\partial^2 u^N}{\partial x^2}(x, t)u^N(x, t)w(x)\mathrm{d}x \geq \frac{1}{4}\int_{-1}^{1}\left[\frac{\partial u^N}{\partial x}(x, t)\right]^2 w(x)\mathrm{d}x \,.$$

Then from (6.1.6) it follows that

$$\frac{1}{2}\frac{\mathrm{d}}{\mathrm{d}t}\sum_{k=0}^{N}\left[u^N(x_k, t)^2\right] w_k + \frac{1}{4}\int_{-1}^{1}\left[\frac{\partial u^N}{\partial x}(x, t)\right]^2 w(x)\mathrm{d}x \leq 0 \,;$$

whence

$$\sum_{k=0}^{N}\left[u^N(x_k, t)^2\right] w_k + \frac{1}{2}\int_0^t\int_{-1}^{1}\left[\frac{\partial u^N}{\partial x}(x, s)\right]^2 w(x)\mathrm{d}x\,\mathrm{d}s \leq \sum_{k=0}^{N}[u_0(x_k)]^2 w_k \,.$$

The sum on the left-hand side represents the *discrete* $L^2$-norm of the solution with respect to the Chebyshev weight. It does not coincide with the continuous $L^2$-norm $\int_{-1}^{1}[u^N(x,t)]^2 w(x)\mathrm{d}x$ since $(u^N)^2$ is a polynomial of degree $2N$. However, as pointed out in Sect. 5.3 (see (5.3.2)), it is uniformly equivalent to this norm, i.e.,

$$\int_{-1}^{1}\left[u^N(x,t)\right]^2 w(x)\mathrm{d}x \le \sum_{k=0}^{N}\left[u^N(x_k,t)\right]^2 w_k \le 2\int_{-1}^{1}\left[u^N(x,t)\right]^2 w(x)\mathrm{d}x \,.$$

On the other hand, the sum on the right-hand side can be bounded, for instance, by twice the square of the maximum of the data on the interval $[-1,1]$. We conclude that, for any $t>0$,

$$\int_{-1}^{1}\left[u^N(x,t)\right]^2 w(x)\mathrm{d}x+\frac{1}{2}\int_0^t\int_{-1}^{1}\left[\frac{\partial u^N}{\partial x}(x,s)\right]^2 w(x)\,\mathrm{d}x\,\mathrm{d}s \le 2\max_{-1\le x\le 1}|u_0(x)|^2 \,.$$

This proves that the Chebyshev collocation scheme is stable. Note that this stability estimate provides a bound for both the weighted $L^2$-norm at any given time and also the weighted "energy" norm integrated over the time interval $(0,t)$.

The convergence of the approximation can be proved by a simple, although crude, argument. Assume the exact solution $u$ to be smooth enough. Its interpolant, $\tilde{u}=I_N u$, defined in Sect. 2.2.3, satisfies the collocation equations

$$\frac{\partial \tilde{u}}{\partial t}(x_k,t)-\frac{\partial^2 \tilde{u}}{\partial x^2}(x_k,t)=r(x_k,t)\,, \qquad t>0\,,\ k=1,\dots,N-1\,,$$

with the truncation error $r=(\partial^2/\partial x^2)(u-\tilde{u})$. Hence, the difference $e=\tilde{u}-u^N$, which is a polynomial of degree $N$ vanishing at the boundary points, satisfies the equations

$$\frac{\partial e}{\partial t}(x_k,t)-\frac{\partial^2 e}{\partial x^2}(x_k,t)=r(x_k,t)\,, \qquad t>0\,,\ k=1,\dots,N-1$$

and the initial condition, $e(x_k,0)=0$, $k=0,\dots,N$. The same analysis previously used yields

$$\begin{aligned}\frac{1}{2}\frac{\mathrm{d}}{\mathrm{d}t}\sum_{k=0}^{N}[e(x_k,t)]^2 w_k &+ \frac{1}{4}\int_{-1}^{1}\left[\frac{\partial e}{\partial x}(x,t)\right]^2 w(x)\mathrm{d}x \\ &\le \sum_{k=0}^{N} r(x_k,t)e(x_k,t)w_k \le \frac{1}{2}\sum_{k=0}^{N}[r(x_k,t)]^2 w_k+\frac{1}{2}\sum_{k=0}^{N}[e(x_k,t)]^2 w_k \,.\end{aligned}$$

Here we have used the Cauchy-Schwarz inequality (see (A.2)). By the Gronwall lemma (see (A.15)) we get

$$\sum_{k=0}^{N}[e(x_k,t)]^2 w_k + \frac{1}{2}\int_0^t \int_{-1}^{1} \left[\frac{\partial e}{\partial x}(x,s)\right]^2 w(x)\mathrm{d}x\,\mathrm{d}s$$
$$\leq \exp(t)\int_0^t \sum_{k=0}^{N}[r(x_k,s)]^2 w_k \mathrm{d}s\ . \tag{6.1.7}$$

If we drop the second term on the left-hand side, we get an estimate of the discrete $L^2$-norm of the error $u - u^N$ at the collocation points:

$$\sum_{k=0}^{N}\left[u(x_k,t) - u^N(x_k,t)\right]^2 w_k \leq \exp(t)\int_0^t \sum_{k=0}^{N}[r(x_k,s)]^2 w_k \mathrm{d}s\ .$$

Hence, the scheme is convergent provided the truncation error vanishes as $N$ tends to infinity. Now we have

$$\begin{aligned}
\sum_{k=0}^{N}[r(x_k,s)]^2 w_k &= \sum_{k=0}^{N}[I_N r(x_k,s)]^2 w_k \leq 2\int_{-1}^{1}[I_N r(x,s)]^2 w(x)\mathrm{d}x \\
&= 2\int_{-1}^{1}\left[\left(I_N\frac{\partial^2 u}{\partial x^2} - \frac{\partial^2}{\partial x^2}(I_N u)\right)(x,s)\right]^2 w(x)\mathrm{d}x \\
&\leq 4\int_{-1}^{1}\left[\left(\frac{\partial^2 u}{\partial x^2} - I_N\frac{\partial^2 u}{\partial x^2}\right)(x,s)\right]^2 w(x)\mathrm{d}x \\
&+ 4\int_{-1}^{1}\left[\frac{\partial^2}{\partial x^2}(u - I_N u)(x,s)\right]^2 w(x)\mathrm{d}x\ ,
\end{aligned}$$

where we have used the equivalence (5.3.2) between discrete and continuous $L^2$-norms. Applying the estimate (5.5.26) in evaluating the right-hand side, we obtain the bound

$$\left(\sum_{k=0}^{N}\left[u(x_k,t) - u^N(x_k,t)\right]^2 w_k\right)^{1/2}$$
$$\leq CN^{3-m}\exp\left(\frac{t}{2}\right)\left(\int_0^t |u(s)|^2_{H_w^{m;N}(-1,1)}ds\right)^{1/2}, \tag{6.1.8}$$

where the norm on the right-hand side is defined in (5.5.8), and $C$ is a constant independent of $N$ and $u$.

Using (6.1.7) once more, one can derive an estimate for the spatial derivative of the error, i.e.,

$$\left(\int_0^t \int_{-1}^1 \left[\left(\frac{\partial u}{\partial x} - \frac{\partial u^N}{\partial x}\right)(x,s)\right]^2 w(x)\mathrm{d}x\,\mathrm{d}s\right)^{1/2} \\ \leq CN^{3-m}\exp\left(\frac{t}{2}\right)\left(\int_0^t |u(s)|^2_{H_w^{m;N}(-1,1)}\mathrm{d}s\right)^{1/2} . \tag{6.1.9}$$

This inequality proves that the approximation is convergent and the error decays faster than algebraically when the solution is infinitely smooth; the issue of the smoothness of the exact solution will be addressed below.

The previous analysis allows us to prove the convergence of the method in square mean norms by a transparent argument, namely, the comparison between the spectral solution and the Chebyshev interpolant of the exact solution at the collocation nodes. However, the rate of decay of the error predicted by this theory is not optimal, i.e., it is slower than the one corresponding to the best approximation. According to the previous estimate, the energy norm of the error decays at least like $N^{3-m}$, while the error of best approximation in the same norm decays like $N^{1-m}$ (see Sect. 5.5.2). Furthermore, the right-hand side of (6.1.9) blows up exponentially in time.

A more careful analysis allows us to state that the error for the collocation approximation considered here is actually asymptotic with the best approximation error, i.e., the following estimate can be obtained:

$$\left(\int_{-1}^1 \left[(u-u^N)(x,t)\right]^2 w(x)\mathrm{d}x\right)^{1/2} + \left(\int_0^t \int_{-1}^1 \left[\left(\frac{\partial u}{\partial x} - \frac{\partial u^N}{\partial x}\right)(x,s)\right]^2 w(x)\mathrm{d}x\,\mathrm{d}s\right)^{1/2} \\ \leq CN^{1-m}\left\{\int_0^t \left(\left|\frac{\partial u}{\partial t}(s)\right|^2_{H_w^{m-2;N}(-1,1)} + |u(s)|^2_{H_w^{m;N}(-1,1)}\right)\mathrm{d}s\right\}^{1/2} . \tag{6.1.10}$$

The details of this analysis are given in Example 3 of Sect. 6.5.1.

The previous estimates show that the rate of convergence of $u^N$ to $u$ as $N \to \infty$ depends on how many times $u$ is differentiable. Since in general the solution is obviously not known explicitly, the issue of deriving the smoothness of $u$ from the smoothness of the initial condition $u_0$ (which is the only nonzero data of our problem) arises in a natural way. For quite general initial- and boundary-value problems, analyzing how the smoothness of the data influences the smoothness of the solution may be a tremendously difficult task. In the present case, however, simple arguments can be used, that nevertheless illuminate a subtlety of the question: the smoothness of $u_0$ is a necessary condition for the smoothness of $u$, yet it is not sufficient. In order for $u(x,t)$ to be $k$-times continuously differentiable in $-1 \leq x \leq 1$ and $t \geq 0$, obviously $u_0(x) = u(x,0)$ has to be $k$-times continuously differentiable in $-1 \leq x \leq 1$. On the other hand, the continuity of $u$ at $x = \pm 1$, $t = 0$ and the boundary condition (6.1.3) force $u_0$ to vanish at $x = \pm 1$. If $u$ is twice continuously differentiable, (6.1.2) and again the fact that $u$ is zero at $x = \pm 1$ for all times, force the second derivative of $u_0$ to vanish at $x = \pm 1$; indeed,

$$\frac{d^2 u_0}{dx^2}(\pm 1) = \frac{\partial^2 u}{\partial x^2}(\pm 1, 0) = \frac{\partial u}{\partial t}(\pm 1, 0) = 0 .$$

If $u$ is four times continuously differentiable, then by differentiating the equation with respect to time we get

$$\frac{\partial^2 u}{\partial t^2} = \frac{\partial^3 u}{\partial t \partial x^2} = \frac{\partial^2}{\partial x^2}\left(\frac{\partial u}{\partial t}\right) = \frac{\partial^2}{\partial x^2}\left(\frac{\partial^2 u}{\partial x^2}\right) = \frac{\partial^4 u}{\partial x^4} ,$$

which, as above, yields $\frac{d^4 u_0}{dx^4}(\pm 1) = 0$. The argument can be iterated to prove that $u$ is infinitely differentiable (for all $x$ and $t$) if so is $u_0$ (for all $x$) *and if* all the even derivatives of $u_0$ vanish at $x = \pm 1$.

This is an example of *compatibility conditions* between the initial data $u_0$ and the boundary data (here, identically zero), that have to be satisfied to guarantee the smoothness of the exact solution. More general initial- and boundary-value problems require more elaborated compatibility conditions. Boyd (1999) provides a detailed discussion of this issue; a thorough mathematical analysis can be found, e.g., in Brezzi and Gilardi (1987).

### 6.1.3 A Legendre Tau Method for the Poisson Equation

In Sect. 1.2.4 we considered the homogeneous Dirichlet problem for the Poisson equation in the square $\Omega = (-1,1) \times (-1,1)$:

$$\begin{aligned} -\Delta u &= f , &\quad& -1 < x ,\ y < 1 , \\ u &= 0 && \text{if } x = \pm 1 \text{ or } y = \pm 1 . \end{aligned}$$

This problem was approximated by the following Legendre tau method. Let $\mathbb{P}_N$ denote the space of polynomials in two variables, $x, y$, of degree at most $N$ in each variable. The spectral solution $u^N$ belongs to $\mathbb{P}_N$ and is defined by

$$-\int_\Omega \Delta u^N \phi \, dx\, dy = \int_\Omega f \phi \, dx\, dy \quad \text{for all } \phi \in \mathbb{P}_{N-2} , \tag{6.1.11}$$

and by the boundary condition

$$u^N(x,y) = 0 \quad \text{if } x = \pm 1 \text{ or } y = \pm 1 . \tag{6.1.12}$$

The last condition was imposed in (1.2.75) in an integral way, i.e., it was translated into a set of linear relations among the Legendre coefficients. Since the problem is intrinsically formulated in a variational way, it is natural to try to derive the stability of the method from an appropriate choice of the test function $\phi$ in (6.1.11). Both the choices $\phi = u^N$ and $\phi = -\Delta u^N$ – which would immediately give stability – are not allowed, since these functions are polynomials of degree higher than $N-2$. They could be projected onto the space $\mathbb{P}_{N-2}$ of the admissible functions for (6.1.11). Instead, we adopt a different strategy. Since $u^N$ vanishes at the boundary of the square $(-1,1)\times(-1,1)$, it can be factored as

$$u^N(x,y) = (1-x^2)(1-y^2)q(x,y) \quad \text{for a } q \in \mathbb{P}_{N-2} \, .$$

We choose $\phi = q$ in (6.1.11). Let us denote by $b(x,y)$ the bubble function $(1-x^2)(1-y^2)$. Applying Green's formula twice (in which $\partial/\partial n$ is the outward normal derivative on the boundary $\partial\Omega$ of the square), we have

$$\begin{aligned} -\int_\Omega \Delta u^N q \, \mathrm{d}x \, \mathrm{d}y &= \int_\Omega \nabla(bq)\cdot\nabla q \, \mathrm{d}x \, \mathrm{d}y - \int_{\partial\Omega} \frac{\partial(bq)}{\partial n} q \, \mathrm{d}\sigma \\ &= \int_\Omega b|\nabla q|^2 \, \mathrm{d}x \, \mathrm{d}y + \frac{1}{2}\int_\Omega \nabla b \cdot \nabla(q^2) \mathrm{d}x \, \mathrm{d}y - \int_{\partial\Omega} \frac{\partial b}{\partial n} q^2 \, \mathrm{d}\sigma \\ &= \int_\Omega b|\nabla q|^2 \, \mathrm{d}x \, \mathrm{d}y - \frac{1}{2}\int_\Omega (\Delta b) q^2 \, \mathrm{d}x \, \mathrm{d}y - \frac{1}{2}\int_{\partial\Omega} \frac{\partial b}{\partial n} q^2 \, \mathrm{d}\sigma \, . \end{aligned} \tag{6.1.13}$$

Each term on the right-hand side is positive. On the other hand, the right-hand side of (6.1.11) can be bounded by the Cauchy-Schwarz inequality as follows:

$$\begin{aligned} &\left|\int_\Omega f q \, \mathrm{d}x \, \mathrm{d}y\right| \\ &= \left|\int_\Omega \frac{f}{\sqrt{|\Delta b|}}\sqrt{|\Delta b|} q \, \mathrm{d}x \, \mathrm{d}y\right| \le \left(\int_\Omega \frac{f^2}{|\Delta b|} \mathrm{d}x \, \mathrm{d}y\right)^{1/2} \left(\int_\Omega |\Delta b| q^2 \, \mathrm{d}x \, \mathrm{d}y\right)^{1/2} \\ &\le \int_\Omega \frac{f^2}{|\Delta b|} \mathrm{d}x \, \mathrm{d}y + \frac{1}{4}\int_\Omega |\Delta b| q^2 \, \mathrm{d}x \, \mathrm{d}y \, . \end{aligned}$$

By (6.1.13) and this inequality one gets

$$\int_\Omega b|\nabla q|^2 \mathrm{d}x \, \mathrm{d}y + \frac{1}{4}\int_\Omega |\Delta b| q^2 \, \mathrm{d}x \, \mathrm{d}y \le \int_\Omega \frac{f^2}{|\Delta b|} \mathrm{d}x \, \mathrm{d}y \, . \tag{6.1.14}$$

The integral on the right-hand side is certainly finite if $f$ is bounded in $\Omega$. Finally, using the identity $\nabla u^N = b\nabla q + q\nabla b$, and noting that $b \le 1$ and $|\nabla b|^2 \le 2|\Delta b|$, we have

$$\begin{aligned} \int_\Omega |\nabla u^N|^2 \mathrm{d}x \, \mathrm{d}y &\le 2\int_\Omega b^2 |\nabla q|^2 \mathrm{d}x \, \mathrm{d}y + 2\int_\Omega |\nabla b|^2 q^2 \, \mathrm{d}x \, \mathrm{d}y \\ &\le 2\int_\Omega b|\nabla q|^2 \mathrm{d}x \, \mathrm{d}y + 4\int_\Omega |\Delta b| q^2 \, \mathrm{d}x \, \mathrm{d}y \, ; \end{aligned}$$

whence, by (6.1.14),

$$\int_\Omega |\nabla u^N|^2 \mathrm{d}x\,\mathrm{d}y \le 16 \int_\Omega \frac{f^2}{|\Delta b|} \mathrm{d}x\,\mathrm{d}y\,. \tag{6.1.15}$$

This proves the stability of the Legendre tau method in the energy norm. Indeed, $(\int_\Omega |\nabla u^N|^2 \mathrm{d}x\,\mathrm{d}y)^{1/2}$ is a norm for $u^N$ since $u^N$ is zero on $\partial\Omega$.

In order to derive the convergence of the scheme, let $\tilde{u}$ denote a polynomial of degree $N$ vanishing on $\partial\Omega$, to be chosen later as a suitable approximation of the exact solution $u$. Then $e = \tilde{u} - u^N$ satisfies

$$-\int_\Omega \Delta e \phi \,\mathrm{d}x\,\mathrm{d}y = \int_\Omega \Delta(u - \tilde{u})\phi \,\mathrm{d}x\,\mathrm{d}y \quad \text{for all } \phi \in \mathbb{P}_{N-2}\,.$$

By the previous argument we get

$$\int_\Omega |\nabla e|^2 \mathrm{d}x\,\mathrm{d}y \le 16 \int_\Omega \frac{|\Delta(u-\tilde{u})|^2}{|\Delta b|} \mathrm{d}x\,\mathrm{d}y\,;$$

whence, by the triangle inequality,

$$\int_\Omega |\nabla(u-u^N)|^2 \mathrm{d}x\,\mathrm{d}y \le 2\int_\Omega |\nabla(u-\tilde{u})|^2 \mathrm{d}x\,\mathrm{d}y + C_1 \|\Delta(u-\tilde{u})\|^2_{L^\infty(\Omega)}\,, \tag{6.1.16}$$

where $C_1 = 32C_0$, and $C_0$ is the value of the integral of $1/|\Delta b|$ over $\Omega$ ($C_0$ is less than 3).

We now use a result about Sobolev spaces, which states that any function in $H^2(\Omega)$, where $\Omega$ is a two-dimensional domain, is bounded; furthermore, there is a constant $C > 0$ such that $\|v\|_{L^\infty(\Omega)} \le C\|v\|_{H^2(\Omega)}$ for all $v \in H^2(\Omega)$. Applying this result, we have $\|\Delta(u-\tilde{u})\|_{L^\infty(\Omega)} \le \|u-\tilde{u}\|_{H^4(\Omega)}$, and we are led to choose as $\tilde{u}$ the polynomial that satisfies (5.8.16) for $l = 4$ and $\lambda = 1$. Then, the square root of the right-hand side of (6.1.16) can be bounded by $CN^{4-m}|u|_{H^{m;N}(\Omega)}$ . This, however, is not the best rate of convergence. A more clever choice of $\tilde{u}$, involving orthogonal projections in Sobolev spaces of high order, yields the estimate, for all real $p > 4$ and $m < N$,

$$\left(\int_\Omega |\nabla(u-u^N)|^2 \mathrm{d}x\,\mathrm{d}y\right)^{1/2} \le CN^{2-m}\left(\int_\Omega |D^m u|^p \mathrm{d}x\,\mathrm{d}y\right)^{1/p}\,. \tag{6.1.17}$$

The details are given in Sacchi-Landriani (1988).

As for the previous example, we remark that the smoothness of $u$, which determines the rate of convergence of the approximation, depends on the smoothness of the internal and boundary data and on certain *compatibility conditions* among them, which also involve the smoothness of the domain. Note for instance that the boundary conditions considered here force the Laplacian of a smooth solution $u$ to vanish at the four corners of the domain, requiring $f$ to vanish there as well. We refer to Grisvard (1985) for a rigorous mathematical treatment of the subject.

## 6.2 Towards a General Theory

In the previous section a mathematical analysis was sketched for the stability and convergence properties of three representative spectral methods. This analysis relied in a fundamental way upon interpreting the schemes as projection methods over suitable subspaces with respect to the appropriate inner products. The projection analysis is certainly natural for the Galerkin, G-NI and tau methods. It appears, however, to be unnatural for the collocation method, which is usually implemented in a pointwise manner. Unfortunately, in all but the simplest cases, the pointwise analysis of collocation methods is not only far more difficult than their projection analysis, it is also less precise, i.e., the error estimates suggest a lower rate of convergence than is achieved in practice. (The mathematical reasons for this are similar to those that make optimal error estimates easier to obtain for finite-element methods than for finite-difference methods.) An additional reason for preferring the projection analysis of collocation methods is that it enables all spectral methods to be discussed in terms of the same general theory.

As we noted in the introduction of Chap. 1, the finite-dimensional space on which the equation is projected is not necessarily the same finite-dimensional space in which the spectral solution lies. Galerkin methods invariably use the same space for both purposes. The Legendre tau approximation discussed in Sect. 6.1.3 is an example of a situation in which the two spaces differ. Many familiar collocation methods also use two different spaces. It follows that a unified approach to the theory must necessarily involve two families of finite-dimensional spaces, one for the trial functions and the other for the test functions.

The most straightforward technique for establishing the stability of the spectral schemes – the so-called energy method – is based on choosing the solution itself as the test function. This approach is successful if the spaces of the trial and test functions coincide, and if the spectral operator is positive with respect to a suitable inner product (as occurred in the first two examples of the previous section). If either of these hypotheses is not satisfied, then the energy method cannot be used. In an alternative strategy, which is often invoked, stability is proven by building up a suitable test function that depends in some way on the spectral solution. This was the strategy employed in the last example of Sect. 6.1. Generally speaking, the inequality that is associated with the energy method and that ensures stability must be replaced by a more general inequality. Mathematically, this inequality amounts to the requirement that the spectral operator be an isomorphism (i.e., a continuous invertible map) between the spaces of trial and test functions, and that a suitable norm of its inverse be bounded independently of the discretization parameter.

The convergence analysis given for the introductory examples of this chapter used the standard technique of systematically comparing the spectral solution with a projection of the exact solution onto the space of the trial functions. This strategy is essentially the same as that used in the proof of the Lax-Richtmyer equivalence theorem (which states that for consistent approximations, stability is equivalent to convergence).

The last two examples in Sect. 6.1 show that the error estimate (i.e., the rate of decay of the error) predicted by this approach is extremely sensitive to the approximation properties of the particular projection of the exact solution that one chooses in this analysis. Both the truncated series and the interpolant of the exact solution appear to be viable candidates for the projection. However, the rates of decay predicted by choosing these functions may be asymptotically worse than the errors of best approximation in the same norms. (This point has already been emphasized in Chap. 5.) Typically, one chooses a projection of the exact solution that yields the same approximation properties as the best approximation. Such projection operators were introduced in Sects. 5.4.2, 5.5.2 and 5.8 and will play a key role in the subsequent convergence analysis.

## 6.3 General Formulation of Spectral Approximations to Linear Steady Problems

Let $\Omega$ be an open bounded domain in $\mathbb{R}^d$, with piecewise smooth boundary $\partial\Omega$. We assume that we want to approximate the boundary-value problem

$$\mathcal{L}u = f \qquad \text{in } \Omega\,, \tag{6.3.1}$$

$$\mathcal{B}u = 0 \qquad \text{on } \partial\Omega_b\,, \tag{6.3.2}$$

where $\mathcal{L}$ is a linear differential operator in $\Omega$, and $\mathcal{B}$ is a set of linear boundary differential operators on a part (or the whole) of $\partial\Omega$ that we call $\partial\Omega_b$.

We assume that there exists a Hilbert space $X$ such that $\mathcal{L}$ is an unbounded operator in $X$ (see (A.1) and (A.3)). We will denote by $(u,v)$ the inner product in $X$ and by $\|u\| = (u,u)^{1/2}$ the associated norm. Typically, $X$ will be a space of real or complex functions defined in $\Omega$ that are square integrable with respect to a suitable *weight* function. Hereafter, by weight function we shall mean a continuous and strictly positive function in $\Omega$ that is properly or improperly integrable. The domain of definition of $\mathcal{L}$, i.e., the subset $D(\mathcal{L})$ of those functions $u$ of $X$ for which $\mathcal{L}u$ is still an element of $X$, is supposed to be a dense subspace of $X$ (see (A.6)). Thus, $\mathcal{L}$ is a linear operator from $D(\mathcal{L})$ to $X$.

The following elementary example will serve as a model for the theoretical presentation. Let us consider the second-derivative operator, $\mathcal{L} = -\mathrm{d}^2/\mathrm{d}x^2$, on the interval $\Omega = (-1,1)$. If $w$ denotes either the Legendre weight, $w(x) = 1$, or the Chebyshev weight, $w(x) = 1/\sqrt{1-x^2}$, we set $X = L^2_w(-1,1) = \{v \mid \int_{-1}^1 v^2(x)w(x)\mathrm{d}x < \infty\}$ with $(u,v) = \int_{-1}^1 u(x)v(x)w(x)\mathrm{d}x$. Then $\mathcal{L}$ is an unbounded operator in $X$ whose domain is

$$D(\mathcal{L}) = \left\{ v \in C^1(-1,1) \;\middle|\; \frac{\mathrm{d}^2 v}{\mathrm{d}x^2} \in L^2_w(-1,1) \right\} ,$$

where the derivative is taken in the sense of distributions (see (A.10)).

We assume that the boundary operators make sense when applied to all the functions of the domain $D(\mathcal{L})$. Prescribing the boundary conditions (6.3.2) amounts to restricting the domain of $\mathcal{L}$ to the subspace $D_B(\mathcal{L})$ of $D(\mathcal{L})$ defined by

$$D_B(\mathcal{L}) = \{ v \in D(\mathcal{L}) \mid \mathcal{B}v = 0 \text{ on } \partial\Omega_b \} ,$$

which again we assume to be dense in $X$. Hence, we consider $\mathcal{L}$ as acting between $D_B(\mathcal{L})$ and $X$,

$$\mathcal{L} : D_B(\mathcal{L}) \subset X \longrightarrow X ,$$

and problem (6.3.1)–(6.3.2) can be written as

$$\begin{aligned} &u \in D_B(\mathcal{L}) , \\ &\mathcal{L}u = f , \end{aligned} \tag{6.3.3}$$

for $f \in X$ (the equality is between two functions in $X$).

In the previous example, the operator $\mathcal{L}$ can be supplemented, for instance, either with Dirichlet boundary conditions, $\mathcal{B}u(\pm 1) = u(\pm 1) = 0$, or with Neumann boundary conditions, $\mathcal{B}u(\pm 1) = u_x(\pm 1) = 0$. Notice that in both cases the boundary conditions make sense, since the functions of $D_B(\mathcal{L})$ are continuous with their first derivative. The density of $D_B(\mathcal{L})$ into $L^2_w(-1,1)$ is a consequence of the density of $\mathscr{D}(-1,1)$ into $L^2_w(-1,1)$ (see (A.9)).

The second condition in (6.3.3) can be equivalently written as

$$(\mathcal{L}u, v) = (f, v) \qquad \text{for all } v \in X .$$

The left-hand side is a bilinear form on $D_B(\mathcal{L}) \times X$ (i.e., it is a real- or complex-valued function that depends linearly on both arguments); we will denote it by $a(u,v)$. Similarly, the right-hand side is a linear form on $X$, that will be denoted by $F(v)$. Thus, (6.3.3) can be written as

$$\begin{aligned} &u \in D_B(\mathcal{L}) , \\ &a(u,v) = F(v) \qquad \text{for all } v \in X . \end{aligned} \tag{6.3.4}$$

The bilinear form $a(u,v)$ can often be given an equivalent expression that is defined on couples of spaces other than $D_B(\mathcal{L}) \times X$, say $W \times V$, that are more appropriate for showing that (6.3.3) is well-posed and for defining a numerical approximation. The space $W$ contains functions that are less regular than those in $D_B(\mathcal{L})$, whereas $V$ contains functions that are more regular than those in $X$. Usually, the equivalent expression is obtained by applying some integration-by-parts and using the boundary conditions. For instance, for the example above with $w(x) = 1$, we have

$$(\mathcal{L}u, v) = \int_{-1}^{1} -\frac{\mathrm{d}^2 u}{\mathrm{d}x^2} v \,\mathrm{d}x = \int_{-1}^{1} \frac{\mathrm{d}u}{\mathrm{d}x}\frac{\mathrm{d}v}{\mathrm{d}x}\,\mathrm{d}x = a(u,v)\,,$$

provided $\mathrm{d}v/\mathrm{d}x$ belongs to $L^2(-1,1)$ and at each endpoint of the interval either $\mathrm{d}u/\mathrm{d}x$ or $v$ vanishes. Once the spaces $W$ and $V$ are introduced, the formulation of the problem is

$$\begin{aligned} &u \in W\,, \\ &a(u,v) = F(v) \qquad \text{for all } v \in V\,. \end{aligned} \tag{6.3.5}$$

Concerning the conditions that guarantee the well-posedness of the problem, the simplest case occurs when the operator $\mathcal{L}$ satisfies a *coercivity condition*. Let us assume that there is a Hilbert space $E$ densely contained in $X$ with norm $\|u\|_E$, for which there exists a positive constant $C$ such that $\|u\| \le C\|u\|_E$ for all $u \in E$; moreover, let $D_B(\mathcal{L})$ be densely contained in $E$. We assume that the bilinear form $a(u,v)$ is defined on $E \times E$ and there exist constants $\alpha > 0$ and $A > 0$ such that

$$\alpha\|u\|_E^2 \le a(u,u) \qquad \text{for all } u \in E\,, \tag{6.3.6}$$

$$|a(u,v)| \le A\|u\|_E\|v\|_E \quad \text{for all } u \in E \text{ and } v \in E\,. \tag{6.3.7}$$

Thus, the spaces $W$ and $V$ mentioned above coincide with $E$, which is the subspace of the functions $u \in X$ having "finite" energy, the energy being precisely given by $\|u\|_E^2$. Inequality (6.3.6) is the coercivity condition for the bilinear form $a(u,v)$; it states that $\mathcal{L}$ supplemented with the prescribed boundary conditions is a positive operator, that is coercive over $E$. On the other hand, (6.3.7) is a continuity condition for $\mathcal{L}$ (in the sense that $(\mathcal{L}u,v)$ depends continuously on $u$ and $v$ in the norm of $E$). Furthermore, the linear form $F(v) = (f,v)$ obviously satisfies the inequality $|F(v)| \le \|f\|\,\|v\| \le C\|f\|\,\|v\|_E$, i.e., there exists a constant $C_F > 0$ such that

$$|F(v)| \le C_F\|v\|_E \qquad \text{for all } v \in E\,. \tag{6.3.8}$$

Under conditions (6.3.6)–(6.3.8), the Lax-Milgram theorem (see (A.5)) assures us that there exists a unique $u$ that is a solution of the problem

$$\begin{aligned} &u \in E\,, \\ &a(u,v) = F(v) \qquad \text{for all } v \in E\,. \end{aligned} \tag{6.3.9}$$

Such a function depends continuously on $f$, namely, one has

$$\|u\|_E \le \frac{C}{\alpha}\|f\|\,, \tag{6.3.10}$$

and actually $u$ solves the original problem (6.3.3).

Going back to the example considered above, let us assume that Dirichlet boundary conditions are prescribed for the operator $\mathcal{L} = -\mathrm{d}^2/\mathrm{d}x^2$. Then conditions (6.3.6) and (6.3.7) are satisfied with $E = H^1_{w,0}(-1,1)$ (see (A.11.c)), which is a Hilbert space for the norm

$$\|u\|_E = \left( \int_{-1}^{1} |u_x|^2 w \, \mathrm{d}x \right)^{1/2} .$$

This result is immediate if the Legendre weight is used, whereas it will be proven in Chap. 7 (see Theorem 7.1) for the Chebyshev weight. Note that all functions in $E$ satisfy the boundary conditions. On the other hand, if Neumann boundary conditions are prescribed for the operator $\mathcal{L} = -\mathrm{d}^2/\mathrm{d}x^2 + I$, and if we choose $w(x) = 1$, then conditions (6.3.6) and (6.3.7) are satisfied with $E = H^1(-1,1)$ (see (A.11.a)), which is a Hilbert space for the norm

$$\|u\|_E = \left( \int_{-1}^{1} (|u|^2 + |u_x|^2) \, \mathrm{d}x \right)^{1/2} .$$

In this case, the functions in $E$ need not satisfy the boundary conditions; however, the solution of (6.3.9) will satisfy them, as it can be seen by counterintegrating by parts and letting $v$ vary in $E$.

The positivity condition (6.3.6) is the most immediate condition that guarantees the well-posedness of problem (6.3.3). However, there are situations for which it is not fulfilled. In such cases, one can resort to a more general condition, known as the *inf-sup condition*, which we now present.

Let $W \subseteq X$ and $V \subseteq X$ be Hilbert spaces, whose norms will be denoted by $\|u\|_W$ and $\|u\|_V$, respectively. We assume that the inclusion of $V$ into $X$ is continuous, in the sense that there exists a suitable constant $C > 0$ such that $\|v\| \le C\|v\|_V$ for all $v \in X$. We suppose that $D_B(\mathcal{L})$ is densely contained in $W$ and that $V$ is densely contained in $X$. Furthermore, we assume that the bilinear form $a(u,v)$ is defined in $W \times V$, and that there exist constants $\alpha > 0$ and $A > 0$ such that

$$0 < \sup_{u \in W} a(u,v) \quad \text{for all } v \in V \,, \quad v \neq 0 \,, \tag{6.3.11}$$

$$\alpha \|u\|_W \le \sup_{\substack{v \in V \\ v \neq 0}} \frac{a(u,v)}{\|v\|_V} \quad \text{for all } u \in W \,, \tag{6.3.12}$$

$$|a(u,v)| \le A\|u\|_W \|v\|_V \quad \text{for all } u \in W \text{ and } v \in V \,. \tag{6.3.13}$$

Using an extended form of the Lax-Milgram theorem (see Nečas (1962)), conditions (6.3.11)–(6.3.13) together with (6.3.8) assure that problem (6.3.5) has a unique solution that depends continuously on the data, i.e.,

$$\|u\|_W \le \frac{C}{\alpha} \|f\| \,.$$

Again, the function $u$ so defined is indeed the solution of (6.3.3).

Note that conditions (6.3.11) and (6.3.12) are implied by the coercivity condition (6.3.6) by choosing $V = W = E$.

As an example, consider a second-order operator of the form $\mathcal{L}u = -(a(x)u_x)_x$ in the interval $\Omega = (-1,1)$, where $a(x)$ is a smooth, strictly positive function. It will be supplemented by homogeneous Dirichlet boundary conditions. The operator $\mathcal{L}$ can still be defined on $X = L^2_w(-1,1)$, its domain of definition once again being $D(\mathcal{L}) = \{v \in C^1(-1,1) \mid v_{xx} \in L^2_w(-1,1)\}$. The coercivity condition (6.3.6) may not be satisfied with $E = H^1_{w,0}(-1,1)$. However, conditions (6.3.11)–(6.3.13) are fulfilled if we take $W = V = H^1_{w,0}(-1,1)$ (see the discussion in Example 3 of Sect. 6.4).

Another example is given by the operator $\mathcal{L}u = -u_{xx} + u$ supplemented with homogeneous Neumann boundary conditions. For this problem, conditions (6.3.11)–(6.3.13) are fulfilled with the choice $V = L^2_w(-1,1)$ and $W = \{u \in H^2_w(-1,1) \mid u_x(\pm 1) = 0\}$ (see Example 4 of Sect. 6.4).

### *Spectral Approximations*

We will describe in general terms the process that leads to the definition of a spectral approximation of problem (6.3.3). The discussion of Galerkin, collocation, G-NI and tau methods given in Sect. 6.4 will be based on the framework we are going to state.

Keeping in mind the formulations (6.3.9) or (6.3.5) of Problem (6.3.3), a spectral approximation will be cast in the form

$$\begin{aligned} &u^N \in X_N\,, \\ &a_N(u^N, v) = F_N(v) \qquad \text{for all } v \in Y_N\,, \end{aligned} \tag{6.3.14}$$

where $X_N$ and $Y_N$ are finite-dimensional subspaces of $X$ having the same dimension, $a_N$ is a bilinear form defined in $X_N \times Y_N$ that approximates the bilinear form $a$, whereas $F_N$ is a linear form on $Y_N$ that approximates the linear form $F$. Depending on how the boundary conditions are enforced, $X_N$ may be contained in $D_B(\mathcal{L})$, i.e., each function of $X_N$ satisfies exactly the prescribed boundary conditions, or not; in the latter case, the spectral solution will satisfy the boundary conditions in an approximate way only.

The *Galerkin method* consists of restricting both the trial and the test function spaces for (6.3.9) or (6.3.5) to a finite-dimensional space $X_N$; thus, a Galerkin scheme is defined by

$$\begin{aligned} &u^N \in X_N \\ &a(u^N, v) = F(v) \qquad \text{for all } v \in X_N\,. \end{aligned} \tag{6.3.15}$$

The *Galerkin with numerical integration (G-NI) method* is obtained from this formulation by replacing all the integrals that appear in the bilinear and linear forms by high-precision (Gaussian) quadrature formulas. Appending a suffix $N$ to the resulting forms, a G-NI scheme can be written as

$$\begin{aligned} &u^N \in X_N\,, \\ &a_N(u^N, v) = F_N(v) \qquad \text{for all } v \in X_N\,. \end{aligned} \tag{6.3.16}$$

Formulations (6.3.15) or (6.3.16) are quite general, as they account for the possible weak imposition of boundary conditions, and even for the weak enforcement of subdomain matching, in the multidomain form of spectral methods discussed in Chaps. 5 and 6.

A more restricted setting often suffices in the single-domain form of spectral methods, considered in this book. Indeed, in such a case all approximating functions are usually smooth in $\Omega$ (e.g., they are polynomials); hence, the operator $\mathcal{L}$ is surely defined on $X_N$. Whenever the trial functions in $X_N$ individually satisfy the boundary conditions, i.e., whenever $X_N \subset D_B(\mathcal{L})$, the bilinear form $a$ can actually be written in the strong form $a(u,v) = (\mathcal{L}u, v)$ for all $u \in X_N$. Then, the Galerkin method reads as

$$\begin{aligned} &u^N \in X_N \,, \\ &(\mathcal{L}u^N, v) = (f, v) \qquad \text{for all } v \in X_N \,. \end{aligned} \tag{6.3.17}$$

The *tau method* is obtained by allowing the test functions to vary in a space $Y_N$ different from $X_N$: it has the same dimension as $X_N$, but its functions need not individually satisfy the boundary conditions, as must those in $X_N$. A tau scheme is usually written as

$$\begin{aligned} &u^N \in X_N \,, \\ &(\mathcal{L}u^N, v) = (f, v) \qquad \text{for all } v \in Y_N \,. \end{aligned} \tag{6.3.18}$$

The *collocation method* can be written in a form similar to (6.3.17), namely,

$$\begin{aligned} &u^N \in X_N \,, \\ &(\mathcal{L}_N u^N, v)_N = (f, v)_N \qquad \text{for all } v \in X_N \,. \end{aligned} \tag{6.3.19}$$

Here, $\mathcal{L}_N$ is an approximation of $\mathcal{L}$, usually obtained by replacing exact derivatives by interpolation derivatives (see Sects. 2.1.3, 2.3.2, and 2.4.2). Furthermore, $(u,v)_N$ is a bilinear form, usually defined through the values of $u$ and $v$ at the collocation points only, that is an inner product in $X_N$. This form is indeed defined on a subspace $Z \subset X$, composed of continuous functions for which the pointwise value is meaningful; obviously, we assume that $\mathcal{L}_N$ maps $X_N$ into $Z$ and that $f \in Z$.

**Remark.** Formulations (6.3.17)–(6.3.19) can be summarized in the abstract form

$$\begin{aligned} &u^N \in X_N \,, \\ &(\mathcal{L}_N u^N - f, v)_N = 0 \qquad \text{for all } v \in Y_N \,, \end{aligned} \tag{6.3.20}$$

with suitable definitions of $\mathcal{L}_N$, $Y_N$ and $(u,v)_N$ which depend on the particular method. This form attests to the fact that a spectral scheme so defined is actually a method of *weighted residuals*. The choice of the space $Y_N$ and the inner product $(u,v)_N$ in $Y_N$ defines the way the residual $\mathcal{L}_N u^N - f$ is minimized.

An equivalent operational form of (6.3.20) is

$$\begin{aligned} & u^N \in X_N \,, \\ & Q_N \mathcal{L}_N u^N = Q_N f \,, \end{aligned} \tag{6.3.21}$$

where $Q_N : Z \subseteq X \longrightarrow Y_N$ satisfies

$$(z - Q_N z, v)_N = 0 \quad \text{for all } v \in Y_N \,, \tag{6.3.22}$$

i.e., it is the orthogonal projection upon $Y_N$ in the inner product $(u, v)_N$. Fig. 6.1 represents the function spaces and the operators involved in the formulation (6.3.21).

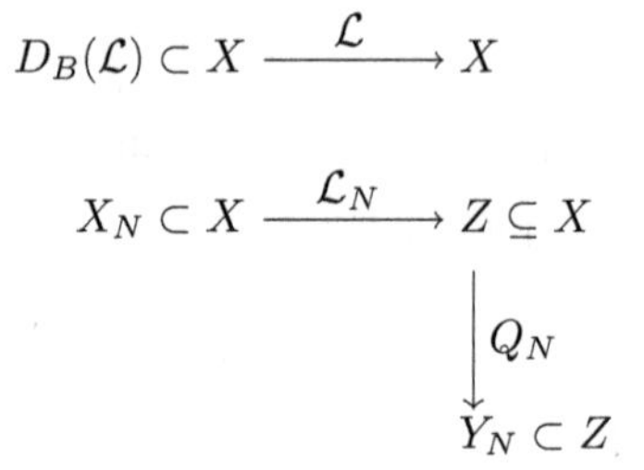

$$\begin{array}{ccc} D_B(\mathcal{L}) \subset X & \xrightarrow{\quad \mathcal{L} \quad} & X \\ & & \\ X_N \subset X & \xrightarrow{\quad \mathcal{L}_N \quad} & Z \subseteq X \\ & & \big\downarrow Q_N \\ & & Y_N \subset Z \end{array}$$

**Fig. 6.1.** The spaces and the operators involved in the abstract formulation (6.3.21) of a spectral method

## 6.4 Galerkin, Collocation, G-NI and Tau Methods

In this section, we will provide a general formulation of the fundamental types of spectral methods. The formulation will be given in a way that fits into the general framework introduced above and at the same time permits the construction of an algorithm for the solution. The essential elements for each method are the spaces of the trial and of the test functions, the bilinear and linear forms $a_N(u, v)$ and $F_N(v)$, or, in a more restricted setting, the projection operator $Q_N$ and the inner product $(u, v)_N$. Several examples of approximations to steady boundary-value problems will be discussed for each method. General theorems will be given that guarantee stability and convergence results for each method. Some of the cumbersome details will be omitted.

Galerkin, collocation, G-NI and tau methods are not the only schemes of spectral type that can be conceived and that are actually used in applications. Indeed, for some problems a method that combines two or more of these schemes may be the most flexible and efficient. An important example is provided by certain algorithms for the incompressible Navier-Stokes equations, that couple a tau discretization of the diffusive term with a different type of spectral method (Galerkin, collocation or pseudospectral) for the convective term. Such combined schemes can be often analyzed using elements of the theory presented separately here for the four fundamental schemes.

***The Space*** $\mathrm{Pol}_N(\Omega)$

In what follows we maintain the same notation used in Sect. 6.3. However, we now specify that the domain $\Omega$ in which the problem (6.3.1) has to be solved, is the product of the intervals $(0, 2\pi)$ or $(-1, 1)$ according to the type of prescribed boundary conditions. Precisely, we set

$$\Omega = \prod_{k=0}^{d} I_k ,$$

where $I_k = (0, 2\pi)$ if periodicity is required in the $x_k$-direction, and $I_k = (-1, 1)$ otherwise. Thus, $\Omega$ may be either the physical domain or the computational domain on which the original problem has been mapped, as is done in many applications (see Sects. 2.5 and 2.8.1).

For each integer $N$ the spectral approximation involves functions that in each variable are either trigonometric or algebraic polynomials of degree up to $N$. We shall denote by $\mathrm{Pol}_N(\Omega)$ the set of these functions. Precisely, $\mathrm{Pol}_N(\Omega)$ *is the space of the continuous functions* $u : \Omega \to \mathbb{C}$ *such that* $u$ *is a trigonometric polynomial of degree* $\leq N$ *in the variables* $x_k$ *for which* $I_k = (0, 2\pi)$*, and an algebraic polynomial of degree* $\leq N$ *in each of the remaining variables.* More generally, $N$ could denote a multi-integer, $N = (N_1, \ldots, N_d)$, collecting the (possibly different) polynomial degrees in each space variable. If there are no directions of periodicity, the functions of $\mathrm{Pol}_N(\Omega)$ will be real-valued.

It will always be assumed that $\mathrm{Pol}_N(\Omega)$ is contained in the domain of definition $D(\mathcal{L})$ of the operator $\mathcal{L}$.

The geometric and functional setting just described is the natural one in classical spectral methods. As noted in Sect. 2.9, spectral expansions in non-Cartesian domains $\Omega$ such as triangles, hexahedra, pyramids and prisms can be defined as well, and indeed they have become popular in recent years. In these cases, the definition of the space $\mathrm{Pol}_N(\Omega)$ should be modified conveniently. For instance, if $\Omega = \mathcal{T}$ is a tetrahedron, $\mathrm{Pol}_N(\Omega)$ is the space $\mathcal{P}_N(\mathcal{T})$ of the algebraic polynomials of total degree $\leq N$ in the variables $x_1$, $x_2$ and $x_3$.

Although we will not present specific examples, the reader is invited to bear in mind that all the subsequent analysis applies to the non-Cartesian situation as well.

### 6.4.1 Galerkin Methods

In this subsection, we confine ourselves to the conceptually simplest version of a Galerkin method, the one in which trial and test functions individually satisfy all prescribed boundary conditions. This is the natural approach for handling the Dirichlet problem for a second-order operator. A different strategy consists of using trial and test functions that fulfill only some, or even none, of the boundary conditions; the remaining ones are incorporated into the weak (integral) formulation of the differential equation by manipulating the boundary terms that appear after integration-by-parts. The latter approach, which allows for a natural treatment, e.g., of the Neumann or mixed Dirichlet/Neumann problem for a second-order operator, can be analyzed by the same tools as for the G-NI methods; hence, we refer to Section 6.4.3 for its study.

Let $X_N$ be the subspace of $\mathrm{Pol}_N(\Omega)$ of the functions that satisfy the boundary conditions, so that $X_N \subset D_B(\mathcal{L})$. Choose a basis $\{\phi_k, k \in J\}$ in $X_N$, where $J$ is a set of indices. The $\phi_k$'s need not be orthogonal in the inner product of $X$. A Galerkin method is defined by the equations

$$\begin{aligned} &u^N \in X_N\,, \\ &(\mathcal{L}u^N, \phi_k) = (f, \phi_k) \qquad \text{for all } k \in J\,. \end{aligned} \tag{6.4.1}$$

Usually, the unknowns are the coefficients $\alpha_k$ in the expansion $u^N = \sum_{k\in J} \alpha_k \phi_k$. Equations (6.4.1) can be equivalently written as

$$\begin{aligned} &u^N \in X_N\,, \\ &(\mathcal{L}u^N, v) = (f, v) \qquad \text{for all } v \in X_N\,, \end{aligned} \tag{6.4.2}$$

which is nothing else than (6.3.17). It follows that with respect to the general formulation (6.3.20), a Galerkin method is defined by the choices $Y_N = X_N$ and $(u, v)_N = (u, v)$, the inner product of $X$. We note that $Q_N$ is the orthogonal projection from $X$ into $X_N$ in the inner product of $X$. Moreover, we have assumed that $\mathcal{L}_N = \mathcal{L}$, as occurs in most applications. A generalization of the Galerkin method is the so-called *Petrov-Galerkin* method. With this method, test functions differ from trial functions, though they individually satisfy the boundary conditions. In this case, we have $X_N \neq Y_N$, and (6.4.2) is replaced by

$$\begin{aligned} &u^N \in X_N\,, \\ &(\mathcal{L}u^N, v) = (f, v) \qquad \text{for all } v \in Y_N\,. \end{aligned}$$

An example is given by Leonard's method for the incompressible Navier-Stokes equations (see CHQZ3, Sect. 3.4).

### *Stability and Convergence*

We are now concerned with the *stability and convergence properties* of Galerkin approximations. The simplest case occurs when the the bilinear form $a(u, v) = (\mathcal{L}u, v)$ satisfies the coercivity condition (6.3.6) and the continuity condition (6.3.7), and each $X_N$ is contained in $E$. Then we have

$$\alpha\|u\|_E^2 \leq (\mathcal{L}u, u) \quad \text{for all } u \in X_N \tag{6.4.3}$$

and

$$|(\mathcal{L}u, v)| \leq A\|u\|_E\|v\|_E \quad \text{for all } u,\ v \in X_N\,. \tag{6.4.4}$$

*If (6.4.3) holds, then the Galerkin approximation (6.4.2) is stable, in the sense that the following estimate holds:*

$$\|u^N\|_E \leq \frac{C}{\alpha}\|f\| \,. \tag{6.4.5}$$

Actually, choosing as test function in (6.4.2) the solution itself, and using the coercivity condition (6.4.3) on the left-hand side and the Cauchy-Schwarz inequality on the right-hand side, one has

$$\alpha\|u^N\|_E^2 \leq (\mathcal{L}u^N, u^N) = (f, u^N) \leq \|f\|\|u^N\| \,.$$

Recalling that $\|u^N\| \leq C\|u^N\|_E$, we have (6.4.5). Note that this inequality is the same as the one satisfied by the exact solution (compare with (6.3.10)). Inequality (6.4.5) also proves that (6.4.2) has a unique solution, since the problem is linear (indeed, the only solution corresponding to $f = 0$ is $u^N = 0$).

When (6.4.3) is satisfied, the stability of the approximation (6.4.2) is achieved by the *energy method* (and (6.4.5) is referred to as an *energy inequality*).

If stability is assured, convergence is a consequence of a consistency hypothesis, according to the Lax-Richtmyer equivalence theorem. In the Galerkin framework, the consistency hypothesis is expressed by the condition that $X$ is well-approximated by the family of the $X_N$'s. More precisely, assume that there exists a dense subspace $\mathcal{W} \subseteq D_B(\mathcal{L})$, ($\mathcal{W}$ will be a space of sufficiently smooth functions), and for all $N > 0$, a projection operator

$$R_N : \mathcal{W} \longrightarrow X_N \,, \tag{6.4.6}$$

such that for $N \to \infty$,

$$\|u - R_N u\|_E \longrightarrow 0 \quad \text{for all } u \in \mathcal{W} \,. \tag{6.4.7}$$

Under this consistency hypothesis, the approximation (6.4.2) is convergent. Actually, $e = u^N - R_N u$ satisfies, by (6.4.2),

$$(\mathcal{L}e, v) = (\mathcal{L}(u - R_N u), v) \quad \text{for all } v \in X_N \,.$$

Then by (6.4.3) and (6.4.4), it follows that

$$\|e\|_E \leq \frac{A}{\alpha}\|u - R_N u\|_E \,.$$

Since $u - u^N = u - R_N u - e$, we deduce the error bound

$$\|u - u^N\|_E \leq \left(1 + \frac{A}{\alpha}\right)\|u - R_N u\|_E \,. \tag{6.4.8}$$

This inequality implies convergence for all $u \in \mathcal{W}$ due to the assumption (6.4.7). (Note that convergence occurs even if $u$ is just a function in $E$, provided $\mathcal{W}$ is dense in $E$.) The above equality states the well-known fact that the error of a Galerkin approximation behaves like the error of best approximation in the norm for which stability is proven (*Céa's lemma*).

In order to check the consistency hypothesis, one could choose as $R_N$ the orthogonal projection operator onto $X_N$ with respect to the inner product $(u, v)$ of $X$. However, such orthogonal projection of an element $u \in E$ is generally less accurate than the best approximation of $u$ in the energy norm among the elements in $X_N$. (This has been noticed throughout Chap. 5.) Thus, this choice (that nevertheless allows us to prove convergence) is not the best possible one from the point of view of the analysis of convergence: *the rate of decay of the error predicted by estimate (6.4.8) with such $R_N u$ is generally slower than the real one.*

To get an optimal error estimate, $R_N u$ is usually chosen as the best approximation of $u$ in $X_N$ with respect to the $E$-norm, or as an element in $X_N$ that asymptotically behaves like the best approximation in the $E$-norm, namely,

$$\|u - R_N u\|_E \le C \inf_{v \in X_N} \|u - v\|_E$$

for a constant $C$ independent of $N$. This error can be bounded according to the estimates presented in Chap. 5. Spectral convergence is then a consequence of the smoothness of the exact solution; as noted at the end of Sect. 6.1.3, this in turn follows from the smoothness of the data and possibly the fulfillment of certain compatibility conditions among them.

We now consider some examples that illustrate the theory presented so far.

### *Examples*

**Example 1.** *The Helmholtz Equation in the Square with Periodic Boundary Conditions.* Let us consider the boundary-value problem

$$\begin{aligned} -\Delta u + \lambda u &= f \qquad \text{in } \Omega = (0, 2\pi) \times (0, 2\pi) \, , \\ u &\text{ periodic in } \Omega \, , \end{aligned}$$

with $\lambda > 0$ and $f \in L^2(\Omega)$. Using the Fourier Galerkin approximation, the solution $u^N$ belongs to $X_N = \text{span}\{e^{i(kx+my)} \mid -N \le k, m \le N-1\}$ and satisfies, for $-N \le k, m \le N-1$,

$$\int_\Omega (-\Delta u^N + \lambda u^N) e^{-i(kx+my)} \mathrm{d}x \, \mathrm{d}y = \int_\Omega f e^{-i(kx+my)} \mathrm{d}x \, \mathrm{d}y \, .$$

Equivalently, the Fourier coefficients $\hat{u}^N_{km}$ of $u^N$ are defined in terms of the Fourier coefficients $\hat{f}_{km}$ of $f$ by the set of linear relations

$$(k^2 + m^2 + \lambda) \hat{u}^N_{km} = \hat{f}_{km} \, , \quad -N \le k, m \le N-1 \, .$$

Thus, $X = L^2(\Omega)$, and $(u, v) = \int_\Omega u(x, y) \overline{v(x, y)} \mathrm{d}x \, \mathrm{d}y$.

Stability is established as follows. Using integration-by-parts and the periodicity condition we have

$$\int_\Omega (-\Delta u + \lambda u)\overline{u}\, dx\, dy = \int_\Omega (|\nabla u|^2 + \lambda |u|^2) dx\, dy$$
$$\geq \min(\lambda, 1) \int_\Omega (|\nabla u|^2 + |u|^2) dx\, dy ,$$

for all $u \in X_N$. The integral on the right-hand side is precisely the square of the norm $\|u\|_{H^1(\Omega)}$ in the Hilbert space $H^1_p(\Omega)$ defined in (A.11.d). Hence, the stability condition (6.4.3) is verified with $E = H^1_p(\Omega)$ and $\alpha = \min(\lambda, 1)$, and the approximation is stable according to (6.4.5). Condition (6.4.4) follows easily by integrating by parts and using the Cauchy-Schwarz inequality. As regards the convergence analysis, the truncation operator $P_N$ defined in (5.8.3) gives the best approximation error in the norm of any $H^m_p(\Omega)$, $m > 0$. Therefore, we can choose this operator as $R_N$ in (6.4.6). Using the estimate (5.8.4), we get the optimal error bound

$$\|u - u^N\|_{H^1(\Omega)} \leq C N^{1-m} |u|_{H^{m;N}(\Omega)}, \quad m \geq 1 . \qquad \square$$

**Example 2.** *The Poisson Equation in the Square with Dirichlet Boundary Conditions.* Let us consider the problem

$$-\Delta u = f \qquad \text{in } \Omega = (-1,1) \times (-1,1) ,$$
$$u = 0 \qquad \text{on } \partial\Omega .$$

Denote by $X_N = \{v \in \mathbb{P}_N \mid v = 0 \text{ on } \partial\Omega\}$ the space of algebraic polynomials of degree at most $N$ in each variable, vanishing on the boundary of the square. A modal basis for $X_N$ is given by

$$\phi_{km}(x,y) = \phi_k(x)\phi_m(y) , \quad 2 \leq k, m \leq N ,$$

where

$$\phi_k(x) = \begin{cases} L_0(x) - L_k(x) , & k \text{ even} , \\ L_1(x) - L_k(x) , & k \text{ odd} , \end{cases} \tag{6.4.9}$$

if the Legendre polynomials introduced in Sect. 2.3 are used (see also (2.3.33) and (4.1.31)), or

$$\phi_k(x) = \begin{cases} T_0(x) - T_k(x) , & k \text{ even} , \\ T_1(x) - T_k(x) , & k \text{ odd} , \end{cases} \tag{6.4.10}$$

if the Chebyshev polynomials introduced in Sect. 2.4 are used instead. With any of these choices, the Galerkin equations to be satisfied by $u^N \in X_N$ are

$$-\int_{\Omega} \Delta u^N \phi_{km} w(x,y) \mathrm{d}x\, \mathrm{d}y = \int_{\Omega} f \phi_{km} w(x,y) \mathrm{d}x\, \mathrm{d}y\ ,$$

where $w(x,y) = w(x)w(y)$, and $w(x)$ is either the Legendre or the Chebyshev weight according to whether (6.4.10) or (6.4.9) is used for the basis. In the present example we choose $X = L^2_w(\Omega)$ and $(u,v) = \int_{\Omega} uvw\, \mathrm{d}x\, \mathrm{d}y$ (see (A.9.h)).

Let us discuss the stability of the approximation. In the Legendre case, all $u \in X_N$ satisfy

$$-\int_{\Omega} \Delta u u\, \mathrm{d}x\, \mathrm{d}y = \int_{\Omega} |\nabla u|^2 \mathrm{d}x\, \mathrm{d}y\ .$$

Since $u$ is zero on $\partial\Omega$, the $L^2$-norm of its gradient controls the norm $\|u\|_{H^1(\Omega)} = \{\int_{\Omega}(|u|^2 + |\nabla u|^2)\mathrm{d}x\, \mathrm{d}y\}^{1/2}$ of $H^1(\Omega)$, according to the Poincaré inequality (A.13). We choose $E$ to be the subspace $H^1_0(\Omega)$ of the functions in $H^1(\Omega)$ that vanish on $\partial\Omega$ (see (A.11.c)). $E$ is a Hilbert space under the same norm as $H^1(\Omega)$. Thus, (6.4.3) is verified, and the scheme is stable.

In order to prove the convergence, $R_N u$ is chosen to be the best approximation of $u$ among the functions in $X_N$ in the norm of $E$. By (5.8.15) and (6.4.8) we conclude that the following optimal error estimate holds:

$$\|u - u^N\|_{H^1(\Omega)} \leq C N^{1-m} |u|_{H^{m;N}(\Omega)}, \quad m \geq 1.$$

In the Chebyshev case, it is not immediate that the quantity

$$-\int_{\Omega} \Delta u u w\, \mathrm{d}x\, \mathrm{d}y = \int_{\Omega} |\nabla u|^2 w\, \mathrm{d}x\, \mathrm{d}y + \int_{\Omega} u \nabla u \nabla w\, \mathrm{d}x\, \mathrm{d}y$$

is positive, due to the presence of the Chebyshev weight. However (see Sect. 7.1), the right-hand side actually controls the norm

$$\|u\|_{H^1_w(\Omega)} = \left\{ \int_{\Omega} (u^2 + |\nabla u|^2) w\, \mathrm{d}x\, \mathrm{d}y \right\}^{1/2}$$

of the weighted Sobolev space $H^1_w(\Omega)$ (defined in (A.11.b)). Thus, we have the same stability and convergence results as above, provided the Chebyshev weight is inserted in all the norms. □

So far we have assumed that the Galerkin approximation (6.4.2) satisfies the discrete coercivity condition (6.4.3). There are cases in which this condition is not fulfilled (see Example 3). Another way of getting stability and convergence results is to check a discrete form of the inf-sup condition (6.3.11) and (6.3.12). This condition is also suitable for the analysis of Petrov-Galerkin methods. We refer to the forthcoming subsection on tau methods for the detailed description of this approach.

### 6.4.2 Collocation Methods

To define a collocation method for approximating (6.3.1)–(6.3.2), one uses as many distinct points

$$x_k \,, \quad k \in J \text{ (a set of indices)}, \tag{6.4.11}$$

in the domain $\Omega$ or on its boundary $\partial\Omega$ as the dimension of the space $\mathrm{Pol}_N(\Omega)$ in which the spectral solution is sought. At a number of these points, located on $\partial\Omega$, the boundary conditions are imposed. The remaining points are used to enforce the differential equation.

We assume that the set $J$ is unisolvent for $\mathrm{Pol}_N(\Omega)$, i.e., for any $k \in J$, there exists a polynomial $\phi_k \in \mathrm{Pol}_N(\Omega)$, necessarily unique, such that

$$\phi_k(x_m) = \begin{cases} 1 & \text{if } k = m \,, \\ 0 & \text{if } k \neq m \,. \end{cases} \tag{6.4.12}$$

This is certainly true in all the applications, where the points (6.4.11) are products of distinct points in each space variable. Consistently with the one-dimensional definition of Chap. 1 (see (1.2.55)), the $\phi_k$'s are called characteristic Lagrange polynomials, or discrete delta-functions. They form a basis for the polynomials of degree $N$, since $v(x) = \sum_{k\in J} v(x_k)\phi_k(x)$ for all $v \in \mathrm{Pol}_N(\Omega)$. A collocation method is obtained by requiring that the differential equation be satisfied at a number of points $\{x_k\}$ (those in the interior of the domain, and possibly some on the boundary) and that the boundary conditions (or, possibly, some of them) be satisfied at the remaining $x_k$'s. To be precise, let $J$ be divided into two disjoint subsets, $J_e$ and $J_b$, such that if $k \in J_b$, the $x_k$'s are on the part $\partial\Omega_b$ of the boundary where the boundary conditions (6.3.2) are prescribed. Moreover, let $\mathcal{L}_N$ be an approximation to the operator $\mathcal{L}$ in which derivatives are taken via interpolation at the points $x_k$'s (see Sects. 2.1.3, 2.3.2, and 2.4.2). The collocation solution is a polynomial $u^N \in \mathrm{Pol}_N(\Omega)$ that satisfies the equations

$$\mathcal{L}_N u^N(x_k) = f(x_k) \qquad \text{for all } k \in J_e \,, \tag{6.4.13}$$

$$\mathcal{B}u^N(x_k) = 0 \qquad \text{for all } k \in J_b \,. \tag{6.4.14}$$

The unknowns in a collocation method are the values of $u^N$ at the points (6.4.11), i.e., the coefficients of $u^N$ with respect to the Lagrange basis (6.4.12). The set $J_b$ is empty in Fourier approximations for periodic problems since the trigonometric polynomials are themselves periodic. However, $J_b$ may be empty even in approximations to nonperiodic problems. In these cases, the boundary conditions are taken into account implicitly in the definition of the operator $\mathcal{L}_N$ (see, e.g., Canuto (1986)), or via a penalty approach (such as the one discussed in Sect. 3.7; see in particular (3.7.7)).

We will now set the collocation method (6.4.13)–(6.4.14) into the framework given in Sect. 6.3. To this end, we introduce a bilinear form $(u,v)_N$ on the space $Z = C^0(\overline{\Omega})$ of the functions continuous up to the boundary of $\Omega$ by fixing a family of weights $w_k > 0$ and setting

$$(u,v)_N = \sum_{k\in J} u(x_k)\overline{v(x_k)}w_k \, . \tag{6.4.15}$$

The existence of the Lagrange basis (6.4.12) ensures that (6.4.15) is an inner product on $\mathrm{Pol}_N(\Omega)$. Consequently, we define a *discrete norm* on $\mathrm{Pol}_N(\Omega)$ as

$$\|u\|_N = \sqrt{(u,u)_N} \quad \text{for } u \in \mathrm{Pol}_N(\Omega) \, . \tag{6.4.16}$$

The basis of the $\phi_k$'s is orthogonal under the *discrete* inner product (6.4.15).

We make the assumption that the nodes $\{x_k\}$ and the weights $\{w_k\}$ are such that

$$(u,v)_N = (u,v) \quad \text{for all } u,\ v \text{ such that } uv \in \mathrm{Pol}_{2N-1}(\Omega) \, . \tag{6.4.17}$$

This means that the discrete inner product (6.4.15) must approximate with enough precision the inner product of $X$. Condition (6.4.17) introduces a constraint in the choice of the collocation points. In all the applications, this assumption is fulfilled since the $x_k$'s are the knots of quadrature formulas of Gaussian type.

Let $X_N$ be the space of the polynomials of degree $\leq N$ that satisfy the boundary conditions (6.4.14), i.e.,

$$X_N = \{v \in \mathrm{Pol}_N(\Omega) \,|\, \mathcal{B}v(x_k) = 0 \text{ for all } k \in J_b\} \, . \tag{6.4.18}$$

Then the collocation method is equivalently written as

$$\begin{aligned} & u^N \in X_N \, , \\ & (\mathcal{L}_N u^N, \phi_k)_N = (f, \phi_k)_N \qquad \text{for all } k \in J_e \, . \end{aligned} \tag{6.4.19}$$

If $Y_N$ is the space spanned by the $\phi_k$'s with $k \in J_e$, i.e.,

$$Y_N = \{v \in \mathrm{Pol}_N(\Omega) \,|\, v(x_k) = 0 \text{ for all } k \in J_b\} \, , \tag{6.4.20}$$

then (6.4.19) can be written as

$$\begin{aligned} & u^N \in X_N \, , \\ & (\mathcal{L}_N u^N, v)_N = (f, v)_N \qquad \text{for all } v \in Y_N \, . \end{aligned} \tag{6.4.21}$$

This is precisely (6.3.20). Equivalently, (6.4.19) can be written in the form

$$Q_N(\mathcal{L}_N u^N - f) = 0 \tag{6.4.22}$$

(see (6.3.21)). For a collocation approximation, $Q_N v$ is the polynomial of degree $N$ matching $v$ at the interior points $\{x_k, k \in J_e\}$ and vanishing at the boundary points $\{x_k, k \in J_b\}$.

Note that in the special case where all the boundary conditions are of Dirichlet type, i.e., if $\mathcal{B}v \equiv v$, one has $X_N = Y_N$. In this case the collocation method can be viewed as a G-NI method, i.e., as a Galerkin method in which the *continuous* inner product $(u, v)$ is replaced by the *discrete* inner product $(u, v)_N$ (compare (6.4.21) with (6.3.17)).

### *Stability and Convergence*

We consider now the *stability and convergence properties* of the collocation approximation (6.4.13)–(6.4.14). As for the Galerkin approximation, the simplest situation occurs when the operator $\mathcal{L}$ satisfies the coercivity condition (6.3.6) and the continuity condition (6.3.7) with respect to a suitable energy space, $E$.

Again we assume that $X_N$ is contained in $E$ for all $N > 0$. Moreover, we assume that for all $u \in X_N$, $\|u\|_N \leq C\|u\|_E$ with $C > 0$ independent of $N$ (see (6.4.16)). A coercivity condition for the approximation (6.4.21), by analogy with condition (6.3.6), is as follows.

*If there exists a constant $\overline{\alpha} > 0$ (independent of $N$) such that*

$$\overline{\alpha}\|u\|_E^2 \leq (Q_N \mathcal{L}_N u, u)_N \quad \text{for all } u \in X_N \ , \tag{6.4.23}$$

*then the approximation is stable, in the sense that the following estimate holds:*

$$\|u^N\|_E \leq \frac{C}{\overline{\alpha}}\|f\|_N \, . \tag{6.4.24}$$

Actually, one has

$$\overline{\alpha}\|u^N\|_E^2 \leq (Q_N \mathcal{L}_N u^N, u^N)_N = (Q_N f, u^N)_N \leq \|Q_N f\|_N \|u^N\|_N \leq C\|f\|_N \|u^N\|_E .$$

We use here the fact that $Q_N$ is the projection operator upon $Y_N$ with respect to the discrete inner product $(u, v)_N$.

We move now to the convergence analysis. Let $R_N$ be a projection operator from a dense subspace $\mathcal{W}$ of $D_B(\mathcal{L})$ upon $X_N$. For each $u \in \mathcal{W}$, we further require $R_N u$ to satisfy the exact boundary conditions, i.e.,

$$R_N : \mathcal{W} \longrightarrow X_N \cap D_B(\mathcal{L}) \, . \tag{6.4.25}$$

The following error bound between the exact and the collocation solutions holds:

$$\|u - u^N\|_E \leq \left(1 + \frac{A}{\overline{\alpha}}\right) \|u - R_N u\|_E + \frac{1}{\overline{\alpha}} \frac{|(\mathcal{L} R_N u, e) - (Q_N \mathcal{L}_N R_N u, e)_N|}{\|e\|_E}$$
$$+ \frac{1}{\overline{\alpha}} \frac{|(f, e) - (Q_N f, e)_N|}{\|e\|_E} \, , \tag{6.4.26}$$

with $e = u^N - R_N u$.

Assume for the moment that (6.4.26) is proven. It follows that *convergence is assured if the following three consistency conditions are fulfilled:*

$$\|u - R_N u\|_E \longrightarrow 0 \tag{6.4.27a}$$

*as* $N \to \infty$, *for all* $u \in \mathcal{W}$;

$$\sup_{\substack{v \in X_N \\ v \neq 0}} \frac{(\mathcal{L} R_N u, v) - (Q_N \mathcal{L}_N R_N u, v)_N}{\|v\|_E} \longrightarrow 0 \tag{6.4.27b}$$

*as* $N \to \infty$, *for all* $u \in \mathcal{W}$;

$$\sup_{\substack{v \in X_N \\ v \neq 0}} \frac{(f, v) - (Q_N f, v)_N}{\|v\|_E} \longrightarrow 0 \tag{6.4.27c}$$

*as* $N \to \infty$, *for all* $f \in \mathcal{Z}$ *smooth enough.*

Proof of (6.4.26). From (6.3.1) and (6.4.22) it follows that, for any $v \in X_N$,

$$(\mathcal{L} u, v) = (f, v) \tag{6.4.28}$$

and

$$(Q_N \mathcal{L}_N u^N, v)_N = (Q_N f, v)_N \, . \tag{6.4.29}$$

On the other hand,

$$(Q_N \mathcal{L}_N e, v)_N = (Q_N \mathcal{L}_N u^N, v)_N - (Q_N \mathcal{L}_N R_N u, v)_N \, .$$

Adding and subtracting $(\mathcal{L} u, v)$ and using (6.4.28) and (6.4.29) we obtain

$$\begin{aligned} (Q_N \mathcal{L}_N e, v)_N = {} & (Q_N f, v)_N - (f, v) + (\mathcal{L}(R_N u - u), v) \\ & + (\mathcal{L} R_N u, v) - (Q_N \mathcal{L}_N R_N u, v)_N \, . \end{aligned}$$

Taking $v = e$ and using the hypotheses (6.3.7) and (6.4.23) it follows that

$$\begin{aligned} \overline{\alpha} \|e\|_E^2 \leq {} & |(Q_N f, e)_N - (f, e)| + A \|R_N u - u\|_E \|e\|_E \\ & + |(\mathcal{L} R_N u, e) - (Q_N \mathcal{L}_N R_N u, e)_N| \, . \end{aligned}$$

Now (6.4.26) follows using the triangle inequality $\|u - u^N\|_E \leq \|u - R_N u\|_E + \|e\|_E$. □

The positivity condition (6.4.23) is the most immediate condition that guarantees the well-posedness of problem (6.4.21). However, there are situations where (6.4.23) is not fulfilled. This occurs, for instance, when the norms involved in the stability and convergence analysis depend on weight functions like the Chebyshev norms presented in Chap. 5. In these cases, the discrete analog of the inf-sup condition provides a more general criterion for checking the stability of the scheme.

Let us assume that the operator $\mathcal{L}$ satisfies conditions (6.3.11) to (6.3.13). Assume that for all $N > 0$, $X_N \subset W$ and $Y_N \subset V$. Moreover, assume that $\|v\|_N \le C\|v\|_V$ for all $v$ in $Y_N$, with $C > 0$ independent of $N$. Then we have the following inf-sup condition for problem (6.4.21).

*If there exists a constant* $\overline{\alpha} > 0$ *independent of* $N$ *such that*

$$\overline{\alpha}\|u\|_W \le \sup_{\substack{v\in Y_N \\ v\neq 0}} \frac{(\mathcal{L}_N u, v)_N}{\|v\|_V} \quad \textit{for all } u \in X_N \,, \tag{6.4.30}$$

*then*

$$\|u^N\|_W \le \frac{C}{\overline{\alpha}}\|f\|_N \,. \tag{6.4.31}$$

The proof of (6.4.31) is a slight modification of the one of (6.4.67) pertaining to tau approximations.

Concerning the *convergence* of the method, one can bound the error $u-u^N$ according to the following formula:

$$\begin{aligned}\|u-u^N\|_W \le \left(1+\frac{A}{\overline{\alpha}}\right)\|u-R_N u\|_W \\ + \frac{1}{\overline{\alpha}} \sup_{\substack{v\in Y_N \\ v\neq 0}} \frac{|(\mathcal{L}R_N u, v) - (\mathcal{L}_N R_N u, v)_N|}{\|v\|_V} \\ + \frac{1}{\overline{\alpha}} \sup_{\substack{v\in Y_N \\ v\neq 0}} \frac{|(f,v)-(f,v)_N|}{\|v\|_V} \,.\end{aligned} \tag{6.4.32}$$

As in the previous case, $R_N$ is a projection operator from a dense subspace $\mathcal{W} \subseteq D_B(\mathcal{L})$ into $X_N \cap D_B(\mathcal{L})$. The proof of (6.4.32) mimics that of (6.4.26). The error, $e = u^N - R_N u$, satisfies

$$(\mathcal{L}_N e, v)_N = (\mathcal{L}(u-R_N u), v) + (\mathcal{L}R_N u, v) - (\mathcal{L}R_N u, v)_N + (f,v)_N - (f,v) \,,$$

for any $v \in Y_N$. We divide both sides by $\|v\|_V$ and take the supremum over all the functions in $Y_N$. Then, (6.4.32) follows from (6.4.30) and (6.3.13).

According to (6.4.32), *the approximation is convergent if the three following conditions hold true*:

$$\|u - R_N u\|_W \longrightarrow 0 \tag{6.4.33a}$$

*as* $N \to \infty$, *for all* $u \in \mathcal{W}$;

$$\sup_{\substack{v\in Y_N \\ v\neq 0}} \frac{(\mathcal{L}R_N u, v) - (\mathcal{L}_N R_N u, v)_N}{\|v\|_V} \longrightarrow 0 \tag{6.4.33b}$$

*as* $N \to \infty$, *for all* $u \in \mathcal{W}$;

$$\sup_{\substack{v \in Y_N \\ v \neq 0}} \frac{(f,v) - (f,v)_N}{\|v\|_V} \longrightarrow 0 \tag{6.4.33c}$$

*as* $N \to \infty$, *for all* $f \in Z$ *smooth enough.*

These are precisely the conditions to be checked in any specific situation in order to prove the convergence and to establish the rate of decay of the error.

We emphasize that the stability and convergence estimates given for the collocation problem include as special cases the ones for the Galerkin and tau approximations, provided that the discrete inner product is replaced by the continuous one. The last two terms appearing in the right-hand side of the convergence estimate (6.4.32) for collocation are precisely due to the use of quadrature formulas in the collocation scheme. Therefore, the conditions (10.4.58) and (10.4.61) are the most general ones that assure stability and convergence for the general spectral approximation (6.3.21).

We want to bring the attention of the reader to the concept of *algebraic stability*, introduced by Gottlieb and Orszag (1977) for approximations by spectral methods.

In both the stability criteria (6.4.23) and (6.4.30) we require that the constant $\overline{\alpha}$ be independent of $N$. This is not necessary for the convergence of the method. The constant $\overline{\alpha}$ may depend on $N$ in an algebraic way, i.e., it may be of the form $\overline{\alpha} = O(N^{-r})$ for a suitable $r > 0$. In this case, convergence is still assured, according to the estimates (6.4.26) and (6.4.32), provided that the exact solution $u$ is so smooth that the deviation $u - R_N u$ vanishes fast enough. Precisely, convergence occurs if $\|u - R_N u\|_E$ (or $\|u - R_N u\|_W$) decays as $O(N^{-r'})$ for an $r' > r$. This is a slightly different form of the concept of algebraic stability presented in Gottlieb and Orszag (1977, Sect. 5).

### *Examples*

We now consider some examples that illustrate the theory presented above for collocation methods.

**Example 3.** *The Dirichlet Problem for a Variable-Coefficient Second-Order Operator in the Interval* $(-1,1)$. We consider the problem

$$\begin{aligned} -(au_x)_x &= f\,, & -1 < x < 1\,, \\ u(-1) &= u(1) = 0\,, \end{aligned}$$

where $a(x)$ is continuously differentiable and satisfies $a(x) \geq \alpha_0 > 0$ in $[-1,1]$, and $f$ is continuous.

For a fixed integer $N > 0$, set $J = \{0, 1, \ldots, N\}$ and choose as points (6.4.11) the nodes $\{x_k, k \in J\}$ of the $(N+1)$-point Gauss-Lobatto quadrature formula with respect to the Legendre or Chebyshev weight. If $\{w_k, k \in J\}$ are the corresponding weights, assumption (6.4.17) is satisfied. Denote by $I_N v$ the polynomial of degree $N$ that interpolates a continuous function $v$ at the points $x_k, k \in J$. The collocation approximation to $u$ is a polynomial $u^N$ of degree $N$ that satisfies the equations

$$\begin{aligned} &-\left[I_N(a u_x^N)\right]_x(x_k) = f(x_k)\,, \qquad k = 1, \dots, N-1\,, \\ &u^N(x_0) = u^N(x_N) = 0\,. \end{aligned} \tag{6.4.34}$$

Thus, the operator $\mathcal{L}u = -(a u_x)_x$ has been approximated by the operator $\mathcal{L}_N u = -[I_N(a u_x)]_x$, in which the outer derivative has been replaced by the interpolation derivative at the collocation points. Problem (6.4.34) corresponds to (6.4.13)–(6.4.14), with $J_e = \{1, \dots, N-1\}$ and $J_b = \{0, N\}$. The space $X$ can be chosen here as $L^2_w(-1,1)$, where $w$ is either the Legendre or the Chebyshev weight function. The spaces $X_N$ and $Y_N$ coincide in this case, and one has

$$X_N = Y_N = \mathbb{P}^0_N(-1,1) = \{v \in \mathbb{P}_N \mid v(-1) = v(1) = 0\}\,,$$

where, as usual, $\mathbb{P}_N$ denotes the space of algebraic polynomials of degree $\le N$ in the variable $x$.

The stability and convergence analysis is easy if the Legendre points are used. In this case, the scheme satisfies a stability condition of the type (6.4.23). To check this result, let us start by observing that

$$(Q_N \mathcal{L}_N u, u)_N = (\mathcal{L}_N u, u)_N \quad \text{for all } u \in X_N\,,$$

since $Q_N$ is now the orthogonal projection onto $X_N$ for the discrete inner product $(u, v)_N$. Furthermore, for all $u \in X_N$,

$$\begin{aligned} (\mathcal{L}_N u, u)_N &= -\int_{-1}^{1} [I_N(a u_x)]_x u \,\mathrm{d}x = \int_{-1}^{1} I_N(a u_x) u_x \,\mathrm{d}x \\ &= \sum_{k=0}^{N} a(x_k)[u_x(x_k)]^2 w_k \ge \alpha_0 \sum_{k=0}^{N} [u_x(x_k)]^2 w_k \\ &= \alpha_0 \int_{-1}^{1} [u_x(x)]^2 \mathrm{d}x\,. \end{aligned}$$

Each change between integral and sum is allowed since the integrands are polynomials of degree at most $2N-1$. Thus, (6.4.23) holds with $E = H^1_0(-1,1)$ due to the Poincaré inequality (see (A.13) and (A.11.c)). We observe that the collocation scheme here considered for the Legendre nodes coincides with a G-NI scheme (see the discussion in Sect. 1.2.3); hence it can also be analyzed as described in Sect. 6.4.3.

Let us consider now the Chebyshev collocation points. If the coefficient $a$ in (6.4.34) is constant, say $a \equiv 1$, the scheme still fulfills the positivity condition (6.4.23), with $E = H^1_{w,0}(-1,1)$ defined in (A.11.c). Actually,

$$(\mathcal{L}_N u, u)_N = -\int_{-1}^{1} u_{xx} u w \, \mathrm{d}x \,,$$

which dominates the norm of $H^1_{w,0}(-1,1)$ as shown in Sect. 7.1 (see (7.1.16)). If $a(x)$ is not constant in the interval $(-1,1)$, the operator $\mathcal{L}_N$ may be indefinite in the inner product $(u,v)_N$. This can be seen by the following heuristic argument (which, however, can be made mathematically rigorous). For $N$ large enough, $(\mathcal{L}_N u, u)_N$ approaches $(\mathcal{L}u, u) = \int_{-1}^{1} a u_x (uw)_x \mathrm{d}x$. Now $u_x(uw)_x$ may be strictly negative in a region excluding the endpoints and the origin, though its average on $(-1,1)$ is positive according to (7.1.16). Thus, if $a$ is large in this region and small elsewhere, $(\mathcal{L}u,u)$ and consequently $(\mathcal{L}_N u, u)_N$ are strictly negative. The argument in turn shows that the coercivity condition (6.4.23) may not be satisfied in this case. However, it is possible to prove that the collocation scheme (6.4.34) is stable according to the more general inf-sup condition (6.4.30), where $W = V = H^1_{w,0}(-1,1)$. More precisely, for any polynomial $u \in X_N$, it is possible to construct a polynomial $v \in X_N$, that depends on $u$ but is different from it, such that $\|v\|_{H^1_w(-1,1)} \le C\|u\|_{H^1_w(-1,1)}$, and $(\mathcal{L}_N u, v)_N \ge \tilde{\alpha}\|u\|^2_{H^1_w(-1,1)}$ for two positive constants $C$ and $\tilde{\alpha}$ independent of $N$. This clearly implies (6.4.34). The proof is rather technical and can be found in Canuto and Quarteroni (1984).

The convergence of the approximation can be proved by checking the conditions (10.4.55) for the Legendre points and the conditions (10.4.61) for the Chebyshev points. In both cases an optimal error estimate is obtained by choosing as $R_N u$ the best polynomial approximation of $u$ in the norm of $H^1_{w,0}$, as defined in (5.4.29) or (5.5.17). The precise result is

$$\|u - u^N\|_{H^1_w(-1,1)} \le CN^{1-m}\left(|u|_{H^{m;N}_w(-1,1)} + |f|_{H^{m-1,N}_w(-1,1)}\right) \,,$$

where the seminorms on the right-hand side are defined in (5.4.10) or (5.5.8). □

**Example 4.** *The Neumann Problem for a Constant-Coefficient Elliptic Operator in the Interval* $(-1,1)$. The problem,

$$\begin{aligned} -u_{xx} + u &= f\,, \qquad -1 < x < 1\,, \\ u_x(-1) &= u_x(1) = 0\,, \end{aligned} \tag{6.4.35}$$

can be approximated by the following collocation method:

$$\begin{aligned} &(-u^N_{xx} + u^N)(x_k) = f(x_k)\,, \qquad 1 \le k \le N-1\,, \\ &u^N_x(-1) = u^N_x(1) = 0\,, \end{aligned} \tag{6.4.36}$$

where $u^N$ is an algebraic polynomial of degree $N$ and $\{x_k \,|\, k \in J\}$ are the points introduced in the previous example. Again, we set $X = L^2_w(-1,1)$, whereas now

$$X_N = \{v \in \mathbb{P}_N \,|\, v_x(-1) = v_x(1) = 0\}$$

and

$$Y_N = \{v \in \mathbb{P}_N \,|\, v(-1) = v(1) = 0\} \,.$$

Each $v \in Y_N$ can be written as $v(x) = z(x)(1-x^2)$ with $z \in \mathbb{P}_{N-2}$. Thus, $Y_N$ can be identified with $\mathbb{P}_{N-2}$, in the sense that $Y_N = (1-x^2)\mathbb{P}_{N-2}$. In this example, the formulation (6.4.21) reads as follows:

$$\sum_{k=0}^{N} \left[-u^N_{xx} + u^N\right](x_k)z(x_k)\left(1-x_k^2\right)w_k = \sum_{k=0}^{N} f(x_k)z\,(x_k)(1-x_k^2)\,w_k$$
$$\text{for all } z \in \mathbb{P}_{N-2} \,. \qquad (6.4.37)$$

Due to the relation (2.2.17), the higher order term on the left-hand side can be integrated exactly, namely,

$$-\sum_{k=0}^{N} u^N_{xx}(x_k)z(x_k)(1-x_k^2)w_k = -\int_{-1}^{1} u^N_{xx}(x)z(x)\eta(x)\mathrm{d}x \,, \qquad (6.4.38)$$

where $\eta(x) = \sqrt{1-x^2}$ is a Jacobi weight on the interval $(-1,1)$. So, one is naturally led to establish the stability of (6.4.36) in a norm depending on the weight $\eta$. Actually, if we choose $z = -u^N_{xx}$ in (6.4.37), then (6.4.38) is precisely the square of the norm of $u^N_{xx}$ in $L^2_\eta(-1,1)$, i.e., $\int_{-1}^{1}[u^N_{xx}(x)]^2\eta(x)\mathrm{d}x$. In view of the inf-sup condition (6.4.30), this observation suggests the choice of the space $W$ as

$$W = \{v \in L^2_\eta(-1,1) \,\big|\, v_{xx} \in L^2_\eta(-1,1)\} \,,$$

with norm

$$\|v\|^2_W = \int_{-1}^{1} \left[v^2(x) + v^2_{xx}(x)\right]\eta(x)\mathrm{d}x \,.$$

The natural norm for the test functions $z$ is the norm of $L^2_\eta(-1,1)$. In terms of the original test functions, $v = (1-x^2)z$, this norm reads as

$$\int_{-1}^{1} z^2(x)\eta(x)\mathrm{d}x = \int_{-1}^{1} \left(\frac{v(x)}{1-x^2}\right)^2 \eta(x)\mathrm{d}x = \|v\|^2_V \,. \qquad (6.4.39)$$

Thus, $V$ will be the space of those functions $v$ for which the right-hand side of (6.4.39) is finite.

Within this framework it can be shown that the stability and convergence conditions (6.4.30) and (10.4.61) hold. The following error estimate can be proven:

$$\|u - u^N\|_W \le CN^{2-m}\left(|u|_{H^{m;N}_w(-1,1)} + |f|_{H^{m-1;N}_w(-1,1)}\right) \,, \qquad m \ge 2 \,.$$

Details can be found in Canuto and Quarteroni (1984). □

### 6.4.3 G-NI Methods

In order to highlight the essential features of a G-NI method, we assume that the operator $\mathcal{L}$ can be represented as $\mathcal{L}u = -\nabla \cdot \mathcal{F} + \mathcal{L}_0 u$, where $\mathcal{L}_0$ is a linear operator, and $\mathcal{F} = \mathcal{F}(u)$ is a vector-valued function depending linearly on $u$, that we call a *flux*. Furthermore, we assume that the boundary conditions, $\mathcal{B}u = 0$, can be split into a set of linear homogeneous conditions acting on $u$, say $\mathcal{B}_0 u = 0$, that are enforced on a part $\Gamma_0$ of $\partial\Omega_b$, and a set of linear conditions acting on the flux $\mathcal{F}$, say $\mathcal{B}_1\mathcal{F} = 0$, that are enforced on $\Gamma_1 = \partial\Omega_b \setminus \Gamma_0$. A typical example is a Dirichlet condition, $u = 0$, on $\Gamma_0$ and a no-flux condition, $\mathbf{n} \cdot \mathcal{F} = 0$, on $\Gamma_1$, where $\mathbf{n}$ denotes the outward normal vector to $\partial\Omega$. We actually admit a more general situation than the one considered in (6.3.2), namely, we allow for an inhomogeneous flux condition, $\mathcal{B}_1\mathcal{F} = g$, on $\Gamma_1$. The case of inhomogeneous conditions on $u$, $\mathcal{B}_0 u = \eta$, can be reduced to the homogeneous case by the change of unknown function $u \to u_0 = u - u_\eta$, where $u_\eta$ is any known function (called lifting or extension of $\eta$) satisfying $\mathcal{B}_0 u_\eta = \eta$.

Denoting by $(u, v)$ the inner product in $X = L^2(\Omega)$ and by $(u, v)_\Gamma$ the $L^2$-inner product on a portion $\Gamma$ of $\partial\Omega$, after application of the divergence theorem, we have (formally)

$$
\begin{aligned}
(\mathcal{L}u, v) &= (-\nabla \cdot \mathcal{F}, v) + (\mathcal{L}_0 u, v) \\
&= (\mathcal{F}, \nabla v) - (\mathbf{n} \cdot \mathcal{F}, v)_{\partial\Omega} + (\mathcal{L}_0 u, v) \\
&= (\mathcal{F}, \nabla v) + (\mathcal{L}_0 u, v) \\
&\quad - (\mathbf{n} \cdot \mathcal{F}, v)_{\Gamma_0} - (\mathbf{n} \cdot \mathcal{F}, v)_{\Gamma_1} - (\mathbf{n} \cdot \mathcal{F}, v)_{\partial\Omega\setminus\partial\Omega_b} .
\end{aligned}
\tag{6.4.40}
$$

Now let us assume that the test functions $v$ satisfy the boundary conditions $\mathcal{B}_0 v = 0$ on $\Gamma_0$; this information can be used to manipulate the boundary term on $\Gamma_0$. Similarly, we can use the prescribed boundary conditions, $\mathcal{B}_1\mathcal{F} = g$, on $\Gamma_1$ to manipulate the boundary term on $\Gamma_1$. For instance, if the boundary conditions are Dirichlet on $\Gamma_0$ and no-flux on $\Gamma_1$, both boundary terms vanish. After these manipulations have been performed, the integral relations, $(\mathcal{L}u, v) = (f, v)$, that enforce the differential equation in $\Omega$ are transformed into a set of relations of the form $a(u, v) = F(v)$. Here, $a(u, v)$ is defined as $(\mathcal{F}, \nabla v) + (\mathcal{L}_0 u, v) +$ the bilinear boundary terms on the right-hand side of (6.4.40) after manipulation, whereas $F(v)$ is defined as $(f, v) +$ the linear boundary term on $\Gamma_1$ depending on $g$ (see (6.4.49) below for an example).

From now on, we assume that $\Gamma_0$ and $\Gamma_1$ are (possibly empty) unions of sides (in 2D) or faces (in 3D) of $\Omega$. Let $X_N$ be the subspace of $\mathrm{Pol}_N(\Omega)$ of the functions satisfying the boundary conditions, $\mathcal{B}_0 v = 0$, on $\Gamma_0$. Then, a Galerkin approximation is defined as follows.

$$
\begin{aligned}
&u^N \in X_N , \\
&a\left(u^N, v\right) = F(v) \qquad \text{for all } v \in X_N .
\end{aligned}
\tag{6.4.41}
$$

Since the boundary conditions $\mathcal{B}_1\mathcal{F} = g$ are not enforced directly on the functions of $X_N$, $u^N$ need not satisfy them exactly. However, as the weak formulation (6.4.41) has been obtained by integration-by-parts incorporating these conditions into the boundary terms, a counter-integration-by-parts usually allows one to show that the flux conditions are satisfied by $u^N$ in an approximate way, i.e., they are satisfied exactly in the limit as $N \to \infty$ if convergence occurs.

The next step is to make the integrals appearing in $a(u,v)$ and $F(v)$ easily computable even in the presence of variable coefficients. To this end, a tensor-product quadrature formula, based on Gaussian points, is introduced to compute the integrals in $\Omega$, and similar formulas are used to compute the boundary terms. Functions in $X_N$ can often be identified by their values at the quadrature points, taking also into account the boundary conditions they satisfy on $\Gamma_0$; this is accomplished by introducing a nodal basis associated with the quadrature points. However, in certain cases a different nodal basis, or a modal basis, is used instead (see Sects. 2.3.3, 2.8 and 2.9). Denoting by $a_N(u,v)$ and $F_N(v)$ the forms obtained from $a(u,v)$ and $F(v)$ by numerical integration, we end up with the following G-NI scheme:

$$\begin{aligned} &u^N \in X_N , \\ &a_N(u^N, v) = F_N(v) \qquad \text{for all } v \in X_N . \end{aligned} \tag{6.4.42}$$

### *Stability and Convergence*

The *stability and convergence analysis* for G-NI approximations is similar to the one given in the previous section for collocation methods. We assume again that the operator $\mathcal{L}$ satisfies the coercivity condition (6.3.6) and the continuity condition (6.3.7) with respect to a suitable energy space $E$; furthermore, we assume that $X_N \subseteq E$ for all $N > 0$. A stability condition for the approximation (6.4.42), by analogy with condition (6.3.6), is as follows.

*If there exists a constant* $\overline{\alpha} > 0$ *(independent of* $N$*) such that*

$$\overline{\alpha}\|v\|_E^2 \leq a_N(v,v) \quad \textit{for all } v \in X_N , \tag{6.4.43}$$

*and if there exists a constant* $C_F > 0$ *(independent of* $N$*) such that*

$$|F_N(v)| \leq C_F \|v\|_E \quad \textit{for all } v \in X_N , \tag{6.4.44}$$

*then the approximation is stable, in the sense that the following estimate holds:*

$$\|u^N\|_E \leq \frac{C_F}{\overline{\alpha}} . \tag{6.4.45}$$

Indeed, it is enough to choose $v = u^N$ in (6.4.42). The result implies existence and uniqueness of the solution of the G-NI scheme, since $X_N$ is finite dimensional.

As for the convergence analysis, let $R_N$ be a projection operator from a dense subspace $\mathcal{W}$ of $E$ upon $X_N$. Setting $e = u^N - R_N u$, the following error bound between the exact and the G-NI solutions holds:

$$\|u - u^N\|_E \leq \left(1 + \frac{A}{\overline{\alpha}}\right) \|u - R_N u\|_E + \frac{1}{\overline{\alpha}} \frac{|a(R_N u, e) - a_N(R_N u, e)|}{\|e\|_E} + \frac{1}{\overline{\alpha}} \frac{|F(e) - F_N(e)|}{\|e\|_E} . \tag{6.4.46}$$

The proof is similar to the one given for (6.4.26). It follows that *convergence is assured if the following three consistency conditions are fulfilled*:

$$\|u - R_N u\|_E \longrightarrow 0 \tag{6.4.47a}$$

*as* $N \to \infty$, *for all* $u \in \mathcal{W}$;

$$\sup_{\substack{v \in X_N \\ v \neq 0}} \frac{a(R_N u, v) - a_N(R_N u, v)}{\|v\|_E} \longrightarrow 0 \tag{6.4.47b}$$

*as* $N \to \infty$, *for all* $u \in \mathcal{W}$;

$$\sup_{\substack{v \in X_N \\ v \neq 0}} \frac{F(v) - F_N(v)}{\|v\|_E} \longrightarrow 0 \tag{6.4.47c}$$

*as* $N \to \infty$, *for all sufficiently smooth data*, $f$ *and* $g$, *appearing in* $F$ *and* $F_N$.

Conditions (6.4.47b) and (6.4.47c) are often called *Strang conditions* and estimate (6.4.46) is known as the *Strang lemma* (see, e.g., Quarteroni and Valli (1994), Theorem 5.5.1).

The following example illustrates the theory described above.

**Example 5.** *A Second-Order Operator in Divergence Form under Mixed Boundary Conditions.* Let $\Omega = (-1,1)^d$ be the square ($d = 2$) or the cube ($d = 3$), and let us partition its boundary $\partial\Omega$ into the open side or face $\Gamma_1 = \{\mathbf{x} = (x_1, \ldots, x_d) \,|\, x_1 = 1, |x_j| < 1 \text{ for } j = 2, \ldots, d\}$ and the remaining part $\Gamma_0 = \partial\Omega \backslash \Gamma_1$. We consider the general second-order equation with mixed Dirichlet and Neumann boundary conditions:

$$\begin{aligned} \mathcal{L}u \equiv -\nabla \cdot \boldsymbol{\mathcal{F}} + \gamma u = f \quad & \text{in } \Omega \, , \\ u = 0 \quad & \text{on } \Gamma_0 \, , \\ \mathbf{n} \cdot \boldsymbol{\mathcal{F}} = g \quad & \text{on } \Gamma_1 \, , \end{aligned} \tag{6.4.48}$$

where the flux is $\boldsymbol{\mathcal{F}} = \boldsymbol{\mathcal{F}}(u) = \boldsymbol{\alpha} \nabla u + \boldsymbol{\beta} u$. The coefficients $\boldsymbol{\alpha}$ (a symmetric and positive-definite matrix), $\boldsymbol{\beta}$ (a vector) and $\gamma$ are smooth functions defined in $\bar{\Omega}$, whereas $f$, $g$ are given data.

Taking the inner product of $\mathcal{L}u$ with a test function $v$ vanishing on $\Gamma_0$ and using the divergence theorem, we get

$$(\mathcal{L}u, v) = \int_\Omega (\nabla v)^T \boldsymbol{\alpha} \nabla u + \int_\Omega u(\boldsymbol{\beta} \cdot \nabla v) + \int_\Omega \gamma uv - \int_{\Gamma_1} \mathbf{n} \cdot (\boldsymbol{\alpha} \nabla u + \boldsymbol{\beta} u)v.$$

Next, in view of the discretization of the problem, we use the *skew-symmetric* decomposition $\boldsymbol{\beta} \cdot \nabla v = \frac{1}{2}\boldsymbol{\beta} \cdot \nabla v + \frac{1}{2}(\nabla \cdot (\boldsymbol{\beta} v) - (\nabla \cdot \boldsymbol{\beta})v)$ and the divergence theorem again to write the second integral on the right-hand side as

$$\int_\Omega u(\boldsymbol{\beta} \cdot \nabla v) = \tfrac{1}{2} \int_\Omega (u(\boldsymbol{\beta} \cdot \nabla v) - (\boldsymbol{\beta} \cdot \nabla u)v) - \tfrac{1}{2} \int_\Omega (\nabla \cdot \boldsymbol{\beta})uv + \tfrac{1}{2} \int_{\Gamma_1} \boldsymbol{\beta} \cdot \mathbf{n} uv\,.$$

It follows that $u$ solves the variational problem

$$\begin{aligned} &\int_\Omega (\nabla v)^T \boldsymbol{\alpha} \nabla u + \tfrac{1}{2} \int_\Omega (u(\boldsymbol{\beta} \cdot \nabla v) - (\boldsymbol{\beta} \cdot \nabla u)v) \\ &\quad + \int_\Omega (-\tfrac{1}{2}\nabla \cdot \boldsymbol{\beta} + \gamma)uv + \tfrac{1}{2} \int_{\Gamma_1} \boldsymbol{\beta} \cdot \mathbf{n} uv = \int_\Omega fv + \int_{\Gamma_1} gv \end{aligned} \tag{6.4.49}$$

for all $v$ smooth enough and vanishing on $\Gamma_0$. This suggests that we define the bilinear form $a(u, v)$ as the left-hand side of (6.4.49), and the linear form $F(v)$ as the right-hand side of (6.4.49). Both forms are naturally defined on the (closed) subspace $E$ of $H^1(\Omega)$ of the functions vanishing on $\Gamma_0$, endowed with the norm $\|v\|_E = \left(\int_\Omega |v|^2 + \int_\Omega |\nabla v|^2\right)^{1/2}$ of $H^1(\Omega)$. More precisely, the form $a(u, v)$ satisfies (6.3.7) whereas the form $F(v)$ satisfies (6.3.8), provided the components of $\boldsymbol{\alpha}$ and $\gamma$ are bounded in $\bar{\Omega}$, the components of $\boldsymbol{\beta}$ and its divergence are bounded in $\bar{\Omega}$, $f$ is square integrable in $\Omega$, and $g$ is square integrable in $\Gamma_1$. This can be seen by repeatedly applying the Cauchy-Schwarz inequality (see (A.2)) and also using the fact that the $H^1$-norm of a function controls the $L^2$-norm of its restriction on $\Gamma_1$, i.e., there exists a constant $C > 0$ such that $\|v\|_{L^2(\Gamma_1)} \le C\|v\|_E$ for all $v \in E$. Concerning the positivity condition (6.3.6), we observe that the second integral on the left-hand side of (6.4.49) vanishes for $v = u$ (indeed, it is skew-symmetric). Thus, we have

$$a(u, u) = \int_\Omega (\nabla u)^T \boldsymbol{\alpha} \nabla u + \int_\Omega (-\tfrac{1}{2}\nabla \cdot \boldsymbol{\beta} + \gamma)u^2 + \tfrac{1}{2} \int_{\Gamma_1} \boldsymbol{\beta} \cdot \mathbf{n} u^2\,.$$

We now assume that the operator $\mathcal{L}$ is uniformly elliptic, i.e., there exists a constant $\alpha_0 > 0$ such that $\boldsymbol{\xi}^T \boldsymbol{\alpha} \boldsymbol{\xi} \ge \alpha_0 |\boldsymbol{\xi}|^2$ in $\Omega$ for all $\boldsymbol{\xi} \in \mathbb{R}^d$; furthermore, we assume that $-\frac{1}{2}\nabla \cdot \boldsymbol{\beta} + \gamma \ge 0$ in $\Omega$ and that $\Gamma_1 \subseteq \partial\Omega_+ = \{\mathbf{x} \in \partial\Omega \mid \boldsymbol{\beta} \cdot \mathbf{n} \ge 0\}$. As mentioned in Sect. A.13, there exists a constant $C > 0$ such that the Poincaré inequality $\|v\|_{L^2(\Omega)} \le C\|\nabla v\|_{(L^2(\Omega))^d}$ holds for all functions $v$ in $H^1(\Omega)$ vanishing on $\Gamma_0$ (this inequality is precisely (A.13.3) when $\Gamma_0 = \partial\Omega$). Then (6.3.6) holds, since

$$a(u, u) \ge \alpha_0 \int_\Omega |\nabla u|^2 \ge \frac{\alpha_0}{C^2 + 1} \|u\|_E^2 \qquad \text{for all } u \in E\,. \tag{6.4.50}$$

This implies the existence and uniqueness in $E$ of the solution of (6.4.49).

We now discretize the problem by the G-NI approach. Let $X_N \subset E$ be the subspace of $\mathrm{Pol}_N(\Omega) = \mathbb{P}_N(-1,1)^d$ of the polynomials vanishing on $\Gamma_0$; note that $X_N$ is obtained by tensorizing polynomials of degree $\leq N$ in the $x_1$-variable, vanishing at $x_1 = -1$, with polynomials of degree $\leq N$ in each of the remaining variables, vanishing at the endpoints of the interval $(-1,1)$. Let $\{x_j, w_j\}_{j=0,\dots,N}$ denote the $N+1$ nodes and weights of the Legendre Gauss-Lobatto quadrature formula in $[-1,1]$; by tensorization, we obtain the $(N+1)^d$ nodes and weights $\{\mathbf{x_k}, w_\mathbf{k}\}_{\mathbf{k}\in J}$ (where $J$ denotes the set of the $d$-dimensional indices $\mathbf{k}$) of the corresponding formula in $\bar{\Omega} = [-1,1]^d$, that satisfies

$$\int_\Omega \varphi(\mathbf{x}) = \sum_{\mathbf{k}\in J} \varphi(\mathbf{x_k}) w_\mathbf{k} \qquad \text{for all } \varphi \in \mathrm{Pol}_{2N-1}(\Omega).$$

Note that a polynomial $v \in \mathrm{Pol}_N(\Omega)$ belongs to $X_N$ if and only if $v(\mathbf{x_k}) = 0$ for all quadrature points $\mathbf{x_k}$ sitting on $\Gamma_0$.

We also need a quadrature formula on $\Gamma_1$. This is obtained in the obvious manner by tensorizing $d-1$ times to get the nodes and weights $\{\mathbf{x}'_{\mathbf{k}'}, w_{\mathbf{k}'}\}_{\mathbf{k}'\in J'}$ in $[-1,1]^{d-1}$, then setting $\mathbf{x}_{\mathbf{k}'} = (1, \mathbf{x}'_{\mathbf{k}'})$ (here $J'$ denotes the set of the $(d-1)$-dimensional indices $\mathbf{k}'$); the resulting formula satisfies

$$\int_{\Gamma_1} \psi(\mathbf{x}) = \sum_{\mathbf{k}'\in J'} \psi(\mathbf{x}_{\mathbf{k}'}) w_{\mathbf{k}'} \qquad \text{for all } \psi \in \mathrm{Pol}_{2N-1}(\Gamma_1).$$

For simplicity, in the sequel we will write $\sum_J \varphi$ in lieu of $\sum_{\mathbf{k}\in J} \varphi(\mathbf{x_k}) w_\mathbf{k}$, as well as $\sum_{J'} \psi$ in lieu of $\sum_{\mathbf{k}'\in J'} \psi(\mathbf{x}_{\mathbf{k}'}) w_{\mathbf{k}'}$. We will also set $\|\varphi\|_{N,\Omega} = (\sum_J \varphi^2)^{1/2}$ and $\|\psi\|_{N,\Gamma_1} = (\sum_{J'} \psi^2)^{1/2}$.

From now on we assume that all coefficients and data appearing in (6.4.49) are continuous functions. The G-NI scheme is obtained from this formulation by replacing integrals with quadrature formulas; precisely, $u^N \in X_N$ is defined as the solution of

$$\begin{aligned} \sum_J (\nabla v)^T \boldsymbol{\alpha} \nabla u + \tfrac{1}{2} \sum_J \left(u^N (\boldsymbol{\beta}\cdot\nabla v) - (\boldsymbol{\beta}\cdot\nabla u^N) v\right) + \sum_J \left(-\tfrac{1}{2}\nabla\cdot\boldsymbol{\beta} + \gamma\right) u^N v \\ + \tfrac{1}{2} \sum_{J'} \boldsymbol{\beta}\cdot\mathbf{n} u^N v = \sum_J f v + \sum_{J'} g v \qquad \text{for all } v \in X_N. \end{aligned} \tag{6.4.51}$$

We denote the left-hand side of (6.4.51) by $a_N(u,v)$ and the right-hand side of (6.4.51) by $F_N(v)$, so that the scheme can be written as (6.4.42).

Let us discuss the stability of the method. To this end, we recall the fundamental equivalence (5.3.2) of discrete and continuous $L^2$-norms of polynomials, that by tensorization yields the following equivalences:

$$c_1 \|\varphi\|_{L^2(\Omega)} \le \|\varphi\|_{N,\Omega} \le c_2 \|\varphi\|_{L^2(\Omega)} \qquad \text{for all } \varphi \in \mathrm{Pol}_N(\Omega)\,, \qquad (6.4.52)$$

and

$$c_1' \|\psi\|_{L^2(\Gamma_1)} \le \|\psi\|_{N,\Gamma_1} \le c_2' \|\psi\|_{L^2(\Gamma_1)} \qquad \text{for all } \psi \in \mathrm{Pol}_N(\Gamma_1)\,, \quad (6.4.53)$$

for suitable constants $c_1, c_2, c_1', c_2' > 0$. Then, (6.4.43) follows by observing that, as for (6.4.50),

$$a_N(u,u) \ge \alpha_0 \sum_J |\nabla u|^2 \ge \alpha_0 c_1^2 \|\nabla u\|^2_{L^2(\Omega)} \ge \frac{\alpha_0 c_1^2}{C^2+1} \|u\|_E^2 \quad \text{for all } u \in X_N .$$

The right-hand side $F_N(v)$ is estimated as follows:

$$\begin{aligned} |F_N(v)| &\le \sum_J |f|\,|v| + \sum_{J'} |g|\,|v| \le \|f\|_{N,\Omega} \|v\|_{N,\Omega} + \|g\|_{N,\Gamma_1} \|v\|_{N,\Gamma_1} \\ &\le c_2 2^{d/2} \|f\|_{L^\infty(\Omega)} \|v\|_{L^2(\Omega)} + c_2' 2^{(d-1)/2} \|g\|_{L^\infty(\Gamma_1)} \|v\|_{L^2(\Gamma_1)}\,, \end{aligned}$$

where we have used the fact that the quadrature formula is exact on the constants. Then, inequality (6.4.44) follows easily. We conclude that (6.4.45) holds, which in turn – as already noted – implies the existence and uniqueness of the solution $u^N$ of the G-NI scheme.

Finally, we establish the convergence of the approximation. Let $\bar{N}$ denote the largest integer $\le N/2$, and let us introduce a projection operator $R_N : E \to X_{\bar{N}} \subset X_N$ that yields an optimal approximation error in the $H^1$-norm, i.e.,

$$\|u - R_N u\|_{H^1(\Omega)} \le C N^{1-m} |u|_{H^{m;\bar{N}}(\Omega)} \qquad (6.4.54)$$

for all $u \in H^m(\Omega) \cap E$, $m \ge 1$; we refer to Sect. 5.8.2 for the construction of such an operator. Then, (6.4.47a) is fulfilled by taking $\mathcal{W} = H^{m_0}(\Omega) \cap E$ for an arbitrary $m_0 > 1$. In order to prove (6.4.47b), we estimate each contribution to the error, $a(R_N u, v) - a_N(R_N u, v)$, separately. The first one comes from the diffusion term

$$\mathcal{E} = \int_\Omega (\nabla v)^T \boldsymbol{\alpha} \nabla R_N u - \sum_J (\nabla v)^T \boldsymbol{\alpha} \nabla R_N u. \qquad (6.4.55)$$

For each component $\alpha$ of $\boldsymbol{\alpha}$, let $\alpha_N \in \mathrm{Pol}_{\bar{N}}(\Omega)$ be an approximation of $\alpha$ such that $\|\alpha - \alpha_N\|_{L^\infty(\Omega)} \to 0$ as $N \to \infty$, provided $\alpha$ is smooth enough. For instance, $\alpha_N$ can be chosen as the best approximation of $\alpha$ in $\mathrm{Pol}_{\bar{N}}(\Omega)$ in the $L^\infty$-norm, for which one has the estimate (that generalizes (5.4.16))

$$\|\alpha - \alpha_N\|_{L^\infty(\Omega)} \le C N^t |\alpha|_{W^{t;\bar{N},\infty}(\Omega)}\,, \qquad t > d/2\,,$$

with $|\alpha|_{W^{t;\bar{N},\infty}(\Omega)} = \max_{\bar{t} \le k \le t} \max_{1 \le i \le d} \|D_i^k u\|^2_{L^\infty(\Omega)}$, $\bar{t} = \min(t, \bar{N}+1)$. Then, we add and substract $\boldsymbol{\alpha}_N$ from each term on the right-hand side of (6.4.55).

Since $(\nabla v)^T \boldsymbol{\alpha}_N \nabla R_N u$ belongs to $\mathrm{Pol}_{2N-1}(\Omega)$, the quadrature formula integrates it exactly; hence,

$$|\mathcal{E}| = \left| \int_\Omega (\nabla v)^T (\boldsymbol{\alpha} - \boldsymbol{\alpha}_N) \nabla R_N u - \sum_J (\nabla v)^T (\boldsymbol{\alpha} - \boldsymbol{\alpha}_N) \nabla R_N u \right|$$

$$\leq \|\boldsymbol{\alpha} - \boldsymbol{\alpha}_N\|_{L^\infty(\Omega)} (\|\nabla R_N u\|_{L^2(\Omega)} \|\nabla v\|_{L^2(\Omega)} + \|\nabla R_N u\|_{N,\Omega} \|\nabla v\|_{N,\Omega}).$$

Using again the exactness of the quadrature formula for $(R_N u)^2 \in \mathrm{Pol}_{2N-2}(\Omega)$ as well as (6.4.52), we get

$$|\mathcal{E}| \leq (1 + c_2) \|\boldsymbol{\alpha} - \boldsymbol{\alpha}_N\|_{L^\infty(\Omega)} \|\nabla R_N u\|_{L^2(\Omega)} \|\nabla v\|_{L^2(\Omega)}.$$

Finally, we note that $\|R_N u\|_{H^1(\Omega)} \leq C \|u\|_{H^1(\Omega)}$ by (6.4.54), so that we obtain the following bound for the error term $\mathcal{E}$:

$$|\mathcal{E}| \leq C \|\boldsymbol{\alpha} - \boldsymbol{\alpha}_N\|_{L^\infty(\Omega)} \|u\|_E \|v\|_E.$$

All other terms appearing in $a(R_N u, v) - a_N(R_N u, v)$ can be handled similarly, provided the coefficients appearing therein are smooth enough. Under this assumption, we obtain (6.4.47b).

To conclude the consistency discussion, we establish (6.4.47c). We apply estimate (5.8.22) to the internal error, $\int_\Omega fv - \sum_J fv$, and the boundary error, $\int_{\Gamma_1} gv - \sum_{J'} gv$, separately. For all $v \in X_N$, we get

$$\begin{aligned} |F(v) - F_N(v)| &\leq C_1 (\|f - P_{N-1} f\|_{L^2(\Omega)} + \|f - I_N f\|_{L^2(\Omega)}) \|v\|_{L^2(\Omega)} \\ &\quad + C_2 (\|g - P'_{N-1} g\|_{L^2(\Gamma_1)} + \|g - I'_N g\|_{L^2(\Gamma_1)}) \|v\|_{L^2(\Gamma_1)}, \end{aligned}$$

where $P'_{N-1}$ and $I'_N$ denote $L^2$-projection and interpolation on $\Gamma_1$, respectively. Recalling that $\|v\|_{L^2(\Omega)} + \|v\|_{L^2(\Gamma_1)} \leq C \|v\|_E$ for all $v \in E$, we obtain

$$\begin{aligned} |F(v) - F_N(v)| \leq C \; &\big(\|f - P_{N-1} f\|_{L^2(\Omega)} + \|f - I_N f\|_{L^2(\Omega)} \\ &+ \|g - P'_{N-1} g\|_{L^2(\Gamma_1)} + \|g - I'_N g\|_{L^2(\Gamma_1)}\big) \|v\|_E. \end{aligned}$$

Using the results of Sect. 5.8.2, we deduce that the right-hand side tends to 0 as $N \to \infty$, provided $f$ and $g$ are smooth enough, yielding (6.4.47c).

An estimate of the G-NI error $\|u - u_N\|_{H^1(\Omega)}$ can be obtained by (6.4.46). If we assume suitable regularity for the solution $u$, the coefficients $\boldsymbol{\alpha}$, $\boldsymbol{\beta}$ and $\gamma$, and the data $f$ and $g$, then the approximation results of Chap. 5 together with the error analysis sketched above allow us to bound each term appearing on the right-hand side of (6.4.46). Precisely, let us assume that $u \in H^m(\Omega)$, $\boldsymbol{\alpha} \in (W^{t,\infty}(\Omega))^{d\times d}$, $\boldsymbol{\beta} \in (W^{\tau,\infty}(\Omega))^d$, $\gamma \in W^{\theta,\infty}(\Omega)$, $f \in H^\mu(\Omega)$ and $g \in H^\nu(\Gamma_1)$. Then, we end up with the following result:

$$\begin{aligned} \|u - u_N\|_{H^1(\Omega)} &\leq C_1 N^{1-m} |u|_{H^{s;\tilde{N}}(\Omega)} + C_2 N^{-t} |\boldsymbol{\alpha}|_{(W^{t;\tilde{N},\infty}(\Omega))^{d\times d}} \\ &\quad + C_3 N^{-\tau} |\boldsymbol{\beta}|_{(W^{\tau;\tilde{N},\infty}(\Omega))^d} + C_4 N^{-\theta} |\gamma|_{W^{\theta;\tilde{N},\infty}(\Omega)} \\ &\quad + C_5 N^{-\mu} |f|_{H^{\mu;N-1}(\Omega)} + C_6 N^{-\nu} |g|_{H^{\nu;N-1}(\Gamma_1)}. \end{aligned}$$

□

### 6.4.4 Tau Methods

Tau methods are mostly used for constant-coefficient, nonperiodic problems. The definition of these methods is particularly simple for problems in one spatial dimension. We begin with this case, and then we consider the general situation.

We assume that the differential problem (6.3.1) is defined in the interval $\Omega = (-1, 1)$, and we recall that $\partial\Omega_b$ is the set of the endpoints where the boundary conditions (6.3.2) are imposed.

Let $\{\phi_k, k = 0, 1, \dots\}$ be a system of algebraic polynomials, orthogonal with respect to the inner product $\int_{-1}^{1} u(x)v(x)w(x)\mathrm{d}x$, where $w > 0$ is a weight function on $(-1, 1)$. We assume that each $\phi_k$ is a polynomial of effective degree $k$. The tau solution is a polynomial of degree $N$, $u^N = \sum_{k=0}^{N} \alpha_k \phi_k$, whose coefficients in the expansion according to this basis are the unknowns of the problem. They are determined in the following way: denote by $\beta$ the number of boundary conditions prescribed at the endpoints of the interval (for instance, $\beta = 2$ if $\mathcal{L}$ is a nondegenerate second-order operator). The differential equation (6.3.1) is projected onto the space of polynomials of degree $N - \beta$,

$$\int_{-1}^{1} \mathcal{L}u^N \phi_k w \, \mathrm{d}x = \int_{-1}^{1} f \phi_k w \, \mathrm{d}x \,, \qquad k = 0, 1, \dots, N - \beta \,, \tag{6.4.56}$$

and the boundary conditions (6.3.2) are imposed exactly on $\partial\Omega_b$:

$$\sum_{k=0}^{N} \alpha_k \mathcal{B} \phi_k = 0 \quad \text{at the points of } \partial\Omega_b \,. \tag{6.4.57}$$

Conditions (6.4.57) are necessary since the basis functions do not automatically satisfy the boundary conditions, unlike the basis used in a Galerkin method as considered in Sect. 6.4.1.

In order to cast a tau method in the framework of Sect. 6.3, we set $X = L^2_w(-1, 1)$,

$$X_N = \{v \in \mathbb{P}_N \mid Bv = 0 \text{ at the points of } \partial\Omega_b\} \,, \tag{6.4.58}$$

and

$$Y_N = \mathbb{P}_{N-\beta} \,. \tag{6.4.59}$$

Then the tau method is equivalent to

$$\begin{aligned} & u^N \in X_N \,, \\ & (\mathcal{L}u^N, v) = (f, v) \qquad \text{for all } v \in Y_N \,. \end{aligned} \tag{6.4.60}$$

With respect to the general setting (6.3.22), in a tau method the projector $Q_N$ is the orthogonal projection operator from $X$ upon $Y_N$ relative to the inner product $(u, v)$ of $X$.

We consider now the $d$-dimensional case. The domain $\Omega$ is the product of $d$ copies of the interval $(-1,1)$ and the functions of $\mathrm{Pol}_N(\Omega)$ are algebraic polynomials in each variable. In the sequel, we will mean by "side" a $(d-1)$-dimensional subset of $\partial\Omega$ characterized by the equation $x_i = c$ for some $i \in \{1,\ldots,d\}$ and $c \in \{-1,1\}$. We assume that on a given side of the boundary the same kind of boundary conditions are given. We exclude, for example, the use of Dirichlet boundary conditions on part of a side and Neumann boundary conditions on the rest of it.

A basis in $\mathrm{Pol}_N(\Omega)$ can be built as a product of the basis functions $\{\phi_k\}$ in each variable. Define the lattice

$$J = \{\mathbf{k} = (k_1,\ldots,k_n) \,|\, k_i \text{ is an integer, } 0 \le k_i \le N \text{ for } i = 1,\ldots,d\} ,$$

and set

$$\phi_{\mathbf{k}}(\mathbf{x}) = \phi_{k_1}(x_1)\cdots\phi_{k_d}(x_d) .$$

Then $\{\phi_{\mathbf{k}},\ \mathbf{k} \in J\}$ is a basis in $\mathrm{Pol}_N(\Omega)$ that is orthogonal for the inner product

$$(u,v) = \int_{-1}^{1} w(x_1)\mathrm{d}x_1 \cdots \int_{-1}^{1} u(\mathbf{x})v(\mathbf{x})w(x_d)\,\mathrm{d}x_d .$$

The solution of a spectral tau scheme is a polynomial in $\mathrm{Pol}_N(\Omega)$ expanded in this basis. Its coefficients in this expansion are determined by two sets of linear equations. The first set is obtained by requiring that the residual $\mathcal{L}_N u^N - f$ be orthogonal to a family of basis functions of reduced degree. The $\phi_{\mathbf{k}}$'s that are retained as test functions are the ones whose degree in each direction is at most $N$ minus the number of boundary conditions prescribed on the sides orthogonal to that direction. More precisely, for each $i = 1,\ldots,d$, denote by $\beta_i$ the total number of boundary conditions prescribed on the sides $x_i = \pm 1$. Define the sublattice

$$J_e = \{\mathbf{k} = (k_1,\ldots,k_d) \in J \,|\, 0 \le k_i \le N - \beta_i \text{ for } i = 1,\ldots,d\} ,$$

where the subscript $e$ stands for equation. (See Fig. 6.2 for an example.) The differential equation is enforced by requiring that the tau solution $u^N \in \mathrm{Pol}_N(\Omega)$ satisfies the set of equations

$$(\mathcal{L}u^N, \phi_{\mathbf{k}}) = (f, \phi_{\mathbf{k}}) \quad \text{for all } \mathbf{k} \in J_e . \tag{6.4.61}$$

The remaining equations are obtained by imposing the boundary conditions. These give a set of algebraic relations involving the coefficients of $u^N$ with respect to the orthogonal basis $\{\phi_{\mathbf{k}} \,|\, \mathbf{k} \in J\}$.

The most direct way of taking into account the boundary conditions in a tau method consists of projecting, separately for each side upon the space of polynomials of degree $N$, the equation to be satisfied at the boundary (see, for instance, Example 1.2.4 in Chap. 1). This method may lead to an overdetermined set of boundary equations due to possible continuity conditions at the corners (in two dimensions) or edges (in three dimensions). In the

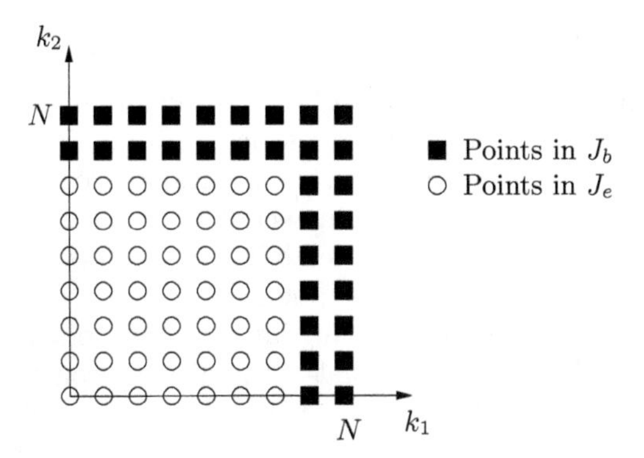


**Fig. 6.2.** The set $J$ in frequency space for the tau approximation to the Dirichlet boundary-value problem for the Laplace equation in the square (Example 8)

quoted example, the number of equations represented by (1.2.77) is $4N+4$, while only $4N$ independent equations have to be added to (1.2.76) in order to determine $u^N$. The rank of the system is only $4N$.

We describe hereafter a mathematically rigorous procedure of boundary projection that leads to the correct number of linearly independent boundary equations. To this end, define the inner product $(u,v)_{\partial\Omega_b}$ between two functions, $u$ and $v$, on $\partial\Omega_b$ as follows. If $S$ is a side of $\partial\Omega_b$ orthogonal to the direction $x_i$, let $\sigma$ be the independent variable on $S$ and let $\tilde{w}(\sigma) = \prod_{\substack{j=1 \\ j\neq i}}^{d} w(x_j)$. Then we set

$$(u,v)_{\partial\Omega_b} = \sum_{\text{sides of } \partial\Omega_b} \int_S u(\sigma)v(\sigma)\tilde{w}(\sigma)\mathrm{d}\sigma \,. \tag{6.4.62}$$

Next, we consider the set of indices $J_b = J - J_e$ and take into account the boundary conditions (6.3.2) by requiring that the tau solution $u^N$ satisfy the set of equations

$$\left(\mathcal{B}u^N, \phi_{\mathbf{k}}\right)_{\partial\Omega_b} = 0 \quad \text{for all } \mathbf{k} \in J_b \,. \tag{6.4.63a}$$

Condition (6.4.63a) involves the traces of the $\phi_{\mathbf{k}}$'s on $\partial\Omega_b$ only, with $\mathbf{k} \in J_b$. These traces are linearly independent on $\partial\Omega_b$, and actually they generate the space $C_0(\partial\Omega_b;N)$ of all the continuous functions on $\partial\Omega_b$ that are polynomials of degree up to $N$ on each side of $\partial\Omega_b$. The proof of this property is not hard, but is rather technical and will be left to the reader. Thus, (6.4.63a) is equivalent to

$$\left(\mathcal{B}u^N, \psi\right)_{\partial\Omega_b} = 0 \quad \text{for all } \psi \in C^0(\partial\Omega_b;N) \,. \tag{6.4.63b}$$

Any convenient basis in $C^0(\partial\Omega_b;N)$ can be used to enforce (6.4.63a), such as a basis whose functions are nonzero on at most $d$ contiguous sides of $\partial\Omega_b$.

We conclude that a multidimensional tau method is represented again by (6.4.60), where now $X = L^2_w(\Omega)$ (see (A.9.h)) and

$$X_N = \{v \in \mathrm{Pol}_N(\Omega) \,|\, (\mathcal{B}v, \phi_{\mathbf{k}})_{\partial\Omega_b} = 0 \text{ for all } \mathbf{k} \in J_b\} \,, \tag{6.4.64}$$

$$Y_N = \mathrm{span}\{\phi_{\mathbf{k}} \,,\ \mathbf{k} \in J_e\} \,. \tag{6.4.65}$$

***Stability and Convergence***

We are next concerned with the problem of *stability and convergence* for the tau approximation (6.4.60). Since the space $X_N$ of basis functions is different from the space $Y_N$ of test functions, the natural approach now is the discrete form of the inf-sup condition given in Sect. 6.3. We assume, therefore, that the operator $\mathcal{L}$ is such that the associated bilinear form $a(u,v) = (\mathcal{L}u, v)$ satisfies (6.3.11)–(6.3.13). Moreover, we assume here that $X_N \subset W$ and $Y_N \subset V$ for all $N > 0$. Then, we have the following inf-sup condition, due to Babuška (see, e.g., Babuška and Aziz (1972)), that is the discrete counterpart of (6.3.12).

*If there exists a constant $\overline{\alpha} > 0$ independent of $N$ such that*

$$\overline{\alpha}\|u\|_W \le \sup_{\substack{v\in Y_N\\ v\neq 0}} \frac{(\mathcal{L}u, v)}{\|v\|_V} \quad \textit{for all } u \in X_N\,, \tag{6.4.66}$$

*then the following estimate holds:*

$$\|u^N\|_W \le \frac{C}{\overline{\alpha}}\|f\|\,, \tag{6.4.67}$$

where the constant $C$, independent of $N$, satisfies $\|v\| \le C\|v\|_V$ for all $v \in V$.

Inequality (6.4.67) implies that (6.4.60) has a unique solution (since $X_N$ and $Y_N$ have the same dimension) and the approximation is stable. The bound (6.4.67) is obtained by dividing each term in (6.4.60) by $\|v\|_V$, then taking the supremum over all $v \in Y_N$ and using (6.4.66) together with the continuity of the inclusion of $V$ into $X$.

Concerning the *convergence* of the method, as for the Galerkin approximation, let $R_N$ be a linear operator from a dense subspace $\mathcal{W} \subseteq D_B(\mathcal{L})$ into $X_N$ such that for $N \to \infty$,

$$\|u - R_N u\|_W \longrightarrow 0 \quad \text{for all } u \in \mathcal{W}\,. \tag{6.4.68}$$

By an argument similar to that used for proving (6.4.8), the following error bound between the solution of (6.3.3) and the tau solution of (6.4.60) can be established:

$$\|u - u^N\|_W \le \left(1 + \frac{A}{\overline{\alpha}}\right)\|u - R_N u\|_W\,. \tag{6.4.69}$$

Thus, the tau method is convergent.

A stability condition of type inf-sup can also be given for Galerkin approximations. Obviously, it is obtainable from (6.4.66) by replacing $Y_N$ with $X_N$. The coercivity condition (6.4.3) is nothing but a particular form of this condition, in which $W = V = E$. Actually, (6.4.66) can be written as

$$\alpha\|u\|_E \le \frac{(\mathcal{L}u, u)}{\|u\|_E} \quad \text{for all } u \in X_N,\ u \neq 0\,, \tag{6.4.70}$$

which is clearly implied by (6.4.3).

A similarity can be established between collocation and tau methods. Indeed, from the tau equations (6.4.61) and (6.4.63) one can obtain the collocation equations (6.4.13) and (6.4.14) formally by replacing the continuous inner product with the discrete one, and taking as $\phi_k$ the characteristic Lagrange polynomials (6.4.12). In both methods, the basis in which the solution is expanded is orthogonal with respect to the inner product involved in the scheme.

### *Examples*

We now consider some examples that illustrate the theory described above.

**Example 6.** *The Dirichlet Problem for a Second-Order Elliptic Operator in the Interval* $(-1,1)$. Consider the problem

$$\begin{aligned} &\mathcal{L}u \equiv -u_{xx} + \lambda^2 u = f\,, \qquad -1 < x < 1\,,\ \lambda \in \mathbb{R}\,, \\ &u(-1) = u(1) = 0\,. \end{aligned}$$

We look for a tau solution $u^N$ expanded in Chebyshev polynomials. Thus, we assume that $f \in L^2_w(-1,1)$ ($w$ being the Chebyshev weight), and we determine the solution $u^N(x) = \sum_{k=0}^{N} \alpha_k T_k(x)$ by the conditions

$$\begin{aligned} &\int_{-1}^{1} \left(-u^N_{xx} + \lambda^2 u^N\right)(x) T_k(x) w(x) \mathrm{d}x \\ &\qquad = \int_{-1}^{1} f(x) T_k(x) w(x) \mathrm{d}x\,, \qquad \text{for } k = 0,1,\dots,N-2\,, \\ &\sum_{k=0}^{N} \alpha_k (-1)^k = \sum_{k=0}^{N} \alpha_k = 0\,. \end{aligned}$$

In the present case, $X_N = \{v \in \mathbb{P}_N \mid v(-1) = v(1) = 0\}$ and $Y_N = \mathbb{P}_{N-2}$.

Let us now discuss the stability and convergence of the approximation. Throughout this and the next example, we will use the simplified notation $\|u\|_{m,w}$ instead of $\|u\|_{H^m_w(-1,1)}$ for $m \geq 0$ (see (5.5.7)). If $u$ is any polynomial of degree $N$ that vanishes on the boundary, then $v = -u_{xx}$ is a polynomial of degree $N-2$, and

$$\begin{aligned} (\mathcal{L}u, v) &= \int_{-1}^{1} (u_{xx})^2 w \,\mathrm{d}x + \lambda^2 \int_{-1}^{1} u_x (uw)_x \,\mathrm{d}x \\ &\geq \|u_{xx}\|^2_{0,w} + \frac{\lambda^2}{4} \|u\|^2_{0,w} \geq C \|u\|^2_{2,w}\,. \end{aligned}$$

We have used (7.1.16) and the Poincaré inequality (A.13) (if $\lambda = 0$ this inequality must be used twice). Therefore, the inf-sup condition (6.4.66) is satisfied if we choose $W = H^2_w(-1,1)$ and $V = L^2_w(-1,1)$, and we have the estimate

$$\|u^N\|_{2,w} \leq C\|f\|_{0,w} \tag{6.4.71}$$

for a constant $C$ independent of $N$ and $\lambda$. The convergence of the method can be established as a consequence of (6.4.69) by defining the projection operator $R_N$ as follows. Let $u$ denote again the exact solution we want to approximate, and let $R_N u$ be an algebraic polynomial of degree $\leq N$ that satisfies (5.5.15) for $l = 2$ and vanishes at the boundary points. It can be easily constructed in the form $R_N u = P_N^2 u - p_1$, where $P_N^2 u$ is the orthogonal projection of $u$ upon $\mathbb{P}_N$ in the $H_w^2$-inner product, which itself satisfies (5.5.15), whereas $p_1$ is the linear polynomial that matches $P_N^2 u$ at the boundary points. Using the inclusion $H_w^1(-1,1) \subset C^0([-1,1])$ (see (A.11.a)), one has

$$\|p_1\|_{2,w} \leq C\|u - P_N^2 u\|_{2,w} \leq CN^{2-m}|u|_{H_w^{m;N}(-1,1)}, \qquad m \geq 2 .$$

Thus, we obtain the optimal convergence estimate

$$\|u - u^N\|_{2,w} \leq CN^{2-m}|u|_{H_w^{m;N}(-1,1)}, \qquad m \geq 2 . \tag{6.4.72}$$

Finally, we note that $v = -u_{xx}^N$ is not the only test function that allows us to obtain a stability estimate for the scheme under consideration. Actually, if $u$ denotes here any polynomial in $X_N$, and if we set $v = P_{N-2}u$, we have

$$(\mathcal{L}u, v) = -\int_{-1}^{1} u_{xx} P_{N-2}uw \,\mathrm{d}x + \lambda^2 \int_{-1}^{1} uP_{N-2}uw \,\mathrm{d}x$$

$$= \int_{-1}^{1} u_x(uw)_x \mathrm{d}x + \lambda^2 \int_{-1}^{1} (P_{N-2}u)^2 w \,\mathrm{d}x .$$

Thus, taking $v = P_{N-2}u^N$ in the tau scheme yields the estimate

$$\frac{1}{2}\|u_x^N\|_{0,w} + \lambda\|P_{N-2}u^N\|_{0,w} \leq C\|f\|_{0,w} . \tag{6.4.73}$$

If $\lambda \gg 1$, (6.4.73) contains the new information that the $L_w^2$-norm of $P_{N-2}u^N$ is $O(1/\lambda)$. This kind of result has been used by Canuto and Sacchi-Landriani (1986) in the analysis of the Kleiser-Schumann method for the Navier-Stokes equations (see CHQZ3, Sect. 3.7). □

**Example 7.** *The Neumann Problem for a Second-Order Elliptic Operator in the Interval* $(-1,1)$. Consider the problem

$$\mathcal{L}u \equiv -u_{xx} + u = f , \qquad -1 < x < 1 ,$$

$$u_x(-1) = u_x(1) = 0 .$$

Again, we look for a tau solution $u^N$ expanded in Chebyshev polynomials. Thus, $u^N(x) = \sum_{k=0}^{N} \alpha_k T_k(x)$ is determined by the conditions

$$
\begin{aligned}
&\int_{-1}^{1} \left(-u_{xx}^N + u^N\right)(x) T_k(x) w(x) \mathrm{d}x \\
&\qquad = \int_{-1}^{1} f(x) T_k(x) w(x) \mathrm{d}x \qquad \text{for } k = 0, 1, \ldots, N-2 \,, \qquad (6.4.74) \\
&\sum_{k=0}^{N-1} \beta_k (-1)^k = \sum_{k=0}^{N-1} \beta_k = 0 \,,
\end{aligned}
$$

where the $\beta_k$'s are the coefficients of the Chebyshev expansion of the derivative $u_x^N$ (see (2.4.22)). We now have $X_N = \{v \in \mathbb{P}_N \,|\, v_x(-1) = v_x(1) = 0\}$ and $Y_N = \mathbb{P}_{N-2}$.

Let us deal with the stability analysis. Note that for all $u \in X_N$, $P_{N-2}\mathcal{L}u = \mathcal{L}u - (u - P_{N-2}u)$, where $P_{N-2}$ is the orthogonal projection operator on $\mathbb{P}_{N-2}$. Hence,

$$
\begin{aligned}
(\mathcal{L}u, P_{N-2}\mathcal{L}u) &= \|\mathcal{L}u\|_{0,w}^2 - (\mathcal{L}u, u - P_{N-2}u) \\
&\geq \|\mathcal{L}u\|_{0,w}^2 - \|\mathcal{L}u\|_{0,w} \|u - P_{N-2}u\|_{0,w} \,.
\end{aligned} \tag{6.4.75}
$$

Now, by (5.5.9), we have $\|u - P_{N-2}u\|_{0,w} \leq C_0 N^{-2} \|u\|_{2,w}$. Moreover, it is possible to prove the *a priori* estimate

$$\|u\|_{2,w} \leq C_1 \|\mathcal{L}u\|_{0,w} \,,$$

for a suitable constant $C_1 > 0$. By (6.4.75) we get

$$(\mathcal{L}u, P_{N-2}\mathcal{L}u) \geq (1 - C_0 C_1 N^{-2}) \|\mathcal{L}u\|_{0,w}^2 \geq (2C_1^2)^{-1} \|u\|_{2,w}^2 \,,$$

provided $N$ is so large that $1 - C_0 C_1 N^{-2} \geq 1/2$. Since $\|P_{N-2}\mathcal{L}u\|_{0,w} \leq C_3 \|u\|_{2,w}$, we conclude that the estimate

$$\frac{(\mathcal{L}u, P_{N-2}\mathcal{L}u)}{\|P_{N-2}\mathcal{L}u\|_{0,w}} \geq \frac{1}{2C_1^2 C_3} \|u\|_{2,w} \,, \tag{6.4.76}$$

holds.

This proves that the scheme (6.4.74) satisfies the stability condition (6.4.66), if we define $W = \{v \in H_w^2(-1,1) \,|\, v_x(-1) = v_x(1) = 0\}$ and $V = L_w^2(-1,1)$.

The convergence analysis is straightforward, in view of (6.4.69). Define the projector $R_N$ onto $X_N$ as

$$(R_N u)(x) = u(-1) + \int_{-1}^{x} \left(P_{N-1}^{1,0} u_x\right)(\xi) \mathrm{d}\xi \,,$$

where $P_{N-1}^{1,0}$ is the operator introduced in (5.5.17). Then it is easy to prove that $\|u - R_N u\|_{2,w} \leq C N^{2-m} |u|_{H_w^{m;N}(-1,1)}$; whence, by (6.4.69) we get the optimal error estimate

$$\|u - u^N\|_{2,w} \leq C N^{2-m} |u|_{H_w^{m;N}(-1,1)} \,.$$

□

**Example 8.** *A Legendre Tau Method for the Poisson equation.* We consider again the tau approximation introduced in Sect. 1.2.4 and analyzed in Sect. 6.1.3. The aim here is to incorporate this scheme in the previous general framework.

The tau solution is expanded into Legendre polynomials, $\phi_{\mathbf{k}}(x,y) = L_{k_1}(x)L_{k_2}(y)$, namely, $u^N(x,y) = \sum_{k=0}^{N}\sum_{m=0}^{N} \hat{u}_{km} L_k(x) L_m(y)$. Thus, the natural choice for the Hilbert space $X$ is the space $L^2(\Omega)$ ($\Omega$ being the square $(-1,1)\times(-1,1)$), with inner product

$$(u,v) = \int_{-1}^{1}\int_{-1}^{1} u(x,y)v(x,y)\mathrm{d}x\,\mathrm{d}y\;.$$

The boundary conditions are prescribed over the whole boundary of $\Omega$; hence $\partial\Omega_b = \partial\Omega$ and the boundary inner product takes the form

$$\begin{aligned}(u,v)_{\partial\Omega} = &\int_{-1}^{1} u(x,-1)v(x,-1)\mathrm{d}x + \int_{-1}^{1} u(x,1)v(x,1)\mathrm{d}x \\ &+ \int_{-1}^{1} u(-1,y)v(-1,y)\mathrm{d}y + \int_{-1}^{1} u(1,y)v(1,y)\mathrm{d}y\;.\end{aligned}$$

Exactly one boundary condition is prescribed on each side of $\Omega$; hence, we have

$$J_e = \{(k_1,k_2)\,|\,0 \le k_1, k_2 \le N-2\}$$

and

$$J_b = \{(k_1,k_2)\,|\,N-1 \le k_i \le N, \text{ for at least one index } i = 1,2\}\;.$$

Thus, equations (1.2.74) are nothing but (6.4.61), while equations (1.2.77) clearly imply (6.4.63b). We look now for a basis of $C^0(\partial\Omega_b;N)$, the space of polynomials of degree $N$ on each side of $\Omega$ that are continuous at the corners. Define for $k \ge 2$, $l_k(x) = L_k(x) - L_{\overline{k}}(x)$, where $\overline{k} = k \pmod 2$ (i.e., $\overline{k} = 0$ if $k$ is even, $\overline{k} = 1$ if $k$ is odd). Thus, $l_k(+1) = l_k(-1) = 0$. Furthermore, set $l_\pm(x) = L_N(x) \pm L_{N-1}(x)$, so that $l_\pm(\pm 1) \ne 0$ and $l_\pm(\mp 1) = 0$. Each of the functions

$$\Psi_{(k,+)}(x,y) = l_k(x)l_+(y)\,, \qquad k \ge 2\,,$$

is a linear combination of basis functions $\phi_{\mathbf{k}}(x,y)$ with $\mathbf{k} \in J_b$; hence, (6.4.63) yields

$$\begin{aligned}(u^N, \Psi_{(k,+)})_{\partial\Omega} &= \int_{-1}^{1} u^N(x,1)l_k(x)\mathrm{d}x \\ &= \sum_{m=0}^{N}\left[\hat{u}_{km} + \hat{u}_{\overline{k},m}\right] = 0\,, \qquad 2 \le k \le N\,.\end{aligned} \tag{6.4.77a}$$

In the same way, the test functions $\Psi_{(k,-)}(x,y) = l_k(x)l_-(y)$ and $\Psi_{(\pm,k)}(x,y) = l_\pm(x)l_k(y)$ yield, respectively, the relations

$$\sum_{m=0}^{N}(-1)^m \left[\hat{u}_{km} - \hat{u}_{\overline{k},m}\right] = 0\,, \qquad 2 \leq k \leq N\,, \tag{6.4.77b}$$

$$\sum_{k=0}^{N} [\hat{u}_{km} + \hat{u}_{k,\overline{m}}] = 0\,, \qquad 2 \leq m \leq N\,, \tag{6.4.77c}$$

$$\sum_{k=0}^{N}(-1)^k [\hat{u}_{km} - \hat{u}_{k,\overline{m}}] = 0\,, \qquad 2 \leq m \leq N\,. \tag{6.4.77d}$$

Finally, the test functions $\Psi_{(\pm,\pm)}(x,y) = l_\pm(x)l_\pm(y)$ give the remaining relations

$$\sum_{m=0}^{N} [\hat{u}_{Nm} + \hat{u}_{N-1,m}] + \sum_{k=0}^{N} [\hat{u}_{kN} + \hat{u}_{k,N-1}] = 0\,, \tag{6.4.78a}$$

$$\sum_{m=0}^{N}(-1)^m [\hat{u}_{Nm} + \hat{u}_{N-1,m}] + \sum_{k=0}^{N} [\hat{u}_{kN} - \hat{u}_{k,N-1}] = 0\,, \tag{6.4.78b}$$

$$\sum_{m=0}^{N} [\hat{u}_{Nm} - \hat{u}_{N-1,m}] + \sum_{k=0}^{N}(-1)^k [\hat{u}_{kN} + \hat{u}_{k,N-1}] = 0\,, \tag{6.4.78c}$$

$$\sum_{m=0}^{N}(-1)^m [\hat{u}_{Nm} - \hat{u}_{N-1,m}] + \sum_{k=0}^{N}(-1)^k [\hat{u}_{kN} - \hat{u}_{k,N-1}] = 0\,. \tag{6.4.78d}$$

Note that the functions $\Psi_{(k,\pm)}$, and $\Psi_{(\pm,k)}$ are nonzero on one side of $\Omega$, while $\Psi_{(\pm,\pm)}$ are nonzero on two contiguous sides of $\Omega$. We conclude that (6.4.77) and (6.4.78) are equivalent to (6.4.63b).

For the present scheme, one has $X_N = \{v \in \mathbb{P}_N \,|\, v \equiv 0 \text{ on } \partial\Omega\}$ and $Y_N = \mathbb{P}_{N-2}$. Here, $\mathbb{P}_N$ is the space of the algebraic polynomials of degree $\leq N$ in each variable.

Let us now discuss the stability and convergence of the method. In Sect. 6.1.3 the test function designed to prove stability was $q(x,y) = u^N(x,y)/[(1-x^2)(1-y)^2]$. This appears to be a natural choice for tau approximations to homogeneous Dirichlet boundary-value problems. Actually, any $u \in X_N$ can be split into the product $u = bq$ where $q$ is a polynomial of the space $Y_N$, and $b$ is a polynomial of minimal degree that vanishes on $\partial\Omega_b$.

If for a suitable choice of Hilbert spaces $W$ and $V$, there exist positive constants $\alpha_1$ and $\alpha_2$ independent of $N$ such that

$$\alpha_1 \|u\|_W^2 \leq (\mathcal{L}u, q) \quad \text{for all } u \in X_N\,, \tag{6.4.79}$$

$$\|q\|_V \leq \alpha_2 \|u\|_W \quad \text{for all } u \in X_N\,, \tag{6.4.80}$$

then (6.4.66) is satisfied with $\overline{\alpha} = \alpha_1/\alpha_2$.

In the current example we set $b(x,y) = (1-x^2)(1-y^2)$, and we define the norms

$$\|u\|_W = \left(\int_\Omega b|\nabla q|^2 \mathrm{d}x\,\mathrm{d}y + \frac{1}{2}\int_\Omega |\Delta b| q^2\,\mathrm{d}x\,\mathrm{d}y\right)^{1/2}, \quad \text{with } q = u/b\,,$$

and

$$\|v\|_V = \left(\int_\Omega |\Delta b| v^2\,\mathrm{d}x\,\mathrm{d}y\right)^{1/2}$$

($W$ and $V$ being defined as the weighted Sobolev spaces of the functions for which these norms, respectively, are finite). In the present example, however, the continuity condition (6.3.13) is not verified. Rather we have, by the Cauchy-Schwarz inequality,

$$\left|\int_\Omega \Delta u v\,\mathrm{d}x\,\mathrm{d}y\right| \le \left(\int_\Omega \frac{|\Delta u|^2}{|\Delta b|}\mathrm{d}x\,\mathrm{d}y\right)^{1/2} \left(\int_\Omega |\Delta b|\, v^2\,\mathrm{d}x\,\mathrm{d}y\right)^{1/2}.$$

Hence, the operator $\mathcal{L}$ turns out to be continuous with respect to a stronger norm than the norm of $W$. More precisely, if we define $\|u\|_{\tilde{W}}$ to be the maximum of the values attained in $\Omega$ by any derivative of $u$ of order up to 2 (mathematically, the Sobolev space $\tilde{W}$ for which such a norm is finite is denoted by $W^{2,\infty}(\Omega)$), we have

$$|(\mathcal{L}u, v)| \le \tilde{A}\|u\|_{\tilde{W}}\|v\|_V \quad \text{for all } u \in \tilde{W} \text{ and } v \in V\,,$$

with $\tilde{A} = (\int_\Omega (1/|\Delta b|)\mathrm{d}x\,\mathrm{d}y)^{1/2} < +\infty$. The convergence estimate (6.4.69) has to be modified into

$$\|u - u^N\|_W \le C\|u - R_N u\|_{\tilde{W}}\,.$$

Using this inequality and a suitable projection operator, one gets the estimate (6.1.17) given in Sect. 6.1. □

## 6.5 General Formulation of Spectral Approximations to Linear Evolution Problems

Our attention now turns to an abstract formulation of spectral approximations to time-dependent problems. It is based on the same mathematical setting introduced in Sect. 6.3. We will retain the same notation here without referring repeatedly to Sect. 6.3.

We will analyze *semi-discrete* approximations only; in particular, the time variable will not be discretized. Some time-marching methods commonly used in combination with spectral approximations are discussed in Appendix D;

an example of analysis of a fully discrete scheme for the heat equation is given in Sect. 7.5.

Consider the initial-boundary-value problem

$$u_t + \mathcal{L}u = f \qquad \text{in } \Omega \times (0, +\infty) , \tag{6.5.1}$$

$$\mathcal{B}u = 0 \qquad \text{on } \partial\Omega_b \times (0, +\infty) , \tag{6.5.2}$$

$$u = u_0 \qquad \text{in } \Omega \text{ for } t = 0 . \tag{6.5.3}$$

The initial value $u_0$ is a function belonging to the space $X$, and the right-hand side $f$ is a continuous function of the variable $t$ with values in $X$, i.e., $f(t) \in X$ for each $t > 0$. A solution for this problem is an $X$-valued function $u(t)$ such that $u$ is continuous for all $t \geq 0$, $du/dt$ exists and is continuous for all $t > 0$, $u(0) = u_0$, $u(t) \in D_B(\mathcal{L})$ for all $t > 0$, and (6.5.1) holds for all $t > 0$. In compact notation:

$$\begin{aligned} & u \in C^1(0, +\infty; X) , \quad u(t) \in D_B(\mathcal{L}) , && \text{for } t > 0 , \\ & \frac{du}{dt}(t) + \mathcal{L}u(t) = f(t) && \text{for } t > 0 , \\ & u(0) = u_0 . \end{aligned} \tag{6.5.4}$$

We assume that problem (6.5.4) is well posed. For a rigorous definition of well-posedness and for conditions assuring the well-posedness, we refer, e.g., to Hille and Philips (1957) or to Richtmyer (1978), Chap. 16.

Any spectral approximation to the time-independent problem (6.3.3), as defined in Sect. 6.3, yields in a natural way a semi-discrete spectral approximation to the evolution problem (6.5.4). In the most general setting, the time-dependent counterpart of (6.3.14) consists of looking for an approximation $u^N(t)$ satisfying

$$\begin{aligned} & u^N \in C^1([0, +\infty); X_N) , \\ & \left( \frac{du^N}{dt}(t), v \right)_N + a_N(u^N(t), v) = (f(t), v)_N && \text{for all } v \in Y_N,\ t > 0 , \\ & u^N(0) = u_0^N , \end{aligned} \tag{6.5.5}$$

where $(u, v)_N$ denotes an approximation of the inner product $(u, v)$ in $X$ and $u_0^N$ is an approximation of $u_0$. Galerkin and G-NI schemes, possibly incorporating the weak enforcement of boundary conditions, can be formulated in this manner.

In the more restricted situation in which $X_N \subset D_B(\mathcal{L})$, the second condition in (6.5.5) can be replaced by

$$\left( \frac{du^N}{dt}(t) + \mathcal{L}_N u^N(t) - f(t), v \right)_N = 0 \qquad \text{for all } v \in Y_N,\ t > 0 , \tag{6.5.6}$$

which is the time-dependent counterpart of (6.3.20), or by its operational form

$$Q_N\left(\frac{\mathrm{d}u^N}{\mathrm{d}t}(t)+\mathcal{L}_N u^N(t)-f(t)\right)=0 \qquad \text{for all } t>0\,, \tag{6.5.7}$$

which corresponds to (6.3.21).

### *Galerkin, Collocation, G-NI and Tau Approximations*

The formulation (6.5.5) summarizes various spectral approximations to the evolution problem (6.5.4). In particular, the Galerkin, collocation, G-NI and tau schemes, that have been defined in Sect. 6.4 for steady problems, apply in the present situation also. The time-derivative term $\mathrm{d}u^N/\mathrm{d}t$ is treated formally in the same way as the right-hand side $f$. Each of these procedures transforms (6.5.7) into a system of ordinary differential equations whose unknowns are the coefficients of $u^N(t)$ with respect to the chosen (modal or nodal) basis in $X_N$. From a mathematical point of view, each of these methods is defined by the same choice of the spaces $X_N$ and $Y_N$, the bilinear and linear forms $a_N(u,v)$ and $F_N(v)$ (or the operator $\mathcal{L}_N$ and the inner product $(u,v)_N$) made in Sect. 6.4. It is therefore straightforward to extend the material of that section to the case of time-dependent problems.

### 6.5.1 Conditions for Stability and Convergence: The Parabolic Case

In order to discuss questions of stability (in space) and convergence for spectral approximations to time-dependent problems, we distinguish between equations of parabolic and hyperbolic type. We start with the parabolic case, which is characterized by the fact that the operator $\mathcal{L}$ is coercive (or weakly coercive) with respect to a norm that is stronger than the one of $X$.

As for time-independent problems, the simplest stability condition arises from an *energy* inequality. We will assume henceforth that all the hypotheses made in Sect. 6.3 hold true; in particular *we assume that the spatial operator* $\mathcal{L}$ *satisfies the continuity condition* (6.3.7) *and the coercivity condition* (6.3.6).

We consider first a *Galerkin approximation*, for which (6.5.5) takes the form

$$\left(\frac{\mathrm{d}u^N}{\mathrm{d}t}(t),v\right)+a(u^N(t),v)=(f(t),v) \quad \text{for all } v\in X_N,\ t>0\,. \tag{6.5.8}$$

Then, taking $v=u^N(t)$, we get, for each $t>0$,

$$\frac{1}{2}\frac{\mathrm{d}}{\mathrm{d}t}\|u^N(t)\|^2+\alpha\|u^N(t)\|_E^2\le(f(t),u^N(t))\,.$$

Now, applying the algebraic inequality $ab \leq (1/4\varepsilon)a^2 + \varepsilon b^2$, with $\varepsilon = \alpha/2$, to the right-hand side, we can find a constant $C$ depending on $\alpha$ but independent of $N$ such that we have, for all $t > 0$,

$$\|u^N(t)\|^2 + \alpha \int_0^t \|u^N(s)\|_E^2 \, \mathrm{d}s \leq \|u_0^N\|^2 + C \int_0^t \|f(s)\|^2 \mathrm{d}s \, . \tag{6.5.9}$$

This proves the *stability* (in space) of the Galerkin approximation.

Concerning its *convergence*, let us set $e(t) = R_N u(t) - u^N(t)$, where $R_N$ is the projection operator introduced in (6.4.6). Then, the error function $e(t)$ satisfies the inequality

$$\frac{1}{2}\frac{\mathrm{d}}{\mathrm{d}t}\|e\|^2 + \alpha\|e\|_E^2 \leq |(u_t - R_N u_t, e) + a(u - R_N u, e)| \, . \tag{6.5.10}$$

For any function $g \in X$, we can define the new norm

$$\|g\|_{E^*} = \sup_{\substack{v \in E \\ v \neq 0}} \frac{(g, v)}{\|v\|_E} \, . \tag{6.5.11}$$

This is the norm of $g$ in the dual space $E^*$ of $E$ (see (A.1.c)). Note that $\|g\|_{E^*} \leq C\|g\|_X$, since $\|v\|_X \leq C\|v\|_E$ for all $v \in E$. Then, using the above definition and the continuity of the operator $\mathcal{L}$ (see (6.3.7)), it follows that

$$|(u_t - R_N u_t, e) + a(u - R_N u, e)| \leq C\{\|u_t - R_N u_t\|_{E^*} + \|u - R_N u\|_E\}\|e\|_E \, .$$

Therefore, for all $t > 0$ the following error bound can be inferred from (6.5.10):

$$\begin{aligned} \|e(t)\|^2 + \alpha \int_0^t \|e(s)\|_E^2 \, \mathrm{d}s &\leq \|e(0)\|^2 \\ &+ C\left\{\int_0^t \|(u_t - R_N u_t)(s)\|_{E^*}^2 \, \mathrm{d}s + \int_0^t \|(u - R_N u)(s)\|_E^2 \, \mathrm{d}s\right\} , \end{aligned} \tag{6.5.12}$$

where $C$ is a constant independent of $N$.

We conclude that *the approximation is convergent if each term on the right-hand side tends to 0 as $N \to \infty$ for $u$, $u_t$, and $u_0$ regular enough.* In particular, this is true if the hypothesis (6.4.7) holds uniformly in $t$ for time-dependent functions $u = u(t)$ and $u_t = u_t(t)$ in a suitable class. The approximation results given in Chap. 5 guarantee this property. As discussed at the end of Sect. 6.1.2, the smoothness of the solution follows from the smoothness of the initial and boundary data and possibly the fulfillment of certain compatibility conditions among them.

G-NI approximations, satisfying the assumptions of Sect. 6.4.3 (see in particular (6.4.43)–(6.4.44) as well as (6.4.47a)–(6.4.47c)), can be analyzed in a similar manner.

**Example 1.** *A Fourier Galerkin Method for the Heat Equation.* Consider the one-dimensional heat equation problem

$$\begin{aligned} &u_t - u_{xx} = f\,, && 0 < x < 2\pi,\ t > 0\,, \\ &u(x,0) = u_0(x)\,, && 0 < x < 2\pi\,, \\ &u(x,t)\ \ 2\pi\text{-periodic in } x && \text{for all } t \geq 0\,. \end{aligned} \tag{6.5.13}$$

Its Galerkin approximation consists of looking for a function $u^N(t) \in S_N$, where $S_N$ is the space of trigonometric polynomials defined in (5.1.1), that satisfies

$$\left(u_t^N - u_{xx}^N - f, v\right) = 0 \quad \text{for all } v \in S_N,\ t > 0\,, \tag{6.5.14}$$

and $u^N(0) = P_N u_0$ (see (2.1.7)). In this case, the operator $\mathcal{L} = -\partial^2/\partial x^2$ satisfies the following energy identity:

$$(\mathcal{L}u, u) = a(u,u) = \int_0^{2\pi} |u_x|^2\,\mathrm{d}x\,.$$

The square root of the right-hand side is just a semi-norm for the space $E = H_p^1(0,2\pi)$ (see (A.11.d)). However, using the change of variable $u^N(t) \rightarrow w^N(t) = e^{-t}u^N(t)$, (6.5.14) becomes

$$\left(w_t^N - w_{xx}^N + w^N - e^t f, v\right) = 0 \quad \text{for all } v \in S_N,\ t > 0\,. \tag{6.5.15}$$

The new operator $\tilde{\mathcal{L}} = -\partial^2/\partial x^2 + I$ satisfies the coercivity estimate (6.3.6); hence, stability and convergence follow by the previous general results. The trick of the above change of variable is used each time the bilinear form $a(u,v)$ associated with the operator $\mathcal{L}$ is only *weakly coercive* on $E$, i.e., it satisfies the inequality

$$\alpha |u|_E^2 \leq a(u,u) \qquad \text{for all } u \in E\,, \tag{6.5.16}$$

where $|u|_E$ is a seminorm on $E$ such that $(\|u\|^2 + |u|_E^2)^{1/2} = \|u\|_E$.

Continuing our analysis, we choose $R_N = P_N$ in (6.5.10), and we observe that for all $v \in H_p^1(0,2\pi)$ we have, by (5.1.9),

$$|(u_t - P_N u_t, v)| = |(u_t - P_N u_t, v - P_N v)| \leq CN^{1-m} |u_t|_{H^{m-2}(0,2\pi)} |v|_{H^1(0,2\pi)}\,,$$

where $|v|_{H^s(0,2\pi)} = \|v^{(s)}\|_{L^2(0,2\pi)}$ is the seminorm of $v$ of order $s$. Hence, $\|u_t - R_N u_t\|_{[H_p^1(0,2\pi)]^*} \leq CN^{1-m}|u_t|_{H^{m-2}(0,2\pi)}$. Thus, we obtain the following error estimate, which holds for all $t > 0$ and $m \geq 1$:

$$\|u(t) - u^N(t)\|_{L^2(0,2\pi)} + \left(\int_0^t \|\left(u - u^N\right)(s)\|_{H^1(0,2\pi)}^2 \mathrm{d}s\right)^{1/2}$$

$$\leq CN^{1-m}\left(\int_0^t |u_t(s)|_{H^{m-2}(0,2\pi)}^2 \mathrm{d}s + \int_0^t |u(s)|_{H^m(0,2\pi)}^2 \mathrm{d}s\right)^{1/2}\,. \qquad \square$$

We consider now *tau approximations* of problem (6.5.4). The tau method has been introduced for steady problems in Sect. 6.4.4. When applied to the evolution problem (6.5.4), it yields the scheme

$$\left(\frac{\mathrm{d}u^N}{\mathrm{d}t}(t) + \mathcal{L}u^N(t) - f(t), v\right) = 0 \quad \text{for all } v \in Y_N,\ t > 0\ . \tag{6.5.17}$$

Thus, *stability* can be obtained provided the following inequality holds:

$$(\mathcal{L}u, Q_N u) \geq \overline{\alpha}\|u\|_E^2 \quad \text{for all } u \in X_N\ , \tag{6.5.18}$$

where $\overline{\alpha}$ is a positive constant, and $Q_N$ is the orthogonal projection upon $Y_N$ in the inner product of $X$. Indeed, choosing $v = Q_N u^N(t)$ as test function, for all $t > 0$, we obtain the following stability result:

$$\|Q_N u^N(t)\|^2 + \overline{\alpha}\int_0^t \|u^N(s)\|_E^2\,\mathrm{d}s \leq \|u_0^N\|^2 + C\int_0^t \|f(s)\|^2\,\mathrm{d}s\ . \tag{6.5.19}$$

Proceeding as done for the Galerkin approximation, the convergence inequality takes now the form

$$\begin{aligned}\|Q_N e(t)\|^2 + \overline{\alpha}\int_0^t \|e(s)\|_E^2\,\mathrm{d}s &\leq \|e(0)\|^2 \\ +C\int_0^t \|Q_N(u_t - R_N u_t)(s)\|_{E^*}^2\,\mathrm{d}s &+ \int_0^t \|Q_N\mathcal{L}(u - R_N u)(s)\|_{E^*}^2\,\mathrm{d}s\ .\end{aligned} \tag{6.5.20}$$

This inequality, together with the approximation results of Chap. 5, allows one to prove the *convergence* of the scheme.

**Example 2.** *A Legendre Tau Method for the Heat Equation.* We consider the initial-boundary-value problem

$$\begin{aligned} &u_t - u_{xx} = f\ , && -1 < x < 1,\ t > 0\ , \\ &u(-1,t) = u(1,t) = 0\ , && t > 0\ , \\ &u(x,0) = u_0(x)\ , && -1 < x < 1\ . \end{aligned}$$

The solution $u^N(x,t)$ of the Legendre tau approximation of this problem is for all $t \geq 0$ a polynomial of degree $N$ in $x$, that is zero at $x = \pm 1$ and satisfies for all $v \in \mathbb{P}_{N-2}$ the equations

$$\begin{aligned} &\int_{-1}^1 \left[u_t^N(x,t) - u_{xx}^N(x,t)\right] v(x)\,\mathrm{d}x = \int_{-1}^1 f(x,t)v(x)\,\mathrm{d}x\ , \qquad t > 0\ , \\ &\int_{-1}^1 \left[u^N(x,0) - u_0(x)\right] v(x)\mathrm{d}x = 0\ . \end{aligned} \tag{6.5.21}$$

It follows that this scheme conforms to the abstract form (6.5.7) if we set $X = L^2(-1,1)$, $X_N = \{u \in \mathbb{P}_N \,|\, u(-1) = u(1) = 0\}$, $Y_N = \mathbb{P}_{N-2}$, $\mathcal{L}_N = \mathcal{L} = -\partial^2/\partial x^2$, and if the projection $Q_N : L^2(-1,1) \to \mathbb{P}_{N-2}$ is the truncation $P_{N-2}$ of the Legendre series.

For all $u \in X_N$ we have

$$-\int_{-1}^{1} u_{xx} P_{N-2} u \,\mathrm{d}x = -\int_{-1}^{1} u_{xx} u \,\mathrm{d}x = \int_{-1}^{1} (u_x)^2 \mathrm{d}x \,.$$

It follows that the stability condition (6.5.18) is verified with $E = H_0^1(-1,1)$ (see (A.11.c)), since $\|u\|_E = (\int_{-1}^{1} (u_x)^2 \mathrm{d}x)^{1/2}$ is a norm for this space (see (A.13)). Hence, the Legendre tau approximation (6.5.21) is stable, and (6.5.19) gives for all $t > 0$ the estimate

$$\begin{aligned} \|P_{N-2} u^N(t)\|^2_{L^2(-1,1)} &+ \int_0^t \|u_x^N(s)\|^2_{L^2(-1,1)} \mathrm{d}s \\ &\le \|u_0\|^2_{L^2(-1,1)} + C \int_0^t \|f(s)\|^2_{L^2(-1,1)} \mathrm{d}s \,. \end{aligned}$$

A bound for the error $u - u^N$ can be derived from the estimate (6.5.20). The operator $R_N$ is chosen as the orthogonal projection on $X_N$ in the norm of $H_0^1(-1,1)$, as defined in (5.4.29). We bound each term on the right-hand side of (6.5.20). The first term is bounded by the square of $C(\|u_0 - P_{N-2}u_0\|_{L^2(-1,1)} + \|u_0 - R_N u_0\|_{L^2(-1,1)})$. Concerning the second term, we have, for each $v \in H_0^1(-1,1)$,

$$\begin{aligned} (P_{N-2}(u_t - R_N u_t), v) &= (u_t - R_N u_t, v) - (u_t - R_N u_t, v - P_{N-2} v) \\ &= ((u_t - R_N u_t)_x, (\phi - R_N \phi)_x) - (u_t - R_N u_t, v - P_{N-2} v) \,, \end{aligned}$$

where $\phi$ is the only function in $H_0^1(-1,1)$ satisfying $-\phi_{xx} = v$. Then, using the approximation results for the operators $P_{N-2}$ and $R_N$ given in (5.4.11) and (5.4.30), respectively, and recalling (6.5.11), we obtain

$$\|P_{N-2}(u_t - R_N u_t)\|_{E^*} \le C N^{1-m} |u_t|_{H^{m-2;N}(-1,1)} \,. \tag{6.5.22}$$

For the last term of (6.5.20) we have, for all $v \in H_0^1(-1,1)$,

$$\begin{aligned} (P_{N-2}(u - R_N u)_{xx}, v) &= -((u - R_N u)_x, v_x) - ((u - R_N u)_{xx}, v - P_{N-2} v) \\ &= -((u - R_N u)_x, v_x) - (u_{xx} - P_{N-2} u_{xx}, v - P_{N-2} v) \,. \end{aligned}$$

Here we have used the fact that both $P_{N-2} u_{xx}$ and $(R_N u)_{xx}$ are orthogonal to $v - P_{N-2} v$. Using the same approximation results as before, we deduce

$$\|P_{N-2}(u - R_N u)_{xx}\|_{E^*} \le C N^{1-m} |u|_{H^{m;N}(-1,1)} \,. \tag{6.5.23}$$

Combining the previous results we obtain the final error estimate, for all $t > 0$ and all $m \geq 2$,

$$\|u(t) - P_{N-2}u^N(t)\|_{L^2(-1,1)} + \left( \int_0^t \left\| \left(u_x - u_x^N\right)(s) \right\|^2_{L^2(-1,1)} \mathrm{d}s \right)^{1/2}$$
$$\leq CN^{1-m} \left( \int_0^t \left( |u_t(s)|^2_{H^{m-2;N}(-1,1)} + |u(s)|^2_{H^{m;N}(-1,1)} \right) \mathrm{d}s \right)^{1/2} . \tag{6.5.24}$$

□

Finally, let us consider *collocation approximations* to (6.5.4). We recall that collocation methods for steady problems have been introduced in Sect. 6.4.2. For simplicity, we assume in (6.5.6) that $Y_N = X_N$, which is the case when the boundary conditions are of Dirichlet type. Moreover, we assume that the discrete operator $\mathcal{L}_N$ satisfies the coercivity inequality

$$(\mathcal{L}_N u, u)_N \geq \overline{\alpha}\|u\|^2_E \quad \text{for all } u \in X_N . \tag{6.5.25}$$

The technique already applied to the other spectral schemes yields, for each $t > 0$, the *stability* inequality

$$\|u^N(t)\|^2_N + \overline{\alpha} \int_0^t \|u^N(s)\|^2_E \, \mathrm{d}s \leq \|u_0^N\|^2_N + C \int_0^t \|f(s)\|^2_N \, \mathrm{d}s . \tag{6.5.26}$$

We recall here that the discrete norm $\|u\|_N = \sqrt{(u,u)_N}$ can be controlled by $C\|u\|_X$ for all $u \in \mathrm{Pol}_N(\Omega)$, with $C$ independent of $N$ (see Sect. 5.3).

Concerning the *convergence* of the approximation, the following estimate, that is the counterpart of estimate (6.4.26) for evolution equations, holds for all $t > 0$:

$$\begin{aligned}
&\|e(t)\|^2_N + 2\overline{\alpha} \int_0^t \|e(s)\|^2_E \, \mathrm{d}s \\
&\quad \leq \|e(0)\|^2_N + C \int_0^t \|u_t - R_N u_t\|^2_{E^*} \, \mathrm{d}s + \int_0^t \|u - R_N u\|^2_E \, \mathrm{d}s \\
&\qquad + C \int_0^t \left( \frac{(R_N u_t, e) - (R_N u_t, e)_N}{\|e\|_E} \right)^2 \mathrm{d}s \\
&\qquad + \int_0^t \left( \frac{(\mathcal{L} R_N u, e) - (\mathcal{L}_N R_N u, e)_N}{\|e\|_E} \right)^2 \mathrm{d}s \\
&\qquad + \int_0^t \left( \frac{(f,e) - (f,e)_N}{\|e\|_E} \right)^2 \mathrm{d}s .
\end{aligned} \tag{6.5.27}$$

This estimate can be obtained by adapting to the present situation the proof of estimate (6.5.20), taking into account the extra errors due to the discrete inner product.

**Example 3.** *A Chebyshev Collocation Method for the Heat Equation with Dirichlet Boundary Conditions.* We consider again the scheme presented in Sect. 1.2.2. This scheme is analyzed in Sect. 6.1.2, where it is actually proven that the stability condition (6.5.25) holds in this case with $E = H^1_{w,0}(-1,1)$ defined in (A.11.b) and $\|u\|_E = (\int_{-1}^1 |u_x|^2 w(x)\,dx)^{1/2}$. Moreover, it is claimed there that the optimal error bound (6.1.10) holds. Indeed, this estimate is an immediate consequence of the general estimate (6.5.27).

We choose as $R_N u$ the orthogonal projection of $u$ upon $\mathbb{P}_{N-1}$ rather than $\mathbb{P}_N$ with respect to the inner product of $H^1_{w,0}(-1,1)$ (see (5.5.17)). Then the three last terms of (6.5.27) are zero in the current situation, while the two remaining ones can be handled as in Example 2. □

### 6.5.2 Conditions for Stability and Convergence: The Hyperbolic Case

The *energy* approach for equations of hyperbolic type takes the following general form. It is assumed that there exists a Hilbert space $E \subset X$ with norm $\|v\|_E$ such that $D_B(\mathcal{L}) \subset E$ and $\|v\| \le C\|v\|_E$ for all $v \in E$. Moreover, it is assumed that there exists a constant $\bar{C} > 0$ such that

$$\|\mathcal{L}v\| \le \bar{C}\|v\|_E \quad \text{for all } v \in D_B(\mathcal{L}) \,, \tag{6.5.28}$$

and that the operator $\mathcal{L}$ satisfies the nonnegativity property

$$0 \le (\mathcal{L}v, v) \quad \text{for all } v \in D_B(\mathcal{L}) \,. \tag{6.5.29}$$

Considering discrete approximations, we refer again to the general setting (6.5.5). Galerkin, G-NI and certain collocation approximations fit into this scheme with the choice $X_N = Y_N$, where $X_N$ is contained in $E$. In such cases the natural discrete counterpart of condition (6.5.29) is

$$0 \le a_N(v, v) \quad \text{for all } v \in X_N \,. \tag{6.5.30}$$

*If this assumption is fulfilled for all $N > 0$, the approximation scheme* (6.5.5) *is stable* (in space) in the norm $\|u^N\|_N$ associated with the inner product $(u, v)_N$. Indeed, taking $v = u^N(t)$ in (6.5.5) and using the Gronwall lemma (see (A.15)), we obtain the following estimate:

$$\|u^N(t)\|_N^2 \le \|u_0^N\|_N^2 + \exp(t) \int_0^t \|f(s)\|_N^2 ds \quad \text{for all } t > 0 \,. \tag{6.5.31}$$

In order to study the convergence of the approximation, we suppose that the discrete and continuous norms are uniformly equivalent on $X_N$, i.e.,

$$C_1\|v\| \le \|v\|_N \le C_2\|v\| \quad \text{for all } v \in X_N \,,$$

with two constants $C_1$ and $C_2$ independent of $N$. This condition is always fulfilled in the cases of interest, as has been shown in Chap. 5 (see Sect. 5.3). Indeed, we recall that $(u,v)_N$ does coincide with the inner product $(u,v)$ in $X$ for Galerkin methods, whereas for G-NI and collocation methods it takes the usual meaning of the discrete inner product defined by a Gaussian quadrature formula.

To get a *convergence* estimate, we set as usual $e(t) = R_N u(t) - u^N(t)$, where $R_N$ is a suitable projection operator defined as in (6.4.6). The equation satisfied by the error function $e(t)$ is easily obtained from (6.5.4), which we write in the equivalent variational form

$$\left(\frac{du}{dt}(t), v\right) + a(u(t),v) = (f(t),v) \qquad \text{for all } v \in E,\ t > 0\,,$$

and from (6.5.5), in which we write $u^N = R_N u - e$. Assumption (6.5.30) together with the Gronwall lemma, allow us to get a bound for $\|e\|_N$, that in turn implies a bound for $\|e\|$. Then, by the triangle inequality, $\|u - u^N\| \leq \|u - R_N u\| + \|e\|$, we obtain the desired convergence estimate, that reads as follows: for all $t > 0$,

$$\begin{aligned}
&\|u(t) - u_N(t)\|^2 \\
&\quad \leq C\Bigg\{ \|u(t) - R_N u(t)\|^2 + \|u_0^N - R_N u_0\|^2 \\
&\qquad + \exp(t)\left[\int_0^t (\|u_t - R_N u_t\|^2 + \|u - R_N u\|_E^2)\, ds \right.\\
&\qquad + \int_0^t \left(\frac{(R_N u_t, e) - (R_N u_t, e)_N}{\|v\|}\right)^2 ds \\
&\qquad \left. + \int_0^t \left(\frac{a(R_N u, e) - a_N(R_N u, e)}{\|e\|}\right)^2 ds + \int_0^t \left(\frac{(f,e)-(f,e)_N}{\|e\|}\right)^2 ds \right]\Bigg\}\,.
\end{aligned} \tag{6.5.32}$$

The three last terms on the right-hand side are absent in a Galerkin approximation; they originate from the quadrature error in a G-NI method, or, equivalently, from the aliasing error in a collocation method. Again, the convergence of the methods is guaranteed if each term on the right-hand side of (6.5.32) vanishes as $N \to \infty$. This can be proven for regular solutions using the approximation results given in Chap. 5.

It is worth noticing that if the bilinear form $a_N(u,v)$ not only satisfies (6.5.30) but also is coercive with respect to the norm of $X$, i.e., if there exists a constant $\bar{\alpha} > 0$ such that $\bar{\alpha}\|v\|^2 \leq a_N(v,v)$ for all $v \in X_N$, then the exponential term in the estimates (6.5.31) and (6.5.32) can be replaced by a constant (in time) depending on $\bar{\alpha}$.

Now we present some examples that illustrate the theory given above.

**Example 4.** *Fourier Galerkin and Collocation Approximations to a Two-Dimensional Advection Equation.* We consider the advection problem in skew-symmetric form

$$\begin{aligned} &u_t + \boldsymbol{\beta}\cdot\nabla u + \nabla\cdot(\boldsymbol{\beta}u) = 0\,, && \mathbf{x}\in\Omega=(0,2\pi)^2,\ \ t>0\,,\\ &u(\mathbf{x},0)=u_0(\mathbf{x})\,, && \mathbf{x}\in\Omega\,,\\ &u(\mathbf{x},t)\ \text{periodic in}\ \mathbf{x}\,, && t>0\,. \end{aligned} \tag{6.5.33}$$

We have set $\mathbf{x}=(x_1,x_2)$, and we assume that $\boldsymbol{\beta}=(\beta_1(\mathbf{x}),\beta_2(\mathbf{x}))$ and $u_0$ are given regular and periodic functions. Denote by $\mathbf{k}=(k_1,k_2)$ any couple of integers (positive or negative). Then $\mathbf{k}\cdot\mathbf{x}=k_1x_1+k_2x_2$ denotes the Euclidean inner product of $\mathbb{R}^2$. Finally we denote by $J$ the set of multi-indexes $\mathbf{k}=(k_1,k_2)$ such that $-N\le k_1\le N-1$ for $i=1,2$.

The Fourier Galerkin approximation to $u$ is the function $u^N(\mathbf{x},t)=\sum_{k\in J}\alpha_{\mathbf{k}}(t)e^{i\mathbf{k}\cdot\mathbf{x}}$ that satisfies the equations

$$\begin{aligned} &\int_\Omega [u_t^N+\mathcal{L}u^N](\mathbf{x},t)e^{-i\mathbf{k}\cdot\mathbf{x}}\,dx = 0 && \text{for }\mathbf{k}\in J,\ t>0\,,\\ &\alpha_{\mathbf{k}}(0)=\frac{1}{2\pi}\int_\Omega u_0(\mathbf{x})e^{-i\mathbf{k}\cdot\mathbf{x}}\,dx && \text{for }\mathbf{k}\in J\,. \end{aligned} \tag{6.5.34}$$

Here $\mathcal{L}u=\boldsymbol{\beta}\cdot\nabla u+\nabla\cdot(\boldsymbol{\beta}u)$ is the linear operator associated to the problem (6.5.33).

Problem (6.5.34) is a particular case of (6.5.5) corresponding to the choice $X_N=Y_N=\mathrm{span}\{e^{i\mathbf{k}\cdot\mathbf{x}},\mathbf{k}\in J\}$, $(u,v)_N=(u,v)=\int_\Omega u(\mathbf{x})\overline{v(\mathbf{x})}\,d\mathbf{x}$, and $a_N(u,v)=a(u,v)=(\mathcal{L}u,v)=(\boldsymbol{\beta}\cdot\nabla u,v)-(u,\boldsymbol{\beta}\cdot\nabla v)$.

The continuity property (6.5.28) holds, taking as $E$ the space $H_p^1(\Omega)$, defined in (A.11.d). Furthermore, we obviously have

$$(\mathcal{L}u,u)=0 \quad \text{for all } u\in H_p^1(\Omega)\,; \tag{6.5.35}$$

hence, (6.5.5) holds. From (6.5.31), it follows that (6.5.34) is a stable approximation to (6.5.33), namely,

$$\|u^N(t)\|_{L^2(\Omega)} \le \|P_Nu_0\|_{L^2(\Omega)} \le \|u_0\|_{L^2(\Omega)}\,, \tag{6.5.36}$$

where $P_N$ denotes the orthogonal projection from $X=L^2(\Omega)$ onto $X_N$. Moreover, taking $R_Nu=P_Nu$, the convergence estimate (6.5.32) gives in the present situation the following inequality for all $t>0$ and $m\ge 1$:

$$\begin{aligned} &\|u(t)-u^N(t)\|_{L^2(\Omega)} \\ &\le CN^{1-m}\exp\left(\frac{t}{2}\right)\left(\int_0^t\left(|u_t(s)|^2_{H^{m-1;N}(\Omega)}+|u(s)|^2_{H^{m;N}(\Omega)}\right)dt\right)^{1/2}. \end{aligned} \tag{6.5.37}$$

Let us now introduce the $4N^2$ collocation points $\mathbf{x}_{jk} = (x_j, x_k)$, $0 \le j$, $k \le 2N-1$, with $x_j = \pi j/N$, and denote by $I_N u \in X_N$ the interpolant of $u$ at these points. The Fourier collocation approximation to $u$ is the function $u^N(\mathbf{x},t) = \sum_{jk} u^N(\mathbf{x}_{jk},t)\varphi_{jk}(\mathbf{x})$ (where $\varphi_{jk}$ are the characteristic Lagrange trigonometric polynomials at the collocation points) satisfying the equations

$$\begin{aligned} \left[u_t^N + \mathcal{L}_N u^N\right](\mathbf{x}_{jk},t) &= 0 \qquad \text{for } t>0 \text{ and } 0 \le j,\ k \le 2N-1\ , \\ u^N(\mathbf{x}_{jk},0) &= u_0(\mathbf{x}_{jk}) \qquad \text{for } 0 \le j,\ k \le 2N-1\ . \end{aligned} \tag{6.5.38}$$

Here $\mathcal{L}_N u = \boldsymbol{\beta}\cdot\nabla u + \nabla\cdot I_N(\boldsymbol{\beta}u)$ for all $u \in X_N$; it represents the interpolation approximation of $\mathcal{L}u$ (see Sect. 2.1.3). This scheme can be written in the general form (6.5.5) by setting

$$(u,v)_N = \left(\frac{\pi}{N}\right)^2 \sum_{0\le j,\,k\le 2N-1} u(\mathbf{x}_{jk})\overline{v(\mathbf{x}_{jk})}$$

(note that $(u,v)_N = (u,v)$ for all $u,v \in X_N$, due to (2.1.33)) and

$$a_N(u,v) = (\mathcal{L}_N u, v)_N = (\boldsymbol{\beta}\cdot\nabla u, v)_N - (u, \boldsymbol{\beta}\cdot\nabla v)_N\ ,$$

which immediately implies $a_N(v,v) = 0$ for all $v \in X_N$. This proves that the collocation scheme is quadratically conservative, as discussed in Sect. 4.5. Moreover, since $\|v\|_N^2 \equiv (v,v)_N = \|v\|^2_{L^2(\Omega)}$ for all $v \in X_N$, the stability estimate (6.5.31) gives

$$\|u^N(t)\|_{L^2(\Omega)} \le \|I_N u_0\|_{L^2(\Omega)} \le \max_{x\in\overline{\Omega}} |u_0(x)|\ .$$

Furthermore, the same convergence estimate as (6.5.37) can be proven for the Fourier collocation solution, taking now $R_N u = I_N u$ in (6.5.32) and using the approximation properties of this operator (see Sect. 5.1.3).

The stability and convergence analysis for the approximation schemes (6.5.34) and (6.5.38) has been given first by Pasciak (1980). □

**Example 5.** *G-NI Approximations to a One-Dimensional Advection-Reaction Equation in the Interval* $(-1,1)$. We consider the one-dimensional, variable-coefficient advection-reaction problem

$$\begin{aligned} u_t + (\beta u)_x + \gamma u &= f\ , \qquad -1 < x < 1,\ t > 0\ , \\ u(x,0) &= u_0(x)\ , \qquad -1 < x < 1\ , \end{aligned} \tag{6.5.39}$$

where $\beta$ and $\gamma$ are given smooth functions in $[-1,1]$; for simplicity, we assume them independent of $t$, although the subsequent analysis can be easily adapted to the most general case. The boundary conditions for this problem must be prescribed at those points of the boundary where the flux, $\mathcal{F}(u) = \beta u$, is entering. Precisely, we introduce the sets $B_\pm = \{x_b \in \{-1,1\} \mid \pm\beta(x_b)n_b > 0\}$, with $n_b = x_b$. The set $B_-$ ($B_+$, resp.) is the *inflow* (*outflow*, resp.) boundary of the domain $(-1,1)$. Then, we prescribe the value of $u$ at the inflow:

$$u(x_b) = 0 \qquad \text{at all } x_b \in B_- \,. \tag{6.5.40}$$

This example generalizes the initial-boundary-value problem (3.7.1) considered in Sect. 3.7.

We set $\mathcal{L}u = (\beta u)_x + \gamma u$, $X = L^2(-1,1)$ and $E = H^1(-1,1)$; then, $D_B(\mathcal{L}) = \{v \in E \,|\, v \text{ satisfies (6.5.40)}\,\}$, and (6.5.28) is easily checked. On the other hand, the following *skew-symmetric* decomposition of the advection term,

$$(\beta u)_x = \tfrac{1}{2}\beta u_x + \tfrac{1}{2}\beta_x u + \tfrac{1}{2}(\beta u)_x \,,$$

implies, after integrating by parts, the relation

$$\begin{aligned}(\mathcal{L}u, v) = \tfrac{1}{2}(\beta u_x, v) - \tfrac{1}{2}(u, \beta v_x) + \left(\left(\tfrac{1}{2}\beta_x + \gamma\right) u, v\right) \\ + \tfrac{1}{2} \sum_{x_b \in \{-1,1\}} \beta(x_b) n_b u(x_b) v(x_b) \qquad \text{for all } u, v \in E \,.\end{aligned} \tag{6.5.41}$$

Thus, taking $u = v \in D_B(\mathcal{L})$, we get

$$(\mathcal{L}v, v) = \int_{-1}^{1} \left(\tfrac{1}{2}\beta_x + \gamma\right) v^2 \, dx + \tfrac{1}{2} \sum_{x_b \in B_+} \beta(x_b) n_b v^2(x_b) \,.$$

The second term on the right-hand side is nonnegative by definition of $B_+$; thus, (6.5.29) is satisfied provided we assume that

$$\tfrac{1}{2}\beta_x + \gamma \geq 0 \qquad \text{in } (-1,1) \,. \tag{6.5.42}$$

(Note that, since $\beta_x$ and $\gamma$ are bounded by assumption, this condition is always fulfilled after applying the change of dependent variable, $u(t) \to w(t) = e^{-ct}u(t)$, for a suitable $c > 0$.)

In view of the numerical approximations, it is convenient to introduce the bilinear form on $E$:

$$\begin{aligned}a(u, v) = \tfrac{1}{2}(\beta u_x, v) - \tfrac{1}{2}(u, \beta v_x) + \left(\left(\tfrac{1}{2}\beta_x + \gamma\right) u, v\right) \\ + \tfrac{1}{2} \sum_{x_b \in B_+} \beta(x_b) n_b u(x_b) v(x_b) \,,\end{aligned} \tag{6.5.43}$$

so that

$$(\mathcal{L}u, v) = a(u, v) + \tfrac{1}{2} \sum_{x_b \in B_-} \beta(x_b) n_b u(x_b) v(x_b) \tag{6.5.44}$$

and

$$(\mathcal{L}u, v) = a(u, v) \quad \text{if } u \text{ or } v \text{ belong to } D_B(\mathcal{L}) \,. \tag{6.5.45}$$

Let us introduce the discrete inner product $(u,v)_N$ built by the quadrature formula that uses the $N$-degree Legendre Gauss-Lobatto points introduced in Sect. 2.2.3. A first G-NI method is obtained by enforcing the boundary conditions exactly, i.e., by choosing $X_N = \{v \in \mathbb{P}_N \mid v \text{ satisfies (6.5.40)}\}$, and by approximating the bilinear form $a(u,v)$ as follows. For all $u, v \in X_N$, we set

$$
\begin{aligned}
a_N(u,v) &= \tfrac{1}{2}(\beta u_x, v)_N - \tfrac{1}{2}(u, \beta v_x)_N + \left(\left(\tfrac{1}{2}\beta_x + \gamma\right) u, v\right)_N \\
&\quad + \tfrac{1}{2} \sum_{x_b \in B_+} \beta(x_b) n_b u(x_b) v(x_b) .
\end{aligned}
\tag{6.5.46}
$$

Taking $u = v$ and using again (6.5.42), we immediately see that (6.5.30) is fulfilled. Thus, the G-NI scheme: find $u^N(t) \in X_N$ such that

$$
\begin{aligned}
&(u_t^N(t), v)_N + a_N(u^N(t), v) = (f(t), v)_N \qquad \text{for all } v \in X_N \ , \ t > 0 \ , \\
&u^N(0) = u_0^N = I_N u_0 \ ,
\end{aligned}
\tag{6.5.47}
$$

is stable, i.e., it satisfies (6.5.31). This G-NI scheme is a particular collocation scheme at the interior Legendre Gauss-Lobatto points and at the boundary points not belonging to $B_-$. Precisely, taking as $v$ in (6.5.47) the characteristic Lagrange polynomials centered at any of these points and counter-integrating by parts, we immediately obtain that the condition

$$
u_t^N + \tfrac{1}{2}\beta u_x^N + \tfrac{1}{2}\left(I_N\left(\beta u^N\right)\right)_x + \left(\tfrac{1}{2}\beta_x + \gamma\right) u^N = f \tag{6.5.48}
$$

holds therein. Note that the exact derivative $(\beta u^N)_x$ has been approximated by the interpolation derivative $(I_N(\beta u^N))_x$ as discussed in Sect. 2.3.2.

An alternative approach, which has already been considered in Sect. 3.7 and which can be easily extended to more general situations, consists of enforcing the boundary conditions in a weak manner. In order to better understand such a treatment, here we extend (6.5.40) to the nonhomogeneous case, i.e., we assume that the boundary conditions are

$$
u(x_b) = u_b \qquad \text{at all } x_b \in B_- \ . \tag{6.5.49}
$$

Now we set $X_N = \mathbb{P}_N$, which in particular implies that we do not require the G-NI solution $u^N(t)$ to satisfy exactly these conditions. The expression of $(\mathcal{L}u, v)$ given by (6.5.44) is approximated by

$$
(\mathcal{L}u, v) \simeq a_N(u,v) + \tfrac{1}{2} \sum_{x_b \in B_-} \beta(x_b) n_b u_b v(x_b) \ , \tag{6.5.50}
$$

where $a_N(u,v)$ is still given by (6.5.46), and we have incorporated the conditions (6.5.49) in the boundary term on $B_-$. The resulting G-NI scheme with weak imposition of the boundary conditions is as follows: find $u^N(t) \in \mathbb{P}_N$ such that

$$(u_t^N(t), v)_N + a_N(u^N(t), v) = (f(t), v)_N + \tfrac{1}{2} \sum_{x_b \in B_-} |\beta(x_b) n_b| u_b v(x_b)$$

$$\text{for all } v \in \mathbb{P}_N \;, \; t > 0 \;,$$

$$u^N(0) = u_0^N = I_N u_0 \;. \tag{6.5.51}$$

Obviously, (6.5.30) is still satisfied with the present choice of $X_N$; hence, (6.5.31) holds (in the homogeneous case $u_b = 0$). The scheme has the following interpretation: at the interior Legendre Gauss-Lobatto points and at the boundary points not belonging to $B_-$, we still have (6.5.48), whereas at the inflow boundary points we have

$$\begin{aligned} u_t^N + \tfrac{1}{2}\beta u_x^N + \tfrac{1}{2}(I_N(\beta u^N))_x + \left(\tfrac{1}{2}\beta_x + \gamma\right) u^N - f & \\ + \tfrac{1}{2w_b} |\beta(x_b) n_b| \left(u^N(x_b) - u_b\right) = 0 \;, & \end{aligned} \tag{6.5.52}$$

where $w_b$ is the Legendre Gauss-Lobatto weight associated with the point $x_b$. As already noted in Sect. 3.7, since $1/w_b \sim cN^2$ as $N \to \infty$, eq. (6.5.52) shows that the boundary condition is indeed enforced by a *penalty* method.

Finally, the convergence analysis of both G-NI schemes closely follows the steps presented in Example 5 of Sect. 6.4.2, to which we refer for more details. Denoting again by $\bar{N}$ the largest integer $\le N/2$, it is convenient to choose as $R_N u$ a polynomial in $\mathbb{P}_{\bar{N}}$ matching the boundary values of $u$ and satisfying

$$\|u - R_N u\|_{H^k(-1,1)} \le C N^{k-m} |u|_{H^{m;N}(-1,1)} \;, \qquad 0 \le k \le 1 \le m \;. \tag{6.5.53}$$

Such an approximation can be built as indicated in Sect. 5.4.2. With this definition of $R_N u$, we can apply the abstract estimate (6.5.32). Note that we have $\|u_t - R_N u_t\| \le C N^{1-m} |u_t|_{H^{m-1;N}(-1,1)}$, whereas $(R_N u_t, v) - (R_N u_t, v)_N = 0$ for all $v \in \mathbb{P}_N$ due to the exactness of the quadrature rule. The error $a(R_N u, v) - a_N(R_N u, v)$ can be estimated as indicated in the Example 5 cited above, i.e., by interlacing an approximation of each coefficient, $\beta$, $\beta_x$ or $\gamma$, in $\mathbb{P}_{\bar{N}}$ with optimal convergence properties in $L^\infty$. In bounding the error $(\beta R_N u, v_x) - (\beta R_N u, v_x)_N$, we make use of the inverse inequality $\|v_x\| \le C N^2 \|v\|$ for all $v \in \mathbb{P}_N$ (see Sect. 5.4.1). Finally, the error $(f, v) - (f, v)_N$ can be estimated by (5.5.29).

The convergence result is as follows. Let us assume that, for all $t \ge 0$, $u \in H^m(-1,1)$ and $u_t \in H^{m-1}(-1,1)$; furthermore, let us assume that $\beta \in W^{\tau,\infty}(-1,1)$, $\gamma \in W^{\vartheta,\infty}(-1,1)$ and $f \in H^\mu(-1,1)$. Then, for both versions of the G-NI scheme considered in the present Example, the following error bound holds, for all $t > 0$:

$$
\begin{aligned}
\|u(t) - u_N(t)\| &\le C_1 N^{-m} \left( |u(t)|_{H^{m;\tilde N}(-1,1)} + |u_0|_{H^{m;\tilde N}(-1,1)} \right) \\
&+ \exp\left(\frac{t}{2}\right) \left[ C_2 N^{1-m} \left( \int_0^t (|u(s)|^2_{H^{m;\tilde N}(-1,1)} + |u_t(s)|^2_{H^{m-1;\tilde N}(-1,1)})\, \mathrm{d}s \right)^{1/2} \right. \\
&+ C_2 N^{2-\tau} |\beta|_{W^{\tau,\infty;\tilde N}(-1,1)} \left( \int_0^t (\|u(s)\|^2_{H^1(-1,1)})\, \mathrm{d}s \right)^{1/2} \\
&+ C_3 N^{-\vartheta} |\gamma|_{W^{\vartheta,\infty;\tilde N}(-1,1)} \left( \int_0^t (\|u(s)\|^2_{L^2(-1,1)})\, \mathrm{d}s \right)^{1/2} \\
&\left. + C_4 N^{-\mu} \left( \int_0^t |f(s)|^2_{H^{\mu;N-1}(-1,1)}\, \mathrm{d}s \right)^{1/2} \right] .
\end{aligned}
\tag{6.5.54}
$$

□

**Example 6.** *A Chebyshev Collocation Approximation to a One-Dimensional Advection-Reaction Equation in the Interval* $(-1,1)$. We consider here the same boundary-value problem as in the previous example, but we focus on Chebyshev collocation approximations.

At first, let us assume that $B_- = \{-1\}$, i.e., the only inflow boundary point is $x_b = -1$. A Chebyshev collocation approximation can be defined as follows. Let

$$
\begin{aligned}
x_j &= \cos\left(-\pi + \frac{2\pi j}{2N+1}\right) , \qquad 0 \le j \le N , \\
w_0 &= \frac{\pi}{2N+1} , \quad w_j = 2w_0, \qquad 1 \le j \le N ,
\end{aligned}
\tag{6.5.55}
$$

be, respectively, the nodes and the weights of the Chebyshev Gauss-Radau quadrature formula having as prescribed boundary node $x_0 = -1$ (see (2.4.13), where the prescribed node is $x = 1$ instead). For all $t \ge 0$, the collocation approximation to $u$ is the polynomial $u^N(t) \in \mathbb{P}_N$ satisfying

$$
\begin{aligned}
&[u^N_t + \mathcal{L}_N u^N](x_j, t) = f(x_j, t) , \qquad && 1 \le j \le N,\ t > 0 , \\
&u^N(x_j, 0) = u_0(x_j) , && 0 \le j \le N, \\
&u^N(x_0, t) = 0 , && t > 0 .
\end{aligned}
\tag{6.5.56}
$$

Here, $\mathcal{L}_N u^N = \frac{1}{2}\{\beta u^N_x + [I_N(\beta u^N)]_x\} + [\frac{1}{2}(I_N\beta)_x + \gamma]u^N$ is the *skew-symmetric interpolation* decomposition of $\mathcal{L}u^N$, since $I_N$ denotes the interpolation operator with respect to the nodes $\{x_j\}$. We set $X_N = \{u \in \mathbb{P}_N \mid u(-1) = 0\}$ and $Y_N = X_N$. Moreover, we define a discrete inner product as follows:

$$
(u, v)_N = \sum_{j=0}^{N} u(x_j) v(x_j) \tilde w_j , \qquad \tilde w_j = (1 - x) w_j . \tag{6.5.57}
$$

Then (6.5.56) can be equivalently written in the form (6.5.6) taking $u_0^N = I_N u_0$. The stability and convergence analysis can be carried out according to the theory of this section, setting

$$X = L^2_{\tilde{w}}(-1,1)\,, \quad \text{where } \tilde{w}(x) = (1-x)\frac{1}{\sqrt{1-x^2}} = \left(\frac{1-x}{1+x}\right)^{1/2} . \tag{6.5.58}$$

The details of the analysis can be found in Canuto and Quarteroni (1982b).

For the other inflow conditions, $B_- = \{+1\}$, or $B_- = \{\pm 1\}$, or $B_- = \emptyset$, the collocation scheme is still defined as in (6.5.56) with the appropriate changes in the last equation. The collocation points are the nodes of the Chebyshev Gauss quadrature formula including those boundary points where boundary conditions are given. In the analysis, the weight $\tilde{w}(x)$ becomes $\tilde{w}(x) = \varepsilon(x)(1/\sqrt{1-x^2})$, where $\varepsilon(x)$ is $(1+x)$, or 1, or $(1-x^2)$, respectively. The same kind of stability and convergence results can be proven. □

Going back to the general theory, we finally consider *tau methods*. They usually assume that $X_N \subset D_B(\mathcal{L})$ and $Y_N \neq X_N$. We set $(u,v)_N = (u,v)$ and $a_N(u,v) = (\mathcal{L}u, v)$ in (6.5.5), or, equivalently, $\mathcal{L}_N = \mathcal{L}$ in (6.5.6). The discrete counterpart of condition (6.5.29) is now

$$0 \le (\mathcal{L}v, Q_N v) \quad \text{for all } v \in X_N \,, \tag{6.5.59}$$

where $Q_N$ denotes the orthogonal projection upon $Y_N$ in the inner product of $X$. Taking $v = Q_N u^N(t)$ in (6.5.6) and using again the Gronwall lemma (see (A.15)), we obtain the following *stability* estimate:

$$\|Q_N u^N(t)\|^2 \le \|u_0^N\|^2 + \exp(t)\int_0^t \|f(s)\|^2 ds \quad \text{for all } t > 0 \,. \tag{6.5.60}$$

If we introduce a suitable approximation operator $R_N$ in $X_N$, and we apply this bound to the error $e(t) = R_N u(t) - u^N(t)$, we obtain a *convergence* estimate, that for tau methods reads as follows. For all $t > 0$,

$$\begin{aligned} &\|u(t) - Q_N u^N(t)\|^2 \\ &\quad \le 2\|u(t) - Q_N R_N u(t)\|^2 + 2\|Q_N(u_0^N - R_N u_0)\|^2 \\ &\qquad + C\exp(t)\int_0^t (\|(u_t - R_N u_t)(s)\|^2 + \|(u - R_N u)(s)\|_E^2) ds \,. \end{aligned} \tag{6.5.61}$$

Again, the exponential on the right-hand sides of (6.5.60) and (6.5.61) can be dropped if (6.5.59) is replaced by the stronger condition $\bar{\alpha}\|v\|^2 \le (\mathcal{L}v, Q_N v)$ for all $v \in X_N$, for a suitable constant $\bar{\alpha} > 0$.

The following examples serve as an illustration of this theory.

**Example 7.** *A Legendre Tau Method for the Equation* $u_t + u_x = f$. We consider the initial-boundary-value problem

$$\begin{aligned} &u_t + u_x = f\,, && -1 < x < 1,\ t > 0\,, \\ &u(-1,t) = 0\,, && t > 0\,, \\ &u(x,0) = u_0(x)\,, && -1 < x < 1\,. \end{aligned} \tag{6.5.62}$$

As usual, let $L_k(x)$ denote the $k$-th Legendre polynomial. The Legendre tau approximate solution, $u^N(x,t) = \sum_{k=0}^N \alpha_k(t) L_k(x)$, to this problem is defined by the set of equations

$$\begin{aligned} &\int_{-1}^{1} \left[u_t^N + u_x^N\right](x,t) L_k(x)\,\mathrm{d}x = \int_{-1}^{1} f(x,t) L_k(x)\,\mathrm{d}x \\ &\qquad\qquad\qquad\qquad\qquad\qquad\qquad \text{for } k = 0,\dots,N-1,\ t > 0\,, \\ &\sum_{k=0}^{N} (-1)^k \alpha_k(t) = 0\,, \qquad\qquad\qquad t > 0\,, \\ &\alpha_k(0) = \left(k+\tfrac{1}{2}\right)\int_{-1}^{1} u_0(x) L_k(x)\,\mathrm{d}x\,, \qquad k = 0,\dots,N-1\,. \end{aligned} \tag{6.5.63}$$

This scheme fits into the general formulation (6.5.6) provided one sets $X = L^2(-1,1)$, $X_N = \{u \in \mathbb{P}_N \mid u(-1) = 0\}$, $Y_N = \mathbb{P}_{N-1}$, $\mathcal{L}_N = \mathcal{L} = \partial/\partial x$ and $(u,v)_N = (u,v) = \int_{-1}^{1} u(x)v(x)\,\mathrm{d}x$. The projection $Q_N$ is the orthogonal projection $P_{N-1}$ over the space of polynomials of degree up to $N-1$ with respect to this inner product (see (2.2.6)). The continuity condition (6.5.28) holds with $E = H^1(-1,1)$ (this space is defined in (A.11.a). Moreover, if $v \in D_B(\mathcal{L})$, one has

$$\int_{-1}^{1} v_x v\,\mathrm{d}x = \tfrac{1}{2} v^2(1)\,,$$

which proves (6.5.29). On the other hand, if $v \in X_N$, then $v_x$ is a polynomial of degree $\leq N-1$; hence, again

$$\int_{-1}^{1} v_x P_{N-1} v\,\mathrm{d}x = \int_{-1}^{1} v_x v\,\mathrm{d}x = \tfrac{1}{2} v^2(1)\,,$$

and (6.5.59) is satisfied. It follows that the scheme is stable, namely, for all $t > 0$, (6.5.60) yields the estimate

$$\|P_{N-1} u^N(t)\|^2_{L^2(-1,1)} \leq \|u_0\|^2_{L^2(-1,1)} + \exp(t) \int_0^t \|f(s)\|^2_{L^2(-1,1)} \mathrm{d}s\,. \tag{6.5.64}$$

We apply now the general convergence estimate (6.5.61) to the present situation. It is convenient to choose $R_N u$ as the best approximation of $u$ in

$X_{N-1} \subset X_N$ with respect to the norm of $E = H^1(-1,1)$. In this case $Q_N R_N u = R_N u$. It is possible to prove an error estimate for $R_N$ similar to (5.4.30), namely,

$$\|u - R_N u\|_{H^k(-1,1)} \leq C N^{k-m} |u|_{H^{m;N}(-1,1)} \,, \quad k = 0 \text{ or } 1 \text{ and } m \geq 1 \,. \tag{6.5.65}$$

Noting that $Q_N(u_0^N - R_N u_0) = P_{N-1}u_0 - R_N u_0$, using (6.5.65) and (5.4.11) we obtain from (6.5.61) that

$$\begin{aligned} &\|u(t) - P_{N-1}u^N(t)\|_{L^2(-1,1)} \\ &\quad \leq C N^{1-m} \exp\left(\frac{t}{2}\right) \left[\int_0^t \left(|u_t(s)|^2_{H^{m-1;N}(-1,1)} + |u(s)|^2_{H^{m;N}(-1,1)}\right) \mathrm{d}s\right]^{1/2} , \end{aligned} \tag{6.5.66}$$

which holds for all $t > 0$ and $m \geq 1$. We have bounded $|u_0|_{H^{m-1;N}(-1,1)}$ and $|u(t)|_{H^{m-1;N}(-1,1)}$ by the last integral on the right-hand side of the previous inequality. This is allowed by classical results of functional analysis (see, e.g., Lions and Magenes (1972)).

The stability and convergence analysis for the scheme (6.5.63) can be also carried out using a test function different from $Q_N u^N$ (or $Q_N e$). Indeed, take $v(t) = (u^N(t))/b$ as test function in (6.5.6) with $b(x) = 1 + x$ and define a new inner product $[u, v] = \int_{-1}^{1} u(x)v(x)(\mathrm{d}x/b(x))$. Then, setting $|\!|\!|v|\!|\!| = [v, v]^{1/2}$, we have

$$\frac{1}{2}\frac{\mathrm{d}}{\mathrm{d}t}|\!|\!|u^N(t)|\!|\!|^2 + [u_x^N(t), u^N(t)] = [f(t), u^N(t)], \quad t > 0 \,.$$

Integrating by parts, we have

$$[u_x^N, u^N] = \frac{1}{2}\int_{-1}^{1} v^2 \,\mathrm{d}x + v^2(1) \,.$$

Moreover,

$$[f, u^N] = \int_{-1}^{1} f v \,\mathrm{d}x \leq \|f\|_{L^2(-1,1)} \|v\|_{L^2(-1,1)} \leq \tfrac{1}{2}\|f\|^2_{L^2(-1,1)} + \tfrac{1}{2}\|v\|^2_{L^2(-1,1)} \,.$$

On the other hand, it is evident that $|\!|\!|u^N(t)|\!|\!|^2 \geq \frac{1}{2}\|u^N(t)\|^2_{L^2(-1,1)}$. Therefore, integrating in time we obtain

$$\|u^N(t)\|^2_{L^2(-1,1)} \leq \|u_0^N\|^2_{L^2(-1,1)} + \int_0^1 \|f(s)\|^2_{L^2(-1,1)} \,\mathrm{d}s, \quad t > 0 \,. \tag{6.5.67}$$

We stress that with this new stability estimate all frequencies of the solution $u^N$ are controlled. Moreover, the bound on the right-hand side of (6.5.67) does not blow up in time, unlike the one in (6.5.64). Concerning convergence, by the usual argument, one can obtain the following error estimate:

$$\|u(t) - u_N(t)\|_{L^2(-1,1)}$$

$$\leq CN^{1-m} \left\{ \int_0^t (|u_t(s)|^2_{H^{m-1;N}(-1,1)} + |u(s)|^2_{H^{m;N}(-1,1)} \, \mathrm{d}s \right\}^{1/2} , \tag{6.5.68}$$

which improves (6.5.66). □

**Example 8.** *A Chebyshev Tau Method for the Equation* $u_t - xu_x = f$. We consider the initial-boundary-value problem

$$\begin{aligned} &u_t + xu_x = f \,, && -1 < x < 1, \ t > 0 \,, \\ &u(-1,t) = u(1,t) = 0 \,, && t > 0 \,, \\ &u(x,0) = u_0(x) \,, && -1 < x < 1 \,. \end{aligned} \tag{6.5.69}$$

The Chebyshev tau solution $u^N(x,t) = \sum_{k=0}^N \alpha_k(t) T_k(x)$ of this problem is defined by the conditions

$$\int_{-1}^1 \left[u_t^N(x,t) - xu_x^N(x,t)\right] T_k(x) w(x) \, \mathrm{d}x = \int_{-1}^1 f(x,t) T_k(x) w(x) \, \mathrm{d}x \,,$$

$$k = 0, \dots, N-2, \text{ and } t > 0 \,,$$

$$\sum_{k=0}^N (-1)^k \alpha_k(t) = \sum_{k=0}^N \alpha_k(t) = 0 \,, \qquad t > 0 \,,$$

$$\alpha_k(0) = \frac{2}{c_k \pi} \int_{-1}^1 u_0(x) T_k(x) w(x) \, \mathrm{d}x \,, \qquad k = 0, \dots, N-2 \,. \tag{6.5.70}$$

Here $T_k(x)$ is the $k$-th Chebyshev polynomial, $w(x) = (1-x^2)^{-1/2}$ is the Chebyshev weight, and the $c_k$'s are defined in (2.4.10).

Problem (6.5.70) can be expressed in the form (6.5.6) by setting $X = L_w^2(-1,1)$, $X_N = \{u \in \mathbb{P}_N \mid u(-1) = u(1) = 0\}$, $Y_N = \mathbb{P}_{N-2}$, $\mathcal{L}_N = \mathcal{L} = -x(\partial/\partial x)$ and $(u,v)_N = (u,v)_w = \int_{-1}^1 u(x)v(x)w(x)\,\mathrm{d}x$. The projection operator $Q_N$ is the orthogonal projection operator $P_{N-2}$ over $\mathbb{P}_{N-2}$ with respect to the Chebyshev inner product $(u,v)_w$.

The positivity condition (6.5.59) takes the form

$$-\int_{-1}^1 xu_x P_{N-2} uw \, \mathrm{d}x \geq 0 \quad \text{for all } u \in X_N \,.$$

It is satisfied in the current example since one has

$$-\int_{-1}^1 xu_x P_{N-2} uw \, \mathrm{d}x = \int_{-1}^1 xu_x (u - P_{N-2} u) w \, \mathrm{d}x + \frac{1}{2} \int_{-1}^1 u^2 (xw)_x \, \mathrm{d}x \,.$$

The last term is positive since $xw(x)$ is an increasing function. The other term, using (2.4.4) and (2.4.22), equals $\frac{1}{2}N\hat{u}_N^2+\frac{1}{2}(N-1)\hat{u}_{N-1}^2$ (where $\hat{u}_N$ and $\hat{u}_{N-1}$ denote the two last Chebyshev coefficients of $u$); hence, it is positive. The convergence analysis follows along the guidelines of the previous example.

A different approach consists of choosing $v = u/b$, where $b(x) = 1 - x^2$, as a test function. A straightforward calculation reveals that

$$\begin{aligned}(\mathcal{L}u, v)_w &= -\int_{-1}^{1} x u_x v w \, dx \\ &= \frac{1}{2}\int_{-1}^{1} v^2 \frac{1}{w} \, dx + \frac{3}{2}\int_{-1}^{1} v^2 x^2 w \, dx \\ &\geq \frac{1}{2}\int_{-1}^{1} u^2 w \, dx + \frac{3}{2}\int_{-1}^{1} v^2 x^2 w \, dx \, .\end{aligned} \tag{6.5.71}$$

Then, proceeding as in the previous example, stability and convergence inequalities like (6.5.67) and (6.5.68) can be proven, relative to the weighted Chebyshev norms. □

## 6.6 The Error Equation

It has been shown in Sects. 6.3 and 6.5 that many spectral schemes are defined through a projection of the differential equation onto a finite-dimensional space of polynomials. For these schemes, the spectral solution is characterized by a set of weighted residual, or weak, equations (see (6.3.20) and (6.5.6)).

It is also useful to characterize a spectral solution as the exact solution of a suitable differential problem. This problem is of the same type as the original problem to be discretized. It only differs in a forcing term that takes into account the error committed by the spectral projection. The new differential equation is called the *error equation* of the method.

The error equation can be exploited in deriving the stability and convergence properties of spectral schemes. It was first used for this purpose by M. Dubiner and by Gottlieb and Orszag (1977). Since the spectral solution satisfies the error equation pointwise over the whole domain, it is also possible to deduce from it local information on the qualitative behavior of the solution, as opposed to the global information produced by variational methods. On the other hand, the analysis based on the error equation is usually confined to simple model problems, such as constant-coefficient problems.

For brevity, our discussion of the error equation will be limited to evolution problems only. However, a similar discussion could be carried out for steady or eigenvalue problems as well. In what follows, we refer for both notation and hypotheses to the abstract formulation of spectral approximations for evolution problems, given in Sect. 6.5 (see (6.5.7)), that extends the steady-state situation described in Sect. 6.3 (see (6.3.21)).

In particular, we recall that for all $t > 0$, the spectral solution $u^N(t)$ belongs to a finite-dimensional space $X_N$ and that the spectral operator $\mathcal{L}_N$ maps $X_N$ into a space $Z$ that is either a space of square-integrable functions or a space of continuous functions on the domain $\Omega$. We have assumed that $X_N \subset Z$ and the data $f(t) \in Z$ for all $t > 0$. Hence, $u_t^N + \mathcal{L}_N u^N - f$ is an element of $Z$ for all $t \geq 0$. By definition, $Q_N(u_t^N + \mathcal{L}_N u^N - f) = 0$ (see (6.5.7)), where $Q_N$ is a projection upon a finite-dimensional space $Y_N$.

The error equation arises from the trivial decomposition

$$w = Q_N w + Q_N^* w \quad \text{for all } w \in Z ,$$

where

$$Q_N^* w = w - Q_N w .$$

Taking into account (6.5.7), one has

$$u_t^N + \mathcal{L}_N u^N - f = Q_N^* \left( u_t^N + \mathcal{L}_N u^N - f \right) , \tag{6.6.1}$$

or equivalently,

$$u_t^N + \mathcal{L}_N u^N = Q_N^* \left( u_t^N + \mathcal{L}_N u^N \right) + Q_N f . \tag{6.6.2}$$

This is precisely the error equation. The right-hand side of (6.6.1) represents the error generated pointwise by the spectral approximation scheme. It is precisely from the analysis of this error that one can infer information about the spectral solution. In all the relevant schemes, the space $Z$ contains the space $\mathrm{Pol}_N(\Omega)$ of the polynomials of degree $N$, introduced at the beginning of Sect. 6.4. Thus, we make here the assumptions that $X_N$ and $Y_N$ are contained in $\mathrm{Pol}_N(\Omega)$ and that the spectral operator $\mathcal{L}_N$ actually maps $X_N$ into $\mathrm{Pol}_N(\Omega) \subset Z$. The last assumption is certainly true if $\mathcal{L}_N$ has constant coefficients.

Under these hypotheses, $Q_N^*(u_t^N + \mathcal{L}_N u^N)$ is a polynomial in $\mathrm{Pol}_N(\Omega)$ for all $t > 0$. Hence, it can be expanded according to any basis $\{\phi_k \mid k \in J\}$, in $\mathrm{Pol}_N(\Omega)$, as

$$Q_N^* \left( u_t^N + \mathcal{L}_N u^N \right) = \sum_{k \in J} \tau_k(t) \phi_k , \qquad t \geq 0 . \tag{6.6.3}$$

This expression takes a simplified form in some relevant cases.

### *Full Fourier Approximations*

If the boundary conditions are all periodic, $\mathrm{Pol}_N(\Omega)$ is a space of trigonometric polynomials, and $X_N = Y_N = \mathrm{Pol}_N(\Omega)$. Thus, $Q_N^* v = 0$ for all $v \in \mathrm{Pol}_N(\Omega)$, and the error equation becomes

$$u_t^N + \mathcal{L}_N u^N = Q_N f . \tag{6.6.4}$$

As a simple example, let us consider the Fourier Galerkin approximation to the heat equation that has been presented in Example 1 of Sect. 6.5. In this case the spectral solution $u^N$ satisfies the following error equation:

$$u_t^N - u_{xx}^N = P_N f\,, \qquad 0 < x < 2\pi,\ t > 0\,,$$

where $P_N f$ is the truncation of order $N$ of the Fourier series of $f$ (see (2.1.7)). For a collocation approximation to the same heat problem (6.5.13), the error equation satisfied by the spectral solution $u^N$ would be

$$u_t^N - u_{xx}^N = I_N f\,, \qquad 0 < x < 2\pi,\ t > 0\,,$$

where now $I_N f$ is the interpolant of $f$ at the collocation points (see (2.1.28)).

### *Collocation and Tau Methods for Nonperiodic Boundary Conditions*

For collocation methods, the natural basis in $\mathrm{Pol}_N(\Omega)$ is the nodal Lagrange basis associated to the collocation points, that has been introduced in (6.4.12). This basis is orthogonal with respect to the inner product $(u,v)_N$. On the other hand, in tau methods, $\mathrm{Pol}_N(\Omega)$ is represented in terms of the modal orthogonal basis with respect to the inner product $(u,v)$ of $X$.

Note that for all $v \in \mathrm{Pol}_N(\Omega)$, $Q_N^* v$ is orthogonal to any polynomial in $Y_N$ in the inner product $(u,v)_N$. This follows from the definition of $Q_N^* v$. Hence $Q_N^* v$ has no components along the elements in $Y_N$. In particular, (6.6.3) becomes

$$Q_N^*\left(u_t^N + \mathcal{L}_N u^N\right) = \sum_{k\in J_b} \tau_k(t)\phi_k\,, \qquad t \ge 0\,. \tag{6.6.5}$$

This expansion, recalling the definition of the set $J_b$, shows that the error on the left-hand side of (6.6.5) arises from the process by which the boundary conditions are taken into account in the spectral scheme.

An explicit representation of the coefficients $\tau_k(t)$ can be derived from (6.6.2) using the orthogonality of the basis functions in $\mathrm{Pol}_N(\Omega)$. One immediately has, for all $t > 0$,

$$\tau_k(t) = \frac{1}{(\phi_k,\phi_k)_N}\left(\frac{\mathrm{d}}{\mathrm{d}t}(u^N,\phi_k)_N + (\mathcal{L}_N u^N,\phi_k)_N\right) \quad \text{for all } k \in J_b\,.$$

As an example, consider Chebyshev approximations to the heat equation problem

$$\begin{aligned} &u_t - u_{xx} = f\,, && -1 < x < 1,\ t > 0\,,\\ &u(-1,t) = u(1,t) = 0\,, && t > 0\,,\\ &u(x,0) = u_0(x)\,, && -1 < x < 1\,. \end{aligned} \tag{6.6.6}$$

The error equation pertaining to the Chebyshev tau approximation of (6.6.6) is

$$u_t^N - u_{xx}^N = \tau_N(t)T_N + \tau_{N-1}(t)T_{N-1} + P_{N-2}f \; . \tag{6.6.7}$$

Here $\tau_k(t) = \mathrm{d}a_k/\mathrm{d}t$ for $k = N, N-1$, where $a_k(t)$ are the Chebyshev coefficients of the expansion of $u^N$, and $P_{N-2}f$ is the truncation of order $N-2$ of the Chebyshev series of $f$ (see (2.2.16)).

The collocation approximation to (6.6.6) has the form

$$u_t^N - u_{xx}^N = \{\tau_0(t)(1+x) + \tau_N(t)(1-x)\}T_N' + I_N f \; , \tag{6.6.8}$$

where

$$\begin{aligned}
\tau_0(t) &= \frac{2}{N^2}u_t^N(1,t) - u_{xx}^N(1,t), \\
\tau_N(t) &= (-1)^N \frac{2}{N^2}u_t^N(-1,t) - u_{xx}^N(-1,t) \; ,
\end{aligned}$$

and $I_N f$ is the interpolant of $f$ at the Chebyshev collocation points.

The error equation has been extensively used to derive stability estimates for constant-coefficient equations in the 1977 monograph by Gottlieb and Orszag (see Sects. 7 and 8) and in several subsequent papers by Gottlieb and coworkers. In this book, examples of analysis based on the error equation are reported in Sect. 7.2, where the stability in the maximum norm for solutions of singular perturbation problems is investigated, and in Sect. 7.6.2, where the tau method for the equation $u_t + u_x = 0$ is considered.

# 7. Analysis of Model Boundary-Value Problems

In this chapter, we apply the techniques for the theoretical analysis of spectral approximations to some differential operators and differential equations that are representative building blocks of the mathematical modelling in continuum mechanics. We first study the Poisson equation, followed by singularly perturbed elliptic equations that model advection-diffusion and reaction-diffusion processes featuring sharp boundary layers. Subsequently, we develop an eigenvalue analysis for several matrices produced by spectral approximations to diffusion, advection-diffusion and pure advection problems. We extend our analysis to the closely related study of the low-order preconditioning of spectral matrices.

In the second part of the chapter, we analyze time-dependent problems. At first we consider the heat equation, and we provide an example of analysis for a fully discrete (in space and time) scheme. Linear scalar hyperbolic equations are analysed next, with a particular emphasis on the issues of spatial stability and the resolution of the Gibbs phenomenon for discontinuous solutions through filtering, singularity detection and spectral reconstruction techniques.

Finally, we provide theoretical results for spectral approximations to nonlinear problems. We describe the mathematical foundation of the spectral viscosity method for scalar conservation laws, and we detail the analysis of the approximation of a non-singular branch of solutions for the steady Burgers equation.

## 7.1 The Poisson Equation

Numerous spectral algorithms for the numerical simulation of physical phenomena require the approximate solution of one or more Poisson equations of the type

$$-\Delta u = f \tag{7.1.1}$$

in a bounded domain $\Omega \subset \mathbb{R}^d$ $(d = 1, 2, 3)$. Here $\Delta = \sum_{i=1}^{d} \partial^2/\partial x_i^2$ denotes the Laplace operator in $d$ space variables, $u$ is the unknown function, and $f$ is the given data.

Among the boundary conditions that are more commonly associated to the Poisson equation (7.1.1) are homogeneous Dirichlet conditions

$$u = 0 \quad \text{on } \partial\Omega \,. \tag{7.1.2}$$

As usual in spectral methods, we assume that the computational domain is the Cartesian product of $d$ copies of the interval $(-1,1)$, i.e., $\Omega = (-1,1)^d$.

In Chap. 6 we discussed from a general point of view conditions which guarantee the convergence of spectral approximations to boundary-value problems. These conditions concern on the one hand the properties of approximation of the space of polynomials chosen to represent the discrete solution, and on the other hand the fulfillment of suitable properties of coercivity by the differential operator and by its spectral approximation.

Several examples have been given in Chap. 6 to illustrate the application of the theory to specific problems. Some of those pertained to the Poisson equation with Dirichlet boundary conditions, in one or more space dimensions.

Hereafter, we collect the most relevant theoretical facts about the Laplace operator submitted to homogeneous Dirichlet boundary conditions, and about its approximations of spectral type. We show that the coercivity conditions of Chap. 6 are fulfilled with a natural choice of the norms.

Nonperiodic boundary-value problems are usually approximated by Legendre or Chebyshev methods. From a theoretical point of view, the analysis of Chebyshev methods is more involved, due to the presence of the singular weight. Thus, it is convenient to treat separately Legendre and Chebyshev methods.

### 7.1.1 Legendre Methods

The natural norms in which to set the analysis of these methods are the norms of the standard (non-weighted) Sobolev spaces $H^m(\Omega)$ (see (A.11.a)). A central role is played by the Hilbert space $H_0^1(\Omega)$, defined in (A.11.c).

The operator $\mathcal{L} = -\Delta$ is a linear unbounded operator in $L^2(\Omega)$ (see (A.3)). Supplemented with homogeneous Dirichlet boundary conditions, its domain of definition is the dense subspace $D_B(\mathcal{L}) = \{v \in H^2(\Omega) : v|_{\partial\Omega} = 0\}$. If $u \in D_B(\mathcal{L})$ and $v \in H_0^1(\Omega)$, integration-by-parts yields

$$-\int_\Omega \Delta u v \, dx = \int_\Omega \nabla u \cdot \nabla v \, dx \,. \tag{7.1.3}$$

The right-hand side, which defines a symmetric bilinear form $a(u,v)$, is precisely the inner product of the Hilbert space $H_0^1(\Omega)$ (see (A.11.c)). It follows that the coercivity and continuity assumptions (6.3.6) and (6.3.7) are satisfied with the choice $E = H_0^1(\Omega)$.

Using (7.1.3) the following weak (or variational) formulation of the boundary-value problem (7.1.1)–(7.1.2) is obtained: one looks for a function $u \in H_0^1(\Omega)$ such that

$$\int_\Omega \nabla u \cdot \nabla v \, dx = \int_\Omega f v \, dx \quad \text{for all } v \in H_0^1(\Omega) . \tag{7.1.4}$$

Here, we have assumed $f \in L^2(\Omega)$. More general data $f \in H^{-1}(\Omega)$ (the dual space of $H_0^1(\Omega)$, (see (A.11.c)) is allowed, in which case the right-hand side has to be replaced by the duality pairing $\langle f, v \rangle$ between $H^{-1}(\Omega)$ and $H_0^1(\Omega)$. By the Riesz representation theorem (see (A.1.d)), there exists a unique solution of problem (7.1.4). If $f \in L^2(\Omega)$, then one can prove that the second derivatives of $u$ are square integrable in $\Omega$. Hence, we conclude that $u \in D_B(\mathcal{L})$.

Now we turn to the numerical approximations. Since the coercivity assumption (6.4.3) is fulfilled, it follows that the Legendre Galerkin method for (7.1.1)–(7.1.2) is stable (hence, convergent) in the $H_0^1(\Omega)$-norm, or equivalently, in the $H^1(\Omega)$-norm. The same conclusion holds for the G-NI method based on the Gauss-Lobatto points (2.3.12) in each space direction. It has been already observed that such a method, when applied with full Dirichlet boundary conditions, coincides with the Legendre collocation method. To study its stability, let us consider the discrete inner product

$$(u, v)_N = \sum_{j \in J} u(x_j) v(x_j) w_j , \tag{7.1.5}$$

where $\{x_j \,|\, j \in J\}$ denotes the tensor product of the one-dimensional Gauss-Lobatto points and $\{w_j \,|\, j \in J\}$ are the corresponding weights. Then if $u \in \mathbb{P}_N(\Omega)$ and $v \in \mathbb{P}_N^0(\Omega)$, the space of polynomials of degree $N$ in each space variable vanishing on $\partial\Omega$, one has

$$(-\Delta u, v)_N = (\nabla u, \nabla v)_N = a_N(u, v) . \tag{7.1.6}$$

This follows by integration-by-parts, since in each direction of differentiation the quadrature rule can be replaced by the exact integral, the integrand being a polynomial of degree at most $2N-1$ in that direction. On the other hand, by (5.3.2), the right-hand side of (7.1.6) is an inner product on $\mathbb{P}_N^0(\Omega)$, which induces a norm equivalent to the $H_0^1(\Omega)$-norm. Thus, (6.4.43), or equivalently (6.4.23), is fulfilled with $E = H_0^1(\Omega)$. The corresponding convergence estimate, based on (6.4.8) or (6.4.46) and the approximation results of Sects. 5.4 and 5.8.2, is

$$\|u - u^N\|_{H^1(\Omega)} \le C N^{1-m} |u|_{H^{m;N}(\Omega)}, \quad m \ge 1, \tag{7.1.7}$$

if $u^N$ is the Galerkin solution, or

$$\|u - u^N\|_{H^1(\Omega)} \le C N^{1-m} \{ |u|_{H^{m;N}(\Omega)} + |f|_{H^{m-1;N}(\Omega)} \}, \quad m > 1 + d/2, \tag{7.1.8}$$

if $u^N$ is the G-NI (or, equivalently in this case, collocation) solution.

For the analysis of the tau approximation to (7.1.1)–(7.1.2) we have to resort to the generalized "inf-sup" condition (6.4.66). In the one-dimensional case we endow $X_N = \mathbb{P}_N^0$ with the norm of $H^2(-1,1)$ and $Y_N = \mathbb{P}_{N-2}$ with the norm of $L^2(-1,1)$ and we choose as test function $v = -u_{xx}$. This yields the stability of the method in the norm of $H^2(-1,1)$. The two-dimensional case has been discussed in Example 8 of Sect. 6.4.4.

### 7.1.2 Chebyshev Methods

Let $w(x) = \prod_{i=1}^d (1-x_i^2)^{-1/2}$ be the Chebyshev weight in dimension $d$. Chebyshev methods are naturally studied in the norms of the weighted Sobolev spaces $H_w^m(\Omega)$ (see (A.11.b)). Here we consider the operator $\mathcal{L} = -\Delta$ as a linear unbounded operator in $L_w^2(\Omega)$. The domain of definition of $\mathcal{L}$ with Dirichlet boundary conditions is the dense subspace $D_B(\mathcal{L}) = \{v \in H_w^2(\Omega) : v|_{\partial\Omega} = 0\}$. This result is immediate in one space dimension, whereas in more space dimensions it requires a complex proof due to Bernardi and Maday (1986).

Let $u \in D_B(\mathcal{L})$ and $v \in H_{w,0}^1(\Omega)$ (see (A.11.c) for the definition of this space). Integrating by parts in a formal manner we get

$$-\int_\Omega \Delta u v w \, dx = \int_\Omega \nabla u \cdot \nabla(vw) \, dx \, . \tag{7.1.9}$$

The right-hand side is nonsymmetric in its arguments $u$ and $v$, due to the presence of the weight $w$. Let us set

$$a(u,v) = \int_\Omega \nabla u \cdot \nabla(vw) \, dx \, . \tag{7.1.10}$$

The bilinear form $a(u,v)$ is defined, continuous and coercive on the product space $H_{w,0}^1(\Omega) \times H_{w,0}^1(\Omega)$, as stated precisely in the following theorem:

**Theorem 7.1.**

(i) *There exists a constant $A > 0$ such that for all $u, v \in H_{w,0}^1(\Omega)$*

$$|a(u,v)| \le A \|u\|_{H_w^1(\Omega)} \|v\|_{H_w^1(\Omega)} \, ; \tag{7.1.11}$$

(ii) *there exists a constant $\alpha > 0$ such that for all $u \in H_{w,0}^1(\Omega)$*

$$\alpha \|u\|_{H_w^1(\Omega)}^2 \le a(u,u) \, . \tag{7.1.12}$$

This result was proved by Canuto and Quarteroni (1981) in one dimension, and was extended to higher space dimensions by Funaro (1981).

Hereafter, we give the proof for the one-dimensional case, since it already contains all the essential elements of the analysis. The bilinear form (7.1.10) becomes

$$a(u,v) = \int_{-1}^1 u_x (vw)_x \, dx \, , \tag{7.1.13}$$

where $w(x) = (1-x^2)^{-1/2}$ is the Chebyshev weight. Let us start with the following inequality.

**Lemma 7.1.** *For all* $u \in H^1_{w,0}(-1,1)$

$$\int_{-1}^{1} u^2(x) w^5(x)\,\mathrm{d}x \le \frac{8}{3} \int_{-1}^{1} u_x^2(x) w(x)\,\mathrm{d}x\,. \tag{7.1.14}$$

PROOF. Let us split the left-hand side as

$$\int_{-1}^{1} u^2(x) w^5(x)\,\mathrm{d}x = \int_{-1}^{0} u^2(x) w^5(x)\,\mathrm{d}x + \int_{0}^{1} u^2(x) w^5(x)\,\mathrm{d}x\,.$$

Since $w(x) \le (1-x)^{-1/2}$ if $0 \le x \le 1$,

$$\begin{aligned}\int_{0}^{1} u^2(x) w^5(x)\,\mathrm{d}x &\le \int_{0}^{1} u^2(x)(1-x)^{-5/2}\mathrm{d}x \\ &= \int_{0}^{1} \left[\frac{1}{1-x}\int_{x}^{1} u_x(s)\,\mathrm{d}s\right]^2 (1-x)^{-1/2}\mathrm{d}x\,.\end{aligned}$$

Now we apply Hardy's inequality (A.14) with $\alpha = -1/2$, and we get

$$\int_{0}^{1} u^2(x) w^5(x)\,\mathrm{d}x \le \frac{8}{3} \int_{0}^{1} u_x^2(x) w(x)\,\mathrm{d}x\,.$$

The same inequality holds over the interval $(-1,0)$, whence the result. □

Let us prove part (i) of Theorem 7.1. Precisely, we will prove that *for all* $u$ *and* $v \in H^1_{w,0}(-1,1)$, *the following inequality holds*:

$$\left|\int_{-1}^{1} u_x (vw)_x \mathrm{d}x\right| \le \left(1 + \sqrt{\frac{8}{3}}\right) \|u_x\|_{L^2_w(-1,1)} \|v_x\|_{L^2_w(-1,1)}\,. \tag{7.1.15}$$

Indeed, by the identity

$$\int_{-1}^{1} u_x (vw)_x \mathrm{d}x = \int_{-1}^{1} u_x v_x w\,\mathrm{d}x + \int_{-1}^{1} u_x (v w_x w^{-1}) w\,\mathrm{d}x\,,$$

and the application of the Cauchy-Schwarz inequality (A.2) to both terms on the right-hand side, one gets

$$|a(u,v)| \le \|u_x\|_{L^2_w(-1,1)} \left\{ \|v_x\|_{L^2_w(-1,1)} + \left(\int_{-1}^{1} v^2 w_x^2 w^{-1} \mathrm{d}x\right)^{1/2} \right\}.$$

Noting that $w_x = x w^3$, it follows using (7.1.14) that

$$\int_{-1}^{1} v^2 w_x^2 w^{-1} \mathrm{d}x \le \int_{-1}^{1} v^2 w^5 \mathrm{d}x \le \frac{8}{3} \|v_x\|^2_{L^2_w(-1,1)}\,,$$

whence the result.

Finally, we prove part (ii) of Theorem 7.1. Precisely, we shall prove that *for all* $u \in H^1_{w,0}(-1,1)$ *the following inequality holds*:

$$\frac{1}{4}\|u_x\|^2_{L^2_w(-1,1)} \leq \int_{-1}^{1} u_x(uw)_x \mathrm{d}x \,. \tag{7.1.16}$$

Then, (7.1.12) will follow from the Poincaré inequality (A.13). (Note that the Poincaré inequality is implied by the inequality (7.1.12), since $w(x) \geq 1$.)

To obtain (7.1.16), one uses partial integration (which is allowed by (7.1.14)) and gets

$$\begin{aligned} a(u,u) &= \int_{-1}^{1} (u_x)^2 w \,\mathrm{d}x + \int_{-1}^{1} u u_x w_x \mathrm{d}x \\ &= \int_{-1}^{1} (u_x)^2 w \,\mathrm{d}x - \frac{1}{2}\int_{-1}^{1} u^2 w_{xx} \mathrm{d}x \,. \end{aligned} \tag{7.1.17}$$

In order to estimate the last integral on the right-hand side, let us use another expression for $a(u,u)$, namely,

$$\begin{aligned} a(u,u) &= \int_{-1}^{1} \left[(u_x)^2 w^2 + u u_x w_x w\right] w^{-1} \,\mathrm{d}x \\ &= \int_{-1}^{1} \left[(u_x w)^2 + 2u_x w u w_x + (u w_x)^2\right] w^{-1} \mathrm{d}x \\ &\quad - \int_{-1}^{1} \left(u u_x w_x + u^2 w_x^2 w^{-1}\right) \mathrm{d}x \\ &= \int_{-1}^{1} [(uw)_x]^2 w^{-1} \mathrm{d}x + \int_{-1}^{1} u^2 \left(\frac{w_{xx}}{2} - w_x^2 w^{-1}\right) \mathrm{d}x \,. \end{aligned} \tag{7.1.18}$$

By the identity $w_{xx} - 2w_x^2 w^{-1} = w^5$ we obtain

$$\frac{1}{2}\int_{-1}^{1} u^2 w^5 \mathrm{d}x \leq a(u,u) \,. \tag{7.1.19}$$

On the other hand, since $w_{xx} = (1+2x^2)w^5$,

$$\int_{-1}^{1} u^2 w_{xx} \leq 3\int_{-1}^{1} u^2 w^5 \mathrm{d}x \leq 6a(u,u) \,.$$

Thus, recalling (7.1.17)

$$a(u,u) \geq \int_{-1}^{1} (u_x)^2 w \,\mathrm{d}x - 3a(u,u) \,,$$

or, equivalently,

$$4a(u,u) \geq \int_{-1}^{1} (u_x)^2 w \, \mathrm{d}x \, ,$$

whence the result.

Let us now turn to the general $d$-dimensional case. Theorem 7.1 essentially states that the Laplace operator with homogeneous Dirichlet boundary conditions fulfills the coercivity and the continuity conditions (6.3.6) and (6.3.7) with respect to the Hilbert space $E = H^1_{w,0}(\Omega)$. In Sect. 6.3 we made the general claim that whenever these conditions apply to a boundary-value problem, its well-posedness can be established. Let us check this statement in the present situation. Problem (7.1.1)–(7.1.2) can be formulated in a weak (or variational) form which involves the Chebyshev weight as follows: One looks for a function $u \in H^1_{w,0}(\Omega)$ such that

$$\int_{\Omega} \nabla u \cdot \nabla (vw) \mathrm{d}x = \int_{\Omega} f v w \, \mathrm{d}x \quad \text{for all } v \in H^1_{w,0}(\Omega) \, . \tag{7.1.20}$$

(The data $f$ is assumed to belong to $L^2_w(\Omega)$.) By Theorem 7.1, we can apply the Lax-Milgram Theorem (see (A.5)) to this problem; this assures the existence of a unique solution. Finally, one can prove (this is technical) that the solution not only belongs to $H^1_{w,0}(\Omega)$, but also it is more regular, i.e., $u \in H^2_w(\Omega)$ (Bernardi and Maday (1986)). Thus, given arbitrary data $f \in L^2_w(\Omega)$, there exists a unique solution in $D_B(\mathcal{L})$ to the problem (7.1.1)–(7.1.2).

Let us now consider the numerical approximation of this problem by Chebyshev methods. The Galerkin method is proven to be stable in the norm of $H^1_{w,0}(\Omega)$ as a direct consequence of Theorem 7.1. Here, we apply it to functions $u$ and $v$, which are polynomials of degree $N$ in each space variable and which vanish on the boundary (i.e., $u, v \in \mathbb{P}^0_N(\Omega)$). Theorem 7.1 ensures that the assumptions (6.4.3) and (6.4.4) are satisfied. The corresponding convergence estimate, based on (6.4.8) and the approximation estimates (5.5.19) or (5.8.32), reads as follows:

$$\|u - u^N\|_{H^1_w(\Omega)} \leq C N^{1-m} |u|_{H^{m;N}_w(\Omega)} \quad \text{for } m \geq 1 \, . \tag{7.1.21}$$

The two-dimensional case has been considered in Example 2 of Sect. 6.4.1.

The stability of the collocation method which uses the Gauss-Lobatto points (2.4.14) for the Chebyshev weight in each space direction follows from a specific version of Theorem 7.1. In dimension one, the stability is actually a direct consequence of Theorem 7.1, since

$$\sum_{j=0}^{N} u_{xx}(x_j) u(x_j) w_j = \int_{-1}^{1} u_{xx} u w \, \mathrm{d}x \quad \text{for all } u \in \mathbb{P}_N \, .$$

Thus, condition (6.4.23) is fulfilled. This result has been applied in the second example of Sect. 6.1. Let us now detail the stability analysis in dimension two. The collocation solution of (7.1.1)–(7.1.2) is a polynomial $u^N \in \mathbb{P}_N^0(\Omega)$ ($\Omega$ is the square $(-1,1) \times (-1,1)$) satisfying

$$-\Delta u^N = f \quad \text{at } \mathbf{x}_{ij} \quad \text{for } 1 \le i\,,\ j \le N-1\,, \tag{7.1.22}$$

where

$$\mathbf{x}_{ij} = \left(\cos\frac{i\pi}{N}, \cos\frac{j\pi}{N}\right)\,, \qquad 0 \le i\,,\ j \le N\,. \tag{7.1.23}$$

Setting

$$(u,v)_N = \sum_{i,j=0}^{N} u(\mathbf{x}_{ij})v(\mathbf{x}_{ij})w_i w_j\,, \tag{7.1.24}$$

let us define the bilinear form on $\mathbb{P}_N^0(\Omega) \times \mathbb{P}_N^0(\Omega)$

$$a_N(u,v) = -(\Delta u, v)_N\,. \tag{7.1.25}$$

Then, (7.1.22) is equivalent to the variational equations

$$a_N\left(u^N, v\right) = (f,v)_N \quad \text{for all } v \in \mathbb{P}_N^0(\Omega)\,. \tag{7.1.26}$$

Using the exactness of the Gauss-Lobatto formula and integration-by-parts, one gets the identity

$$a_N(u,v) = (\nabla u, \nabla(vw)w^{-1})_N \tag{7.1.27}$$

for all $u$ and $v \in \mathbb{P}_N^0(\Omega)$. Note that $\nabla(vw)w^{-1} \in \mathbb{P}_N(\Omega)^2$. Although $a_N(u,v)$ does not equal $a(u,v)$ (the form defined in (7.1.10)) for all $u, v \in \mathbb{P}_N^0(\Omega)$, it nonetheless retains the same continuity and coercivity properties of the form $a(u,v)$. Precisely, the following result has been proved by Funaro (1981):

**Theorem 7.2.**

(i) *There exists a constant $\tilde{A} > 0$ independent of $N$ such that for all $u, v \in \mathbb{P}_N^0(\Omega)$*

$$|a_N(u,v)| \le \tilde{A}\|u\|_{H_w^1(\Omega)}\|v\|_{H_w^1(\Omega)}\,; \tag{7.1.28}$$

(ii) *there exists a constant $\tilde{\alpha} > 0$ independent of $N$ such that for all $u \in \mathbb{P}_N^0(\Omega)$*

$$\tilde{\alpha}\|u\|_{H_w^1(\Omega)}^2 \le a_N(u,u)\,. \tag{7.1.29}$$

It follows that the stability condition (6.4.23) is satisfied with $E = H_{w,0}^1(\Omega)$. Thus, the energy method of Sect. 6.4.2 can be applied to obtain the stability and the convergence of the scheme (7.1.22) (see (6.4.24) and (6.4.26)). Moreover, the approximation results of Sects. 5.5 and 5.8 yield the following error estimate:

$$\|u - u^N\|_{H_w^1(\Omega)} \le CN^{1-m}\left\{|u|_{H_w^{m;N}(\Omega)} + |f|_{H_w^{m-1;N}(\Omega)}\right\} \tag{7.1.30}$$

provided $m > 2$.

### 7.1.3 Other Boundary-Value Problems

So far we have discussed the Dirichlet boundary-value problem for the Poisson equation. The analysis can be extended to cover other boundary conditions (such as Neumann or Robin conditions) as well as more general second-order elliptic operators.

Legendre Galerkin and G-NI approximations are based on the classical weak formulation of these problems, in which non-Dirichlet boundary conditions are accounted for in the boundary integral terms. Consequently, the "energy method", corresponding to the stability conditions (6.4.3) or (6.4.43), is still the most appropriate tool for their analysis. On the other hand, for most collocation or tau approximations, this method turns out to be inadequate, and one has to resort to the more general coercivity condition of the type (6.4.66) or (6.4.30).

Examples of analysis for different elliptic boundary-value problems have been given throughout Chap. 6. Example 3 of Sect. 6.4.2 contains a discussion of the Dirichlet boundary-value problem for a second-order elliptic operator in dimension one, with variable coefficients in the higher order term. The Neumann problem with the strong enforcement of the boundary conditions is considered in the subsequent Example 4. Both examples concern collocation methods. A Chebyshev tau approximation to the one-dimensional Neumann problem is analyzed in Example 7 of Sect. 6.4.4.

## 7.2 Singularly Perturbed Elliptic Equations

In this section, we provide some mathematical insight on the behavior of spectral schemes for the approximation of second-order singular perturbation problems. We consider the model boundary-layer problem

$$\begin{aligned} &-\nu u_{xx} + \mathcal{L}u = 0 \,, \qquad -1 < x < 1 \,, \quad \nu > 0 \,, \\ &u(-1) = 0 \,, \; u(1) = 1 \,, \end{aligned} \tag{7.2.1}$$

where $\mathcal{L}u = u$ (Helmholtz equation) or $\mathcal{L}u = u_x$ (advection-diffusion equation). Both choices of the operator $\mathcal{L}$ are directly relevant to fluid dynamics applications. The solution of Helmholtz problems like (7.2.1) is a major component of several methods for the spectral simulation of an incompressible flow in a channel (see CHQZ3, Sect. 3.4). On the other hand, the advection-diffusion problem is a simple model of viscous flow near a wall.

The function $u(x) = \sinh((x+1)/\sqrt{\nu})/\sinh(2/\sqrt{\nu})$ is the exact solution of the Helmholtz problem; it has a boundary layer of width $O(\sqrt{\nu})$ near $x = 1$ as $\nu \to 0$. The exact solution of the advection-diffusion problem is $u(x) = (\mathrm{e}^{(x-1)/\nu} - \mathrm{e}^{-2/\nu})/(1-\mathrm{e}^{-2/\nu})$, which again has a boundary layer near $x = 1$, but now of width $O(\nu)$. Obviously, if we fix $\nu$ and let $N$ tend to infinity, any spectral approximation $u^N$ to (7.2.1) will eventually exhibit exponential convergence. For instance, in the Legendre Galerkin case, by applying the abstract error estimate (6.4.8), one gets the bound

$$\|u - u^N\|_{H^1(0,1)} \leq \frac{C}{\sqrt{\nu}} \inf \|u - v^N\|_{H^1(0,1)} ,$$

where the infimum is taken over all polynomials in $\mathbb{P}_N$ that satisfy the same boundary conditions as $u$. Next, applying an approximation error bound such as (5.4.13) to the right-hand side, one gets the error bound

$$\|u - u^N\|_{H^1(0,1)} \leq \frac{C}{\sqrt{\nu}} N^{-s} \|D_x^{s+1} u\|_{L^2_{s+1}(0,1)} ,$$

for all $s \geq 1$; here, $C$ is a constant depending on $s$ but independent of $\nu$ and $N$, while the norm on the right-hand side is defined in (5.4.14). A simple argument based on writing the equation as $u_{xx} = \nu^{-1}\mathcal{L}u$ and successively differentiating this relation, taking also into account that outside the boundary-layer region $u$ is basically zero whereas inside it the slope is inversely proportional to the width of the region, proves that $\|D_x^{s+1} u\|_{L^2_{s+1}(0,1)}$ scales as $C'\nu^{-s/4}$ for the Helmholtz equation, and as $C'\nu^{-s/2}$ for the advection-diffusion equation. Therefore, we obtain the convergence estimate

$$\|u - u^N\|_{H^1(0,1)} \leq \frac{C''}{\sqrt{\nu}} \left( \frac{1}{\nu^{1/4} N} \right)^s \tag{7.2.2}$$

in the former case, and

$$\|u - u^N\|_{H^1(0,1)} \leq \frac{C''}{\sqrt{\nu}} \left( \frac{1}{\nu^{1/2} N} \right)^s \tag{7.2.3}$$

in the latter case. This proves the claimed result and, in particular, that spectral convergence is achieved as soon as the boundary layer can be fully resolved by the numerical scheme. From an alternative perspective, the behavior of the error becomes similar to that observed for the approximation of a pure second-order, Poisson problem. This reflects the fact that, from a mathematical point of view, the leading term in (7.2.1) is the second-order, diffusion term, while the first- or zeroth-order terms are merely compact perturbations of it.

For the analysis of singular perturbation problems, the focus is upon results which hold for any values of $\nu$ and $N$. (*Uniformity*, or *robustness*, of the estimates with respect to the singular perturbation parameter is desired.) Such results describe the behavior of the spectral solution also in those regimes in which the boundary layer is either fully unresolved (this is of academic interest only, unless the boundary layer does not affect the essential physics of the problem) or marginally resolved. (In this case, the full understanding of the phenomenon may provide insight that enables the design of numerical devices that enhance the performance of spectral methods).

Since the differential operator in (7.2.1) has constant coefficients, the approximations of Galerkin, collocation (equivalent to G-NI) and tau type can be investigated by the error equation technique described in Sect. 6.6; the analysis has been provided by Canuto (1988). In order to give a concrete illustration of this analysis, we will consider Chebyshev schemes, and we will represent the corresponding spectral solutions as $u^N(x) = \sum_{k=0}^N \hat{u}_k T_k(x)$; however, results similar to the forthcoming ones hold for Legendre discretizations as well.

Let us consider the Helmholtz problem first. The error equation (see Sect. 6.6) satisfied by $u^N$ is

$$-\nu u_{xx}^N + u^N = \lambda \Phi_N + \mu \Phi_{N-1} , \qquad -1 < x < 1 , \tag{7.2.4}$$

where $\lambda, \mu$ are suitable constants depending on $\nu$ and $N$ and determined by the boundary conditions, whereas $\Phi_n$ ($n = N$ or $N-1$) are polynomials depending only on the discretization method. More precisely, we have $\Phi_N = T'_{N+1}$ and $\Phi_{N-1} = T'_N$ for the Galerkin method, $\Phi_N = xT'_N$ and $\Phi_{N-1} = T'_N$ for the collocation method, and $\Phi_N = T_N$ and $\Phi_{N-1} = T_{N-1}$ for the tau method. A careful analysis of (7.2.4) shows that for all $\nu > 0$ and all $N > 0$ the Chebyshev coefficients of $u^N$ satisfy the bounds

$$0 < \hat{u}_k < \tfrac{1}{2} , \qquad 0 \le k \le N . \tag{7.2.5}$$

This property can be viewed as a sort of "maximum principle" in transform space, in the sense that all the Chebyshev coefficients of $u^N$ are strictly positive. Note that the usual maximum principle in physical space (which states that $0 \le u(x) \le 1$ for $-1 \le x \le 1$) does not hold for the spectral solutions to (7.2.1), as reflected by the onset of a Gibbs phenomenon near $x = 1$ when $\nu$ becomes small compared with $N^{-1}$.

An important implication of (7.2.5) is that $u^N$ is uniformly bounded in the interval $[-1, 1]$, *independently of* $N$ *and* $\nu$. In fact,

$$|u^N(x)| \le \sum_{k=0}^N \hat{u}_k |T_k(x)| \le \sum_{k=0}^N \hat{u}_k T_k(1) = u^N(1) = 1 . \tag{7.2.6}$$

Thus, the spectral solutions, although possibly highly oscillatory, are stable in the maximum norm.

The error equation for the advection-diffusion problem is

$$-\nu u_{xx}^N + u_x^N = \eta \Phi_{N-1} , \qquad -1 < x < 1 , \tag{7.2.7}$$

with $\Phi_{N-1}$ as before (note that the Galerkin and collocation schemes coincide due to the precision of the Gauss-Lobatto quadrature formula). For the Galerkin scheme, it is proven in Canuto (1988) that, for all $\nu > 0$ and $N > 0$,

$$\hat{u}_k > 0 \quad \text{for } k = 1, \ldots, N , \tag{7.2.8}$$

which implies the bound $u^N(x) \leq 1$ for $-1 \leq x \leq 1$, with the same proof as (7.2.6). Interestingly enough, in the "unresolved" regime (i.e., if $\nu N^2 \to 0$), the asymptotic behavior of $u^N$ depends on the parity of $N$. If $N$ is odd, the first coefficient $\hat{u}_0$ is strictly positive, too, implying the uniform bound $|u^N(x)| \leq 1$ for $-1 \leq x \leq 1$; more precisely, the analysis yields

$$u^N \simeq \tfrac{1}{2} + \tfrac{1}{2} T_N \qquad \text{in } [-1, 1] \, . \tag{7.2.9}$$

If $N$ is even, $\hat{u}_0$ is negative and one has

$$u^N \simeq \hat{u}_0 + \hat{u}_N T_N \qquad \text{in } [-1, 1] \tag{7.2.10}$$

with $|\hat{u}_0| \simeq \hat{u}_N \simeq C(\nu N^2)^{-1}$. Hence, in this case $u^N$ is not bounded from below independently of $\nu$. An illustration of these effects is provided in Fig. 7.1, which displays solutions to (7.2.1) for the advection-diffusion case ($\mathcal{L}u = u_x$). The dominant highest and lowest frequency components of the numerical solution are apparent in the figure, along with the striking difference between solutions with odd $N$ and even $N$.

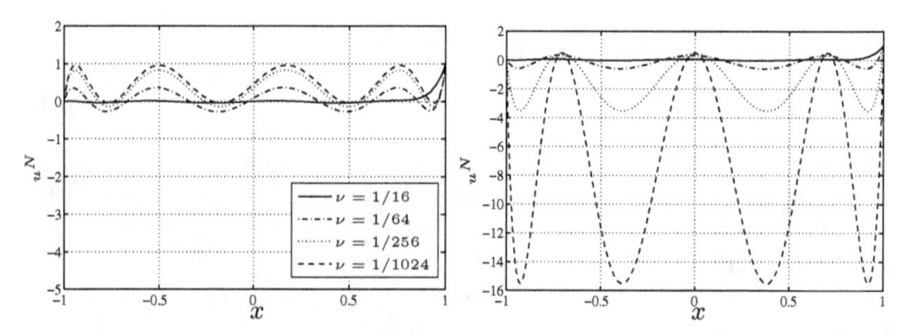


**Fig. 7.1.** Galerkin solutions $u^N$ of the advection-diffusion problem $-\nu u_{xx} + u_x = 0$, $-1 < x < 1$, $u(-1) = 0$, $u(1) = 1$, for several values of $\nu$ for odd $N$ $(= 9)$ (*left*) and even $N$ $(= 8)$ (*right*)

For the tau scheme, one has the asymptotic expansion

$$u^N \simeq \hat{u}_{N-2} T_{N-2} + \hat{u}_N T_N \qquad \text{in } [-1, 1] \tag{7.2.11}$$

with $|\hat{u}_{N-2}| \simeq \hat{u}_N \simeq CN$ if $N$ is odd, and $\hat{u}_{N-2} \simeq |\hat{u}_N| \simeq C(\nu N)^{-1}$ if $N$ is even. Again, $u^N$ is not bounded independently of $\nu$.

Another consequence of the error equation analysis concerns the limit behavior of $u^N$ as $\nu \to 0$ and $N \to \infty$. It can be shown that the maximum error $\|u - u^N\|_{L^\infty(-1,1)}$ between the exact solution $u$ of (7.2.1) and *any* spectral approximation $u^N$ satisfies an estimate of the form

$$\|u - u^N\|_{L^\infty(-1,1)} \leq C \min\left(1, \frac{1}{\nu N^4}\right) \tag{7.2.12}$$

for the Helmholtz equation, and of the form

$$\|u - u^N\|_{L^\infty(-1,1)} \leq \frac{C}{\nu N^2} \tag{7.2.13}$$

for the advection-diffusion equation. Here $C > 0$ is a constant independent of $\nu$ and $N$. A gain of a factor $N^{-1/2}$ occurs if the maximum norm is replaced by the $L^2_w$-norm. This proves again that the scaling of the resolution requirements for a spectral method to accurately resolve a boundary layer is that the number of modes be inversely proportional to the square root of the boundary-layer width.

The highly localized structure of the solution to (7.2.1) calls naturally for a multidomain strategy (see CHQZ3, Chaps. 5–6), in order to enhance the resolution within the boundary layer and to avoid the propagation of spurious oscillations in the outer region. We refer to Sect. 3.4 in Schwab (1998) for a thorough investigation of such a strategy. By extending the Shishkin-mesh approach (see, e.g., Roos, Stynes and Tobiska (1995)) to spectral methods, it is shown that placing one domain in the outer region and one domain in the boundary layer, with a properly chosen position of the interface point, guarantees exponential decay of the error as $\nu \to 0$ and $N \to \infty$ (since the solution of the linear, constant-coefficient boundary-layer problem in one dimension is obviously analytic). Schwab also provides additional robust and sharp estimates on the global polynomial approximation to problem (7.2.1) in a single domain.

### 7.2.1 Stabilization of Spectral Methods

The spurious oscillations that affect a spectral approximation to a singularly perturbed problem in the "under-resolved" regime are globally spread over the domain (or the subdomain, in a multidomain method). They are associated with the highest frequency components in the truncated expansions, as clearly documented by the asymptotic expansions (7.2.9)–(7.2.11) for the model problems considered above; note, however, that in some cases, such as (7.2.10), the lowest frequency component is affected too (see Fig. 4.17).

Several strategies can be invoked to cure such instabilities by controlling the onset of spurious oscillations. A simple approach consists of applying a filtering procedure (see Sect. 2.1.4), which damps the highest components of the spectrum of the discrete solution. Obviously, this approach is particularly well-suited for those methods that use a modal basis for the expansion; the need to match the boundary conditions after filtering suggests indeed a boundary-adapted basis (see Sect. 2.3.3). If a nodal basis is used instead, as for collocation and G-NI methods, then one incurs the extra cost of transferring from physical to frequency space and back. In all cases, some *a priori* knowledge of the structure of the spurious oscillations, which may be difficult to obtain except for model problems, seems required to properly tune the filter.

Other stabilization techniques are inspired by procedures that originated in the framework of low-order schemes, such as the $h$-version of finite elements. For advection-diffusion problems, the use of a Petrov-Galerkin approach – in which test functions are different from trial functions, the former being biased by the advection (or stream) direction – dates back to the mid 1970's (Christie et al. (1976)). Melenk and Schwab (1999) consider spectral and spectral-element versions ($p$- and $hp$-versions in their terminology) of this approach. Another strategy for stabilization, quite popular in the finite-element community, is known as the *SUPG-stabilization* (where SUPG stands for *streamline upwind Petrov-Galerkin (method)*) after Brooks and Hughes (1982), or – lately – as the bubble stabilization after Brezzi, Bristeau, Franca, Mallet and Rogé (1992) and Brezzi and Russo (1994). The adaptation of these ideas for spectral methods has been investigated by Canuto (1994), Pasquarelli and Quarteroni (1994) and Canuto and Puppo (1994).

In order to illustrate the SUPG strategy, let us consider the model Dirichlet problem for the advection-diffusion-reaction equation

$$\begin{aligned} &\mathcal{L}u \equiv -\nu u_{xx} + \beta u_x + \gamma u = f \;, \qquad -1 < x < 1 \;, \\ &u(-1) = 0 \;,\; u(1) = 0 \;, \end{aligned} \tag{7.2.14}$$

where $\nu > 0$ is a constant, and the coefficients $\beta$, $\gamma$ are smooth functions satisfying $-\frac{1}{2}\beta_x + \gamma \geq \sigma_0$ in $(-1,1)$ for some constant $\sigma_0 \geq 0$. The associated bilinear form $a(u,v) = \int_{-1}^{1}(\nu u_x v_x + \beta u_x v + \gamma u v)\,dx$ is continuous and coercive in $H_0^1(-1,1)$; precisely we have $a(v,v) \geq \nu\|v_x\|^2_{L^2(-1,1)} + \sigma_0\|v\|^2_{L^2(-1,1)}$ for all $v \in H_0^1(-1,1)$. However, for the advection-dominated case – $\beta$ is $O(1)$ but $\nu \ll 1$ – only a poor control on the gradient of $v$ is to be expected. At the discrete level, this is reflected by the fact that the Legendre Galerkin approximation of $u$, i.e., the function $u^N$ belonging to a polynomial subspace $V_N$ of $H_0^1(-1,1)$ and satisfying

$$a(u^N, v) = (f, v) \qquad \text{for all } v \in V_N \;, \tag{7.2.15}$$

may be polluted by spurious oscillations. Indeed, taking $v = u^N$, one gets the stability estimate (using the Cauchy-Schwarz and Poincaré inequalities)

$$\nu\|u_x^N\|^2_{L^2(-1,1)} + \sigma_0\|u^N\|^2_{L^2(-1,1)} \leq \min(1/\sigma_0, 4/\nu)\,\|f\|^2_{L^2(-1,1)} \;. \tag{7.2.16}$$

A similar situation occurs for the G-NI approximation. (The subsequent discussion will be based on the Legendre Galerkin method, but the implementation of the resulting stabilized schemes in a G-NI setting is straightforward.)

The structure of the discrete solution, in the limit $\nu \to 0$, $N$ fixed, can be easily understood in the constant-coefficient case $\beta = 1$, $\gamma = 0$ (for which the Galerkin and G-NI approximations coincide). Indeed, $u^N$ can be written as $u^N(x) = 1 + x - 2\tilde{u}^N(x)$, where $\tilde{u}^N \in \mathbb{P}_N$ is the discrete approximation to (7.2.1) with $\mathcal{L}u = u_x$. The results of the theory developed above apply to such a function. In particular, if $N$ is even, similarly to (7.2.10), we have $\tilde{u}^N(x) \simeq \hat{u}_0 + \hat{u}_N L_N(x)$ in $[-1,1]$, with $|\hat{u}_0| \simeq \hat{u}_N \simeq C(\nu N^2)^{-1}$. Therefore, $u^N$ is not bounded from above independently of $\nu$, and its highest and lowest frequency components are predominant. Some plots of $u^N$, for different values of $\nu$ and $N$ are shown in Fig. 4.17: the spurious dominant components of $u^N$ are apparent.

A tighter control on the variation of the discrete solution is obtained by modifying (7.2.15) in a strongly consistent way, i.e., by requiring that $u^N$ satisfy

$$a(u^N, v) + (\mathcal{L}u^N, \beta v_x)_\tau = (f, v) + (f, \beta v_x)_\tau \qquad \text{for all } v \in V_N\ , \qquad (7.2.17)$$

where $(g, w)_\tau$ denotes a weighted $L^2$-inner product, with nonnegative weight $\tau$ depending on the discretization but virtually independent of $\nu$ in the singular perturbation limit. Formulation (7.2.17) is the prototype of any SUPG-like stabilization method. It is strongly consistent, in the sense that the exact solution $u$ of (7.2.14) fulfills it; this implies, in particular, that the formal infinite order of accuracy of the spectral Legendre method is preserved. The added value over the standard Galerkin formulation comes from the extra control on the quantity $\|\beta u_x^N\|_\tau^2 = (\beta u_x^N, \beta u_x^N)_\tau$: it appears on the left-hand side of an estimate like (7.2.16) after choosing as usual $v = u^N$ in (7.2.17) and performing some manipulation. The norm $\|\beta u_x^N\|_\tau$ is called the *SUPG-norm* of the streamline derivative $\beta u_x^N$.

Let us now detail one particular realization of (7.2.17), proposed in Canuto (1994), which marries the accuracy of global polynomial expansions with the flexibility of local low-order finite elements. Let $-1 = x_0 < x_1 < \ldots < x_N = 1$ denote the Legendre Gauss-Lobatto points, and let $\Lambda_j = [x_{j-1}, x_j]$, $j = 1, \ldots, N$, be the "elements", or "cells", of size $h_j = x_j - x_{j-1}$ defined by two consecutive Gauss-Lobatto points. We introduce two finite-element spaces on the decomposition $\Lambda = \{\Lambda_j\}_j$ of the domain $[-1,1]$: $S_h^{(0)}$ is the space of the piecewise-constant functions on $\Lambda$, whereas $S_h^{(1)}$ is the space of the continuous, piecewise-linear functions on $\Lambda$. Correspondingly, we introduce a projection operator $J_h$ from $L^2(-1,1)$ or $C^0([-1,1])$ onto $S_h^{(0)}$, such as either the $L^2$-orthogonal projection operator or the interpolation operator at one selected point in each cell; furthermore, we introduce the interpolation operator $I_h$ from $C^0([-1,1])$ onto $S_h^{(1)}$ at the nodes of the decomposition $\Lambda$. Given any polynomial $v^N \in \mathbb{P}_N(-1,1)$, let $v^h = I_h v^N \in S_h^{(1)}$ be its piecewise-linear interpolant; the mapping $I_h : \mathbb{P}_N \to S_h^{(1)}$ is obviously bijective. Remarkably, the two polynomial functions $v^N$ and $v^h$ stay uniformly close to each other; indeed, there exist constants $C_i > 0$ independent of $N$ such that, for all $v^N \in \mathbb{P}_N$,

$$C_1 \|v^N\|_{L^2(-1,1)} \leq \|v^h\|_{L^2(-1,1)} \leq C_2 \|v^N\|_{L^2(-1,1)} \qquad (7.2.18)$$

and

$$C_3 \|v_x^N\|_{L^2(-1,1)} \le \|v_x^h\|_{L^2(-1,1)} \le C_4 \|v_x^N\|_{L^2(-1,1)} \,. \tag{7.2.19}$$

We term this property the *uniform low-order/high-order interpolation property*. An equivalent statement is that the bijection $I_h : \mathbb{P}_N \to S_h^{(1)}$ is an isomorphism in both the $L^2$-norm and the $H^1$-norm. The proof, given in Canuto (1994), exploits the property that each Gauss-Lobatto weight $w_j$ is a good approximation to the local spacing $\frac{1}{2}(h_j + h_{j-1})$ of the mesh, uniformly in $j$ and $N$ (see (7.4.5)–(7.4.6) below).

After choosing stabilization parameters $\tau_j > 0$ in each cell according to strategies that will be detailed shortly, we arrive at the following modified form of (7.2.17): find $u^N \in V_N = \mathbb{P}_N^0(-1,1)$ such that

$$\begin{aligned} a(u^N, v^N) &+ \sum_{j=1}^N \tau_j \int_{\Lambda_j} (\mathcal{L}u^N)_h \beta_h v_x^h \\ &= (f, v^N) + \sum_{j=1}^N \tau_j \int_{\Lambda_j} f_h \beta_h v_x^h \qquad \text{for all } v^N \in V_N \end{aligned} \tag{7.2.20}$$

(here, for notational simplicity, given any function $g$, such as $f$ and $\beta$, we set $g_h \equiv J_h g$). Note that it is fundamental that the *same* projection operator $J_h$ be applied to both $\mathcal{L}u^N$ and $f$ in order to preserve spectral accuracy.

We study the stability of this approximation under the assumption that $J_h$ is the $L^2$-orthogonal projection upon $S_h^{(0)}$; furthermore, we assume that $\sigma_0 > 0$ (we refer to Canuto and Puppo (1994) for the case $\sigma_0 = 0$). For convenience, we denote by $\mathcal{D}_\beta u^h$ the piecewise-constant function $\beta_h u_x^h$. Taking $v^N = u^N$ in (7.2.20), in each cell we have

$$\begin{aligned} \int_{\Lambda_j} (\mathcal{L}u^N)_h \mathcal{D}_\beta u^h &= \int_{\Lambda_j} \mathcal{L}u^N \mathcal{D}_\beta u^h \\ &= -\nu \int_{\Lambda_j} u_{xx}^N \mathcal{D}_\beta u^h + \int_{\Lambda_j} \beta_h u_x^N \mathcal{D}_\beta u^h + \int_{\Lambda_j} (\beta - \beta_h) u_x^N \mathcal{D}_\beta u^h + \int_{\Lambda_j} \gamma u^N \mathcal{D}_\beta u^h \\ &= S_{1j} + S_{2j} + S_{3j} + S_{4j} \,. \end{aligned}$$

The terms $S_{2j}$ give the desired extra control; indeed, recalling the definition of $u^h$, we have

$$\begin{aligned} S_{2j} &= \mathcal{D}_\beta u^h \, \beta_h \int_{\Lambda_j} u_x^N = \mathcal{D}_\beta u^h \, \beta_h \left(u^N(x_j) - u^N(x_{j-1})\right) \\ &= \mathcal{D}_\beta u^h \, \beta_h \left(u^h(x_j) - u^h(x_{j-1})\right) = \left(\mathcal{D}_\beta u^h\right)^2 h_j = \int_{\Lambda_j} \left(\mathcal{D}_\beta u^h\right)^2 \,. \end{aligned}$$

By defining the SUPG-norm as $\|w\|_\tau^2 = \sum_j \tau_j \|w\|_{L^2(\Lambda_j)}^2$, we thus have

$$\sum_{j=1}^N \tau_j S_{2j} = \|\mathcal{D}_\beta u^h\|_\tau^2 \,.$$

Using among others the inequality

$$\sum_{j=1}^{N} h_j^2 \, \|v_x^N\|_{L^2(\Lambda_j)}^2 \leq C \|v^N\|_{L^2(-1,1)}^2 \qquad \text{for all } v^N \in \mathbb{P}_N \,, \tag{7.2.21}$$

where $C > 0$ is a constant independent of $N$, the other integral terms can be bounded as follows (see (Canuto and Puppo (1994) for the details):

$$\left| \sum_{j=1}^{N} \tau_j \left(S_{1j} + S_{3j} + S_{4j}\right) \right| \leq C \|\mathcal{D}_\beta u^h\|_\tau \Big\{ \max_j (\tau_j^{1/2} \nu^{1/2} h_j^{-1}) \nu^{1/2} \|u_x^N\|_{L^2(-1,1)}$$

$$+ \max_j \left[\tau_j^{1/2} \left(\|\beta_x\|_{L^\infty(\Lambda_j)} + \|\gamma\|_{L^\infty(\Lambda_j)}\right)\right] \|u^N\|_{L^2(-1,1)} \Big\} \,.$$

We conclude that if the stabilization parameters $\tau_j$ are chosen in such a way that the quantities

$$\max_j (\tau_j^{1/2} \nu^{1/2} h_j^{-1}) \quad \text{and} \quad \max_j \left(\tau_j^{1/2} (\|\beta_x\|_{L^\infty(\Lambda_j)}) + \|\gamma\|_{L^\infty(\Lambda_j)})\right) \tag{7.2.22}$$

are small enough, we obtain the stability estimate

$$\nu \|u_x^N\|_{L^2(-1,1)}^2 + \|\mathcal{D}_\beta u^h\|_\tau^2 + \sigma_0 \|u^N\|_{L^2(-1,1)}^2 \leq C \|f\|_{L^2(-1,1)}^2 \tag{7.2.23}$$

for a constant $C$ independent of $\nu$ and $N$.

We now discuss the choice of the weights $\tau_j$. The classical SUPG recipe proposed by Franca, Frey and Hughes (1992) gives

$$\tau_j = \min \left( \frac{h_j}{2\|\beta\|_{L^\infty(\Lambda_j)}}, \frac{h_j^2}{12\nu} \right) \,; \tag{7.2.24}$$

a tuning parameter $c_0$ can be placed in front of such an expression to enforce the smallness of (7.2.22). (See also Pasquarelli and Quarteroni (1994) for similar choices in the context of approximations like (7.2.17) using either the SUPG method or the GaLS (Galerkin Least Squares) method.)

A different strategy of selection comes from identifying the SUPG-stabilized scheme (7.2.20) as the one produced by a standard Galerkin method in which the trial-test space $V_N$ is augmented by a space $B_h$ of "bubbles" (a bubble is a function which is nonzero only in one cell), and then the bubble components are eliminated from the resulting block $2 \times 2$-system. To be precise, set $W_N = V_N \oplus B_h$ and split any $w^N \in W_N$ as $w^N = v^N + v^b$. The standard Galerkin discretization of problem (7.2.14) based on the space $W_N$ can be formulated in the split form: find $u^N \in V_N$ and $u^b \in B_h$ such that

$$a(u^N, v^N) + a(u^b, v^N) = (f, v^N) \qquad \text{for all } v^N \in V_N \,,$$

$$a(u^N, v^b) + a(u^b, v^b) = (f, v^b) \qquad \text{for all } v^b \in B_h \,.$$

These equations are then modified as follows:

$$a(u^N, v^N) + a_h(u^b, v^h) = (f, v^N) \qquad \text{for all } v^N \in V_N \; , \tag{7.2.25}$$

$$(J_h(\mathcal{L}u^N), v^b) + a_h(u^b, v^b) = (J_h f, v^b) \qquad \text{for all } v^b \in B_h \; , \tag{7.2.26}$$

where $a_h(u,v) = \nu(u_x, v_x) + (\beta_h u_x, v)$ (for simplicity, we consider now a pure advection-diffusion problem). Next, we compute the bubble contributions from (7.2.26) and eliminate them from (7.2.25). By virtue of the properties of bubbles, we can accomplish this with a cell-by-cell procedure. Denoting by $B_{h,j}$ the space of bubbles on $\Lambda_j$ and setting $u_j^b = u^b_{|\Lambda_j}$, (7.2.26) yields

$$a_h(u_j^b, v_j^b) = (J_h(f - \mathcal{L}u^N)_{|\Lambda_j}, v_j^b) \qquad \text{for all } v_j^b \in B_{h,j} \; .$$

Since $J_h(f - \mathcal{L}u^N)_{|\Lambda_j}$ is constant, we can introduce the bubble $b_j \in B_{h,j}$ satisfying

$$a_h(b_j, v_j^b) = (1, v_j^b) \qquad \text{for all } v_j^b \in B_{h,j} \; , \tag{7.2.27}$$

which allows us to write $u_j^b = J_h(f - \mathcal{L}u^N)_{|\Lambda_j}\, b_j$. Substituting this expression into (7.2.25) and working out some algebra (the complete details can be found, e.g., in Canuto and Puppo (1994)), we end up precisely with (7.2.20) with $\tau_j$ given by

$$\tau_j = \frac{\left(\int_{\Lambda_j} b_j\right)^2}{h_j \nu \int_{\Lambda_j} b_{j,x}^2} \; . \tag{7.2.28}$$

Thus, the determination of the stabilization parameter in each cell is reduced to the determination of the bubble function satisfying (7.2.27). The *residual-free bubble* strategy (Brezzi and Russo (1994)) consists of choosing as $B_{h,j}$ not just a finite-dimensional space, but the largest admissible bubble space, i.e., the infinite-dimensional space $H_0^1(\Lambda_j)$. With such a choice, (7.2.27) is nothing but the constant-coefficient advection-diffusion problem

$$\begin{cases} -\nu b_{j,xx} + \beta_{h,j} b_{j,x} = 1 \; , & x_{j-1} < x < x_j \; , \\ b_j(x_{j-1}) = 0 \; , \; b_j(x_j) = 0 \; , & \end{cases} \tag{7.2.29}$$

with $\beta_{h,j} = \beta_{h|\Lambda_j}$. The solution to this problem satisfies $\nu \int_{\Lambda_j} b_{j,x}^2 = \int_{\Lambda_j} b_j$, as can easily be seen by multiplying the equation by $b_j$ and integrating over $\Lambda_j$; thus, (7.2.28) simplifies to

$$\tau_j = \frac{1}{h_j} \int_{\Lambda_j} b_j \; , \tag{7.2.30}$$

and the stabilization parameter can be obtained in all regimes by integrating $b_j$ (exactly or in an approximate way) over the cell. In the present one-dimensional situation, we actually have the analytical expression

$$b_j(x) = \frac{1}{\beta_{h,j}}(x - x_{j-1}) - \frac{h_j}{\beta_{h,j}} \frac{\mathrm{e}^{\beta_{h,j}(x-x_j)/\nu} - \mathrm{e}^{-\beta_{h,j}h_j/\nu}}{1 - \mathrm{e}^{-\beta_{h,j}h_j/\nu}} .$$

In the singular perturbation limit $\nu << |\beta_{h,j}|$, one has $b_j(x) \simeq (x-x_{j-1})/\beta_{h,j}$ if $\beta_{h,j} > 0$ or $b_j(x) \simeq (x_j - x)/\beta_{h,j}$ if $\beta_{h,j} < 0$; whence we obtain $\tau_j \simeq h_j/(2\beta_{h,j})$, consistent with (7.2.24).

With the prescribed choice of the stabilization parameters $\tau_j$, the term $\|\mathcal{D}_\beta u^h\|_\tau$ appearing on the left-hand side of (7.2.23) provides a uniform control on the variation of $u^N$ at the Gauss-Lobatto points, thereby preventing the onset of spurious oscillations. Fig. 7.2 provides an example of the results produced by the stabilization. Most of the spurious oscillations which would affect the pure Galerkin solution are absent; yet, the extent of the boundary layer is correctly confined to one cell. The values of the stabilized solution at the LGL nodes are spectrally accurate, while a simple post-processing, consisting of piecewise-linearly interpolating such values, suffices to produce a graphically correct approximation of the true solution. An additional feature of the method is that it allows a natural definition of a preconditioner for use in an iterative solution procedure for the resulting algebraic system. Indeed, it is enough to take the linear finite-element scheme, set on the same LGL mesh and stabilized by the same SUPG strategy. The spectra of the resulting preconditioned operators, varying $\nu$ and $N$, are uniformly close to the segment $[0.5, 1]$ on the real axis. We refer to Fig. 4.28 for an example; further results are given in Canuto and Puppo (1994).

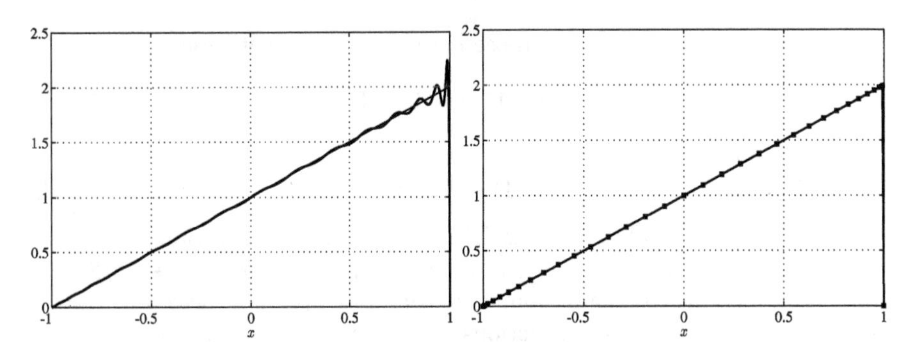


**Fig. 7.2.** Solution of the advection-diffusion problem $-\nu u_{xx} + u_x = 1$, $-1 < x < 1$, $u(-1) = u(1) = 0$, for $\nu = 10^{-4}$ and $N = 32$, by the stabilized scheme (7.2.20). (*left*) Spectral solution $u^N$, (*right*) finite-element interpolant, $u_h = I_h u^N$, of the spectral solution

The extension of the stabilized schemes described so far to the multidimensional case is rather straightforward. The tensor-product domain $\Omega$ is split into cells that are tensor products of intervals $\Lambda_j$. In each cell, $J_h$ is still an orthogonal projection over the constants, whereas $I_h$ is the multilinear interpolant at the vertices of the cell. The residual-free bubble strategy yields again the stabilization parameter as the average of the bubble over the cell, as in (7.2.30). However, the multidimensional analog of problem (7.2.29) can no longer be solved analytically; an efficient procedure for computing approximate values of the stabilization parameter in each cell, based on a preprocessing stage followed by extrapolation, can be found in Canuto, Russo and van Kemenade (1998). Applications to the full Navier-Stokes equations are also considered therein.

It is worth mentioning that a discretization-independent, functional interpretation of the norm $\|\mathcal{D}_\beta u^h\|_\tau$ can be given, relating it to the anisotropic smoothness of fractional order 1/2 along the streamlines of the discrete solution; we refer to Canuto and Tabacco (2001) for the details.

## 7.3 The Eigenvalues of Some Spectral Operators

We shall give a brief theoretical discussion of the qualitative behavior of the eigenvalues of some relevant spectral approximations to the following differential operators: the pure second-derivative operator $\mathcal{L}u = -u_{xx}$, the advection-diffusion operator $\mathcal{L}u = -\nu u_{xx} + \beta u_x$, and the first-order hyperbolic operator $\mathcal{L}u = u_x$. All the operators are associated with nonperiodic boundary conditions.

### 7.3.1 The Discrete Eigenvalues for $\mathcal{L}u = -u_{xx}$

The boundary conditions we impose here are of Dirichlet type

$$u(-1) = u(1) = 0 \tag{7.3.1a}$$

or of Neumann type

$$u_x(-1) = u_x(1) = 0\ . \tag{7.3.1b}$$

The exact eigenvalues for the Dirichlet boundary conditions are $\lambda_m = (\pi m/2)^2$, $m = 1, 2, \dots$, with eigenfunctions $u_m(x) = \sin(\pi m(x+1)/2)$; the eigenvalues for the Neumann boundary conditions include these plus $\lambda_0 = 0$, and the corresponding eigenfunctions are $u_m(x) = \cos(\pi m(x+1)/2)$.

We will first consider the *collocation* method that uses the Gauss-Lobatto points $x_j$, $j = 0, \dots, N$, (see Sect. 2.2.3) with respect to the Chebyshev or the Legendre weight $w(x)$. The corresponding eigenvalues $\lambda^N$ are defined by the relations

$$-u^N_{xx}(x_j) = \lambda^N u^N(x_j)\ , \qquad j = 1, \dots, N-1\ , \tag{7.3.2}$$

where $u^N$ is a non-trivial polynomial of degree $N$ which satisfies the boundary conditions (7.3.1a) or (7.3.1b).

It has been proved by Gottlieb and Lustman (1983) that for the Chebyshev points $x_j = \cos(j\pi)/N$, the eigenvalues are all real, nonnegative, and distinct. Gottlieb and Lustman actually prove their result for a wider class of boundary conditions than (7.3.1), namely for

$$\alpha u(1) + \beta u_x(1) = 0\,, \qquad \gamma u(-1) + \delta u_x(-1) = 0\,, \tag{7.3.3}$$

with $\alpha, \beta, \gamma > 0$ and $\delta < 0$ (these conditions can be relaxed to allow $\alpha$ and $\gamma$, or $\beta$ and $\delta$ to be zero). Starting from the error equation associated with (7.3.2) (see Sect. 6.6), their method consists of finding an explicit expression for the characteristic polynomial of the collocation matrix. Next, they prove that this polynomial satisfies an algebraic condition which implies that its roots are real, nonnegative and simple. The method can be used to prove the same kind of result when the collocation points are the Legendre Gauss-Lobatto points defined in (2.3.12).

For the Dirichlet boundary conditions, it is easy to derive an upper- and a lower-bound for the eigenvalues of the collocation operator. Multiplying each equation (7.3.2) by $u^N(x_j)w_j$ – where $w_j$ is the $j$-th weight of the Gauss-Lobatto formula (2.2.17) – and summing up we get

$$-\sum_{j=0}^{N} u_{xx}^N(x_j)u^N(x_j)w_j = \lambda^N \sum_{j=0}^{N} \left(u^N(x_j)\right)^2 w_j\,.$$

By the exactness of the quadrature rule over $\mathbb{P}_{2N-1}$, we then have

$$\lambda^N = \frac{-\int_{-1}^{1} u_{xx}^N u^N w \, dx}{\|u^N\|_N^2}\,. \tag{7.3.4}$$

Here $\|u^N\|_N$ denotes the discrete $L_w^2$-norm of $u^N$ (see (2.2.24)), which is uniformly equivalent to the standard norm $\|u^N\|_{L_w^2(-1,1)}$ (see (5.3.2)). Integrating by parts the numerator of (7.3.4) if $w$ is the Legendre weight, or using inequalities (7.1.11) and (7.1.12) if $w$ is the Chebyshev weight, we obtain the bounds

$$\bar{c}_1 \frac{\|u_x^N\|_{L_w^2(-1,1)}^2}{\|u^N\|_{L_w^2(-1,1)}^2} \le \lambda^N \le \bar{c}_2 \frac{\|u_x^N\|_{L_w^2(-1,1)}^2}{\|u^N\|_{L_w^2(-1,1)}^2}$$

for two constants $\bar{c}_1$ and $\bar{c}_2$ independent of $N$. Using the Poincaré inequality (A.13.2) on the left-hand side, and the inverse inequality (5.4.5) or (5.5.4) (with $p = 2$) on the right-hand side, we conclude that there exist two positive constants $c_1$, $c_2$ independent of $N$ such that

$$0 < c_1 \le \lambda^N \le c_2 N^4\,. \tag{7.3.5}$$

This estimate is optimal, as can be observed from the results in Fig. 4.6. From the theoretical point of view, having a smaller exponent of $N$ in (7.3.5) would imply a smaller exponent in the inverse inequality (5.4.5) or (5.5.4), which is not possible.

We now consider the Legendre *Galerkin* or *G-NI methods*. The discrete eigenvalue problem consists of finding non-trivial polynomials $u^N \in V_N$, where $V_N = \mathbb{P}_N^0(-1,1)$ or $\mathbb{P}_N(-1,1)$, depending on whether Dirichlet or Neumann conditions are enforced, such that

$$(u_x^N, v_x)_{L^2(-1,1)} = \lambda^N (u^N, v)_* \qquad \text{for all } v \in V_N , \tag{7.3.6}$$

where $(u,v)_* = (u,v)_{L^2(-1,1)}$ in the Galerkin case or $(u,v)_* = (u,v)_N$ as defined in (2.2.24) in the G-NI case.

At the algebraic level, the problem is formulated as the generalized eigenvalue problem

$$K\mathbf{u} = \lambda^N M\mathbf{u} ,$$

where $K$ is the stiffness matrix (symmetric and positive definite for Dirichlet boundary conditions, symmetric and positive semi-definite for Neumann conditions), whereas $M$ is the mass matrix associated with the inner product $(u,v)_*$ (invariably symmetric and positive definite). (See Sects. 2.3.3 and 3.8 for more details on these matrices.) Thus, the discrete eigenvalues are all real and strictly positive, except for the zero eigenvalue when Neumann boundary conditions are applied. Taking $v = u^N$ in (7.3.6), we get

$$\lambda^N = \frac{\|u_x^N\|^2_{L^2(-1,1)}}{\|u^N\|_*^2} ,$$

which leads to the same conclusions as for the collocation scheme discussed above; in particular, a bound of the type (7.3.5) holds for all nonzero eigenvalues.

The extreme eigenvalues of the stiffness matrix $K$ itself, defined by the relation

$$K\mathbf{u} = \lambda^N \mathbf{u} , \tag{7.3.7}$$

are also of interest, particularly in the solution of the linear systems generated by Galerkin and G-NI methods. Indeed, as discussed in Chap. 4, the 2-norm condition number $\kappa_2(K) = \lambda^N_{max}/\lambda^N_{min}$, which coincides with the iterative condition number $\mathcal{K}(K)$ (see (C.1.10)) for the present symmetric problems, influences both the sensitivity to round-off errors of a direct solution method, and the rate of convergence of an iterative solution scheme.

We aim at bounding from below and from above the eigenvalues of (7.3.7). We consider homogeneous Dirichlet boundary conditions. Let $u^N \in \mathbb{P}_N^0(-1,1)$ be the unique polynomial satisfying $u^N(x_j) = u_j$, $j = 1, \dots, N-1$, where $\mathbf{u} = (u_j)$. From (7.3.7) we obtain

$$\lambda^N = \frac{\mathbf{u}^T K\mathbf{u}}{\mathbf{u}^T\mathbf{u}} = \frac{\|u_x^N\|^2_{L^2(-1,1)}}{\mathbf{u}^T\mathbf{u}} . \tag{7.3.8}$$

For any $j$, we have $u_j = \int_{-1}^{x_j} u_x^N(s)\,\mathrm{d}s$; whence by the Cauchy-Schwarz inequality we have

$$|u_j| \le \left(\int_{-1}^{x_j} |u_x^N(s)|^2\,\mathrm{d}s\right)^{1/2} \left(\int_{-1}^{x_j} \mathrm{d}s\right)^{1/2} \le \sqrt{2}\,\|u_x^N\|_{L^2(-1,1)}\,.$$

Thus, $\mathbf{u}^T\mathbf{u} = \sum_{j=1}^{N-1} u_j^2 \le 2(N-1)\|u_x^N\|^2_{L^2(-1,1)}$; inserting this inequality into (7.3.8), we find the lower bound

$$\frac{1}{2(N-1)} \le \lambda^N\,. \tag{7.3.9}$$

In order to get an upper bound, we invoke the inverse inequality (5.4.7) to obtain

$$\|u_x^N\|^2_{L^2(-1,1)} \le CN^2 \int_{-1}^{1} \frac{|u^N(x)|^2}{1-x^2}\,\mathrm{d}x = CN^2 \sum_{j=1}^{N-1} \frac{|u^N(x_j)|^2}{1-x_j^2}\,w_j\,,$$

where we have used the exactness of the LGL quadrature formula, which is permissible since $(u^N)^2/(1-x^2) \in \mathbb{P}_{2N-2}$. The asymptotic behavior of the nodes and weights of the LGL formula, (2.3.15) and (2.3.16), yields the bound $w_j/(1-x_j^2) \le C$ for all $j$; whence $\|u_x^N\|^2_{L^2(-1,1)} \le CN^2\mathbf{u}^T\mathbf{u}$. Therefore, we obtain the upper bound

$$\lambda^N \le CN^2\,. \tag{7.3.10}$$

It is possible to prove that both bounds (7.3.9) and (7.3.10) are sharp, i.e., $\lambda^N_{min} \sim cN^{-1}$ and $\lambda^N_{max} \sim c'N^2$ as $N \to \infty$. Thus, in particular,

$$\kappa_2(K) = \mathcal{K}(K) \sim CN^3 \tag{7.3.11}$$

(see Bernardi and Maday (1997); see also Schwab (1998)). Similar results hold for the homogeneous Neumann boundary conditions (after removing the null eigenvalue). Such an asymptotic behavior is clearly documented in Fig. 4.7.

Let us finally consider the *tau* approximation for the second-derivative operator. The corresponding eigenvalues are defined by

$$-\hat{u}_k^{(2)} = \lambda^N \hat{u}_k\,, \qquad k = 0, \dots, N-2\,, \tag{7.3.12}$$

where $\hat{u}_k$ and $\hat{u}_k^{(2)}$ denote respectively the $k$-th coefficient of $u^N$ and of $u^N_{xx}$ in the expansion according to the Chebyshev or the Legendre basis. As usual, the two highest coefficients of $u^N$ are determined by the boundary conditions (7.3.1). An equivalent formulation of (7.3.12) is

$$-\int_{-1}^{1} u^N_{xx} v w\,\mathrm{d}x = \lambda^N \int_{-1}^{1} u^N v w\,\mathrm{d}x \quad \text{for all } v \in \mathbb{P}_{N-2}\,. \tag{7.3.13}$$

For the Chebyshev method, the technique of Gottlieb and Lustman (1983) can be adapted to prove that the eigenvalues of (7.3.12) and (7.3.1) are real, nonnegative, and distinct. For the Dirichlet boundary conditions, the positivity of the eigenvalues is an easy consequence of their being real, since one can choose $v = -u^N_{xx}$ in (7.3.13) and use inequalities (7.1.11)–(7.1.12) to get the estimate

$$\bar{c}_1 \frac{\int_{-1}^{1} |u^N_{xx}|^2 w \, \mathrm{d}x}{\int_{-1}^{1} |u^N_x|^2 w \, \mathrm{d}x} \le \lambda^N \le \bar{c}_2 \frac{\int_{-1}^{1} |u^N_{xx}|^2 w \, \mathrm{d}x}{\int_{-1}^{1} |u^N_x|^2 w \, \mathrm{d}x} .$$

Since $u^N$ is a polynomial vanishing at $x = \pm 1$, its first derivative $u^N_x$ vanishes for at least one point in the interval $(-1, 1)$. Thus, we can apply to the function $u^N_x$ the Poincaré inequality (A.13.2) and the inverse inequality (5.5.4) (with $p = 2$ and $r = 1$) to get an estimate of the type (7.3.5). For both Dirichlet and Neumann boundary conditions, the largest computed eigenvalue grows asymptotically as $N^4$.

The theory is instead very easy for the Legendre method. By choosing $v = -\bar{u}^N_{xx}$ in (7.3.13) and integrating by parts, one proves that $\lambda^N$ has to be real and positive. The inverse inequality (5.4.5) ensures again that $\lambda^N$ can grow at most as $O(N^4)$. Furthermore, for Dirichlet boundary conditions, $\lambda^N$ is uniformly bounded away from 0.

The particular constants appearing in the $O(N^4)$ asymptotic growth of the largest eigenvalues, for various discretization methods, are reported in Table 4.2.

### 7.3.2 The Discrete Eigenvalues for $\mathcal{L}u = -\nu u_{xx} + \beta u_x$

We assume that $\nu$ is a strictly positive constant, while $\beta$ is a smooth real function of $x$. Hereafter, we shall submit $u$ to the Dirichlet boundary conditions (7.3.1a).

The exact eigenvalues of this operator are, in general, complex due to the presence of the first-order, advection term. Moreover, multiplying the equation $\mathcal{L}u = \lambda u$ by $\bar{u}$ and using a standard integration-by-parts argument one gets

$$\mathrm{Re}(\lambda) = \frac{\nu \int_{-1}^{1} |u_x|^2 \mathrm{d}x - \frac{1}{2} \int_{-1}^{1} \beta_x |u|^2 \mathrm{d}x}{\int_{-1}^{1} |u|^2 \mathrm{d}x} . \tag{7.3.14}$$

This shows that the real part of the eigenvalues need not be positive, whenever $\nu$ is small and $b_x$ is strictly positive. However, the Poincaré inequality (A.13.1) yields $\int_{-1}^{1} |u_x|^2 \mathrm{d}x \ge c \int_{-1}^{1} |u|^2 \mathrm{d}x$ for all $u \in H^1_0(-1, 1)$, with $c = \frac{1}{4}\pi^2$ by direct computation. Hence,

$$\mathrm{Re}(\lambda) \ge \frac{\pi^2}{4}\nu - \frac{1}{2}\beta_1 ,$$

with $\beta_1 = \max\{\beta_x(x), -1 \le x \le 1\}$. This implies that only a finite number of eigenvalues have negative real parts. In particular, $\mathrm{Re}(\lambda) > 0$ if $\beta$ is constant.

Let us discuss first the behavior of the eigenvalues of the spectral *Galerkin* operator. They are defined by the existence of a non-trivial polynomial $u^N \in \mathbb{P}_N^0(-1,1)$ such that

$$\int_{-1}^{1} (-\nu u_{xx}^N + \beta u_x^N) v w \, dx = \lambda^N \int_{-1}^{1} u^N v w \, dx \quad \text{for all } v \in \mathbb{P}_N^0(-1,1) \,. \tag{7.3.15}$$

An estimate for $\mathrm{Re}(\lambda^N)$ can be obtained by choosing $v = \overline{u}^N$. In the Legendre case, we get exactly (7.3.14) satisfied by $\lambda^N$ and $u^N$; whence

$$\mathrm{Re}(\lambda^N) \geq \frac{\pi^2}{4}\nu - \frac{1}{2}\beta_1 \,, \tag{7.3.16}$$

as for the exact eigenvalues. In the Chebyshev case, we have, by (7.1.16),

$$-\mathrm{Re}\left(\int_{-1}^{1} u_{xx}^N \overline{u}^N w \, dx\right) \geq \frac{1}{4}\int_{-1}^{1} |u_x^N|^2 w \, dx \,,$$

while by the Cauchy-Schwarz inequality, assuming $u^N$ to be normalized by $\int_{-1}^{1} |u^N|^2 w \, dx = 1$, we have

$$\left|\mathrm{Re}\left(\int_{-1}^{1} \beta u_x^N \overline{u}^N w \, dx\right)\right| \leq \beta_0 \left(\int_{-1}^{1} |u_x^N|^2 w \, dx\right)^{1/2} ,$$

where $\beta_0 = \max\{|\beta(x)|, -1 \leq x \leq 1\}$. Hence, $\mathrm{Re}(\lambda^N) \geq (\nu/4)\|u_x^N\|^2_{L^2_w(-1,1)} - \beta_0 \|u_x^N\|_{L^2_w(-1,1)}$ which implies

$$\mathrm{Re}(\lambda^N) \geq -\frac{\beta_0}{\nu} \,. \tag{7.3.17}$$

This proves that the real parts of the eigenvalues of the Galerkin method are uniformly bounded from below.

For both the Legendre and the Chebyshev methods a bound for $|\lambda^N|$ is obtained by choosing again $v = \overline{u}^N$ in (7.3.15) and taking the modulus of both sides. One gets

$$|\lambda^N| \leq \frac{c\nu \|u_{xx}^N\|^2_{L^2_w(-1,1)} + \beta_0 \|u_x^N\|_{L^2_w(-1,1)} \|u^N\|_{L^2_w(-1,1)}}{\|u^N\|_{L^2_w(-1,1)}} ,$$

whence, by the inverse inequality (5.4.5) or (5.5.4),

$$|\lambda^N| \leq \nu O(N^4) + \beta_0 O(N^2) \,. \tag{7.3.18}$$

The eigenvalues of the *collocation* operator for the advection-diffusion problem are defined by the relation

$$-\nu u_{xx}^N(x_j) + \beta(x_j) u_x^N(x_j) = \lambda^N u^N(x_j) \,, \qquad j = 1, \ldots, N-1 \,, \tag{7.3.19}$$

where again $u^N$ is a non-trivial polynomial of degree $N$, zero at $x = \pm 1$.

Equivalently, we have

$$-\nu \int_{-1}^{1} u_{xx}^N v w \,\mathrm{d}x + (\beta u_x^N, v)_N = \lambda^N (u^N, v)_N \qquad \text{for all } v \in \mathbb{P}_N^0(-1,1)\,, \tag{7.3.20}$$

where $(u,v)_N$ is defined in (2.2.24).

The theoretical estimates (7.3.17) and (7.3.18) derived above hold for the eigenvalues of (7.3.19) as well. It is enough to adapt the arguments previously used, taking into account the exactness of the quadrature formula related to the collocation nodes (see (2.2.25)), and the uniform equivalence of the continuous and discrete norms over $\mathbb{P}_N$ (see Sect. 5.3). In the Legendre case, we obtain the bound (7.3.16) instead of (7.3.18) if the collocation method is implemented using the skew-symmetric form of the advection term (see Example 5 of Sect. 6.5.2):

$$\begin{aligned}-\nu u_{xx}^N(x_j) + \tfrac{1}{2}\beta(x_j)u_x^N(x_j) + \tfrac{1}{2}\left(I_N(\beta u^N)\right)_x(x_j)\\ - \tfrac{1}{2}\beta_x(x_j)u^N(x_j) = \lambda^N u^N(x_j)\,, \qquad j = 1,\dots,N-1\,.\end{aligned}$$

Numerical experiments for collocation approximations to the operators

$$\mathcal{L}u = -\nu u_{xx} + u_x \qquad \text{and} \qquad \mathcal{L}u = -\nu u_{xx} + x u_x$$

support the estimates (7.3.16) or (7.3.17), as well as (7.3.18). In the former case for Legendre approximations, all the eigenvalues have nonnegative real parts, whereas for Chebyshev approximations, there are some eigenvalues with negative real parts when $\nu$ and $N$ are small. In the latter case for Legendre approximations, the real parts of the eigenvalues are bounded from below by $-\frac{1}{2}$, whereas for Chebyshev approximations, the real parts of the eigenvalues can have quite large negative values when $\nu$ and $N$ are small.

At last, we consider the *tau* scheme, which reads as (7.3.15) except that the test functions lie in $\mathbb{P}_{N-2}$. The estimate (7.3.17) on $\mathrm{Re}(\lambda^N)$ can be obtained in the same manner as for the Galerkin scheme, using now $v = P_{N-2}\overline{u}^N$ as a test function. For the Legendre tau method, it is possible to obtain a lower bound as close to (7.3.16) as desired, provided that $N$ is large enough. Indeed,

$$\begin{aligned}\mathrm{Re}\Big(\int_{-1}^{1} \beta u_x^N P_{N-2}\overline{u}^N \,\mathrm{d}x\Big)\\
&= \mathrm{Re}\left(\int_{-1}^{1} \beta u_x^N \overline{u}^N \,\mathrm{d}x\right) - \mathrm{Re}\left(\int_{-1}^{1} \beta u_x^N (\overline{u}^N - P_{N-2}\overline{u}^N)\mathrm{d}x\right)\\
&= -\frac{1}{2}\int_{-1}^{1} \beta_x |u^N|^2 \mathrm{d}x - \mathrm{Re}\left(\int_{-1}^{1} \beta u_x^N (\overline{u}^N - P_{N-2}\overline{u}^N)\mathrm{d}x\right)\\
&= -\frac{1}{2}\int_{-1}^{1} \beta_x |P_{N-2}u^N|^2 \mathrm{d}x - \frac{1}{2}\int_{-1}^{1} \beta_x \left[|u^N|^2 - |P_{N-2}u^N|^2\right] \mathrm{d}x\\
&\quad -\mathrm{Re}\left(\int_{-1}^{1} \beta u_x^N (\overline{u}^N - P_{N-2}\overline{u}^N)\mathrm{d}x\right)\,.\end{aligned}$$

The last two integrals on the right-hand side are easily shown to be bounded by $CN^{-1}\|u_x^N\|^2_{L^2(-1,1)}$, according to the estimate (5.4.11). Hence,

$$\mathrm{Re}(\lambda^N) \geq \frac{\left(\nu - \frac{C}{N}\right)\|u_x^N\|^2_{L^2(-1,1)} - \beta_1/2\|u^N\|^2_{L^2(-1,1)}}{\|u^N\|^2_{L^2(-1,1)}} \, .$$

By the Poincaré inequality, we conclude that

$$\mathrm{Re}(\lambda^N) \geq \left(\frac{\pi^2}{4}\nu - \frac{1}{2}\beta_1\right) - \frac{C}{N}$$

for a constant $C > 0$ depending on $\beta$ but independent of $\nu$.

For both the Legendre and the Chebyshev methods, the bound (7.3.18) on $|\lambda^N|$ is obtained by choosing $v = -\overline{u}^N_{xx}$ as a test function in the tau scheme and taking the modulus of both sides. One gets

$$|\lambda^N| \leq c\frac{\nu\|u^N_{xx}\|^2_{L^2_w(-1,1)} + \beta_0\|u^N_x\|_{L^2_w(-1,1)}\|u^N_{xx}\|_{L^2_w(-1,1)}}{\|u^N_x\|_{L^2_w(-1,1)}}$$

and concludes again by the inverse inequality (5.4.5) or (5.5.4).

### 7.3.3 The Discrete Eigenvalues for $\mathcal{L}u = u_x$

We associate to this operator the boundary condition

$$u(1) = 0 \, , \tag{7.3.21}$$

instead of $u(-1) = 0$, to conform to the discussion and the numerical results given in Sect. 4.3.2.

At first, we consider *collocation* methods. We choose here the collocation points to be the Gauss-Lobatto points $\{x_j\}_{j=0}^N$ for the Chebyshev or the Legendre weight, as defined in Sect. 2.2.3. Other choices of collocation points are possible. The eigenvalues of the collocation operator are defined by the set of equations

$$\begin{aligned} &u_x^N(x_j) = \lambda^N u^N(x_j) \, , \qquad j = 0, \ldots, N-1 \, , \\ &u^N(x_N) = 0 \, , \end{aligned} \tag{7.3.22}$$

provided $u^N$ is a non-trivial polynomial of degree $N$ (we assume here that in both cases the nodes are ordered left-to-right).

The eigenvalues of (7.3.22) are complex numbers, whose real parts are all nonpositive. For the Chebyshev points, this sign property follows from a stability result, due to Gottlieb and Turkel (1985), for the associated time-dependent problem

$$\begin{aligned} &u_t^N(x_j,t) = u_x^N(x_j,t) , && j=0,\dots,N-1 , \ t>0 , \\ &u^N(x_N,t) = 0 , && t>0 , \\ &u^N(x_j,0) = u_0(x_j) , && j=0,\dots,N . \end{aligned} \tag{7.3.23}$$

They prove that, for each $N > 0$, there exists a spatial norm of $u^N$ which remains bounded for all times $t > 0$. This clearly implies that the eigenvalues of the spatial operator in (7.3.23) have nonpositive real parts. Moreover, an estimate of the form

$$|\lambda^N| \le O(N^2) \tag{7.3.24}$$

for each eigenvalue follows easily from the identity

$$\int_{-1}^{1} u_x^N \overline{u}^N w \,\mathrm{d}x = \lambda^N \sum_{j=0}^{N} |u^N(x_j)|^2 w_j , \tag{7.3.25}$$

taking into account (5.5.4) and (5.3.2). This identity is obtained in the usual way by multiplying the $j$-th equation in (7.3.22) by $\overline{u}^N(x_j)w_j$, summing over $j = 0,\dots,N$ and using (2.2.25). One can prove that estimate (7.3.24) is sharp, as confirmed by numerical experiments (see Fig. 4.8).

The analysis for the Legendre collocation operator is easier. Equation (7.3.25) (where now $w \equiv 1$) implies the nonpositivity of $\mathrm{Re}(\lambda^N)$ (since $\mathrm{Re}\left(\int_{-1}^{1} u_x^N \overline{u}^N \,\mathrm{d}x\right) = -\frac{1}{2}|u^N(-1)|^2 \le 0$), as well as the growth estimate (7.3.24).

Next, we consider the *Legendre G-NI* method, with the weak imposition of the boundary conditions, that we have already considered in Sect. 3.7 (see also Example 5 in Sect. 6.5.2). The discrete form of the eigenvalue problem reads

$$(u_x^N, v)_N - u^N(1)v(1) = \lambda^N (u^N, v)_N \qquad \text{for all } v \in \mathbb{P}_N , \tag{7.3.26}$$

where $u^N \in \mathbb{P}_N$ and $(\cdot,\cdot)_N$ is the LGL inner product in $(-1,1)$. Choosing $v = \overline{u}^N$ and using the exactness of the LGL quadrature formula, we get

$$\mathrm{Re}\left( (u_x^N, \overline{u}^N)_N \right) = \tfrac{1}{2}|u^N(1)|^2 - \tfrac{1}{2}|u^N(-1)|^2 ;$$

whence

$$\mathrm{Re}\left(\lambda^N\right) = -\frac{\frac{1}{2}\left(|u^N(1)|^2 + |u^N(-1)|^2\right)}{\|u^N\|_N^2} \le 0 .$$

This proves the nonpositivity of the real parts of the discrete eigenvalues. From (7.3.26) again with $v = \overline{u}^N$, we also have

$$|\lambda^N| \le \frac{\|u_x^N\|_{L^2(-1,1)} \|u^N\|_{L^2(-1,1)} + |u^N(1)|^2}{\|u^N\|_N^2} ;$$

applying the inverse inequalities (5.4.5) with $r = 1$ and $p = 2$, and (5.4.3) with $p = 2$ and $q = \infty$, to bound the numerator, and the equivalence (5.3.2) of discrete and continuous $L^2$-norms in the denominator, we obtain again a bound of the form (7.3.24) on the modulus of the eigenvalues.

Finally, the eigenvalues arising from the *tau* approximation of the advection operator $\mathcal{L}$ are defined by the existence of a non-trivial polynomial $u^N$ of degree $N$ vanishing at $x = 1$ and such that

$$\hat{u}_k^{(1)} = \lambda^N \hat{u}_k \;, \qquad k = 0, \ldots, N-1 \;, \tag{7.3.27}$$

where $\hat{u}_k$ and $\hat{u}_k^{(1)}$ denote respectively the $k$-th coefficient of $u^N$ and of $u_x^N$ in the expansion according to the Chebyshev or the Legendre basis. Equation (7.3.27) is equivalent to the variational form

$$(u_x^N, v)_w = \lambda^N (u^N, v)_w \quad \text{for all } v \in \mathbb{P}_{N-1} \,. \tag{7.3.28}$$

The real parts of the eigenvalues of (7.3.27) are all strictly negative. In order to show this result, let us consider first the Chebyshev method. Equation (7.3.28) yields the error equation (see Sect. 6.6)

$$u_x^N = \lambda^N u^N + \alpha^N T_N \;, \qquad -1 < x < 1 \;; \tag{7.3.29}$$

by equating the coefficients of $T_N$ on both sides we get

$$\alpha^N = -\lambda^N \hat{u}_N \frac{\pi}{2} \,.$$

Let us multiply equation (7.3.29) by $(1+x)\overline{u}_x^N(x)w(x)$ and integrate over $(-1,1)$. It is easily checked using (2.4.22) that the $N$-th Chebyshev coefficient of the function $(1+x)\overline{u}_x^N$ is $N\overline{\hat{u}}_N$. Thus, setting $\tilde{w}(x) = (1+x)w(x)$ we have

$$\int_{-1}^{1} |u_x^N|^2 \tilde{w}(x) \mathrm{d}x = \lambda^N \left[ \int_{-1}^{1} u^N \overline{u}_x^N \tilde{w}(x) \mathrm{d}x - N \frac{\pi}{2} |\hat{u}_N|^2 \right] .$$

Note that $\mathrm{Re}\left( \int_{-1}^{1} u^N \overline{u}_x^N \tilde{w} \mathrm{d}x \right) = -\frac{1}{2} \int_{-1}^{1} |u^N|^2 \tilde{w}_x \mathrm{d}x < 0$; whence it follows that $\mathrm{Re}(\lambda^N) < 0$.

A bound for the modulus of $\lambda^N$ can be obtained by setting $v = P_{N-1}\overline{u}^N$ in (7.3.28) and using the Cauchy-Schwarz inequality to get

$$|\lambda^N| \le \frac{\|u_x^N\|_{L^2_w(-1,1)}}{\|P_{N-1}u^N\|_{L^2_w(-1,1)}} \,.$$

One can prove (following the argument used in Canuto and Quarteroni (1982a) to obtain the inverse inequality (5.5.4)) that there exists a constant $C > 0$ independent of $N$ such that

$$\|v_x\|_{L^2_w(-1,1)} \le C N^2 \|P_{N-1} v\|_{L^2_w(-1,1)} \quad \text{for all } v \in \mathbb{P}_N \text{ such that } v(1) = 0 \,.$$

Thus, one obtains again the estimate (7.3.24).

For the eigenvalues of the Legendre tau method, the nonpositivity of the real parts follows immediately setting $v = \overline{u}_x^N$ in (7.3.28), since $\mathrm{Re}\left(\int_{-1}^{1} u^N \overline{u}_x^N \, dx\right) = -\frac{1}{2}|u^N(-1)|^2 \leq 0$. On the other hand, the eigenvalues of the Legendre tau method differ qualitatively from those of the Chebyshev tau method, in that their largest modulus satisfies an estimate of the form

$$|\lambda^N| \leq O(N) \tag{7.3.30}$$

instead of (7.3.24). This rather surprising fact was proved by Dubiner (1991a), using an asymptotic analysis. On the other hand, when the Legendre tau method is applied to a system of hyperbolic equations, the corresponding eigenvalues grow again at the rate of $O(N^2)$, as predicted by the inverse inequality (5.4.5).

## 7.4 The Preconditioning of Spectral Operators

In this section, we review some of the theoretical results on the preconditioning of spectral operators by low-order finite-difference or finite-element operators.

The case of periodic boundary conditions is investigated in Sect. 4.4.2 for the most significant one-dimensional constant-coefficient operators. The preconditioning properties of several low-order operators are easily derived from the available analytical expression of the eigenvalues and eigenvectors of the corresponding matrices. The one-dimensional results can be immediately extended to the multidimensional case by exploiting the tensor-product structure of both the spectral and the finite-order operators, as indicated below; obviously, this approach presumes that the exact preconditioner is applied.

For the case of nonperiodic boundary conditions, the analytical expression of the eigenvalues of the preconditioned matrices is seldom available; an example is given by (4.4.37). In general, one must be content either with empirical results, such as those presented in Sect. 4.4, or with theoretical bounds on the spectra of the preconditioned matrices, which show, e.g., that their eigenvalues are bounded away from 0 and $\infty$ uniformly in $N$.

Results of the latter type can be easily provided for the Laplacian submitted to Dirichlet boundary conditions, when it is discretized by the Legendre Galerkin or G-NI methods that use the nodal basis at the Legendre Gauss-Lobatto nodes (thus, the G-NI method coincides with the collocation method at these points) and preconditioned by (multi-)linear finite elements based at the same points. The core of the analysis is furnished by the *uniform low-order/high-order interpolation property*, expressed by the two equivalence estimates (7.2.18)–(7.2.19); they state that interpolating a set of data given at a Legendre Gauss-Lobatto grid in $(-1, 1)$ either by a global polynomial or by a piecewise-linear function yields interpolants which have uniformly equivalent $L^2$- and $H^1$- norms. For the analysis of the G-NI method, another tool is the uniform equivalence between the continuous and discrete $L^2$-norms of a polynomial, given by (5.3.2).

In order to illustrate these results, let us start with the one-dimensional problem

$$-\frac{\mathrm{d}^2 u}{\mathrm{d}x^2} = f \,, \qquad -1 < x < 1 \,,$$
$$u(-1) = u(1) = 0 \,.$$

The Legendre Galerkin method which uses the nodal basis of $\mathbb{P}_N^0(-1,1)$ at the Gauss-Lobatto nodes yields the system

$$K\mathbf{u} = M\mathbf{f} \qquad (\mathbf{u} \in \mathbb{R}^{N-1}) \,,$$

where $K$ ($M$, resp.) is the stiffness (mass, resp.) matrix associated with this basis (for its general form, see (3.8.16)). The linear finite-element approximation of the same problem leads to a stiffness (mass, resp.) matrix denoted by $K_{FE}$ ($M_{FE}$, resp.); we refer to Sect. 4.4.2 for their definitions. Restricting (7.2.18)–(7.2.19) to functions $v^N \in \mathbb{P}_N^0(-1,1)$, these inequalities are equivalently written as

$$c_1 \mathbf{v}^T M_{FE} \mathbf{v} \le \mathbf{v}^T M \mathbf{v} \le c_2 \mathbf{v}^T M_{FE} \mathbf{v} \,, \tag{7.4.1}$$

and

$$c_3 \mathbf{v}^T K_{FE} \mathbf{v} \le \mathbf{v}^T K \mathbf{v} \le c_4 \mathbf{v}^T K_{FE} \mathbf{v} \,, \tag{7.4.2}$$

for all $\mathbf{v} \in \mathbb{R}^{N-1}$, for suitable constants independent of $N$.

If we consider the G-NI method, we rather have the system

$$K_{GNI} \mathbf{u} = M_{GNI} \mathbf{f} \,,$$

with $K_{GNI} = K$ due to the exactness of the LGL inner product, whereas $M_{GNI}$ is the diagonal matrix of the LGL weights (see (3.8.11)). By the equivalence of norms (5.3.2), we have

$$c_5 \mathbf{v}^T M \mathbf{v} \le \mathbf{v}^T M_{GNI} \mathbf{v} \le c_6 \mathbf{v}^T M \mathbf{v} \,, \tag{7.4.3}$$

for all $\mathbf{v} \in \mathbb{R}^{N-1}$. We will also consider the lumped finite-element mass matrix $M_{FE,d}$ introduced in Sect. 4.4.2. The explicit calculation of the matrix elements and the application to the matrix $(M_{FE,d})^{-1} M_{FE}$ of the classical Gerschgorin theorem on the localization of eigenvalues yield

$$c_7 \mathbf{v}^T M_{FE} \mathbf{v} \le \mathbf{v}^T M_{FE,d} \mathbf{v} \le c_8 \mathbf{v}^T M_{FE} \mathbf{v} \,, \tag{7.4.4}$$

for all $\mathbf{v} \in \mathbb{R}^{N-1}$; again, all constants in the previous estimates are independent of $N$.

Relation (7.4.2) immediately tells us that the eigenvalues of the symmetric and positive-definite matrix $K_{FE}^{-1/2} K K_{FE}^{-1/2}$ all lie in the interval $[c_3, c_4]$. Since this matrix is similar to the preconditioned matrix $P_1 = K_{FE}^{-1} K_{GNI} = K_{FE}^{-1} K$, considered in (4.4.45) and corresponding to the weak form of the finite-element preconditioning, we deduce that its iterative condition number (see (C.1.10)), $\mathcal{K}(P_1)$, is bounded by $c_4/c_3$ uniformly in $N$.

The eigenvalue analysis of the matrix $P_2 = (M_{FE}^{-1} K_{FE})^{-1} M_{GNI}^{-1} K_{GNI}$, introduced in (4.4.46) and corresponding to the strong form of the finite-element preconditioning, is less immediate. It relies again upon (7.4.2); in addition, one needs a bound for the ratios

$$r_j = \frac{(M_{GNI})_{jj}}{(M_{FE,d})_{jj}} = \frac{w_j}{\frac{1}{2}(h_j + h_{j-1})}, \qquad (h_j = x_{j+1} - x_j), \tag{7.4.5}$$

between the LGL weights and the local spacing of the LGL grid. Note that the numerator and denominator, respectively, are the elements of the diagonal spectral and (lumped) finite-element mass matrices, $M_{GNI}$ and $M_{FE,d}$, respectively, introduced in Sect. 4.4.2. The asymptotic expressions (2.3.15) and (2.3.16) easily show that $r_j \sim 1$ for all $j$ and $N$; Parter (2001a) proves indeed the sharp estimates

$$0.9 \le r_j \le 1 \tag{7.4.6}$$

for all ratios. Using these results, Parter (2001b) proves that the real parts of the eigenvalues of $P_2$ are uniformly positively bounded away from 0, and that the eigenvalues are uniformly bounded in modulus. By the same technique, one can easily prove that the eigenvalues of the matrix $P_3$ defined in (4.4.46) have the same properties. The same results hold for the finite-difference preconditioning of the collocation matrix, $H_{FD}^{-1} L_{coll}$ (see Parter and Rothman (1995) and Parter (2001a)). Similar results hold for the Chebyshev rather than the Legendre method; they have been obtained by Kim and Parter (1997). The extension from the Poisson equation to the Helmholtz equation $-\Delta u + \gamma u = f$, with a (possibly non-constant) coefficient $\gamma \ge 0$, poses no extra difficulty to the analysis.

The extension of the previous results to the multidimensional case

$$\begin{aligned} -\Delta u &= f \qquad && \text{in } \Omega = (-1,1)^d , \\ u &= 0 && \text{on } \partial\Omega , \end{aligned}$$

relies upon the tensor-product structure of the matrices at hand. Considering for instance the two-dimensional case, it is easily seen that the spectral mass matrix $M$ can be expressed as $M = M_x \otimes M_y$, where $M_x$, $M_y$ denote the one-dimensional mass matrices in each direction. Similarly, the bilinear finite-element mass matrix $M_{FE}$ has (with obvious choice of notation) the structure $M_{FE} = M_{FE,x} \otimes M_{FE,y}$. The spectral and finite-element stiffness matrices for the Laplacian operator, $K$ and $K_{FE}$, can be written as

$$K = K_x \otimes M_y + M_x \otimes K_y \quad \text{and} \quad K_{FE} = K_{FE,x} \otimes M_{FE,y} + M_{FE,x} \otimes K_{FE,y} .$$

The matrices involving the diagonal (lumped) versions of the mass matrices have an analogous form.

Now, if $A_i$, $B_i$, $i = 1, 2$, are symmetric and positive-definite matrices of order $n$ satisfying

$$\mathbf{v}^T A_i \mathbf{v} \le c_i^* \, \mathbf{v}^T B_i \mathbf{v} \qquad \text{for all } \mathbf{v} \in \mathbb{R}^n ,$$

then one has

$$\mathbf{v}^T (A_1 \otimes A_2)\mathbf{v} \le c_1^* c_2^* \, \mathbf{v}^T (B_1 \otimes B_2)\mathbf{v} \qquad \text{for all } \mathbf{v} \in \mathbb{R}^{n \times n} .$$

Using this property and the relations (7.4.1) and (7.4.2) in each direction, we get, for all $\mathbf{v} \in \mathbb{R}^{n \times n}$,

$$c_1 c_3 \, \mathbf{v}^T (K_{FE,x} \otimes M_{FE,y})\mathbf{v} \le \mathbf{v}^T (K_x \otimes M_y)\mathbf{v} \le c_2 c_4 \, \mathbf{v}^T (K_{FE,x} \otimes M_{FE,y})\mathbf{v}$$

and a similar sequence of inequalities in which the roles of $x$ and $y$ are interchanged; summing up the corresponding terms in the two sequences, we obtain the two-dimensional version of (7.4.2), with left constant $c_1 c_3$ and right constant $c_2 c_4$. As in the one-dimensional case, this immediately yields that the eigenvalues of the preconditioned stiffness matrix $P_0 = K_{FE}^{-1} K$ lie in the interval $[c_1 c_3, c_2 c_4]$, i.e., $\mathcal{K}(P_0) \le c_2 c_4 / c_1 c_3$.

Replacing the exact mass matrices by their lumped diagonal approximations, we prove in the same way that the iterative condition numbers of the matrices $P_1 = K_{FE}^{-1} K_{GNI}$ and $P_3 = K_{FE,app}^{-1} K_{GNI}$, introduced in (4.4.64) and (4.4.66), respectively, are uniformly bounded in $N$. We note that the tight bounds (7.4.6) on the elements of the one-dimensional diagonal matrix $M_{FE,d}^{-1} M_{GNI}$ tell us that $M_{FE,d}$ is an extremely good approximation to $M_{GNI}$, better in fact than the exact mass matrix $M_{FE}$. This explains why the condition number of $P_3$, which involves the former finite-element mass matrix, is smaller than that of $P_1$, which involves the latter matrix instead; the behavior is clearly documented in Table 4.9. At last, concerning the preconditioned matrix $P_2$, defined in (4.4.65), Parter (2001b) proves, as in the one-dimensional case, that the real parts of its eigenvalues are uniformly positively bounded away from 0, and that the eigenvalues are uniformly bounded in modulus. The same results can be proven for the matrix $P_4$ defined in (4.4.67).

## 7.5 The Heat Equation

Semi-discrete (discrete in space, continuous in time) approximations to this equation, submitted to Dirichlet, Neumann or Robin boundary conditions, can be analyzed by the energy method presented in Sect. 6.5.1. The three examples of that section illustrate its application to one-dimensional schemes.

The stability and convergence analysis of spectral schemes for the multidimensional heat equation can be established in a similar manner, using the continuity and coercivity results of Sect. 7.1 for the exact and discrete bilinear forms associated with Laplace's operator.

The aim of this section is to provide the reader with one example of analysis of a fully discrete approximation. For simplicity, we consider the one-dimensional heat equation submitted to Dirichlet boundary conditions:

$$\begin{aligned} &u_t - u_{xx} = 0\,, && -1 < x < 1\,, \quad t > 0\,, \\ &u(-1,t) = u(1,t) = 0\,, && t > 0\,, \\ &u(x,0) = u_0(x)\,, && -1 < x < 1\,. \end{aligned} \tag{7.5.1}$$

The analysis for the two-dimensional equation can be found in Bressan and Quarteroni (1986). We will deal with the Chebyshev collocation method in space; as usual, Fourier or Legendre methods would pose fewer difficulties from the technical point of view. On the other hand, the time variable will be discretized by a *$\theta$-method*, defined in (D.2.11). This family of methods includes, among others, both the forward and backward Euler methods (for $\theta = 0$ and $\theta = 1$, respectively), and the Crank-Nicolson method (for $\theta = \frac{1}{2}$). A $\theta$-method is explicit for $\theta = 0$, implicit for all other values of $\theta$.

Let $\Delta t > 0$ be the time-step, let $t^k = k\Delta t$, and let $\phi_j^k$ denote the value of the function $\phi$ for $x = x_j$ and $t = t^k$, where $x_j = \cos \pi j/N$. The fully discrete approximation to (7.5.1) reads as follows:

For any $k \geq 0$, $u^{N,k}$ is a polynomial of degree $N$ which satisfies

$$\begin{aligned} &u_j^{N,k+1} - u_j^{N,k} - \Delta t\left\{\theta (u_{xx}^N)_j^{k+1} + (1-\theta)(u_{xx}^N)_j^k\right\} = 0\,, && 1 \leq j \leq N-1\,, \\ &u_0^{N,k+1} = u_N^{N,k+1} = 0, && \\ &u_j^{N,0} = u_0(x_j)\,, && 0 \leq j \leq N\,. \end{aligned} \tag{7.5.2}$$

The absolute stability region of a $\theta$-method, as a function of $\theta$, is described in Sect. D.2.3. From the eigen-analysis of Sect. 7.3 (see, in particular, (7.3.5)), it follows that for $\theta < \frac{1}{2}$ the method has a severe stability restriction on the time-step $\Delta t$ of the form $\Delta t \leq C_\theta / N^4$, where $C_\theta$ is a positive constant monotonically increasing with $\theta$. The more restrictive condition is for the explicit backward Euler method ($\theta = 0$); whereas the condition is more and more alleviated as $\theta$ approaches $\frac{1}{2}$. To avoid any restriction, from now on $\theta$ will be chosen to satisfy $\frac{1}{2} \leq \theta \leq 1$, since in this case the method is $A$-stable (see Appendix D). In general, implicit time-discretization methods are customary for the heat equation.

By standard arguments, (7.5.2) can be restated as follows: for all $k \geq 0$, $u^{N,k} \in \mathbb{P}_N^0$ satisfies, for all $v \in \mathbb{P}_N^0$,

$$\left(u^{N,k+1} - u^{N,k}, v\right)_N + \Delta t\; a\left(\theta u^{N,k+1} + (1-\theta) u^{N,k}, v\right) = 0\,, \tag{7.5.3}$$

where $a(u,v)$ is defined in (7.1.13) and coincides with $-(u_{xx},v)_N$ when $u$ and $v$ are elements of $\mathbb{P}_N^0$. Furthermore, $u^{N,0} = I_N u_0$ is the interpolant of $u_0$ at the $(N+1)$ Legendre-Gauss-Lobatto points.

For convenience of notation we denote here by $\|v\|_0$ the norm of $v$ in $L^2_w(-1,1)$. To prove stability, let us take $v = \theta u^{N,k+1} + (1-\theta)u^{N,k}$ in (7.5.3). By (7.1.16), we get

$$\begin{aligned}\theta\|u^{N,k+1}\|_N^2 + (1-2\theta)\left(u^{N,k+1},u^{N,k}\right)_N - (1-\theta)\|u^{N,k}\|_N^2 \\ + \frac{\Delta t}{4}\|\theta u_x^{N,k+1} + (1-\theta)u_x^{N,k}\|_0^2 \le 0\,.\end{aligned} \tag{7.5.4}$$

Since $1-2\theta \le 0$, the Cauchy-Schwarz inequality gives

$$(1-2\theta)\left(u^{N,k+1},u^{N,k}\right)_N \ge \left(\tfrac{1}{2}-\theta\right)\left(\|u^{N,k+1}\|_N^2 + \|u^{N,k}\|_N^2\right)\,.$$

Then from (7.5.4) it follows that

$$\|u^{N,k+1}\|_N^2 + \frac{\Delta t}{2}\|\theta u_x^{N,k+1} + (1-\theta)u_x^{N,k}\|_0^2 \le \|u^{N,k}\|_N^2\,, \tag{7.5.5}$$

and thus, for all $k \ge 0$,

$$\|u^{N,k}\|_N^2 + \frac{\Delta t}{2}\sum_{j=0}^{k-1}\|\theta u_x^{N,j+1} + (1-\theta)u_x^{N,j}\|_0^2 \le \|u_0\|_N^2\,. \tag{7.5.6}$$

This shows that the scheme (7.5.2) is *unconditionally stable* if $\theta \in [\frac{1}{2},1]$. From (7.5.5) we deduce that $\|u^{N,k+1}\|_N \le \|u^{N,k}\|_N$ for all $k \ge 0$, which means that the scheme is contractive (see (D.1.8)). The same conclusion could be derived from the fact that the absolute stability region of the $\theta$-method for $\theta \ge \frac{1}{2}$ includes the negative real axis, and the eigenvalues of the spatial operator are real and negative. The energy (or variational) argument used above to prove (7.5.5) provides an alternative method of investigation, which yields the richer information about the spatial derivative of the quantity $u^{N,j+1} + (1-\theta)u^{N,j}$.

We prove now that certain norms of the error $u(t_k) - u^{N,k}$ tend to zero as both $\Delta t$ and $1/N$ tend to zero. In the sequel, given a function $v = v(x,t)$, we will denote by $v(t)$ the function of $x$ such that $(v(t))(x) = v(x,t)$; furthermore, we will set $v(t_k) = v^k$ to soften the notation. Using the function $\tilde{u}(t) = \Pi_N u(t) \in \mathbb{P}_N^0$, a projection defined by the condition $a(u(t) - \Pi_N u(t), v) = 0$ for all $v \in \mathbb{P}_N^0$ (see (5.5.21)), we get

$$\left(\theta\tilde{u}_t^{k+1} + (1-\theta)\tilde{u}_t^k, v\right)_w + a\left(\theta\tilde{u}^{k+1} + (1-\theta)\tilde{u}^k, v\right) = \left(\delta^k, v\right)_w\,,$$

for all $v \in \mathbb{P}_N^0$ and $k \ge 0$, where $\delta^k = \theta(\tilde{u}-u)_t^{k+1} + (1-\theta)(\tilde{u}-u)_t^k$. Then setting $e^k = u^{N,k} - \tilde{u}^k$ and using (7.5.3), we obtain

$$\begin{aligned}&\frac{1}{\Delta t}\left(u^{N,k+1} - u^{N,k}, v\right)_N - \left(\theta\tilde{u}_t^{k+1} + (1-\theta)\tilde{u}_t^k, v\right)_N + a\left(\theta e^{k+1} + (1-\theta)e^k, v\right) \\ &\quad = -\left(\delta^k, v\right)_w - E\left(\theta\tilde{u}_t^{k+1} + (1-\theta)\tilde{u}_t^k, v\right)\,,\end{aligned} \tag{7.5.7}$$

where the bilinear form $E$ is defined as $E(\phi,\psi) = (\phi,\psi)_w - (\phi,\psi)_N$.

Using the standard approximation results (7.1.21) and (5.3.4b), we obtain

$$\left|(\delta^k, v)_w + E\left(\theta \tilde{u}_t^{k+1} + (1-\theta)\tilde{u}_t^k, v\right)\right| \leq C_1 \|\gamma^k(u)\|_0 \|v\|_N \,, \tag{7.5.8}$$

where $\gamma^k(u) = \|\delta^k\|_0 + \|\theta \tilde{u}_t^{k+1} + (1-\theta)\tilde{u}_t^k\|_0$ and

$$|\gamma^k(u)| \leq C_2 N^{-r}\left(|u_t^k|_{H_w^{r;N}(-1,1)} + |u_t^{k+1}|_{H_w^{r;N}(-1,1)}\right), \quad r \geq 1 \,. \tag{7.5.9}$$

Now let $z = z(t)$ be any continuously differentiable function in the semi-infinite interval $(0, +\infty)$, and define

$$\varepsilon^k(z) = \frac{1}{\Delta t}\left(z^{k+1} - z^k\right) - \left(\theta z_t^{k+1} + (1-\theta) z_t^k\right).$$

If $z \in C^2(0,+\infty)$, then using the Taylor formula with the integral form of the remainder gives

$$\varepsilon^k(z) = \frac{1}{\Delta t}\int_{t^k}^{t^{k+1}} \left(s - (1-\theta)t^{k+1} - \theta t^k\right) z_{tt}(s) ds \,;$$

whence

$$|\varepsilon^k(z)| \leq \max(\theta, 1-\theta)\int_{t^k}^{t^{k+1}} |z_{tt}(s)| ds \leq \int_{t^k}^{t^{k+1}} |z_{tt}(s)| ds \,. \tag{7.5.10}$$

If $\theta = 1/2$ and $z \in C^3(0,+\infty)$, then a better estimate is obtained from a higher-order Taylor formula, namely

$$\varepsilon^k(z) = \frac{1}{2\Delta t}\int_{t^k}^{t^{k+1}} (t^k - s)(t^{k+1} - s) z_{ttt}(s) ds \,;$$

whence

$$|\varepsilon^k(z)| \leq \frac{\Delta t}{8}\int_{t^k}^{t^{k+1}} |z_{ttt}(s)| ds \,. \tag{7.5.11}$$

From (7.5.7) we obtain, using the above definition of $\varepsilon^k$,

$$\begin{aligned}
&\frac{1}{\Delta t}\left(e^{k+1} - e^k, v\right)_N + a\left(\theta e^{k+1} + (1-\theta)e^k, v\right) \\
&\quad = -\left\{(\varepsilon^k(\tilde{u}), v)_N + (\delta^k, v)_w + E(\theta \tilde{u}_t^{k+1} + (1-\theta)\tilde{u}_t^k, v)\right\}.
\end{aligned}$$

Taking $v = \theta e^{k+1} + (1-\theta)e^k$, proceeding in a manner similar to the stability proof, using (7.5.8) and the Cauchy-Schwarz inequality, we obtain

$$\begin{aligned}
&\|e^{k+1}\|_N^2 - \|e^k\|_N^2 + \frac{\Delta t}{2}\|\theta e_x^{N,k+1} + (1-\theta)e_x^{N,k}\|_0^2 \\
&\quad \leq 2\Delta t\left(\|\varepsilon^k(\tilde{u})\|_N + C_1|\gamma^k(u)|\right)\|\theta e^{k+1} + (1-\theta)e^k\|_N \,.
\end{aligned} \tag{7.5.12}$$

By the Poincaré inequality (see (A.13.2)) and the equivalence of continuous and discrete norms (see (5.3.2)), there exists a constant $c_P > 0$ such that $\|v_x\|_0 \geq c_P \|v\|_N$ for all $v \in \mathbb{P}_N^0$. Hence, using the Young inequality $ab \leq \frac{1}{2\eta}a^2 + \frac{\eta}{2}b^2$ for all $a, b \in \mathbb{R}$ and arbitrary $\eta > 0$, we get

$$\begin{aligned}\|e^{k+1}\|_N^2 &+ \frac{c_P}{4}\Delta t \|\theta e_x^{N,k+1} + (1-\theta)e_x^{N,k}\|_0^2 \\ &\leq \|e^k\|_N^2 + \frac{4}{c_P}\Delta t \left(\|\varepsilon^k(\tilde{u})\|_N + C_1|\gamma^k(u)|\right)^2 ;\end{aligned}$$

applying the above estimate recursively yields

$$\begin{aligned}\|e^k\|_N^2 &+ \frac{c_P}{4}\Delta t \sum_{j=0}^{k-1} \|\theta e_x^{N,j+1} + (1-\theta)e_x^{N,j}\|_0^2 \\ &\leq \|e^0\|_N^2 + \frac{4}{c_P}\Delta t \sum_{j=0}^{k-1} \left(\|\varepsilon^j(\tilde{u})\|_N + C_1|\gamma^j(u)|\right)^2 \equiv RHS\,.\end{aligned} \tag{7.5.13}$$

Since $e^0 = u^{N,0} - \tilde{u}(0) = I_N u_0 - \Pi_N u_0$ from (7.1.21) and (5.5.22), it follows using (5.3.2) that

$$\|e^0\|_N \leq 2\|e^0\|_0 \leq C_2 N^{-r} |u_0|_{H_w^{r;N}(-1,1)}\,, \quad r \geq 1\,. \tag{7.5.14}$$

We are going now to estimate the term $\|\varepsilon^j(\tilde{u})\|_N$ in the case $\theta = 1/2$. Since $\varepsilon^j(\tilde{u}) \in \mathbb{P}_N^0$, using (5.3.2), (7.5.11) and the Cauchy-Schwarz inequality yields

$$\begin{aligned}\|\varepsilon^j(\tilde{u})\|_N^2 \leq 4\|\varepsilon^j(\tilde{u})\|_0^2 &\leq \frac{1}{16}\Delta t^2 \int_{-1}^1 \left(\int_{t^j}^{t^{j+1}} |\tilde{u}_{ttt}(x,s)|\,\mathrm{d}s\right)^2 w(x)\,\mathrm{d}x \\ &\leq \frac{\Delta t^3}{16}\int_{t^j}^{t^{j+1}} \|\tilde{u}_{ttt}(s)\|_0^2\,\mathrm{d}s\,;\end{aligned}$$

whence

$$\sum_{j=0}^{k-1} \|\varepsilon^j(\tilde{u})\|_N^2 \leq \frac{\Delta t^3}{16}\int_0^{t^k} \|\tilde{u}_{ttt}(s)\|_0^2\,\mathrm{d}s\,. \tag{7.5.15}$$

Finally, from (7.5.9), (7.5.14) and (7.5.15), we see that the right-hand side of (7.5.13) can be estimated as

$$\begin{aligned}RHS \leq N^{-2r} &\left(C_2|u_0|_{H_w^{r;N}(-1,1)}^2 + C_3\sum_{j=0}^{k}|u_t^j|_{H_w^{r;N}(-1,1)}^2\right) \\ &+ C_4\Delta t^4 \int_0^{t^k} \|\tilde{u}_{ttt}(s)\|_0^2\mathrm{d}s\,.\end{aligned}$$

We now recall that $u(t_k) - u^{N,k} = \big(u(t_k) - \Pi_N u(t_k)\big) + e^k$, and we use the triangle inequalities for both the $L^2_w$- and the $H^1_w$-norms. This yields

$$\|u(t_k) - u^{N,k}\|_{L^2_w(-1,1)}$$
$$\leq N^{-r}\left(C_2|u_0|^2_{H^{r;N}_w(-1,1)} + C_5|u(t_k)|^2_{H^{r;N}_w(-1,1)} + C_3\Delta t\sum_{j=0}^{k}|u^j_t|^2_{H^{r;N}_w(-1,1)}\right)^{1/2}$$
$$+ \Delta t^2\left(C_4\int_0^{t^k}\|\tilde{u}_{ttt}(s)\|^2_{L^2_w(-1,1)}\mathrm{d}s\right)^{1/2}$$

and

$$\left(\Delta t\sum_{j=0}^{k-1}\|\theta(u(t_j) - u^{N,j}) + (1-\theta)(u(t_{j+1}) - u^{N,j+1})\|^2_{H^1_w(-1,1)}\right)^{1/2}$$
$$\leq N^{1-r}\left(C_6|u_0|^2_{H^{r-1;N}_w(-1,1)} + C_7|u(t_k)|^2_{H^{r;N}_w(-1,1)} + C_8\Delta t\sum_{j=0}^{k}|u^j_t|^2_{H^{r-1;N}_w(-1,1)}\right)^{1/2}$$
$$+ \Delta t^2\left(C_4\int_0^{t^k}\|\tilde{u}_{ttt}(s)\|^2_{L^2_w(-1,1)}\mathrm{d}s\right)^{1/2}.$$

We finally note that since $\tilde{u}_t = \Pi_N u_t$, the time derivatives of $\tilde{u}$ can be replaced with those of $u$, using (7.5.1) in a straightforward way.

The above convergence analysis has been carried out for $\theta = 1/2$. If $\theta \in (\frac{1}{2}, 1]$, it is easily seen, using (7.5.10) instead of (7.5.11), that the previous estimates still hold provided one replaces $\Delta t^2$ by $\Delta t$ and $\tilde{u}_{ttt}$ by $\tilde{u}_{tt}$ on the right-hand side.

The previous analysis can be adapted to cover the case of a full second-order parabolic equation

$$u_t - \nu u_{xx} + \beta u_x + \gamma u = f\,, \tag{7.5.16}$$

when the bilinear form $a(u,v)$ associated with the spatial part of the operator satisfies the coercivity condition $a(v,v) \geq \alpha\|v\|^2_{H^1_w(-1,1)}$ for all $v \in H^1_w(-1,1)$, for some $\alpha > 0$. Proceeding as above, one obtains similar results, in which the coercivity constant $\alpha$ appears in the denominator on the right-hand side. Should $\alpha$ be small, as in a singular perturbation problem, one can get estimates which do not depend explicitly on $\alpha$, by replacing the bound (7.5.12) with the bound

$$\|e^{k+1}\|^2_N - \|e^k\|^2_N + 2\alpha\Delta t\|\theta e^{N,k+1}_x + (1-\theta)e^{N,k}_x\|^2_0$$
$$\leq 4\Delta t\left(\|\varepsilon^k(\tilde{u})\|_N + C_1|\gamma^k(u)|\right)^2 + 2\Delta t\left(\|e^{k+1}\|^2_N + \|e^k\|^2_N\right)\,,$$

and then proceeding with the discrete form of the Gronwall lemma (see Sect. A.15). In this case, an exponential term $e^{\sigma t}$ (for some $\sigma > 0$) multiplies the norms on the right-hand sides of the final estimates. Obviously, these estimates become of little interest when the equation is integrated over long time intervals.

## 7.6 Linear Hyperbolic Equations

In this section, we present the numerical analysis of a number of spectral methods for linear hyperbolic problems. The discussion will be mainly confined to the one-dimensional case. We will consider the model *scalar* problem

$$\begin{aligned} & u_t + a(x)u_x = 0 \qquad \text{for } t > 0 , \\ & u(x,0) = u_0(x) , \end{aligned} \tag{7.6.1}$$

in a suitable space interval, supplemented with proper boundary conditions. The real functions $a$ and $u_0$ are assumed to be smooth. (Note that elsewhere in the book the velocity coefficient $a$ was indicated by $\beta$; we prefer to adopt the alternative symbol here in order to conform to a classical notation in the context of pure hyperbolic equations.) As in the previous section, for each $t$ we denote by $u(t)$ the function of $x$ such that $(u(t))(x) = u(x,t)$. Since both periodic and nonperiodic boundary conditions are relevant in applications, but require different techniques in the analysis, they will be considered in separate subsections. We also review some theoretical results about the resolution of the Gibbs phenomenon; subsequently, we deal with the challenge of recovering the exponential decay of the error from spectral approximations to discontinuous solutions of hyperbolic equations.

Spectral discretizations of hyperbolic *systems* of equations will be considered in CHQZ3, Sect. 4.2. The investigation of their mathematical properties, focused on the assumptions on the boundary conditions which ensure the stability of the approximations, is therefore deferred to CHQZ3, Sect. 4.2.4.

### 7.6.1 Periodic Boundary Conditions

In (7.6.1), $u, u_0$ and $a$ are supposed to be $2\pi$-periodic functions. Let us first recall that the solution $u$ is defined by the formula

$$u(x,t) = u_0(X(0;x,t)) , \tag{7.6.2}$$

where $X(\tau;x,t)$ denotes the solution of the backward initial-value problem

$$\begin{aligned} & \frac{\mathrm{d}X}{\mathrm{d}\tau} = a(X) , \qquad 0 \le \tau \le t , \\ & X(t) = x . \end{aligned} \tag{7.6.3}$$

According to (7.6.2), the maximum norm of $u$ on the interval $(0, 2\pi)$ (see (A.9.f)) is constant in time, i.e.,

$$\|u(t)\|_{L^\infty(0,2\pi)} = \|u_0\|_{L^\infty(0,2\pi)} \qquad \text{for all } t > 0\,. \tag{7.6.4}$$

On the other hand, the $L^2$-norm of $u$, although finite for all $t > 0$, may grow exponentially in time with respect to its value at $t = 0$ (i.e., the ratio $\|u(t)\|_{L^2(0,2\pi)}/\|u_0\|_{L^2(0,2\pi)}$ may grow exponentially). Indeed, multiplying (7.6.1) by $u$ and integrating by parts over $(0, 2\pi)$, we get

$$\frac{\mathrm{d}}{\mathrm{d}t}\int_0^{2\pi} u^2 \mathrm{d}x - \int_0^{2\pi} a_x u^2 \mathrm{d}x = 0\;;$$

whence, setting $\alpha = \max\limits_{0\le x\le 2\pi} a_x(x)$, we obtain

$$\|u(t)\|^2_{L^2(0,2\pi)} \le e^{\alpha t}\|u_0\|^2_{L^2(0,2\pi)}\,, \qquad t > 0\,. \tag{7.6.5}$$

This estimate is sharp in describing the behavior of the $L^2$-norm of the solution on a finite time interval. Take for instance the case $a(x) = x$ and choose the initial data $u_0$ such that $u_0(x) = 1$ if $|x| \le \eta$, $u_0(x) = 0$ elsewhere (this example is nonperiodic, but if $\eta$ is chosen small enough compared to $t$, it is equivalent to a periodic problem; furthermore, $u_0$ is not smooth, but one can easily regularize it.) A direct computation yields $u(x,t) = 1$ if $|x| \le \eta e^t$, $u(x,t) = 0$ elsewhere, whence $\|u(t)\|^2_{L^2(\mathbb{R})} = e^t\|u_0\|^2_{L^2(\mathbb{R})}$.

However, the $L^2$-norm of $u$ is bounded independently of $t$ when $a$ is of one sign. In fact, in this case (7.6.1) is equivalent to

$$\frac{1}{a}u_t + u_x = 0\,,$$

which, by multiplication by $u$ and integration-by-parts, yields

$$\frac{\mathrm{d}}{\mathrm{d}t}\int_0^{2\pi} \frac{1}{a(x)}u^2(x,t)\mathrm{d}x = 0\,,$$

and therefore

$$\|u(t)\|^2_{L^2(0,2\pi)} \le \frac{\max\limits_{0\le x\le 2\pi}|a(x)|}{\min\limits_{0\le x\le 2\pi}|a(x)|}\|u_0\|^2_{L^2(0,2\pi)}\,. \tag{7.6.6}$$

Finally, we recall that if the functions $a$ and $u_0$ are globally smooth, then so is $u$; this follows from (7.6.2)–(7.6.4). Nevertheless, $u$ develops gradients (in space) which grow exponentially in time at each point $\xi$ where $a$ changes sign with strictly negative derivative. Indeed, let us differentiate (7.6.2) at $x = \xi$ by the chain rule, using the facts that $X(\tau;\xi,t) = \xi$ for all $\tau$ and that $Y = \partial X/\partial x$ is the solution of the backward initial-value problem

$$\frac{\mathrm{d}Y}{\mathrm{d}\tau} = a_x(X)Y\,, \qquad 0 \le \tau \le t\,,$$
$$Y(t) = 1\,,$$

obtained by differentiating (7.6.3) with respect to $x$. We arrive at the formula

$$u_x(\xi, t) = e^{-a_x(\xi)t} u_{0,x}(\xi)\,, \tag{7.6.7}$$

which demonstrates the exponential steepening of the solution near these special points. Such a behavior poses a difficulty for any numerical approximation of (7.6.1).

Let us now consider spectral methods for this problem. A semi-discrete Fourier approximation $u^N(t)$ is a trigonometric polynomial of degree $N$ in $x$, i.e., $u_N(t) \in S_N$ where $S_N$ is defined in (5.1.1). It can be defined by a *Galerkin method*:

$$\begin{aligned} &\hat{u}_{k,t} + \left(a\, u_x^N\right)_k^\wedge = 0\,, \qquad && -N \le k \le N-1\,,\ t > 0\,, \\ &\hat{u}_k(0) = \hat{u}_{0,k}\,, && -N \le k \le N-1\,. \end{aligned} \tag{7.6.8}$$

Here $\hat{u}_k$ denotes the $k$-th Fourier coefficient of $u^N$. Another way of defining $u^N$ is by a *collocation method*:

$$\begin{aligned} &u_t^N(x_j, t) + a(x_j) u_x^N(x_j, t) = 0\,, \qquad && j = 0, \ldots, 2N-1\,,\ t > 0\,, \\ &u^N(x_j, 0) = u_0(x_j)\,, && j = 0, \ldots, 2N-1\,, \end{aligned} \tag{7.6.9}$$

where $x_j = j\pi/N$.

We discuss now the stability and convergence properties of these methods. The Galerkin solution satisfies, by (7.6.8),

$$\begin{aligned} &\left(u_t^N + a u_x^N, v\right) = 0 \qquad && \text{for all } v \in S_N\,,\ t > 0\,, \\ &u^N(0) = P_N u_0\,, \end{aligned} \tag{7.6.10}$$

where $(u, v) = \int_0^{2\pi} u\bar{v}\,\mathrm{d}x$ and $P_N$ is the $L^2$-projection operator upon $S_N$. Setting $v = u^N$ we obtain

$$\frac{\mathrm{d}}{\mathrm{d}t}\int_0^{2\pi} |u^N|^2 \mathrm{d}x - \int_0^{2\pi} a_x |u^N|^2 \mathrm{d}x = 0\,;$$

whence

$$\|u^N(t)\|^2_{L^2(0,2\pi)} \le e^{\alpha t} \|u_0\|^2_{L^2(0,2x)}\,, \qquad t > 0\,. \tag{7.6.11}$$

This estimate is the same as the one for the exact solution of (7.6.1) (see (7.6.5)). Thus, the $L^2$-norm of the Fourier Galerkin solution is bounded independently of $N$ on every finite time interval $[0, T]$. On the other hand, for each fixed $N$ the $L^2$-norm of $u^N$ is allowed to grow exponentially as $t \to \infty$, precisely as may the $L^2$-norm of the exact solution, according to (7.6.5).

There are examples in which $\|u^N(t)\|_{L^2(0,2\pi)}$ does grow exponentially in time as $t \to \infty$. This happens, e. g., for the equation $u_t + \sin(\delta x - \gamma)u_x = 0$, as reported in Gottlieb (1981), Sect. 3. Such a phenomenon is attributed (see Gottlieb (1981), Gottlieb, Orszag and Turkel (1981)) to the eventual insufficient resolution of the numerical scheme (for a fixed $N$), which surfaces as soon as excessively steep gradients are developed in the solution. According to the mechanism described by (7.6.7), oscillations which grow in time are produced in the numerical solution. However, if resolution is improved, i.e., if $N$ is increased, then the growth with time of $\|u^N(t)\|_{L^2(0,2\pi)}$ is retarded.

The fact that oscillations are bounded independently of $N$ on every fixed time interval can also be established by investigating the behavior of higher order Sobolev norms of the spectral solution. Setting $v = -u^N_{xx}$ in (7.6.10) we get

$$\frac{1}{2}\frac{\mathrm{d}}{\mathrm{d}t}\int_0^{2\pi} |u_x^N|^2 \mathrm{d}x - \int_0^{2\pi} a\ u_x^N \bar{u}_{xx}^N \mathrm{d}x = 0\ ;$$

whence

$$\|u_x^N(t)\|^2_{L^2(0,2\pi)} \le e^{\alpha t}\|u_{0,x}\|^2_{L^2(0,2\pi)}\ . \tag{7.6.12}$$

This estimate together with (7.6.11) proves that $u^N(x,t)$ is bounded independently of $N$ for all fixed intervals $0 \le t \le T$.

Finally, the convergence theory established in Sect. 6.5.2 and the approximation estimate (5.1.10) allow us to derive the following error estimate from the stability bound (7.6.11):

$$\|u(t)-u^N(t)\|_{L^2(0,2\pi)} \le Ce^{\alpha t/2}N^{1-m} \max_{0\le\tau\le t} \|u^{(m)}(\tau)\|_{L^2(0,2\pi)}\ , \tag{7.6.13}$$

provided $u(\tau) \in H_p^m(0,2\pi)$ for $0 \le \tau \le t$, with $m \ge 1$.

We turn now to the Fourier collocation method (7.6.9). If $a(x)$ does not vanish in $[0,2\pi]$, then (7.6.9) can be written as

$$\frac{1}{a(x_j)} u_t^N(x_j,t) + u_x^N(x_j,t) = 0\ , \qquad j = 0,\dots,2N-1\ .$$

Let us multiply each equation by $\bar{u}^N(x_j,t)(\pi/N)$, and sum up over $j$. By the exactness of the trapezoidal rule, based on the points $x_j$, for all trigonometric polynomials of degree $\le 2N$ (see Sect. 2.1.2), and by the skew-symmetry of the spatial operator, we get

$$\frac{\mathrm{d}}{\mathrm{d}t}\sum_{j=0}^{2N-1} \frac{1}{a(x_j)}|u^N(x_j,t)|^2 \frac{\pi}{N} = 0\ ;$$

whence

$$\|u^N(t)\|^2_{L^2(0,2\pi)} \le \frac{\max_{0\le x\le 2\pi} |a(x)|}{\min_{0\le x\le 2\pi} |a(x)|} \|I_N u_0\|^2_{L^2(0,2\pi)}\ , \tag{7.6.14}$$

where $I_N u_0$ is the trigonometric interpolant of $u_0$ at the collocation nodes. This proves the stability of the method, provided that the initial data is continuous or of bounded variation. Such a result was first established by Gottlieb (1981). Again, the convergence of the method can be inferred using the technique described in Sect. 6.5.2.

The analysis becomes much more involved when the coefficient $a(x)$ changes sign in the domain. Note that the equation degenerates into $\frac{\partial u}{\partial t} = 0$ at points where $a$ vanishes, leading to vertical characteristic lines and to the decoupling of the problem set in $(-1,1)$ into independent subproblems set in subintervals. A stability result such as (7.6.11), possibly with $u_0$ replaced by $I_N u_0$ in the collocation case, cannot hold. Indeed, when a fixed resolution (i.e., a fixed $N$) is used in the approximation of a solution in which steeper and steeper gradients develop in time, then aliasing effects may eventually become significant and adversely affect the stability. By carefully examining the interplay between aliasing, resolution and stability, Goodman, Hou and Tadmor (1994) proved that the standard Fourier collocation method for a general coefficient $a$ is only algebraically stable (in the sense of Gottlieb and Orszag (1977)), or weakly unstable, i. e., it satisfies

$$\|u^N(t)\|^2_{L^2(0,2\pi)} \leq C(t) N \|I_N u_0\|^2_{L^2(0,2\pi)} , \qquad t > 0 . \qquad (7.6.15)$$

However, the weak instability stems only from the high, unresolved modes through aliasing. In practice, well enough resolved computations keep the aliasing error below the truncation error, and results appear as if they were produced by a stable method.

$L^2$-stability can be rigorously proven for two variants of the collocation method – the skew-symmetric version and the filtered version. We begin by considering a Fourier collocation approximation of (7.6.1) in which the spatial term is discretized in a skew-symmetric way (see Gottlieb and Orszag (1977), Kreiss and Oliger (1979), Pasciak (1980)). Since $au_x$ can be decomposed as

$$au_x = \tfrac{1}{2}\left[au_x + (au)_x\right] - \tfrac{1}{2} a_x u ,$$

one considers the scheme

$$u_t^N(x_j,t) + \tfrac{1}{2}\left[a u_x^N + \mathcal{D}_N\left(a u^N\right)\right](x_j,t) - \tfrac{1}{2} a_x(x_j) u^N(x_j,t) = 0 , \qquad j = 0,\dots,2N-1 , \qquad (7.6.16)$$

where $\mathcal{D}_N$ represents the interpolation derivative operator at the collocation points (see (2.1.44)). Since, by (2.1.33),

$$\mathrm{Re}\left(\left(\mathcal{D}_N(au^N), u^N\right)_N\right) = -\mathrm{Re}\left(\left(au^N, u_x^N\right)_N\right) = -\mathrm{Re}\left(\left(au_x^N, u^N\right)_N\right) ,$$

we obtain, by multiplying (7.6.14) by $\bar{u}^N(x_j,t)(\pi/N)$ and summing over $j$,

$$\frac{\mathrm{d}}{\mathrm{d}t}\|u^N(t)\|^2_{L^2(0,2\pi)} \leq \alpha \|u^N(t)\|^2_{L^2(0,2\pi)}, \quad t > 0 ,$$

where again $\alpha = \max_{0 \le x \le 2\pi} a_x(x)$. Thus,

$$\|u^N(t)\|^2_{L^2(0,2\pi)} \le e^{\alpha t} \|I_N u_0\|^2_{L^2(0,2\pi)} , \tag{7.6.17}$$

which proves stability. Again by the methods of Sect. 6.5.2 one can prove the following convergence estimate (Pasciak (1980)):

$$\|u(t) - u^N(t)\|_{L^2(0,2\pi)} \le C e^{\alpha t/2} N^{1-m} \|u_0^{(m)}\|_{L^2(0,2\pi)}, \quad m \ge 1 . \tag{7.6.18}$$

(The two-dimensional version of this scheme is discussed in Sect. 6.5.2, Example 4.)

The skew-symmetric decomposition costs twice as much as a standard collocation method. Furthermore, although it provides an $L^2$-stable solution, it does not prevent the onset of oscillations near the points where sharp gradients are developed. Alternatively put, stability is not guaranteed in norms that yield control over the gradient of the spectral solution. Since oscillations, as well as the possible instability of the numerical solution, are due to the growth of the higher order modes, an attractive alternative to the skew-symmetric decomposition consists of inserting into the scheme (7.6.9) a filtering or smoothing mechanism. This can be accomplished by using the scheme

$$u_t^N(x_j, t) + a(x_j) \left(\mathcal{S}_N u_x^N\right)(x_j, t) = 0 , \qquad j = 0, \dots, 2N-1 , \tag{7.6.19}$$

where $\mathcal{S}_N : S_N \to S_N$ is a smoothing operator acting in transform space (see Sect. 2.1.4). The computational effort required by this process is generally relatively modest.

The class of filters proposed by Kreiss and Oliger (1979) offers the theoretical advantage of facilitating the derivation of a stability estimate in the $L^2$-norm. Here is a short description of their method. Fix three real, strictly positive constants $m, s$ and $j$. Let $M$ denote the largest integer $\le (1-(1/m))N$. For each $u = \sum_{k=-N}^{N} \hat{u}_k e^{ikx} \in S_N$, define $u_M \in S_M$ to be the truncation of $u$ of order $M$, i.e., $u_M = \sum_{|k| \le M} \hat{u}_k e^{ikx}$. Then, the smoothing operator $\mathcal{S}_N$ is defined as $\mathcal{S}_N u = \sum_{k=-N}^{N} \sigma_k \hat{u}_k e^{ikx}$, where

$$\sigma_k = \begin{cases} 1 \quad \text{if } |k| \le M \text{ or } |\hat{u}_k| \le \dfrac{\gamma \|u_M\|_{L^2(0,2\pi)}}{|2\pi k|^s} , \\ \dfrac{\gamma \|u_M\|_{L^2(0,2\pi)}}{|2\pi k|^s |\hat{u}_k|} \quad \text{otherwise} . \end{cases} \tag{7.6.20}$$

Note that $\mathcal{S}_N$ is bounded in the $L^2$-norm, i.e., $\|\mathcal{S}_N u\|_{L^2(0,2\pi)} \le \|u\|_{L^2(0,2\pi)}$ for all $u \in S_N$, and it leaves unchanged the lower portion of the spectrum, i.e., $\mathcal{S}_N u_M = u_M$. Moreover, $\mathcal{S}_N$ leaves unchanged the functions in $S_N$ which are "sufficiently smooth", in the sense that

$$\left\| \frac{\mathrm{d}^s u}{\mathrm{d}x^s} \right\|_{L^2(0,2\pi)} \leq \delta \|u\|_{L^2(0,2\pi)} \quad \text{for a suitable constant } \delta > 0\,,$$

provided that $m$ and $j$ are properly chosen as functions of $\delta$ (see Kreiss and Oliger (1979), Lemma 4.2).

The operator $\mathcal{S}_N$ prescribes a minimal rate of decay of the higher order coefficients, since $|\sigma_k \hat{u}_k| \leq O(|k|^{-s})$. Thus, according to (5.1.7), $\mathcal{S}_N$ enforces a minimal smoothing on the high-frequency component of $u$. This suggests that the choice of the actual value of the parameter $s$ should be based upon *a priori* information on the regularity of the exact solution of (7.6.1).

Kreiss and Oliger prove that with their filter the solution of (7.6.19) satisfies the estimate

$$\frac{\mathrm{d}}{\mathrm{d}t}\|u^N(t)\|^2_{L^2(0,2\pi)} \leq \left\{ \max_{0 \leq x \leq 2\pi} |(I_N a)_x| + O(N^{2-s}) \right\} \|u^N(t)\|^2_{L^2(0,2\pi)}\,, \tag{7.6.21}$$

provided that the $k$-th Fourier coefficient of $a$ decays at least as fast as $|k|^{-s}$. Thus, if $s > 2$, (7.6.21) implies that the $L^2$-norm of $u^N(t)$ is bounded independently of $N$ on every finite time interval.

Smoothing operators other than Kreiss and Oliger's can be used in (7.6.19) in order to stabilize the computation: for instance, those generated by the class of filters introduced in Sect. 2.1.4, which include the exponential filter considered by Majda, McDonough and Osher (1978) (see Sect. 7.6.4). As for the skew-symmetric scheme, there are no practical examples which indicate that the use of these filtering methods produces for linear problems more stable results than the straightforward collocation method.

### 7.6.2 Nonperiodic Boundary Conditions

We now assume that (7.6.1) holds in the interval $-1 < x < 1$, and that the value of $u$ is prescribed for $t > 0$ at the inflow boundary points. This means that $u$ is required to satisfy the conditions

$$\begin{aligned} u(-1,t) &= g_-(t) \quad \text{if } a(-1) > 0\,, \\ u(1,t) &= g_+(t) \quad \text{if } a(1) < 0\,, \end{aligned} \qquad t > 0\,, \tag{7.6.22}$$

where $g_\pm$ are smooth data. Under these boundary conditions, problem (7.6.1), (7.6.22) is well-posed in the $L^2$-norm, since by multiplication of (7.6.1) by $u$ and partial integration we have

$$\frac{\mathrm{d}}{\mathrm{d}t}\int_{-1}^{1} u^2 \mathrm{d}x - \int_{-1}^{1} a_x u^2 \mathrm{d}x + \sigma_+ a(1) g_+^2 - \sigma_- a(-1) g_-^2 \leq 0\,,$$

where

$$\sigma_- = \begin{cases} 1 & \text{if } a(-1) < 0\,, \\ 0 & \text{if } a(-1) \geq 0\,, \end{cases} \qquad \sigma_+ = \begin{cases} 0 & \text{if } a(+1) \leq 0\,, \\ 1 & \text{if } a(+1) > 0\,. \end{cases}$$

It follows that, setting $\alpha = \max_{-1\leq x\leq 1} a_x(x)$, one has

$$\|u(t)\|^2_{L^2(0,2\pi)} \leq e^{\alpha t}\|u_0\|^2_{L^2(0,2\pi)} + \int_0^t e^{\alpha(t-s)}\left\{-\sigma_+ a(1) g_+^2(s) + \sigma_- a(-1) g_-^2(s)\right\} \mathrm{d}s \,. \tag{7.6.23}$$

This result predicts that the stability analysis of Legendre discretization methods can be naturally accomplished by resorting to the energy approach. This is indeed the case. Spectral Legendre methods for hyperbolic problems have been introduced in Sect. 3.7; various strategies of enforcement of the boundary conditions are discussed therein, and their $L^2$-stability is established. The complete stability and convergence analysis of the Legendre G-NI scheme is detailed in the Example 5 of Sect. 6.5.2.

When we move to the analysis of Chebyshev methods, our road goes immediately uphill. The most natural norm in which to seek the stability of Chebyshev approximations seems to be the one involving the Chebyshev weight $w(x) = (1-x^2)^{-1/2}$. However, as pointed out by Gottlieb and Orszag (1977) and Gottlieb and Turkel (1985), the initial-boundary-value problem (7.6.1), (7.6.22) need not be well-posed in such a norm. A simple counter-example (Gottlieb and Orszag (1977)) is provided by the constant-coefficient problem

$$u_t + u_x = 0\,, \qquad u(-1,t) = 0\,, \tag{7.6.24}$$

with the initial condition

$$u(x,0) = u_0^\varepsilon(x) = \begin{cases} 1 - \dfrac{|x|}{\varepsilon} & \text{if } |x| \leq \varepsilon\,, \\ 0 & \text{if } |x| > \varepsilon\,. \end{cases} \tag{7.6.25}$$

It is easily seen that the $L^2_w$-norm of the solution satisfies the relations

$$\|u_0^\varepsilon\|_{L^2_w(-1,1)} = O(\varepsilon^{1/2}) \qquad \text{but} \qquad \|u(1)\|_{L^2_w(-1,1)} = O(\varepsilon^{1/4})\,.$$

Since $\varepsilon$ is arbitrarily small, the problem is not stable in the $L^2_w$-norm.

Greater freedom in the choice of the weighted norm in which to seek stability is obtained by allowing the weight function $\tilde{w}$ to be of the form $\tilde{w}(x) = r(x)w(x)$ with $r(x) = (1-x)^\lambda(1+x)^\mu$; the exponents $\lambda$ and $\mu$ equal 0 or 1 in such a way that $r(x)$ vanishes at the outflow boundary points for (7.6.1) (see Gottlieb and Orszag (1977), Gottlieb (1981), Canuto and Quarteroni (1982b)). When the boundary conditions are homogeneous, the stability in the $L^2_{\tilde{w}}$-norm follows from the identity

$$\frac{\mathrm{d}}{\mathrm{d}t}\int_{-1}^1 u^2\tilde{w}\mathrm{d}x - \int_{-1}^1 \left\{a_x + \tilde{w}^{-1}(a\tilde{w}_x)\right\} u^2\tilde{w}\mathrm{d}x = 0$$

by observing that the term in braces is bounded from above by a finite constant. Note that now waves always propagate toward boundary points where the weight vanishes. In the case of nonhomogeneous boundary conditions, stability can be inferred from the homogeneous case, provided that $g_{\pm}(t)$ are differentiable functions.

An account of the stability results for several Chebyshev schemes for problem (7.6.1) and (7.6.22), under particular assumptions on the coefficient $a$, can be found in Sect. 12.1.2 of Canuto et al. (1988).

A compromise between the efficiency of Chebyshev methods (related to the use of fast transform algorithms) and the ease of analysis of Legendre methods are the so-called Chebyshev-Legendre methods introduced by Don and Gottlieb (1994). The Chebyshev nodes are used to represent the discrete solution, but the differential equation is enforced at the Legendre nodes.

### 7.6.3 The Resolution of the Gibbs Phenomenon

Sect. 2.1.4 is devoted to the Gibbs phenomenon, which occurs in the approximation of discontinuous functions by spectral (and high-order) methods. Therein we have investigated its structure, and we have discussed several cures based on simple filtering (or smoothing) techniques. Hereafter, we complete the treatment of those filters by reporting some theoretical results. Furthermore, we review more sophisticated techniques that allow the reconstruction of the function with spectral accuracy away from the discontinuities, from the knowledge of Gibbs-oscillating discrete approximations. These methods have a wide application in the general field of signal and image processing. They are relevant to the matter of the present chapter, as they can be applied in a post-processing stage to the output of spectral discretizations of hyperbolic problems with discontinuous solutions. This particular issue will be discussed in the next subsection.

#### *Filters*

An axiomatic definition of filters of order $p \geq 2$ in Fourier space is given in (2.1.79). Several results are known about the convergence to $u$ of the smoothed Fourier series $\mathcal{S}_N u$ defined in (2.1.66) (see Vandeven (1991), Gottlieb and Shu (1997)). An example of such results is as follows. Let $u$ be a $2\pi$-periodic function which is piecewise infinitely differentiable, i.e., there exist $r$ singularity points $0 \leq x_0 < x_1 < \cdots < x_{r-1} < 2\pi$ such that in each interval $[x_{m-1}, x_m]$ (with $x_r = x_0 + 2\pi$) $u$ can be extended to a $C^\infty$-function up to the boundary. Given a point $x \in [0, 2\pi)$ different from each $x_m$, let $d(x) > 0$ denote the distance between $x$ and the nearest singularity point (taking into account the periodicity). Then, if $\mathcal{S}_N u$ is defined as in (2.1.70)–(2.1.71) through a filter $\sigma$ (see (2.1.79)) of order $p$, there exists a constant $C_\sigma > 0$ independent of $u$, $x$ and $N$ such that

$$|u(x) - \mathcal{S}_N u(x)| \le C_\sigma N^{1-p} d(x)^{1-p} |||u|||_p \,, \tag{7.6.26}$$

where $|||u|||_p$ is the so-called broken Sobolev norm

$$|||u|||_p = \left( \sum_{m=1}^{r} \|u\|^2_{H^p(x_{m-1},x_m)} \right)^{1/2} .$$

This shows that the error decays at least as fast as $N^{1-p}$ at each point of smoothness for $u$.

Among all filters of order $p$, the Vandeven filter (2.1.84) is optimal, in the sense that it minimizes the $L^2$-norm of the $p$-th derivative of $\sigma$ in $[0, \pi]$. This norm enters the estimate

$$|K_N(\xi)| \le C'_\sigma N^{1-p} \|\sigma^{(p)}\|_{L^2(0,\pi)} \{\xi^{-p} + (2\pi - \xi)^{-p}\} \tag{7.6.27}$$

of the decay of the smoothing kernel (2.1.71) away from the origin (mod $2\pi$). Obviously, the more concentrated the kernel, the better its approximation properties.

The effects of various filters on the square wave (2.1.21) were illustrated in Figs. 2.6, 2.8 and 2.9. Figure 7.3 now illustrates the convergence of the pointwise errors for this function at the points $x = 0.51\pi$, $x = 0.6\pi$ and $x = \pi$, which are at increasing distance from the nearest discontinuity (at $x = \pi/2$). (For ready comparison with the related figures from Chapter 2, the Chapter 1-4 convention for $N$ is employed on the abscissas.) The benefit of a higher order filter is clearly in evidence.

Three straight lines are provided for each filtered result, with slopes one order less, equal to, and one order greater than the order of the filter. Evidently, for larger values of $N$ the convergence is more rapid than the above estimate (7.6.26). Moreover, the convergence behavior is more regular the further one is from the discontinuity.

The "half-sine" function (2.1.22) exhibits more regular convergence, as illustrated in Fig. 7.4 . But once again, the convergence estimate appears overly pessimistic for large $N$.

Finally, we mention that if $u$ is not only piecewise infinitely differentiable but also piecewise analytic, an exponentially convergent approximation of $u(x)$ at any regular point $x$ can be recovered, by letting the order, $p$, of the Vandeven filter to grow with $N$. The precise relation is $p = cN^{\beta/4}$ where $c$ is a constant independent of $N$, whereas $\beta$ is such that $d(x) > N^{-1+\beta}$.

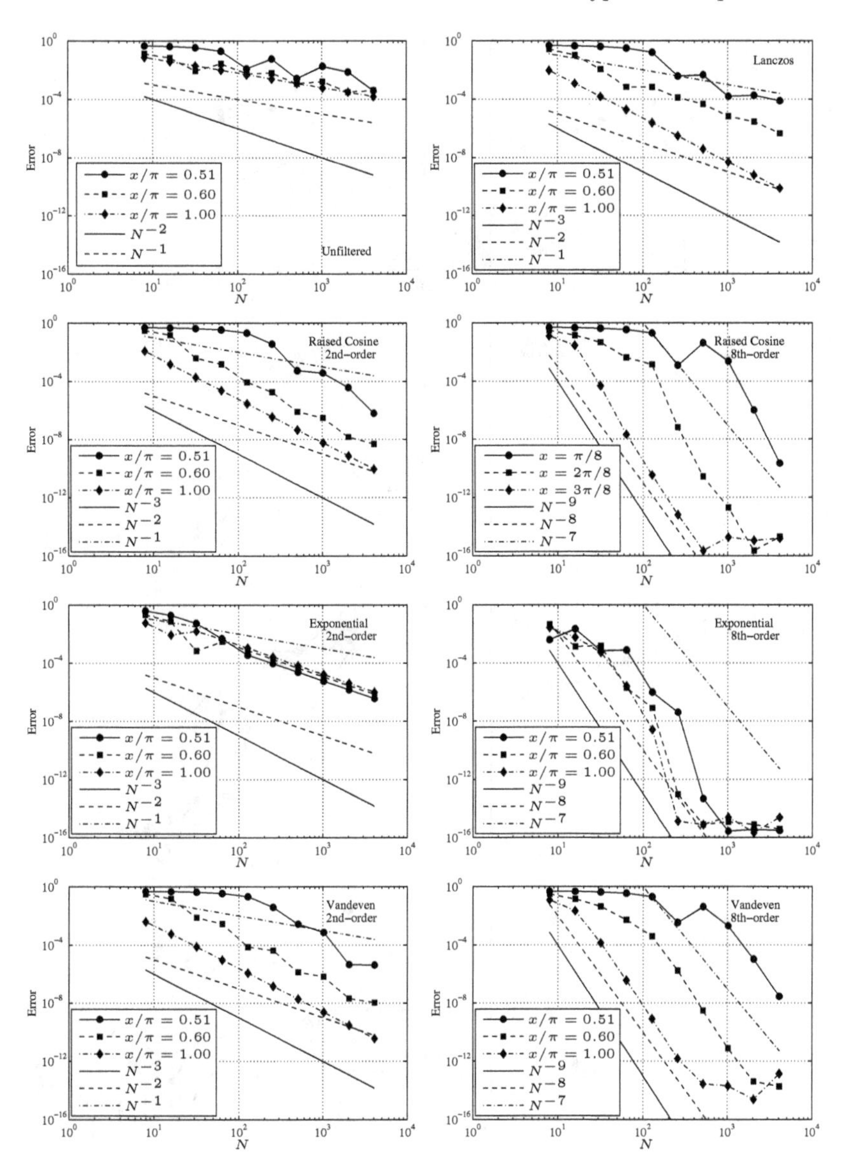


**Fig. 7.3.** Pointwise convergence for various filters applied to the square wave

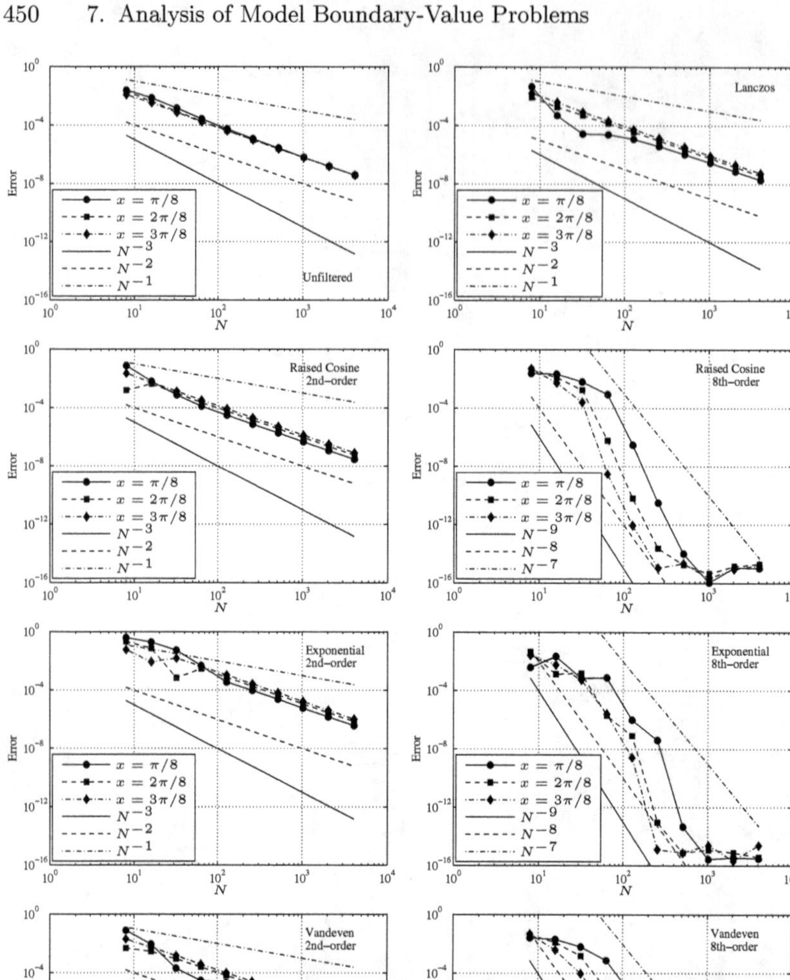


**Fig. 7.4.** Pointwise convergence for various filters applied to $\sin(x/2)$

### *Spectral Reconstruction Methods*

Several methods have been proposed that allow the reconstruction of a piecewise-smooth function up to the singularities starting from its truncated Fourier series (or, more generally, from a truncated expansion in orthogonal polynomials). Early results in spectral methods date back to Gottlieb, Lustman and Orszag (1981) and Abarbanel, Gottlieb and Tadmor (1986); they were based on subtracting the truncated expansions of suitably chosen step functions from the truncated expansion of the function of interest, and then applying a filter to the difference. An account of subsequent reconstruction methods can be found, e.g., in Gelb and Tanner (2006).

We briefly describe the class of reprojection methods initially proposed by Gottlieb et al. (1992) and subsequently developed by Gottlieb and Shu (1997, 1998). The idea underlying this approach is that the slowly convergent truncated series of the discontinuous function can be re-projected, in each interval of smoothness, onto an appropriate truncated orthogonal expansion in this interval. If the new basis guarantees spectral accuracy in the expansion of analytic functions, and if the projection of the high modes in the original basis upon the low modes of the new basis is exponentially small, then the reprojected expansion is spectrally convergent to the restriction of the original function to the interval.

For the sake of definiteness, assume that $u$ is a $2\pi$-periodic, piecewise-analytic function, whose truncated Fourier series is $P_N u = \sum_{k=-N}^{N} \hat{u}_k \phi_k$, with $\varphi_k(x) = e^{ikx}$. Let $[a, b]$ be an interval in which $u$ is analytic, which is mapped onto the reference interval $[-1, 1]$ by the transformation

$$\xi(x) = -1 + 2\frac{x-a}{b-a} .$$

For any value of a parameter $\lambda \geq 0$, let $\Psi^\lambda = \{\psi_m^\lambda(\xi)\}_{m\geq 0}$ be an orthogonal system for the inner product

$$(f, g)_\lambda = \int_{-1}^{1} f(\xi) g(\xi)\, w_\lambda(\xi) \mathrm{d}\xi ,$$

where $w_\lambda$ is a weight function in $(-1, 1)$, depending on $\lambda$; we assume that the orthogonal system is a basis in $L^2_{w_\lambda}(-1, 1)$.

The family of bases $\Psi^\lambda$ is termed a *Gibbs complement* for the basis $\Phi = \{\phi_k\}_{k\in\mathbb{Z}}$ if the two following conditions are satisfied:

i) Let

$$P_\lambda v = \sum_{m=0}^{\lambda} \frac{1}{\gamma_m^\lambda} (v, \psi_m^\lambda)_\lambda \psi_m^\lambda , \qquad \text{with } \gamma_m^\lambda = (\psi_m^\lambda, \psi_m^\lambda)_\lambda ,$$

be the *diagonal* orthogonal projection operator associated with the family of bases $\Psi^\lambda$ (diagonal means that both the truncation index and the weight index are varied at the same time). If $v$ is analytic in $[-1,1]$, then $P_\lambda v$ tends to $v$ exponentially fast as $\lambda \to \infty$, i.e.,

$$\|v - P_\lambda v\|_{L^\infty(-1,1)} \leq C_1 e^{-c_2\lambda} .$$

ii) Given a function $v$ defined in $[a,b]$, denote by $\tilde{v}$ the function $\tilde{v}(\xi) = v(x(\xi))$ defined in $[-1,1]$. Then, there exists $\alpha$, $\beta < 1$ such that for $\lambda = \beta N$

$$\left| \frac{1}{\gamma_m^\lambda} (\tilde{\phi}_k, \psi_m^\lambda)_\lambda \right| \|\psi_m^\lambda\|_{L^\infty(-1,1)} \leq \left( \frac{\alpha N}{|k|} \right)^\lambda \qquad \text{for all } |k| > N,\ m \leq \lambda .$$

Under these assumptions Gottlieb and Shu prove the following result: If $u$ is analytic in a disk of the complex plane containing the real interval $[a,b]$, and if $\lambda = \beta N$, then $P_\lambda(\widetilde{P_N u})$ tends to $\tilde{u}$ exponentially fast as $N \to \infty$, i.e.,

$$\|\tilde{u} - P_\lambda(\widetilde{P_N u})\|_{L^\infty(-1,1)} \leq C_3 e^{-c_4 N} . \qquad (7.6.28)$$

Note that the condition $\lambda \to \infty$ as $N \to \infty$ is necessary to achieve the exponential convergence in (7.6.28); choosing $\lambda$ constant or uniformly bounded with respect to $N$ would only yield a finite-order convergence.

An example of a Gibbs complement (not only for the Fourier system but also for the Legendre and Chebyshev systems) is provided by the Gegenbauer polynomials $\psi_m^\lambda(\xi) = C_m^\lambda(\xi)$ defined in (2.5.10). Boyd (2005) and Gelb and Tanner (2006) propose alternatives to them with better numerical properties.

### *Singularity detection*

The reconstruction methods mentioned above require the knowledge of the location of the singularities of the piecewise-smooth function $u$ in order to perform the reconstruction. Gottlieb, Lustman and Orszag (1981) were the first to incorporate a singularity detection device into a spectral scheme for hyperbolic problems; they locate each discontinuity and determine its strength by comparing the spectrum of the numerical solution with the spectrum of a step function. Since then, several techniques have been developed to detect these points starting from the truncated Fourier series of $u$; they find application in the more general field of *edge detection* in signal and image processing.

Gelb and Tadmor (2000a) elaborate a general framework for this problem, which incorporates most of the techniques proposed in the literature. Assume that the $2\pi$-periodic function $u$ is smooth except for a finite number of jump discontinuities located at $x_j$, where the one-sided limits $u(x_j^\pm) = \lim_{x \to x_j^\pm} u(x)$ exist and are finite. Define the jump function

$$[u](x) = u(x^+) - u(x^-) ,$$

whose knowledge is equivalent to the knowledge of the position and the strength of the jumps of $u$. An approximation of $[u]$ is obtained by taking the convolution, $K_N^\sigma * u = K_N^\sigma * P_N u$, between the truncated Fourier series of $u$ and a so-called *concentration* kernel $K_N^\sigma$. This is defined as

$$K_N^\sigma(t) = -\sum_{k=1}^{N} \sigma\left(\frac{k}{N}\right) \sin kt \,, \tag{7.6.29}$$

where the concentration factors $\sigma(\xi) = \xi\eta(\xi)$ satisfy $\eta \in C^2([0,1])$ and the normalization condition $\int_0^1 \eta(\xi)\,\mathrm{d}\xi = 1$. Then, setting

$$K_N^\sigma * u(x) = K_N^\sigma * P_N u(x) = i\pi \sum_{|k|\le N} \operatorname{sign}(k)\, \sigma\left(\frac{|k|}{N}\right) \hat{u}_k e^{ikx} \,,$$

one has

$$\| K_N^\sigma * u - [u] \|_{L^\infty(0,2\pi)} \le C \frac{\log N}{N} \,,$$

with $C$ depending on $\sigma$ and $u$ but not on $N$.

Examples of concentration factors are the trigonometric factors $\sigma(\xi) = \mathrm{Si}(\pi)^{-1} \sin \pi\xi$, with $\mathrm{Si}(\pi) = \int_0^\pi \frac{\sin t}{t}\, dt$, the polynomial factors $\sigma(\xi) = \alpha\xi^\alpha$, and the exponential factors $\sigma(\xi) = C\xi \exp(1/(\xi(\xi-1)))$.

A nonlinear enhancement of the singularity detection is also possible. Let us set $\varepsilon = \log N/N$; the above result shows that

$$K_N^\sigma * u \sim \begin{cases} \mathcal{O}(\varepsilon) & \text{at the smoothness points of } u \,, \\ [u] & \text{at the jumps of } u \,. \end{cases}$$

Thus,

$$\varepsilon^{-p/2} |K_N^\sigma * u|^p \sim \begin{cases} \mathcal{O}(\varepsilon^{p/2}) & \text{at the smoothness points of } u \,, \\ \big|[u]\big|^p \varepsilon^{-p/2} & \text{at the jumps of } u \,. \end{cases}$$

Letting $p > 1$ increase yields a better and better separation of scales. Introducing a critical threshold $J$, one can then select those jumps satisfying $\big|[u]\big| > J^{1/p}\sqrt{\varepsilon}$ by setting

$$K_{N,J}^\sigma u(x) = \begin{cases} K_N^\sigma * u(x) & \text{if } \varepsilon^{-p/2} |K_N^\sigma * u(x)|^p > J \,, \\ 0 & \text{otherwise} \,. \end{cases}$$

Gelb and Tadmor (2000a) provide further theoretical results, as well as a wealth of numerical demonstrations.

A different approach to edge detection, based on the *minmod* function typically used in numerical conservation laws to reduce spurious oscillations, is proposed by Archibald, Gelb and Yoon (2005).

### 7.6.4 Spectral Accuracy for Non-Smooth Solutions

The convergence results presented in the previous subsections are meaningful under the assumption that the exact solution be smooth enough, in the sense that it belongs to a Sobolev space of sufficiently high order. In hyperbolic problems, however, discontinuities in the data are propagated toward the interior of the domain, and if the operator is nonlinear, discontinuities can even develop in a finite time starting from smooth data.

If global convergence at a spectral rate is unattainable in such cases, at least one can hope to achieve spectral accuracy in those regions where the solution is smooth. The results of the analysis by Majda, McDonough and Osher (1978) indicate that it is not realistic to expect spectral accuracy directly in the numerical solution obtained by a standard collocation scheme. They consider problem (7.6.1) with $a = 1$, under periodic boundary conditions and a discontinuous initial condition with a single jump discontinuity located at a collocation point. If the Fourier collocation method is applied in the conventional manner, then in a region which excludes the discontinuity, the maximum error, for any $t > 0$, decays as $N^{-2}$. However, it is possible to achieve a convergence rate of infinite order by a proper filtering of the initial condition. This filtering is applied to the *continuous* Fourier coefficients of $u_0(x)$. The application of this filtering to the *discrete* Fourier coefficients of $u_0(x)$ still leads to second-order convergence. The filter has the structure (2.1.79), and it is of infinite order, i.e., is perfectly flat in a neighborhood of $\theta = 0$ and $\theta = \pi$.

The possibility of generating a spectrally accurate approximation of the discontinuous solution of a hyperbolic problem relies on the following property: a discretization method of spectral type (i.e., which guarantees spectral accuracy on smooth solutions) produces a very accurate approximation of *a projection* of the exact solution upon the finite-dimensional trial function space, rather than of the solution itself. Using the information encoded in this projection, one can accurately reconstruct the solution itself through a post-processing stage.

For instance, we will prove below that any Fourier coefficient of a discontinuous, periodic solution is approximated within spectral accuracy by a Fourier Galerkin method (obviously, provided the cut-off parameter is large enough). This means that the discrete solution is an accurate approximation of the truncated Fourier series of the exact solution. This, in turn, is a poorly convergent approximation of the solution itself (whenever discontinuities exist). Yet, the techniques described in the previous subsection allow one to detect the jumps and to reconstruct the solution up to the singularities, starting from such information.

As opposed to global post-processing, one can post-process a collocation or G-NI solution by a local smoothing, in order to recover spectral accuracy. The idea is based on the observation that while the pointwise convergence of a high-order polynomial approximation to a discontinuous solution is very slow, the convergence in a weighted mean – the weight being a smooth function – is very fast because oscillations kill each other on the average. Local smoothing can be carried out by a convolution in physical space with a localized function, and hence by a weighted mean which approximates exceedingly well the exact value of the solution.

From a rigorous mathematical point of view, the convergence in the mean can he measured in terms of a *Sobolev norm of negative order*. For simplicity, let us confine ourselves to the case of periodic functions. Each function $f \in L^2(0, 2\pi)$ defines a continuous linear form on the space $H_p^s(0, 2\pi)$ $(s \geq 0)$ (introduced in (A.11.d)), given by the mapping $\phi \to (f, \phi) = \int_0^{2\pi} f(x)\overline{\phi(x)}\mathrm{d}x$. Thus, $f$ can be identified with an element in the dual space of $H_p^s(0, 2\pi)$, here denoted by $H_p^{-s}(0, 2\pi)$ (see (A.1.c)). Its norm in this space is given by

$$\|f\|_{-s} = \sup_{\phi \in H_p^s(0,2\pi)} \frac{|(f, \phi)|}{\|\phi\|_s} . \tag{7.6.30}$$

For the remainder of this subsection $\|\phi\|_s$ denotes the norm of $\phi$ in $H_p^s(0, 2\pi)$. As usual, let $P_N f \in S_N$ be the symmetric truncation of the Fourier series of $f$ to $2N + 1$ modes. We want to estimate the error $f - P_N f$ in a negative Sobolev norm. By definition of $P_N$ we have, for all $\phi \in H_p^s(0, 2\pi)$,

$$(f - P_N f, \phi) = (f - P_N f, \phi - P_N \phi) .$$

Hence,

$$\begin{aligned} |(f - P_N f, \phi)| &\leq \|f - P_N f\|_0 \|\phi - P_N \phi\|_0 \\ &\leq C N^{-s} \|\phi\|_s \|f\|_0 . \end{aligned}$$

Here we have used (5.1.9). Thus, we obtain the estimate

$$\|f - P_N f\|_{-s} \leq C N^{-s} \|f\|_0 , \quad s \geq 0 . \tag{7.6.31}$$

Note that even though $f$ is merely square integrable, the truncation error in a negative Sobolev norm decays at a rate which depends solely upon the order of the norm.

As first pointed out by Mercier (1981), the previous argument can be extended to get an estimate in negative norms for the error between the exact and the spectral solutions to a linear hyperbolic problem. Let $\mathcal{L}$ be a linear, first-order hyperbolic operator with smooth periodic coefficients such that $(\mathcal{L}u, u) \geq 0$ for all $u \in H_p^1(0, 2\pi)$. Denote by $u = u(t)$ the solution of the following initial-boundary-value problem:

$$\begin{aligned} &u_t + \mathcal{L}u = 0 , && 0 < x < 2\pi , \quad t > 0 , \\ &u \;\; 2\pi\text{-periodic in } x , && \\ &u(0) = u_0 \in L^2(0, 2\pi) . && \end{aligned} \tag{7.6.32}$$

Let $u^N = u^N(t) \in S_N$ be the solution of the following Galerkin approximation of (7.6.32):

$$\begin{aligned} &(u_t^N + \mathcal{L}u^N, v) = 0 && \text{for all } v \in S_N\,, \quad t > 0\,, \\ &(u^N(0) - u_0, v) = 0 && \text{for all } v \in S_N\,. \end{aligned} \tag{7.6.33}$$

We want to estimate the quantity $(u(t) - u^N(t), \phi)$, where $\phi \in H_p^s(0, 2\pi)$. To this end, let $\mathcal{L}^*$ be the adjoint of $\mathcal{L}$, i.e., $(\mathcal{L}^* w, v) = (w, \mathcal{L}v)$ for all $v$ and $w \in H_p^1(0, 2\pi)$. Define $w = w(t)$ to be the solution of the hyperbolic problem

$$\begin{aligned} &w_t + \mathcal{L}^* w = 0\,, && 0 < x < 2\pi\,, \quad t > 0\,, \\ &w \;\; 2\pi\text{-periodic in } x\,, \\ &w(0) = \phi\,. \end{aligned} \tag{7.6.34}$$

Next, consider the corresponding Galerkin approximation $w^N = w^N(t) \in S_N$, which satisfies

$$\begin{aligned} &(w_t^N + \mathcal{L}^* w^N, v) = 0 && \text{for all } v \in S_N\,,\; t > 0\,, \\ &(w^N(0) - \phi, v) = 0 && \text{for all } v \in S_N\,. \end{aligned} \tag{7.6.35}$$

For a fixed $t > 0$ we have

$$\begin{aligned} (u(t) - u^N(t), \phi) &= (u(t) - u^N(t), w(0)) \\ &= (u(t), w(0)) - (u^N(t), w^N(0))\,. \end{aligned}$$

Set $\tilde{w}(s) = w(t - s)$. Then, for $0 < s < t$,

$$\frac{\mathrm{d}}{\mathrm{d}s}(u(s), \tilde{w}(s)) = (u_s, \tilde{w}) + (u, \tilde{w}_s) = -(\mathcal{L}u, \tilde{w}) + (u, \mathcal{L}^* \tilde{w}) = 0\,.$$

Thus,

$$(u(t), w(0)) = (u_0, w(t))\,. \tag{7.6.36}$$

Similarly,

$$(u^N(t), w^N(0)) = (u^N(0), w^N(t)) = (u_0, w^N(t))\,. \tag{7.6.37}$$

It follows from (7.6.36) that

$$(u(t) - u^N(t), \phi) = (u_0, w(t) - w^N(t))\,. \tag{7.6.38}$$

Under the assumptions on $\mathcal{L}$, if $\phi$ belongs to $H_p^s(0, 2\pi)$, then the solution to (7.6.34) belongs to $H_p^s(0, 2\pi)$ for all times and $\|w(t)\|_s \le C\|\phi\|_s$ (see, e.g., Taylor (1981)). Moreover, the theory of Sect. 6.5.2 yields the error estimate

$$\|w(t) - w^N(t)\|_0 \le CN^{-s}\|\phi\|_s\,. \tag{7.6.39}$$

Thus, we obtain the error estimate in negative Sobolev norm

$$\|u(t) - u^N(t)\|_{-s} \leq CN^{-s}\|u_0\|_0 \,, \quad s \geq 0 \,. \tag{7.6.40}$$

The previous proof can be suitably adapted to cover the case of a Fourier collocation approximation.

A slight modification of the arguments above yield the desired estimate for the Fourier coefficients. Take $\phi = \phi_k = e^{ikx}$ in (7.6.38), so that $\left(u(t) - u^N(t), \phi_k\right) = 2\pi(\hat{u}_k(t) - \hat{u}_k^N(t))$. Using (7.6.38), we get

$$|\hat{u}_k(t) - \hat{u}_k^N(t)| \leq CN^{-s}\|\phi_k\|_s\|u_0\|_0 \,.$$

Since $\|\phi_k\|_s \leq C|k|^s$, we conclude that

$$|\hat{u}_k(t) - \hat{u}_k^N(t)| \leq C\left(\frac{|k|}{N}\right)^s \|u_0\|_0 \,. \tag{7.6.41}$$

This proves the spectral convergence of each Fourier coefficient of the Galerkin solution to the corresponding coefficient of the exact solution, even in the case in which $u_0$ (and consequently $u$) is a discontinuous function.

Finally, we are going to use (7.6.40) in order to show that it is possible to use the information contained in $u^N(t)$ to approximate $u(t)$ with spectral accuracy at each point where $u$ is smooth. The idea, already sketched in Mercier (1981), has been developed independently by Gottlieb and coworkers, both theoretically and computationally (see Gottlieb (1985), Gottlieb and Tadmor (1985), Abarbanel, Gottlieb and Tadmor (1986)).

Let us drop the dependence upon time in all the functions which appear hereafter. Assume that at time $t > 0$ the solution $u$ of (7.6.32) is infinitely smooth in an open neighborhood $J$ of a point $x_0 \in [0, 2\pi]$. Let us choose an infinitely differentiable, periodic function $\rho = \rho(x)$ such that $\rho$ is identically zero outside $J$, $\rho$ is nonnegative everywhere, and $\rho(x_0) = 1$. Thus, the function $\rho u$ is everywhere smooth and $(\rho u)(x_0) = u(x_0)$. For each fixed $M > 0$, the maximum error between $\rho u$ and its Fourier truncation $P_M(\rho u)$ can be estimated according to (5.1.12)–(5.1.14):

$$\|\rho u - P_M(\rho u)\|_{L^\infty(0,2\pi)} \leq C(1 + \log M)M^{-s}\|\rho u\|_{s,\infty} \,, \quad (s \geq 0) \,.$$

The norm $\|\rho u\|_{s,\infty}$ is the maximum modulus over $(0, 2\pi)$ of all the derivatives of $\rho u$ of order up to $s$. Such a quantity can be bounded by a constant, depending upon $\rho$, times the maximum modulus over $J$ of the derivatives of $u$ of order up to $s$. This latter quantity is finite by assumption and will be denoted by $\|u\|_{s,\infty,J}$. Thus,

$$|u(x_0) - P_M(\rho u)(x_0)| \leq C(1 + \log M)M^{-s}\|u\|_{s,\infty,J} \,. \tag{7.6.42}$$

On the other hand, we have the following representation of $P_M(\rho u)$ as a convolution integral (see (2.1.55)):

$$P_M(\rho u)(x_0) = \frac{1}{2\pi}\int_0^{2\pi} D_M(x_0 - y)\rho(y)u(y)\mathrm{d}y \,, \tag{7.6.43}$$

where $D_M$ is the Dirichlet kernel, used here with the classical notation (2.1.56) (with $N$ replaced by $2M$). For a fixed $M$, the function $\phi(y) = D_M(x_0 - y)\rho(y)$ is an infinitely smooth, periodic function. Thus, we can apply (7.6.40) and get

$$\left| \int_0^{2\pi} D_M(x_0 - y)\rho(y)u(y)\mathrm{d}y - \int_0^{2\pi} D_M(x_0 - y)\rho(y)u^N(y)\mathrm{d}y \right| \\ \leq CN^{-s}\|u_0\|_0\|\phi\|_s \, . \tag{7.6.44}$$

The norm $\|\phi\|_s$ can be bounded by $C(1+M)^{s+1}\|\rho\|_s$. Finally, we choose $M$ as an increasing function of $N$ satisfying $M(N) < N$, and we denote by

$$Ru^N(x_0) = \frac{1}{2\pi}\int_0^{2\pi} D_M(x_0, y)\rho(y)u^N(y)\mathrm{d}y \, , \tag{7.6.45}$$

the regularized value of $u^N$ at the point $x_0$. Note that this value can only be evaluated exactly once the Fourier coefficients of $\phi$ are known; in practice, in order to evaluate the integral in (7.6.45) one can use a trapezoidal rule with sufficiently many points. If we choose $M = N^\beta$ with $0 < \beta < 1$, then by (7.6.42) and (7.6.44) we obtain the following error estimate:

$$\left|u(x_0) - Ru^N(x_0)\right| \leq C_1\left(1 + \log N\right) N^{-s\beta} + C_2 N^{-s+\beta(1+s)} \, , \tag{7.6.46}$$

where the constants depend upon Sobolev norms of $\rho$ and $u$ over the interval $J$. We conclude that $u(x_0)$ can he approximated with spectral accuracy starting from the knowledge of the Galerkin approximation $u^N$. An asymptotic balance of the errors in (7.6.46) is achieved, up to the logarithmic factor, by choosing $\beta = 1/2$. Tadmor and Tanner (2002) suggest instead the choice $M = \theta N$, where $\theta \in (0,1)$ is proportional to the size of the interval $J$.

A number of generalizations of the previous results are possible. First, one can consider a collocation approximation, in which case the integral in (7.6.45) is replaced by the trapezoidal rule, and only the values of $u$ in physical space are needed. An extra error term due to aliasing is added, but the asymptotic behavior of the error is the same. The extension of the above results to the two-dimensional case is considered by Gelb and Tadmor (2002). Next one can consider Legendre or Chebyshev methods for nonperiodic problems. An integral representation of the truncation operator, similar to (7.6.43), is still available. The Dirichlet kernel has to be replaced by the kernel

$$K_M(\xi) = \sum_{k=0}^{M} \left(k + \tfrac{1}{2}\right) L_k(\xi)L_k(0) \tag{7.6.47}$$

in a Legendre method, and by the kernel

$$K_M(\xi) = \frac{2}{\pi}\sum_{k=0}^{M} T_k(\xi)T_k(0) \tag{7.6.48}$$

**Table 7.1.** Results of smoothing of the spectral approximation of $u(x)$ (from Gottlieb and Tadmor (1985))

| $x_\nu = \frac{\pi\nu}{8}$ | $\lvert u(x_\nu) - u^N(x_\nu)\rvert$ | | $\lvert u(x_\nu) - Ru^N(x_\nu)\rvert$ | |
|---|---|---|---|---|
| $\nu$ equals | $N = 64$ | $N = 128$ | $N = 64$ | $N = 128$ |
| 2 | 6.4(−3) | 3.2(−3) | 4.8(−6) | 5.8(−10) |
| 3 | 1.0(−2) | 5.2(−3) | 5.9(−6) | 7.9(−10) |
| 4 | 1.5(−2) | 7.8(−3) | 7.7(−6) | 6.3(−10) |
| 5 | 2.3(−2) | 1.1(−2) | 12.9(−6) | 1.1(−10) |

in a Chebyshev method. For the details we refer to Gottlieb and Tadmor (1985) and Abarbanel, Gottlieb and Tadmor (1986).

From a computational point of view, one has to choose a proper cut-off function $\rho$ whose support is in the region of smoothness of the solution, and also choose a value for $\beta$. As usual, the method may require a fine-tuning of the parameters for the problem at hand. Gottlieb and Tadmor (1985) consider the piecewise-$C^\infty$ function

$$u(x) = \begin{cases} \sin\dfrac{x}{2}\,, & 0 \le x < \pi\,, \\ -\sin\dfrac{x}{2}\,, & \pi \le x < 2\pi\,, \end{cases} \tag{7.6.49}$$

and use an exponential cut-off function. Denoting by $u^N$ the truncation of the Fourier series, the results listed in Table 7.1 have been reported. The unsmoothed error decays linearly in $N^{-1}$, whereas spectral accuracy is clearly documented for the regularized approximation.

## 7.7 Scalar Conservation Laws

So far, our analysis has been confined to linear hyperbolic problems. Now we consider a scalar nonlinear equation in the form of a conservation law

$$\frac{\partial u}{\partial t} + \frac{\partial \mathcal{F}(u)}{\partial x} = 0\,, \tag{7.7.1}$$

where $\mathcal{F} = \mathcal{F}(u)$ is a flux smoothly depending on the real variable $u$. An example is the inviscid Burgers equation, corresponding to the choice $\mathcal{F}(u) = \frac{1}{2}u^2$. (Recall that the viscous Burgers equation is discussed in Sects. 3.1 and 3.3.) We assume $2\pi$-periodic boundary conditions (which call for Fourier discretization methods), although all the subsequent discussion can be extended to the case of inflow boundary conditions (with Legendre or Chebyshev discretizations). An initial condition $u(0) = u_0$ completes the problem.

The most striking difference between the linear model (7.6.1) and the current nonlinear model is that even if $u_0$ is a smooth periodic function, the solution $u$ may develop jump discontinuities, called shocks, at a finite time. This occurs whenever two characteristic curves, which are defined by the condition

$$\frac{\mathrm{d}x}{\mathrm{d}t} = \mathcal{F}'(u(x,t))$$

and which carry a constant value of $u$, intersect. For the Burgers equation, the characteristics satisfy $\frac{\mathrm{d}x}{\mathrm{d}t} = u(x,t)$, and since $u$ is constant on them, they are straight lines. When a shock appears, the strong form (7.7.1) of the conservation law becomes meaningless, and one has to resort to the weak form

$$\int_0^\infty \int_0^{2\pi} \left( u \frac{\partial \varphi}{\partial t} + \mathcal{F}(u) \frac{\partial \varphi}{\partial x} \right) \mathrm{d}x \, \mathrm{d}t = \int_0^{2\pi} u_0(x) \varphi(0,t) \, \mathrm{d}x \, , \qquad (7.7.2)$$

valid for all smooth functions $\varphi$, $2\pi$-periodic in $x$ and vanishing for $t$ large enough. From this form one can derive the speed $s$ of the propagation of a shock, given by the Rankine-Hugoniot condition

$$s = \frac{[\mathcal{F}(u)]}{[u]} \, . \qquad (7.7.3)$$

The naive application of a spectral method to the discretization of a conservation law brings good news and bad news. The good news is that if the sequence of discrete solutions $u^N$ produced by a Galerkin or a collocation method is bounded and converges almost everywhere, as $N \to \infty$, to a limit $u$, then $u$ is a weak solution of the conservation law, i.e., it satisfies (7.7.2); consequently, any shocks that are present are propagated with the correct speed. This result was proven, even for nonperiodic problems, by Gottlieb, Lustman and Orszag (1981), following an argument due to Lax and Wendroff (1960).

The somewhat bad news is that as soon as the solution $u$ develops steep gradients (and eventually shocks), the spectral solution $u^N$ exhibits a Gibbs phenomenon, i.e., spurious oscillations appear. But the (partially offsetting) good news is that the transition between the pre-shock and the post-shock states always occurs within one mesh interval. Thus, a very accurate shock position is inherent in a spectral solution. Furthermore, the spurious oscillations are not in themselves insurmountable, for according to a result of Lax (1978) and as discussed in Sect. 7.6.3, they contain sufficient information to permit the reconstruction of an exact solution within spectral accuracy.

The really bad news is that a naively generated spectral solution fails to fulfill the *entropy condition*, which is appended to the conservation law in order to select the physically relevant solution (the so called *entropy solution*). Such a failure (which however is not peculiar to spectral methods), can be easily seen by considering the Fourier Galerkin discretization of (7.7.1):

$$u^N \in S_N \ : \quad (u_t^N + (\mathcal{F}(u^N))_x, v) = 0 \qquad \text{for all } v \in S_N \ , \ t > 0 \ , \quad (7.7.4)$$

where $(u,v) = \int_0^{2\pi} u(x)\overline{v(x)}\,\mathrm{d}x$ is the $L^2$-inner product. Choosing $v = u^N$ and assuming all variables to be real, we get

$$\frac{1}{2}\frac{\mathrm{d}}{\mathrm{d}t}\int_0^{2\pi} |u^N|^2\,\mathrm{d}x + \int_0^{2\pi} (\mathcal{G}(u^N))_x\,\mathrm{d}x = 0 \ ,$$

where $\mathcal{G} = \mathcal{G}(u)$ satisfies $\mathcal{G}'(u) = u\mathcal{F}'(u)$ for all real $u$. Applying the periodic boundary conditions, the second integral vanishes. Thus, the *energy functional* $E(u^N)$, defined as

$$(E(u^N))(t) = \frac{1}{2}\int_0^{2\pi} |u^N(x,t)|^2\,\mathrm{d}x \ ,$$

is conserved for all times. This contrasts with the behavior of the entropy solution because, assuming $\mathcal{F}$ convex, the entropy condition

$$\frac{\partial}{\partial t}\left(\frac{1}{2}|u|^2\right) + \frac{\partial \mathcal{G}(u)}{\partial x} \leq 0$$

forces the entropy function $\frac{1}{2}|u|^2$, and consequently the exact energy $E(u)$, to decay as soon as a shock is developed. The above discussion shows that a dissipative mechanism should be inserted into the spectral scheme in order to approximate the entropy solution; simply applying a post-processing at the end of the computation will not suffice.

Viscosity is the paramount dissipative mechanism. Indeed, a result by Kružkov (1970) states that any entropy solution is the limit of a sequence of viscous solutions (solutions of the conservation laws augmented by a viscous term) as the viscosity parameter tends to zero. Adding numerical (or artificial) viscosity is a classical device for stabilizing low-order numerical schemes, as well as to guarantee in the limit the fulfillment of the entropy condition. However, in spectral methods such a device must be applied very delicately, for the injected dissipation should not destroy the potential spectral accuracy for smooth solutions. The *spectral viscosity method* was introduced by Tadmor (1989) to provide a satisfactory answer to this issue: it damps the higher order modes of the discrete solution while leaving the lower order modes unchanged. A comparable effect, in practice, can be achieved by properly filtering the discrete solution during the time evolution (say, at regular time intervals) using a spectral filter which leaves the lower portion of the spectrum unaffected. However, the spectral viscosity approach offers the advantage of allowing a rigorous stability and convergence analysis.

The Fourier Galerkin version of the spectral viscosity method is as follows:

$$\left(u_t^N + \left(\mathcal{F}(u^N)\right)_x, v\right) + \varepsilon_{N,s}\left(Q_M D_x^s u^N, D_x^s v\right) = 0 \qquad \text{for all } v \in S_N \ , \ t > 0 \ . \quad (7.7.5)$$

Here, $s \geq 1$ defines the order, $2s$, of the added (super-)viscosity, whereas $\varepsilon_{N,s} > 0$ is the artificial (super-)viscosity coefficient, which scales as $\varepsilon_{N,s} \sim \frac{C_s}{N^{2s-1}}$, for a suitable positive constant $C_s$. Classical second-order viscosity corresponds to the choice $s = 1$; superviscosity effects (see below) are obtained for $s > 1$. The low-pass filter operator $Q_M v$ is defined as

$$Q_M v(x) = \sum_{M<|k|\leq N} q_k \hat{v}_k e^{ikx} ,$$

where the cut-off parameter $M < N$ is linked to $N$ by the relations

$$M \sim N^{\vartheta} \qquad \text{for } \vartheta < \frac{2s-1}{2s} , \tag{7.7.6}$$

whereas the smoothing factors satisfy

$$1 - \left(\frac{M}{|k|}\right)^{\frac{2s-1}{\vartheta}} \leq q_k \leq 1 , \qquad |k| > N . \tag{7.7.7}$$

Note that the extra term appearing in (7.7.5), with respect to the standard Fourier Galerkin formulation (7.7.4), only depends on $u^N - P_M u^N$. This, together with relation (7.7.6), guarantees that spectral accuracy is preserved for smooth solutions.

The choice of the superviscosity dissipation, as opposed to the standard viscosity, is motivated by the aim of concentrating the viscosity effects on the higher modes, since in the former case $M$ can be chosen closer to $N$ (see (7.7.6)). This leads to sharper profiles near shocks. It is even possible to choose $s$ depending on $N$, as $s \sim N^{\mu}$, $\mu < \frac{1}{2}$ (see Tadmor (1998)).

The analysis, initiated by Tadmor (1989) and Maday and Tadmor (1989) (see also Tadmor (1998) for a review of the subject), establishes the entropy dissipation bound

$$\|u^N(t)\|^2_{L^2(0,2\pi)} + \varepsilon_{N,s} \int_0^t \|D_x^s u^N(\tau)\|^2_{L^2(0,2\pi)} \, d\tau \leq C \|u_0\|^2_{L^2(0,2\pi)} \, . \tag{7.7.8}$$

Using a compensated compactness argument due to Murat (1978), this estimate allows one to prove that if the sequence $u^N$ is uniformly bounded in the $L^\infty$-norm (this property can be rigorously proven for $s = 1$), then it converges to the unique entropy solution of the conservation law in the $L^p$-norm on any bounded set in the $(x,t)$-plane for any $p < +\infty$.

Let us stress the practical implication of this convergence result. If the solution $u$ is not smooth, then the convergence of $u^N$ to $u$ is slow. In this case, however, the formal spectral accuracy assured by the spectral viscosity method manifests itself in the fact that the convergence of $u^N$ to the truncation $P_N u$ is fast. Consequently, a post-processing stage, such as the one described in Sect. 7.6.3, can be successfully applied to reconstructing an accurate approximation of $u$ from the knowledge of $u^N$. An example is given by Gelb and Tadmor (2000b).

The spectral viscosity method has been extended to the nonperiodic case by Maday, Ould Kaber and Tadmor (1993) using Legendre expansions; see also Tadmor (1998). The low-pass filter $Q_M$ which appears in the artificial viscosity term now becomes

$$Q_M v(x) = \sum_{M<k\leq N} q_k \hat{v}_k L_k(x) ,$$

where the smoothing factors satisfy

$$1 - \left(\frac{M}{k}\right)^4 \leq q_k \leq 1 , \qquad k > M .$$

Note the power 4 as opposed to the power close to 2 appearing in the equivalent Fourier condition (7.7.7) for $s = 1$; the difference can be understood by recalling the inverse inequalities in Sobolev norms for Legendre and trigonometric expansions; see (5.4.5) and (5.1.5). Furthermore, since $Q_M$ operates in transform space, the enforcement of the boundary condition(s) is more conveniently accomplished via a weak, or penalty, formulation, as described in Sect. 3.7.1. Several forms of the dissipative term are possible; see Guo, Ma and Tadmor (2001).

The spectral viscosity method in the multidimensional case has been introduced by Chen, Du and Tadmor (1993). Further developments and numerical results are discussed in Tadmor (1998).

The spectral viscosity method is not the only technique of spectral type developed for conservation laws. The *spectral cell-averaging method* (Cai, Gottlieb and Shu (1989), Cai, Gottlieb and Harten (1990)) is based on the cell-averaged, or finite-volume, formulation of the conservation law. A reconstruction stage produces the fluxes at the cell interfaces from the spectrally accurate cell averages. This stage is accomplished via a global interpolating polynomial, using reconstruction techniques similar to those described in Sect. 7.6.3. A short account of the method can be found, e.g., in Bernardi and Maday (1997), Sect. 30. (This method uses a staggered grid; for some discussions of a staggered grid in the context of spectral methods see Sects. 3.7.1 and 4.4.2 and CHQZ3, Chap. 4.)

## 7.8 The Steady Burgers Equation

We consider here the nonlinear problem

$$\begin{aligned} -\nu u_{xx} + u u_x &= f , \qquad -1 < x < 1 , \\ u(-1) = u(1) &= 0 . \end{aligned} \tag{7.8.1}$$

We intend to show that Chebyshev (Galerkin and collocation) approximations to this problem are stable and convergent for all positive values of $\nu$. This is the simplest example of the rigorous results that can be obtained for nonlinear problems. We choose to outline the analysis in the general framework that has been used for more difficult nonlinear problems such as the Navier-Stokes equations. Legendre approximations, including those produced by the G-NI approach, can be analysed along the same lines, with the benefit of avoiding the technical difficulties related to the Chebyshev weight. For this reason, we focus on Chebyshev approximations.

We assume that $f \in L^2_w(-1,1)$, where $w$ is, as usual, the Chebyshev weight. Let $a(u,v)$ denote again the bilinear form (7.1.13) defined on the product space $H^1_{w,0}(-1,1) \times H^1_{w,0}(-1,1)$ and associated with the second-derivative operator with Dirichlet boundary conditions. Moreover, let us set

$$\lambda = \nu^{-1}, \quad \mathcal{G}(\lambda, u) = \lambda(uu_x - f) . \tag{7.8.2}$$

Each $u \in H^1_{w,0}(-1,1)$ is bounded in $[-1,1]$ (see (A.11.a)); hence, $\mathcal{G}(\lambda,u) \in L^2_w(-1,1)$. Thus, we can consider the following weak formulation of problem (7.8.1):

$$\begin{aligned} &u \in H^1_{w,0}(-1,1) , \\ &a(u,v) + (\mathcal{G}(\lambda,u),v)_w = 0 \quad \text{for all } v \in H^1_{w,0}(-1,1) . \end{aligned} \tag{7.8.3}$$

Here $(z,v)_w$ denotes the inner product in $L^2_w(-1,1)$. For each positive $\lambda$ and each $u \in H^1_{w,0}(-1,1)$, the linear form $v \to (\mathcal{G}(\lambda,u),v)_w$ is continuous on $H^1_{w,0}(-1,1)$. Hence, $\mathcal{G}(\lambda,u)$ can be regarded as an element of the dual space $H^{-1}_w(-1,1)$ of $H^1_{w,0}(-1,1)$ (see (A.1.c)), so that $(\mathcal{G}(\lambda,u),v)_w = \langle \mathcal{G}(\lambda,u), v\rangle$ for all $v \in H^1_{w,0}(-1,1)$. (The symbol $\langle \cdot,\cdot \rangle$ denotes the duality pairing between $H^{-1}_w(-1,1)$ and $H^1_{w,0}(-1,1)$.)

Let $\mathcal{T} : H^{-1}_w(-1,1) \to H^1_{w,0}(-1,1)$ be the linear operator which associates to an element $g \in H^{-1}_w(-1,1)$ the solution $\mathcal{T}g \in H^1_{w,0}(-1,1)$ of the problem

$$a(\mathcal{T}g, v) = \langle g, v\rangle \quad \text{for all } v \in H^1_{w,0}(-1,1) . \tag{7.8.4}$$

This problem has indeed a unique solution since the bilinear form $a(u,v)$ satisfies the assumptions of the Lax-Milgram Theorem (A.5), as shown in Sect. 7.1. It follows that problem (7.8.3) can be written equivalently in the form

$$\begin{aligned} &u \in H^1_{w,0}(-1,1) , \\ &\mathcal{E}(\lambda,u) \equiv u + \mathcal{T}\mathcal{G}(\lambda,u) = 0 . \end{aligned} \tag{7.8.5}$$

Many nonlinear problems depending upon a parameter can be formulated in a manner similar to (7.8.5). A remarkable instance is provided by the Navier-Stokes equations for steady viscous incompressible flows (see CHQZ3, Sect. 3.1), in which case $\lambda$ is the inverse of the kinematic viscosity $\nu$.

In general, the linear operator $\mathcal{T}$ acts between the dual space $V'$ of a Banach space $V$ (see (A.1)) and the Banach space $V$ itself, i.e.,

$$\mathcal{T} : V' \longrightarrow V \,. \tag{7.8.6}$$

It represents the inverse of the linear part of the differential problem (for instance, the inverse of the Stokes operator in the steady incompressible Navier-Stokes equations). The operator $\mathcal{G}$ maps $\mathbb{R} \times V$ into the dual space $V'$ in a continuously differentiable way:

$$\mathcal{G} : \mathbb{R} \times V \longrightarrow V' \,, \tag{7.8.7}$$

and represents the nonlinear part of the problem. The full problem can be written as a nonlinear equation in $V$, in the form

$$\begin{aligned} &u(\lambda) \in V \,, \\ &\mathcal{E}(\lambda, u(\lambda)) \equiv u(\lambda) + \mathcal{T}\mathcal{G}(\lambda, u(\lambda)) = 0 \,. \end{aligned} \tag{7.8.8}$$

Here we have stressed the dependence of the solution upon the parameter $\lambda$, which is usually restricted to vary in a closed, bounded interval $\Lambda$ of the real line.

Let us make the technical assumption that there exists a Banach space $W \subset V'$ such that

$$\mathcal{G}(\cdot,\cdot) \textit{ is a continuous mapping from } \mathbb{R}^+ \times V \textit{ into } W \,, \tag{7.8.9}$$

and

$$\mathcal{T} \textit{ is a compact operator } \text{(see (A.3))} \textit{ from } W \textit{ into } V \,. \tag{7.8.10}$$

For the Burgers problem (7.8.1), these hypotheses are fulfilled, for instance, with the choice $W = L^2_w(-1,1)$. In fact, if $g \in L^2_w(-1,1)$, the solution $\xi = \mathcal{T}g$ of (7.8.4) (i.e., the solution of the boundary-value problem $-\xi_{xx} = g$ in $-1 < x < 1$, with $\xi(-1) = \xi(1) = 0$) belongs to $H^2_w(-1,1)$, which is compactly imbedded in $H^1_w(-1,1)$.

We shall confine our analysis to the case of a *nonsingular branch* of solutions $\{(\lambda, u(\lambda)) : \lambda \in \Lambda\}$ of (7.8.8), i.e., to a branch of solutions along which the Fréchet derivative (see (A.4)) $D_u\mathcal{E}(\lambda, u)$ of the map $\mathcal{E}$ with respect to the variable $u$ is invertible. More precisely, we assume that there exists a positive constant $\alpha > 0$ such that

$$\|v + \mathcal{T}D_u\mathcal{G}(\lambda, u(\lambda))v\|_V \geq \alpha \|v\|_V \quad \text{for all } v \in V \text{ and all } \lambda \in \Lambda \,. \tag{7.8.11}$$

Here the symbol $D_u\mathcal{G}(\lambda_0, u_0)$ denotes the Fréchet derivative of $\mathcal{G}(\lambda, u)$ with respect to the variable $u$, computed at the point $(\lambda_0, u_0)$. For problem (7.8.1), condition (7.8.11) amounts to the requirement that for all $g \in H^{-1}_w(-1,1)$ and all $\lambda \in \Lambda$, the problem

$$\begin{aligned} & v \in H^1_{w,0}(-1,1) \,, \\ & -v_{xx} + \lambda(u(\lambda)v_x + u_x(\lambda)v) = g \end{aligned} \tag{7.8.12}$$

has a unique solution, which satisfies the inequality

$$\|v\|_{H^1_{w,0}(-1,1)} \leq C\|g\|_{H^{-1}_w(-1,1)} \,.$$

We are going now to introduce a general approximation to any problem which can be written in the form (7.8.8), provided the assumptions (7.8.9)–(7.8.11) are satisfied. Further, we shall state a general theorem to be used for the analysis of stability and convergence of such approximations. As a particular case, this theorem will be used to infer stability and convergence of both Galerkin and collocation Chebyshev approximations to the Burgers problem (7.8.1), which was previously written in the form (7.8.5).

For any integer $N$, let $V_N$ be a finite-dimensional subspace of $V$, and let $\mathcal{G}_N : \mathbb{R}^+ \times V_N \to V'$ be a suitable approximation to $\mathcal{G}$. Further, let $\mathcal{T}_N : V' \to V_N$ be a linear operator which approximates $\mathcal{T}$. The following is a finite-dimensional approximation to problem (7.8.8):

$$\begin{aligned} & u^N(\lambda) \in V_N \,, \\ & \mathcal{E}_N(\lambda, u^N(\lambda)) \equiv u^N(\lambda) + \mathcal{T}_N \mathcal{G}_N(\lambda, u^N(\lambda)) = 0 \,. \end{aligned} \tag{7.8.13}$$

The next theorem, due to Maday and Quarteroni (1982), is concerned with the convergence of the discrete solutions $\{(\lambda, u^N(\lambda)), \lambda \in \Lambda\}$ (problem (7.8.13)) to the nonsingular branch of the exact solutions $\{(\lambda, u(\lambda)), \lambda \in \Lambda\}$ (problem (7.8.8)).

**Theorem 7.3.** *Assume that (7.8.9)–(7.8.11) hold. Moreover, assume that for some integer $m \geq 2$, $\mathcal{G} : \Lambda \times V \to W$ is a $C^m$ mapping, and $D^m\mathcal{G}$ is bounded over any bounded subset of $\Lambda \times V$. Concerning the discrete problem, we assume that*

$$\lim_{N\to\infty} \|\mathcal{T} - \mathcal{T}_N\|_{\mathcal{L}(W,V)} = 0 \,. \tag{7.8.14}$$

(See (A.3) for the definition of the norm of a linear operator.) *About $\mathcal{G}_N$, we assume that it is a $C^m$ mapping from $\Lambda \times V_N \to V'$, and that there exists a positive function $K : \mathbb{R}^+ \to \mathbb{R}^+$ such that*

$$\|D^l\mathcal{G}_N(\lambda, v)\|_{\mathcal{L}_l(\Lambda\times V_N, W)} \leq K(|\lambda| + \|v\|_V) \,, \quad l = 1, \ldots, m \,. \tag{7.8.15}$$

(See again (A.3) for the definition of the norm of a multilinear operator.) *Further, we assume that there exists a projection operator $\Pi_N : V \to V_N$ satisfying*

$$\lim_{N\to\infty} \|v - \Pi_N v\|_V = 0 \quad \textit{for all } v \in V \,, \tag{7.8.16}$$

*and such that*

$$\lim_{N\to\infty} \sup_{\lambda\in\Lambda} \|D_u(\mathcal{G} - \mathcal{G}_N)(\lambda, \Pi_N u(\lambda))\|_{\mathcal{L}(V_N, V')} = 0 \,. \tag{7.8.17}$$

*Then there exists a neighborhood $\Theta$ of the origin in $V$, and for $N$ large enough, a unique $C^m$ mapping $\lambda \in \Lambda \to u^N(\lambda) \in V_N$ such that for all $\lambda \in \Lambda$*

$$\mathcal{E}_N\left(\lambda, u^N(\lambda)\right) = 0\,, \qquad u^N(\lambda) - u(\lambda) \in \Theta\,, \tag{7.8.18}$$

*and the following estimate holds:*

$$\begin{aligned}\|u(\lambda) - u^N(\lambda)\|_V \leq & C(\|u(\lambda) - \Pi_N u(\lambda)\|_V + \|(\mathcal{T} - \mathcal{T}_N)\mathcal{G}(\lambda, u(\lambda))\|_V \\ & + \|\mathcal{T}_N(\mathcal{G} - \mathcal{G}_N)(\lambda, \Pi_N u(\lambda))\|_V)\,,\end{aligned} \tag{7.8.19}$$

*with a positive constant $C$ independent of $\lambda$ and $N$.*

A qualitative interpretation of this theorem is in order. There are several assumptions on the approximations to the linear and nonlinear components of the problem. Assumption (7.8.16) means that $V$ is well-approximated by the sequence of subspaces $V_N$, and (7.8.14) means that the linear operator $\mathcal{T}$ is well-approximated by the sequence of operators $\mathcal{T}_N$. Naturally enough, stricter requirements are placed on the approximation to the nonlinear operator $\mathcal{G}$. Assumption (7.8.15) means that the derivatives of $\mathcal{G}_N$ up to order $m$ are locally Lipschitz continuous, and (7.8.17) states that the Fréchet derivative of $\mathcal{G}_N$ approximates that of $\mathcal{G}$ as $N \to \infty$.

The first conclusion, (7.8.18), is that, for fixed $N$, there is a unique branch of nonsingular solutions and that these solutions are bounded uniformly with respect to $N$. Finally, inequality (7.8.19) exhibits the dependence of the error on the approximation properties of $\Pi_N$, $\mathcal{T}_N$ and $\mathcal{G}_N$.

### *Chebyshev Galerkin Approximation*

We return now to problem (7.8.1), and its equivalent formulation (7.8.5), with $\mathcal{G}$ and $\mathcal{T}$ defined in (7.8.2), (7.8.4). For any $\lambda \in \mathbb{R}^+$, we look for a polynomial $u^N(\lambda) \in V_N = \{v \in \mathbb{P}_N(-1,1) | v(\pm 1) = 0\}$ which satisfies

$$a\left(u^N(\lambda), v\right) + \left(\mathcal{G}\left(\lambda, u^N(\lambda)\right), v\right)_w = 0 \quad \text{for all } v \in V_N\,. \tag{7.8.20}$$

This is a Chebyshev Galerkin approximation. We define the operator $\mathcal{T}_N : V' \to V_N$ by

$$a(\mathcal{T}_N g, v) = \langle g, v\rangle \quad \text{for all } v \in V_N\,. \tag{7.8.21}$$

Then it follows that $\mathcal{T}_N = \Pi_N \mathcal{T}$, where $\Pi_N : V \to V_N$ is the operator defined in (5.5.21), namely:

$$a(\Pi_N u - u, v) = 0 \quad \text{for all } v \in V_N\,. \tag{7.8.22}$$

Owing to (7.8.21), the Chebyshev Galerkin approximation to (7.8.1) can be restated as follows:

$$\begin{aligned}& u^N(\lambda) \in V_N\,, \\ & \mathcal{E}_N\left(\lambda, u^N(\lambda)\right) \equiv u^N(\lambda) + \mathcal{T}_N \mathcal{G}\left(\lambda, u^N(\lambda)\right) = 0\,.\end{aligned} \tag{7.8.23}$$

This is precisely the form (7.8.13); in the current situation, however, $\mathcal{G}_N \equiv \mathcal{G}$. To apply Theorem 7.3, we need to check that the assumptions (7.8.14)–(7.8.17) are fulfilled.

Property (7.8.16) follows from the fact that each function $v \in H^1_{w,0}(-1,1)$ can be approximated in the norm of $H^1_w(-1,1)$ by a sequence of more regular functions $\tilde{v}_n \in H^m_w(-1,1) \cap H^1_{w,0}(-1,1)$, with $m > 1$. Then one applies to each such $\tilde{v}_n$ the convergence estimate (7.1.21) for the Chebyshev Galerkin approximation (where $\tilde{v}_n^N$ is indeed $\Pi_N \tilde{v}_n$). In order to check (7.8.14), let us choose $W = L^2_w(-1,1)$. Recalling that $\mathcal{T}_N = \Pi_N \mathcal{T}$ we have

$$\|\mathcal{T} - \mathcal{T}_N\|_{\mathcal{L}(W,V)} = \sup_{g \in L^2_w(-1,1)} \frac{\|\mathcal{T}g - \Pi_N \mathcal{T}g\|_{H^1_w(-1,1)}}{\|g\|_{L^2_w(-1,1)}} .$$

Using again (7.1.21) and the definition of the operator $\mathcal{T}$, we have

$$\|\mathcal{T}g - \Pi_N \mathcal{T}g\|_{H^1_w(-1,1)} \le CN^{-1} \|\mathcal{T}g\|_{H^2_w(-1,1)} \le C' N^{-1} \|g\|_{L^2_w(-1,1)} ;$$

whence (7.8.14) follows. Moreover, both (7.8.15) and (7.8.17) are trivially verified for all integers $m \ge 0$.

By (7.8.18) and (7.8.19) we conclude that for any branch $\{(\lambda, u(\lambda)), \lambda \in \Lambda\}$, $\Lambda \subset \mathbb{R}^+$, of nonsingular solutions of(7.8.1), there exists a $C^\infty$ mapping: $\lambda \in \Lambda \to u^N(\lambda) \in V_N$, such that $u^N(\lambda)$ is the only solution of the Chebyshev Galerkin approximation (7.8.20) in a neighborhood of $u(\lambda)$. Moreover, one has the estimate

$$\begin{aligned} \|u(\lambda) - u^N(\lambda)\|_{H^1_w(-1,1)} \le C \|u(\lambda) - \Pi_N u(\lambda)\|_{H^1_w(-1,1)} \\ + \|\mathcal{T}\mathcal{G}(\lambda, u(\lambda)) - \Pi_N \mathcal{T}\mathcal{G}(\lambda, u(\lambda))\|_{H^1_w(-1,1)} . \end{aligned}$$

Noting that from (7.8.5), $\mathcal{T}\mathcal{G}(\lambda, u(\lambda)) = -u(\lambda)$, and using again (7.1.21) we get the convergence estimate

$$\|u(\lambda) - u^N(\lambda)\|_{H^1_w(-1,1)} \le CN^{1-m} |u(\lambda)|_{H^m_w(-1,1)}, \quad m \ge 1 , \qquad (7.8.24)$$

for a constant $C$ which depends only upon the parameter interval $\Lambda$.

### *Chebyshev Collocation Approximation*

Let $x_j = \cos(\pi j/N)$, $j = 0, \dots, N$ , be the Chebyshev Gauss-Lobatto points (see (2.4.14)), and let $I_N v$ be the interpolant of $v$ at these points (see Sect. 2.2.3). We look now for a polynomial $u^N = u^N(\lambda)$ of degree $N$ which satisfies

$$\begin{aligned} &-u^N_{xx} + \lambda \left( \tfrac{1}{2} \left( I_N \left(u^N\right)^2 \right)_x - f \right) \quad \text{at } x = x_j \ , \ 1 \le j \le N-1 \\ &u^N(x_0) = u^N(x_N) = 0 \ . \end{aligned} \qquad (7.8.25)$$

Introducing the discrete inner product $(u, v)_N$ associated with the Chebyshev points $\{x_j\}$ (see (2.2.24)), we can restate this collocation problem as follows:

$$\begin{aligned} &u^N \in V_N \ , \\ &a\left(u^N, v\right) + \lambda \left(\tfrac{1}{2}\left(I_N(u^N)^2\right)_x - f, v\right)_N = 0 \quad \text{for all } v \in V_N \ . \end{aligned} \tag{7.8.26}$$

We have used (2.2.25) to replace $-(u^N_{xx}, v)_N$ by $a(u^N, v)$. We define the operator $\mathcal{G}_N : \mathbb{R}^+ \times V_N \to V'$ by setting

$$\langle \mathcal{G}_N(\lambda, v), \phi \rangle = \lambda \left(\tfrac{1}{2}\left(I_N(v^2)\right)_x - f, \phi\right)_N \quad \text{for all } \phi \in V \ .$$

Note that again by (2.2.25) we have

$$\langle \mathcal{G}_N(\lambda, v), \phi \rangle = \lambda \left\{ \left(\tfrac{1}{2}\left(I_N(u^N)^2\right)_x, \phi\right)_w - (f, \phi)_N \right\} \quad \text{for all } \phi \in V_N \ .$$

If we define $\mathcal{T}_N : V' \to V_N$ as in (7.8.21), then problem (7.8.25) fits into the general form (7.8.13).

The assumptions of Theorem 7.3 can be checked by very technical arguments, which will not be reported here. The interested reader can refer to the paper by Maday and Quarteroni (1982). The conclusion of the analysis is that there exists a $C^\infty$ mapping $\lambda \in \Lambda \to u^N(\lambda) \in V_N$ such that $u^N(\lambda)$ is the only solution of the Chebyshev collocation approximation (7.8.25) in a neighborhood of $u(\lambda)$, and such that the error estimate (7.8.19) holds.

Let us briefly work out this estimate in our particular case. The first two terms on the right-hand side can be handled as they were for the Chebyshev Galerkin approximation; we concentrate on the last term. For the sake of (notational) simplicity, we drop the dependence of $u$ on $\lambda$. Moreover, let us set $\phi = \Pi_N u$ and $\psi = \mathcal{T}_N(\mathcal{G} - \mathcal{G}_N)(\lambda, \Pi_N u)$. By (7.8.21) and (7.1.16) we have

$$\begin{aligned} \gamma \|\psi\|^2_{H^1_w(-1,1)} &\le a(\psi, \psi) = \langle (\mathcal{G} - \mathcal{G}_N)(\lambda, \phi), \psi \rangle \\ &= \tfrac{\lambda}{2} \left( \left[\phi^2 - I_N(\phi^2)\right]_x, \psi \right)_w + \lambda \left[ (f, \psi) - (f, \psi)_N \right] \ . \end{aligned} \tag{7.8.27}$$

Integrating by parts and using the Cauchy-Schwarz inequality together with inequality (7.1.14) yields

$$\begin{aligned} |([\phi^2 - I_N(\phi^2)]_x, \psi)_w| &= \left| \int_{-1}^{1} [\phi^2 - I_N(\phi^2)](\psi w)_x \mathrm{d}x \right| \\ &\le C \|\phi^2 - I_N(\phi^2)\|_{L^2_w(-1,1)} \|\psi\|_{H^1_w(-1,1)} \ . \end{aligned}$$

Now, by the triangle inequality,

$$\begin{aligned} \|(I - I_N)(\phi^2)\|_{L^2_w(-1,1)} &\le \|(I - I_N)(u^2)\|_{L^2_w(-1,1)} \\ &\quad + \|(I - I_N)(u^2 - \phi^2)\|_{L^2_w(-1,1)} \ . \end{aligned}$$

Assuming that $f \in H^{m-1}_w(-1,1)$ for some $m \ge 2$, it is easily seen using equation (7.8.1) that $u \in H^m_w(-1,1)$. Thus, by (5.5.23) we have

$$\|(I - I_N)(u^2)\|_{L^2_w(-1,1)} \le C N^{-m} |u|^2_{H^{m;N}_w(-1,1)} \ ,$$

while again by (5.5.23) and the estimate (7.1.21) for $u^N = \Pi_N u = \phi$ it follows that

$$\begin{aligned}\|(I - I_N)(u^2 - \phi^2)\|_{L^2_w(-1,1)} &\leq C_1 N^{-1} \|u^2 - \phi^2\|_{H^1_w(-1,1)} \\ &\leq C_1 N^{-1} \|u + \Pi_N u\|_{H^1_w(-1,1)} \|u - \Pi_N u\|_{H^1_w(-1,1)} \\ &\leq C_2 N^{-m} \|u\|_{H^1_w(-1,1)} |u|_{H^{m;N}_w(-1,1)} \,.\end{aligned}$$

Finally, the error on the forcing term in (7.8.27) can be handled as shown in Sect. 5.3 (see formula (5.3.4b)), to give

$$|(f,\psi)_w - (f,\psi)_N| \leq C N^{1-m} |f|_{H^{m-1;N}_w(-1,1)} \|\psi\|_{L^2_w(-1,1)} \,.$$

The final result of the convergence analysis, here just summarized, is the following error estimate for the Chebyshev collocation approximation (7.8.25):

$$\|u(\lambda) - u^N(\lambda)\|_{H^1_w(-1,1)} \leq C N^{1-m} \left\{ |u(\lambda)|^2_{H^{m;N}_w(-1,1)} + |f|_{H^{m-1;N}_w(-1,1)} \right\} , \tag{7.8.28}$$

for a constant $C$ which depends only upon the parameter interval $\Lambda$.

# Appendix A. Basic Mathematical Concepts

## A.1 Hilbert and Banach Spaces

### (a) Hilbert Spaces

Let $X$ be a real vector space. An *inner product* on $X$ is a function $X \times X \to \mathbb{R}$, denoted by $(u, v)$, that satisfies the following properties:

(i) $(u, v) = (v, u)$ for all $u, v \in X$;
(ii) $(\alpha u + \beta v, w) = \alpha(u, w) + \beta(v, w)$ for all $\alpha, \beta \in \mathbb{R}$ and all $u, v, w \in X$;
(iii) $(u, u) \geq 0$ for all $u \in X$;
(iv) $(u, u) = 0$ implies $u = 0$.

Two elements $u, v \in X$ are said to be *orthogonal* in $X$ if $(u, v) = 0$. The inner product $(u, v)$ defines a *norm* on $X$ by the relation

$$\|u\| = (u, u)^{1/2} \quad \text{for all } u \in X .$$

The *distance* between-two elements $u, v \in X$ is the positive number $\|u - v\|$. A *Cauchy sequence* in $X$ is a sequence $\{u_k \mid k = 0, 1, \dots\}$ of elements of $X$ that satisfies the following property:

> for each positive number $\varepsilon > 0$, there exists an integer $N = N(\varepsilon) > 0$ such that the distance $\|u_k - u_m\|$ between any two elements of the sequence is smaller than $\varepsilon$ provided both $k$ and $m$ are larger than $N(\varepsilon)$.

A sequence in $X$ is said to *converge* to an element $u \in X$ if the distance $\|u_k - u\|$ tends to 0 as $k$ tends to $\infty$.

A *Hilbert space* is a vector space equipped with an inner product for which all the Cauchy sequences are convergent.

***Examples***

(i) $\mathbb{R}^n$ endowed with the Euclidean product

$$(\mathbf{u}, \mathbf{v}) = \sum_{i=1}^{n} u_i v_i$$

is a finite-dimensional Hilbert space.

(ii) If $[a,b] \subset \mathbb{R}$ is an interval, the space $L^2(a,b)$ (see (A.9.f)) is an infinite-dimensional Hilbert space for the inner product

$$(u,v) = \int_a^b u(x)v(x)\,\mathrm{d}x \,.$$

If $X$ is a complex vector space, the inner product on $X$ will be a complex-valued function. Then condition (i) has to be replaced by

(i′) $$(u,v) = \overline{(v,u)} \quad \text{for all } u,v \in X \,.$$

## (b) Banach Spaces

The concept of Banach space extends that of Hilbert space. Given a vector space $X$, a *norm* on $X$ is a function $X \to \mathbb{R}$, denoted by $\|u\|$, that satisfies the following properties:

$$\begin{aligned}
\|u+v\| &\le \|u\| + \|v\| && \text{for all } u,v \in X \,; \\
\|\lambda u\| &= |\lambda|\|u\| && \text{for all } u \in X \,, \text{ and all } \lambda \in \mathbb{R} \,; \\
\|u\| &\ge 0 && \text{for all } u \in X \,; \\
\|u\| &= 0 && \text{if and only if } u = 0 \,.
\end{aligned}$$

A *Banach space* is a vector space equipped with a norm for which all the Cauchy sequences are convergent.

***Examples***

(i) $\mathbb{R}^n$ endowed with the norm

$$\|\mathbf{u}\| = \left( \sum_{i=1}^n |u_i|^p \right)^{1/p}$$

(with $1 \le p < +\infty$) is a finite-dimensional Banach space.

(ii) If $[a,b] \subset \mathbb{R}$ is an interval and $1 \le p < +\infty$, the space $L^p(a,b)$ (see (A.9.f)) is an infinite-dimensional Banach space for the norm

$$\|u\| = \left( \int_a^b |u(x)|^p \mathrm{d}x \right)^{1/p} .$$

## (c) Dual Spaces

Let $X$ be a Hilbert or a Banach space. A linear form $F : X \to \mathbb{R}$ is said to be *continuous* if there exists a constant $C > 0$ such that

$$|F(u)| \le C\|u\| \quad \text{for all } u \in X \,.$$

The set of all the linear continuous forms on $X$ is a vector space. We can define a norm on this space by setting

$$\|F\| = \sup_{\substack{u \in X \\ u \neq 0}} \frac{F(u)}{\|u\|} .$$

The vector space of all the linear continuous forms on $X$ is called the *dual space* of $X$ and is denoted by $X'$. Endowed with the previous norm, it is itself a Banach space.

The bilinear form from $X' \times X$ into $\mathbb{R}$ defined by

$$\langle F, u \rangle = F(u)$$

is called the *duality pairing* between $X$ and $X'$.

### (d) The Riesz Representation Theorem

If $X$ is a Hilbert space, the dual space $X'$ can be canonically identified with $X$ (hence, it is a Hilbert space). In fact, the Riesz representation theorem states that for each linear continuous form $F$ on $X$, there exists a unique element $u \in X$ such that

$$\langle F, v \rangle = (u, v) \quad \text{for all } v \in X .$$

Moreover, $\|F\|_{X'} = \|u\|_X$.

## A.2 The Cauchy-Schwarz Inequality

Let $X$ be a Hilbert space, endowed with the inner product $(u, v)$ and the associated norm $\|u\|$ (see (A.1.a)). The Cauchy-Schwarz inequality states that

$$|(u, v)| \leq \|u\| \, \|v\| \quad \text{for all } u, v \in X .$$

Of particular importance in the analysis of numerical methods for partial differential equations is the Cauchy-Schwarz inequality in the weighted Lebesgue spaces $L^2_w(\Omega)$, where $\Omega$ is a domain in $\mathbb{R}^n$ and $w = w(\mathbf{x})$ is a weight function (see (A.9.h)). The previous inequality becomes:

$$\left| \int_\Omega u(\mathbf{x}) v(\mathbf{x}) w(\mathbf{x}) \, \mathrm{d}\mathbf{x} \right| \leq \left( \int_\Omega u^2(\mathbf{x}) w(\mathbf{x}) \, \mathrm{d}\mathbf{x} \right)^{1/2} \left( \int_\Omega v^2(\mathbf{x}) w(\mathbf{x}) \, \mathrm{d}\mathbf{x} \right)^{1/2}$$

for all functions $u, v \in L^2_w(\Omega)$.

## A.3 Linear Operators Between Banach Spaces

Let $X$ and $Y$ be Banach spaces (see (A.1.b)). A linear operator $L$ defined on $X$ and taking values in $Y$, $L : X \to Y$, is said to be *bounded*, or *continuous*, if there exists a constant $C > 0$ such that

$$\|Lv\|_Y \leq C\|v\|_X \quad \text{for all } v \in X \,.$$

The smallest constant $C$ for which the inequality holds is denoted by $\|L\|$, i.e.,

$$\|L\| = \sup_{\substack{v \in X \\ v \neq 0}} \frac{\|Lv\|_Y}{\|v\|_X} \,.$$

The vector space of all the linear bounded operators between $X$ and $Y$ is denoted by $\mathcal{L}(X,Y)$. It is a Banach space for the norm $\|L\|$ just defined.

In the formulation of differential problems, it may be convenient to consider linear operators that are only defined on a subset of a Banach space $X$ (say, with values in $X$). The *domain* $D(L)$ of a linear operator $L : X \to X$ is the largest subset of $X$ on which $L$ is defined, i.e., $v \in D(L)$ if and only if there exists $g \in X$ such that $Lv = g$. We say that $L$ is an *unbounded* operator if

$$\sup_{\substack{v \in D(L) \\ v \neq 0}} \frac{\|Lv\|_X}{\|v\|_X} = +\infty \,.$$

***Example*** Consider the linear differential operator $Lv = \mathrm{d}^2v/\mathrm{d}x^2$, where $v$ is a function on the interval $(a,b)$ of the real line. $L$ can be considered as a *bounded* operator between the Banach spaces $X = C^2([a,b])$ and $Y = C^0([a,b])$ (see (A.7)), or as an *unbounded* operator in $X = C^0([a,b])$. In the former case, the numerator is $\|Lv\|_Y$, which measures the second derivative of $v$, and the denominator is $\|v\|_X$, which measures all the derivatives of $v$ up to order 2. The ratio of these norms is bounded. In the latter case, the domain of $L$ is $D(L) = C^2([a,b])$, considered now as a subspace of $C^0([a,b])$. Here the numerator is again the maximum norm of the second derivative, but the denominator is the weaker norm which measures only the function itself. Taking bounded, but rapidly oscillatory functions, this ratio can be arbitrarily large.

A linear continuous operator $L : X \to Y$ is said to be *compact* if for each sequence $\{v_n \in X \mid n = 0, 1, \dots\}$ such that $\|v_n\|_X \leq C$, one can find a subsequence $\{v_{n_k} \mid k = 0, 1, \dots\}$ and an element $v \in X$ such that

$$\|Lv_{n_k} - Lv\|_Y \longrightarrow 0 \quad \text{as } n_k \longrightarrow \infty \,.$$

Finally an operator $L : X^l \to Y$ is said to be *multilinear* if it is linear in each of its variables. A multilinear operator $L$ is continuous if the quantity

$$\|L\| = \sup_{v_1,\dots,v_l \in X} \frac{\|L(v_1,\dots,v_l)\|_Y}{\|v_1\|_X \cdots \|v_l\|_X}$$

is finite. The space of the multilinear operators $L : X^l \to Y$ is denoted by $\mathcal{L}_l(X,Y)$ and is a Banach space for the norm just introduced.

## A.4 The Fréchet Derivative of an Operator

Let $A$ be a mapping between a Banach space $X$ and a Banach space $Y$, i.e., $A : X \to Y$. We say that $A$ is *Fréchet differentiable* at a point $u_0 \in X$ if there exists a linear continuous operator $L \in \mathcal{L}(X,Y)$ such that

$$\lim_{\substack{w \in X \\ \|w\|_X \to 0}} \frac{\|A(u_0+w) - A(u_0) - Lw\|_Y}{\|w\|_X} = 0 \,.$$

If this happens, the linear operator $L$ is unique. It is termed the *Fréchet derivative* of $A$ at the point $u_0$, and is denoted by $A'(u_0)$.

## A.5 The Lax-Milgram Theorem

Let $V$ be a real Hilbert space (see (A.1.a)). Let $a : V \times V \to \mathbb{R}$ be a bilinear continuous form on $V$, i.e., $a$ satisfies

(i) $a(\lambda u + \mu v, w) = \lambda a(u,w) + \mu a(v,w)$ and
$a(u, \lambda v + \mu w) = \lambda a(u,v) + \mu a(u,w)$
for all $u, v, w \in V$ and all $\lambda, \mu \in \mathbb{R}$;

(ii) there exists a constant $\beta > 0$ such that

$$|a(u,v)| \le \beta \|u\|_V \|v\|_V \quad \text{for all } u, v \in V \,.$$

(iii) there exists a constant $\alpha > 0$ such that

$$a(u,u) \ge \alpha \|u\|_V^2 \quad \text{for all } u \in V \,,$$

i.e., the form $a$ is $V$-*coercive*, or $V$-*elliptic.*

Then for each form $F \in V'$ (the dual space of $V$, see (A.1.c)), there exists a unique solution $u \in V$ to the variational problem

$$a(u,v) = F(v) \quad \text{for all } v \in V \,.$$

Moreover, the following inequality holds:

$$\|u\|_V \le \frac{\beta}{\alpha} \|F\|_{V'} \,.$$

Note that the Riesz representation theorem (A.1.d) follows from the Lax-Milgram theorem applied to the inner product $(u,v)$. This is indeed a symmetric bilinear form, for which (ii) is nothing but the Cauchy-Schwarz inequality (A.2), and (iii) follows from the definition of Hilbertian norm.

## A.6 Dense Subspace of a Normed Space

Let $X$ be a Hilbert or a Banach space with norm $\|v\|$. Let $S \subset X$ be a subspace of $X$. $S$ is said to be *dense* in $X$ if for each element $v \in X$ there exists a sequence $\{v_n \mid n = 0, 1, \dots\}$ of elements $v_n \in S$, such that

$$\|v - v_n\| \longrightarrow 0 \quad \text{as } n \longrightarrow \infty\,.$$

Thus, each element of $X$ can be approximated arbitrarily well by elements of $S$, in the distance induced by the norm of $X$.

For example, the subspace $C^0([a,b])$ of the continuous functions on a bounded, closed interval $[a,b]$ of the real line, is dense in $L^2(a,b)$, the space of the measurable square-integrable functions on $(a,b)$. Indeed, for each function $v \in L^2(a,b)$ and each $n > 0$, one can find a continuous function $v_n \in C^0([a,b])$ such that

$$\int_a^b |v(x) - v_n(x)|^2 \mathrm{d}x \leq \frac{1}{n^2}\,.$$

## A.7 The Spaces $C^m(\overline{\Omega})$, $m \geq 0$

Let $\Omega = (a,b)^d \subset \mathbb{R}^d$, with $d = 1, 2$ or 3. Let us denote by $\overline{\Omega}$ the closure of $\Omega$, i.e., the closed poly-interval $[a,b]^d$. For each multi-index $\alpha = (\alpha_1, \dots, \alpha_d)$ of nonnegative integers, set $|\alpha| = \alpha_1 + \dots + \alpha_d$ and $D^\alpha v = \partial^{|\alpha|} v / \partial x_1^{\alpha_1} \dots \partial x_d^{\alpha_d}$.

We denote by $C^m(\overline{\Omega})$ the vector space of the functions $v : \overline{\Omega} \to \mathbb{R}$ such that for each multi-index $\alpha$ with $0 \leq |\alpha| \leq m$, $D^\alpha v$ exists and is continuous on $\Omega$. Since a continuous function on a closed, bounded (poly)-interval is bounded there, one can set

$$\|v\|_{C^m(\overline{\Omega})} = \sup_{0 \leq |\alpha| \leq m} \sup_{\mathbf{x} \in \overline{\Omega}} |D^\alpha v(\mathbf{x})|\,.$$

This is a norm for which $C^m(\overline{\Omega})$ is a Banach space (see (A.1.b)).

The space $C^\infty(\overline{\Omega})$ is the space of the infinitely differentiable functions on $\overline{\Omega}$. Thus, a function $v$ belongs to $C^\infty(\overline{\Omega})$ if and only if it belongs to $C^m(\overline{\Omega})$ for all $m > 0$.

## A.8 Functions of Bounded Variation and the Riemann(-Stieltjes) Integral

Let $[a,b] \subset \mathbb{R}$ be a bounded interval of the real line, and let $u : [a,b] \to \mathbb{R}$ be a given function. The *total variation* of $u$ on $[a,b]$ is defined by

$$V(u) = \sup_{a = x_0 < x_1 < \dots < x_n = b} \sum_{i=1}^{n} |u(x_i) - u(x_{i-1})|\,,$$

where the supremum is taken over all the partitions of $[a,b]$ by a finite number of points, i.e. over all the sets of $n+1$ points such that $a = x_0 < x_1 < \cdots < x_n = b$, $n$ being arbitrary.

A function is said to be of *bounded variation* in $[a,b]$ if $V(u)$ is finite. Note that a function of bounded variation is certainly bounded.

A continuously differentiable function $u$ in $[a,b]$ is of bounded variation; its total variation can be equivalently expressed as

$$V(u) = \int_a^b |u'(x)| \, \mathrm{d}x \, .$$

The same is true for an absolutely continuous function in $[a,b]$, i.e., a continuous function that admits an integrable derivative in the sense of distributions (see (A.10.b)). However, a function of bounded variation need not be continuous. For instance, the step function

$$u(x) = \begin{cases} 0 & \text{if } x < 0 \, , \\ 1 & \text{if } x \geq 0 \, , \end{cases}$$

is of bounded variation on each interval $[a,b]$ of the real line. On the contrary, $u(x) = x \sin(1/x)$ is an example of a continuous function that is not of bounded variation in any interval containing the origin.

A function $u$ of bounded variation can be split into the difference

$$u(x) = \alpha(x) - \beta(x) \, ,$$

where $\alpha$ and $\beta$ are monotonically increasing functions. This property makes possible the definition of the *Riemann-Stieltjes* integral with respect to a function

$$\int_a^b f(x) \, \mathrm{d}u(x) \, .$$

of bounded variation. We start by defining the Riemann-Stieltjes integral of a bounded function on $[a,b]$ with respect to a monotonically increasing function $\alpha(x)$. Given a partition $\mathcal{P} = \{a = x_0 < x_1 < \cdots < x_n = b\}$, let us set $M_i = \sup\{f(x) | x_{i-1} \leq x \leq x_i\}$ and $m_i = \inf\{f(x) | x_{i-1} \leq x \leq x_i\}$. Next we define

$$\overline{\int_a^b} f(x) \, \mathrm{d}\alpha = \inf_{\mathcal{P}} \sum_{i=1}^n M_i (\alpha(x_i) - \alpha(x_{i-1}))$$

and

$$\underline{\int_a^b} f(x) \, \mathrm{d}\alpha = \sup_{\mathcal{P}} \sum_{i=1}^n m_i (\alpha(x_i) - \alpha(x_{i-1})) \, ,$$

the infimum and the supremum being taken over all the partitions $\mathcal{P}$ of $[a,b]$. If the two numbers just defined are equal, we denote their common value by

$$\int_a^b f(x)\,\mathrm{d}\alpha\,,$$

and we say that $f$ is *Riemann-Stieltjes integrable* with respect to $\alpha$.

If $\alpha(x) \equiv x$, the previous integral coincides with the classical *Riemann integral*.

The Riemann-Stieltjes integral of a bounded function on $[a, b]$ with respect to a function of bounded variation $u$ is defined as

$$\int_a^b u(x)\,\mathrm{d}u = \int_a^b u(x)\,\mathrm{d}\alpha - \int_a^b u(x)\,\mathrm{d}\beta\,,$$

where $u = \alpha - \beta$ is any decomposition of $u$ into the difference of two monotonically increasing functions. This definition is independent of the particular decomposition.

The following integration-by-parts rule for functions of bounded variation holds. Let $u$ and $v$ be continuous functions of bounded variation on $[a, b]$. Then,

$$\int_a^b u(x)\,\mathrm{d}v = u(b)v(b) - u(a)v(a) - \int_a^b v(x)\,\mathrm{d}u\,.$$

## A.9 The Lebesgue Integral and $L^p$-Spaces

Let us start with a schematic account of the Lebesgue measure on a hounded interval $(a, b)$ of the real line. A complete introduction to the Lebesgue integration theory can be found, e.g., in Royden (1968) or Rudin (1966).

### (a) The Lebesgue (Outer) Measure

Each set $A$ contained in $(a, b)$ can be covered by a countable union of open intervals $I_N$, i.e. $A \subset \bigcup_{n=0}^{\infty} I_n$. Taking into account this property, *the Lebesgue outer measure* $\mu(A)$ of the set $A$ is defined as

$$\mu(A) = \inf \sum_n |I_n|\,,$$

where $|I_n|$ denotes the length of the interval $I_n$, and the infimum is taken over all the coverings of $A$ by open intervals. Note that the measure of an interval is its length. Each countable set has zero measure.

### (b) Measurable Sets

For each set $A \subseteq (a, b)$, let $\widetilde{A}$ denote the complementary set of $A$ in $(a, b)$, i.e. $\widetilde{A} = \{x \in (a, b) : x \notin A\}$.

A set $A \subseteq (a, b)$ is said to be measurable if

$$\mu(A) + \mu(\widetilde{A}) = \mu((a, b)) = b - a\,.$$

In Lebesgue's measure theory only measurable sets are of interest.

### (c) Simple Measurable Functions

A function $s : (a, b) \to [0, +\infty)$ is a *simple measurable function* if it assumes only a finite number of values $\{s_0, \dots, s_n\}$, and if each set $A_i = \{x \in (a, b) : s(x) = s_i\}$ is measurable.

### (d) Measurable Functions

A positive function $u : (a, b) \to [0, +\infty)$ is *measurable* if it is the pointwise limit of simple measurable functions – more precisely, if there exist simple measurable functions $s^{(k)}$ such that

(i) $0 \le s^{(1)} \le s^{(2)} \le \dots \le u$

(ii) $s^{(k)}(x) \to u(x)$ as $k \to \infty$, for all $x \in (a, b)$.

A real function $u : (a, b) \to \mathbb{R}$ is measurable if both its positive and negative parts, $u^+ = \max\{u, 0\}$ and $u^- = \max\{-u, 0\}$, are measurable.

### (e) The Lebesgue Integral

If $s$ is a simple measurable function on $(a, b)$, we set

$$\int_a^b s \,\mathrm{d}\mu = \sum_{i=0}^{n} s_i \mu(A_i) .$$

If $u$ is a positive measurable function on $(a, b)$, we set

$$\int_a^b u \,\mathrm{d}\mu = \sup \int_a^b s \,\mathrm{d}\mu ,$$

the supremum being taken over all the simple measurable functions such that $0 \le s \le u$. The value of the right-hand side is a nonnegative number or $+\infty$. We call it the *Lebesgue integral* of $u$ on $(a, b)$.

A positive measurable function $u$ is said to be *Lebesgue integrable on* $(a, b)$ if

$$\int_a^b u \,\mathrm{d}\mu < +\infty .$$

A real measurable function $u$ on $(a, b)$ is said to be Lebesgue integrable if both its positive and negative parts, $u^+$ and $u^-$, are Lebesgue integrable. In this case we define the Lebesgue integral of $u$ on $(a, b)$ as

$$\int_a^b u \,\mathrm{d}\mu = \int_a^b u^+ \,\mathrm{d}\mu - \int_a^b u^- \,\mathrm{d}\mu .$$

### (f) The Spaces $L^p(a,b)$, $1 \le p \le \infty$

Let us now define several spaces of integrable functions in the sense of Lebesgue. Hereafter we will use the more conventional notation $\int_a^b u(x)\mathrm{d}x$, $\int_\Omega u(\mathbf{x})\,\mathrm{d}(\mathbf{x})$, etc. to denote Lebesgue integrals. Since two integrable functions that differ on a set of zero measure have the same integral, they can be identified from the point of view of the Lebesgue integration theory, i.e., they belong to the same equivalence class. This identification is always presumed here and in the sequel.

Let $(a,b)$ be a bounded interval of $\mathbb{R}$, and let $1 \le p < +\infty$. We denote by $L^p(a,b)$ the space of the measurable functions $u : (a,b) \to \mathbb{R}$ such that $\int_a^b |u(x)|^p \mathrm{d}x < +\infty$. Endowed with the norm

$$\|u\|_{L^p(a,b)} = \left( \int_a^b |u(x)|^p \mathrm{d}x \right)^{1/p} ,$$

it is a Banach space (see (A.1.b)).

For $p = +\infty$, $L^\infty(a,b)$ is the space of the measurable functions $u : (a,b) \to \mathbb{R}$ such that $|u(x)|$ is bounded outside a set of measure zero. If $M$ denotes the smallest real number such that $|u(x)| \le M$ outside a set of measure zero, we define a norm on $L^\infty(a,b)$ by setting

$$\|u\|_{L^p(a,b)} = \operatorname*{ess\,sup}_{x\in(a,b)} |u(x)| = M .$$

(If $u$ is continuous on $[a,b]$, then $\|u\|_{L^\infty(a,b)}$ is the maximum of the absolute value of $u$ on $[a,b]$.) Again $L^\infty(a,b)$ is a Banach space.

The index $p = 2$ is of special interest because $L^2(a,b)$ is not only a Banach space but also a Hilbert space (see (A.1.a)). The inner product is

$$(u,v) = \int_a^b u(x)v(x)\mathrm{d}x ,$$

which induces the norm

$$\|u\|_{L^2(a,b)} = \left( \int_a^b |u(x)|^2 \mathrm{d}x \right)^{1/2} .$$

It is also possible to define $L^p$-spaces of complex measurable functions. The previous definitions and norms hold unchanged provided the absolute value of $u$ is replaced by the modulus of $u$. The inner product of the complex $L^2(a,b)$-space is

$$(u,v) = \int_a^b u(x)\overline{v(x)}\mathrm{d}x .$$

### (g) The Weighted Spaces $L^p_w(-1,1)$, $1 \leq p \leq +\infty$

Let $w(x)$ be a weight function on the interval $(-1,1)$, i.e., a continuous, strictly positive and integrable function on $(-1,1)$. For $p < +\infty$, we denote by $L^p_w(-1,1)$ the Banach space of the measurable functions $u : (a,b) \to \mathbb{R}$ such that $\int_a^b |u(x)|^p w(x) \mathrm{d}x < +\infty$. It is endowed with the norm

$$\|u\|_{L^p_w(-1,1)} = \left( \int_a^b |u(x)|^p w(x) \mathrm{d}x \right)^{1/p} .$$

For $p = \infty$ we set $L^\infty_w(-1,1) = L^\infty(-1,1)$.

The space $L^2_w(-1,1)$ is a Hilbert space for the inner product

$$(u,v)_w = \int_a^b u(x)v(x)w(x) \mathrm{d}x ,$$

which induces the weighted norm

$$\|u\|_{L^2_w(a,b)} = \left( \int_a^b |u(x)|^2 w(x) \mathrm{d}x \right)^{1/2} .$$

### (h) The Spaces $L^p(\Omega)$ and $L^p_w(\Omega)$, $1 \leq p \leq +\infty$

The previous definitions can be extended in a straightforward way to more than one space dimension. Let $\Omega$ denote a bounded, open domain in $\mathbb{R}^d$, for $d = 2$ or $3$ (for instance, $\Omega = (0,2\pi)^d$ or $\Omega = (-1,1)^d$), and let $\mathrm{d}\mathbf{x}$ be the Lebesgue measure on $\mathbb{R}^d$.

For $p < +\infty$, we denote by $L^p(\Omega)$ the space of the measurable functions $u : \Omega \to \mathbb{R}$ such that $\int_\Omega |u(\mathbf{x})|^p \mathrm{d}\mathbf{x} < +\infty$. It is a Banach space for the norm

$$\|u\|_{L^p(\Omega)} = \left( \int_\Omega |u(\mathbf{x})|^p \mathrm{d}\mathbf{x} \right)^{1/p} .$$

$L^\infty(\Omega)$ is the Banach space of the measurable functions $u : \Omega \to \mathbb{R}$ that are bounded outside a set of measure zero, equipped with the norm

$$\|u\|_{L^\infty(\Omega)} = \operatorname*{ess\,sup}_{\mathbf{x} \in \Omega} |u(\mathbf{x})| .$$

The space $L^2(\Omega)$ is a Hilbert space for the inner product

$$(u,v) = \int_\Omega u(\mathbf{x})v(\mathbf{x}) \mathrm{d}\mathbf{x} ,$$

which induces the norm

$$\|u\|_{L^2(\Omega)} = \left( \int_\Omega |u(\mathbf{x})|^2 \mathrm{d}\mathbf{x} \right)^{1/2} .$$

Again one can consider $L^p(\Omega)$ spaces of complex functions in a straightforward manner.

If $w(\mathbf{x})$ denotes a weight function on $\Omega$, the weighted spaces $L^p_w(\Omega)$ can be defined, by analogy to $L^p_w(a,b)$, as the Banach spaces of the measurable functions $u : \Omega \to \mathbb{R}$ such that the function $\mathbf{x} \to |u(\mathbf{x})|^p w(\mathbf{x})$ is Lebesgue integrable on $\Omega$. In particular, the space $L^2_w(\Omega)$ is a Hilbert space for the inner product

$$(u,v)_w = \int_\Omega u(\mathbf{x})v(\mathbf{x})w(\mathbf{x})\mathrm{d}\mathbf{x} ,$$

which induces the weighted norm

$$\|u\|_{L^2_w(\Omega)} = \left( \int_\Omega |u(\mathbf{x})|^2 w(\mathbf{x}) \mathrm{d}\mathbf{x} \right)^{1/2} .$$

## A.10 Infinitely Differentiable Functions and Distributions

Let $\Omega$ be a bounded, open domain in $\mathbb{R}^d$, for $d = 1, 2$ or 3. If $\alpha = (\alpha_1, \ldots, \alpha_d)$ is a multi-index of nonnegative integers, let us set

$$D^\alpha v = \frac{\partial^{\alpha_1 + \cdots + \alpha_d} v}{\partial x_1^{\alpha_1} \cdots \partial x_d^{\alpha_d}} .$$

We denote by $\mathscr{D}(\Omega)$ the vector space of all the infinitely differentiable functions $\phi : \Omega \to \mathbb{R}$, for which there exists a closed set $K \subset \Omega$ such that $\phi \equiv 0$ outside $K$.

We say that a sequence of functions $\phi_n \in \mathscr{D}(\Omega)$ *converges in* $\mathscr{D}(\Omega)$ to a function $\phi \in \mathscr{D}(\Omega)$ as $n \to \infty$, if there exists a common closed set $K \subset \Omega$ such that all the $\phi_n$ vanish outside $K$, and $D^\alpha \phi_n \to D^\alpha \phi$ uniformly on $K$ as $n \to \infty$, for all nonnegative multi-indices $\alpha$.

### (a) Distributions

Let $T$ be a linear form on $\mathscr{D}(\Omega)$, i.e., a linear mapping $T : \mathscr{D}(\Omega) \to \mathbb{R}$. We shall denote the value of $T$ on the element $\phi \in \mathscr{D}(\Omega)$ by $\langle T, \phi \rangle$. $T$ is said to be *continuous* if for each sequence $\phi_n \in \mathscr{D}(\Omega)$ that converges in $\mathscr{D}(\Omega)$ to a function $\phi \in \mathscr{D}(\Omega)$ as $n \to \infty$, one has

$$\langle T, \phi_n \rangle \longrightarrow \langle T, \phi \rangle \quad \text{as } n \longrightarrow \infty .$$

A *distribution* is a linear continuous form on $\mathscr{D}(\Omega)$. The set of all the distributions on $\Omega$ is a vector space denoted by $\mathscr{D}'(\Omega)$.

*Examples*

(i) Each integrable function $f \in L^1(\Omega)$ (see (A.9.f)) can be identified with the distribution $T_f$ defined by

$$\langle T_f, \phi \rangle = \int_\Omega f(\mathbf{x})\phi(\mathbf{x})\mathrm{d}\mathbf{x} \quad \text{for all } \phi \in \mathscr{D}(\Omega) .$$

(ii) Let $\mathbf{x}_0 \in \Omega$. The linear form on $\mathscr{D}(\Omega)$,

$$\langle \delta_{\mathbf{x}_0}, \phi \rangle = \phi(\mathbf{x}_0) \quad \text{for all } \phi \in \mathscr{D}(\Omega) ,$$

is a distribution, which is commonly (but improperly) called the "Dirac function".

We notice that if $T_1$ and $T_2$ are two distributions, then they are "equal in the sense of distributions" if

$$\langle T_1, \phi \rangle = \langle T_2, \phi \rangle \quad \text{for all } \phi \in \mathscr{D}(\Omega) .$$

## (b) Derivative of Distributions

Let $\alpha$ be a nonnegative multi-index and set $m = \alpha_1 + \cdots + \alpha_d$. For each distribution $T \in \mathscr{D}'(\Omega)$ let us consider the linear form on $\mathscr{D}(\Omega)$:

$$\langle D^\alpha T, \phi \rangle = (-1)^m \langle T, D^\alpha \phi \rangle \quad \text{for all } \phi \in \mathscr{D}(\Omega) .$$

This linear form is continuous on $\mathscr{D}(\Omega)$; hence, it is a distribution, which is called the $\alpha$-distributional *derivative* of $T$.

It follows that each integrable function $u \in L^1(\Omega)$ is infinitely differentiable in the sense of distributions, and the following Green's formula holds:

$$\langle D^\alpha u, \phi \rangle = (-1)^m \int_\Omega u(\mathbf{x}) D^\alpha \phi(\mathbf{x})\mathrm{d}\mathbf{x} \quad \text{for all } \phi \in \mathscr{D}(\Omega) .$$

If $u$ is $m$-times continuously differentiable in $\Omega$, then the $\alpha$-distributional derivative of $u$ coincides with the classical derivative of index $\alpha$. In general, a distributional derivative of an integrable function can be an integrable function or merely a distribution. We say that the $\alpha$-distributional derivative of an integrable function $u \in L^1(\Omega)$ is an *integrable function* if there exists $g \in L^1(\Omega)$ such that

$$\langle D^\alpha u, \phi \rangle = \int_\Omega g(\mathbf{x})\phi(\mathbf{x})\mathrm{d}\mathbf{x} \quad \text{for all } \phi \in \mathscr{D}(\Omega) .$$

*Examples*

(i) Consider the function $u(x) = \frac{1}{2}|x|$ in the interval $(-1,1)$. Note that $u$ is not classically differentiable at the origin. The first derivative of $u$ in the distributional sense is represented by the step function

$$v(x) = \begin{cases} 1/2 & \text{if } x > 0 \,, \\ -1/2 & \text{if } x < 0 \,. \end{cases}$$

(ii) Consider the function $v$ now defined. Note that the classical derivative is zero at all the points $x \neq 0$. The first derivative of $v$ in the sense of distributions is the "Dirac function" $\delta_0$ at the origin. This distribution cannot be represented by an integrable function.

Functions having a certain number of distributional derivatives that can be represented by integrable functions play a fundamental role in the modern theory of partial differential equations. The spaces of these functions are named Sobolev spaces (see (A.11)).

### (c) Periodic Distributions

Let $\Omega = (0, 2\pi)^d$, for $d = 1, 2$ or 3. We define the space $C_p^\infty(\overline{\Omega})$ as the vector space of the functions $u : \overline{\Omega} \to \mathbb{C}$ that have derivatives of any order continuous in the closure $\overline{\Omega}$ of $\Omega$, and $2\pi$-periodic in each space direction. A sequence $\phi_n \in C_p^\infty(\overline{\Omega})$ *converges in* $C_p^\infty(\overline{\Omega})$ to a function $\phi \in C_p^\infty(\overline{\Omega})$ if $D^\alpha \phi_n \to D^\alpha \phi$ uniformly on $\overline{\Omega}$, as $n \to \infty$ for all nonnegative multi-indices $\alpha$.

A *periodic distribution* is a linear form $T : C_p^\infty(\overline{\Omega}) \to \mathbb{C}$ that is continuous, i.e., such that

$$\langle T, \phi_n \rangle \longrightarrow \langle T, \phi \rangle \quad \text{as } n \longrightarrow \infty \,,$$

whenever $\phi_n \to \phi$ in $C_p^\infty(\overline{\Omega})$.

The *derivative* of index $\alpha$ of a periodic distribution $T$ is the periodic distribution $D^\alpha T$ defined by

$$\langle D^\alpha T, \phi \rangle = (-1)^m \langle T, D^\alpha \phi \rangle \quad \text{for all } \phi \in C_p^\infty(\overline{\Omega})$$

(where $m = \alpha_1 + \cdots + \alpha_d$).

Note that each function in $\mathscr{D}(\Omega)$ also belongs to $C_p^\infty(\overline{\Omega})$. Thus, it is easily seen that each periodic distribution is indeed a distribution in the sense of (A.10.a).

## A.11 Sobolev Spaces and Sobolev Norms

We introduce hereafter some relevant Hilbert spaces, which occur in the numerical analysis of boundary-value problems. They are spaces of square-integrable functions (see (A.9)), which possess a certain number of derivatives (in the sense of distributions, see (A.10.b)) representable as square-integrable functions.

### (a) The Spaces $H^m(a,b)$ and $H^m(\Omega)$, $m \geq 0$

Let $(a,b)$ be a bounded interval of the real line, and let $m \geq 0$ be an integer.

We define $H^m(a,b)$ to be the vector space of the functions $v \in L^2(a,b)$ such that all the distributional derivatives of $u$ of order up to $m$ can be represented by functions in $L^2(a,b)$. In short,

$$H^m(a,b) = \left\{ v \in L^2(a,b) : \text{ for } 0 \leq k \leq m, \frac{\mathrm{d}^k u}{\mathrm{d}x^k} \in L^2(a,b) \right\} .$$

$H^m(a,b)$ is endowed with the inner product

$$(u,v)_m = \sum_{k=0}^{m} \int_a^b \frac{\mathrm{d}^k u}{\mathrm{d}x^k}(x) \frac{\mathrm{d}^k v}{\mathrm{d}x^k}(x) \mathrm{d}x$$

for which $H^m(a,b)$ is a Hilbert space. The associated norm is

$$\|v\|_{H^m(a,b)} = \left( \sum_{k=0}^{m} \left\| \frac{\mathrm{d}^k v}{\mathrm{d}x^k} \right\|^2_{L^2(a,b)} \right)^{1/2} .$$

The Sobolev spaces $H^m(a,b)$ form a hierarchy of Hilbert spaces, in the sense that $\ldots H^{m+1}(a,b) \subset H^m(a,b) \subset \cdots \subset H^0(a,b) \equiv L^2(a,b)$, each inclusion being continuous (see (A.3)). Clearly, if a function $u$ has $m$ classical continuous derivatives in $[a,b]$, then $u$ belongs to $H^m(a,b)$ – in other words, $C^m([a,b]) \subset H^m(a,b)$ with continuous inclusion. Conversely, if $u$ belongs to $H^m(a,b)$ for $m \geq 1$, then $u$ has $m-1$ classical continuous derivatives in $[a,b]$, i.e., $H^m(a,b) \subset C^{m-1}([a,b])$ with continuous inclusion. This is an example of the so-called "Sobolev imbedding theorems". As a matter of fact, $H^m(a,b)$ can be equivalently defined as

$$H^m(a,b) = \left\{ v \in C^{m-1}([a,b]) : \frac{\mathrm{d}}{\mathrm{d}x} v^{(m-1)} \in L^2(a,b) \right\} ,$$

where the last derivative is in the sense of distributions.

Functions in $H^m(a,b)$ can be approximated arbitrarily well by infinitely differentiable functions in $[a,b]$, in the distance induced by the norm of $H^m(a,b)$. In other words,

$$C^\infty([a,b]) \text{ is dense in } H^m(a,b)$$

(see (A.6) for the definition of density of a subspace).

Set now $\Omega = (a,b)^d$, for $d = 2$ or $3$. Given a multi-index $\alpha = (\alpha_1, \ldots, \alpha_d)$ of nonnegative integers, we set $|\alpha| = \alpha_1 + \cdots + \alpha_d$ and

$$D^\alpha v = \frac{\partial^{|\alpha|} v}{\partial x_1^{\alpha_1} \cdots \partial x_d^{\alpha_d}} .$$

The previous definition of Sobolev spaces can be extended to higher space dimensions as follows. We define

$$H^m(\Omega)=\{v\in L^2(\Omega) : \text{for each nonnegative multi-index } \alpha \text{ with } |\alpha| \le m, \text{ the distributional derivative } D^\alpha v \text{ belongs to } L^2(\Omega)\}.$$

This is a Hilbert space for the inner product

$$(u,v)_m = \sum_{|\alpha|\le m} \int D^\alpha u(\mathbf{x}) D^\alpha v(\mathbf{x}) \mathrm{d}\mathbf{x} ,$$

which induces the norm

$$\|v\|_{H^m(\Omega)} = \left( \sum_{|\alpha|\le m} \|D^\alpha v\|^2_{L^2(\Omega)} \right)^{1/2} .$$

Functions in $H^m(\Omega)$ for $m \ge 1$ need not have the derivatives of order $m-1$ continuous in $\Omega$. However, the weaker Sobolev inclusion $H^m(\Omega) \subset C^{m-2}(\overline{\Omega})$ ($m \ge 2$) holds. On the other hand, as in the one-dimensional case

$$C^\infty(\overline{\Omega}) \text{ is dense in } H^m(\Omega) .$$

### (b) The Spaces $H^m_w(-1,1)$ and $H^m_w(\Omega)$, $m \ge 0$

In the definition of a Sobolev space, one can require that the function as well as its distributional derivatives be square integrable with respect to a weight function $w$ (see (A.9)). This is the most natural framework in dealing with Chebyshev methods.

Let now $(a,b)$ be the interval $(-1,1)$. We choose the weight function $w$ to be the Chebyshev weight $w(x) = (1-x^2)^{-1/2}$ (although the following definitions can be given for an arbitrary weight function). We set

$$H^m_w(-1,1) = \left\{ v \in L^2_w(-1,1) : \text{for } 0 \le k \le m, \text{ the distributional derivative } \frac{\mathrm{d}^k u}{\mathrm{d}x^k} \text{ belongs to } L^2_w(-1,1) \right\} .$$

$H^m_w(-1,1)$ is a Hilbert space for the inner product

$$(u,v)_{m,w} = \sum_{k=0}^m \int_{-1}^1 \frac{\mathrm{d}^k u}{\mathrm{d}x^k}(x) \frac{\mathrm{d}^k v}{\mathrm{d}x^k}(x) \frac{\mathrm{d}x}{\sqrt{1-x^2}} ,$$

which induces the norm

$$\|u\|_{H^m_w(-1,1)} = \left( \sum_{k=0}^m \left\| \frac{\mathrm{d}^k v}{\mathrm{d}x^k} \right\|^2_{L^2_w(-1,1)} \right)^{1/2} .$$

For $\Omega = (-1,1)^d$ ($d = 2$ or 3) and $w = w(\mathbf{x}) = \prod_{i=1}^d (1 - x_i^2)^{-1/2}$ (the $d$-dimensional Chebyshev weight), we define $H_w^m(\Omega)$ by analogy to $H^m(\Omega)$. Precisely we set

$$H_w^m(\Omega) = \{v \in L_w^2(\Omega) : \text{for each nonnegative multi-index } \alpha \text{ with } |\alpha| < m, \text{ the distributional derivative } D^\alpha v \text{ belongs to } L_w^2(\Omega)\} .$$

This space is endowed with the Hilbertian inner product

$$(u, v)_{m,w} = \sum_{|\alpha| \leq m} \int_\Omega D^\alpha u(\mathbf{x}) D^\alpha v(\mathbf{x}) w(\mathbf{x}) \mathrm{d}\mathbf{x}$$

and the associated norm

$$\|v\|_{H_w^m(\Omega)} = \left( \sum_{|\alpha| \leq m} \|D^\alpha v\|_{L_w^2(\Omega)}^2 \right)^{1/2} .$$

The properties of inclusion and density previously recalled for $H^m(a,b)$ and $H^m(\Omega)$ hold for $H_w^m(-1,1)$ and $H_w^m(\Omega)$ as well. Moreover, we note that $H_w^m(\Omega) \subset H^m(\Omega)$ for all $m \geq 0$.

### (c) The Spaces $H_0^1(a,b)$, $H_{w,0}^1(-1,1)$ and $H_0^1(\Omega)$, $H_{w,0}^1(\Omega)$

Dirichlet conditions are among the simplest and most common boundary conditions to be associated with a differential operator. Therefore, the subspaces of the Sobolev spaces $H^m$ spanned by the functions satisfying homogeneous Dirichlet boundary conditions play a fundamental role.

Since the functions of $H^1(a,b)$ are continuous up to the boundary by the Sobolev imbedding theorem, it is meaningful to introduce the following subspace of $H^1(a,b)$:

$$H_0^1(a,b) = \{v \in H^1(a,b) : v(a) = v(b) = 0\} .$$

This is a Hilbert space for the same inner product of $H^1(a,b)$. It is often preferable to endow $H^1(a,b)$ with a different, although equivalent, inner product. This is defined as

$$[u, v] = \int_a^b \frac{\mathrm{d}u}{\mathrm{d}x}(x) \frac{\mathrm{d}v}{\mathrm{d}x}(x) \mathrm{d}x .$$

By the Poincaré inequality (A.13), it is indeed an inner product on $H_0^1(a,b)$. The associated norm, denoted by

$$\|v\|_{H_0^1(a,b)} = \left( \int_a^b \left| \frac{\mathrm{d}v}{\mathrm{d}x} \right|^2 \mathrm{d}x \right)^{1/2} ,$$

is equivalent to the $H_0^1(a,b)$-norm, in the sense that there exists a constant $C > 0$ such that, for all $v \in H_0^1(a,b)$,

$$C\|v\|_{H^1(a,b)} \le \|v\|_{H_0^1(a,b)} \le \|v\|_{H^1(a,b)} \,.$$

Again, this follows from the Poincaré inequality.

The subspace $H_{w,0}^1(-1,1)$ of $H_w^1(-1,1)$ is defined similarly, namely, we set

$$H_{w,0}^1(-1,1) = \{v \in H_w^1(-1,1) : v(-1) = v(1) = 0\} \,.$$

Again, it can be endowed with the weighted inner product

$$[u,v]_w = \int_{-1}^{1} \frac{\mathrm{d}u}{\mathrm{d}x}(x)\, \frac{\mathrm{d}v}{\mathrm{d}x}(x) \frac{\mathrm{d}x}{\sqrt{1-x^2}} \,.$$

The associated norm

$$\|v\|_{H_{w,0}^1(-1,1)} = \left( \int_{-1}^{1} \left| \frac{\mathrm{d}v}{\mathrm{d}x} \right|^2 \frac{\mathrm{d}x}{\sqrt{1-x^2}} \right)^{1/2}$$

is equivalent to the norm of $H_w^1(-1,1)$, due to the Poincaré inequality.

The functions of $H_0^1(a,b)$ can be approximated arbitrarily well in the norm of this space not only by infinitely differentiable functions on $[a,b]$, but also by infinitely differentiable functions that vanish identically in a neighborhood of $x = a$ and $x = b$. In other words,

$$\mathscr{D}((a,b)) \text{ is dense in } H^1(a,b)$$

(see (A.10) and (A.6)). A similar result holds for $H_{w,0}^1(-1.1)$, i.e.,

$$\mathscr{D}((-1,1)) \text{ is dense in } H_{w,0}^1(-1,1) \,.$$

We turn now to more space dimensions. If $\Omega$ is the Cartesian product of $d$ intervals ($d = 2$ or 3), the functions of $H^1(\Omega)$ need not be continuous on the closure of $\Omega$. Thus, their pointwise values on the boundary $\partial\Omega$ of $\Omega$ need not be defined. However, it is possible to extend the trace operator $v \mapsto v|_{\partial\Omega}$ (classically defined for functions $v \in C^0(\overline{\Omega})$) so as to be a linear continuous mapping between $H^1(\Omega)$ and $L^2(\Omega)$, the space of the square-integrable functions on $\partial\Omega$ (see Lions and Magenes (1972), Chapter 1, for the rigorous definition of the trace of a function $v \in H^1(\Omega)$). With this in mind, it is meaningful to define $H_0^1(\Omega)$ as the subspace of $H^1(\Omega)$ of the functions whose trace at the boundary is zero. Precisely we set

$$H_0^1(\Omega) = \{v \in H^1(\Omega) : v|_{\partial\Omega} = 0\} \,.$$

This is a Hilbert space for the inner product of $H^1(\Omega)$, or for the inner product

$$[u,v] = \int_{\Omega} \nabla u(\mathbf{x}) \cdot \nabla v(\mathbf{x})\, \mathrm{d}\mathbf{x} \,.$$

The associated norm is denoted by

$$\|v\|_{H_0^1(\Omega)} = \left( \int_\Omega |\nabla v|^2 \mathrm{d}\mathbf{x} \right)^{1/2}$$

and is equivalent to the $H^1(\Omega)$-norm, by the Poincaré inequality (A.13).

In a completely similar manner we introduce the space

$$H_{w,0}^1(\Omega) = \{v \in H_w^1(\Omega) : v|_{\partial\Omega} \equiv 0\}$$

endowed with the inner product

$$[u, v]_w = \int_\Omega \nabla u(\mathbf{x}) \cdot \nabla v(\mathbf{x}) w(\mathbf{x}) \mathrm{d}\mathbf{x}$$

and the norm

$$\|v\|_{H_{w,0}^1(\Omega)} = \left( \int_\Omega |\nabla v|^2 w(\mathbf{x}) \mathrm{d}\mathbf{x} \right)^{1/2} .$$

Concerning the approximation of the functions of $H_0^1(\Omega)$ by infinitely smooth functions, the following result holds:

$$\mathscr{D}(\Omega) \text{ is dense in } H_0^1(\Omega) \text{ (respectively in } H_{w,0}^1(\Omega)) .$$

The dual spaces (see (A.1.c)) of the Hilbert spaces of type $H_0^1$ now defined are usually denoted by $H^{-1}$. Thus, $H^{-1}(a, b)$ is the dual space of $H_0^1(a, b)$. $H_w^{-1}(-1, 1)$ is the dual space of $H_{w,0}^1(-1, 1)$, and so on.

Finally let us mention that for $m \geq 2$, one can define the subspaces $H_0^m(a, b)$ of $H^m(a, b)$ (and similarly for $H_w^m(-1, 1)$, etc.) of the functions of $H^m(a, b)$ whose derivatives of order up to $m - 1$ vanish on the boundary of the domain of definition. Again, these spaces are Hilbert spaces for the inner product of $H^m(a, b)$, or for an equivalent inner product that only involves the derivatives of order $m$.

### (d) The Spaces $H_p^m(0, 2\pi)$ and $H_p^m(\Omega)$, $m \geq 0$

In the analysis of Fourier methods, the natural Sobolev spaces are those of periodic functions. In this framework, functions are complex valued, and their derivatives are taken in the sense of the periodic distributions (see (A.10.c)). We set

$$H_p^m(0, 2\pi) = \Big\{ v \in L^2(0, 2\pi) : \text{ for } 0 \leq k \leq m, \text{ the derivative } \frac{\mathrm{d}^k v}{\mathrm{d}x^k} \text{ in the sense of periodic distribution belongs to } L^2(0, 2\pi) \Big\} .$$

$H_p^m(0,2\pi)$ is a Hilbert space for the inner product

$$(u,v)_m = \sum_{k=0}^{m} \int_0^{2\pi} \frac{\mathrm{d}^k u}{\mathrm{d}x^k}(x) \overline{\frac{\mathrm{d}^k v}{\mathrm{d}x^k}}(x)\mathrm{d}x ,$$

whose associated norm is

$$\|v\|_{H_p^m(0,2\pi)} = \left( \sum_{k=0}^{m} \left\| \frac{\mathrm{d}^k v}{\mathrm{d}x^k} \right\|^2_{L^2(0,2\pi)} \right)^{1/2} .$$

The space $H_p^m(0,2\pi)$ coincides with the space of the functions $v : [0,2\pi] \to \mathbb{C}$ that have $m-1$ continuously differentiable, $2\pi$-periodic derivatives on $[0,2\pi]$, and such that the periodic distributional derivative $(\mathrm{d}/\mathrm{d}x)v^{(m-1)}$ can be represented by a function of $L^2(0,2\pi)$.

The space $C_p^\infty([0,2\pi])$ introduced in (A.10.c) is dense in $H_p^m(0,2\pi)$. If $\Omega = (0,2\pi)$ for $d = 2$ or 3, we set

$$\begin{aligned} H_p^m(\Omega) = \{v \in L^2(\Omega) : & \text{ for each integral multi-index } \alpha \text{ with } |\alpha| \le m, \\ & \text{ the derivative } D^\alpha v \text{ in the sense of periodic} \\ & \text{ distributions belongs to } L^2(\Omega)\}. \end{aligned}$$

This is a Hilbert space for the inner product

$$(u,v)_m = \sum_{|\alpha| \le m} \int_\Omega D^\alpha u(\mathbf{x}) \overline{D^\alpha v(\mathbf{x})} \mathrm{d}\mathbf{x} ,$$

with associated norm

$$\|v\|_{H_p^m(\Omega)} = \left( \sum_{|\alpha| \le m} \|D^\alpha v\|^2_{L^2(\Omega)} \right)^{1/2} .$$

The space $C_p^\infty(\overline{\Omega})$ is dense in $H_p^m(\Omega)$. Note that since a periodic distribution is also a distribution (see (A.10.c)), each space $H_p^m(0,2\pi)$ (resp. $H_p^m(\Omega)$) is a subspace of the space $H^m(0,2\pi)$ (resp. $H^m(\Omega)$).

## A.12 The Sobolev Inequality

Let $(a,b) \subset \mathbb{R}$ be a bounded interval of the real line. For each function $u \in H^1(a,b)$ (see (A.11.a)) the following inequality holds:

$$\|u\|_{L^\infty(a,b)} \le \left( \frac{1}{b-a} + 2 \right)^{1/2} \|u\|^{1/2}_{L^2(a,b)} \|u\|^{1/2}_{H^1(a,b)} .$$

## A.13 The Poincaré Inequality

Let $v$ be a function of $H^1(a,b)$ (see (A.11.a)). We know that $v$ is continuous on $[a,b]$. Assume that at a point $x_0 \in [a,b]$, $v_0(x_0) = 0$. The Poincaré inequality states that there exists a constant $C$ (depending upon the interval length $b-a$) such that

$$\|v\|_{L^2(a,b)} \le C\|v'\|_{L^2(a,b)} , \tag{A.13.1}$$

i.e., the $L^2$-norm of the function is bounded by the $L^2$-norm of the derivative. The Poincaré inequality applies to functions belonging to $H_0^1(a,b)$ (see (A.11.c)), for which $x_0 = a$ or $b$, and also to functions of $H^1(a,b)$ that have zero average on $(a,b)$, since necessarily such functions change sign in the domain.

A similar inequality holds if we replace $H^1(a,b)$ with $H_w^1(a,b)$ (see (A.11.b)). Precisely, there exists a constant $C > 0$ such that, for all $v \in H_w^1(a,b)$ vanishing at a point $x_0 \in [a,b]$,

$$\|v\|_{L_w^2(a,b)} \le C\|v'\|_{L_w^2(a,b)} . \tag{A.13.2}$$

In space dimension $d \ge 2$, the functions to which the Poincaré inequality applies must vanish on a manifold of dimension $d-1$. Confining ourselves to the case of functions vanishing on the boundary $\partial\Omega$ of the domain of definition $\Omega$, one has

$$\|v\|_{L^2(\Omega)} \le C\|\nabla v\|_{(L^2(\Omega))^d} \quad \text{for all } v \in H_0^1(\Omega) \tag{A.13.3}$$

and

$$\|v\|_{L_w^2(\Omega)} \le C\|\nabla v\|_{(L_w^2(\Omega))^d} \quad \text{for all } v \in H_{w,0}^1(\Omega) . \tag{A.13.4}$$

(See (A.11.c) for the definition of the spaces $H_0^1(\Omega)$ and $H_{w,0}^1(\Omega)$.) The same results hold if the domain $\Omega$ is simply connected and $v$ only vanishes on a portion of $\partial\Omega$ of positive measure.

## A.14 The Hardy Inequality

Let $a < b$ be two real numbers, and let $\alpha < 1$ be a real constant. The following inequalities hold for all measurable functions $\phi$ on $(a,b)$:

$$\int_a^b \left[\frac{1}{t-a}\int_a^t \phi(s)\mathrm{d}s\right]^2 (t-a)^\alpha \mathrm{d}t \le \frac{4}{1-\alpha}\int_a^b \phi^2(t)(t-a)^\alpha \mathrm{d}t$$

and, similarly,

$$\int_a^b \left[\frac{1}{b-t}\int_t^b \phi(s)\mathrm{d}s\right]^2 (b-t)^\alpha \mathrm{d}t \le \frac{4}{1-\alpha}\int_a^b \phi^2(t)(b-t)^\alpha \mathrm{d}t .$$

## A.15 The Gronwall Lemma

Let $\phi = \phi(t)$ be a continuous function in the interval $[0, t^*]$ that is differentiable on $(0, t^*)$. If there exists a constant $\alpha \in \mathbb{R}$ and a continuous function $g(t)$ such that for $0 < t < t^*$, $\phi$ satisfies the inequality

$$\phi'(t) \leq \alpha\phi(t) + g(t)$$

(or equivalently,

$$\phi(t) \leq \phi(0) + \int_0^t [\alpha\phi(s) + g(s)]\mathrm{d}s) \ ,$$

then $\phi$ satisfies the inequality

$$\phi(t) \leq e^{\alpha t}\phi(0) + \int_0^t g(s)e^{\alpha(t-s)}\mathrm{d}s \ .$$

# Appendix B. Fast Fourier Transforms

### *Basics*

The Fast Fourier Transform (FFT) is a recursive algorithm for evaluating the discrete Fourier transform and its inverse. The FFT is conventionally written for the evaluation of

$$\tilde{u}_k = \sum_{j=0}^{N-1} u_j e^{+2\pi ijk/N} , \qquad k = 0, 1, \ldots, N-1 , \tag{B.1.a}$$

$$\tilde{u}_k = \sum_{j=0}^{N-1} u_j e^{-2\pi ijk/N} , \qquad k = 0, 1, \ldots, N-1 , \tag{B.1.b}$$

where $u_j$, $j = 0, 1, \ldots, N-1$, are a set of complex data. The FFT quickly became a widely used tool in signal processing after its description by Cooley and Tukey (1965). (As noted later by Cooley, Lewis and Welch (1969), most essential components of the FFT date back to the 1920s.) The Cooley-Tukey algorithm enables the sums in (B.1) to be evaluated in $5N \log_2 N$ real operations (when $N$ is a power of 2), instead of the $8N^2$ real operations required by the straightforward sum. Moreover, calculation of (B.1) via the FFT incurs less error due to round-off than the direct summation method (Cooley, Lewis and Welch (1969)).

Many versions of the FFT are now in existence. The review by Temperton (1983) contains an especially clear description of a simple yet efficient one. It allows $N$ to be of the form

$$N = 2^p 3^q 4^r 5^s 6^t \tag{B.2}$$

and has the operation count

$$N(5p + 9\tfrac{1}{3}q + 8\tfrac{1}{2}r + 13\tfrac{3}{5}s + 13\tfrac{1}{3}t - 6) . \tag{B.3}$$

No additional flexibility is gained by the inclusion of the factors 4 and 6. The algorithm is, however, more efficient when these factors are included. Not only is the operation count lower – for example, by 15% when $N = 64$ – but, due to the higher ratio of arithmetic operations to memory accesses, most Fortran compilers generate more efficient code for the larger factors. For the sake of simplicity, however, throughout this book we shall use $(5 \log_2 N - 6)N$ as the operation count for the complex FFT; moreover, the lower order term linear in $N$ will usually be omitted.

We should also mention the book by Brigham (1974) which is devoted entirely to the Fast Fourier Transform and the FFTW package by Frigo and Johnson (2005), which received the 1999 Wilkinson Prize for Numerical Software. (The FFTW software is available at http://www.fftw.org/.)

### *Use in Spectral Methods*

In applications of Fourier spectral methods, the sums that one must evaluate are

$$\tilde{u}_k = \frac{1}{N} \sum_{j=0}^{N-1} u_j e^{-2\pi ijk/N} , \qquad k = -\frac{N}{2}, -\frac{N}{2}+1, \ldots, \frac{N}{2}-1 , \tag{B.4}$$

and

$$u_j = \sum_{k=-N/2}^{N/2-1} \tilde{u}_k e^{2\pi ijk/N} , \qquad j = 0, 1, \ldots, N-1 \tag{B.5}$$

(see (2.1.25) and (2.1.27)). From (B.4) it is apparent that, for integers $p$ and $k$,

$$\tilde{u}_{k+pN} = \tilde{u}_k . \tag{B.6}$$

When the array $(u_0, u_1, \ldots, u_{N-1})$ is fed into a standard FFT for evaluating (B.1.b) it returns, in effect, the array

$$(N\tilde{u}_0, N\tilde{u}_1, \ldots, N\tilde{u}_{N/2-1}, N\tilde{u}_{-N/2}, N\tilde{u}_{-N/2+1}, \ldots, N\tilde{u}_{-1}) .$$

Conversely, when this array (without the factor $N$) is fed into the standard FFT for evaluating (B.1.a) (with the plus sign), the array $(u_0, u_1, \ldots, u_{N-1})$ is returned.

In most applications of spectral methods the direct use of the complex FFT (B.1) is needlessly expensive. This is true, for example if the function $u_j$ is real or if a cosine transform (for a Chebyshev spectral method) is desired. These issues have been addressed by Orszag (1971a, Appendix II) and by Brachet et al. (1983, Appendix C). A summary of some of the relevant transformations follows.

### *Real Transforms*

The simplest case occurs when many real transforms are desired at once, as arises for multidimensional problems. They can be computed pairwise. Suppose that $u_j^1$ and $u_j^2$, $j = 0, 1, \ldots, N-1$, are two sets of real data.

Then one can define

$$v_j = u_j^1 + iu_j^2 \tag{B.7}$$

and compute $\tilde{v}_k$ according to (B.4) by the standard $N$-point complex FFT. Then the transforms $\tilde{u}_k^1$ and $\tilde{u}_k^2$ can be extracted according to

$$\begin{aligned} \tilde{u}_k^1 &= \frac{1}{2}(\tilde{v}_k + \overline{\tilde{v}}_{-k}) \\ \tilde{u}_k^2 &= -\frac{i}{2}(\tilde{v}_k - \overline{\tilde{v}}_{-k}) \end{aligned} , \qquad k = 0, 1, \ldots, \frac{N}{2} - 1 . \tag{B.8}$$

(The Fourier coefficients of real data for negative $k$ are related to those for positive $k$ by $\tilde{u}_{-k} = \overline{\tilde{u}}_k$.) This process is readily reversed. In fact, if one is performing a Fourier collocation derivative, one need not even bother with the separation (B.8) in Fourier space, since

$$\left.\frac{du^1}{dx}\right|_j + i \left.\frac{du^2}{dx}\right|_j = \sum_{k=-N/2}^{N/2-1} ik\tilde{v}_k . \tag{B.9}$$

If only a single real transform is desired, then one may follow the prescription given by Orszag (1971a). Let $M = N/2$ and define

$$v_j = u_{2j} + iu_{2j+1} , \qquad j = 0, 1, \ldots, M-1 . \tag{B.10}$$

Then take an $M$-point transform of $v_j$, set $\tilde{v}_M = \tilde{v}_0$, and extract the desired coefficients via

$$\tilde{u}_k = \frac{1}{2}(\tilde{v}_k + \overline{\tilde{v}}_{M-k}) - \frac{i}{2}e^{2\pi ik/N}(\tilde{v}_k - \overline{\tilde{v}}_{M-k}) , \qquad k = 0, 1, \ldots, M-1 . \tag{B.11}$$

For both of these approaches the cost of a single, real-to-half-complex transform is essentially $(5/2)N \log_2 N$.

### *Chebyshev Transforms*

The discrete Chebyshev transforms based on the Gauss-Lobatto points (2.4.14) are given by

$$\tilde{u}_k = \frac{2}{N\bar{c}_k} \sum_{j=0}^{N} \frac{1}{\bar{c}_j} u_j \cos\frac{\pi jk}{N} , \qquad k = 0, 1, \ldots, N , \tag{B.12}$$

(see (2.2.22) and (2.4.15)) and

$$u_j = \sum_{k=0}^{N} \tilde{u}_k \cos\frac{\pi jk}{N} , \qquad j = 0, 1, \ldots, N \tag{B.13}$$

(see (2.2.21) and (2.4.17)). Suppose that the transform (B.12) is desired for two real sets of data $u_j^1$ and $u_j^2$. Then define the complex data $v_j$ by

$$v_j = \begin{cases} u_j^1 + iu_j^2 \ , & j = 0, 1, \ldots, N \ , \\ v_{2N-j} \ , & j = N+1, N+2, \ldots, 2N-1 \ , \end{cases} \tag{B.14}$$

and by periodicity (with period $2N$) for other integers $j$. Next, define $\tilde{v}_k$, $k = 0, 1, \ldots, N$, by (B.12) and define $\tilde{V}_k$, $k = 0, 1, \ldots, 2N-1$, by (B.1.a) with $N$ replaced by $2N$. It is readily shown that

$$\tilde{V}_k = \frac{1}{N\bar{c}_k}\tilde{v}_k \ , \qquad k = 0, 1, \ldots, N \ , \tag{B.15}$$

and that

$$\tilde{V}_k = \sum_{l=0}^{N-1} v_{2l} e^{2\pi i k l/N} + e^{\pi i k/N} \sum_{l=0}^{N-1} v_{2l+1} e^{2\pi i k l/N} \ . \tag{B.16}$$

Now, define $w_j$ by

$$w_j = v_{2j} + i(v_{2j+1} - v_{2j-1}) \ , \qquad j = 0, 1, \ldots, N-1 \ , \tag{B.17}$$

and compute $\tilde{w}_k$ according to the complex FFT (B.1.a). We have

$$\begin{aligned} \tilde{w}_k &= \sum_{l=0}^{N-1} v_{2l} e^{2\pi i k l/N} + i(1 - e^{2\pi i k/N}) \sum_{l=0}^{N-1} v_{2l+1} e^{2\pi i k l/N} \ , \\ \tilde{w}_{N-k} &= \sum_{l=0}^{N-1} v_{2l} e^{2\pi i k l/N} - i(1 - e^{2\pi i k/N}) \sum_{l=0}^{N-1} v_{2l+1} e^{2\pi i k l/N} \ . \end{aligned} \tag{B.18}$$

Consequently,

$$\begin{aligned} \tilde{v}_0 &= \frac{1}{N} \sum_{j=0}^{N} \frac{1}{\bar{c}_j} v_j \ , \\ \tilde{v}_k &= \frac{1}{N} \left[ \left( \frac{1}{2} + \frac{1}{4\sin\frac{\pi k}{N}} \right) \tilde{w}_k + \left( \frac{1}{2} - \frac{1}{4\sin\frac{\pi k}{N}} \right) \tilde{w}_{N-k} \right] \ , \\ \tilde{v}_N &= \frac{1}{N} \sum_{j=0}^{N} (-1)^j \frac{1}{\bar{c}_j} v_j \ . \end{aligned} \tag{B.19}$$

The desired real coefficients $\tilde{u}_k^1$ and $\tilde{u}_k^2$ are the real and imaginary parts, respectively, of the $\tilde{v}_k$. Thus, the discrete Chebyshev transform (B.12) can be computed in $\frac{5}{2}N\log_2 N + 4N$ real operations per transform, assuming that a large number of such transforms are computed. The inverse discrete

Chebyshev transform (B.13) can be evaluated with only minor modifications to the algorithm given by (B.14), (B.17) and (B.19).

Discrete sine transforms can be handled in a similar manner: (B.14) (with $v_{2N-j}$ replaced by $-v_{2N-j}$) and (B.17) are retained as is the central equation in (B.19) with the coefficient of $\tilde{w}_{N-k}$ having the opposite sign; the entire $\tilde{v}_k$ term is multiplied by $i$ and one sets $\tilde{v}_0 = \tilde{v}_N = 0$. Swarztrauber (1986) described how real cosine and sine transforms can be computed without the pre- and post-processing costs incurred by (B.17) and (B.19).

### *Other Cosine Transforms*

In some applications, such as the use of a staggered grid in Navier-Stokes calculations (see CHQZ3, Sect. 3.4) and in simulations of flows with special symmetries (Brachet et al. (1983)), discrete Chebyshev transforms with respect to the Gauss points (see (2.4.12) but with $N-1$ in place of $N$) are required. Consider

$$\tilde{u}_k = \frac{2}{N} \sum_{j=0}^{N-1} u_j \cos \frac{(2j+1)\pi k}{2N} , \qquad k = 0, 1, \ldots, N-1 . \tag{B.20}$$

Brachet et al. (1983) have provided prescriptions for computing efficiently this and related sums. Put

$$v_j = \begin{cases} u_{2j} , & j = 0, 1, \ldots, \frac{N}{2} - 1 , \\ u_{2N-2j-1} , & j = \frac{N}{2}, \frac{N}{2} + 1, \ldots, N-1 , \end{cases} \tag{B.21}$$

and compute $\tilde{v}_k$ according to (B.1.a). Then $\tilde{u}_k$ may be extracted via

$$\tilde{u}_k = \frac{1}{N} \left[ e^{2\pi i k/2N} \tilde{v}_k + e^{-2\pi i k/2N} \tilde{v}_{N-k} \right] , \qquad k = 0, 1, \ldots, N-1 . \tag{B.22}$$

The corresponding inverse Chebyshev transform

$$u_j = \sum_{k=0}^{N-1} \tilde{u}_k \cos \frac{(2j+1)\pi k}{2N} \tag{B.23}$$

can be evaluated by reversing these steps.

For some problems the Chebyshev expansion may be over the interval $[0,1]$ instead of $[-1,1]$. Moreover, it may also be useful to use only the odd (or even) polynomials (Spalart (1984); see also Sect. 2.7.1). Spalart (1986, private communication) explained how to employ the FFT for an expansion over $[0,1]$ in terms of just the odd Chebyshev polynomials. The collocation points are

$$x_j = \cos \frac{(2j+1)\pi}{2N} , \qquad j = 0, 1, \ldots, N-1 , \tag{B.24}$$

the series expansion is

$$u^N(x) = \sum_{k=0}^{N-1} \tilde{u}_k T_{2k+1}(x) , \tag{B.25}$$

and the discrete transforms are

$$\tilde{u}_k = \frac{2}{N} \sum_{j=0}^{N-1} u_j \cos \frac{(2k+1)(2j+1)\pi}{4N} , \qquad k = 0, 1, \ldots, N-1 , \tag{B.26}$$

and

$$u_j = \sum_{k=0}^{N-1} \tilde{u}_k \cos \frac{(2k+1)(2j+1)\pi}{4N} , \qquad j = 0, 1, \ldots, N-1 . \tag{B.27}$$

(In order for a half-interval Chebyshev expansion to be spectrally accurate, one needs $u(x)$ and all of its derivatives to vanish at $x = 0$.) Spalart's trick for evaluating (B.27) is to define

$$\tilde{v}_k = \frac{\tilde{u}_k + \tilde{u}_{k-1}}{2 \cos \left( \dfrac{k\pi}{2N} \right)} , \qquad k = 0, 1, \ldots, N , \tag{B.28}$$

where $\tilde{u}_{-1} = \tilde{u}_N = 0$, to compute $v_j$ according to (B.13), and then to extract $u_j$ via

$$u_j = \frac{\tilde{v}_j + \tilde{v}_{j+1}}{2 \cos \dfrac{(2j+1)\pi}{4N}} , \qquad j = 0, 1, \ldots, N-1 . \tag{B.29}$$

(Note however, that this transform is not suitable for use with the Gauss-Lobatto points.)

# Appendix C. Iterative Methods for Linear Systems

In this appendix, we review some of the most important iterative methods for the solution of a linear system of the same form,

$$L\mathbf{u} = \mathbf{f} \,, \tag{C.0.1}$$

as the one considered in (4.8). The discussion will be at a tutorial level. For an extensive presentation and a thorough analysis the reader may refer to Golub and Van Loan (2003), Saad (1996), Greenbaum (1997), Van der Vorst (2003), and to the ample literature cited therein.

## C.1 A Gentle Approach to Iterative Methods

A particularly simple iterative scheme is the *Richardson* (1910) *method*. Given an initial guess $\mathbf{v}^0$ to $\mathbf{u}$, subsequent approximations are obtained via

$$\mathbf{v}^{n+1} = \mathbf{v}^n + \omega \mathbf{r}^n \,, \tag{C.1.1}$$

where $\omega$ is a relaxation parameter and

$$\mathbf{r}^n = \mathbf{f} - L\mathbf{v}^n \tag{C.1.2}$$

is the residual associated with $\mathbf{v}^n$. The error obeys the relation

$$\left(\mathbf{v}^{n+1} - \mathbf{u}\right) = G\left(\mathbf{v}^n - \mathbf{u}\right) \,, \tag{C.1.3}$$

where the iteration matrix $G$ of the Richardson scheme is given by

$$G = I - \omega L \,. \tag{C.1.4}$$

The iterative scheme is convergent if the spectral radius $\rho$ of $G$ is less than 1. In the case of the Richardson scheme this condition is equivalent to

$$|1 - \omega\lambda| < 1 \,, \tag{C.1.5}$$

for all the eigenvalues $\lambda$ of $L$. The simultaneous fulfilment of these inequalities is possible only if all the eigenvalues of $L$ have nonzero real parts of constant sign. A particularly relevant case is that of a matrix with all real and strictly positive eigenvalues; symmetric and positive-definite matrices enjoy this property, but these are not necessary conditions. For example, the matrices generated by Chebyshev or Legendre collocation discretizations of second-order problems have all real and strictly positive eigenvalues. In such a situation, we have $0 < \lambda_{\min} \leq \lambda_{\max}$, where $\lambda_{\min}$ and $\lambda_{\max}$ are the extreme eigenvalues of $L$. The convergence condition (C.1.5) is satisfied for $0 < \omega < \omega_{\max}$, where

$$\omega_{\max} = 2/\lambda_{\max} \, . \tag{C.1.6}$$

The best choice of $\omega$ is that which minimizes $\rho$. It is obtained from the relation

$$(1 - \omega\lambda_{\max}) = -(1 - \omega\lambda_{\min}) \, , \tag{C.1.7}$$

for then the largest values of $1-\omega\lambda$ are equal in magnitude and have opposite sign (see Fox and Parker (1968), Quarteroni and Valli (1994)). The optimal relaxation parameter is thus

$$\omega_{\text{opt}} = \frac{2}{\lambda_{\max} + \lambda_{\min}} \, . \tag{C.1.8}$$

It produces the spectral radius

$$\rho = \frac{\lambda_{\max} - \lambda_{\min}}{\lambda_{\max} + \lambda_{\min}} \, . \tag{C.1.9}$$

Note that the dependence upon the extreme eigenvalues enters only in the combination

$$\mathcal{K} = \frac{\lambda_{\max}}{\lambda_{\min}} \, . \tag{C.1.10}$$

We shall call this ratio the *iterative condition number* of $L$ to distinguish it from the spectral condition number defined in (4.3.2). Obviously, for a symmetric and positive-definite matrix $L$, the iterative and spectral condition numbers coincide. However, for some nonsymmetric discretization matrices that have real positive eigenvalues, such as those mentioned above, the spectral and the iterative condition numbers might differ. In terms of this ratio, (C.1.9) becomes

$$\rho = \frac{\mathcal{K} - 1}{\mathcal{K} + 1} \, . \tag{C.1.11}$$

Define the *rate of convergence* $\mathcal{R}$ to be

$$\mathcal{R} = -\log \rho \, , \tag{C.1.12}$$

and denote its reciprocal by $\mathcal{J}$. The latter quantity measures the number of iterations required to reduce the error by a factor of $e$. This immediately follows from the error bound

$$\|\mathbf{v}^n - \mathbf{u}\|_L \leq \rho^n \|\mathbf{v}^0 - \mathbf{u}\|_L ,$$

which holds with $\|\mathbf{v}\|_L = (\mathbf{v}^T L \mathbf{v})^{1/2}$. Clearly, the larger the convergence rate that a method has for a problem, the fewer iterations that are required to obtain a solution to a given accuracy. For the Richardson method described above, the number of iterations increases as

$$\mathcal{J} \cong \frac{1}{2}\mathcal{K} . \tag{C.1.13}$$

The basic Richardson method (C.1.1) can be improved and extended in several ways. The discussion thus far concerned only the *stationary* Richardson method. In a *non-stationary* Richardson method, the parameter $\omega$ in (C.1.1) is allowed to depend on $n$, i.e. to change in the course of iterations, in order to speed up the convergence.

For a *static* non-stationary Richardson (NSR) method one cycles through a fixed number $k$ of parameters. Using the minimax property of Chebyshev polynomials, one derives the following expressions for the optimal parameters (Young (1954)):

$$\omega_j = \frac{2/\lambda_{\min}}{(\mathcal{K}-1)\cos\dfrac{(2j-1)\pi}{2k} + (\mathcal{K}+1)} , \qquad j = 1, \dots, k , \tag{C.1.14}$$

and the effective spectral radius

$$\rho = \frac{1}{\left[T_k\left(\dfrac{\mathcal{K}+1}{\mathcal{K}-1}\right)\right]^{1/k}} . \tag{C.1.15}$$

Both $\omega_j$ (for all $j$) and $\rho$ depend on $\mathcal{K}$. However, this approach suffers from the same limitation as the basic Richardson method – information must be available on the eigenvalues of $L$ in order to compute $\mathcal{K}$.

A broad family of *dynamic* non-stationary Richardson methods are based on an optimality strategy that does not require the knowledge of the extreme eigenvalues. We address dynamic non-stationary Richardson methods in Sects. C.2 and 4.5.2.

The primary cause of the inefficiency of the Richardson method is that the convergence rate decreases as the iterative condition number increases; in spectral methods, the condition number typically increases with the approximation parameter $N$. This can be alleviated by *preconditioning* the problem, in effect solving

$$H^{-1}L\mathbf{u} = H^{-1}\mathbf{f}$$

rather than (C.0.1). (This is called *left preconditioning.* Other options are available as well, such as *right preconditioning* or *symmetric preconditioning*; see (C.2.15) and (C.2.18), respectively.)

A preconditioned version of (C.1.1) is

$$H\left(\mathbf{v}^{n+1}-\mathbf{v}^{n}\right)=\omega\mathbf{r}^{n}\ . \tag{C.1.16}$$

One obvious requirement for $H$ is that this equation can be solved inexpensively, i.e., in fewer operations than are required to evaluate $L\mathbf{v}^n$. The effective iteration matrix is now

$$G=I-\omega H^{-1}L\ . \tag{C.1.17}$$

The second requirement on the preconditioning matrix is that $H^{-1}$ be a good approximation to $L^{-1}$, i.e., that the new iterative condition number $\mathcal{K}(H^{-1}L)$ be much smaller than $\mathcal{K}(L)$. In such circumstances, the new spectral radius $\rho$ is much smaller than that of the non-preconditioned Richardson method. This property can be rigorously justified whenever $L$ and $H$ are both symmetric and positive definite. Indeed, denoting by $H^{1/2}$ the square root of $H$, (C.1.16) can be written equivalently as

$$\mathbf{w}^{n+1}=\mathbf{w}^{n}+\omega(H^{-1/2}\mathbf{f}-H^{-1/2}LH^{-1/2}\mathbf{w}^{n})$$

with $\mathbf{w}^n=H^{1/2}\mathbf{v}^n$, showing that (C.1.16) is nothing but a Richardson iteration applied to the symmetric and positive-definite matrix $H^{-1/2}LH^{-1/2}$. Since this matrix is similar to $H^{-1}L$, we have

$$\mathcal{K}(H^{-1/2}LH^{-1/2})=\mathcal{K}(H^{-1}L).$$

The discussion so far has presumed that the eigenvalues of $H^{-1}L$ are confined to the interval $[\lambda_{\min},\lambda_{\max}]$ on the positive real axis. However, the Richardson iteration schemes can work on problems for which the eigenvalues are complex but have positive real parts. If we still use a real $\omega$, then it should obey the following restriction for convergence:

$$\omega<2\frac{\mathrm{Re}\,(\lambda_i)}{|\lambda_i|^2}\,,$$

for all eigenvalues $\lambda_i$ of $H^{-1}L$ (see, e.g., Quarteroni and Valli (1994), Sect. 2.4). One could also use a complex $\omega$, in which case the iterations can be performed entirely in real arithmetic according to

$$\mathbf{v}^{n+1}=\mathbf{v}^{n}+2\,\mathrm{Re}\{\omega\}H^{-1}\mathbf{r}^{n}+|\omega|^{2}H^{-1}LH^{-1}\mathbf{r}^{n}\ . \tag{C.1.18}$$

The value of the optimal parameter $\omega_{opt}$ is obtained by solving a minimax problem in complex arithmetic.

## C.2 Descent Methods for Symmetric Problems

Unlike the stationary Richardson method discussed previously, descent methods have no parameters such as $\omega$ that require knowledge of the extreme eigenvalues $\lambda_{\min}$ and $\lambda_{\max}$ of the matrix $L$ or of $H^{-1}L$, where $H$ is a suitable preconditioner. The principle is to adjust the current guess $\mathbf{v}^n$ via

$$H(\mathbf{v}^{n+1} - \mathbf{v}^n) = \alpha_n \mathbf{r}^n \,, \tag{C.2.1}$$

where $\mathbf{r}^n = \mathbf{f} - L\mathbf{v}^n$ is the residual, and the scalar $\alpha_n$ – the dynamic relaxation parameter – is chosen according to some optimality criterium, as described below. In this section we will assume that both $L$ and $H$ are symmetric and positive-definite (but the reader should be aware that these iterative methods may work even if this condition is not satisfied).

The most natural option for defining $\alpha_n$ is to minimize the Euclidean norm of the new residual $\mathbf{r}^{n+1}$; another option is to minimize the so-called $H$-norm of the new preconditioned residual $\mathbf{p}^{n+1} = H^{-1}\mathbf{r}^{n+1}$, i.e., the quantity $\|\mathbf{p}^{n+1}\|_H = (H\mathbf{p}^{n+1}, \mathbf{p}^{n+1})^{1/2} = \|\mathbf{r}^{n+1}\|_{H^{-1}}$. Both options are referred to as *preconditioned minimum residual Richardson* (PMRR) *methods* and will be denoted by PMRR$_2$ and PMRR$_H$, respectively. An additional option is to minimize the $L$-norm of the new error $\mathbf{e}^{n+1} = \mathbf{u} - \mathbf{v}^{n+1}$, i.e., the quantity $\|\mathbf{e}^{n+1}\|_L = (L\mathbf{e}^{n+1}, \mathbf{e}^{n+1})^{1/2}$. This is referred to as a *preconditioned steepest descent Richardson* (PSDR) *method.*

The corresponding algorithms can be written compactly as follows:

***Preconditioned Richardson Methods***

Initialize

$$\mathbf{v}^0, \quad \mathbf{r}^0 = \mathbf{f} - L\mathbf{v}^0, \quad H\mathbf{p}^0 = \mathbf{r}^0 \,.$$

Iterate

$$\begin{aligned}
&\alpha_n \text{ defined according to one of the rows of table C.1} \,, \\
&\mathbf{v}^{n+1} = \mathbf{v}^n + \alpha_n \mathbf{p}^n \,, \\
&\mathbf{r}^{n+1} = \mathbf{r}^n - \alpha_n L\mathbf{p}^n \,, \\
&H\mathbf{p}^{n+1} = \mathbf{r}^{n+1} \,.
\end{aligned} \tag{C.2.2}$$

Note that for non-preconditioned iterations, then $H = I$ and $\mathbf{p}^n = \mathbf{r}^n$ in Table C.1. (In particular, PMRR$_2$ and PMRR$_H$ coincide if $P = I$.)

For PMRR$_H$ iterations the following estimate holds for the preconditioned residual:

$$\|\mathbf{p}^n\|_H \leq \left(\frac{\mathcal{K} - 1}{\mathcal{K} + 1}\right)^n \|\mathbf{p}^0\|_H \,, \tag{C.2.3}$$

**Table C.1.** The three different strategies for Richardson iterations (PMRR )

| Name of method | Acceleration parameter | Method minimizes |
|---|---|---|
| $\text{PMRR}_2$ | $\alpha_n = \dfrac{(\mathbf{r}^n, L\mathbf{p}^n)}{(L\mathbf{p}^n, L\mathbf{p}^n)}$ | $\lVert \mathbf{r}^{n+1} \rVert$ |
| $\text{PMRR}_H$ | $\alpha_n = \dfrac{(\mathbf{p}^n, L\mathbf{p}^n)}{(L\mathbf{p}^n, H^{-1}L\mathbf{p}^n)}$ | $\lVert \mathbf{p}^{n+1} \rVert_H$ |
| PSDR | $\alpha_n = \dfrac{(\mathbf{p}^n, \mathbf{r}^n)}{(\mathbf{p}^n, L\mathbf{p}^n)}$ | $\lVert \mathbf{e}^{n+1} \rVert_L$ |

where $\mathcal{K}$ still denotes the iterative condition number of $H^{-1}L$, while for PSDR iterations we have

$$\lVert \mathbf{e}^n \rVert_L \leq \left( \frac{\mathcal{K}-1}{\mathcal{K}+1} \right)^n \lVert \mathbf{e}^0 \rVert_L \tag{C.2.4}$$

(see Quarteroni and Valli (1994), Sect. 2.4). Note that when $H = I$ (no preconditioning), the PSDR method reduces to the classical steepest descent (or gradient) algorithm. Also note that, in both cases, the number of iterations required for convergence is proportional to

$$\mathcal{J} = \frac{1}{2}\mathcal{K} . \tag{C.2.5}$$

When the eigenvalues of the preconditioned matrix $H^{-1}L$ are complex but with dominant real parts, a surrogate for $\mathcal{K}$ that is still representative of the convergence behavior of the Richarson iterations is

$$\mathcal{K}^* = \frac{\max_j |\lambda_j|}{\min_j |\lambda_j|} . \tag{C.2.6}$$

A substantial improvement in convergence rate can be achieved by using *conjugate direction methods* in place of PMRR or PSDR. The two most common conjugate direction methods are known as the conjugate gradient method and the conjugate residual method. These methods were proposed by Hestenes and Stiefel (1952) as a direct method for solving symmetric and positive-definite linear systems. For such problems the conjugate direction methods produce the exact answer (in the absence of round-off errors) in a finite number of steps. In the late 1960s and early 1970s these methods began to be considered seriously as iterative, rather than direct, solution schemes that can produce a very accurate result in a small number of iterations. The papers by Reid (1971) and by Concus, Golub and O'Leary (1976) were particularly influential.

In a non-preconditioned conjugate direction method the update of the iterate is generalized from (C.2.1) to

$$\mathbf{v}^{n+1} = \mathbf{v}^n + \alpha_n \mathbf{p}^n \,. \tag{C.2.7}$$

In the conjugate gradient version, the directions satisfy the orthogonality property

$$\left(\mathbf{p}^{n+1}, L\mathbf{p}^n\right) = 0 \,. \tag{C.2.8}$$

The scheme is initialized with an initial guess $\mathbf{v}^0$. The initial direction vector is chosen to be $\mathbf{p}^0 = \mathbf{r}^0$, where $\mathbf{r}^0$ is the initial residual. Subsequent iterations are made according to the following formulas:

***Conjugate Gradient (CG) Method***

$$\begin{aligned}
\alpha_n &= \frac{(\mathbf{r}^n, \mathbf{r}^n)}{(\mathbf{p}^n, L\mathbf{p}^n)} \,, \\
\mathbf{v}^{n+1} &= \mathbf{v}^n + \alpha_n \mathbf{p}^n \,, \\
\mathbf{r}^{n+1} &= \mathbf{r}^n - \alpha_n L\mathbf{p}^n \,, \\
\beta_n &= \frac{\left(\mathbf{r}^{n+1}, \mathbf{r}^{n+1}\right)}{(\mathbf{r}^n, \mathbf{r}^n)} \,, \\
\mathbf{p}^{n+1} &= \mathbf{r}^{n+1} + \beta_n \mathbf{p}^n \,.
\end{aligned} \tag{C.2.9}$$

In (C.2.9) the formula for the familiar scalar $\alpha_n$ results from the requirement that $\mathbf{v}^{n+1}$ minimize the energy norm of the error, and the formula for the additional scalar $\beta_n$ follows from the requirement (C.2.8).

The following orthogonality properties hold:

$$\left(\mathbf{r}^k, \mathbf{r}^l\right) = 0, \quad \left(\mathbf{p}^k, L\mathbf{p}^l\right) = 0 \quad \text{for } k \neq l \,. \tag{C.2.10}$$

The first of these implies that $\mathbf{r}^m = 0$ for some $m \leq nd$, where $nd$ is the order of the matrix $L$. (Here we use the symbol $nd$ to denote the dimension of the linear system (C.0.1) instead of $n$ as done in Chap. 4, given that $n$ is a natural symbol for the iteration index.) This explains the claim that the exact solution is obtained in a finite number of iterations. However, the presence of rounding errors leads to some contamination of the residual and direction vectors. The second orthogonality relation shows that the CG method does far more than the original requirement (C.2.8); indeed, we say that the directions $\{\mathbf{p}^k\}$ are *L-conjugated.*

The favorable convergence properties of this method are reflected by the estimate for the energy error (which improves the one in (C.2.4)):

$$\|\mathbf{e}^n\|_L \leq 2 \left(\frac{\sqrt{\mathcal{K}} - 1}{\sqrt{\mathcal{K}} + 1}\right)^n \|\mathbf{e}^0\|_L \,. \tag{C.2.11}$$

The number of iterations required for convergence is therefore proportional to

$$\mathcal{J} = \frac{1}{2}\sqrt{\mathcal{K}}\,. \tag{C.2.12}$$

This is a decided improvement over the result (C.2.5). Of course, the CG method is more costly per iteration, both in CPU time and storage.

The *conjugate residual method* is similar, but now the orthogonality property is

$$\left(L\mathbf{p}^{n+1}, L\mathbf{p}^{n}\right) = 0\,, \tag{C.2.13}$$

and the requirement on $\mathbf{v}^{n+1}$ is that it minimize the Euclidean norm of the residual.

Let us now include a symmetric preconditioning, denoted as usual by $H$, in these descent methods. It is tempting to write (C.0.1) as either

$$\tilde{L}\mathbf{u} = \tilde{\mathbf{f}} \quad \text{with} \quad \tilde{L} = H^{-1}L \quad \text{and} \quad \tilde{\mathbf{f}} = H^{-1}\mathbf{f} \tag{C.2.14}$$

or

$$\tilde{L}\tilde{\mathbf{u}} = \mathbf{f}, \quad \text{where} \quad \tilde{L} = LH^{-1} \quad \text{and} \quad \tilde{\mathbf{u}} = H\mathbf{u}, \tag{C.2.15}$$

and then apply the preceding formulas to either (C.2.14) or (C.2.15). However, $\tilde{L}$ is not necessarily symmetric and positive definite (unless $L$ and $H^{-1}$ commute). We can, however, choose $Q$ such that

$$H = QQ^T\,, \tag{C.2.16}$$

and use

$$\tilde{L}\tilde{\mathbf{u}} = \tilde{\mathbf{f}}\,, \tag{C.2.17}$$

with

$$\tilde{L} = Q^{-1}LQ^{-T}, \quad \tilde{\mathbf{f}} = Q^{-1}\mathbf{f}, \quad \tilde{\mathbf{u}} = Q^T\mathbf{u}\,. \tag{C.2.18}$$

We also use

$$\tilde{\mathbf{v}} = Q^T\mathbf{v}, \quad \tilde{\mathbf{p}} = Q^T\mathbf{p}, \quad \tilde{\mathbf{r}} = Q^{-1}\mathbf{r}\,. \tag{C.2.19}$$

This ensures that the matrix $\tilde{L}$ is symmetric and positive definite. After inserting (C.2.18) into the preceding schemes and then manipulating the expressions into computationally convenient forms, we arrive at the following:

### *Preconditioned Conjugate Gradient (PCG) Method*

Initialize

$$\mathbf{v}^0, \quad \mathbf{r}^0 = \mathbf{f} - L\mathbf{v}^0, \quad H\mathbf{z}^0 = \mathbf{r}^0, \quad \mathbf{p}^0 = \mathbf{z}^0\,.$$

Iterate

$$
\begin{aligned}
\alpha_n &= \frac{(\mathbf{r}^n, \mathbf{z}^n)}{(\mathbf{p}^n, L\mathbf{p}^n)} , \\
\mathbf{v}^{n+1} &= \mathbf{v}^n + \alpha_n \mathbf{p}^n , \\
\mathbf{r}^{n+1} &= \mathbf{r}^n - \alpha_n L\mathbf{p}^n , \\
H\mathbf{z}^{n+1} &= \mathbf{r}^{n+1} , \\
\beta_n &= \frac{(\mathbf{r}^{n+1}, \mathbf{z}^{n+1})}{(\mathbf{r}^n, \mathbf{z}^n)} , \\
\mathbf{p}^{n+1} &= \mathbf{z}^{n+1} + \beta_n \mathbf{p}^n .
\end{aligned}
\tag{C.2.20}
$$

***Preconditioned Conjugate Residual (PCR) Method***

Initialize

$$
\mathbf{v}^0, \quad \mathbf{r}^0 = \mathbf{f} - L\mathbf{v}^0, \quad H\mathbf{z}^0 = \mathbf{r}^0, \quad \mathbf{p}^0 = \mathbf{z}^0 .
$$

Iterate

$$
\begin{aligned}
\alpha_n &= \frac{(\mathbf{r}^n, L\mathbf{p}^n)}{(L\mathbf{p}^n, L\mathbf{p}^n)} , \\
\mathbf{v}^{n+1} &= \mathbf{v}^n + \alpha_n \mathbf{p}^n , \\
\mathbf{r}^{n+1} &= \mathbf{r}^n - \alpha_n L\mathbf{p}^n , \\
H\mathbf{z}^{n+1} &= \mathbf{r}^{n+1} , \\
\beta_n &= -\frac{(L\mathbf{z}^{n+1}, L\mathbf{p}^n)}{(L\mathbf{p}^n, L\mathbf{p}^n)} , \\
\mathbf{p}^{n+1} &= \mathbf{z}^{n+1} + \beta_n \mathbf{p}^n . \\
L\mathbf{p}^{n+1} &= L\mathbf{z}^{n+1} + \beta_n L\mathbf{p}^n .
\end{aligned}
\tag{C.2.21}
$$

The preconditioned conjugate gradient method minimizes the $L$-norm of the error; thus, the associated error satisfies (C.2.11). However, now the relevant condition number is that of $Q^{-1}LQ^{-T}$ (which coincides with that of $H^{-1}L$) rather than that of $L$.

For the CG and CR methods, their orthogonality properties are lost when applied to nonsymmetric problems. In this case they are more properly called the *truncated conjugate gradient* (TCG) and *truncated conjugate residual* (TCR) *methods*. Their preconditioned versions are abbreviated as the PTCG and PTCR methods, and they are given by (C.2.20) and (C.2.21), respectively.

Although the descent methods described in this section may work for nonsymmetric problems, the methods in the following section are usually preferable for the general case.

## C.3 Krylov Methods for Nonsymmetric Problems

The subject of iterative schemes for nonsymmetric problems has received much attention since the 1980's. The descent methods that we discuss in this subsection are but a small subset of the schemes that have been proposed.

Since the matrix $L$ is not symmetric, we can use either one of the transformations (C.2.14)–(C.2.15) or (C.2.16)–(C.2.19). The preconditioned matrix $\tilde{L}$ determines the performance of Krylov methods.

When the Richardson method (C.1.1) is applied to the solution of the linear system (C.0.1), the residual, $\mathbf{r}^n = \mathbf{f} - L\mathbf{v}^n$, at the $n$-th iteration can be related to the initial residual as

$$\mathbf{r}^n = \prod_{j=0}^{n-1}(I - \omega_j L)\mathbf{r}^0 = \mathbf{p}_n(L)\mathbf{r}^0 , \tag{C.3.1}$$

where $\omega_j$ is the relaxation parameter at the $j$-th step, while $\mathbf{p}_n(L)$ indicates a polynomial in $L$ of degree $n$.

Let us introduce the space

$$K_m(L;\mathbf{w}) = span\{\mathbf{w}, L\mathbf{w}, \dots, L^{m-1}\mathbf{w}\} , \qquad m \geq 1 , \tag{C.3.2}$$

called the *Krylov space of order* $m$ associated with the matrix $L$ and the vector $\mathbf{w}$. Then, $\mathbf{r}^n \in K_{n+1}(L;\mathbf{r}^0)$. From (C.1.1) we obtain

$$\mathbf{v}^n = \mathbf{v}^0 + \sum_{j=1}^{n-1} \omega_j \mathbf{r}^j ;$$

thus,

$$\mathbf{v}^n - \mathbf{v}^0 \in K_n(L;\mathbf{r}^0)$$

and

$$\mathbf{v}^n - \mathbf{v}^0 = p_{n-1}(L)\mathbf{r}^0 .$$

More generally, methods can be devised in such a way that

$$\mathbf{v}^n - \mathbf{v}^0 = q_{n-1}(L)\mathbf{r}^0 , \tag{C.3.3}$$

where $q_{n-1}$ is a polynomial chosen so that $\mathbf{v}^n$ represents the "best" approximation of the solution $\mathbf{u}$ in $\tilde{K}_n = \mathbf{v}^0 + K_n(L;\mathbf{r}^0)$. Any such method is called a *Krylov method.*

For any fixed $m \geq 1$, an orthonormal basis $\{\mathbf{w}_i\}$ for $K_m(L;\mathbf{w})$ can be computed using the so-called *Arnoldi algorithm.* Setting $\mathbf{w}_1 = \mathbf{v}/\|\mathbf{w}\|$, we apply the Gram-Schmidt procedure: for $k \geq 1$,

$$g_{ik} = \mathbf{w}_i^T L\mathbf{w}_k , \qquad i = 1, \dots, k , \tag{C.3.4}$$

$$\mathbf{z}_k = L\mathbf{w}_k - \sum_{i=1}^{k} g_{ik}\mathbf{w}_i , \tag{C.3.5}$$

$$g_{k+1,k} = \|\mathbf{z}_k\| . \tag{C.3.6}$$

Should $\mathbf{z}_k = \mathbf{0}$ the process terminates, and we say that a *breakdown* of the algorithm has occurred. Otherwise, we set

$$\mathbf{w}_{k+1} = \frac{\mathbf{z}_k}{\|\mathbf{z}_k\|} , \tag{C.3.7}$$

and the algorithm continues, incrementing $k$ by 1.

If the algorithm terminates at the step $m$, then $\{\mathbf{w}_1, \dots, \mathbf{w}_m\}$ forms a basis for $K_m(L; \mathbf{v})$. In such a case, denoting by $W_m \in \mathbb{R}^{n\times m}$ the matrix whose columns are the vectors $\mathbf{w}_i$, we obtain

$$W_m^T L W_m = G_m, \quad W_{m+1}^T L W_m = \hat{G}_m , \tag{C.3.8}$$

where $\hat{G}_m \in \mathbb{R}^{(m+1)\times m}$ is an upper-Hessenberg matrix whose entries are the $g_{ij}$, while $G_m \in \mathbb{R}^{m\times m}$ is the restriction of $\hat{G}_m$ to the first $m$ rows and $m$ columns. In our application the Krylov space will be invariably constructed for $\mathbf{v} = \mathbf{r}^0$.

This algorithm for generating an orthonormal basis for a Krylov space of any order is the foundation for solving the linear system (C.0.1) by a Krylov method. The most natural approach would be to search for $\mathbf{v}^n$ as the vector that minimizes the error $\|\mathbf{v}^n - \mathbf{u}\|$ in $\tilde{K}_n$. However, since $\mathbf{u}$ is unknown, this method would not work in practice. Two alternative strategies that are workable are

1. Compute $\mathbf{v}^n$ by enforcing that the residual $\mathbf{r}^n$ be orthogonal to any vector in $K_n(L; \mathbf{r}^0)$, i.e.,

$$\mathbf{v}^T(\mathbf{f} - L\mathbf{v}^n) = 0 \qquad \forall \mathbf{v} \in K_n(L; \mathbf{r}^0) . \tag{C.3.9}$$

   This leads to the so-called *full orthogonalization method* (FOM).

2. Compute $\mathbf{v}^n \in \tilde{K}_n$ by minimizing the norm of the residual $\mathbf{r}^n$, i.e.,

$$\|\mathbf{f} - L\mathbf{v}^n\| = \min_{\mathbf{v}\in\tilde{K}_n} \|\mathbf{f} - L\mathbf{v}\| , \tag{C.3.10}$$

   which yields the *generalized minimum residual method* (GMRES).

Note that

$$\mathbf{v}^n = \mathbf{v}^0 + W_n \mathbf{q}^n, \tag{C.3.11}$$

where $\mathbf{q}^n$ has to be chosen according to the selected optimality criterion ((C.3.9) or (C.3.10)).

Then,

$$\mathbf{r}^n = \mathbf{r}^0 - LW_n\mathbf{q}^n ,$$

since $\mathbf{r}^0 = \mathbf{w}_1\|\mathbf{r}_0\|$. From (C.3.8) it follows that

$$\mathbf{r}^n = W_{n+1}(\|\mathbf{r}^0\|\mathbf{e}_1 - \hat{G}_n\mathbf{q}^n) , \tag{C.3.12}$$

where $\mathbf{e}_1$ is the first unit vector of the canonical basis of $\mathbb{R}^{n+1}$. Thus, in the GMRES method the solution at step $n$ is computed through (C.3.11) where

$$\mathbf{q}^n \text{ minimizes } \| \, (\|\mathbf{r}^0\|\mathbf{e}_1 - \hat{G}_n\mathbf{q}) \, \| \text{ with respect to } \mathbf{q}. \tag{C.3.13}$$

Note that the matrix $W_{n+1}$ appearing in (C.3.12) does not change the value of $\|\mathbf{r}^0\|$ since it is an orthogonal matrix.

Clearly, the GMRES method will be the more effective the smaller the number of iterations, particularly since at each step one has to solve a least-squares problem (C.3.13). The GMRES method in exact arithmetic enjoys the so-called finite-termination property, i.e., it terminates after at most $nd$ iterations, where again $nd$ denotes the order of the matrix $L$. Premature stops are due to a breakdown in the Arnoldi orthonormalization algorithm. This breakdown occurs only if the computed solution $\mathbf{v}^n$ coincides with the exact solution $\mathbf{u}$ for some $n < nd$. However, unless acceptable convergence is reached after just a few iterations, the GMRES method requires prohibitive computational costs for the orthogonalization and excessive storage for the retention of the Krylov subspace bases.

A popular variant consists of restarting GMRES after each $m$ iteration steps. This algorithm is referred to as GMRES($m$); the nonrestarted version is sometimes called *full GMRES*. As pointed out in van der Vorst (2003), there is no simple rule to determine a suitable value of $m$; in fact, the speed of convergence of GMRES($m$) may vary drastically for nearby values of $m$. In some cases, a superlinear convergence behaviour of the full GMRES iterations is observed.

The convergence analysis of GMRES is not trivial, and we report just some of the more elementary results here. If $L$ is positive definite, i.e., its symmetric part $L_S$ has positive eigenvalues, then the $n$-th residual decreases according to the following bound:

$$\|\mathbf{r}^n\| \leq \sin^n(\beta)\|\mathbf{r}^0\| \,, \tag{C.3.14}$$

where $\cos(\beta) = \lambda_{\min}(L_S)/\|L\|$ with $\beta \in [0, \pi/2)$. As usual, $\| \cdot \|$ denotes the Euclidean vector or matrix norm. Moreover, GMRES($m$) converges for all $m \geq 1$. In order to obtain a bound on the residual at a step $n \geq 1$, let us assume that the matrix $L$ is diagonalizable:

$$L = T\Lambda T^{-1} \,,$$

where $\Lambda$ is the diagonal matrix of eigenvalues, $\{\lambda_j\}_{j=1,\dots,nd}$, and $T = (\boldsymbol{\omega}^1, \dots, \boldsymbol{\omega}^{\mathbb{N}^d})$ is the matrix whose columns are the right eigenvectors of $L$. Under these assumptions, the residual norm after $n$ steps of GMRES satisfies

$$\|\mathbf{r}^n\| \leq \kappa_2(T)\delta\|\mathbf{r}^0\| \,,$$

where $\kappa_2(T) = \|T\|_2\|T^{-1}\|_2$ is the condition number of $T$ defined in (4.3.2), and

$$\delta = \min_{p \in \mathbb{P}_n, p(0)=1} \; \max_{1 \leq i \leq nd} |p(\lambda_i)| \,.$$

Moreover, suppose that the initial residual is dominated by $m$ eigenvectors, i.e., $\mathbf{r}^0 = \sum_{j=1}^m \alpha_j \boldsymbol{\omega}^j + \mathbf{e}$, with $\|\mathbf{e}\|$ small in comparison to $\|\sum_{j=1}^m \alpha_j \boldsymbol{\omega}^j\|$, and assume that if some complex $\boldsymbol{\omega}^j$ appears in the previous sum, then its conjugate $\overline{\boldsymbol{\omega}}^j$ appears as well. Then

$$\|\mathbf{r}^n\| \leq \kappa_2(T) c_n \|\mathbf{e}\| \ ,$$

$$c_n = \max_{p>n} \prod_{j=1}^{n} \left| \frac{\lambda_p - \lambda_j}{\lambda_j} \right| \ .$$

Very often, $c_n$ is of order one; hence, $n$ steps of GMRES reduce the residual norm to the order of $\|\mathbf{e}\|$ provided that $\kappa_2(T)$ is not too large.

In general, as highlighted from the previous estimate, the eigenvalue information alone is not enough, and information on the eigensystem is also needed. If the eigensystem is orthogonal, as for normal matrices, then $\kappa_2(T) = 1$, and the eigenvalues are descriptive for convergence. Otherwise, upper bounds for $\|\mathbf{r}^n\|$ can be provided in terms of both spectral and pseudospectral information, as well as the so-called *field of values* of $L$:

$$\mathcal{F}(L) = \{\mathbf{v}^* L \mathbf{v} \mid \|\mathbf{v}\| = 1\}.$$

If $0 \notin \mathcal{F}(L)$, then the estimate (C.3.14) can be improved by replacing $\lambda_{\min}(L_S)$ with $\mathrm{dist}(0, \mathcal{F}(L))$.

An extensive discussion of convergence of GMRES and GMRES($m$) can be found in Saad (1996), Embree (1999) and van der Vorst (2003).

The GMRES method can of course be implemented for a preconditioned system. We provide here an implementation of the preconditioned GMRES method with a left preconditioner $H$.

### *Preconditioned GMRES (PGMRES) Method*

$\mathbf{v}^0$, $H\mathbf{r}^0 = \mathbf{f} - L\mathbf{v}^0$, $\beta = \|\mathbf{r}^0\|$, $\mathbf{v}^1 = \mathbf{r}^0/\beta$.

Iterate

$$\begin{aligned}
&\textit{For } j = 1, \ldots, n \textit{ Do} \\
&\quad \text{Compute } H\mathbf{w}^j = L\mathbf{v}^j \\
&\quad \textit{For } i = 1, \ldots, j \textit{ Do} \\
&\qquad g_{ij} = (\mathbf{v}^i)^T \mathbf{w}^j \\
&\qquad \mathbf{w}^j = \mathbf{w}^j - g_{ij} \mathbf{v}_i \\
&\quad \textit{End Do} \\
&\quad g_{j+1,j} = \|\mathbf{w}^j\| \\
&\quad (\textit{if } g_{j+1,j} = 0 \text{ set } n = j \text{ and } \textit{Goto } (1)) \\
&\quad \mathbf{v}^{j+1} = \mathbf{w}^j / g_{j+1,j} \\
&\textit{End Do} \\
&W_n = [\mathbf{v}^1, \ldots, \mathbf{v}^n],\ \hat{G}_n = \{g_{ij}\},\ 1 \leq j \leq n,\ 1 \leq i \leq j+1;
\end{aligned} \tag{C.3.15}$$

(1) Compute $\mathbf{q}^n$ , the minimizer of $\|\beta \mathbf{e}_1 - \hat{G}_n \mathbf{q}\|$
Set $\mathbf{v}^n = \mathbf{v}^0 + W_n \mathbf{q}^n$

More generally, as proposed by Saad (1996), a variable preconditioner $H_n$ can be used at the $n$-th iteration, yielding the so-called *flexible GMRES* method. The use of a variable preconditioner is especially interesting in those situations where the preconditioner is not explicitly given, but implicitly defined, for instance, as an approximate Jacobian in a Newton iteration or by a few steps of an inner iteration process. Another meaningful case is the one of domain decomposition preconditioners (of either Schwarz or Schur type) where the preconditioning step involves one or several substeps of local solves in the subdomains (see CHQZ3, Chap. 6).

Several considerations for the practical implementation of GMRES, its relation with FOM, how to restart GMRES, and the Householder version of GMRES can be found in Saad (1996).

A different approach to iterative methods for nonsymmetric matrices consists of generalizing the conjugate gradient method through a specific characterization of the properties satisfied by the residual.

The property that the residual vectors $\mathbf{r}^n$ generated by the CG method satisfy a three-term recurrence is lost when $L$ is not symmetric. The *bi-conjugate gradient* (Bi-CG) *method* introduced by Fletcher (1976) constructs a residual $\mathbf{r}^k$ orthogonal to another row of vectors $\tilde{\mathbf{r}}^0, \tilde{\mathbf{r}}^1, \ldots, \tilde{\mathbf{r}}^{n-1}$, and, vice versa, $\tilde{\mathbf{r}}^n$ is orthogonal with respect to $\mathbf{r}^0, \mathbf{r}^1, \ldots, \mathbf{r}^{n-1}$. This method enjoys the finite-termination property, but there is no minimization property as in CG or GMRES for the intermediate steps. When this method converges, both $\{\mathbf{r}^n\}$ and $\{\tilde{\mathbf{r}}^n\}$ converge towards zero but only the convergence of the $\{\tilde{\mathbf{r}}^n\}$ is exploited. Based on this observation, Sonneveld (1989) proposed a modification called the *conjugate gradient-squared* (CGS) *method* that focuses more strongly on the $\{\mathbf{r}^n\}$ vectors. CGS generates residual vectors $\mathbf{r}^n$ given by

$$\mathbf{r}^n = p_n^2(L)\mathbf{r}^0 ,$$

where $p_n(L)$ is that $n$-th degree polynomial in $L$ for which $p_n(L)\mathbf{r}^0$ is equal to the residual at the $n$-th step obtained by means of the Bi-CG method.

In the *Bi-CGStab method*, introduced by van der Vorst (1992), instead of simply squaring the Bi-CG polynomial, as in CGS, the more general form

$$\mathbf{r}^n = q_n(L)p_n(L)\mathbf{r}^0 , \tag{C.3.16}$$

is used, where now $q_n(x) = \prod_{i=1}^n (1 - \omega_i x)$, and $\omega_i$ are suitable constants chosen in such a way that $\|\mathbf{r}^n\|$ is minimized with respect to $\omega_i$.

The preconditioned algorithm can be described as follows:

***Preconditioned Bi-CGStab (PBi-CGStab) Method***

Initialize

$$\mathbf{v}^0, \quad \mathbf{r}^0 = \mathbf{f} - L\mathbf{v}^0, \quad \text{choose } \tilde{\mathbf{r}}^0 \text{ s.t. } (\tilde{\mathbf{r}}^0, \mathbf{r}^0) \neq 0, \ (\text{e.g., } \tilde{\mathbf{r}}^0 = \mathbf{r}^0)$$

Iterate

$$\begin{aligned}
&\rho_{n-1} = (\mathbf{r}^{n-1}, \tilde{\mathbf{r}}^0) \\
&\mathit{if}\ \rho_{n-1} = 0 \\
&\quad \mathit{then} \text{ the method fails} \\
&\mathit{end\ if} \\
&\mathit{if}\ n = 1 \\
&\quad \mathit{then}\ \mathbf{p}^n = \mathbf{r}^{n-1} \\
&\quad \mathit{else}\ \beta_{n-1} = (\rho_{n-1}/\rho_{n-2})(\alpha_{n-1}/\omega_{n-1}) \\
&\qquad\quad \mathbf{p}^n = \mathbf{r}^{n-1} + \beta_{n-1}(\mathbf{p}^{n-1} - \omega_{n-1}\mathbf{w}^{n-1}) \\
&\mathit{end\ if} \\
&H\hat{\mathbf{p}} = \mathbf{p}^n \\
&\mathbf{w}^n = L\hat{\mathbf{p}} \\
&\alpha_n = \rho_{n-1}/(\mathbf{w}^n, \tilde{\mathbf{r}}^0) \\
&\mathbf{s} = \mathbf{r}^{n-1} - \alpha_n \mathbf{w}^n \\
&\mathit{if}\ \|\mathbf{s}\| \text{ small enough} \\
&\quad \mathit{then}\ \mathbf{v}^n = \mathbf{v}^{n-1} + \alpha_n \hat{\mathbf{p}};\ \text{quit} \\
&\mathit{end\ if} \\
&H\hat{\mathbf{s}} = \mathbf{s} \\
&\mathbf{t} = L\hat{\mathbf{s}} \\
&\omega_n = (\mathbf{t}, \mathbf{s})/(\mathbf{t}, \mathbf{t}) \\
&\mathbf{v}^n = \mathbf{v}^{n-1} + \alpha_n \hat{\mathbf{p}} + \omega_n \hat{\mathbf{s}} \\
&\mathit{if}\ \mathbf{v}^n \text{ is accurate enough} \\
&\quad \mathit{then} \text{ quit} \\
&\mathit{end\ if} \\
&\mathbf{r}^n = \mathbf{s} - \omega_n \mathbf{t} \\
&\text{For continuation it is necessary that } \omega_n \neq 0.
\end{aligned} \tag{C.3.17}$$

For an unfavorable choice of $\tilde{\mathbf{r}}^0$, $\rho_n$ or $(\mathbf{w}^n, \tilde{\mathbf{r}}^0)$ can be 0 or very small. In this case one has to restart, e.g., with $\tilde{\mathbf{r}}^0$ and $\mathbf{v}^0$ given by the last available values of $\mathbf{r}^n$ and $\mathbf{v}^n$. In exact arithmetic, Bi-CGStab is also a finite-termination method (i.e., $\mathbf{v}^n = \mathbf{u}$ for some $n \leq nd$). Its theoretical convergence properties are similar to those of CGS; however, it converges more smoothly, i.e., the oscillations of the residuals (with $n$) of Bi-CGStab are in general less pronounced than those of CGS.

It is clear from the previous algorithm description that a weakness of Bi-CGStab is that a breakdown occurs if an $\omega_n$ is equal to zero (but also a very small $\omega_n$ may be troublesome).

Another non-ideal property is that the $q_n$ polynomial in (C.3.16) has only real roots by construction, whereas optimal reduction polynomials for matrices with complex eigenvalues may also have complex roots. These considerations have led to the introduction of a variant, called Bi-CGStab(2), in which $q_n$ is constructed as a product of quadratic factors. For its derivation and analysis the reader is referred, e.g., to van der Vorst (2003).

Unfortunately, for a general nonsymmetric matrix, Krylov methods are not guaranteed to converge. But neither are any other known iterative methods. As noted earlier, GMRES(m) does have a convergence guarantee if $L_S$ has positive eigenvalues.

# Appendix D. Time Discretizations

In this appendix we will make some general comments about time discretizations, survey standard methods for ODEs and their stability regions, discuss integrating factors for Fourier spatial discretizations, and highlight some low-storage time-discretization formulas that have been widely used in conjunction with spectral methods.

## D.1 Notation and Stability Definitions

The typical evolution equation can be written

$$\begin{aligned} \frac{\partial u}{\partial t} &= f(u,t) \ , \quad t > 0, \\ u(0) &= 0 \ , \end{aligned} \tag{D.1.1}$$

where the (generally) nonlinear operator $f$ contains the spatial part of the PDE. Following the general formulation of Chap. 6, the semi-discrete version is

$$Q_N \frac{\mathrm{d}u^N}{\mathrm{d}t} = Q_N f_N(u^N, t) \ ,$$

where $u^N$ is the spectral approximation to $u$, $f_N$ denotes the spectral approximation to the operator $f$, and $Q_N$ is the spatial projection operator which characterizes the scheme. Let us denote by $\mathbf{u}(t)$ the vector of the spatial unknowns which determine $u^N(t)$. For example, in a collocation method for a Dirichlet boundary-value problem, $\mathbf{u}(t)$ represents the set of the interior grid-point values of $u^N(t)$. Then the previous discrete problem can be written in the form

$$\begin{aligned} \frac{\mathrm{d}\mathbf{u}}{\mathrm{d}t} &= \mathbf{f}(\mathbf{u},t) \ , \qquad t > 0, \\ \mathbf{u}(0) &= \mathbf{u}_0 \ , \end{aligned} \tag{D.1.2}$$

where $\mathbf{f}$ is the vector-valued function governing the semi-discrete problem. For Galerkin and G-NI methods, $\mathbf{f}$ may incorporate the matrix $M^{-1}$, where $M$

denotes the mass matrix which expresses the projection $Q_N \frac{du^N}{dt}$ algebraically as $M\frac{d\mathbf{u}}{dt}$. For time-dependent, linear PDEs, (D.1.2) reduces to

$$\frac{d\mathbf{u}}{dt} = -L\mathbf{u} + \mathbf{b}\,, \qquad t > 0, \\ \mathbf{u}(0) = \mathbf{u}_0\,, \tag{D.1.3}$$

where $L$ is the matrix representing the spatial discretization by the chosen spectral method. (The use of the negative sign in front of $L$ in (D.1.3) is consistent with the notation of Chap. 4 – see (4.8) – for describing the discretization of a time-independent boundary-value problem. In that chapter, we introduced and analyzed some representative spectral discretization matrices.) This is also called a method-of-lines approach or a continuous-in-time discretization. In describing the time discretizations, we denote the time-step by $\Delta t$, the $n$-th time-level by $t_n = n\Delta t$, the approximate solution at time-step $n$ by $\mathbf{u}^n$, and use $\mathbf{f}^n = \mathbf{f}(\mathbf{u}^n, t^n)$.

The corresponding (linear, scalar) model problem is

$$\frac{du}{dt} = \lambda u\,, \tag{D.1.4}$$

where $\lambda$ is a complex number, which for (D.1.2) is "representative" of the partial derivative of $f$ with respect to $u$ (in the scalar case) or of the eigenvalues of the Jacobian matrix $(\partial f_i/\partial u_j)_{i,j}$ in the vector case, and which for (D.1.3) is representative of the eigenvalues of $-L$.

In most applications of spectral methods to partial differential equations the spatial discretization is spectral but the temporal discretization uses conventional finite differences. (See, however, Morchoisne (1979, 1981) and Tal-Ezer (1986a, 1989) for some exploratory work on methods using spectral discretizations in both space and time. See also Schötzau and Schwab (2000) for high-order discontinuous Galerkin methods in time, albeit coupled with the *hp*-version of finite elements rather than with spectral methods.) Some standard references from the extensive literature on numerical methods for ODEs are the books by Gear (1971), Lambert (1991), Shampine (1994), Hairer, Norsett and Wanner (1993), Hairer and Wanner (1996), and Butcher (2003).

If the spatial discretization is presumed fixed, then we use the term stability in its ODE context. The time discretization is said to be *stable* (sometimes called *zero-stable*) if there exist positive constants $\delta$, $\varepsilon$ and $C(T)$, independent of $\Delta t$, such that, for all $T > 0$ (perhaps limited by a maximal $T_{max}$ depending on the problem) and for all $0 \le \Delta t < \delta$,

$$\|\mathbf{u}^n - \mathbf{v}^n\| \le C(T)\|\mathbf{u}^0 - \mathbf{v}^0\| \qquad \text{for } 0 \le t_n \le T \tag{D.1.5}$$

provided that $\|\mathbf{u}^0 - \mathbf{v}^0\| < \varepsilon$, where $\|\mathbf{u}^n\|$ is some spatial norm of $\mathbf{u}^n$. The constant $C(T)$ is permitted to grow with $T$. Here, $\mathbf{v}^n$ is the solution obtained by the same numerical method corresponding to a (perturbed) initial data $\mathbf{v}^0$. On a linear problem (hence in particular, for the problems (D.1.3) or (D.1.4)), (D.1.5) can be equivalently replaced by

$$\|\mathbf{u}^n\| \leq C(T)\|\mathbf{u}^0\| \qquad \text{for } 0 \leq t_n \leq T\,. \tag{D.1.6}$$

For many problems involving integration over long time intervals, a method which admits the temporal growth allowed by the estimate (D.1.5) is undesirable. As one example, take a problem of the form (D.1.2) for which $(\partial f/\partial u)(w,t)$ is negative for all $w$ and $t$, or more generally, for which $f$ satisfies the *right Lipschitz condition*: there exists $\mu < 0$ such that

$$\langle f(u,t) - f(v,t), u - v\rangle \leq \mu\|u - v\|^2 \qquad \text{for all } u, v, t,$$

where $\langle\cdot,\cdot\rangle$ is a suitable scalar product and $\|\cdot\|$ its associated norm. In these cases,

$$\|u(t) - v(t)\| \leq e^{\mu t}\|u(0) - v(0)\|\,.$$

(Such problems are referred to as dissipative Cauchy problems in the ODE literature.) The ODEs resulting from spectral spatial discretizations of the heat equation (with homogeneous boundary data and zero source term) fall into this category. In this case one desires that the time discretization be *asymptotically stable*, i.e., that instead of (D.1.5) it satisfy the stronger requirement

$$\|\mathbf{u}^n - \mathbf{v}^n\| \to 0 \quad \text{as} \quad t_n \to +\infty, \tag{D.1.7}$$

or that it be *contractive* (or *B-stable*):

$$\|\mathbf{u}^n - \mathbf{v}^n\| \leq C\|\mathbf{u}^{n-1} - \mathbf{v}^{n-1}\| \qquad \text{for all } n \geq 1, \tag{D.1.8}$$

for a suitable constant $C < 1$ independent of $n$.

As another example for which the above notion of stability is too weak, consider ODEs resulting from the spatial discretization of linear, spatially periodic, purely hyperbolic systems. For these problems, asymptotic stability for the time discretization is undesirable since the exact solution is undamped in time. Instead we rather desire a time discretization which is *temporally stable*, for which we merely require that

$$\|\mathbf{u}^n\| \leq \|\mathbf{u}^0\| \qquad \text{for all } n \geq 1\,. \tag{D.1.9}$$

The notion of *weak instability* is sometimes used in a loose sense for schemes which admit solutions to periodic hyperbolic problems which grow with time, but for which the growth rate decreases with $\Delta t$. For example, the constant $C(T)$ in (D.1.5) might have the form

$$C(T) = e^{\alpha(\Delta t)^p T}\,,$$

where $\alpha > 0$ and $p$ is a positive integer. For such weakly unstable schemes, the longer the time interval of interest, i.e., the larger is $T$, the smaller must

$\Delta t$ be chosen to keep the spurious growth of the solution within acceptable bounds.

Another notion that is relevant to periodic, hyperbolic problems is that of *reversible* (or *symmetric*) time discretizations. These are schemes for which the solution may be marched forward from $t^n$ to $t^{n+1}$ and then backwards to $t^n$ with the starting solution at $t^n$ recovered exactly (except for round-off errors).

Two final definitions are in order for our subsequent discussion. The *absolute stability region* (often referred to just as the *stability region*), say $\mathcal{A}$, of a numerical method is customarily defined for the scalar model problem (D.1.4) to be the set of all complex numbers $\alpha = \lambda\Delta t$ such that any sequence $\{u^n\}$ generated by the method with such $\lambda$ and $\Delta t$ satisfies $\|u^n\| \leq C$ as $t_n \to \infty$, for a suitable constant $C$. Furthermore, a method is called *A-stable* if the region of absolute stability includes the region $\mathrm{Re}(\lambda\Delta t) < 0$. We warn the reader that in some books the absolute stability region is defined as the set of all $\lambda\Delta t$ such that $\|u^n\| \to 0$ as $t_n \to \infty$. This new region, say $\mathcal{A}^0$, would not necessarily coincide with $\mathcal{A}$. In general, if $\mathcal{A}^0$ is non-empty, $\mathcal{A}$ is its closure. However, there are cases for which $\mathcal{A}^0$ is empty (e.g., the midpoint or leap-frog method) and $\mathcal{A}$ is not ($\mathcal{A} = \{z = \alpha i,\ -1 \leq \alpha \leq 1\}$ for the midpoint method). Finally, we note that zero-stable methods are those for which $\mathcal{A}$ contains the origin $z = 0$ of the complex plane.

As noted by Reddy and Trefethen (1990, 1992), having the eigenvalue scaled by the time-step $\Delta t$ falling within the absolute stability region of the ODE method is not always sufficient for stability of the computation. They present a stability criterion utilizing $\epsilon$-pseudospectra. As noted in Sect. 4.3.2, first-derivative (indeed, any odd-order derivative) matrices for nonperiodic problems are nonnormal. However, as discussed by Trefethen (2000, Chapter 10), in almost all cases the "rule-of-thumb" condition involving the standard eigenvalues is acceptable.

On the other hand, we may be interested in the behavior of the computed solution as both the spatial and temporal discretizations are refined. We now define stability by an estimate of the form (D.1.5) where $C$ is independent of $\Delta t$, $\varepsilon$ and the spatial discretization parameter $N$, the norm is independent of $N$, but $\delta$ will in general be a function of $N$. The functional dependence of $\delta$ upon $N$ which is necessary to obtain an estimate of the form (D.1.5) is termed the *stability limit* of the numerical method. If $\delta$ is in fact independent of $N$, then the method is called *unconditionally stable*. Clearly, a necessary condition for the fully discrete problem to be stable is that the semi-discrete problem be stable in the sense discussed in Sect. 6.5. Likewise, a *temporal* stability limit for the fully discrete scheme for a hyperbolic system is the functional dependence of $\delta$ upon $N$ which is necessary to obtain an estimate of the form (D.1.9).

## D.2 Standard ODE Methods

In this section we furnish as a convenience the basic formulas and diagrams for the absolute stability regions for those time discretizations of (D.1.2) that are most commonly used in conjunction with spectral discretizations in space. Among the factors which influence the choice of a time discretization are the accuracy, stability, storage requirements, and work demands of the methods. The storage and work requirements of a method can be deduced in a straight-forward manner from the definition of the method and the nature of the PDE. The accuracy of a method follows from a truncation error analysis and the stability for a given problem is intimately connected with the spectrum of the spatial discretization. In this section we will describe some of the standard methods for ODEs and relate their stability regions to the spectra of the advection and diffusion operators. Bear in mind that in many problems different time discretizations are used for different spatial terms in the equation. The illustrations of the spectra of the spectral differentiation, mass and stiffness matrices furnished in Sect. 4.3 combined with the stability diagrams in this section suffice for general conclusions to be drawn on appropriate choices of time-discretization methods and time-step limits for temporal stability.

For the reader's convenience, Table D.1 provides the numerical values of the intersections of the absolute stability regions with the negative real axis and the positive imaginary axis for all methods discussed in this section.

### D.2.1 Leap Frog Method

The *leap frog* (LF) *method* (also called the *midpoint method*) is a second-order, two-step scheme given by

$$\mathbf{u}^{n+1} = \mathbf{u}^{n-1} + 2\Delta t \mathbf{f}^n \ . \tag{D.2.1}$$

This produces solutions of constant norm for the model problem provided that $\lambda \Delta t$ is on the imaginary axis and that $|\lambda \Delta t| \leq 1$ (see Table D.1). Thus, leap frog is a suitable explicit scheme for problems with purely imaginary eigenvalues. It also is a reversible, or symmetric, method. However, since it is only well-behaved on a segment in the complex $\lambda \Delta t$-plane for the model problem, extra care is needed in practical situations.

The most obvious application is to periodic advection problems, for the eigenvalues of the Fourier approximation to $\mathrm{d}/\mathrm{d}x$ are imaginary. The difficulty with the leap frog method is that the solution is subject to a temporal oscillation with period $2\Delta t$. This arises from the extraneous (spurious) solution to the temporal difference equations. The oscillations can be controlled by every so often averaging the solution at two consecutive time-levels.

Leap frog is quite inappropriate for problems whose spatial eigenvalues have nonzero real parts. This certainly includes diffusion operators. Leap frog is also not viable for advection operators with nonperiodic boundary conditions. The figures in Sect. 4.3.2 indicate clearly that the discrete spectra of Chebyshev and Legendre approximations to the standard advection operator have appreciable real parts.

### D.2.2 Adams-Bashforth Methods

This is a class of explicit multistep methods which includes the simple *forward Euler* (FE) *method*

$$\mathbf{u}^{n+1} = \mathbf{u}^n + \Delta t \mathbf{f}^n \ , \tag{D.2.2}$$

the popular *second-order Adams-Bashforth* (AB2) *method*

$$\mathbf{u}^{n+1} = \mathbf{u}^n + \tfrac{1}{2}\Delta t \left[3\mathbf{f}^n - \mathbf{f}^{n-1}\right] \ , \tag{D.2.3}$$

the still more accurate *third-order Adams-Bashforth* (AB3) *method*

$$\mathbf{u}^{n+1} = \mathbf{u}^n + \tfrac{1}{12}\Delta t \left[23\mathbf{f}^n - 16\mathbf{f}^{n-1} + 5\mathbf{f}^{n-2}\right] \ , \tag{D.2.4}$$

and the *fourth-order Adams-Bashforth* (AB4) *method*

$$\mathbf{u}^{n+1} = \mathbf{u}^n + \tfrac{1}{24}\Delta t \left[55\mathbf{f}^n - 59\mathbf{f}^{n-1} + 37\mathbf{f}^{n-2} - 9\mathbf{f}^{n-3}\right] \ . \tag{D.2.5}$$

These methods are not reversible.

The stability regions $\mathcal{A}$ of these methods are shown in Fig. D.1 (left) and the stability boundaries along the axes are given in Table D.1. Note that the size of the stability region decreases as the order of the method increases. Note also that except for the origin, no portion of the imaginary axis is included in the stability regions of the first- and second-order methods, whereas the third- and fourth-order versions do have some portion of the imaginary axis included in their stability regions. Nevertheless, the AB2 method is weakly unstable, i.e., for a periodic, hyperbolic problem the acceptable $\Delta t$ decreases at $T$ increases.

As is evident from Fig. D.1 (left), higher order AB methods are temporally stable for Fourier approximations to periodic advection problems. Let the upper limit of the absolute stability region along the imaginary axis be denoted by $c$. Then the temporal stability limit is

$$\frac{N}{2}\Delta t \le c$$

or

$$\Delta t \le \frac{c}{\pi}\Delta x \ . \tag{D.2.6}$$

The limit on $\Delta t$ is smaller by a factor of $\pi$ than the corresponding limit for a second-order finite-difference approximation in space. The Fourier spectral approximation is more accurate in space because it represents the high-frequency components much more accurately than the finite-difference method. The artificial damping of the high-frequency components which is produced by finite-difference methods enables the stability restriction on the time-step to be relaxed.

Chebyshev and Legendre approximations to advection problems appear to be temporally stable under all Adams-Bashforth methods for sufficiently small $\Delta t$; precisely, for $\Delta t \leq CN^{-2}$ for a suitable constant $C$. As discussed in Sect. 4.3.2, the spatial eigenvalues all have negative real parts. Thus, the failure of the AB2 method to include the imaginary axis in its absolute stability region does not preclude temporal stability.

The temporal stability limits for Adams-Bashforth methods for Fourier, Chebyshev and Legendre approximations to diffusion equations are easy to deduce since their spatial eigenvalues (i.e., the eigenvalues of the matrix $-L$, where $L = B^{-1}A$ is the matrix considered in Sect. 4.3.1) are real and negative (limited in modulus as indicated in Table 4.2), and the stability bounds along the negative real axis are provided in Table D.1. In this case, $\Delta t$ should be limited by a constant times $N^{-2}$ for Fourier approximations, by a constant times $N^{-4}$ for Chebyshev or Legendre collocation approximations, by a constant times $N^{-3}$ for Legendre G-NI approximations. This follows from the eigenvalue analysis that is carried out in Chap. 4.

### D.2.3 Adams-Moulton Methods

A related set of implicit multistep methods are the Adams-Moulton methods. They include the *backward Euler* (BE) *method*

$$\mathbf{u}^{n+1} = \mathbf{u}^n + \Delta t \mathbf{f}^{n+1} , \tag{D.2.7}$$

the *Crank-Nicolson* (CN) *method*

$$\mathbf{u}^{n+1} = \mathbf{u}^n + \tfrac{1}{2}\Delta t[\mathbf{f}^{n+1} + \mathbf{f}^n] , \tag{D.2.8}$$

the *third-order Adams-Moulton* (AM3) *method*

$$\mathbf{u}^{n+1} = \mathbf{u}^n + \tfrac{1}{12}\Delta t[5\mathbf{f}^{n+1} + 8\mathbf{f}^n - \mathbf{f}^{n-1}] , \tag{D.2.9}$$

and the *fourth-order Adams-Moulton* (AM4) *method*

$$\mathbf{u}^{n+1} = \mathbf{u}^n + \tfrac{1}{24}\Delta t[9\mathbf{f}^{n+1} + 19\mathbf{f}^n - 5\mathbf{f}^{n-1} + \mathbf{f}^{n-2}] . \tag{D.2.10}$$

Forward Euler (FE) (see D.2.2), backward Euler (BE) and Crank-Nicolson (CN) methods are special cases of *θ-methods*, defined as

$$\mathbf{u}^{n+1} = \mathbf{u}^n + \Delta t[\theta \mathbf{f}^{n+1} + (1-\theta)\mathbf{f}^n] , \tag{D.2.11}$$

for $0 \leq \theta \leq 1$. Precisely, they correspond to the choice $\theta = 0$ (FE), $\theta = 1$ (BE) and $\theta = 1/2$ (CN). All $\theta$-methods except for FE are implicit. All $\theta$-methods are first-order accurate, except for CN, which is second-order. For each $\theta < \frac{1}{2}$,

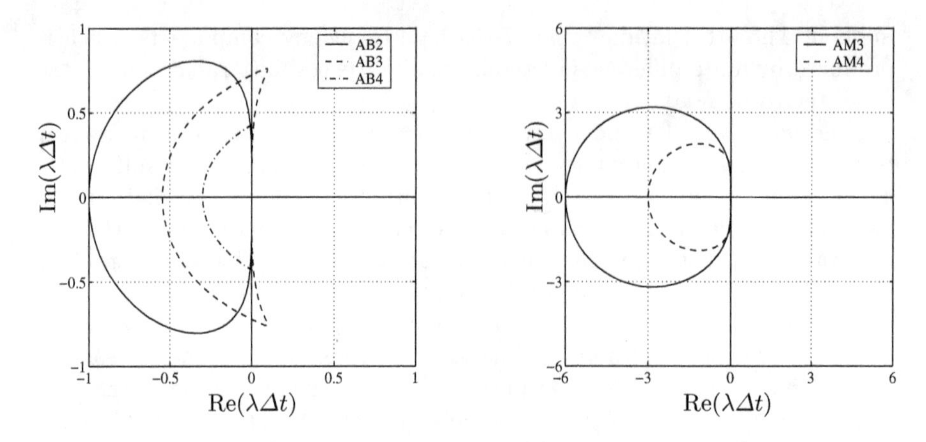


**Fig. D.1.** Absolute stability regions of Adams-Bashforth (*left*) and Adams-Moulton (*right*) methods

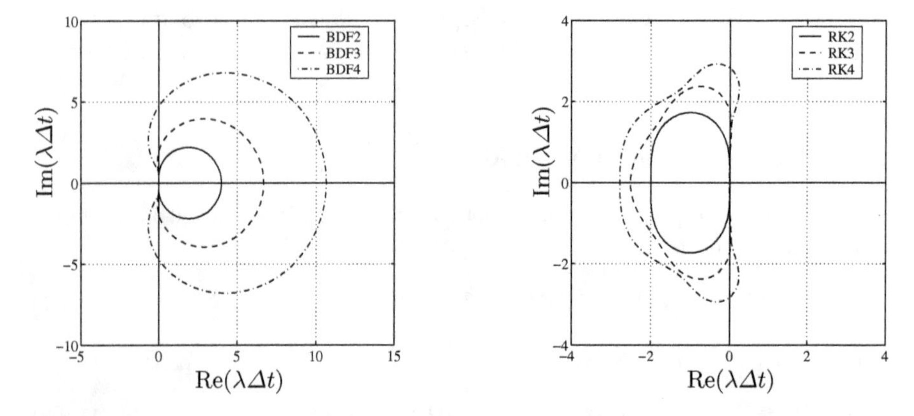


**Fig. D.2.** Absolute stability regions of backwards-difference formulas (*left*) and Runge-Kutta methods (*right*). The BDF methods are absolutely stable on the exteriors (and boundaries) of the regions enclosed by the curves, whereas the RK methods are absolutely stable on the interiors (and boundaries) of the regions enclosed by the curves

the absolute stability region is the circle in the left half-plane $\mathrm{Re}(\lambda\Delta t) \leq 0$ with center $z = (2\theta - 1)^{-1}$ and radius $r = (1 - 2\theta)^{-1}$. The stability region of the CN method coincides with the half-plane $\mathrm{Re}(\lambda\Delta t) \leq 0$. For each $\theta > \frac{1}{2}$, the absolute stability region is the exterior of the open circle in the right half-plane $\mathrm{Re}(\lambda\,\Delta t) > 0$ with center $z = (2\theta - 1)^{-1}$ and radius $r = (2\theta - 1)^{-1}$. Thus, all $\theta$-methods for $\frac{1}{2} \leq \theta \leq 1$ are A-stable.

The absolute stability regions of the third- and fourth-order Adams-Moulton methods are displayed in Fig. D.1 (right) and the stability boundaries along the axes are given in Table D.1. In comparison with the explicit

**Table D.1.** Intersections of absolute stability regions with the negative real axis (left) and with the positive imaginary axis (right)

| Method | $\mathcal{A} \cap \mathbb{R}_-$ | $\mathcal{A} \cap i\mathbb{R}_+$ |
|---|---|---|
| Leap frog (midpoint) | $\{0\}$ | $[0,1]$ |
| Forward Euler | $[-2,0]$ | $\{0\}$ |
| Crank-Nicolson | $(-\infty,0]$ | $[0,+\infty)$ |
| Backward Euler | $(-\infty,0]$ | $[0,+\infty)$ |
| $\theta$-method, $\theta < 1/2$ | $[2/(2\theta-1),0]$ | $\{0\}$ |
| $\theta$-method, $\theta \geq 1/2$ | $(-\infty,0]$ | $[0,+\infty)$ |
| AB2 | $(-1,0]$ | $\{0\}$ |
| AB3 | $[-6/11,0]$ | $[0,0.723]$ |
| AB4 | $[-3/10,0]$ | $[0,0.43]$ |
| AM3 | $[-6,0]$ | $\{0\}$ |
| AM4 | $[-3,0]$ | $\{0\}$ |
| BDF2 | $(-\infty,0]$ | $[0,+\infty)$ |
| BDF3 | $(-\infty,0]$ | $[0,1.94)$ |
| BDF4 | $(-\infty,0]$ | $[0,4.71)$ |
| RK2 | $[-2,0]$ | $\{0\}$ |
| RK3 | $[-2.51,0]$ | $[0,1.73]$ |
| RK4 | $[-2.79,0]$ | $[0,2.83]$ |

Adams-Bashforth method of the same order, an Adams-Moulton method has a smaller truncation error (by factors of five and nine for second and third-order versions), a larger stability region, and requires one fewer levels of storage. However, it does require the solution of an implicit set of equations. The CN method is reversible; the others are not.

The CN method is commonly used for diffusion problems. In Navier-Stokes calculations, it is frequently applied to the viscous and pressure gradient components. Although CN is absolutely stable for the former and temporally stable for the latter, it has the disadvantage that it damps high-frequency components very weakly, whereas in reality these components decay very rapidly. Deville, Kleiser and Montigny-Rannou (1984) have noted that this is undesirable in Navier-Stokes applications for which the solution itself decays rapidly. One remedy is to resort to BE – it damps the high frequency components rapidly. An alternative approach is to use the $\theta$-method (D.2.11) for $\theta = 1/2+\alpha\Delta t$, where $\alpha$ is a small positive constant. This method damps all components of the solution, and although it is formally first-order accurate in time (if $\alpha > 0$) it is "effectively" second order if $\alpha \ll 1$.

The Adams-Moulton methods of third and higher order are only conditionally stable for advection and diffusion problems. The stability limits implied by Fig. D.1 indicate that the stability limit of a high-order Adams-Moulton method is roughly ten times as large for a diffusion problem as the stability limit of the corresponding Adams-Bashforth method. In addition, AM3 and AM4 are weakly unstable for Fourier approximations to advection problems, since the origin is the only part of the imaginary axis which is included in their absolute stability regions.

### D.2.4 Backwards-Difference Formulas

Another class of implicit time discretizations is based upon backwards-difference formulas . These include the *first-order backwards-difference scheme* (BDF1), which is identical to backward Euler, the *second-order backwards-difference scheme* (BDF2)

$$\mathbf{u}^{n+1} = \tfrac{1}{3}[4\mathbf{u}^n - \mathbf{u}^{n-1}] + \tfrac{2}{3}\Delta t\mathbf{f}^{n+1} \,, \tag{D.2.12}$$

the *third-order backwards-difference scheme* (BDF3)

$$\mathbf{u}^{n+1} = \tfrac{1}{11}[18\mathbf{u}^n - 9\mathbf{u}^{n-1} + 2\mathbf{u}^{n-2}] + \tfrac{6}{11}\Delta t\mathbf{f}^{n+1} \,, \tag{D.2.13}$$

and the *fourth-order backwards-difference scheme* (BDF4)

$$\mathbf{u}^{n+1} = \tfrac{1}{25}[48\mathbf{u}^n - 36\mathbf{u}^{n-1} + 16\mathbf{u}^{n-2} - 3\mathbf{u}^{n-3}] + \tfrac{12}{25}\Delta t\mathbf{f}^{n+1} \,. \tag{D.2.14}$$

The absolute stability regions of these methods are displayed in Fig. D.2 (left) and the stability boundaries along the axes are given in Table D.1. The stability regions are much larger than those of the corresponding AM methods, but for orders higher than 2, BDF methods are unstable in a (small) region to the left of the imaginary axis.

### D.2.5 Runge-Kutta Methods

Runge-Kutta methods are single-step, but multistage, time discretizations. The modified Euler version of a *second-order Runge-Kutta* (RK2) *method* can be written

$$\mathbf{u}^{n+1} = \mathbf{u}^n + \frac{1}{2}\Delta t[\mathbf{f}(\mathbf{u}^n, t^n) + \mathbf{f}(\mathbf{u}^n + \Delta t\mathbf{f}(\mathbf{u}^n, t^n), t^n + \Delta t)] \,. \tag{D.2.15}$$

A popular *third-order Runge-Kutta* (RK3) *method* is

$$\begin{aligned}
\mathbf{k}_1 &= \mathbf{f}(\mathbf{u}^n, t_n) \\
\mathbf{k}_2 &= \mathbf{f}(\mathbf{u}^n + \tfrac{1}{2}\Delta t\, \mathbf{k}_1, t_n + \tfrac{1}{2}\Delta t) \\
\mathbf{k}_3 &= \mathbf{f}(\mathbf{u}^n + \tfrac{3}{4}\Delta t\, \mathbf{k}_2, t_n + \tfrac{3}{4}\Delta t) \\
\mathbf{u}^{n+1} &= \mathbf{u}^n + \tfrac{1}{9}\Delta t[2\mathbf{k}_1 + 3\mathbf{k}_2 + 4\mathbf{k}_3] \,.
\end{aligned} \tag{D.2.16}$$

The classical *fourth-order Runge-Kutta* (RK4) *method* is

$$\begin{aligned}
\mathbf{k}_1 &= \mathbf{f}(\mathbf{u}^n, t_n) \\
\mathbf{k}_2 &= \mathbf{f}(\mathbf{u}^n + \tfrac{1}{2}\Delta t\, \mathbf{k}_1, t_n + \tfrac{1}{2}\Delta t) \\
\mathbf{k}_3 &= \mathbf{f}(\mathbf{u}^n + \tfrac{1}{2}\Delta t\, \mathbf{k}_2, t_n + \tfrac{1}{2}\Delta t) \\
\mathbf{k}_4 &= \mathbf{f}(\mathbf{u}^n + \Delta t\, \mathbf{k}_3, t_n + \Delta t) \\
\mathbf{u}^{n+1} &= \mathbf{u}^n + \tfrac{1}{6}\Delta t[\mathbf{k}_1 + 2\mathbf{k}_2 + 2\mathbf{k}_3 + \mathbf{k}_4] \,.
\end{aligned} \tag{D.2.17}$$

All Runge-Kutta methods of a given order have the same stability properties. The absolute stability regions are given in Fig. D.2 (right) and the stability boundaries along the axes are given in Table D.1. Note that the stability region expands as the order increases. Note also that RK2 methods are afflicted with the same weak instability as the AB2 scheme. When storage is not an issue, then the classical RK4 method is commonly used. Otherwise, the low-storage versions of third- and fourth-order methods, such as those described in Sect. D.4, have been preferred.

In the event that $\mathbf{f}$ contains no explicit dependence upon $t$, the following formulation, due to Jameson, Schmidt and Turkel (1981) applies:

$$\begin{aligned}
&\textit{Set} \\
&\qquad \mathbf{u} = \mathbf{u}^n \\
&\textit{For } k = s,\ 1,\ -1 \\
&\qquad \mathbf{u} \leftarrow \mathbf{u}^n + \frac{1}{k}\Delta t \mathbf{f}(\mathbf{u}) \\
&\textit{End For} \\
&\qquad \mathbf{u}^{n+1} = \mathbf{u} \,.
\end{aligned} \tag{D.2.18}$$

It yields a Runge-Kutta method of order $s$ (for linear problems) and requires at most three levels of storage.

## D.3 Integrating Factors

For some applications of spectral methods the use of an *integrating-factor technique* is attractive. The Burgers equation (3.1.1) with periodic boundary conditions will serve here as a simple illustration of handling constant-coefficient linear terms via integrating factors. The semi-discrete Fourier Galerkin formulation of this is given by (3.3.3), which we write here as

$$\frac{\mathrm{d}\hat{u}_k}{\mathrm{d}t} + \hat{g}_k(\hat{\mathbf{u}}) + \nu k^2 \hat{u}_k = 0 \,, \qquad k = -\frac{N}{2}, \dots, \frac{N}{2} - 1 \,, \tag{D.3.1}$$

where $\hat{g}_k(\hat{\mathbf{u}})$ is given by the right-hand side of (3.3.4). Equation (D.3.1) can be written

$$\frac{\mathrm{d}}{\mathrm{d}t}[e^{\nu k^2 t}\hat{u}_k] = -e^{\nu k^2 t}\hat{g}_k(\hat{\mathbf{u}}) .$$

The forward Euler approximation reduces to

$$\hat{u}_k^{n+1} = e^{-\nu k^2 \Delta t}[\hat{u}_k^n - \Delta t \hat{g}_k(\hat{\mathbf{u}}^n)] . \tag{D.3.2}$$

The treatment of the linear term is both unconditionally stable and exact. The accuracy and stability restrictions of the overall time-integration method arise solely from the nonlinear term.

The Fourier collocation method can be handled in a similar, but not equivalent manner:

$$\mathbf{u}^{n+1} = C^{-1}\Lambda C\mathbf{u}^n - \Delta t\mathbf{g}(\mathbf{u}^n) , \tag{D.3.3}$$

where $\mathbf{u}$ represents the vector of unknowns at the collocation points, $C$ represents the discrete Fourier transform matrix (see (2.1.25) and (4.1.9)), $\mathbf{g}(\mathbf{u})$ represents the nonlinear advection term, and

$$\Lambda = \mathrm{diag}\left\{e^{-\nu k^2 \Delta t}\right\} . \tag{D.3.4}$$

This approach was used by Fornberg and Whitham (1978) on the Korteweg-de Vries equation

$$\frac{\partial u}{\partial t} + u\frac{\partial u}{\partial x} + \frac{\partial^3 u}{\partial x^3} = 0 \tag{D.3.5}$$

in their Fourier collocation-leap frog calculations. In this application, exact integration enables the stability limit to be increased from $\Delta t < (1/\pi^3)\Delta x^3$ to $\Delta t < (3/2\pi^2)\Delta x^3$. This is a fivefold increase. Note, however, that the $O(\Delta x^3)$ limit does not disappear entirely in favor of an $O(\Delta x)$ limit, as it would for a Fourier Galerkin method applied in conjunction with exact integration. Chan and Kerkhoven (1985) discuss alternative time discretizations of the Korteweg-de Vries equations. They show that, with the leap frog method for the advection term and the Crank-Nicolson method for the linear term, the stability limit is independent of $\Delta x$ for any finite time interval.

The integrating-factor technique has found extensive use in Fourier Galerkin simulations of homogeneous turbulence (Rogallo (1977); see also CHQZ3, Sect. 3.3) and has also been used for the horizontal diffusion terms in calculations of parallel boundary layers (Spalart (1986); see also CHQZ3, Sect. 3.4.5). The integrating factors are especially useful in these Navier-Stokes applications because they do not suffer from the weak or nonexistent damping of the high-frequency components that arise in backward Euler or Crank-Nicolson discretizations of the viscous terms. Maday, Patera and Rønquist (1990) developed an integrating factor technique that is particularly useful in splitting methods. See CHQZ3, Sect. 3.2.3 for additional discussion.

## D.4 Low-Storage Schemes

When high-order discretization schemes such as spectral methods are employed in space, the primary contributor to the error in the fully discrete approximation is usually the temporal discretization error unless the time discretization itself is at least third order or the time-step is very small. When computations are constrained by memory limitations, a premium is placed on minimizing storage demands. This has made special low-storage Runge-Kutta methods very attractive for large-scale problems. Several popular low-storage Runge-Kutta methods are available that permit third-order or fourth-order temporal accuracy to be obtained with only two levels of storage. Such economies are not available for multistep methods.

We shall note here some of the low-storage Runge-Kutta methods that have been widely used for large-scale spectral computations. The description shall be given for the ODE

$$\frac{d\mathbf{u}}{dt} = \mathbf{g}(\mathbf{u}, t) + \mathbf{l}(\mathbf{u}, t) \, . \tag{D.4.1}$$

where $\mathbf{g}(\mathbf{u}, t)$ is treated with a low-storage Runge-Kutta method and $\mathbf{l}(\mathbf{u}, t)$ is treated implicitly with the Crank-Nicolson method. Such mixed explicit/ implicit time discretizations are very common for incompressible Navier-Stokes computations, for which $\mathbf{g}(\mathbf{u}, t)$ represents (nonlinear) advection and $\mathbf{l}(\mathbf{u}, t)$ (linear) diffusion.

The general representation of a low-storage Runge-Kutta/Crank-Nicolson method requiring only 2 levels of storage (for $\mathbf{u}$ and $\mathbf{h}$) is

$$\begin{aligned}
&\mathbf{h} = \mathbf{0} \\
&\mathbf{u} = \mathbf{u}^n \\
&For\ k = 1\ to\ K \\
&\qquad t^k = t^n + \alpha_k \Delta t \\
&\qquad t^{k+1} = t^n + \alpha_{k+1} \Delta t \\
&\qquad \mathbf{h} \leftarrow \mathbf{g}(\mathbf{u}, t^k) + \beta_k \mathbf{h} \\
&\qquad \mu = \frac{1}{2} \Delta t (\alpha_{k+1} - \alpha_k) \\
&\qquad \mathbf{v} - \mu \mathbf{l}(\mathbf{v}, t^{k+1}) = \mathbf{u} + \gamma_k \Delta t \mathbf{h} + \mu \mathbf{l}(\mathbf{u}, t^k) \\
&\qquad \mathbf{u} \leftarrow \mathbf{v} \\
&End\ For \\
&\mathbf{u}^{n+1} = \mathbf{u}
\end{aligned} \tag{D.4.2}$$

(note that the penultimate instruction in the loop indicates that $\mathbf{v}$ is the solution of the implicit equation on the left-hand side).

Table D.2 lists the values of these parameters for one third-order scheme, due to Williamson (1980), and one fourth-order scheme from Carpenter and Kennedy (1994). The stability limits (on the imaginary axis) for these schemes are 1.73 for the third-order scheme and 3.34 for the fourth-order scheme. Both of these have been widely used for the time discretization in applications of spectral methods. Both references contain a family of low-storage methods. Another low-storage family popular in the spectral methods community originated with A. Wray (unpublished), and was extended by Spalart, Moser and Rogers (1993).

**Table D.2.** Coefficients of low-storage Runge-Kutta/Crank-Nicolson schemes

| | Williamson 3rd-order | Carpenter-Kennedy 4th-order |
|---|---|---|
| $\alpha_1$ | 0 | 0 |
| $\alpha_2$ | 1/3 | 0.1496590219993 |
| $\alpha_3$ | 3/4 | 0.3704009573644 |
| $\alpha_4$ | 1 | 0.6222557631345 |
| $\alpha_5$ | – | 0.9582821306748 |
| $\alpha_6$ | – | 1 |
| $\beta_1$ | 0 | 0 |
| $\beta_2$ | -5/9 | -0.4178904745 |
| $\beta_3$ | -153/128 | -1.192151694643 |
| $\beta_4$ | – | -1.697784692471 |
| $\beta_5$ | – | -1.514183444257 |
| $\gamma_1$ | 1/3 | 0.1496590219993 |
| $\gamma_2$ | 15/16 | 0.3792103129999 |
| $\gamma_3$ | 8/15 | 0.8229550293869 |
| $\gamma_4$ | – | 0.6994504559488 |
| $\gamma_5$ | – | 0.1530572479681 |

# References

S. Abarbanel, D. Gottlieb, E. Tadmor (1986): 'Spectral Methods for Discontinuous Problems'. In: *Numerical Methods for Fluid Dynamics. II*, ed. by K.W. Morton, M.J. Baines (Oxford Univ. Press, London) pp. 129–153

M. Abramowitz, I.A. Stegun (Eds.) (1972): *Handbook of Mathematical Functions with Formulas, Graphs, and Mathematical Tables* (Gov. Printing Office, Washington, D.C.)

L.V. Ahlfors (1979): *Complex Analysis* (McGraw-Hill, New York)

E. Anderson, Z. Bai, C. Bischof, S. Blackford, J. Demmel, J. Dongarra, J. Du Croz, A. Greenbaum, S. Hammarling, A. McKenney, D. Sorensen (1999): LAPACK Users' Guide, 3rd edn. (SIAM, Philadelphia)

R. Archibald, A. Gelb, J. Yoon (2005): Polynomial fitting for edge detection in irregularly sampled signals and images. SIAM J. Numer. Anal. **43**, 259–279

Axelsson (1994): *Iterative Solution Methods* (Cambridge Univ. Press, Cambridge)

A.Y. Aydemir, D.C. Barnes (1984): Three-dimensional nonlinear incompressible MHD calculations. J. Comput. Phys. **53**, 100–123

I. Babuška, A.K. Aziz (1972): 'Survey Lectures on the Mathematical Foundations of the Finite Element Method'. In: *The Mathematical Foundations of the Finite Element Method with Application to Partial Differential Equations*, ed. by A.K. Aziz (Academic Press, New York) pp. 3–359

I. Babuška, B.A. Szabó, I.N. Katz (1981): The $p$-version of the finite element method. SIAM J. Numer. Anal. **18**, 515–545

R. Baltensperger, J.P. Berrut (1999): The errors in calculating the pseudospectral differentiation matrices for Čebyšev-Gauss-Lobatto points. Comput. Math. Appl. **37**(1), 41–48. Errata: Comput. Math. Appl. **38**(1), 119

R. Baltensperger, J.P. Berrut (2001): The linear rational collocation method. J. Comput. Appl. Math. **134**, 243–258

R. Baltensperger, J.P. Berrut, Y. Dubey (2003): The linear rational pseudospectral method with preassigned poles. Num. Algorithms **33**, 53–63

V.A. Barker, L.S. Blackford, J. Dongarra, J. Du Croz, S. Hammarling, M. Marinova, J. Was'niewski, P. Yalamov (2001): LAPACK95 Users' Guide (SIAM, Philadelphia)

R.H. Bartels, G.W. Stewart (1972): Solution of the matrix equation $AX + XB = C$. Comm. ACM **15**, 820–826

Z. Belhachmi, C. Bernardi, S. Deparis, F. Hecht (2006): A truncated Fourier/finite element discretization of the Stokes equations in an axisymmetric domain. Math. Models Methods Appl. Sci., **16**, to appear

M. Benzi (2002): Preconditioning techniques for large linear systems: a survey. J. Comput. Phys. **182**(2), 418–477

M.J. Berger, P. Colella (1989): Local adaptive mesh refinement for shock hydrodynamics. J. Comput. Phys. **82**, 64–84

J. Bergh, J. Löfström (1976): *Interpolation Spaces. An Introduction* (Springer, Berlin)

S. Berlin, M. Wiegel, D.S. Henningson (1999): Numerical and experimental investigations of oblique boundary layer transition. J. Fluid Mech. **393**, 23–57

C. Bernardi, Y. Maday (1986): Propriétés de quelques espaces de Sobolev aved poids et application á la collocation de Tchebycheff. C. R. Acad. Sci. Paris **303**, Serie I, 829–832

C. Bernardi, Dauge, Y. Maday (1999): *Spectral Methods for Axisymmetric Domains* (North-Holland, Amsterdam)

C. Bernardi, Y. Maday (1992a): Polynomial interpolation results in Sobolev spaces. J. Comput. Appl. Math. **43**, 53–80

C. Bernardi, Y. Maday (1992b): *Approximations Spectrales de Problèmes aux Limites Elliptiques* (Springer, Paris)

C. Bernardi, Y. Maday (1997): 'Spectral methods'. In: *Handbook of Numerical Analysis, Vol. 5: Techniques of Scientific Computing*, ed. by P.J. Ciarlet and J.L. Lions (North Holland, Amsterdam) pp. 209–486

C. Bernardi, Y. Maday, B. Métivet (1987): Spectral approximation of the periodic/nonperiodic Navier-Stokes equations. Numer. Math. **44**, 251–264

J.P. Berrut, H. Mittelmann (2001): The linear rational pseudospectral method with iteratively optimized poles for two-point boundary value problems. SIAM J. Sci. Comput. **23**(3), 961–975

F.P. Bertolotti, Th. Herbert, P.R. Spalart (1992): Linear and nonlinear stability of the Blasius boundary layer. J. Fluid Mech. **242**, 441–474

G.A. Blaisdell, N.N. Mansour, W.C. Reynolds (1993): Compressibility effects on the growth and structure of homogeneous turbulent shear flows. J. Fluid Mech. **256**, 443–485

G.A. Blaisdell, O. Zeman (1992): 'Investigation of the Dilatational Dissipation in Compressible homogeneous Shear Flow'. In: *Studying Turbulence Using Numerical Simulation Databases – IV, Proceedings of the 1992 Summer Program, Center for Turbulence Research* (Stanford University/NASA Ames Research Center) pp. 231–245

G. Blatter, M.V. Feigelman, V.B. Geshkenbein, A.I. Larkin, V.M. Vinokur (1994): Vortices in high-temperature superconductors. Rev. Modern Phys. **66**, 1125–1388

D.T. Blackstock (1966): Convergence of the Keck-Boyer perturbation solution for plane waves of finite amplitude in a viscous fluid. J. Acoust. Soc. Am. **39**, 411–413

E.N. Blinova (1944): 'Hydrodynamic theory of pressure and temperature waves and center of action of the atmosphere'. Trans. No. 113 (Regional Control Office, Second Weather Region, Patterson Field, OH)

L. Bos, M.A. Taylor, B.A. Wingate (2001): Tensor-product Gauss-Lobatto nodes are Fekete points for the cube. Math. Comput. **70**, 1543–1547

J.-P. Bouchaud, M. Mezard, G. Parisi (1995): Scaling and intermittency in Burgers turbulence. Phys. Rev. E. **52**, 3656–74

J.P. Boyd (1982): The optimization of convergence for Chebyshev polynomial methods in an unbounded domain. J. Comput. Phys. **45**, 43–79

J.P. Boyd (1984): Asymptotic coefficients of Hermite function series. J. Comput. Phys. **54**, 382–410

J.P. Boyd (1987): Spectral methods using rational basis functions on an infinite interval. J. Comput. Phys. **69**, 112–142

J.P. Boyd (1989): *Chebyshev and Fourier Spectral Methods* (Springer, New York)

J.P. Boyd (2001): *Chebyshev and Fourier Spectral Methods*, 2nd edn. (Dover, New York)

J.P. Boyd (2005): Trouble with Gegenbauer reconstruction for defeating Gibbs' phenomenon: Runge phenomenon in the diagonal limit of Gegenbauer polynomial approximations. J. Comput. Phys. **204**, 253–264

J.P. Boyd, N. Flyer (1999): Compatibility conditions for time-dependent partial differential equations and the rate of convergence of Chebyshev and Fourier spectral methods. Comput. Methods Appl. Mech. Engrg. **175**, 281–309

M.E. Brachet, D.I. Meiron, S.A. Orszag, B.G. Nickel, R.H. Morf, U. Frisch (1983): Small-scale structure of the Taylor-Green vortex. J. Fluid Mech. **130**, 411–452

A. Brandt, S.R. Fulton, G.D. Taylor (1985): Improved spectral multigrid methods for periodic elliptic problems. J. Comput. Phys. **58**, 96–112

N. Bressan, A. Quarteroni (1986): Analysis of Chebyshev collocation methods for parabolic equations. SIAM J. Numer. Anal. **23**, 1138–1154

K.S. Breuer, R.M. Everson (1992): On the errors incurred calculating derivatives using Chebyshev polynomials. J. Comput. Phys. **99**, 56–67

F. Brezzi, M.O. Bristeau, L.P. Franca, M. Mallet, G. Rogé (1992): A relationship between stabilized finite element methods and the Galerkin method with bubble functions. Comput. Methods Appl. Mech. Engrg. **96**, 117–129

F. Brezzi, G. Gilardi (1987): 'Fundamentals of P.D.E. for Numerical Analysis'. In: *Finite Element Handbook*, ed. by H. Kardestuncer (McGraw-Hill, New York)

F. Brezzi, A. Russo (1994): Choosing bubbles for advection-diffusion problems. Math. Models Methods Appl. Sci. **4**, 571–587

E.O. Brigham (1974): *The Fast Fourier Transform* (Prentice-Hall, Englewood Cliffs, NJ)

A.N. Brooks, T.J.R. Hughes (1982): Streamline upwind/Petrov-Galerkin formulations for convection dominated flows with particular emphasis on the incompressible Navier-Stokes equations. Comput. Methods Appl. Mech. Engrg. **32**, 199–259

P.L. Butzer, R.J. Nessel (1971): *Fourier Analysis and Approximation* (Birkhäuser, Basel)

J.M. Burgers (1948): A mathematical model illustrating the theory of turbulence. Adv. Appl. Mech. **1**, 171–199

J.M. Burgers (1974): *The Nonlinear Diffusion Equation* (Reidel, Boston)

W. Cai, D. Gottlieb, A. Harten (1992): Cell averaging Chebyshev methods for hyperbolic problems. Comput. Math. Appl. **24**, 37–49

W. Cai, D. Gottlieb, C.-W. Shu (1989): Essentially nonoscillatory spectral Fourier methods for shock wave calculations. Math. Comput. **52**, 389–410

W. Cai, C.-W. Shu (1993): Uniform high-order spectral methods for one- and two-dimensional Euler equations, J. Comput. Phys. **104**, 427–443

A.B. Cain, J.H. Ferziger, W.C. Reynolds (1984): Discrete orthogonal function expansions for non-uniform grids using the fast Fourier transform. J. Comput. Phys. **56**, 272–286

C. Canuto (1986): Boundary conditions in Legendre and Chebyshev methods. SIAM J. Numer. Anal. **23**, 815–831

C. Canuto (1988): Spectral methods and a maximum principle. Math. Comput. **51**, 615–629

C. Canuto (1994): Stabilization of spectral methods by finite element bubble functions. Comput. Methods Appl. Mech. Engrg. **116**, 13–26

C. Canuto, P. Gervasio, A. Quarteroni (2006): Preconditioning G-NI spectral methods in simple- and multi-domains. In preparation

C. Canuto, M.Y. Hussaini, A. Quarteroni, T.A. Zang (1988): *Spectral Methods in Fluid Dynamics* (Springer, New York)

C. Canuto, M.Y. Hussaini, A. Quarteroni, T.A. Zang (2007): *Spectral Methods. Evolution to Complex Domains and Applications to Fluid Dynamics* (Springer, New York)

C. Canuto, Y. Maday, A. Quarteroni (1982): Analysis for the combined finite element and Fourier interpolation. Numer. Math. **39**, 205–220

C. Canuto, G. Puppo (1994): Bubble stabilization of spectral Legendre methods for the advection-diffusion equation. Comput. Methods Appl. Mech. Engrg. **118**, 239–263

C. Canuto, A. Quarteroni (1981): Spectral and pseudo-spectral methods for parabolic problems with nonperiodic boundary conditions. Calcolo **18**, 197–218

C. Canuto, A. Quarteroni (1982a): Approximation results for orthogonal polynomials in Sobolev spaces. Math. Comput. **38**, 67–86

C. Canuto, A. Quarteroni (1982b): Error estimates for spectral and pseudo-spectral approximations of hyperbolic equations. SIAM J. Numer. Anal. **19**, 629–642

C. Canuto, A. Quarteroni (1984): 'Variational Methods in the Theoretical Analysis of Spectral Approximations'. In: *Spectral Methods for Partial Differential Equations*, ed. by R.G. Voigt, D. Gottlieb, M.Y. Hussaini (SIAM-CBMS, Philadelphia) pp. 55–78

C. Canuto, A. Quarteroni (1985): Preconditioned minimal residual methods for Chebyshev spectral calculations. J. Comput. Phys. **60**, 315–337

C. Canuto, A. Russo, V. van Kemenade (1998): Stabilized spectral methods for the Navier-Stokes equations: residual-free bubbles and preconditioning. Comput. Methods Appl. Mech. Engrg. **166**, 65–83

C. Canuto, G. Sacchi-Landriani (1986): Analysis of the Kleiser-Schumann method. Numer. Math. **50**, 217–243

C. Canuto, A. Tabacco (2001): An anisotropic functional setting for convection-diffusion problems. East-West J. Numer. Math. **9**, 199–231

L. Carleson (1966): On convergence and growth of partial sums of Fourier series. Acta Math. **116**, 135–157

M.H. Carpenter, D. Gottlieb, S. Abarbanel (1993): The stability of numerical boundary treatments for compact high-order finite-difference schemes. J. Comput. Phys. **108**, 272–295.

M.H. Carpenter, C. Kennedy (1994): 'Fourth-order 2N-storage Runge-Kutta Schemes'. NASA TM 109111.

G.F. Carrier, M. Krook, C.E. Pearson (1966): *Functions of a Complex Variable* (McGraw-Hill, New York)

T.F. Chan, T. Kerkhoven (1985): Fourier methods with extended stability intervals for the Korteweg-de Vries equation. SIAM J. Numer. Anal. **22**, 441–454

Q. Chen, I. Babuška (1995): Approximate optimal points for polynomial interpolation of real functions in an interval and in a triangle. Comput. Methods Appl. Mech. Engrg. **128**, 405–417

Q. Chen, I. Babuška (1996): The optimal symmetrical points for polynomial interpolation of real functions in the tetrahedron. Comput. Methods Appl. Mech. Engrg. **137**, 89–94

G.-Q. Chen, Q. Du, E. Tadmor (1993): Spectral viscosity approximations to multidimensional scalar conservation laws. Math. Comp. **61**, 629–643

E.W. Cheney (1966): *Introduction to Approximation Theory* (McGraw-Hill, New York)

I. Christie, D.F. Griffiths, A.R. Mitchell, O.C. Zienkiewicz (1976): Finite element methods for second order differential equations with significant first derivatives. Internat. J. Numer. Methods Engrg. **10**, 1389–1396

Ph.G. Ciarlet (2002): *The Finite Element Method for Elliptic Problems* (SIAM, Philadelphia)

C.W. Clenshaw (1957): The numerical solution of linear differential equations in Chebyshev series. Proc. Cambridge Philos. Soc. **53**, 134–149

C.W. Clenshaw, H.J. Norton (1963): The solution of nonlinear ordinary differential equations in Chebyshev series. Comput. J. **6**, 88–92

G.C. Cohen (2002): *Higher-order Numerical Methods for Transient Wave Equations* (Springer, Berlin)

J.D. Cole (1951): On a quasilinear parabolic equation occurring in aerodynamics. Q. Appl. Math. **9**, 225–236

L. Collatz (1966): *The Numerical Treatment of Differential Equations* (Springer, Berlin)

P. Concus, G.H. Golub (1973): Use of fast direct methods for the efficient numerical solution of nonseparable elliptic equations. SIAM J. Numer. Anal **10**(6), 1103–1120

P. Concus, G.H. Golub, D.P. O'Leary (1976): 'A Generalized Conjugate Gradient Method for the Numerical Solution of Elliptic Partial Differential Equations'. In: *Sparse Matrix Computations*, ed. by J.R. Bunch, D.J. Rose (Academic Press, New York) pp. 309–332

J.W. Cooley, P.A.W. Lewis, P.D. Welch (1969): The fast Fourier transform and its applications. IEEE Trans. Educ. **12**, 27–34

J.W. Cooley, J.W. Tukey (1965): An algorithm for the machine calculation of complex Fourier series. Math. Comput. **19**, 297–301

R. Cools (2003): An encyclopaedia of cubature formulas. Numerical integration and its complexity. J. Complexity **19**, 445–453

O. Coulaud, D. Funaro, O. Kavian (1990): Laguerre spectral approximation of elliptic problems in exterior domains. Comput. Methods Appl. Mech. Engrg. **80**, 451–458

R. Courant, D. Hilbert (1953): *Methods of Mathematical Physics*, Vol. 1 (Wiley-Interscience, New York)

J.P. Dahlburg (1985): 'Turbulent Disruptions from the Strauss Equations'. PhD Thesis, College of William and Mary, Williamsburg, VA

T.A. Davis, I.S. Duff (1999): A combined unifrontal/multifrontal method for unsymmetric sparse matrices. ACM Trans. Math. Software **25**, 1–20

P.J. Davis, P. Rabinowitz (1984): *Methods of Numerical Integration*, 2nd edn. (Academic Press, London, New York)

T.A. Davis (2004): UMFPACK Version 4.3, Univ. of Florida, Gainesville, FL

B.J. Debusschere, H.N. Najm, P.P. Pébay, O.M. Knio, R.G. Ghanem, O.P. LeMaître (2004): Numerical challenges in the use of polynomial chaos representations for stochastic processes. SIAM J. Sci. Comput. **26**(2), 698–719

S.C.R. Dennis, L. Quartapelle (1985): Spectral algorithms for vector elliptic equations in a spherical gap. J. Comput. Phys. **61**, 218–241

M.O. Deville, P.F. Fischer, E.H. Mund (2002): *High-Order Methods for Incompressible Fluid Flow* (Cambridge Univ. Press, Cambridge)

M. Deville, L. Kleiser, F. Montigny-Rannou (1984): Pressure and time treatment of Chebyshev spectral solution of a Stokes problem. Int. J. Numer. Meth. Fluids **4**, 1149–1163

M. Deville, E. Mund (1985): Chebyshev pseudospectral solution of second-order elliptic equations with finite element preconditioning. J. Comput. Phys. **60**, 517–533

M. Deville, E. Mund (1990): Finite element preconditioning for pseudospectral solutions of elliptic problems. SIAM J. Sci. Stat. Comput. **11**, 311–342

G.S Dietachmayer, K.K. Droegemeier (1992): Application of continuous dynamic grid adaption techniques to meteorological modeling. Part I: Basic formulation and accuracy. Mon. Weather Rev. **120**(8), 1675-1706

C.D. Dimitropoulos, A.N. Beris (1997): An efficient and robust spe ctral solver for nonseparable elliptic equations. J. Comput. Phys. **133**, 186–191

W.S. Don, D. Gottlieb (1994): The Chebyshev-Legendre method: implementing Legendre methods on Chebyshev points. SIAM J. Numer. Anal. **31**, 1519–1534

W.S. Don, A. Solomonoff (1995): Accuracy and speed in computing the Chebyshev collocation derivative. SIAM J. Sci. Comput. **16** (6) 1253–1268

C.C. Douglas (1995): Madpack: A family of abstract multigrid or multilevel solvers. Comput. Appl. Math. **14**, 3–20

J. Douglas, J.E. Gunn (1964): A general formulation of alternating direction methods. Numer. Math. **6**, 428–453

M. Dubiner (1991a): Asymptotic analysis of spectral methods. J. Sci. Comput. **2**, 3–31

M. Dubiner (1991b): Spectral methods on triangles and other domains. J. Sci. Comput. **6**, 345–390

U. Ehrenstein, R. Peyret (1989): A Chebyshev collocation method for the Navier-Stokes equations with application to double-diffusive convection. Int. J. Numer. Meth. Fluids **9**, 427–452

E. Eliasen, B. Machenhauer, E. Rasmussen (1970): 'On a Numerical Method for Integration of the Hydrodynamical Equations with a Spectral Representation of the Horizontal Fields'. Rep. No. 2 (Institut for Teoretisk Meteorologi, Univ. Copenhagen)

M. Embree (1999): Convergence of Krylov subspace methods for non-normal matrices. PhD Thesis, Oxford Univ. Computing Laboratories, Oxford

M. Embree, L.N. Trefethen (2005): *Spectra and Pseudospectra: The Behavior of Nonnormal Matrices and Operators* (Princeton Univ. Press, Princeton)

P. Erdös (1961): Problems and results on the theory of interpolation. Acta Math. Acad. Sci. Hungar. **44**, 235–244

G. Erlebacher, T.A. Zang, M.Y. Hussaini (1987): 'Spectral Multigrid Methods for the Numerical Simulation of Turbulence'. In: *Multigrid Methods, Theory, Applications, and Supercomputing*, ed. by S.F. McCormick (Marcel Dekker, New York) pp. 177–194

D.J. Evans (ed.) (1983): *Preconditioning Methods: Analysis and Applications* (Gordon and Breach, New York)

M.V. Feigelman (1980): One-dimensional periodic structures in a weak random potential. Sov. Phys. JETP **52**(3), 555–561 (translated from Zh. Eksper. Teoret. Fiz. **79**(3), 1095–1107)

L. Fejér (1932): Bestimmung derjenigen Abszissen eines Intervalles für welche die Quadratsumme der Grundfunktionen der Lagrangeschen Interpolation im Intervalle $[-1, 1]$ ein möglichst kleines Maximum besitzt. Ann. Scuola Norm. Sup. Pisa **12**, 263–276

B.A. Finlayson, L.E. Scriven (1966): The method of weighted residuals – a review. Appl. Mech. Rev. **19**, 735–748

R. Fletcher (1976): 'Conjugate Gradient Methods for Indefinite Systems'. In: *Lecture Notes in Math., Vol. 506* (Springer, Berlin) pp. 73–89

B. Fornberg (1996): *A Practical Guide to Pseudospectral Methods* (Cambridge Univ. Press, New York)

B. Fornberg, G.B. Whitman (1978): A numerical and theoretical study of certain nonlinear wave phenomena. Philos. Trans. Soc. London **289**, 373–404

L. Fox, I.B. Parker (1968): *Chebyshev Polynomials in Numerical Analysis* (Oxford Univ. Press, London)

L.P. Franca, S.L. Frey, T.J.R. Hughes (1992): Stabilized finite element methods. I. Application to the advective-diffusive model. Comput. Methods Appl. Mech. Engrg. **95**, 253–276

P. Francken, M. Deville, E. Mund (1990): On the spectrum of the iteration operator associated to the finite-element preconditioning of the Chebyshev collocation calculations. Comput. Meth. Appl. Mech. Eng. **80**, 295–3404

R.A. Frazer, W.P. Jones, S.W. Skan (1937): *Approximation to Functions and to the Solution of Differential Equations.* Rep. and Mem. 1799 (Aeronautical Research Council, London)

M. Frigo, S.G. Johnson (2005): The design and implementation of FFTW3. Proc. IEEE **92**(2), 216–231

D. Funaro (1981): 'Numerical Approximation of Parabolic and Hyperbolic Problems by Spectral Methods'. Thesis, Univ. Pavia (in Italian)

D. Funaro (1987): A preconditioning matrix for the Chebyshev differencing operator. SIAM J. Numer. Anal. **24**, 1024–1031

D. Funaro (1988): Computing the inverse of the Chebyshev collocation derivative. SIAM J. Sci. Stat. Comput. **9**, 1050–1057

D. Funaro (1991): 'Estimates of Laguerre spectral projectors in Sobolev spaces'. In: *Orthogonal Polynomials and their Applications*, ed. by C. Brezinski, L. Gori, A. Ronveaux (IMACS, New Brunswick)

D. Funaro (1992): *Polynomial Approximation of Differential Equations* (Springer, Berlin)

D. Funaro (1997): *Spectral Elements for Transport-dominated Equations* (Springer, Berlin)

D. Funaro, D. Gottlieb (1988): A new method of imposing boundary conditions in pseudospectral approximations of hyperbolic equations. Math. Comp. **51**, 519–613

D. Funaro, D. Gottlieb (1991): Convergence results for pseudospectral approximations of hyperbolic systems by a penalty-type boundary treatment. Math. Comp. **57**, 585–596

D. Funaro, O. Kavian (1990): Approximation of some diffusion evolution equations in unbounded domains by Hermite functions. Math. Comput. **57**, 597-619

B. Galerkin (1915): Rods and plates: Series occurring in various questions concerning the elastic equilibrium of rods and plates. Vestn. Inzhen. **19**, 897–908

C.W. Gear (1971): *Numerical Initial Value Problems in Ordinary Differential Equations* (Prentice-Hall, Englewood Cliffs, NJ)

A. Gelb, E. Tadmor (2000a): Detection of edges in spectral data II. Nonlinear enhancement. SIAM J. Numer. Anal. **38**, 1389–1408

A. Gelb, E. Tadmor (2000b): Enhanced spectral viscosity approximations for conservation laws. Appl. Numer. Math. **33**, 3–21

A. Gelb, E. Tadmor (2002): Spectral reconstruction of piecewise smooth functions from their discrete data. Math. Model. Numer. Anal. **36**, 155–175

A. Gelb, J. Tanner (2006): Robust reprojection methods for the resolution of the Gibbs phenomenon. J. Sci. Comput. (special issue dedicated to D. Gottlieb), in press

A. George, J.W.H. Liu (1981) *Computer Solution of Large Sparse Positive Definite Systems* (Prentice-Hall, Englewood Cliffs, NJ)

R.G. Ghanem, P. D. Spanos (1991): *Stochastic Finite Elements: A Spectral Approach* (Springer, New York)

N. Gilbert, L. Kleiser (1990): 'Near-wall Phenomena in Transition to Turbulence'. In: *Near-Wall Turbulence: 1988 Zoran Zaric Memorial Conference*, ed. by S.J. Kline, N.H. Afgan (Hemisphere, Washington) pp. 7–27

J.R. Gilbert, C. Moler, R. Schreiber (1992): Sparse matrices in MATLAB: design and implementation. SIAM J. Matrix Anal. Appl. **13**(1), 333–356.

G.H. Golub, C.F. Van Loan (1996): *Matrix Computations*, 3rd edn. (John Hopkins Univ. Press, Baltimore)

J. Goodman, T. Hou, E. Tadmor (1994): On the stability of the unsmoothed Fourier method for hyperbolic equations. Numer. Math. **67**, 93–129

W.J. Gordon, C.A. Hall (1973a): Construction of curvilinear coordinate systems and their applications to mesh generation. Int. J. Numer. Meth. Eng. **7**, 461–477

W.J. Gordon, C.A. Hall (1973b): Transfinite element methods: blending-function interpolation over arbitrary curved element domains. Numer. Math. **21**, 109–129

D. Gottlieb (1981): The stability of pseudospectral Chebyshev methods. Math. Comput. **36**, 107–118

D. Gottlieb (1985): 'Spectral Methods for Compressible Flow Problems'. In: *Proc. 9th Int Conf. Numerical Methods in Fluid Dynamics*, ed. by Soubbarameyer, J.P. Boujot (Springer, Heidelberg) pp. 48–61

D. Gottlieb, J.S. Hesthaven (2001): Spectral methods for hyperbolic problems. J. Comput. Appl. Math. **128**, 83–131

D. Gottlieb, M.Y. Hussaini, S.A. Orszag (1984): 'Theory and Applications of Spectral Methods'. In: *Spectral Methods for Partial Differential Equations*, ed. by R.G. Voigt, D. Gottlieb, M.Y. Hussaini (SIAM-CBMS, Philadelphia) pp. 1–54

D. Gottlieb, L. Lustman (1983): The spectrum of the Chebyshev collocation operator for the heat equation. SIAM J. Numer. Anal. **20**, 909–921

D. Gottlieb, L. Lustman, S.A. Orszag (1981): Spectral calculations of one-dimensional inviscid compressible flow. SIAM J. Sci. Statist. Comput. **2**, 296–310

D. Gottlieb, L. Lustman, S.A. Orszag (1987): Convergence of spectral methods for hyperbolic initial-boundary value systems. SIAM J. Numer. Anal. **24**, 532–537

D. Gottlieb, S.A. Orszag (1977): *Numerical Analysis of Spectral Methods: Theory and Applications* (SIAM-CBMS, Philadelphia)

D. Gottlieb, S.A. Orszag, E. Turkel (1981): Stability of pseudospectral and finite difference methods for variable coefficient problems. Math. Comput. **37**, 293–305

D. Gottlieb, C.-W. Shu (1997): On the Gibbs phenomenon and its resolution. SIAM Rev. **39**, 644–668

D. Gottlieb, C.-W. Shu (1998): 'A General Theory for the Resolution of the Gibbs Phenomenon'. In: *Tricomi's Ideas and Contemporary Applied Mathematics*, Atti dei Convegni Lincei **147** (Accademia Nazionale dei Lincei, Rome) pp. 39–48

D. Gottlieb, C.-W. Shu, A. Solomonoff, H. Vandeven (1992): On the Gibbs phenomenon I: recovering exponential accuracy from the Fourier partial sums of a nonperiodic analytic function. J. Comput. Appl. Math. **43**, 81–92

D. Gottlieb, E. Tadmor (1985): 'Recovering Pointwise Values of Discontinuous Data Within Spectral Accuracy'. In: *Progress and*

*Supercomputing in Computational Fluid Dynamics*, ed. by E.M. Murman, S.S. Abarbanel (Birkhäuser, Boston) pp. 357–375

D. Gottlieb, E. Turkel (1985): 'Topics in Spectral Methods for Time Dependent Problems'. In: *Numerical Methods in Fluid Dynamics*, ed. by F. Brezzi (Springer, Heidelberg) pp. 115–155

H.L. Grant, R.W. Stewart, A. Moilliet (1962): Turbulence spectra from a tidal channel. J. Fluid Mech. **12**, 241–268

R. Grauer, C. Marliani (1995): Numerical and analytical estimates for the structure functions in two-dimensional magnetohydrodynamic flows. Phys. Plasmas, **2**(1), 41–47

A. Greenbaum (1997): 'Iterative Methods for Solving Linear Systems'. In *Frontiers in Applied Mathematics* **17** (SIAM, Philadelphia)

L. Greengard (1991): Spectral integration and two-point boundary value problems. SIAM J. Numer. Anal. **28**(4), 1071–1080

P. Grisvard (1985): *Elliptic Problems in Nonsmooth Domains* (Pitman, Boston)

C.E. Grosch, S.A. Orszag (1977): Numerical solution of problems in unbounded regions: coordinate transformations. J. Comput. Phys. **25**, 273–296

H. Guillard, J.-A. Desideri (1990): Iterative methods with spectral preconditioning for elliptic equations. Comput. Meth. Appl. Mech. Eng. **80**, 305–312

B.-Y. Guo (1998): *Spectral Methods and their Applications* (World Scientific, Singapore)

B.-Y. Guo (1999): Error estimation of Hermite spectral method for nonlinear partial differential equations. Math. Comput. **68**, 1067-1078

B.-Y. Guo, H.-P. Ma, E. Tadmor (2001): Spectral vanishing viscosity method for nonlinear conservation laws. SIAM J. Numer. Anal. **39**, 1254–1268

B.-Y. Guo, J. Shen (2000): Laguerre-Galerkin method for nonlinear partial differential equations on a semi-infinite interval. Numer. Math. **86**, 635–654

B.-Y. Guo, J. Shen (2003): Spectral and pseudo-spectral approximations using Hermite functions: application to the Dirac equation. Adv. Comput. Math. **19**, 35–55

B.-Y. Guo, L.-L. Wang (2004): Jacobi approximations in non-uniformly Jacobi weighted Sobolev spaces. J. Approx. Th. **128**, 1–41

B.-Y. Guo, L.-L. Wang (2006): Error analysis of spectral methods on a triangle. Adv. Comput. Math., to appear

B.-Y. Guo, C.-L. Xu (2000): Hermite pseudospectral method for nonlinear partial differential equations. Math. Model. Numer. Anal. **34**, 859–872

V. Gurarie, A. Migdal (1996): Instantons in the Burgers equation. Phy. Rev. E **54**, 4908–4914

W. Hackbusch (1985): *Multigrid Methods and Applications* (Springer, Heidelberg)

A. Hageman, D.M. Young (1981): *Applied Iterative Methods* (Academic Press, New York)

D.B. Haidvogel (1977): 'Quasigeostrophic Regional and General Circulation Modelling: an Efficient Pseudospectral Approximation Technique.' In: *Computing Methods in Geophysical Mechanics.* Vol. 25, ed. by R.P. Shaw (ASME, New York)

D.B. Haidvogel, T.A. Zang (1979): The accurate solution of Poisson's equation by expansion in Chebyshev polynomials. J. Comput. Phys. **30**, 167–180

E. Hairer, S.P. Norsett, G. Wanner (1993) *Solving Ordinary Differential Equations I. Nonstiff Problems* (Springer, Heidelberg)

E. Hairer, G. Wanner (1996) *Solving Ordinary Differential Equations II. Stiff and Differential–Algebraic Problems* (Springer, Heidelberg)

P. Haldenwang, G. Labrosse, S. Abboudi, M. Deville (1984): Chebyshev 3-D spectral and 2-D pseudospectral solvers for the Helmholtz equation. J. Comput. Phys. **55**, 115–128

F.R. Hama, J. Nutant (1963): 'Detailed flowfield observations in the transition process in a thick boundary layer'. In: *Proc. 1963 Heat Transfer and Fluid Mechanics Institute* (Stanford Univ. Press, Palo Alto) pp. 77–93

W. Heinrichs (1988): Line relaxation for spectral multigrid methods. J. Comput. Phys. **77**, 166–182

W. Heinrichs (1991): A 3D spectral multigrid method. Appl. Math. Comp. **41**, 117–128

W. Heinrichs (1993): Finite element preconditioning for spectral multigrid methods. Appl. Math. Comp. **59**, 19–40

M.R. Hestenes, E. Stiefel (1952): Methods of conjugate gradients for solving linear systems. J. Res. Natl. Bur. Stand. **49**, 409–436

J.S. Hesthaven (1998): From electrostatics to almost optimal nodal sets for polynomial interpolation in a simplex. SIAM J. Numer. Anal. **35**(2), 655–676

J.S. Hesthaven (2000): Spectral penalty methods. Appl. Numer. Math. **33**(1–4), 23–41

J.S. Hesthaven, C.H. Teng (2000): Stable spectral methods on tetrahedral elements. SIAM J. Sci. Comput. **21**, 2352–2380

J.S. Hesthaven, T. Warburton (2002): Nodal high-order methods on unstructured grids. J. Comput. Phys. **181**, 186–221

E. Hille, R.S. Phillips (1957): *Functional Analysis and Semi-Groups*, Am. Math. Soc. (Providence, Rhode Island)

J.O. Hinze (1975): *Turbulence* (McGraw-Hill, New York)
E. Hopf (1950): The partial differential equation $u_t + uu_x = \mu u_{xx}$. Commun. Pure Appl. Math. **3**, 201–230
W. Huang, R.D. Russell (1997): Analysis of moving mesh partial differential equations with spatial smoothing. SIAM J. Numer. Anal., **34**, 1106–1126
T.J.R. Hughes (2000): *The Finite Element Method: Linear Static and Dynamic Finite Element Analysis* (Dover Publications, Mineola, NY)
D. Jackson (1930): *The Theory of Approximation*, Vol. 11 (AMS Colloquium Publications, New York)
A. Jameson, H. Schmidt, E. Turkel (1981): 'Numerical Solutions of the Euler Equations by Finite Volume Methods Using Runge-Kutta Time Stepping Schemes'. AIAA Paper No. 81–1259
Y. Kaneda, T. Ishihara (2006): High-resolution direct numerical simulation of turbulence. J. Turbulence, to appear
L.V. Kantorovic (1934): On a new method of approximate solution of partial differential equations. Dokl. Akad. Nauk SSSR **4**, 532–536 (in Russian)
G.E. Karniadakis, M. Israeli, S.A. Orszag (1991): High-order splitting methods for the incompressible Navier-Stokes equations. J. Comput. Phys. **97**, 414–443
G.E. Karniadakis, S.J. Sherwin (1999): *Spectral/hp Element Methods for Computational Fluid Dynamics* (Oxford Univ. Press, New York) [2nd edn., Oxford Univ. Press, 2005]
J. Kim, P. Moin, R.D. Moser (1987): Turbulent statistics in fully developed turbulent channel flow at low Reynolds number. J. Fluid Mech. **177**, 133–166
S.D. Kim, S.V. Parter (1997): Preconditioning Chebyshev spectral collocation by finite difference operators. SIAM J. Numer. Anal. **34**, 939–958
L. Kleiser, U. Schumann (1980): 'Treatment of Incompressibility and Boundary Conditions in 3-D Numerical Spectral Simulations of Plane Channel Flows'. In: *Proc. 3rd GAMM Conf. Numerical Methods in Fluid Mechanics*, ed. by E.H. Hirschel (Vieweg, Braunschweig) pp. 165–173
L. Kleiser, U. Schumann (1984): 'Spectral Simulation of the Laminar-Turbulent Transition Process in Plane Poiseuille Flow'. In: *Spectral Mathods for Partial Differential Equations*, ed. by R.G. Voigt, D. Gottlieb, M.Y. Hussaini (SIAM-CBMS, Philadelphia) pp. 141–163
T. Koornwinder (1975): 'Two-variable analogues of the classical orthogonal polynomials'. In: *Theory and Application of Special*

*Functions*, ed. by R.A. Askey (Academic Press, New York) pp. 435–495

D.A. Kopriva, J.H. Kolias (1996): A conservative staggered-grid Chebyshev multidomain method for compressible flows. J. Comput. Phys. **125**, 244–261

K.Z. Korczak, A.T. Patera (1986): Isoparametric spectral element method for solution of the Navier-Stokes equations in complex geometry. J. Comput. Phys. **62**, 361–382

D. Kosloff, H. Tal-Ezer (1993): A modified Chebyshev pseudospectral method with an $\mathcal{O}(N^{-1})$ time step restriction. J. Comput. Phys. **104**, 457–469

H.-O. Kreiss, J. Oliger (1972): Comparison of accurate methods for the integration of hyperbolic equations. Tellus **24**, 199–215

H.-O. Kreiss, J. Oliger (1979): Stability of the Fourier method. SIAM J. Numer. Anal. **16**, 421–433

J. Krug, H. Sophn (1992): 'Kinetic Roughening of Growing Surfaces'. In: *Solids Far from Equililbrium* ed. by C. Godreche (Cambridge Univ. Press, Cambridge) pp. 3738–3742

S.N. Kruškov (1970): First order quasilinear equations in several independent variables. Math. USSR Sbornik **10**, 217–243

J.D. Lambert (1991): *Numerical Methods for Ordinary Differential Systems: The Initial Value Problem* (John Wiley and Sons, New York)

C. Lanczos (1938): Trigonometric interpolation of empirical and analytical functions. J. Math. Phys. **17**, 123–199

P.D. Lax (1978): 'Accuracy and Resolution in the Computation of Solutions of Linear and Nonlinear Equations'. In: *Recent Advances in Numerical Analysis* (Academic Press, New York) pp. 107–117

P.D. Lax, B. Wendroff (1960): Systems of conservation laws. Commun. Pure Appl. Math. **13**, 217–237

S.K. Lele (1992): Compact finite difference schemes with spectral-like resolution. J. Comput. Phys. **103**, 16–42

M.J. Lighthill (1956): 'Viscosity Effects in Sound Waves of Finite Amplitude'. In: *Surveys in Mechanics*, ed. by G.K. Batchelor, R. Davies (Cambridge Univ. Press, Cambridge)

J.-L. Lions, E. Magenes (1972): *Nonhomogeneous Boundary Value Problems and Applications*, Vol. 1 (Springer, Heidelberg)

A. Lundbladh, D.S. Henningson, A.V. Johansson (1992): 'An efficient spectral integration method for the solution of the Navier-Stokes equations'. FFA-TN 1992-28, Aeronautical Research Institute of Sweden, Bromma

L. Lustman (1986): The time evolution of spectral discretizations of hyperbolic systems. SIAM J. Numer. Anal. **23**, 1193–1198

R.E. Lynch, J.R. Rice, D.H. Thomas (1964): Direct solution of partial difference equations by tensor product methods. Numer. Math. **6**, 185–199

Y. Maday (1990): Analysis of spectral projectors in one dimensional domains. Math. Comput. **55**, 537–562

Y. Maday, S.M. Ould Kaber, E. Tadmor (1993): Legendre pseudospectral viscosity method for nonlinear conservation laws. SIAM J. Numer. Anal. **30**, 321–342

Y. Maday, B. Pernaud-Thomas, H. Vandeven (1985): Reappraisal of Laguerre type spectral methods. Rech. Aerosp. **6**, 13–35

Y. Maday, A. Quarteroni (1981): Legendre and Chebyshev spectral approximations of Burgers' equation. Numer. Math. **37**, 321–332

Y. Maday, A. Quarteroni (1982): Approximation of Burgers' equation by pseudospectral methods. R.A.I.R.O. Anal. Numer. **16**, 375–404

Y. Maday, E.M. Rønquist (1990): Optimal error analysis of spectral methods with emphasis on nonconstant coefficients and deformed geometries. Comput. Methods Appl. Mech. Engrg. **80**, 91–115

Y. Maday, E. Tadmor (1989): Analysis of the spectral viscosity method for periodic conservation laws. SIAM J. Numer. Anal. **26**, 854–870

A. Majda, J. McDonough, S. Osher (1978): The Fourier method for nonsmooth initial data. Math. Comput. **32**, 1041–1081

M.R. Malik, T.A. Zang, M.Y. Hussaini (1985): A spectral collocation method for the Navier-Stokes equations. J. Comput. Phys. **61**, 64–88

C.E. Mavriplis (1994): Adaptive mesh strategies for spectral element method. Comp. Meth. in Appl. Mech. Eng. **116**, 77–86

J.A. Meijerink, H.A. Van der Vorst (1981): Guidelines for the usage of incomplete decompositions in solving sets of linear equations as they occur in practical problems. J. Comput. Phys. **44**, 134–155

D.I. Meiron, S.A. Orszag, M. Israeli (1981): Applications of numerical conformal mapping. J. Comput. Phys. **40**, 345–360

J.M. Melenk, Ch. Schwab (1999): An *hp* finite element method for convection-diffusion problems in one dimension. IMA J. Numer. Anal. **19**, 425–453

B. Mercier (1981): 'Analyse Numérique des Méthodes Spectrales'. Note CEA-N-2278 (Commissariat a l'Énergie Atomique Centre d'Études de Limeil, 94190 Villeneuve-Saint Georges) [English translation published as: *An Introduction to the Numerical Analysis of Spectral Methods* (Springer, Heidelberg 1989)]

B. Mercier, G. Raugel (1982): Resolution d'un problème aux limites dans un ouvert axisymétrique par éléments finis en $r$, $z$ et séries de Fourier en $\theta$. R.A.I.R.O. Anal. Numer. **16**, 405–461

P.G. Mestayer, C.H. Gibson, M.F. Coantic, A.S. Patel (1970): Local anisotropy in heated and cooled turbulent boundary layers. Phys. Fluid **19**, 1279–1287

L.M. Milne-Thomson (1966): *Theoretical Aerodynamics* (MacMillan, New York)

G. Mastroianni, G. Monegato (1997): Nyström interpolants based on zeros of Laguerre polynomials for some Wiener-Hopf equations. IMA J. Numer. Anal. **17**, 621–642

Y. Morchoisne (1979): Resolution of Navier-Stokes equations by a space-time pseudospectral method. Rech. Aerosp. **1979–5**, 293–306

Y. Morchoisne (1981): 'Pseudo-spectral Space-Time Calculations of Incompressible Viscous Flows'. AIAA Pap. No. 81–0109

F. Murat (1978): Compacité par compensation. Ann. Scuola Norm. Sup. Pisa **5**, 489–507.

J.W. Murdock (1977): A numerical study of nonlinear effects on boundary-layer stability. AIAA J. **15**, 1167–1173

I. Natanson (1965): *Constructive Function Theory.* Vol. III (Ungar, New York)

J. Nečas (1962): Sur une méthode pour resoudre les equations aux derivées partielles du type elliptique, voisine de la variationnelle. Ann. Sc. Norm. Sup. Pisa **16**, 305–326

S.M. Nikolskii (1951): Inequalities for entire functions of finite degree and their application to the theory of differentiable functions of several variables. Dokl. Akad. Nauk. SSSR **58**, 244–278 (in Russian)

S.M. Nikolskii (1951): *Approximation of Functions of Several Variables and Imbedding Theorems* (Springer, Berlin)

M. Nishioka, M. Asai, S. Iida (1980): 'An Experimental Investigation of the Secondary Instability in Laminar-Turbulent Transition'. In: *Laminar-Turbulent Transition*, ed. by R. Eppler, H. Fasel (Springer, Heidelberg) pp. 37–46

S.A. Orszag (1969): Numerical methods for the simulation of turbulence. Phys. Fluids Suppl. II. **12**, 250–257

S.A. Orszag (1970): Transform method for calculation of vector coupled sums: Application to the spectral form of the vorticity equation. J. Atmosph. Sci. **27**, 890–895

S.A. Orszag (1971a): Numerical simulation of incompressible flows within simple boundaries: I. Galerkin (spectral) representations. Stud. Appl. Math. **50**, 293–327

S.A. Orszag (1971b): Accurate solution of the Orr-Sommerfeld stability equation. J. Fluid Mech. **50**, 689–703

S.A. Orszag (1972): Comparison of pseudospectral and spectral approximations. Stud. Appl. Math. **51**, 253–259

S.A. Orszag (1980): Spectral methods for problems in complex geometries. J. Comput. Phys. **37**, 70–92

S.A. Orszag, L.C. Kells (1980): Transition to turbulence in plane Poiseuille flow and plane Couette flow. J. Fluid Mech. **96**, 159–205

S.A. Orszag, A.T. Patera (1983): Secondary instability of wall-bounded shear flows. J. Fluid Mech. **128**, 347–385

S.A. Orszag, G.S. Patterson, Jr. (1972): Numerical simulation of three dimensional homogeneous isotropic turbulence. Phys. Rev. Lett. **28**, 76–79

R.G. Owens (1998): Spectral approximations on the triangle. Proc. R. Soc. London A **454**, 857–872

S.V. Parter (2001a): Preconditioning Legendre spectral collocation methods for elliptic problems I: finite difference operators. SIAM J. Numer. Anal. **39**, 330–347

S.V. Parter (2001b): Preconditioning Legendre spectral collocation methods for elliptic problems II: finite element operators. SIAM J. Numer. Anal. **39**, 348–362

S.V. Parter, E.E. Rothman (1995): Preconditioning Legendre spectral collocation approximations to elliptic problems. SIAM J. Numer. Anal. **32**, 333-385.

J.E. Pasciak (1980): Spectral and pseudospectral methods for advection equations. Math. Comput. **35**, 1081–1092

F. Pasquarelli, A. Quarteroni (1994): Effective spectral approximations of convection-diffusion equations. Comput. Methods Appl. Mech. Engrg. **116**, 39–51

R. Pasquetti, L. Pavarino, F. Rapetti, E. Zampieri (2006): 'Overlapping Schwarz Preconditioners for Fekete Spectral Elements'. In: *Domain Decomposition Methods in Science and Engineeing*, ed. by O.B. Widlund et al. (Springer, Heidelberg)

R. Pasquetti, F. Rapetti (2004): Spectral element methods on triangles and quadrilaterals: comparisons and applications. Comput. Phys. **198**, 349–362

R. Pasquetti, F. Rapetti (2006): Spectral element methods on unstructured meshes: comparisons and recent advances. J. Sci. Comp. (special issue for ICOSAHOM'04, Brown University), in press

A.T. Patera (1984): A spectral element method for fluid dynamics: laminar flow in a channel expansion. J. Comput. Phys. **54**, 468–488

G.S. Patterson, Jr., S.A. Orszag (1971): Spectral calculations of isotropic turbulence: Efficient removal of aliasing interactions. Phys. Fluids **14**, 2538–2541

R. Peyret (1986): *Introduction to Spectral Methods.* Von Karman Institute Lecture Series 1986–04 (Rhode-Saint Genese, Belgium)

R. Peyret (2002): *Spectral Methods for Incompressible Viscous Flow* (Springer, Heidelberg)

T.N. Phillips, T.A. Zang, M.Y. Hussaini (1986): Preconditioners for the spectral multigrid method. IMA J. Numer. Anal. **6**, 273–292

A.M. Polyakov (1995): Turbulence without pressure. Phys. Rev. E **52**(6), 6183–6188

J. Proriol (1957): Sur une famille de polynômes à deux variables orthogonaux dans un triangle. C. R. Acad. Sci. Paris **257**, 2459–2461

A. Quarteroni (1984): Some results of Bernstein and Jackson type for polynomial approximation in $L_p$-spaces. Jpn. J. Appl. Math. **1**, 173–181

A. Quarteroni (1987): Blending Fourier and Chebyshev interpolation. J. Approx. Theory **51**, 115-126

A. Quarteroni, R. Sacco, F. Saleri (2000): *Numerical Mathematics* (Springer, Heidelberg)

A. Quarteroni, A. Valli (1994): *Numerical Approximations of Partial Differential Equations* (Springer, Heidelberg)

A. Quarteroni, E. Zampieri (1992): Finite element preconditioning for Legendre spectral collocation approximations to elliptic equations and systems. SIAM J. Numer. Anal. **29**, 917–936

S.C. Reddy, L.N. Trefethen (1990): Lax-stability of fully discrete spectral methods via stability regions and pseudo-eigenvalues. Comp. Methods Appl. Mech. Eng. **80**, 147–164

S.C. Reddy, L.N. Trefethen (1992): Stability of the method of lines. Numer. Math. **62**, 235–267

J.K. Reid (1971): 'On the Method of Conjugate Gradients for the Solution of Large Sparse Systems of Linear Equations'. In: *Large Sparse Sets of Linear Equations*, ed. by J.K. Reid (Academic Press, New York) pp. 231–254

M. Renardy, R.C. Rogers (1993): *An Introduction to Partial Differential Equations* (Springer, New York)

L.F. Richardson (1910): The approximate solution by finite differences of physical problems involving differential equations, with an application to the stresses in a masonry dam. Philos. Trans. R. Soc. London Ser. A **210**, 307–357

R.D. Richtmyer (1978): *Principles of Advanced Mathematical Physics*, Vol. 1 (Springer, New York)

T.J. Rivlin (1974): *The Chebyshev Polynomials* (John Wiley and Sons, New York)

R.S. Rogallo (1977): 'An ILLIAC Program for the Numerical Simulation of Homogeneous, Incompressible Turbulence'. NASA TM-73203

M.M. Rogers, R.D. Moser (1992): The three-dimensional evolution of a plane mixing layer: the Kelvin-Helmholtz rollup. J. Fluid Mech. **243**, 183–226

H.-G. Roos, M. Stynes, L. Tobiska (1995): *Numerical Methods for Singularly Perturbed Differential equations. Convection-Diffusion and Flow Problems* (Springer, Berlin)

H.L. Royden (1968): *Real Analysis* (McMillan, New York)

W. Rudin (1966): *Real and Complex Analysis* (McGraw-Hill, New York)

Y. Saad (1996): *Iterative Methods for Sparse Linear Systems* (PWS Publishing Company, Boston)

G. Sacchi-Landriani (1988): Spectral tau approximation of the two-dimensional Stokes problem. Numer. Math. **52**, 683–699

P. Sagaut (2005): *Large Eddy Simulation for Incompressible Flows* (Springer, Heidelberg)

D. Schnack, J. Killeen (1980): Nonlinear two-dimensional magnetohydrodynamic calculations. J. Comput. Phys. **35**, 110–145

C. Schneider, W. Werner (1986): Some new aspects of rational interpolation. Math. Comp. **47**, 285–299

D. Schötzau, Ch. Schwab (2000): Time discretization of parabolic problems by the hp-version of the discontinuous Galerkin finite element method. SIAM J. Numer. Anal. **38**(3), 837–875

Ch. Schwab (1998): *p- and hp-Finite Element Methods* (Oxford Univ. Press, Oxford)

L. Schwartz (1966): *Théorie des Distributions* (Hermann, Paris)

A. Scotti, U. Piomelli (2001): Numerical simulation of pulsating turbulent channel flow. Phys. Fluids **13**, 1367–1384

L.F. Shampine (1994): *Numerical Solution of Ordinary Differential Equations* (Chapman and Hall, New York)

S.F. Shandarin, Ya.B. Zeldovich (1989): The large-scale structure of the universe: turbulence, intermittency, structures in a self-gravitating medium. Rev. Modern Phys. **61**(2), 185–220

J. Shen (1994): Efficient spectral-Galerkin method I. Direct solvers for second- and fourth-order equations by using Legendre polynomials. SIAM J. Sci. Comput. **15**, 1489–1505

J. Shen (1995): Efficient spectral-Galerkin method II. Direct solvers for second- and fourth-order equations by using Chebyshev polynomials. SIAM J. Sci. Comput. **16**, 74–87

S.J. Sherwin, G.E. Karniadakis (1995): A new triangular and tetrahedral basis for high-order finite element methods. Int. J. Num. Meth. Engng. **38**, pp. 3775–3802

I. Silberman (1954): Planetary waves in the atmosphere. J. Meteorol. **11**, 27–34

J.C. Slater (1934): Electronic energy bands in metal. Phys. Rev. **45**, 794–801

P. Sonneveld (1989): CGS, a fast Lanczos-type solver for nonsymmetric linear systems. SIAM J. Sci. Statist. Comput. **10**(1), 36–52

P.R. Spalart (1984): A spectral method for external viscous flows. Contemp. Math. **28**, 315–335

P.R. Spalart (1986): 'Numerical Simulation of Boundary Layers, Part 1: Weak Formulation and Numerical Method'. NASA TM–88222

P.R. Spalart (1988): 'Direct Numerical Study of Leading Edge Contamination'. In *Fluid Dynamics of Three-Dimensional Turbulent Shear Flows and Transition.* AGARD–CP–438, pp. 5.1–5.13

P.R. Spalart, R.D. Moser, M.M. Rogers (1991): Spectral methods for the Navier-Stokes equations with one infinite and two periodic directions. J. Comput. Phys. **96**, 297–324

T.J. Stieltjes (1885): Sur les polynômes de Jacobi. C. R. Acad. Sci. Paris **100**, 620–622

J. Strain (1994): Fast spectrally-accurate solution of variable-coefficient elliptic problems. Proc. Amer. Math. Soc. **122**(3), 843–850

G. Strang, T. Nguyen (1996): *Wavelets and Filter Banks* (Wellesley-Cambridge, Wellesley)

C.L. Streett, T.A. Zang, M.Y. Hussaini (1985): Spectral multigrid methods with applications to transonic potential flow. J. Comput. Phys. **57**, 43–76

A. H. Stroud (1971): *Approximate Calculation of Multiple Integrals* (Prentice-Hall, Englewood Cliffs)

K. Stuben, U. Trottenberg (1982): 'Multigrid Methods: Fundamental Algorithms, Model Problem Analysis and Applications'. In: *Multigrid Methods*, ed. by W. Hackbusch, U. Trottenberg (Springer, Heidelberg) pp. 1–176

P.N. Swarztrauber (1977): The methods of cyclic reduction, Fourier analysis and the FACR algorithm for the discrete solution of Poisson's equation on a rectangle. SIAM Rev. **19**, 490–501

P.N. Swarztrauber (1986): Symmetric FFTs. Math. Comput. **47**, 323–346

G. Szegö (1939): *Orthogonal Polynomials*, Vol. 23 (AMS Coll. Publ., New York)

E. Tadmor (1986): The exponential accuracy of Fourier and Chebyshev differencing methods. SIAM J. Numer. Anal. **23**, 1–10

E. Tadmor (1989): The convergence of spectral methods for nonlinear conservation laws. SIAM J. Numer. Anal. **26**, 30–44

E. Tadmor (1998): 'Approximate Solutions of Nonlinear Conservation Laws'. In: *Advanced Numerical Approximation of Nonlinear*

*Hyperbolic Equations*, Lect. Notes in Math. 1697, ed. by A. Quarteroni (Springer, Heidelberg), pp. 1–150

E. Tadmor, J. Tanner (2002): Adaptive mollifiers – High resolution recovery of piecewise smooth data from its spectral information. Foundat. Comput. Math. **2**, 155–189

H. Tal-Ezer (1986a): Spectral methods in time for hyperbolic equations. SIAM J. Numer. Anal. **23**, 11–26

H. Tal-Ezer (1986b): A pseudospectral Legendre method for hyperbolic equations with an improved stability condition. J. Comput. Phys. **67**, 145–172

H. Tal-Ezer (1989): Spectral methods in time for parabolic problems. SIAM J. Numer. Anal. **26**, 1–11

T. Tang, M.R. Trummer (1996): Boundary layer resolving pseudospectral methods for singular perturbation problems. SIAM J. Sci. Comput. **17**, 430–438

A.E. Taylor (1958): *Introduction to Functional Analysis* (John Wiley and Sons, New York)

M. Taylor (1981): *Pseudodifferenlial Operators* (Princeton Univ. Press, Princeton)

M.A. Taylor, B.A. Wingate, (2000): A generalized diagonal mass matrix spectral element method for non-quadrilateral elements. Appl. Numer. Math. **33**, 259–265

M.A. Taylor, B.A. Wingate, R.E. Vincent (2000): An algorithm for computing Fekete points in the triangle. SIAM J. Numer. Anal. **38**, 1707–1720

C. Temperton (1983): Self-sorting mixed-radix fast Fourier transforms. J. Comput. Phys. **52**, 1–23

H. Tennekes, J.L. Lumley (1972): *A First Course in Turbulence* (Massachusetts Inst. Technology, Cambridge, MA)

V. Theofilis (2000): 'Global linear instability in laminar separated boundary layer flow'. In: *Laminar-Turbulent Transition: IUTAM Symposium*, Sedona, AZ Sept. 13–17, 1999, ed. by H.F. Fasel, W.S. Saric (Springer, New York) pp. 663–668

A.F. Timan (1963): *Theory of Approximation of Functions of a Real Variable* (Pergamon, Oxford)

E.C. Titchmarsh (1962): *Eigenfunction Expansions* (Oxford Univ. Press, London)

L.N. Trefethen (1980): Numerical computation of the Schwarz-Christoffel transformation. SIAM J. Sci. Stat. Comput. **1**, 82–102

L.N. Trefethen (1992): 'Pseudospectra of Matrices'. In: *Numerical Analysis 1991*, ed. by D.F. Griffith, G.A. Watson (Longman Sci. Tech., Harlow, UK) pp. 234–266

L.N. Trefethen (1997): Pseudospectra of linear operators. Siam Rev. **39**(3), 383–406

L.N. Trefethen (2000): *Spectral Methods in MATLAB* (SIAM, Philadelphia)

L.N. Trefethen, M.R. Trummer (1987): An instability phenomenon in spectral methods. SIAM J. Numer. Anal. **24**(5), 1008–1023

H. Vandeven (1990): On the eigenvalues of second-order spectral differentiation operators. Comput. Methods Appl. Mech. Engrg. **80**(1-3), 313–318

H. Vandeven (1991): Family of spectral filters for discontinuous problems. J. Sci. Comput. **6**, 159–192

H.A. van der Vorst (1992): Bi-CGSTAB: a fast and smoothly converging variant of Bi-CG for the solution of nonsymmetric linear systems. SIAM J. Sci. Stat. Comput. **13**(2), 631–644

H.A. van der Vorst (2003): *Iterative Krylov Methods for Large Linear Systems* (Cambridge Univ. Press, Cambridge)

R.S. Varga (1962): *Matrix Iterative Analysis* (Prentice-Hall, Englewood Cliffs, NJ.)

M. Vergassola, B. Dubrulle, U. Frisch, A. Noullez (1994): Burgers equation, devils staircases and the mass distribution for large-scale structures. Astron. Astrophys. **280**, 325–356

J.V. Villadsen, W.E. Stewart (1967): Solution of boundary value problems by orthogonal collocation. Chem. Eng. Sci. **22**, 1483–1501

R.G. Voigt, D. Gottlieb, M.Y. Hussaini (Eds.) (1984): *Spectral Methods for Partial Differential Equations* (SIAM, Philadephia)

T. Warburton, L. Pavarino, J.S. Hesthaven (2000): A pseudo-spectral scheme for the incompressible Navier-Stokes equations using unstructured nodal elements. J. Comput. Phys. **164**, 1–21

G.W. Wei, Y. Gu (2002): Conjugate filter approach for solving Burgers equation. J. Compt. Appl. Math. **149**(2), 439–456

B.D. Welfert (1997): Generation of pseudospectral differentiation matrices I. SIAM J. Numer. Anal. **34**, 1640–1657

P. Wesseling (2004): *An Introduction to Multigrid Methods* (R.T. Edwards, Philadelphia)

N. Wiener (1938): The homogeneous chaos. Amer. J. Math. **60**, 897–936

J.H. Wilkinson (1965): *The Algebraic Eigenvalue Problem* (Clarendon Press, Oxford)

J.H. Williamson (1980): Low-storage Runge-Kutta schemes. J. Comput. Phys. **35**, 48–56

Y.S. Wong, T.A. Zang, M.Y. Hussaini (1986): Efficient iterative techniques for the solution of spectral equations. Comput. Fluids **14**, 85–95

K. Wright (1964): Chebyshev collocation methods for ordinary differential equations. Computer J. **6**, 358–365

T.G. Wright, L.N. Trefethen (2001): Large-scale computation of pseudospectra using ARPACK and eigs. SIAM J. Sci. Comp., **23**(2), 591–605

D. Xiu, G.E. Karniadakis (2002): The Wiener-Askey polynomial chaos for stochastic differential equations. SIAM J. Sci. Comput. **24**, 619–644

C.-L. Xu, B.-Y. Guo (2002): Laguerre pseudospectral method for nonlinear partial differential equations. J. Comput. Math. **20**, 413–428

M. Yokokawa, K. Itakura, A. Uno, T. Ishihara, Y. Kaneda (2002): '16.4-Tflops Direct Numerical Simulation of Turbulence by a Fourier Spectral Method on the Earth Simulator'. In: *Proc. IEEE/ ACM SC2002 Conf.*, Baltimore, 2002; `http://www.sc-2002.org/paperpdfs/pap.pap273.pdf`

D.M. Young (1954): On Richardson's method for solving linear systems with positive definite matrices. J. Math. Phys. **22**, 243–255

D.M. Young (1971): *Iterative Solution of Large Linear Systems* (Academic Press, New York)

T.A. Zang, M.Y. Hussaini (1986): On spectral multigrid methods for the time-dependent Navier-Stokes equations. Appl. Math. Comput. **19**, 359–372

T.A. Zang, M.Y. Hussaini (1987): 'Numerical Simulation of Nonlinear Interactions in Channel and Boundary-layer Transition'. In: *Nonlinear Wave Interaction, in Fluids.* AMD–87, ed. by R.W. Miksad, T.R. Akylas, Th. Herbert, (ASME, New York) pp. 131–145

T.A. Zang, S.E. Krist, G. Erlebacher, M.Y. Hussaini (1987): 'Nonlinear Structures in the Later Stages of Transition'. AIAA Pap. No. 87–1204

T.A. Zang, Y.-S. Wong, M.Y. Hussaini (1982): Spectral multigrid methods for elliptic equations. J. Comput. Phys. **48**, 485–501

T.A. Zang, Y.-S. Wong, M.Y. Hussaini (1984): Spectral multigrid methods for elliptic equations II. J. Comput. Phys. **54**, 489–507

A. Zebib (1984): A Chebyshev method for the solution of boundary value problems. J. Comput Phys. **53**, 443–455

S. Zhao, M. Yedlin (1994): A new iterative Chebyshev spectral method for solving the elliptic equation $\nabla \cdot (\sigma \nabla u) = f$. J. Comput. Phys. **113**, 215–223

O.C. Zienkiewicz, Y.K. Cheung (1967): *The Finite Element Method in Structural and Continuum Mechanics* (McGraw-Hill, London)

A. Zygmund (1959): *Trigonometric Series* (Cambridge Univ. Press, London) [3rd edn., Cambridge Univ. Press, 2003]

# Index

# Scientific Computation

**A Computational Method in Plasma Physics**
F. Bauer, O. Betancourt, P. Garabechan

**Implementation of Finite Element Methods for Navier-Stokes Equations**
F. Thomasset

**Finite-Different Techniques for Vectorized Fluid Dynamics Calculations**
Edited by D. Book

**Unsteady Viscous Flows**
D. P. Telionis

**Computational Methods for Fluid Flow**
R. Peyret, T. D. Taylor

**Computational Methods in Bifurcation Theory and Dissipative Structures**
M. Kubicek, M. Marek

**Optimal Shape Design for Elliptic Systems**
O. Pironneau

**The Method of Differential Approximation**
Yu. I. Shokin

**Computational Galerkin Methods**
C. A. J. Fletcher

**Numerical Methods for Nonlinear Variational Problems**
R. Glowinski

**Numerical Methods in Fluid Dynamics**
Second Edition M. Holt

**Computer Studies of Phase Transitions and Critical Phenomena** O. G. Mouritsen

**Finite Element Methods in Linear Ideal Magnetohydrodynamics**
R. Gruber, J. Rappaz

**Numerical Simulation of Plasmas**
Y. N. Dnestrovskii, D. P. Kostomarov

**Computational Methods for Kinetic Models of Magnetically Confined Plasmas**
J. Killeen, G. D. Kerbel, M. C. McCoy, A. A. Mirin

**Spectral Methods in Fluid Dynamics**
Second Edition
C. Canuto, M. Y. Hussaini, A. Quarteroni, T. A. Zang

**Computational Techniques for Fluid Dynamics 1** Fundamental and General Techniques Second Edition
C. A. J. Fletcher

**Computational Techniques for Fluid Dynamics 2** Specific Techniques for Different Flow Categories Second Edition
C. A. J. Fletcher

**Methods for the Localization of Singularities in Numerical Solutions of Gas Dynamics Problems**
E. V. Vorozhtsov, N. N. Yanenko

**Classical Orthogonal Polynomials of a Discrete Variable**
A. F. Nikiforov, S. K. Suslov, V. B. Uvarov

**Flux Coordinates and Magnetic Filed Structure: A Guide to a Fundamental Tool of Plasma Theory**
W. D. D'haeseleer, W. N. G. Hitchon, J. D. Callen, J. L. Shohet

**Monte Carlo Methods in Boundary Value Problems**
K. K. Sabelfeld

**The Least-Squares Finite Element Method**
Theory and Applications in Computational Fluid Dynamics and Electromagnetics
Bo-nan Jiang

**Computer Simulation of Dynamic Phenomena**
M. L. Wilkins

**Grid Generation Methods**
V. D. Liseikin

**Radiation in Enclosures**
A. Mbiock, R. Weber

**Higher-Order Numerical Methods for Transient Wave Equations**
G. C. Cohen

**Fundamentals of Computational Fluid Dynamics**
H. Lomax, T. H. Pulliam, D. W. Zingg

**The Hybrid Multiscale Simulation Technology** An Introduction with Application to Astrophysical and Laboratory Plasmas A. S. Lipatov

# Scientific Computation

**Computational Aerodynamics and Fluid Dynamics** An Introduction J.-J. Chattot

**Nonclassical Thermoelastic Problems in Nonlinear Dynamics of Shells** Applications of the Bubnov–Galerkin and Finite Difference Numerical Methods
J. Awrejcewicz, V. A. Krys'ko

**A Computational Differential Geometry Approach to Grid Generation** V. D. Liseikin

**Stochastic Numerics for Mathematical Physics** G. N. Milstein, M. V. Tretyakov

**Conjugate Gradient Algorithms and Finite Element Methods**
M. Křížek, P. Neittaanmäki, R. Glowinski, S. Korotov (Eds.)

**Finite Element Methods and Their Applications** Z. Chen

**Mathematics of Large Eddy Simulation of Turbulent Flows**
L. C. Berselli, T. Iliescu, W. J. Layton

**Large Eddy Simulation for Incompressible Flows** An Introduction Third Edition
P. Sagaut

**Spectral Methods** Fundamentals in Single Domains
C. Canuto, M. Y. Hussaini, A. Quarteroni, T. A. Zang

Erratum

# Spectral Methods

## Fundamentals in Single Domains

C. Canuto · M.Y. Hussaini · A. Quarteroni · T.A. Zang



Due to a technical error the caption of Figure 1.6 on page 29 and the content of pages 311 and 312 were reproduced in non-final form. Please find the corrected pages below. On pages 311 and 312 the changes are highlighted in red.

many results obtained from their high-resolution simulations was convincing evidence that the scaled energy spectrum (where the wavenumber is scaled by the inverse of the Kolmogorov length scale $\eta = (\nu^3/\bar{\epsilon})^{1/4}$, with $\nu$ the viscosity and $\bar{\epsilon}$ the average dissipation rate) is not the classical Kolmogorov result of $k^{-5/3}$, but rather $k^{-m}$ with $m \simeq 5/3 - 0.10$.

**Fig. 1.6.** Direct numerical simulation of incompressible isotropic turbulence on a $2048^3$ grid by Y. Kaneda and T. Ishihara (2006): High-Resolution Direct Numerical Simulation of Turbulence. Journal of Turbulence **7**(20), 1–17. The figure shows the regions of intense vorticity in a subdomain with 1/4 the length in each coordinate direction of the full domain [Reprinted with kind permission by the authors and the publisher Taylor & Francis Ltd., http://www.tandf.co.uk/journals]

Rogallo (1977) developed a transformation that permits Fourier spectral methods to be used for homogeneous turbulence flows, such as flows with uniform shear. Blaisdell, Mansour and Reynolds (1993) used the extension of this transformation to the compressible case to simulate compressible, homoge-

$$\|v\|_{H^m_w(\mathbb{R}_+)} = \left( \sum_{j=0}^{m} \|v^{(j)}\|^2_{L^2_w(\mathbb{R}_+)} \right)^{1/2} .$$

A related family of weighted Sobolev spaces is useful, namely,

$$H^m_{w;\alpha}(\mathbb{R}_+) = \left\{ v \in L^2_w(\mathbb{R}_+) \mid (1+x)^{\alpha/2} v \in H^m_w(\mathbb{R}_+) \right\} , \qquad m \geq 0 , \tag{5.7.3}$$

equipped with the natural norm $\|v\|_{H^m_{w;\alpha}(\mathbb{R}_+)} = \|(1+x)^{\alpha/2} v\|_{H^m_w(\mathbb{R}_+)}$.

For each $u \in L^2_w(\mathbb{R}_+)$, let $P_N u \in \mathbb{P}_N$ be the truncation of its Laguerre series, i.e., the orthogonal projection of $u$ upon $\mathbb{P}_N$ with respect to the inner product of $L^2_w(\mathbb{R}_+)$:

$$\int_{\mathbb{R}_+} (u - P_N u)\, \phi\, e^{-x}\, dx = 0 \quad \text{for all } \phi \in \mathbb{P}_N .$$

The following error estimate holds for any $m \geq 0$ and $0 \leq k \leq m$:

$$\|u - P_N u\|_{H^k_w(\mathbb{R}_+)} \leq C N^{k - \frac{m}{2}} \|u\|_{H^m_{w;m}(\mathbb{R}_+)} . \tag{5.7.4}$$

For the orthogonal projection $P^1_N$ upon $\mathbb{P}_N$ in the norm of $H^1_w(\mathbb{R}_+)$, the following estimate holds for $m \geq 1$, $1 \leq k \leq m$:

$$\|u - P^1_N u\|_{H^k_w(\mathbb{R}_+)} \leq C N^{k + \frac{1}{2} - \frac{m}{2}} \|u\|_{H^m_{w;m-1}(\mathbb{R}_+)} ; \tag{5.7.5}$$

the same result holds for the projection $P^{1,0}_N$ upon $\mathbb{P}^0_N$ (Guo and Shen (2000)).

Concerning interpolation, let us consider the $N+1$ Gauss-Radau points $x_j$, $j = 0, \dots , N$, where $x_0 = 0$ and $x_j$, for $j = 1, \dots , N$, are the zeros of $l'_{N+1}(x)$, the derivative of the $(N+1)$-th Laguerre polynomial. For each continuous function $u$ on $\mathbb{R}_+$, let $I_N u \in \mathbb{P}_N$ be the interpolant of $u$ at the points $x_j$. Then, for any integer $m \geq 1$, $0 \leq k \leq m$ and $0 < \epsilon < 1$, one has

$$\|u - I_N u\|_{H^k_w(\mathbb{R}_+)} \leq C_\epsilon N^{k + \frac{1}{2} + \epsilon - \frac{m}{2}} \|u\|_{H^m_{w;m}(\mathbb{R}_+)} \tag{5.7.6}$$

(see Xu and Guo (2002), where additional approximation results can be found). The result stems from the error analysis given by Mastroianni and Monegato (1997) in the family of norms ($r \geq 0$ real)

$$\|v\|_{H^r_{w;*}(\mathbb{R}_+)} = \left( \sum_{k=0}^{\infty} (1+k)^r\, \hat{v}_k^2 \right)^{1/2} ,$$

where $\hat{v}_k = (v, l^{(0)}_k)_{L^2_w(\mathbb{R}_+)}$ are the Laguerre coefficients of $v$. For such norms, one has $\|v\|_{H^r_{w;*}(\mathbb{R}_+)} \leq c \|v\|_{H^r_{w;r}(\mathbb{R}_+)}$ for any integer $r$. Examples of applications to spectral Laguerre discretizations of boundary-value problems in $\mathbb{R}_+$ are provided in the above references. Usually, an appropriate change of unknown function is needed to cast the differential problem into the correct functional setting based on Laguerre-weighted Sobolev spaces.

Hermite approximations can be studied in a similar manner. The basic weighted space $L^2_w(\mathbb{R})$ involves the norm

$$\|v\|_{L^2_w(\mathbb{R})} = \left( \int_{\mathbb{R}} v^2(x) e^{-x^2} dx \right)^{1/2} .$$

The Sobolev spaces $H^m_w(\mathbb{R})$ are defined as above, with respect to this norm. The $L^2_w$-orthogonal projection operator $P_N$ upon $\mathbb{P}_N$ satisfies the estimate

$$\|u - P_N u\|_{H^k_w(\mathbb{R})} \leq C N^{\frac{k}{2} - \frac{m}{2}} \|u\|_{H^m_w(\mathbb{R})} \tag{5.7.7}$$

for all $m \geq 0$ and $0 \leq k \leq m$ (Guo (1999)). Interestingly, all $H^\ell_w$-orthogonal projection operators $P^\ell_N$ upon $\mathbb{P}_N$, for $\ell \geq 0$, coincide with $P_N$, due to property (2.6.12) of Hermite polynomials. For the interpolation operator $I_N$ at the Hermite-Gauss nodes in $\mathbb{R}$, Guo and Xu (2000) proved the estimate

$$\|u - I_N u\|_{H^k_w(\mathbb{R})} \leq C N^{\frac{1}{3} + \frac{k}{2} - \frac{m}{2}} \|u\|_{H^m_w(\mathbb{R})} , \tag{5.7.8}$$

for $m \geq 1$ and $0 \leq k \leq m$.

When dealing with the unbounded intervals $\mathbb{R}_+$ and $\mathbb{R}$, an alternative to polynomials as approximating functions is given by functions that are the product of a polynomial times the natural weight for the interval. Thus, one uses the Laguerre functions $\psi(x) = \phi(x)e^{-x}$ in $\mathbb{R}_+$ or the Hermite functions $\psi(x) = \phi(x)e^{-x^2}$ in $\mathbb{R}$, where $\phi$ is any polynomial in $\mathbb{P}_N$. The behavior at infinity of the function to be approximated may suggest such a choice. We refer, e.g., to Funaro and Kavian (1990) and to Guo and Shen (2003) for the corresponding approximation results and for applications.

## 5.8 Approximation in Cartesian-Product Domains

We shall now extend to several space dimensions some of the approximation results we presented in the previous sections for a single spatial variable. The three expansions of Fourier, Legendre and Chebyshev will be considered. However, we will only be concerned with those Sobolev-type norms that are most frequently applied to the convergence analysis of spectral methods.

### 5.8.1 Fourier Approximations

Let us consider the domain $\Omega = (0, 2\pi)^d$ in $\mathbb{R}^d$, for $d = 2$ or 3, and denote an element of $\mathbb{R}^d$ by $\mathbf{x} = (x_1, \ldots , x_d)$. The space $L^2(\Omega)$, as well as the Sobolev spaces $H^m_p(\Omega)$ of periodic functions, are defined in Appendix A (see (A.9.h)